Lecture Notes in Artificial Intelligence   8933

Subseries of Lecture Notes in Computer Science

LNAI Series Editors

Randy Goebel
University of Alberta, Edmonton, Canada
Yuzuru Tanaka
Hokkaido University, Sapporo, Japan
Wolfgang Wahlster
DFKI and Saarland University, Saarbrücken, Germany

LNAI Founding Series Editor

Joerg Siekmann
DFKI and Saarland University, Saarbrücken, Germany

T0215735

Xudong Luo   Jeffrey Xu Yu   Zhi Li (Eds.)

Advanced Data Mining and Applications

10th International Conference, ADMA 2014
Guilin, China, December 19-21, 2014
Proceedings

 Springer

Volume Editors

Xudong Luo
Sun Yat-sen University
Guangzhou, P.R. China
E-mail: luoxd3@mail.sysu.edu.cn

Jeffrey Xu Yu
Chinese University of Hong Kong
Hong Kong, SAR China
E-mail: yu@se.cuhk.edu.hk

Zhi Li
Guangxi Normal University
Guilin, P.R. China
E-mail: zhili@gxnu.edu.cn

ISSN 0302-9743   e-ISSN 1611-3349
ISBN 978-3-319-14716-1   e-ISBN 978-3-319-14717-8
DOI 10.1007/978-3-319-14717-8
Springer Cham Heidelberg New York Dordrecht London

Library of Congress Control Number: 2014959177

LNCS Sublibrary: SL 7 – Artificial Intelligence

Typesetting: Camera-ready by author, data conversion by Scientific Publishing Services, Chennai, India

Printed on acid-free paper

Springer is part of Springer Science+Business Media (www.springer.com)

Preface

This volume contains the papers presented at ADMA 2014: the 10th International Conference on Advanced Data Mining and Applications held during December 19–21, 2014, in Guilin.

There were 90 submissions. Each submission was reviewed by up to four, and on average 2.9, Program Committee (PC) members and additional reviewers. The committee decided to accept 48 main conference papers plus ten workshop papers.

Since its start in 2005, ADMA has been an important forum on theoretical and application aspects of data mining and knowledge discovery in the world. The previous ADMA conferences were held in Wuhan (2005), Xi'an (2006), Harbin (2007), Chengdu (2008), Beijing (2009 and 2011), Chongqing (2010), Nanjing (2012), and Hangzhou (2013).

Program co-chairs Xudong Luo and Jeffrey Xu Yu put in a lot of effort for the difficult and highly competitive selection of research papers. ADMA 2014 attracted 90 papers from 18 countries and regions. Honorary Chair Ruqian Lu, local Organization Chair Zhi Li, and Steering Committee Chair Xue Li all contributed significantly to the paper review process. We thank Hui Xiong and Osmar R. Zaiane for their keynotes, which were the highlights of the conference.

So many people worked hard to make ADMA 2014 successful. The excellent conference and banquet places were managed by local Organization Chair Zhi Li; Publicity Chair Xianquan Zhang promoted the conference. Treasurer Guoquan Chen played an important role in the financial management.

Different to the previous nine ADMA conferences, we held two important activities. One was to grant the best paper award to three submissions to ADMA 2014. Another was to grant the 10-year best paper award to three papers published in previous ADMA conferences. These papers were selected by Steering Committee Chair Xue Li, PC Co-chairs Xudong Luo and Jeffrey Xu Yu, as well as PC chairs of past conferences. We must thank the Guangxi Key Lab of Multiple Data Mining and Safety at the College of Computer Science and Information Technology at Guangxi Normal University for offering the honorarium for these best papers.

This year, in addition to the conference, we organized for papers to be published in a special issue in the renowned international journal *Knowledge-Based Systems*, which were selected and coordinated by PC Co-chairs Xudong Luo and Jeffrey Xu Yu and Steering Committee Chair Xue Li. The authors of the top 11 papers were invited to extend their contributions for possible publication in *Knowledge-Based Systems*.

Furthermore, we would like to thank the many individuals, including Dingrong Yuan, Xiaofeng Zhu, Zhenjun Tang, and some Master students at the College of Computer Science and Information Technology at Guangxi Normal University, who took care of various aspects of the conference.

We hope that you enjoy these proceedings.

October 2014

Shichao Zhang
Xianxian Li
Changan Yuan

Organization

Program Committee

Liang Chen	Zhejiang University, China
Shu-Ching Chen	Florida International University, USA
Zhaohong Deng	Jiangnan University, China
Xiangjun Dong	Shandong Polytechnic University, China
Stefano Ferilli	University of Bari, Italy
Philippe Fournier-Viger	Centre Universitaire De Moncton, Canada
Yanglan Gan	Gonghua University, China
Yang Gao	Nanjing University, China
Gongde Guo	Fujian Normal University, China
Manish Gupta	University of Illinois at Urbana-Champaign, USA
Sung Ho Ha	Kyungpook National University, Korea
Qing He	University of Chinese Academy of Sciences, China
Wei Hu	Nanjing University, China
Daisuke Ikeda	Kyushu University, Japan
Xin Jin	University of Illinois at Urbana-Champaign, USA
Daisuke Kawahara	Kyoto University, Japan
Sanghyuk Lee	Xi'an Jiaotong-Liverpool University, Japan
Gang Li	Deakin University, Australia
Jinjiu Li	University of Technology Sydney, Australia
Xuelong Li	Chinese Academy of Sciences, China
Zhi Li	Guangxi Normal University, China
Zhixin Li	Guangxi Normal University, China
Bo Liu	Guangdong University of Technology, China
Guohua Liu	Yanshan University, China
Yubao Liu	Sun Yat-sen University, China
Xudong Luo	Sun Yat-sen University, China
Marco Maggini	University of Siena, Italy
Tomonari Masada	Nagasaki University, Japan
Toshiro Minami	Kyushu University, Japan
Yasuhiko Morimoto	Hiroshima University, Japan
Feiping Nie	University of Texas at Arlington, USA
Shirui Pan	University of Technology Sydney, Australia
Tieyun Qian	Wuhan University, China
Faizah Shaari	Polytechnic Sultan Salahuddin Abdul Aziz Shah, Malaysia

Dharmendra Sharma	University of Canberra, Australia
Bin Shen	Zhejiang University, China
Michael Sheng	The University of Adelaide, Australia
Guojie Song	Peking University, China
Yin Song	University of Technology Sydney, Australia
Masashi Sugiyama	Tokyo Institute of Technology, Japan
Zhen-Jun Tang	Guangxi Normal University, China
Hanghang Tong	Carnegie Mellon University, USA
Carlos Toro	Vicomtech, Spain
Eiji Uchino	Yamaguchi University, Japan
Hongzhi Wang	Harbin Institute of Technology, China
Shuliang Wang	Wuhan University, China
Wei Wang	Fudan University, China
Zhihui Wang	Fudan University, China
Jia Wu	University of Technology Sydney, Australia
Zhiang Wu	Nanjing University of Finance and Economics, China
Zhipeng Xie	Fudan University, China
Guandong Xu	University of Technology Sydney, Australia
Bing Xue	Victoria University of Wellington, New Zealand
Zijiang Yang	York University, Canada
Min Yao	Zhejiang University, China
Dayong Ye	University of Wollongong, Australia
Tetsuya Yoshida	Hokkaido University, Japan
Chao Yu	Dalian University of Technology, China
Fusheng Yu	Beijing Normal University, China
Jeffrey X. Yu	Chinese University of Hong Kong, SAR China
Qi Yu	Rochester Institute of Technology, USA
Mengjie Zhang	Victoria University of Wellington, New Zealand
Shichao Zhang	Guangxi Normal University, China
Nenggan Zheng	Zhejiang University, China
Yong Zheng	DePaul University, USA
Sheng Zhong	Nanjing University, China
Haofeng Zhou	University of Texas at Austin, USA
Rong Zhu	Zhejiang University, China
Xiaofeng Zhu	Guangxi Normal University, China
Yi Zhuang	Zhejiang Gongshang University, China

Additional Reviewers

Bu, Zhan
Gomariz Peñalver, Antonio
He, Qing
Hlosta, Martin

Kawakubo, Hideko
Li, Fangfang
Li, Xin
Noh, Yung-Kyun

Pang, Ran
Pazienza, Andrea
Ruan, Wenjie
Shemshadi, Ali
Shie, Bai-En
Siddique, Md. Anisuzzaman
Tangkaratt, Voot
Tao, Haicheng
Tian, Haiman
Vu, Huy Quan

Wang, De
Wang, Huihui
Wang, Youquan
Wu, Cheng Wei
Xiong, Ping
Yuan, Jianjun
Zhai, Tingting
Zhang, Hao
Zhang, Kaibing
Zhu, Tianqing

Table of Contents

Recommend System

Database

Dimensionality Reduction

Advance Machine Learning Techniques

Classification

Big Data and Applications

Clustering Methods

Machine Learning

Data Mining and Database

A New Improved Apriori Algorithm
Based on Compression Matrix

Taoshen Li and Dan Luo

School of Computer, Electronics and Information,
Guangxi University, 530004 Nanning, China
tshli@gxu.edu.cn, 360081437@qq.com

Abstract. The existing Apriori algorithm based on matrix still has the problems that the candidate itemsets are too large and matrix takes up too much memory space. To solve these problems, an improved Apriori algorithm based on compression matrix is proposed. The improvement ideas of this algorithm are as follows: (1) reducing the times of scanning matrix set during compressing by adding two arrays to record the counts of 1 in the row and column; (2) minimizing the scale of matrix and improving space utilization by deleting the itemsets which cannot be connected and the infrequent itemsets in compressing matrix; (3) decreasing the errors of the mining result by changing the condition of deleting the unnecessary transaction column;(4) reducing the cycling number of algorithm by changing the stopping condition of program. Instance analysis and experimental results show that the proposed algorithm can accurately and efficiently mines all frequent itemsets in transaction database, and improves the efficiency of mining association rules.

Keywords: Data mining, Frequent itemsets, Apriori algorithm, Association rules, Compressed matrix.

1 Introduction

Association rule mining is an important research field in data mining. It discovers interesting relationship between transaction records by mining frequent itemsets in database. With the growth of the data sets in the size and complexity, association rule mining is facing with enormous challenge. How to improve the efficiency of discovering the frequent itemsets for large-scale data sets is a major problem of association rule mining.

Apriori algorithm is one of the most efficient association rule mining algorithm which has been brought in 1993 by Agrawal[1]. It follows a generate-and-test methodology for finding frequent item sets, generating successively longer candidate item sets from shorter ones that are known to be frequent. At the kth iteration (for $k>1$), it forms frequent ($k+1$)-itemset candidates based on the frequent k-itemsets, and scans the database once to find the complete set of frequent ($k+1$)-itemsets, L_{k+1}. It can mine association rules accurately and efficiently. However, this algorithm have to scan the transaction database every time when it wants to know the k-frequent item

X. Luo, J.X. Yu, and Z. Li (Eds.): ADMA 2014, LNAI 8933, pp. 1–15, 2014.

sets, and generates huge candidate itemsets. It will require a lot of I/O load and waste huge running time and larger memory space. In order to solve the existed problems of Apriori algorithm, some improved algorithms are proposed, such as FP-Growth algorithm[2],Hash-based algorithm[3], Partitioning-based algorithm[4], Sampling-based algorithm[5], Dynamic Itemsets Counting-based algorithm[6], Scalable algorithm[7], et al. In these improved algorithm, some algorithm is designed too complicated and not practical. Although some algorithms can improve the shortcomings of the Apriori algorithm, but they may cause other problems to increase running cost and affect the operation efficiency.

In [8], Huang et al. first proposed the Apriori algorithm based on matrix, which is called BitMatrix algorithm. The algorithm first converts the transaction database into Boolean Matrix, and then uses the "AND" operation to calculate the candidate support count to find the frequent itemsets. It can find all 1-frequent item sets in the first time scanning and only needs to scan the database once. The following operations are carried on the matrix. In recent years, many researchers have made improvement to the Apriori algorithm based on matrix. Yuan and Huang[9] propose a matrix algorithm for efficient generating large frequent candidate sets. This algorithm first generates a matrix which entries 1 or 0 by passing over the cruel database only once, and then the frequent candidate sets are obtained from the resulting matrix. Zhang et al.[10] present a fast algorithm for mining association rules based on boolean Matrix, which is based on the concept lattice and can certify all association rules with only one scan of databases. Wand and Li [11] put forward an improved Apriori algorithm based on the matix. This algorithm first makes use of the matrix to find out frequent k-itemsets with most items, and finds out all frequent sub-itemsets from frequent k-itemsets and deletes these sub-itemsets from candidate itemset to decrease duplicated scan. And then it finds the remaining frequent itemsets from frequent 2-itemsets to frequent $(k$-1)itemsets. In case of smaller support, this algorithm's performance is better, but its calculation process of the maximum frequent itemsets is too complex.

In order to adapt to the different needs of mining, people has improved Apriori algorithm based on matrix. In [12], a quantitative association rules mining algorithm based on matrix is presented. This algorithm firstly transformed quantitative database into Boolean matrix, and then used Boolean "AND" operation to generate frequent item sets on matrix vector, thereby reducing the database scanning frequency. Khar et al.[13] put forward a Boolean matrix based algorithm for mining multidimensional association rules from relational databases. The algorithm adopts Boolean relational calculus to discover frequent predicate sets, and uses Boolean logical operations to generate the association rules. Luo et al. [14] present an improved Apriori algorithm based on matrix. In order to increase algorithm's running time, this algorithm transforms the event database into matrix database so as to get the matrix item set of maximum item set. When finding the frequent k-item set from the frequent k-item set, only its matrix set is found. Krajca et al. [15] propose an algorithm for computing factorizations based on approximate Boolean matrix decomposition, which first utilizes frequent closed itemsets as candidates for factor-items and then computes the remaining factor-items by incrementally joining items. This algorithm can reduce the dimensionality of data while keeping high approximation degree. In [16], a new algorithm by extracting maximum frequent itemsets based on a Boolean matrix is presented. In [17], a matrix-based and sorting index association rules algorithm is proposed. This algorithm scans database only once, and can directly find the frequency k-itemsets. Especially when frequent item sets are higher or need to have a date mining update, the algorithm has higher efficiency and feasibility. In [18], an improved Apriori algorithm based on MapReduce mode is described, which can handle massive datasets with a large number of nodes on Hadoop platform. In [19], authors realize the parallelization of Apriori algorithm for the massive Web log mining, and verify the efficiency of Apriori algorithm which has been improved by parallelization.

Compared with Apriori algorithm, the Apriori algorithm based on matrix can reduce the number of scanning the database and save the time of computing the support count. But it also exist some problems: (1) scan matrix for many times in the process of computing support count; (2) increase the complex calculation process to reduce the number of frequent candidate itemsets; (3) store a lot of elements unrelated to generate frequent itemsets in matrix due to compression of the matrix is not thorough; (4) decrease the accuracy of generated frequent itemsets because compression transaction delete too many margin transactions; (5)due to the design of algorithms is too complicated, for high intensive and a large quantity of transaction database, algorithm needs to cast a large of times for processing matrix and itemsets. Aiming at these problems, this paper proposes a new improved Apriori algorithm based on compressed matrix, which is called NCMA algorithm. The goal of this NCMA algorithm is to accurately and efficiently mine all frequent itemsets from transaction database.

2   Problem Description

2.1   Problem Description

Let $I=\{I_1,I_2,...,I_m\}$ be a set of item. Let $D=\{D_1,D_2,...,D_n\}$ be a set of database transactions where each transaction T is a set of items such that $T \subset I$. Each transaction is associated with an identifier, called TID. An association rule is an implication of the form $A \Rightarrow B$, where $A \subset I, B \subset I$, $A \cap B=\Phi$. The *support* is the percentage of transactions in D that contain $A \cup B$ (i.e., both A and B). This is taken to be the probability, $P(A \cup B)$. The *confidence* is in the transaction set D is the percentage of transactions in D containing A that also contain B.

A set of items is referred to as an itemset. An itemset that contains k items is a k-itemset. The occurrence frequency of an itemset is the number of transactions that contain the itemset. That is also known, simple, as the frequency, support count of the iemset. If occurrence frequency of an itemset is greater than or equal to product of minimum support(*min_sup*) and the total number of transaction in D, then it satisfies minimum support. An temset is a frequent itemset if it satisfies pre-determined minimum support. The set of frequent k-itemset is commonly denoted an L_k.

Rules that satisfy both a minimum support threshold(min_sup) and a minimum confidence threshold(min_conf) are called strong. Association rule mining aims to find strong rules in the transaction database that satisfy both the minimum support and minimum confidence given by user. It is a two-step process:(1)find all frequent itemsets, (2)generate strong association rules from the frequent itemsets.

In Apriori algorithm based on matrix, the Boolean matrix is used to describe the transaction database. And the number '1' and '0' are used to replace the corresponding items. Firstly, the transaction database needs to be converted into Boolean matrix. If the transaction database contained m items and n transactions the Boolean matrix will have $m+1$ row and $n+2$ columns. The first column registers "items" and the first row records "TID" of the transactions. The last column is used to record the support of the itemsets. Secondly, the minimum support(min_sup) is compared with the support of the itemsets. If the support of the itemsets is smaller than the $min\text{-}sup$, the row of the itemsets will be deleted. By doing this the new Boolean matrix just contains the one-frequent itemsets. And if it wants to know the k-frequent item sets, the "AND" operation will just be carried out on the k rows. Finally all the frequent itemsets can be found out.

The NCMA algorithm is faced with the question: 1) how to reduce the number of scanning the matrix; 2) how to efficiently mine all frequent itemsets from transaction database; 3) delete itemsets to be independent of generated frequent itemset,

2.2   Relevant Property and Theorem

In order to better explain algorithm, some relevant fundamental properties, theorems and inferences are describes as follows:

Property 1: All nonempty subsets of a frequent itemset must also be frequent, and all supersets of a infrequent itemset must also be infrequent.

Inference 1[13]: If L_k is frequent k-itemset in D, then the number of frequent $(k-1)$-itemset that is contained in L_k will be k.

Inference 2[12]: If a frequent k-itemset can also generate frequent $(k+1)$-itemset, then the number of frequent k-itemset must be larger than k.

Property 2: Transaction that does not contains any frequent k-itemset cannot contain any frequent $(k+1)$-itemset.

Theorem 1[20]: If the length of a transaction in D is k, the transaction cannot contain any frequent itemset that number of item is larger than k.

Theorem 2[21]: Assuming that items in itemset are sorted in the dictionary order. While $(k+1)$-itemset is generated by k-itemset, if previous $k-1$ item of two itemset is different, then the connection operation between the two itemset will be give up because the generated itemsets either are duplicated itemsets or infrequent itemsets. At this point, the two itemsets is considered to be not connected.

Inference 3[22]: Assuming that each transaction and itemsets in transaction are sorted in order of the dictionary. If two frequent k-itemsets I_x and I_y are unable to connect, then all itemsets after I_x and I_y are not connected.

3   An Improved Algorithm Based on Compressed Matrix

3.1   Ideas of Improvement

Aiming at existing problems of Apriori algorithm based on matrix, the improved ideas of NCMA algorithm proposed in this paper are as follows.

(1) Improving matrix storage method. In order to reduce the size and the number of scanning matrix, we added two array . The weight array w is used to record the count for the same transaction. Using array w, each column in the matrix only stores one or more transaction information, and no duplicate column. Array m is used to record the number of 1 in each column. When compressing matrix, algorithm can obtain the transaction length by scanning the array m. So algorithm can avoid scanning the entire matrix, reduces scanning time and compresses the matrix quickly and effectively.

(2) Improving method of matrix compression. This improvement is to remove useless transaction for finding frequent itemsets, so that the matrix is compressed to be smaller. Improve matrix compression includes the row (itemsets) compression and column (transaction) compression. The matrix row compression needs to utilize theorem 2 and inference 3, and the process of row compression is as follows: 1) scan matrix, if an itemset is unable to connect with its neighbor itemsets, then the row vector corresponding to this itemset will be deleted; 2) update the values of array m accordingly; 3) the remaining row vectors are combined into a new matrix in original order. The matrix column compression has to use property 2 and theorem 1, and the process of column compression is as follows: 1) after row compression, scan matrix m, if there are value that less than or equal to the 1, then delete column vector corresponding to this value. 2)the remaining column vectors are combined into a new matrix in original order.

(3) Calculating support count. By improving method of calculating support count, the algorithm can reduce the number of scanning the matrix row and more simply and quickly find the frequent itemsets. In process of calculating support count, the "AND" operation will be carried out on the row vectors corresponding to two connected itemsets, and multiplied by the corresponding weight in array w. Its weight sum is the support count of itemset after connected. If the support count of an itemset is greater than or equal to *min_sup*, then the row vector corresponding to this itemset will be saved, otherwise rejected. When an itemset cannot be connected with the next itemset, the connection operation of next itemsets and behind itemsets will be executed, until all the itemsets scanning is completed.

(4) Updating end condition of algorithm to reduce cycle times of algorithm execution. This improvement needs to utilize inference 1 and 2. According to inference 2, when the number of frequent k-itemsets is less than $k+1$(i.e. the row of the matrix is less than $k+1$), the algorithm does not need to find frequent $(k+1)$-itemsets (k+1). At this time, the algorithm can be terminated.

3.2   Algorithm's Description

The NCMA algorithm is described as follows:

Input：transaction database D，minimum support: min_sup

Output：Completed frequent itemset

Method：

```
for(i=0; i<M; i++){   //M is the number of item
    for(j=0; j<N; j++){   // N is the number of non-duplicated transaction
        if(I_i in T_j)   D[i][j] =1;
        else   D[i][j] =0;
    }
}  // Constructing Boolean matrix D
for (i=0; i<N; i++)      // For same transaction count, the count is saved to array w
    if(T_i has duplicated transaction)   w[i]++;
for each l_i in D   {
```

$$support\_count = \sum_{i=1}^{N} l_i;$$

```
    if(support_counts≥min_support×the number of transaction)
        add    l_i to L_1;
}  // Calculating frequent 1-itemset
Output L_1;
delete the column vector appending to infrequent item in D;
D_1←QuickSort(D)        // Quick sorting D in ascending order of support count
for(i=0; i<M_1; i++){       //M_1 is the row number of D_1
    for(j=i+1; j<M_1; j++){
        for(q=0; q<N_1; q++)   {      // N_1 is the column number of D_1
            support_counts+=D_1[i][q]×D_1[j][q]×w[q];
        }
        if(support_counts≥min_support×the number of transaction)   {
            add this itemset to L_2;
            accumulate the row vectors appending to this itemset to array m by bits;
        }
    }
}//Calculating frequent 2-itemset
Output L_2;
D_2←the row vectors appending to L_2;
for(k=2; |L_k|≥k+1; k++){
    for each itemset l in L_k   {
        if(l is unable to carry out join operation with adjacent itemsets)   {
            array m bitwise minus the value of row vector appending to l;
            delete the column vector appending to l;
        }
    } // Row(itemset) compression
    arrange the remaining rows of D_k in order;
    for(i=0; i<N_k; i++){   // N_k is the number of column in D_k
        if(m[i]≤1)
```

```
        delete i-th column and w[i];
} // Column (transaction) compression
arrange the remaining columns of  D_k  in order
clear(m);        // Clear array m
for(i=0; i<M_k'; i++) {        //M_k' is the row number of D_k'
   for(j=i+1; j<M_k'; j++)   {
      if(((l_i[1]=l_j[1])∧(l_i[2]=l_j[2])∧...∧(l_i[k-1]=l_j[k-1])∧(l_i[k]<l_j[k]))   {
         for(q=0; q<N_k'; q++)
            support_counts+=D_k'[i][q]×D_k'[j][q]×w[q];
         if(support_counts≥min_support×the number of transactions) {
            add this itemset to L_{k+1};
            accumulate the row vectors appending to this itemset to array m by bits;
         }
      }
      else break;
   }
}
Output L_{k+1};
D_{k+1}←the row vectors appending to L_{k+1};
}
```

4   Instance Analysis and Experiment of Algorithm

4.1   Instance Analysis

Let's look at an example as shown in Table 1. There are ten transactions in this database, and six itemsets(i.e. $I=\{I_1,I_2,I_3,I_4,I_5,I_6\}$). Suppose that the minimum support count requested is 2(i.e. $min\text{-}sup=2/10=20\%$).

Table 1. A example of transaction database

TID	List of item_IDs
T_1	I_1, I_2, I_4, I_5, I_6
T_2	I_2, I_3, I_5, I_6
T_3	I_2, I_3
T_4	I_3, I_4, I_5
T_5	I_3, I_5, I_6
T_6	I_2, I_3, I_5, I_6
T_7	I_3
T_8	I_2, I_4, I_5, I_6
T_9	I_3, I_4, I_5
T_{10}	I_5, I_6

The execution process of the NCMA algorithm is described as follows:

(1) Scan transaction database D and generate Boolean matrix array m. As seen from Table 1, the transaction T3 and T10 contain the same project, and T5 and T7

also contain the same project. Then, the weights for third and fifth column of the matrix are 2, the weights of rest column are 1. D and w are as follows:

$$w = \begin{bmatrix} 1 & 1 & 2 & 1 & 2 & 1 & 1 & 1 \end{bmatrix}$$

$$D = \begin{matrix} I_1 \\ I_2 \\ I_3 \\ I_4 \\ I_5 \end{matrix} \begin{bmatrix} 1 & 0 & 0 & 1 & 1 & 0 & 1 & 1 \\ 1 & 1 & 1 & 1 & 0 & 1 & 1 & 1 \\ 0 & 0 & 1 & 0 & 1 & 1 & 1 & 1 \\ 0 & 1 & 0 & 1 & 0 & 1 & 0 & 1 \\ 1 & 0 & 0 & 0 & 0 & 0 & 1 & 1 \end{bmatrix}$$

(2) Determine the set of frequent 1-itemsets, L_1. It consist of the candidate set of 1-itemsets satisfying minimum support. The transaction in D are scanned and the support count of all itemsets is accumulated. $support(I_1)=6$, $support(I_2)=8$, $support(I_3)=7$, $support(I_4)=4$, $support(I_5)=3$. And then, the saved row vectors are arranged in ascending order of support count(i.e. $\{I_5, I_4, I_1, I_3, I_2\}$). The initial matrix D_1 generated by recombination is as follows.

$$w = \begin{bmatrix} 1 & 1 & 2 & 1 & 2 & 1 & 1 & 1 \end{bmatrix}$$

$$D_1 = \begin{matrix} I_5 \\ I_4 \\ I_1 \\ I_3 \\ I_2 \end{matrix} \begin{bmatrix} 1 & 0 & 0 & 0 & 0 & 0 & 1 & 1 \\ 0 & 1 & 0 & 1 & 0 & 1 & 0 & 1 \\ 1 & 0 & 0 & 1 & 1 & 0 & 1 & 1 \\ 0 & 0 & 1 & 0 & 1 & 1 & 1 & 1 \\ 1 & 1 & 1 & 1 & 0 & 1 & 1 & 1 \end{bmatrix}$$

(3) Determine the set of frequent 2-itemsets, L_2. Each row of D_1 carries out weighted vector inner product operation with other row in order. For example, $I_5 \wedge I_4 = 00000001, support\_count(I_5I_4) = 0\times1+0\times1+0\times2+0\times1+0\times2+0\times1+0\times1+1\times1=1$. Because the support count of itemset $\{I_5I_4\}$ is less than minimum support count, the row vector appending to itemset is removed. After calculation, a set of $\{I_5I_1, I_5I_3, I_5I_2, I_4I_1, I_4I_3, I_4I_2, I_1I_3, I_1I_2, I_3I_2\}$ is the itemsets which support count is greater than or equal to the minimum support count. The row vectors appending to this itemset are accumulated to array m by bits, and the itemsets appending to the saved row vector is L_2. After recombining D_1 in itemsets order, we can get the new matrix D_2 as shown follows:

$$w = \begin{bmatrix} 1 & 1 & 2 & 1 & 2 & 1 & 1 & 1 \end{bmatrix}$$

$$D_2 = \begin{matrix} I_5I_1 \\ I_5I_3 \\ I_5I_2 \\ I_4I_1 \\ I_4I_3 \\ I_4I_2 \\ I_1I_3 \\ I_1I_2 \\ I_3I_2 \end{matrix} \begin{bmatrix} 1 & 0 & 0 & 0 & 0 & 0 & 1 & 1 \\ 0 & 0 & 0 & 0 & 0 & 0 & 1 & 1 \\ 1 & 0 & 0 & 0 & 0 & 0 & 1 & 1 \\ 0 & 0 & 0 & 1 & 0 & 0 & 0 & 1 \\ 0 & 0 & 0 & 0 & 0 & 1 & 0 & 1 \\ 0 & 1 & 0 & 1 & 0 & 1 & 0 & 1 \\ 0 & 0 & 0 & 0 & 1 & 0 & 1 & 1 \\ 1 & 0 & 0 & 1 & 0 & 0 & 1 & 1 \\ 0 & 0 & 1 & 0 & 0 & 1 & 1 & 1 \end{bmatrix}$$

$$m = \begin{bmatrix} 3 & 1 & 1 & 3 & 1 & 3 & 6 & 9 \end{bmatrix}$$

(4) Compressing matrix D_2. At first, the transactions in D_2 are scanned and the row vectors corresponding to the itemset $\{I_3 I_2\}$ that is unable to connect with its neighbor itemsets are removed. The array m needs to subtract the vector value corresponding to $\{I_3 I_2\}$. The remaining row vectors are recombined into a new D_2 in original order. Secondly, matrix m is scanned. The columns which value is less than or equal to the 1 will be removed, such as 2, 3 and 5 column. The remaining column vectors are recombined into a new D_2' in order. D_2 and D_2' are shown as follows.

$$
w = \begin{bmatrix} 1 & 1 & 2 & 1 & 2 & 1 & 1 & 1 \end{bmatrix}
$$

$$
D_2 = \begin{array}{c} I_5I_1 \\ I_5I_3 \\ I_5I_2 \\ I_4I_1 \\ I_4I_3 \\ I_4I_2 \\ I_1I_3 \\ I_1I_2 \end{array}
\begin{bmatrix}
1 & 0 & 0 & 0 & 0 & 0 & 1 & 1 \\
0 & 0 & 0 & 0 & 0 & 0 & 1 & 1 \\
1 & 0 & 0 & 0 & 0 & 0 & 1 & 1 \\
0 & 0 & 0 & 1 & 0 & 0 & 0 & 1 \\
0 & 0 & 0 & 0 & 0 & 1 & 0 & 1 \\
0 & 1 & 0 & 1 & 0 & 1 & 0 & 1 \\
0 & 0 & 0 & 0 & 1 & 0 & 1 & 1 \\
1 & 0 & 0 & 1 & 0 & 0 & 1 & 1
\end{bmatrix}
$$

$$
m = \begin{bmatrix} 3 & 1 & 0 & 3 & 1 & 2 & 5 & 8 \end{bmatrix}
$$

$$
w = \begin{bmatrix} 1 & 1 & 1 & 1 & 1 \end{bmatrix}
$$

$$
D_2' = \begin{array}{c} I_5I_1 \\ I_5I_3 \\ I_5I_2 \\ I_4I_1 \\ I_4I_3 \\ I_4I_2 \\ I_1I_3 \\ I_1I_2 \end{array}
\begin{bmatrix}
1 & 0 & 0 & 1 & 1 \\
0 & 0 & 0 & 1 & 1 \\
1 & 0 & 0 & 1 & 1 \\
0 & 1 & 0 & 0 & 1 \\
0 & 0 & 1 & 0 & 1 \\
0 & 1 & 1 & 0 & 1 \\
0 & 0 & 0 & 1 & 1 \\
1 & 1 & 0 & 1 & 1
\end{bmatrix}
$$

$$
m = \begin{bmatrix} 3 & 3 & 2 & 5 & 8 \end{bmatrix}
$$

(5) Determine the set of frequent 3-itemsets, L_3. The array m is cleared. Each itemset in D_2' that is able to join carries out weighted vector inner product operation in order. For example, $\{I_5I_1,\ I_5I_3\}$ may be joined to itemset $\{I_5I_1I_3\}$, $I_5 \wedge I_1 \wedge I_3 = I_5I_1 \wedge I_5I_3 = 00011$, $support\_count(I_5I_1I_3) = 0 \times 1 + 0 \times 1 + 0 \times 1 + 1 \times 1 + 1 \times 1 = 2$. Because the support count of itemset $\{I_5I_1I_3\}$ is equal to minimum support count, the row vector appending to the itemset will be retained. Similarly, the row vectors appending to $\{I_5I_1I_2, I_5I_3I_2, I_4I_1I_2, I_4I_3I_2, I_1I_3I_2\}$ will also be retained. At the same time, these row vectors are accumulated to array m by bits. The row vector appending to $\{I_4I_1I_3\}$ will be removed. The itemsets appending to the saved row vector is L_3. After recombination in itemsets order, we can get the matrix D_3 as shown follows. Due to the row number of D_3 is 6 and is larger than 4, the operation will enter the next round of computing.

$$
w = \begin{bmatrix} 1 & 1 & 1 & 1 & 1 \end{bmatrix}
$$

$$
D_3 = \begin{array}{c} I_5I_1I_3 \\ I_5I_1I_2 \\ I_5I_3I_2 \\ I_4I_1I_2 \\ I_4I_3I_2 \\ I_1I_3I_2 \end{array}
\begin{bmatrix}
0 & 0 & 0 & 1 & 1 \\
1 & 0 & 0 & 1 & 1 \\
0 & 0 & 0 & 1 & 1 \\
0 & 1 & 0 & 0 & 1 \\
0 & 0 & 1 & 0 & 1 \\
0 & 0 & 0 & 1 & 1
\end{bmatrix}
$$

$$
m = \begin{bmatrix} 1 & 1 & 1 & 4 & 6 \end{bmatrix}
$$

(6) Compressing matrix D_3. At first, the transactions in D_3 are scanned and the row vectors corresponding to the $\{I_5I_3I_2, I_4I_1I_2, I_4I_3I_2, I_1I_3I_2\}$ that is unable to connect with its neighbor itemsets are removed. When removing one row, the value of array m needs to be updated correspondingly. The remaining row vectors are recombined into a new D_3 in order. Secondly, matrix m is scanned. The columns which its value is less than or equal to the 1 will be removed, such as 1, 2 and 3 column. The remaining column vectors are recombined into a new D_3' in order. D_3 and D_3' are shown as follows.

$$w = \begin{bmatrix} 1 & 1 & 1 & 1 & 1 \end{bmatrix}$$
$$D_3 = \begin{matrix} I_5I_1I_3 \\ I_5I_1I_2 \end{matrix} \begin{bmatrix} 0 & 0 & 0 & 1 & 1 \\ 1 & 0 & 0 & 1 & 1 \end{bmatrix}$$
$$m = \begin{bmatrix} 1 & 0 & 0 & 2 & 2 \end{bmatrix}$$

$$w = \begin{bmatrix} 1 & 1 \end{bmatrix}$$
$$D_3' = \begin{matrix} I_5I_1I_3 \\ I_5I_1I_2 \end{matrix} \begin{bmatrix} 1 & 1 \\ 1 & 1 \end{bmatrix}$$
$$m = \begin{bmatrix} 2 & 2 \end{bmatrix}$$

(7) Determine the set of frequent 4-itemsets, L_4. The array m is cleared again. $\{I_5I_1I_3, I_5I_1I_2\}$ is joined to $\{I_5I_1I_2I_3\}$, and the weighted vector inner product operation is carried out. $I_5 \wedge I_1 \wedge I_3 \wedge I_2 = I_5I_1I_3 \wedge I_5I_1I_2 = 11$, $support\_count(I_5I_1I_3I_2) = 1 \times 1 + 1 \times 1 = 2$. Because the support count of itemset $\{I_5I_1I_3I_2\}$ is equal to minimum support count, the row vector appending to this itemset will be retained. At the same time, this row vectors are accumulated to array m by bits. The itemsets appending to saved row vector is L_4. After recombination in order, we can get the matrix D_4 as shown follows. Due to the row number of D_4 is 2 and is less than 5, the L_5 has not to be determined. At this time, maximal frequent itemsets is frequent 4-itemsets, and running of algorithm is stopped.

$$w = \begin{bmatrix} 1 & 1 \end{bmatrix}$$
$$D_4 = I_5I_1I_3I_2 \begin{bmatrix} 1 & 1 \end{bmatrix}$$
$$m = \begin{bmatrix} 1 & 1 \end{bmatrix}$$

Before ordering L_1, generated L_2 is $\{I_1I_2, I_1I_3, I_1I_4, I_1I_5, I_2I_3, I_2I_4, I_2I_5, I_3I_4, I_3I_5\}$, correspondent L_3 is $\{I_1I_2I_3, I_1I_2I_4, I_1I_2I_5, I_1I_3I_4, I_1I_3I_5, I_1I_4I_5, I_2I_3I_4, I_2I_3I_5, I_2I_4I_5, I_3I_4I_5\}$, L_4 is $\{I_1I_2I_3I_4, I_1I_2I_3I_5, I_1I_2I_4I_5, I_2I_3I_4I_5\}$. After ordering L_1, L_3 is $\{I_5I_1I_3, I_5I_1I_2, I_5I_3I_2, I_4I_1I_3, I_4I_1I_2, I_4I_3I_2, I_1I_3I_2\}$, L_4 is $\{I_5I_1I_3I_2\}$. Compared with the L_3 and L_4 generated before ordering, the frequent itemset number of L_3 and L_4 generated after ordering decrease 3 respectively, thus reducing the three vector inner product operations respectively. Obviously, after the NCMA has sorting frequent 1-itemsets in increasing order, the number of 3 candidate frequent 3-itemsets and later candidate frequent itemsets that are generated by algorithm are decreased. For the frequent itemsets generated in later, the effect of sorting will be getting better and better. So, NCMA algorithm can greatly improve the operation efficiency of running algorithm.

4.2   Experimental Results and Analysis

In order to analyze algorithm's performance, Apriori algorithm, CM_Appriori_1 algorithm proposed in [24] and NCMA algorithm proposed in this paper are compared in experiments. In experiment, three algorithms are programmed by Java programming language. The developmental tool is Eclipse SDK. The configuration of hardware ia as follows: CPU 2.50GHz, 4GB memory, 64-bit Windows 7.

IBM Quest Market-Basket Synthetic Data Generator is common a tool of artificial data synthesis in the research of association rule mining. In experiment, we use this data generator to generate three different experiment databases which have the different number of transactions and items, different average length of transactions.

Table 2 and Figure 1 give the runtime comparison results under different number of transactions for three algorithms. In experiment, the number of items is 40, the average length of transactions is 24, and the support is 30%. The efficiency of the algorithms is measured by the algorithm's running time (seconds).

Table 2. The results under different number of transactions' database

The number of transaction	Appriori algorithm	CM_Appriori_1 algorithm	NCMA algorithm
500	350	6.301	3.245
1000	680	10.686	4.699
1500	1090	16.248	7.591
2000	1420	21.612	10.622
2500	1720	24.32	11.435
3000	2110	28.553	14.424
3500	2460	35.454	16.39

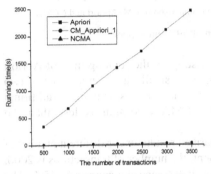   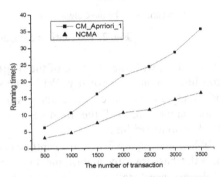

(a) Comparison of the three algorithms   (b) Comparison of CM_Apriori and NCMA

Fig. 1. The runtime comparison of different number of transactions

As can be seen from Figure 1, the running time of the traditional Apriori algorithm is increased with the increase of the number of transactions, and its running time is much larger than the other two Apriori algorithms based on compression matrix. Compared with CM_Apriori algorithm, the running time of NCMA algorithm is smaller, which shows that the NCMA algorithm has a good performance.

Table 3 and Figure 2 show the runtime comparison results of different support under the same database for three algorithms. In experiment, the number of transactions is 40, the average length of transaction is 24, and the support is 30%. The efficiency of the algorithms is measured by the algorithm's running time (seconds).

Table 3. The results under different supports from the same database

Support	Appriori algorithm	CM_Appriori_1 algorithm	NCMA algorithm
15	1177	33.495	15.301
20	415	11.969	5.837
25	176.4	5.034	2.387
30	82.6	2.392	1.28
35	42	1.343	0.78
40	22	0.774	0.468
45	12	0.482	0.359

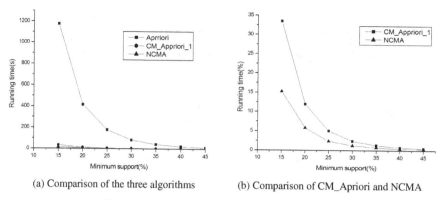

(a) Comparison of the three algorithms   (b) Comparison of CM_Apriori and NCMA

Fig. 2. The runtime comparison of different supports

Obviously, with the increase of the degree of support, the running time of Apriori algorithm is reduced gradually. When the support is small, the running time of two Apriori algorithms based on compression matrix is far less than the traditional Apriori algorithm, and the running time of NCMA algorithm is less than the CM_Apriori algorithm.

Table 4 and Figure 3 give the runtime comparison results under different number of item's database for three algorithms. In experiment, the number of transactions is 2000, and the average length proportion of transaction is 0.6, and the support is 45%. The efficiency of the algorithms is measured by the algorithm's running time (seconds).

Table 4. The results under different number of items' databases

The number of transaction	Appriori algorithm	CM_Appriori_1 algorithm	NCMA algorithm
20	1.655	0.189	0.156
30	13.873	0.492	0.311
40	126.403	2.308	1.186
50	709.824	8.349	3.934
60	4555.225	51.841	22.294

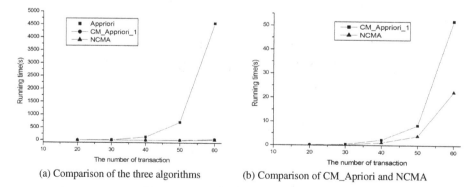

(a) Comparison of the three algorithms   (b) Comparison of CM_Apriori and NCMA

Fig. 3. The runtime comparison of different number of item's databases

As can be seen from the Figure 3, with the increase of item numbers, the running time of the traditional Apriori algorithm is exponential growth, and its running time is much larger than the other two Apriori algorithms based on compression matrix. With the increase of item numbers, the running time of NCMA algorithm is less than CM_ Apriori algorithm, which shows that NCMA algorithm has good operating efficiency.

Table 5 and Figure 4 show the runtime comparison results under different density's database for three algorithms. In experiment, the number of transactions is 1000, the number of items is 50, and the support is 45%. The efficiency of the algorithms is measured by the algorithm's running time (seconds).

Table 5. The results under different density's database

The average number of transaction	Appriori algorithm	CM_Appriori_1 algorithm	NCMA algorithm
10	0.483	0.111	0.093
15	1.717	0.153	0.124
20	15.447	0.348	0.219
25	169.889	2.969	1.22
30	4264.339	70.428	39.137

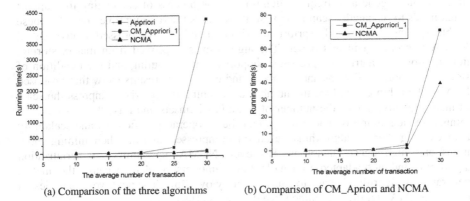

(a) Comparison of the three algorithms   (b) Comparison of CM_Apriori and NCMA

Fig. 4. The runtime comparison of different database density

From the Figure 4, with the increase of database density, the running time of traditional Apriori algorithm is increased exponentially. When the database density is more than 0.5, the running time of Apriori is much larger than CM_Apriori algorithm and NCMA algorithm. The running time of NCMA algorithm is smaller than CM_Apriori algorithm.

According to above experimental results, we can conclude the analysis conclusion.

(1) When dealing with high density and high correlation data, because the Appriori algorithm needs to scan database many times and will generates huge candidate itemsets, its running time is higher than the Apriori algorithm based on compression matrix. The NCMA algorithm only scans the database once, and the sorting of frequent 1-itemsets effectively reduce the number of candidate itemsets produced by the compression matrix. At the same time, the algorithm removes a large number of

useless transactions by using compression matrix, and can quickly compute support count through logic "and" operation.

(2) For more items and high intensive database, in the case of increased number of transactions and dropped support, the performance of NCMA algorithm is much better than that of CM_Apriori algorithm. Because the number of frequent itemsets mined in more items and high dense database is huge, the matrix will become bigger and bigger with the increasing of row number. In this case, the improvement steps of NCMA algorithm in compression matrix have more obvious advantage. Because the algorithm eliminates the itemsets that is unable to be connected when compressing matrix, the scanning matrix generated by compressing is smaller. In addition, because of changing the end condition of the algorithm, the cycle number of NCMA algorithm is less than or equal to CM_Apriori algorithm. As a result, the running time of NCMA algorithm is less than CM_Appriori algorithm, and its running efficiency is also better than CM_Appriori algorithm.

5   Conclusion

The existing Appriori algorithms based on matrix and its improved algorithms still have some problems: (1) the candidate itemsets is too large and matrix takes up too much memory space; (2) because of only compressing the transaction or itemset in the process of compression, compression matrix still stores a lot of elements that are independent of generated frequent itemset; (3) the process of calculating the support count needs to repeatedly scan matrix, so it increases the running time. To solve these problem, a new improved Apriori algorithm based on compressed matrix called NCMA is proposed in this paper. The algorithm is improved from matrix storing, itemsets sorting, matrix compressing, support count computing and the condition of stopping algorithm. The Instance analysis and experiment results show that proposed NCMA algorithm can reduce the number of scanning the matrix, compress the scale of matrix greatly, reduce the number of candidate itemsets and raise the efficiency of mining frequent itemsets. The algorithm has better operation efficiency and scalability than existing Apriori algorithms based on compressed matrix when mining more items and high dense databases. However, NCMA algorithm is a serialization algorithm, and not adapted to handle huge amounts of data because the matrix generated by algorithm is easy to cause memory overflow. The next work is to design parallel NCMA algorithm to mine large databases effectively.

References

1. Agrawal, R., Imielinaki, T., Swami, A.: Mining association rules between sets of items in large databases. In: Proceedings of the ACM SIGMOD Conference on Management of Date, pp. 207–216. ACM press, Washington, D.C (1993)
2. Han, J., Pei, J., Yin, Y.: Mining Frequent Patterns without Candidate Generations. In: Proceedings of the 2000 ACM SIGMOD International Conference on Management of Data, pp. 1–12. ACM Press, Dallas (2000)
3. Park, J.S., Chen, M.S., Yu, P.S.: An effective hash-based algorithms for mining association rules. In: Proceedings of the 1995 ACM SIGMOD International Conference on Management of Data, pp. 175–186. ACM Press, San Jose (1995)
4. Savasere, A., Omiecinski, E., Navathe, S.: An Efficient Algorithm for Mining Association rules. In: Proceedings of the 21st International Conference on Very Large Database, pp. 432–444. ACM Press, New York (1995)

5. Tovionen, H.: Sampling large databases for association rules. In: 22th International Conference on Very Large Database, Bombay, India, pp. 1–12 (1996)
6. Brin, S., Motwan, R.I., Ullman, J.D., et al.: Dynamic Itemset Counting and Implication Rules for Market Basket Data. In: 1997 ACM SIGMOD Conference on Management of Data, pp. 255–264. ACM Press, New York (1997)
7. Zaki, M.J.: Scalable Algorithms for Association Mining. IEEE Transactions on Knowledge and Data Engineering 12(3), 372–390 (2000)
8. Huang, L.S., Chen, H.P., Wang, X., et al.: A Fast Algorithm for Mining Association Rules. Journal of Computer Science and Technology 15(6), 619–624 (2000)
9. Yuan, Y., Huang, T.: A Matrix Algorithm for Mining Association Rules. In: Huang, D.-S., Zhang, X.-P., Huang, G.-B. (eds.) ICIC 2005. LNCS, vol. 3644, pp. 370–379. Springer, Heidelberg (2005)
10. Zhang, Z.L., Liu, J., Zhang, J.: A Fast Algorithm for Mining association Rules Based on Boolean Matrix. In: 2008 International Conference on Wireless Communications, Networking and Mobile Computing, Dalian, China, pp. 1–3 (2008)
11. Wand, F., Li, Y.H.: An improved Apriori algorithm based on the matix. In: 2008 International Seminar on Future BioMedical Information Engineering, Wuhan, China, pp. 152–155 (2008)
12. Liu, H.Z., Dai, S.P., Jiang, H.: Quantitative association rules mining algorithm based on matrix. In: 2009 International Conference on Computational Intelligence and Software Engineering, pp. 1–4. IEEE Computer Society, Wuhan (2009)
13. Khare, N., Adlakha, N., Pardasani, K.R.: An Algorithm for Mining Multidimensional Association Rules using Boolean Matrix. In: 2010 International Conference on Recent Trends in Information, Telecommunication and Computing, pp. 95–99. IEEE Computer Society, Kochi (2010)
14. Luo, X.W., Wang, W.Q.: Improved Algorithms Research for Association Rule Based on Matrix. In: 2010 International Conference on Intelligent Computing and Cognitive Informatics, pp. 415–419. IEEE Computer Society, Kuala Lumpur (2010)
15. Krajca, P., Outrata, J., Vychodil, V.: Using frequent closed itemsets for data dimensionality reduction. In: 11th IEEE Internatinal Conference on Data Mining, pp. 1128–1133. Institute of Electrical and Electronics Engineers Inc, Vancouver (2011)
16. Chen, J.M., Lin, G.F., Yang, Z.H.: Extracting spatial association rules from the maximum frequent itemsets based on Boolean matrix. In: 19th International Conference on Geoinformatics, pp. 1–5. IEEE Computer Society, Shanghai (2011)
17. Zhou, Z.P., Wang, J.F.: An improved matrix sorting index association rule data mining algorithm. In: 33rd Chinese Control Conference, pp. 500–505. IEEE Computer Society, Nanjing (2014)
18. Yang, X.Y., Zhen, L., Fu, Y.: MapReduce as a programming model for association rules algorithm on Hadoop. In: 3rd International Conference on Information Sciences and Interaction Sciences, pp. 99–102. IEEE Computer Society, Chengdu (2010)
19. Wang, Z.Q., Li, H.L.: Research of Massive Web Log Data Mining Based on Cloud Computing. In: Fifth International Conference on Computational and Information Sciences, pp. 591–594. IEEE Computer Society, Shiyang (2013)
20. Mannila, H., Toivonen, H., Verkamo, A.I.: Efficient algorithms for discovering association rules. In: AAAI Workshop on Knowledge Discovery in Databases (KDD 1994), pp. 181–192 (1994)
21. Liu, B.Z.: Improved appriori mining frequent items algorithm. Application Research of computation 29(2), 475–477 (2012) (in Chinese)
22. Li, R., Kang, L.Y., Geng, H.: New optimization association rule algorithm based on array. Science Technology and Engineering 8(21), 5846–5849 (2008) (in Chinese)
23. Fu, S., Liao, M.H., Song, D.: An improved Appriori algorithm based on compression matrix approach. Microelectronics and Computer 29(6), 28–32 (2010) (in Chinese)

FHN: Efficient Mining of High-Utility Itemsets with Negative Unit Profits

Philippe Fournier-Viger

Dept. of Computer Science, University of Moncton, Canada
philippe.fournier-viger@umoncton.ca

Abstract. High utility itemset (HUI) mining is a popular data mining task. It consists of discovering sets of items generating high profit in a transaction database. Several efficient algorithms have been proposed for this task. But few can handle items with negative unit profits despite that such items occurs in many real-life transaction databases. Mining HUIs in a database where items have positive and negative unit profits is a very computationally expensive task. To address this issue, we present an efficient algorithm named FHN (Faster High-Utility itemset miner with Negative unit profits). FHN discovers HUIs without generating candidates and introduces several strategies to handle items with negative unit profits efficiently. Experimental results with six real-life datasets shows that FHN is up to 500 times faster and can use up to 250 times less memory than the state-of-the-art algorithm HUINIV-Mine.

Keywords: frequent pattern mining, high-utility itemset mining, negative unit profit, negative weights, transaction database.

1 Introduction

Frequent Itemset Mining (FIM) [2] is a popular data mining task that is essential to a wide range of applications. Given a transaction database, FIM consists of discovering frequent itemsets. i.e. groups of items (itemsets) appearing frequently in transactions [2]. However, an important limitation of FIM is that it assumes that each item cannot appear more than once in each transaction and that all items have the same importance (weight, unit profit or value). These assumptions often do not hold in real applications. For example, consider a database of customer transactions containing information about the quantities of items in each transaction and the unit profit of each item. FIM algorithms would discard this information and may thus discover many frequent itemsets generating a low profit and fail to discover less frequent itemsets that generate a high profit. To address this issue, the problem of FIM has been redefined as *High-Utility Itemset Mining* (HUIM) to consider the case where items can appear more than once in each transaction and where each item has a weight (e.g. unit profit). The goal of HUIM is to discover high utility itemsets (HUIs), i.e. itemsets generating a high profit. HUIM has a wide range of applications such as website click stream analysis, cross-marketing in retail stores and biomedical applications [3,9,12]. HUIM

X. Luo, J.X. Yu, and Z. Li (Eds.): ADMA 2014, LNAI 8933, pp. 16–29, 2014.

has also inspired several important data mining tasks such as high-utility sequential pattern mining [15,16], high-utility episode mining [14] and high-utility stream mining [11].

The problem of HUIM is widely recognized as more difficult than the problem of FIM. In FIM, the *downward-closure property* states that the support of an itemset is anti-monotonic, that is the supersets of an infrequent itemset are infrequent and subsets of a frequent itemset are frequent. This property is very powerful to prune the search space. In HUIM, the utility of an itemset is neither monotonic or anti-monotonic, that is a high utility itemset may have a superset or subset with lower, equal or higher utility [2]. Thus, techniques to prune the search space developed in FIM cannot be directly applied in HUIM.

Many studies have been carried to develop efficient HUIM algorithms [3,8,4,9,10,12,13]. However, these algorithms are not designed to handle items having negative weights/unit profits, despite that such items occur in many real-life transaction databases. For example, it is common that retail stores sell items at a loss (e.g. printers) to stimulate the sale of other related items (e.g. proprietary printer cartridges). It was demonstrated that if classical HUIM algorithms are applied on databases containing items with negative unit profits, they can generate an incomplete set of HUIs [1]. The reason is that these algorithms over-estimate the utility of itemsets to prune the search space. But, when items with negative unit profits are considered, these estimations may become underestimations, and thus HUIs may be pruned. The state-of-the-art algorithm for mining HUIs while considering negative unit profits is HUINIV-Mine [1]. However, mining HUIs with negative unit profits remains very costly in terms of execution time and memory. Therefore, an important challenge is to design a more efficient algorithm for this task.

In this paper, we address this challenge. We present a novel algorithm named FHN (Fast High-utility itemset miner with Negative unit profits) to mine HUIs while considering both positive and negative unit profits. It extends the current fastest HUI mining algorithm named FHM [4] so that it can handle negative unit profits efficiently. We compare the performance of FHN and HUINIV-Mine on six real-life datasets. Results show that FHN is up to 500 times faster than HUINIV-Mine and consumes up to 250 times less memory. The rest of this paper is organized as follows. Section 2, 3, 4 and 5 respectively presents the problem definition and related work, the FHN algorithm, the experimental evaluation and the conclusion.

2   Problem Definition and Related Work

We first introduce important preliminary definitions.

Definition 1 (transaction database). Let I be a set of items (symbols). A *transaction database* is a set of transactions $D = \{T_1, T_2, ..., T_n\}$ such that for each transaction T_c, $T_c \in I$ and T_c has a unique identifier c called its Tid. Each item $i \in I$ is associated with a positive or negative number $p(i)$, called its external utility (e.g. unit profit). For each transaction T_c such that $i \in T_c$, a positive number $q(i, T_c)$ is called the internal utility of i (e.g. purchase quantity).

Example 1. Consider the database of Fig. 1 (left), which will be used as our running example. This database contains five transactions ($T_1, T_2...T_5$). Transaction T_2 indicates that items a, c, e and g appear in this transaction with an internal utility of respectively 2, 6, 2 and 5. Fig. 1 (right) indicates that the external utility of these items are respectively -5, 1, 3 and 1. Thus, item a is sold at loss.

Definition 2 (utility of an item/itemset in a transaction). The utility of an item i in a transaction T_c is denoted as $u(i, T_c)$ and defined as $p(i) \times q(i, T_c)$. The utility of an itemset X (a group of items $X \subseteq I$) in a transaction T_c is denoted as $u(X, T_c)$ and defined as $u(X, T_c) = \sum_{i \in X} u(i, T_c)$.

TID	Transactions
T_1	(a,1)(c,1)(d,1)
T_2	(a,2)(c,6)(e,2)(g,5)
T_3	(a,1)(b,2)(c,1)(d,6),(e,1),(f,5)
T_4	(b,4)(c,3)(d,3)(e,1)
T_5	(b,2)(c,2)(e,1)(g,2)

Item	a	b	c	d	e	f	g
Profit	-5	2	1	2	3	1	1

Fig. 1. A transaction database (left) and external utility values (right)

Example 2. The utility of item e in T_2 is $u(e, T_2) = 3 \times 2 = 6$. The utility of the itemset $\{c, e\}$ in T_2 is $u(\{c, e\}, T_2) = u(c, T_2) + u(e, T_2) = 1 \times 6 + 3 \times 2 = 12$.

Definition 3 (utility of an itemset in a database). The utility of an itemset X is denoted as $u(X)$ and defined as $u(X) = \sum_{T_c \in g(X)} u(X, T_c)$, where $g(X)$ is the set of transactions containing X.

Example 3. The utility of the itemset $\{c, e\}$ is $u(\{c, e\}) = (u(c, T_2) + u(e, T_2)) + (u(c, T_3) + u(e, T_3)) + (u(c, T_4) + u(e, T_4)) + (u(c, T_5) + u(e, T_5)) = (6+6) + (1+3) + (3+3) + (2+3) = 27$. The utility of the itemset $\{a, d, f\}$ is $u(\{a, d, f\}) = (u(a, T_3) + u(d, T_3)) + u(f, T_3) = -5 + 12 + 5 = 12$.

Definition 4 (problem of HUI mining with/without negative unit profits). Let *minutil* be a threshold set by the user. An itemset X is a *high-utility itemset* if $u(X) \geq minutil$. Otherwise, X is a *low-utility itemset*. The *problem of high-utility itemset mining* is to discover all high-utility itemsets in a database where external utility values are positive. The *problem of high-utility itemset mining with negative unit profits* is to discover all high-utility itemsets in a database where external utility values may be positive or negative.

Example 4. If *minutil* = 20, twenty HUIs should be found in the database. They are $\{a, b, c, d, e, f\}$:20, $\{b, d, f\}$:21, $\{b, d, e, f\}$:24, $\{b, c, d, e, f\}$:25, $\{b, c, d, f\}$:22, $\{d, e, f\}$:20, $\{c, d, e, f\}$:21, $\{c, e, g\}$:24, $\{d\}$:20, $\{b, d\}$:30, $\{b, d, e\}$:36, $\{b, c, d, e\}$:40, $\{b, c, d\}$:34, $\{d, e\}$:24, $\{c, d, e\}$:28, $\{c, d\}$:25, $\{b, e\}$:25, $\{b, c, e\}$:31, $\{b, c\}$:22 and $\{c, e\}$:27, where the number following each itemset is its utility.

Two known properties of HUIs with respect to items having negative unit profits are the following.

Property 1 (a HUI may contain items having negative external utilities). It can be clearly seen from the example that a HUI may contain items having a negative external utility value. For example, $\{a, b, c, d, e, f\}$ contains a, which has an external utility of -5.

Property 2 (a HUI must contain at least an item having a positive external utility [1]). Although a HUI may or may not contain items having negative external utility values, a HUI need to contain at least an item having a positive external utility value (otherwise its utility would be negative and it would not be a HUI).

It can be demonstrated that the utility measure is not monotonic or anti-monotonic. In other words, an itemset may have a utility lower, equal or higher than the utility of its subsets. Therefore, the strategies that are used in FIM to prune the search space based on the anti-monotonicity of the support cannot be directly applied to discover high-utility itemsets. Several HUIM algorithms circumvent this problem by overestimating the utility of itemsets using a measure called the Transaction-Weighted Utilization (TWU) [3,10,12], which is anti-monotonic. The TWU measure assumes that all items have positive external utility values. The TWU measure is defined as follows.

Definition 5 (transaction utility). The *transaction utility* (TU) of a transaction T_c is the sum of the utility of the items from T_c in T_c. i.e. $TU(T_c) = \sum_{x \in T_c} u(x, T_c)$.

Definition 6 (transaction weighted utilization). The *transaction-weighted utilization* (TWU) of an itemset X is defined as the sum of the transaction utility of transactions containing X, i.e. $TWU(X) = \sum_{T_c \in g(X)} TU(T_c)$.

Example 5. Consider the database of the running example and that the external utility value of item a is 5 rather than -5 ($p(a) = 5$). The TU of transactions T_1, T_2, T_3, T_4 and T_5 are respectively 8, 27, 30, 20 and 11. The TWU of items a, b, c, d, e, f and g are 65, 61, 96, 58, 88, 30 and 38. Consider item b. $TWU(\{b\}) = TU(T_3) + TU(T_4) + TU(T_5) = 30 + 20 + 11 = 61$.

The TWU measure has three important properties that are used to prune the search space. These properties only hold if external utility values of items are positive [1].

Property 3 (overestimation). The TWU of an itemset X is higher than or equal to its utility, i.e. $TWU(X) \geq u(X)$ [10].

Property 4 (antimonotonicity). The TWU measure is anti-monotonic. Let X and Y be two itemsets. If $X \subset Y$, then $TWU(X) \geq TWU(Y)$ [10].

Property 5 (pruning). Let X be an itemset. If $TWU(X) < minutil$, then the itemset X is a low-utility itemset as well as all its supersets [10].

Most algorithms for high utility mining that only handle positive external utility values (e.g. Two-Phase [10], IHUP [3] and UPGrowth [12]) utilizes Property 5 to prune the search space. They operate in two phases. In Phase 1, they identify candidate high-utility itemsets by calculating their TWU. In Phase 2, they scan the database to calculate the exact utility of all candidates found in Phase 1 to eliminate low-utility itemsets. However, if such algorithms are applied on databases containing negative unit profits, some HUIs may not be output. For example, consider the running example. If item a has an external utility of -5 rather than 5 as we previously considered, $TWU(\{c, e, g\}) = TWU(T_2) = 7$ and $u(\{c, e, g\}) = 17$, which would be a violation of Property 5 stating that the TWU of an itemset is an overestimation of its utility. The consequence is that if $minutil = 10$, the itemset $\{c, e, g\}$ would not be output by algorithms relying on that property even though this itemset is a HUI (it would be pruned because $TWU(\{c, e, g\}) < minutil$).

To mine high utility itemsets while considering both positive and negative unit profits and output the full set of HUIs, the state-of-the-art algorithm is HUINIV-Mine [1]. It is an extension of the Two-Phase algorithm [10]. To avoid the aforementioned problem, HUINIV-Mine redefines the notion of transaction utility as follows (and thus the TWU measure).

Definition 7 (redefined transaction utility). The *redefined transaction utility* of a transaction T_c is the sum of the utility of the items from T_c having positive external utilities. i.e. $TU(T_c) = \sum_{x \in T_c \wedge p(x) > 0} u(x, T_c)$.

Example 6. Fig. 2 (left) shows the redefined TU of transactions T_1, T_2, T_3, T_4, T_5 for the running example. Fig. 2 (right) shows the TWU of single items based on the redefined transaction utility values. Consider itemsets $\{c, e, g\}$ and $\{e, g\}$. The values $TWU(\{c, e, g\})$ and $TWU(\{e, g\})$ are equal to 17, which are overestimations of $u(\{c, e, g\}) = 17$ and $u(\{e, g\}) = 11$.

Using the redefined transaction utility restores Property 5. This is what allows HUINIV-Mine to find the complete set of HUIs. However, a major problem is that the task of mining HUIs while considering both positive and negative unit profits remain computationally very expensive both in terms of execution time and memory, especially for datasets containing dense or long transactions. It is thus a challenge to build more efficient algorithms.

To address this issue, in this paper, we propose an algorithm named FHN that is a variation of the FHM algorithm [4]. FHM is a recently proposed algorithm for HUI mining, which is designed to handle only positive external utility values. FHM provides the benefit of mining HUIs in a single phase, thus avoiding the candidate generation step of other HUI mining algorithms such as Two-Phase, UPGrowth and IHUP. FHM utilizes the depth-first search procedure and *utility-list* structure recently introduced in HUI-Miner [9] to explore the search space of itemsets, but also provides an efficient optimization named EUCP (Estimated Utility Co-occurrence Pruning) that makes FHM up to 6 times faster than HUI-Miner. FHM associates a *utility-list* [9] to each pattern. Utility-lists allow calculating the utility of a pattern quickly by making join operations with utility-lists of smaller patterns. Utility-lists are defined as follows.

Definition 8 (Utility-list). Let $\succ$ be any total order on items from I. The *utility-list* of an itemset X in a database D is a set of tuples such that there is a tuple $(tid, iutil, rutil)$ for each transaction T_{tid} containing X. The *iutil* element of a tuple is the utility of X in T_{tid}. i.e., $u(X, T_{tid})$. The *rutil* element of a tuple is defined as $\sum_{i \in T_{tid} \wedge i \succ x \forall x \in X} u(i, T_{tid})$.

Example 7. Assume that $\succ$ is the alphabetical order and that the external utility of item a is 5 rather than -5. The utility-list of $\{a\}$ is $\{(T_1, 5, 3), (T_2, 10, 17), (T_3, 5, 25)\}$. The utility-list of $\{d\}$ is $\{(T_1, 2, 0), (T_3, 12, 8), (T_4, 6, 3)\}$. The utility-list of $\{a, d\}$ is $\{(T_1, 7, 0), (T_3, 17, 8)\}$.

FHM discovers HUIs by performing a single database scan to create utility-lists of patterns containing single items. Then, longer patterns are obtained by performing the join operation of utility-lists of shorter patterns. The join operation for single items is performed as follows. Consider two items x, y such that $x \succ y$, and their utility-lists $ul(\{x\})$ and $ul(\{y\})$. The utility-list of $\{x, y\}$ is obtained by creating a tuple $(ex.tid, ex.iutil + ey.iutil, ey.rutil)$ for each pairs of tuples $ex \in ul(\{x\})$ and $ey \in ul(\{y\})$ such that $ex.tid = ey.tid$. The join operation for two itemsets $P \cup \{x\}$ and $P \cup \{y\}$ such that $x \succ y$ is performed as follows. Let $ul(P)$, $ul(\{x\})$ and $ul(\{y\})$ be the utility-lists of P, $\{x\}$ and $\{y\}$. The utility-list of $P \cup \{x, y\}$ is obtained by creating a tuple $(ex.tid, ex.iutil + ey.iutil - ep.iutil, ey.rutil)$ for each set of tuples $ex \in ul(\{x\})$, $ey \in ul(\{y\})$, $ep \in ul(P)$ such that $ex.tid = ey.tid = ep.tid$. Calculating the utility of an itemset using its utility-list and pruning the search space is done as follows.

Property 6 (Calculating utility of an itemset using its utility-list). The utility of an itemset is the sum of *iutil* values in its utility-list [9].

Property 7 (Pruning search space using utility-lists). Let X be an itemset. Let the *extensions* of X be the itemsets that can be obtained by appending an item y to X such that $y \succ i$, $\forall i \in X$. If the sum of *iutil* and *rutil* values in $ul(X)$ is less than *minutil*, then X and its extensions are low utility [9].

Before presenting our algorithm, we demonstrate with an example that the pruning property used by FHM and HUI-Miner is invalid if both negative and positive external utility values appears in a database. Consider that item a and d in the running example have respectively external utility values of 5 and -2, and that $minutil = 10$. The utility-list of $\{a, b\}$ would thus contains a single element, which is $\{(T_3, 9, -3)$. According to Property 7, because the sum of *iutil* and *rutil* values in $ul(\{a, b\})$ is 6, which is less than *minutil*, $\{a, b\}$ and its extensions are low utility. However, this is not the case since itemset $u(\{a, b, f\}) = 14$. Because of this, FHM and HUI-Miner would not find this HUI.

TID	TU
T_1	3
T_2	17
T_3	25
T_4	20
T_5	11

Item	TWU
a	45
b	56
c	77
d	48
e	74
f	25
g	28

Item	a	b	c	d	e	f
b	25					
c	45	56				
d	28	45	48			
e	42	56	74	45		
f	25	25	25	25	25	
g	17	11	28	0	28	0

Fig. 2. Transaction utilities (left), TWU values of single items (center) and EUCS (right)

3   The FHN Algorithm

In this section, we present our proposal, the FHN algorithm. We first describe the main procedure, which is inspired by the FHM [4] algorithm. This procedure can only handle positive external utility values. We then explain how it is adapted to handle negative unit profits without missing any HUIs. We call this new algorithm the FHN algorithm.

3.1   Main Procedure

The main procedure (Algorithm 1) takes as input a transaction database with utility values and the *minutil* threshold. The algorithm first scans the database to calculate the TWU of each item. Then, the algorithm identifies the set I^* of all items having a TWU no less than *minutil* (other items are ignored since they cannot be part of a high-utility itemset by Property 3). The TWU values of items are then used to establish a total order $\succ$ on items, which is the order of ascending TWU values (as suggested in [9]). A second database scan is then performed. During this database scan, items in transactions are reordered according to the total order $\succ$, the utility-list of each item $i \in I^*$ is built and a structure named EUCS (Estimated Utility Co-Occurrence Structure) is built [4]. This latter structure stores the TWU of all pairs of items $\{a, b\}$ such that $u(\{a, b\}) \neq 0$. As suggested in FHM, the EUCS is implemented as a hashmap of hashmaps since in practice a limited number of pairs of items co-occurs in transactions (see [4] for more details). Building the EUCS is very fast (it is performed with a single database scan) and occupies a small amount of memory, bounded by $|I^*| \times |I^*|$, although in practice the size is much smaller because a limited number of pairs of items co-occurs in transactions. After the construction of the EUCS, the depth-first search exploration of itemsets starts by calling the recursive procedure *Search* with the empty itemset $\emptyset$, the set of single items I^*, *minutil* and the EUCS.

The *Search* procedure (Algorithm 2) takes as input (1) an itemset P, (2) extensions of P having the form Pz meaning that Pz was previously obtained by appending an item z to P, (3) *minutil* and (4) the EUCS. The search procedure operates as follows. For each extension Px of P, if the sum of the *iutil* values of the utility-list of Px is no less than *minutil*, then Px is a high-utility itemset

---

Algorithm 1. The FHN algorithm

---

input : D: a transaction database, $minutil$: a user-specified threshold
output: the set of high-utility itemsets

1 Scan D to calculate the TWU of single items;
2 $I^* \leftarrow$ each item i such that $TWU(i) \geq minutil$;
3 Let $\succ$ be the total order of TWU ascending values on I^*;
4 Scan D to built the utility-list of each item $i \in I^*$ and build the $EUCS$ structure;
5 Search $(\emptyset, I^*, minutil, EUCS)$;

---

and it is output (cf. Property 4). Then, if the sum of $iutil$ and $rutil$ values in the utility-list of Px are no less than $minutil$, it means that extensions of Px should be explored (cf. Property 7). This is performed by merging Px with all extensions Py of P such that $y \succ x$ and $TWU(\{x, y\}) \geq minutil$, to form extensions of the form Pxy containing $|Px| + 1$ items. The utility-list of Pxy is then constructed as in FHM by calling the *Construct* procedure (cf. Algorithm 3) to join the utility-lists of P, Px and Py. This latter procedure is the same as in FHM [4] and is thus not detailed here. Then, a recursive call to the *Search* procedure with Pxy is done to calculate its utility and explore its extension(s). Since the *Search* procedure starts from single items, it recursively explores the search space of itemsets by appending single items and it only prunes the search space based on Property 7. It can be easily seen based on Properties 6 and 7 that this procedure is correct and complete to discover all high-utility itemsets.

3.2   Modifying the Algorithm to Handle Negative Item Unit Profits

We next explain how the algorithm is modified to handle negative unit profits. Let the term "positive items" and "negative items" denote items respectively having positive and negative external utility values. To be able to transform the algorithm described in the previous subsection into an algorithm that outputs all HUIs when both negative and positive items are used, we first make a few novel and very important observations that were not done or used in HUINIV-Mine.

It is well-known in HUI mining that appending a positive item z to an itemset X will produce an itemset $X \cup \{z\}$ that may have a utility that is higher, equal or less than X [10]. However, what happens if a negative item z is appended to an itemset X? The following property holds:

Property 8 (downward closure of extensions with negative items). Let X be an itemset and z be a negative item such that $z \notin X$. It follows that $u(X \cup \{z\}) < u(X)$. **Rationale.** Since item z is appended to X, the resulting itemset $X \cup \{z\}$ can only appear in as much or less transactions as X. Thus, the utility provided by each item $e \in X$ to $u(X)$ can only be the same or less. Furthermore, appending z to X can only decrease the overall utility of X since z is by definition a negative item.

Algorithm 2. The *Search* procedure

---

input   : P: an itemset, *ExtensionsOfP*: a set of extensions of P, the *minutil* threshold, the *EUCS* structure

output: the set of high-utility itemsets

```
 1 foreach itemset Px ∈ ExtensionsOfP do
 2 │   if SUM(Px.utilitylist.iutils) ≥ minutil then
 3 │   │   output Px;
 4 │   end
 5 │   if SUM(Px.utilitylist.iutils)+SUM(Px.utilitylist.rutils) ≥ minutil then
 6 │   │   ExtensionsOfPx ← ∅;
 7 │   │   foreach itemset Py ∈ ExtensionsOfP such that y ≻ x do
 8 │   │   │   if TWU({x,y}) ≥ minutil) then
 9 │   │   │   │   Pxy ← Px ∪ Py;
10 │   │   │   │   Pxy.utilitylist ← Construct (P, Px, Py);
11 │   │   │   │   ExtensionsOfPx ← ExtensionsOfPx ∪ Pxy;
12 │   │   │   end
13 │   │   end
14 │   │   Search (Px, ExtensionsOfPx, minutil);
15 │   end
16 end
```

A second key observation is that the previous property can be generalized for successive extensions of an itemset with negative items. However, this generalization is only possible if we define the total order $\succ$ such that negative items always succeed all positive items. In other words, the property can be generalized only if the algorithm always appends positive items before appending negative items to extend itemsets. The generalized property is defined as follows and is very powerful for pruning the search space.

Property 9 (downward closure of transitive extensions of an itemset with negative items). Let X be an itemset. Transitive extensions of X can only have a utility lower than X if $\forall y \in X \ \forall e \in I \setminus X, e \succ y$ and e is a negative item. **Rationale.** For each negative item e that can extend X, extending X with e will result in an itemset with a utility lower than X by Property 8. Since only negative items can be added to X according to the total order $\succ$, the utility of transitive extensions of X can also only be lower than $u(X)$.

Property 10 (pruning condition for itemsets containing negative items based on the total order $\succ$). Let X be an itemset such that $u(X) <$ *minutil* and only negative items can be used to extend X based on the total order $\succ$. Therefore all transitive extensions of X with these items will be low utility and can be pruned. **Rationale.** This pruning condition directly follows from the previous property.

Based on the above properties, the FHN algorithm is obtained by making the following modifications. First, instead of calculating the original TWU, the

---

Algorithm 3. The Construct procedure

> **input** : P: an itemset, Px: the extension of P with an item x, Py: the
> extension of P with an item y
> **output**: the utility-list of Pxy

1 $UtilityListOfPxy \leftarrow \emptyset$;
2 **foreach** tuple $ex \in Px.utilitylist$ **do**
3   **if** $\exists ey \in Py.utilitylist$ and $ex.tid = exy.tid$ **then**
4   **if** $P.utilitylist \neq \emptyset$ **then**
5   Search element $e \in P.utilitylist$ such that $e.tid = ex.tid$.;
6              $exy \leftarrow (ex.tid, ex.iutil + ey.iutil - e.iutil, ey.rutil)$;
7   **end**
8   **else**
9              $exy \leftarrow (ex.tid, ex.iutil + ey.iutil, ey.rutil)$;
10   **end**
11          $UtilityListOfPxy \leftarrow UtilityListOfPxy \cup \{exy\}$;
12   **end**
13 **end**
14 **return** $UtilityListPxy$;

---

redefined TWU is used to avoid underestimating the utility of HUIs containing positive items (similarly to HUINIV-Mine presented in section 2). Fig. 2 shows the redefined transaction utility values (left), TWU of single items (center) and EUCS (right), when this modification is done. Second, utility-lists are redefined such that only utility values of positive items are included in $rutil$ values of utility-lists. The reason is that the algorithm can miss some HUIs if $rutil$ values of negative items are included in utility lists as we have demonstrated in the last paragraph of Section 2. Third, the $\succ$ total order is defined such that all negative items succeed positive items (as previously explained). Fourth, the TWU pruning condition $TWU(\{x, y\}) < minutil$ using the EUCS structure is only used for positive items.

We now discuss the correctness of these modifications for finding all HUIs when positive and negative items are used. This explanation can be broken down into two parts (1) the algorithm first extends an itemset by appending positive items and (2) the algorithm then appends negative items (based on $\succ$).

During the first part, FHN is correct since it behaves as a regular HUI mining algorithm for discovering HUIs containing only positive items. This is true because negative items are always appended after positive items (thus negative items are not considered when forming HUIs containing only positive items). Furthermore, the pruning condition that the sum of $rutil$ and $iutil$ values must be higher than $minutil$ remains correct since $rutil$ values of negative items are not considered in utility-lists when the algorithm is generating HUIs containing only positive items. The pruning condition that an extension Pxy should not be explored if $TWU(\{x, y\} < minutil$ also remains correct since the redefined TWU is used.

To show that the FHN algorithm is also correct in the second part (when negative items are appended), we need to show that the two pruning conditions that are used hold for extensions with negative items. This would ensure that no HUIs containing negative items are missed by the algorithm. The first pruning condition is that extensions of an itemset should not be explored if the sum of *iutil* and *rutil* values are less than *minutil*. It can be observed that in the second part, the sum of *rutil* values will be equal to zero since only positive items are considered in *rutil* values of utility-lists. Thus, the second term of the pruning condition is zero, and the pruning condition is simplified as follows. An extension of an itemset will be pruned if the sum of *iutil* values in its utility-list is less than *minutil*. This condition is correct since it is equivalent to Property 10. The second pruning condition is that an extension Pxy should not be explored if $TWU(\{x, y\}) < minutil$. This does not influence the correctness of the FHN algorithm since this condition is not used for negative items (it is only used for pairs of positive items x and y, as we have previously explained).

4   Experimental Study

We evaluated the performance of the proposed FHN algorithm. Experiments were performed on a computer with a third generation 64 bit Core i5 processor running Windows 7 and 5 GB of free RAM. We compared the performance of FHN with the state-of-the-art algorithm HUINIV-Mine for high-utility itemset mining with negative unit profit. All memory measurements were done using the Java API. Experiments were carried on six real-life datasets having varied characteristics.

- *mushroom* is a dense dataset with 120 distinct items, 8,124 transactions, and an average transaction length of 23 items.
- The *retail* dataset contains 88,162 transactions with 16,470 distinct items and an average transaction length of 10,30 items.
- *kosarak* is a dataset that contains 41,270 distinct items, 990,000 transactions, and transaction have an average length of 8.09 items.
- The *chess* dataset contains 3,196 transactions with 75 distinct items and an average transaction length of 35 items.
- The *psumb* dataset contains 49,046 transactions with 7,116 distinct items and an average transaction length of 74 items.
- The *accidents* dataset contains 340,183 transactions having an average length of 33.80 items, and 468 distinct items.

For all datasets, external utilities for items are generated between -1,000 and 1,000 by using a log-normal distribution and quantities of items are generated randomly between 1 and 5, similarly to the settings of [3,4,9,12]. The source code of all algorithms and datasets can be downloaded from the SPMF data mining library (http://www.philippe-fournier-viger.com/spmf/).

For each dataset, we ran the FHN and HUINIV-Mine algorithms, while decreasing the *minutil* threshold until the algorithms became too long to execute,

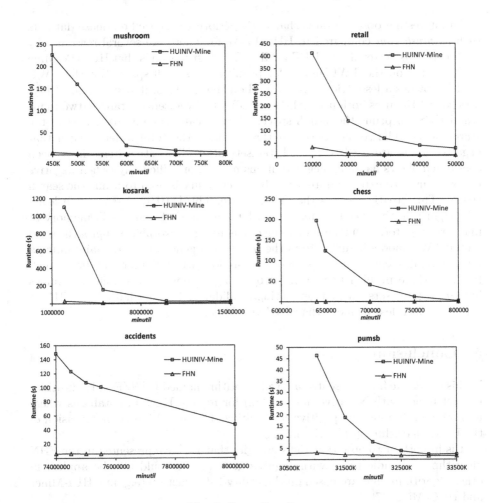

Fig. 3. Execution times

ran out of memory or a clear winner was observed. For each dataset, we recorded the execution time and the maximum memory usage. The comparison of execution times is shown in Fig. 3 for all datasets. For *mushroom, retail, kosarak, chess, psumb* and *accidents*, FHN was respectively up to 42 times faster, 18 times faster, 38 times faster, 500 times, 15 times and 25 times faster than HUINIV-Mine.

In terms of memory usage, FHN uses much less memory than HUINIV-Mine. For the *mushroom* dataset and *minutil* = 450000, HUINIV-Mine used up to 4.97 GB while FHN was using only up to 250 MB. On *kosarak, chess, psumb* and *accidents* HUINIV-Mine ran out of memory under our 5 GB memory limit while FHN was respectively using 20 MB, 1179 MB, 100 MB and 350 MB for the lowest *minutil* values. Lastly, for the *retail* dataset, the memory usage of FHN was about five times less than HUINIV-Mine. Overall, FHN used up to 250 times less memory than HUINIV-Mine.

An interesting observation is that FHN performs very well on dense datasets such as *mushroom* compared to HUINIV-Mine. There are several reasons why FHN performs better than HUINIV-Mine. The first reason is that HUINIV-Mine strictly relies on the TWU model for pruning the search space. But the TWU model provides a less strict upper bound on the utility of itemsets than utility-lists [4,9]. FHN uses both utility-lists and TWU of itemsets containing two items (the EUCS) to prune the search space. Thus, it can prune a larger part of the search space. The second reason is that defining the total order $\succ$ such that negative items are used to extend itemsets after all positive items have been considered allows handling negative items much more efficiently (when negative items are used to extend an itemset, the exact utility is used to prune the search space). This greatly reduces the search space. Third, HUINIV-Mine is a level-wise algorithm that need to maintain a large amount of itemsets in memory to find larger patterns. Furthermore, since it is using a two-phase approach based on the TWU model it suffers from the problem of generating and maintaining a huge amount of candidates in memory before low-utilty itemsets can be pruned. In FHN, these problems are avoided by using a depth-first search that mines high-utility itemsets without generating candidates. This is what allows FHN to consume much less memory than HUINIV-Mine.

5   Conclusion

In this paper, we have presented a novel algorithm named FHN (Fast High-utility itemset miner with Negative unit profits) for mining HUIs in databases where item unit profits may be positive or negative. The algorithm is an extension of the FHM algorithm [4] for HUI mining.

It is important to note that the strategies that we have presented in the FHN algorithm to handle items with negative unit profits could also be applied in other algorithms that are based on the utility-list structure (e.g. in GHUI-Miner and HUG-Miner [7]).

We have performed an extensive experimental study on six real-life datasets to compare the performance of FHN with the state-of-the-art algorithm HUINIV-Mine. Results show that FHN is up to 500 times faster and can use up to 250 times less memory than HUINIV-Mine, and was shown to perform very well on dense datasets. The source code of all algorithms and datasets used in our experiments can be downloaded as part of the SPMF data mining library http://www.philippe-fournier-viger/spmf/. For future work, we are interested in exploring other interesting problems involving utility mining in itemset mining and sequential pattern mining [5,6].

Acknowledgement. This work is financed by a National Science and Engineering Research Council (NSERC) of Canada research grant.

References

1. Chu, C.-J., Tseng, V.S., Liang, T.: An efficient algorithm for mining high utility itemsets with negative item values in large databases. Applied Math. Comput. 215, 767–778 (2009)
2. Agrawal, R., Srikant, R.: Fast algorithms for mining association rules in large databases. In: Proc. Int. Conf. Very Large Databases, pp. 487–499 (1994)
3. Ahmed, C.F., Tanbeer, S.K., Jeong, B.-S., Lee, Y.-K.: Efficient Tree Structures for High-utility Pattern Mining in Incremental Databases. IEEE Trans. Knowl. Data Eng. 21(12), 1708–1721 (2009)
4. Fournier-Viger, P., Wu, C.-W., Zida, S., Tseng, V.S.: FHM: Faster High-Utility Itemset Mining using Estimated Utility Co-occurrence Pruning. In: Andreasen, T., Christiansen, H., Cubero, J.-C., Raś, Z.W. (eds.) ISMIS 2014. LNCS, vol. 8502, pp. 83–92. Springer, Heidelberg (2014)
5. Fournier-Viger, P., Gomariz, A., Campos, M., Thomas, R.: Fast vertical mining of sequential patterns using co-occurrence information. In: Tseng, V.S., Ho, T.B., Zhou, Z.-H., Chen, A.L.P., Kao, H.-Y. (eds.) PAKDD 2014, Part I. LNCS, vol. 8443, pp. 40–52. Springer, Heidelberg (2014)
6. Fournier-Viger, P., Wu, C.-W., Gomariz, A., Tseng, V.S.: VMSP: Efficient Vertical Mining of Maximal Sequential Patterns. In: Sokolova, M., van Beek, P. (eds.) Canadian AI. LNCS, vol. 8436, pp. 83–94. Springer, Heidelberg (2014)
7. Fournier-Viger, P., Wu, C.-W., Tseng, V.S.: Novel Concise Representations of High Utility Itemsets using Generator Patterns. In: Luo, X., Yu, J.X., Li, Z. (eds.) ADMA 2014. LNCS, vol. 8933, pp. 30–43. Springer, Heidelberg (2014)
8. Li, Y.-C., Yeh, J.-S., Chang, C.-C.: Isolated items discarding strategy for discovering high utility itemsets. Data & Knowledge Engineering 64(1), 198–217 (2008)
9. Liu, M., Qu, J.: Mining High Utility Itemsets without Candidate Generation. In: Proceedings of CIKM 2012, pp. 55–64 (2012)
10. Liu, Y., Liao, W., Choudhary, A.: A two-phase algorithm for fast discovery of high utility itemsets. In: Ho, T.-B., Cheung, D., Liu, H. (eds.) PAKDD 2005. LNCS (LNAI), vol. 3518, pp. 689–695. Springer, Heidelberg (2005)
11. Shie, B.-E., Cheng, J.-H., Chuang, K.-T., Tseng, V.S.: A One-Phase Method for Mining High Utility Mobile Sequential Patterns in Mobile Commerce Environments. In: Jiang, H., Ding, W., Ali, M., Wu, X. (eds.) IEA/AIE 2012. LNCS, vol. 7345, pp. 616–626. Springer, Heidelberg (2012)
12. Tseng, V.S., Shie, B.-E., Wu, C.-W., Yu, P.S.: Efficient Algorithms for Mining High Utility Itemsets from Transactional Databases. IEEE Trans. Knowl. Data Eng. 25(8), 1772–1786 (2013)
13. Wu, C.-W., Fournier-Viger, P., Yu., P.S., Tseng, V.S.: Efficient Mining of a Concise and Lossless Representation of High Utility Itemsets. In: Proceedings of ICDM 2011, pp. 824–833 (2011)
14. Wu, C.-W., Lin, Y.-F., Yu, P.S., Tseng, V.S.: Mining High Utility Episodes in Complex Event Sequences. In: Proceedings of ACM SIG KDD 2013, pp. 536–544 (2013)
15. Yin, J., Zheng, Z., Cao, L.: USpan: An Efficient Algorithm for Mining High Utility Sequential Patterns. In: Proceedings of ACM SIG KDD 2012, pp. 660–668 (2012)
16. Yin, J., Zheng, Z., Cao, L., Song, Y., Wei, W.: Efficiently Mining Top-K High Utility Sequential Patterns. In: Proceedings of ICDM 2013, pp. 1259–1264 (2013)

Novel Concise Representations of High Utility Itemsets Using Generator Patterns

Philippe Fournier-Viger[1], Cheng-Wei Wu[2], and Vincent S. Tseng[2]

[1] Dept. of Computer Science, University of Moncton, Canada
[2] Dept. of Comp. Sci. and Info. Eng., National Cheng Kung University, Taiwan
philippe.fournier-viger@umoncton.ca, silvemoonfox@hotmail.com,
tseng@mail.ncku.edu.tw

Abstract. Mining *High Utility Itemsets* (*HUIs*) is an important task with many applications. However, the set of HUIs can be very large, which makes HUI mining algorithms suffer from long execution times and huge memory consumption. To address this issue, concise representations of HUIs have been proposed. However, no concise representation of HUIs has been proposed based on the concept of *generator* despite that it provides several benefits in many applications. In this paper, we incorporate the concept of generator into HUI mining and devise two new concise representations of HUIs, called *High Utility Generators* (*HUGs*) and *Generator of High Utility Itemsets* (*GHUIs*). Two efficient algorithms named *HUG-Miner* and *GHUI-Miner* are proposed to respectively mine these representations. Experiments on both real and synthetic datasets show that proposed algorithms are very efficient and that these representations are up to 36 times smaller than the set of all HUIs.

Keywords: pattern mining, high utility itemset mining, concise representation, high utility generator, generator of high utility itemsets.

1 Introduction

High Utility Itemset Mining (*HUIM*) [2,4,6,13] is an important research topic in data mining. As opposed to *Frequent itemset Mining* (*FIM*) [1], HUIM considers the importance of items (e.g. unit profit) and their quantities in transactions. Therefore, it can be used to discover itemsets having a high utility (e.g. high profit), that is *High Utility Itemsets* (*HUIs*). HUIM has a wide range of applications such as cross-marketing and click stream analysis and biomedical applications [2,4,6,13]. The problem of HUIM is widely recognized as more difficult than that of FIM. In FIM, the *downward-closure property* states that the support (frequency) of an itemset is *anti-monotonic* [1], that is the supersets of an infrequent itemset are infrequent. This property is very powerful to prune the search space. In HUIM, the utility of an itemset is neither monotonic or anti-monotonic. That is, a HUI may have a superset or subset with lower, equal or higher utility [2,4,6,13]. Thus techniques to prune the search space developed in FIM cannot be directly applied to HUIM.

X. Luo, J.X. Yu, and Z. Li (Eds.): ADMA 2014, LNAI 8933, pp. 30–43, 2014.

Many studies have been carried to develop efficient HUIM algorithms [2,4,6,7,13]. However, a crucial problem of HUIM is that the set of HUIs generated by these algorithms can be very large. This makes HUIM algorithms suffer from long execution times and even fail to run due to huge memory consumption or lack of storage space. Moreover, it is very inconvenient for a user to analyze a very large set of HUIs. To address this issue, it was proposed to mine concise representations of HUIs rather than all HUIs. *GUIDE* [9] is an approximate algorithm that integrates the concept of *maximal pattern* from FIM to mine *maximal HUIs* (HUIs having no proper supersets that are high utility). Another work is *CHUD* [14], which adapts the concept of *closed pattern* from FIM to discover *closed HUIs* (HUIs having no proper supersets that are HUIs and have the same support). Although these representations are useful, no work has been done yet on integrating the concept of *generator pattern* (or simply called *generators*) in HUIM despite that generators have shown to provides several benefits over all/closed/maximal patterns in many applications [5,8,10,11,12].

A generator is an itemset that has no proper subset having the same support. Generators provide the following benefits. First, when generators are combined with closed patterns, they give additional information that closed patterns alone cannot provide, for example to generate minimal rules between patterns with a minimal antecedent (generator) and a maximal consequent (closed pattern) [8]. Second, generators can provide higher classification accuracy and are more useful for model selection than using all or only closed patterns [8]. Third, generators are preferable according to the *Minimum Description Length* principle to closed/maximal patterns, since generators are the minimal members of equivalence classes rather than the maximal ones [5,8,11,12]. Lastly, mining generators is generally more efficient than discovering all patterns because it is generally a very small subset of all patterns [5].

Therefore, several interesting questions are raised: How to integrate the concept of generator in HUIM to define meaningful concise representations of HUIs? How much reduction can be achieved by these representations? Can efficient algorithms be developed to mine these representations? What are the pros and cons of these representations? To answer these questions, we investigate the properties of generators in the context of HUIM and devise two alternative concise representations of HUIs using generators, respectively called *High Utility Generators* (HUGs) and *Generator of High Utility Itemsets* (*GHUIs*). We propose two efficient algorithms named *HUG-Miner* and *GHUI-Miner* to respectively mine these representations, and we analyze their respective advantages in terms of speed and number of patterns found. Experimental results on both real and synthetic datasets show that the proposed algorithms are very efficient and that the mined representations are up to 36 times smaller than the set of HUIs.

The rest of this paper is organized as follows. Section 2, 3, 4 and 5 respectively presents the problem definition and related work, the proposed algorithms, the experimental evaluation and the conclusion.

TID	Transactions
T_0	(a,1), (b,5), (c,1), (d,3), (e,1)
T_1	(b,4), (c,3), (d,3), (e,1)
T_2	(a,1), (c,1), (d,1)
T_3	(a,2), (c,6), (e,2)
T_4	(b,2),(c,2),(e,1)

Item	a	b	c	d	e
Profit	5	2	1	2	3

Fig. 1. A transaction database (left) and external utility values (right)

2   Problem Definition and Related Work

2.1   High Utility Itemset Mining

Let I be a set of items. A *transaction database* is a set of transactions $D = \{T_0, T_1, ..., T_n\}$ such that for each transaction T_c, $T_c \in I$ and T_c has a unique identifier c called its *Tid*. Each item $i \in I$ is associated with a positive number $p(i)$, called its *external utility* (e.g. unit profit). For each transaction T_c such that $i \in T_c$, a positive number $q(i, T_c)$ is called the *internal utility* of i (e.g. purchase quantity). For example, Fig. 1 show a transaction database containing five transactions $(T_0, T_1...T_4)$. External utilities of items a, c, and e in T_2 are $q(a, T_2) = 2$, $q(c, T_2) = 6$ and $q(e, T_2) = 2$. Fig. 1 (right) indicates that the external utility of a, b, e are respectively $p(a) = 5$, $p(c) = 1$ and $p(e) = 3$.

Definition 1 (Utility of an itemset). The *utility of an item i* in a transaction T_c is denoted as $u(i, T_c)$ and defined as $p(i) \times q(i, T_c)$. An *itemset* is a set of items. The *utility of an itemset X* in a transaction T_c is defined as $u(X, T_c) = \sum_{i \in X} u(i, T_c)$. The set of transactions containing X is denoted as $g(X)$. The *utility of X in a database* is defined as $u(X) = \sum_{T_c \in g(X)} u(X, T_c)$.

Example 1. The utility of the itemset $\{a, e\}$ in the transaction T_0 is $u(\{a, e\}, T_0) = u(\{a\}, T_0) + u(\{e\}, T_0) = 1 \times 5 + 1 \times 3 = 8$. The utility of $\{a, e\}$ is $u(\{a, e\}, T_0) + u(\{a, e\}, T_3) = 8 + 16 = 24$.

Definition 2 (Problem of HUI mining). An itemset X is a *high utility itemset* if its utility is no less than a user-specified *minimum utility threshold minutil* given by the user. Otherwise, X is a *low utility itemset*. The *problem of high utility itemset mining* is to discover all high utility itemsets in the database.

Example 2. If $minutil = 25$, the complete set of HUIs is $\{a, c\} : 28$, $\{a, c, e\} : 31$, $\{a, b, c, d, e\} : 25$, $\{b, c\} : 28$, $\{b, c, d\} : 34$, $\{b, c, d, e\} : 40$, $\{b, c, e\} : 37$, $\{b, d\} : 30$, $\{b, d, e\} : 36$, $\{b, e\} : 31$ and $\{c, e\} : 27$, where each HUI is annotated with its utility.

HUIM is harder than FIM because the utility measure is not monotonic or anti-monotonic [2,7,13], i.e., an itemset may have a utility lower, equal or higher than the utility of its subsets. Thus, strategies used in FIM to prune the search space based on the anti-monotonicity of the support cannot be directly transfered to HUIM. Several HUIM algorithms circumvent this problem by overestimating the utility of itemsets using the *Transaction-Weighted Utilization (TWU)* measure [2,7,13], which is anti-monotonic, and defined as follows.

Definition 3 (Transaction weighted utilization). The *transaction utility* (TU) of a transaction T_c is the sum of the utility of all the items in T_c. i.e. $TU(T_c) = \sum_{x \in T_c} u(x, T_c)$. The *transaction-weighted utilization* (TWU) of an itemset X is defined as the sum of the transaction utility of transactions containing X, i.e. $TWU(X) = \sum_{T_c \in g(X)} TU(T_c)$.

Example 3. The TUs of T_0, T_1, T_2, T_3 and T_4 are respectively 25, 20, 8, 22 and 9. The TWU of single items a, b, c, d, e are respectively 55, 54, 84, 53 and 76. $TWU(\{c,d\}) = TU(T_0) + TU(T_1) + TU(T_2) = 25 + 20 + 8 = 53$.

Property 1 (Pruning search space using the TWU). Let X be an itemset, if $TWU(X) < minutil$, then X and its supersets are low utility. [7]

Algorithms such as *Two-Phase* [7], *IHUP* [2] and *UP-Growth* [13] utilizes the above property to prune the search space. They operate in two phases. In the first phase, they identify candidate high utility itemsets by calculating their TWUs. In the second phase, they scan the database to calculate the exact utility of all candidates found in the first phase to eliminate low utility itemsets. Recently, an alternative approach called *HUI-Miner*[6] was proposed to mine HUIs directly using a single phase. A faster algorithm named FHM was then proposed. FHM is to our knowledge the fastest algorithm for mining HUIs[4]. It utilizes the depth-first search procedure of HUI-Miner [6] to explore the search space of HUIs but introduces an additional optimization named EUCP [4] that makes FHM up to 6 tims faster than HUI-Miner. In FHM, each itemset is associated with a structure named *utility-list* [4,6]. Utility-lists allow calculating the utility of an itemset quickly by making join operations with utility-lists of shorter patterns. utility-lists a re defined as follows.

Definition 4 (Utility-list). Let $\succ$ be any total order on items from I. The *utility-list* of an itemset X in a database D is a set of tuples such that there is a tuple $(tid, iutil, rutil)$ for each transaction T_{tid} containing X. The *iutil* element of a tuple is the utility of X in T_{tid}. i.e., $u(X, T_{tid})$. The *rutil* element of a tuple is defined as $\sum_{i \in T_{tid} \wedge i \succ x \forall x \in X} u(i, T_{tid})$.

Example 4. Assume that $\succ$ is the alphabetical order. The utility-list of $\{a\}$ is $\{(T_0, 5, 20), (T_2, 5, 3), (T_3, 10, 12)\}$. The utility-list of $\{d\}$ is $\{(T_0, 6, 3), (T_1, 6, 3), (T_2, 2, 0)\}$. The utility-list of $\{a, d\}$ is $\{(T_0, 11, 3), (T_2, 7, 0)\}$.

To discover HUIs, FHM performs a single database scan to create utility-lists of patterns containing single items. Then, longer patterns are obtained by performing the join operation of utility-lists of shorter patterns. The join operation for single items is performed as follows. Consider two items x, y such that $x \succ y$, and their utility-lists $ul(\{x\})$ and $ul(\{y\})$. The utility-list of $\{x, y\}$

is obtained by creating a tuple $(ex.tid, ex.iutil + ey.iutil, ey.rutil)$ for each pairs of tuples $ex \in ul(\{x\})$ and $ey \in ul(\{y\})$ such that $ex.tid = ey.tid$. The join operation for two itemsets $P \cup \{x\}$ and $P \cup \{y\}$ such that $x \succ y$ is performed as follows. Let $ul(P)$, $ul(\{x\})$ and $ul(\{y\})$ be the utility-lists of P, $\{x\}$ and $\{y\}$. The utility-list of $P \cup \{x, y\}$ is obtained by creating a tuple $(ex.tid, ex.iutil + ey.iutil - ep.iutil, ey.rutil)$ for each set of tuples $ex \in ul(\{x\})$, $ey \in ul(\{y\})$, $ep \in ul(P)$ such that $ex.tid = ey.tid = ep.tid$. Calculating the utility of an itemset using its utility-list and pruning the search space is done as follows.

Property 2 (Calculating utility of an itemset using its utility-list). The utility of an itemset is the sum of *iutil* values in its utility-list [6].

Property 3 (Pruning search space using utility-lists). Let X be an itemset. Let the *extensions* of X be the itemsets that can be obtained by appending an item y to X such that $y \succ i$, $\forall i \in X$. If the sum of *iutil* and *rutil* values in $ul(X)$ is less than *minutil*, X and its extensions are low utility [6].

FHM is very efficient. However, it can generate a huge amount of HUIs. This can make the algorithm run out of storage space and fail to terminate. Furthermore, it is very inconvenient for a user to analyze a large set of HUIs.

2.2   Concise Representations of High Utility Itemsets

To discover small and representative subsets of all HUIs, concise representations of HUIs such as *closed HUIs*[14] and *maximal HUIs*[9] have been proposed, which are defined as follows.

Definition 5 (Closed HUIs and maximal HUIs). The *support* of an itemset X in a database D is denoted as $sup(X)$ and defined as $|g(X)|$. A HUI X is a *closed HUI (CHUI)* [14] iff there exists no HUI Y such that $X \subset Y$ and $sup(X) = sup(Y)$. A HUI X is a *maximal HUI (MHUI)* [9] iff there exists no HUI Y, such that $X \subset Y$.

Although these representations are useful in some applications, to our knowledge, no work has been done on integrating the concept of generator pattern from FIM in HUIM, despite that generators provides several benefits over closed and maximal patterns [5,8,11,12]. In FIM, an itemset X is a *generator* (a.k.a *key pattern* or *minimal pattern*) iff there is no itemset Y such that $Y \subset X$ and $sup(X) = sup(Y)$ [5,10,11]. A generator pattern X is the generator of an itemset Y if $X \subseteq Y$ and $sup(X) = sup(Y)$. The concept of generator pattern is directly related to the concept of closed pattern [10,11]. An alternative and equivalent definition of generator patterns and closed patterns is the following. Let an *equivalence class* be the set of all itemsets supported by the same set of transactions. Generator patterns are the minimal members of each equivalence class, while closed patterns are the maximal members of each equivalence class. For example, consider the equivalence class $\{\{a, e\}, \{a, c, e\}\}$ of itemsets appearing in T_0 and T_2. {a,e} is a generator and {a,c,e} is the closed pattern. Each equivalence class contains one or more generators and a single closed pattern [11] called their *closure*. Thus, the *closure of an itemset* X is the unique closed pattern Y such that $sup(X) = sup(Y) \land X \subseteq Y$. Algorithms for mining generator patterns uses the following property to prune the search space [10,11].

Property 4 (Downward closure for generator patterns). An itemset X is not a generator pattern if there exists a strict subset of X that is not a generator [11].

2.3   Integrating the Concept of Generator in HUIM

To understand how the concept of generator can be applied to HUIM, consider the equivalence classes shown in Fig. 2 for the running example. Each equivalence class is represented as a Hasse diagram inside a rectangle and is labelled with the supporting transactions and support of its itemsets. For example, the equivalence class of T_0, T_1 and T_4 contains itemsets $\{b\}$, $\{b,c\}$, $\{b,e\}$ and $\{b,c,e\}$, which have a support of 3. In this figure, HUIs are represented as shapes with a solid line. The first important observation is that some equivalence classes (not shown in the figure) do not contain any HUIs. We name generators appearing in those equivalence classes *l-generators* (*LGs*). It is clear that LGs should not be mined. For example, $\{d\}$ should not be mined since its equivalence class do not contain any HUIs. The second important observation is that some equivalence classes contain at least one HUI. We name generators appearing in these equivalence classes *Generators of High Utility Itemsets* (*GHUIs*). It can be observed that some GHUIs are HUIs and other are not HUIs. We use the term *High Utility Generator* (*HUGs*) to refer to those that are HUIs and the term *Low Utility Generator* (*LUGs*) to refer to those that are not HUIs. LUGs, HUGs, GHUIs and CHUIs for the running example are illustrated in Fig. 2.

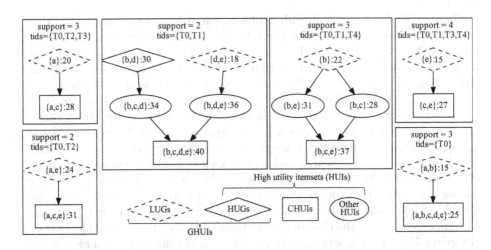

Fig. 2. HUIs and their equivalence classes (represented as Hasse diagrams)

Based on these definitions, it can be clearly seen that the set of all LUGs and the set of all HUGs are disjoint, and are subsets of the set of all GHUIs. Moreover, the set of all GHUIs is disjoint from the set of LGs. Furthermore, three important properties are:

Property 5 (Utility of itemsets in an equivalence class). Consider an equivalence class Q. Given that all itemsets in Q appear in the same set of transactions, it can be demonstrated that if $X \subset Y$ then $u(X) < u(Y)$, $\forall X, Y \in Q$.

Property 6 (utility of supersets of a HUG). In each equivalence class, supersets of a HUG are HUIs.

Property 7 (Highest utility in an equivalence class). In each equivalence class, the closure has the highest utility among all itemsets [14].

In the following, we are interested in mining the concise representations of all GHUIs and that of all HUGs. The set of GHUIs is meaningful following the *Minimum Description Length* principle since GHUIs describe each set of transactions (equivalence class) containing HUIs using minimal sets of items. However, as we will show mining LUGs is more expensive than mining HUGs because LUGs may have a very low utility. For this reason and because HUGs provides the interesting property that all supersets of a HUG are high utility, we also consider mining only HUGs.

3   The GHUI-Miner and HUG-Miner Algorithms

In this section, we first propose an algorithm named GHUI-Miner for mining all GHUIs. Then we explain how a variation named HUG-Miner is devised to mine only HUGs.

Mining GHUIs is a very difficult problem. For the reader, it may seems that GHUIs could be obtained by simply adding a generator checking mechanism used in FIM to a HUIM algorithm. However, doing so would result in an incomplete algorithm. The reason is that HUIM algorithms attempts to only explore HUIs. Thus, only HUGs would be found. To mine all GHUIs, it is thus necessary to design an algorithm that considers a larger search space, that avoids pruning LUGs. Moreover, another important challenge is that when a low utility generator is considered, checking if it is a LUGs requires to verify if there is a HUI in its equivalence class. However, state-of-the-art algorithms for generator mining [10,11] uses a depth-first search and do not keep equivalence classes into memory (maintaining them would be inefficient). Thus, it is not possible to compare an itemset with all the other members of its equivalence class efficiently. An idea that we propose to circumvent this challenge is to only compare a generator with its closure. Since the closure has the highest utility in each equivalence class (Property 7), it is sufficient to only consider the closure to determine if a low utility generator is a LUG. However, even with this solution, a problem is how to obtain the closure of a generator. In FIM, no depth-first search algorithm compute closure and generators at the same time and it cannot be done easily since closed patterns mining algorithms do not visit all generators. In fact, the state-of-the-art algorithm for mining generators and their closures [12] is a compound algorithm that mines closed pattern and generators separately by applying two algorithms and then links generators to closed patterns by post-processing. This approach is not a viable solution for GHUI mining since it would require keeping too many candidate generators in memory. To address this issue, we initially attempted to compute the closure of each found low utility generator on-the-fly. This can be done by intersecting transactions containing the generator. However, even on moderately large datasets, this approach cannot terminate in reasonable time due to the high cost of closure calculation. A better solution that we use in GHUI-Miner is to first mine CHUIs by using a CHUI mining algorithm and then to mine generators using a modified FHM [4] search procedure. The procedure is modified to only explore generator patterns by adding an efficient generator checking mechanism. Furthermore, the procedure is modified to only explores itemsets that are subsets of CHUIs, thus greatly pruning the search space. When the algorithm produces a generator X, if X is high utility, it is output. If X is low utility, its corresponding CHUI is retrieved to quickly determine if X is a LUG that should be output. Furthermore, several pruning conditions are added to further prune the search space.

Algorithm 1. The GHUI-Miner algorithm

---

    input : D: a transaction database, $minutil$: a user-specified threshold,
              $CHUIs$: the set of CHUIs
    output: the set of GHUIs

1   $I_{closed} \leftarrow$ items appearing in CHUIs.;
2   Scan D to calculate the TWU of items in I_{closed};
3   Let $\succ$ be the total order of TWU ascending values on I_{closed};
4   Scan D to build the utility-list of each item $i \in I_{closed}$ and the E structure;
5   **if** $\exists C \in CHUIs$ such that $sup(C) = |D|$ **then** Output $(\emptyset)$;
6   **foreach** $\{i\} \in I_{closed}$ **do**
7   **if** IsGenerator $(\{i\}) = false \vee SUM(i.utilitylist.iutil) +$
          $SUM(i.utilitylist.rutil) < minutil$ **then** $I_{closed} \leftarrow I_{closed} \setminus \{\{i\}\}$;
8   **else if** $SUM(i.utilitylist.iutil) \geq minutil$ **then**
9   Output $(\{i\})$;
10   **if** IsNotStrictSubsetOfACHUI $(\{i\}, CHUIs)$ **then**
11         $I_{closed} \leftarrow I_{closed} \setminus \{\{i\}\}$;
12   **else if** GetClosureOf $(\{i\}, CHUIs) \neq null$ **then** Output $(\{i\})$
13 **end**
14 SearchGHUI $(I_{closed}, minutil, E, CHUIs)$;

Main Procedure of GHUI-Miner. The main procedure of GHUI-Miner (Algorithm 1) takes as input a transaction database D, the $minutil$ threshold and the set of CHUIS. The set of CHUIs need to be first mined using an algorithm for mining CHUIs (CHUD [14] in our implementation). GHUI-Miner first select the set of items appearing in CHUIs (I_{closed}) because only these items may appear in GHUIs. Then, the algorithm scans D to calculate the TWU of each of those items. Note that considering only items in I_{closed} for TWU calculation reduces the TWU values (an thus provide a more tight upper bounds on the utility). The TWU values of items are then used to establish a total order $\succ$ on items, which is the order of ascending TWU values (as suggested in [6]). A second database scan is then performed. During this database scan, items in transactions are re-ordered according to the total order $\succ$, the utility-list of each item $i \in I_{closed}$ is built, and a structure named E is created. The E structure stores the TWU of all pairs of items $\{a, b\}$ such that $u(\{a, b\}) \neq 0$. The E structure could be implemented as a triangular matrix. But, for memory efficiency, we have implemented it as a hashmap of hashmaps since in practice a limited number of pairs of items co-occurs in transactions. After the construction of E, the empty set is output if there is a CHUIs with its support equal to $|D|$ (meaning that the empty set is a LUG). Then, a loop is performed to consider each item $\{i\} \in I_{closed}$. A call to the method $IsGenerator$ is made to verify if $\{i\}$ is a generator (this method will be explained later). If $\{i\}$ is not a generator, then it is removed from I_{closed} because any superset of a non generator cannot be a generator (Property 4). If the sum of $iutil$ and $rutil$ values in the utility-list of $\{i\}$ is less than $minsup$, $\{i\}$ and all its supersets are low utility, $\{i\}$ is discarded (Property 3). Then, the algorithm checks if $\{i\}$ is a GHUI. If the sum of $iutil$ values in the utility-list of $\{i\}$ is no less than $minutil$, $\{i\}$ is a HUG and it is output. If $\{i\}$ is a HUG and $\{i\}$ is not a strict subset of a CHUI, it is removed from I_{closed} because none of its supersets can be a GHUI. If $\{i\}$ is not a HUG, an attempt is made to retrieve the closure of $\{i\}$ from $CHUIs$ (i.e. to retrieve a set Z such that $\{i\} \subseteq Z$ $\land sup(\{i\}) = sup(Z)$. If the closure is found, $\{i\}$ is a LUG and it is output. Then, a recursive depth-first search exploration of generators having more than 1 item starts by calling the procedure $SearchGHUI$ with the set of single items I_{closed}, $minutil$ and the E structure.

The $SearchGHUI$ Procedure. The $SearchGHUI$ procedure (Algorithm 2) takes as input (1) a set of itemsets called $ExtensionsOfP$ having the form Pw meaning that Pw was previously obtained by appending an item w to an itemset P, (2) $minutil$, (3) the E structure and (4) the set of CHUIs. A loop is first performed over each itemset $Px \in ExtensionsOfP$ to explore its extensions. This is performed by merging Px with all extensions Py of $ExtensionsOfP$ such that $y \succ x$ to form extensions of the form Pxy containing $|Px| + 1$ items. For each extension Pxy, several checks are performed to prune the search space. First, if $TWU(\{x, y\}) < minutil$ according to the E structure, then Pxy and all its supersets are low utility and it can be discarded (Property 1). Second, if Pxy is not a subset of a CHUI, it can be discarded since none of its supersets can be GHUIs. Third, if the support of Pxy is equal to the support of Px or Py, Pxy is not a generator and its supersets cannot be generators (Property 4). Thus, Pxy is discarded. Fourth, if the sum of $iutil$ and $rutil$ values in Pxy utility-lists are less than $minsup$, Pxy and all its supersets are low utility and Pxy is thus discarded (Property 3). Fifth, if Pxy is not a generator (which is checked using the $IsGenerator$ method to be described later), it is also discarded. Then, the

---

Algorithm 2. The *SearchGHUI* procedure

---

input : *ExtensionsOfP*: a set of itemsets having a common prefix P, the
minutil threshold, the E structure, *CHUIs*: the set of CHUIs
output: the set of high utility itemsets

```
1  foreach  itemset Px ∈ ExtensionsOfP do
2      ExtensionsOfPx ← ∅;
3      foreach  itemset Py ∈ ExtensionsOfP such that y ≻ x do
4          Pxy ← Px ∪ Py;
5          if TWU(x,y) ≥ minutil according to E ∧ IsSubsetOfACHUI
             (Pxy,CHUIs) ∧ |Pxy.utilitylist| ≠ 0 ∧ |Pxy.utilitylist| ≠
             |Px.utilitylist| ∧ |Pxy.utilitylist| ≠ |Py.utilitylist| ∧
             SUM(i.utilitylist.iutil) + SUM(i.utilitylist.rutil) ≥ minutil ∧
             IsGenerator (Pxy) then
6              if SUM(Pxy.utilitylist.iutil) ≥ minsup then Output (Pxy);
7              else if GetClosureOf ( Pxy, CHUIs) ≠ null then  Output (Pxy) ;
8                  ExtensionsOfPx ← ExtensionsOfPx ∪ {Pxy}
9          end
10     end
11     SearchGHUI (ExtensionsOfPx, minutil, CHUIs);
12 end
```

---

algorithm check if the sum of *iutil* values in the utility-list of Pxy is no less than *minutil*. If yes, then Pxy is a HUG and it is output. Otherwise, an attempt is made to retrivee the closure of Pxy from the set of CHUIs. If the closure is found, Pxy is a LUG and it is output. Furthermore, if Pxy is a generator, it is added to a set *ExtensionsOfPx* for storing itemsets that should be considered for further extensions (Property 4). Finally, a recursive call to the *SearchGHUI* procedure with *ExtensionsOfPx* is done to explore extension(s) of its itemsets.

Since the *SearchGHUI* procedure starts from single items, it recursively explore the search space of generators by appending single items to find all GHUIs. Though, we do not have space to provide the proofs of completeness and correctness, it can be easily seen that this main procedure is correct and complete for finding GHUIs based on previous definitions and properties.

Updating Utility-Lists. In FHM, the *rutil* values in utility-lists are updated assuming that no item m will be added to extend an itemset X if $m \succ k$ for an item $k \in X$. This is a correct assumption for FHM for pruning the search space using Property 3 since items are added to itemsets following the total order $\succ$. However, in GHUI-Miner, we need to consider a larger search space that avoids pruning an itemset X if its closure is a HUI (to keep LUGs), and the closure may contains such an item m. To solve this problem, the *rutil* element of a tuple for an itemset X and a tid *tid* is redefined as $\sum_{i \in T_{tid} \wedge i \notin X} u(i, T_{tid})$. Moroever, utility-lists for the join of two itemsets Px and Py is done as follows. Let $ul(\{x\})$ and $ul(\{y\})$ be the utility-lists of Px and Py. The utility-list of $P \cup \{x, y\}$ is obtained by creating a tuple $(ex.tid, ex.iutil + ey.iutil - ep.iutil, ex.rutil - ey.iutil)$ for each set of tuples $ex \in ul(\{x\})$, $ey \in ul(\{y\})$, $ep \in ul(P)$ such that $ex.tid = ey.tid$.

The *IsGenerator* **Procedure.** GHUI-Miner integrates an efficient mechanism to determine if an itemset is a generator on-the-fly, that is, without having to compare an itemset with its subsets. The mechanism is inspired by the one used in the DefMe algorithm [10], although this later is desgined for a different search procedure. To describe the generator checking mechanism in GHUI-Miner, we introduce the concept of critical transactions. For an itemset X, the *critical transactions* of an item e in X are denoted as $crit(X, e)$ and defined as $g(X \backslash \{e\}) \backslash g(e)$, where $g(X)$ is the set of transactions containing X. Based on this definition, it can be easily seen that a necessary and sufficient condition for an itemset X to be a generator is that $\forall e \in X$, $crit(X, e) \neq \emptyset$. Performing generator checking using this condition requires to be able to compute critical transactions for any itemset efficiently. Fortunately, it can be done by modifying the GHUI-Miner search procedure. Let X be an itemset. It can be demonstrated that for any items a, b, $crit(X \cup \{b\}, a) = crit(X, a) \cap g(\{b\})$. Consider the join of a pair of itemsets $P \cup \{x\}$ and $P \cup \{y\}$ by GHUI-Miner to generate $P \cup \{x, y\}$. The critical objects of $P \cup \{x, y\}$ with respect to an item $z \in P$ can be calculated using $crit(P \cup \{x\}, z)$ and $g(y)$. Thus critical transactions of an itemset can be calculated using critical transactions of the pairs of itemsets that were joined to obtain the new itemset. Note that for an itemset $\{i\}$ containing a single item, $crit(\{i\}, i) = g(\emptyset) \backslash g(i)$. This generator checking mechanism is efficient. In GHUI-Miner, critical transactions are represented as bitsets for memory-efficiency. Moreover, another optimization is to stop generator checking for an itemset as soon as the itemset is determined to not be a generator (a set of critical transactions is found to be empty).

The *IsSubsetOfACHUI* **and** *GetClosure* **Procedures.** These procedures are implemented efficiently by storing CHUIs in a structure that indexes itemsets by their size and their support (a list of lists in our implementation). Thus, when searching for the closure of an itemset X, only itemsets of size greater or equal to $|X|$ and having a support equal to $sup(X)$ are considered. Similarly, when searching for a (strict) superset that is CHUI, only itemsets of size greater than (or equal to) $|X|$ and having a support smaller or equal to $sup(X)$ are considered.

The *HUG – Miner* **Algorithm.** We also propose a variation of GHUI-Miner named HUG-Miner to mine only HUGs, which does not includes instructions specific to LUGs. More precisely, HUG-Miner does not take the set of CHUIs as parameter and it considers the set of items I instead of I_{closed}. In Algorithm 1, lines 5, 10 and 11 are removed. In Algorithm 2, the call to *IsSubsetOfACHUI* and line 7 are removed. Moreover, HUG-Miner updates utility-list as described in section 2.1 since it only need to find HUIs.

4   Experimental Study

We performed an experiment to assess the performance of GHUI-Miner and HUG-Miner, and analyze their respective advantages. The experiment was performed on a computer with a third generation 64 bit Core i5 processor running Windows 7 and 5 GB of free RAM. We compared the performance of

GHUI-Miner and HUG-Miner with the state-of-the-art algorithm FHM for high utility itemset mining. CHUD [14] was used to generate the CHUIs needed by GHUI-Miner. All memory measurements were done using the Java API. The experiment was carried on four real-life datasets commonly used in the HUIM litterature: *mushroom, retail, kosarak* and *foodmart*. These datasets have varied characteristics and represents the main types of data typically encountered in real-life scenarios (dense, sparse and long transactions). Let $|I|$, $|D|$ and A represents the number of transactions, distinct items and average transaction length. *mushroom* is a dense dataset ($|I| = 16,470$, $|D| = 88,162$, $A = 23$). *kosarak* is a dataset that contains many long transactions ($|I| = 41,270$, $|D| = 990,000$, $A = 8.09$). *retail* is a sparse dataset with many different items ($|I| = 16,470$, $|D| = 88,162$, $A = 10,30$). *foodmart* is a sparse dataset ($|I| = 1,559$, $|D| = 4,141$, $A = 4.4$). *foodmart* contains real external and internal utility values. For the other datasets, external utilities for items are generated between 1 and 1,000 by using a log-normal distribution and quantities of items are generated randomly between 1 and 5, as the settings of [2,6,13]. The source code of all algorithms and datasets can be downloaded from http://www.philippe-fournier-viger.com/spmf/.

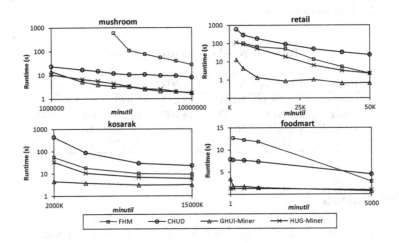

Fig. 3. Execution times

Algorithms were run on each dataset, while decreasing the *minutil* threshold until they became too long to execute, ran out of memory or a clear trend was observed. In fig. 3, we show the execution times of GHUI-Miner, HUG-Miner, CHUD and FHM. In Table 1, we show the number of HUIs, CHUIs, GHUIs, HUGs and LUGs for the lowest *minutil* value for each dataset.

It can first be observed that mining HUGs using HUG-Miner is much faster than mining all GHUIs by running CHUD and then GHUI-Miner. There are two reasons. First, GHUI-Miner needs to explore a larger search space to avoid pruning LUGs. In some cases, LUGs can have a very low utility, which makes them very expensive to mine. Second, mining the set of CHUIs that is needed by GHUI-Miner to verify if generators are LUGs is expensive. In the figure, it can be clearly seen that CHUD is in general the main cost to obtain GHUIs. Thus, an interesting possibility to improve the performance of mining GHUIs is to eventually replace CHUD with a faster algorithm.

A second observation is that mining HUGs is up to 100 times faster than mining all HUIs or CHUIs. On the other hand, mining GHUIs is faster than mining all HUIs using FHM on *mushroom* and *foodmart*. The reason why the combination of CHUD and GHUI-Miner is sometimes slower than FHM is the cost of closure computation and the larger search space to avoid pruning LUGs. The reason why the combination of CHUD and GHUI-Miner is up to 46 times faster than FHM on *mushroom* is that a small proportion of itemsets are generators in that dataset. It is also a well-known fact that algorithms for mining a concise representation of patterns perform better on dense datasets [14].

A third observation is that the set of GHUIs and the set of HUGs are up to 36 times smaller than the set of HUIs. This shows that these concise representations of HUIs are very compact. Furthermore, it can be observed that the set of LUGs is often very small. Thus, if one mines only HUGs using HUG-Miner, it may obtain an interesting compromise by missing a small portion of GHUIs but having a short execution time.

Table 1. Number of patterns found

Dataset	$|HUIs|$	$|CHUIs|$	$|GHUIs|$	$|HUGs|$	$|LUGs|$
mushroom	3,538,181	10,311	98,315	5,979	92,336
retail	14,314	14,090	14,697	14,090	607
kosarak	56	56	67	30	37
foodmart	233,231	6,680	40,178	40,178	0

5   Conclusion

This paper proposes a new framework for mining concise representations of *high utility itemsets using generators*. We investigate the properties of generators and incorporate the concept of generator into HUI mining. We explore two new concise representations of HUIs, called High Utility Generator (HUG) and Generator of High Utility Itemsets (GHUIs). Two efficient algorithms named *HUG-Miner* and *GHUI-Miner* are proposed to respectively mine these representations. The algorithms provide different trade-offs between execution time and completeness. GHUI-Miner captures the complete set of GHUIs but spends more time because it needs to consider generators that are not HUIs. On the other hand, HUG-Miner is over 100 times faster than GHUI-Miner but misses GHUIs that are LUGs. Experimental results on both real-life and synthetic datasets show that the proposed algorithms are very efficient and achieve a massive reduction in terms of number of patterns found. Moreover, HUG-Miner is up to two orders of magnitude faster than the state-of-the-art algorithms for CHUI mining and HUI mining. Source codes of all algorithms and datasets can be downloaded as part of the SPMF pattern mining library http://www.philippe-fournier-viger.com/spmf/. For future work, we will consider utility mining problems in sequential pattern mining [3]

References

1. Agrawal, R., Srikant, R.: Fast algorithms for mining association rules in large databases. In: Proc. Int. Conf. Very Large Databases, pp. 487–499, (1994)
2. Ahmed, C. F., Tanbeer, S. K., Jeong, B.-S., Lee, Y.-K.: Efficient Tree Structures for high utility Pattern Mining in Incremental Databases. In: IEEE Trans. Knowl. Data Eng. 21(12), pp. 1708–1721 (2009)
3. Fournier-Viger, P., Gomariz, A., Campos, M., Thomas, R.: Fast Vertical Sequential Pattern Mining Using Co-occurrence Information. In: Proc. PAKDD 2014, pp. 40-52. (2014)
4. Fournier-Viger, P., Wu, C.-W., Zida, S., Tseng, V. S.: FHM: Faster High-Utility Itemset Mining using Estimated Utility Co-occurrence Pruning. In: Proc. 21st Intern. Symp. Methodologies Intell. Systems, Springer, pp. 83-92 (2014)
5. Gao, C., Wang, J., He, Y., Zhou, L.: Efficient mining of frequent sequence generators. In: Proc. 17th Intern. Conf. World Wide Web:, pp. 1051–1052 (2008)
6. Liu, M., Qu, J.:Mining High Utility Itemsets without Candidate Generation. In Proceedings of CIKM12, pp. 55–64 (2012)
7. Liu, Y., Liao, W., Choudhary, A.: A two-phase algorithm for fast discovery of high utility itemsets. In: Proc. PAKDD 2005, pp. 689–695 (2005)
8. Pham, T.-T., Luo, J., Hong, T.-P., Vo, B..: MSGPs: a novel algorithm for mining sequential generator patterns. In: Proc. 4th Intern. Conf. Computational Collective Intelligence, pp. 393-401 (2012)
9. Shie, B.-E., Yu, P.S., Tseng, V.S.: Efficient algorithms for mining maximal high utility itemsets from data streams with different models. Expert Syst. Appl. 39(17), pp. 12947-12960 (2012)
10. Soulet, A., Rioult, F.: Efficiently Depth-First Minimal Pattern Mining. In: Proc. 18th Pacific-Asia Conf. Knowledge Discovery and Data Mining, pp. 28–39 (2014)
11. Szathmary, L., Valtchev, P., Napoli, A., Godin., R.: Efficient vertical mining of frequent closures and generators. In: Proc. 8th Intern. Symp. Intelligent Data Analysis, August 31 - September 2, Lyon, France, pp. 393–404 (2009)
12. Szathmary, L. et al.: A fast compound algorithm for mining generators, closed patterns, and computing links between equivalence classes. In: Ann. Math. Artif. Intell. 70(1-2), pp. 81–105 (2014)
13. Tseng, V. S., Shie, B.-E., Wu, C.-W., Yu., P. S.: Efficient Algorithms for Mining High Utility Itemsets from Transactional Databases. In: IEEE Trans. Knowl. Data Eng. 25(8), pp. 1772–1786 (2013)
14. Wu, C.-W., Fournier-Viger, P., Yu., P. S, Tseng, V. S.: Efficient Mining of a Concise and Lossless Representation of High Utility Itemsets. In: Proceedings of ICDM11, pp. 824–833 (2011)

Incrementally Updating High-Utility Itemsets with Transaction Insertion

Jerry Chun-Wei Lin[1,2] , Wensheng Gan[1], Tzung-Pei Hong[3,4], and Jeng-Shyang Pan[1,2]

[1] Innovative Information Industry Research Center (IIIRC)
[2] Shenzhen Key Laboratory of Internet Information Collaboration
School of Computer Science and Technology
Harbin Institute of Technology Shenzhen Graduate School
HIT Campus Shenzhen University Town, Xili, Shenzhen, P.R. China
[3] Department of Computer Science and Information Engineering
National University of Kaohsiung, Kaohsiung, Taiwan, R.O.C.
[4] Department of Computer Science and Engineering
National Sun Yat-sen University, Kaohsiung, Taiwan, R.O.C.
`jerrylin@ieee.org`, {`wsgan001,jengshyangpan`}`@gmail.com`,
`tphong@nuk.edu.tw`

Abstract. High-utility itemsets mining (HUIM) is designed to solve the limitations of association-rule mining by considering both the quantity and profit measures. Most algorithms of HUIM are designed to handle the static database. Fewer research handles the dynamic HUIM with transaction insertion, thus requiring the computations of database rescan and combination explosion of pattern-growth mechanism. In this paper, an efficient incremental algorithm with transaction insertion is designed to reduce computations without candidate generation based on the utility-list structures. The enumeration tree and the relationships between 2-itemsets are also adopted in the proposed algorithm to speed up computations. Several experiments are conducted to show the performance of the proposed algorithm in terms of runtime, and memory consumption.

Keywords: utility mining, transaction insertion, utility-list, enumeration tree, dynamic databases.

1 Introduction

Association-rule mining (ARM) [2, 3, 4, 8] is a fundamental task for revealing the relationships among items. Many algorithms have been respectively proposed to efficiently mine the association rules based on whether the level-wise or pattern-growth mechanism [3, 8]. Both the level-wise or pattern-growth approaches can only handle the static database in batch mode. When transactions are changed in the database, new information may arise and old ones may become invalid. The updated database is required to be processed to mine the updated information in batch mode, which is not suitable in practical applications. To solve the limitations of mining algorithms in batch mode [11, 17], Cheung et al. proposed the Fast-UPdated (FUP)

X. Luo, J.X. Yu, and Z. Li (Eds.): ADMA 2014, LNAI 8933, pp. 44–56, 2014.

algorithm [9] to maintain and update the discovered information with transaction insertion. When the itemsets are small in the original database (support ratio is lower than minimum support threshold) but large (support ratio is larger than or equal to the minimum support threshold) in the inserted database, the original database is still required to be rescanned to find the actual occurrence frequencies of the small itemsets in the original database.

For ARM, the implicit factors such as profit or quantity are not considered. High-utility itemsets mining (HUIM) [7, 20] was thus proposed to partially solve the limitations of ARM. It may be thought of as an extension of frequent-itemset mining by considering the sold quantities and profits of the items. Several algorithms have been proposed to mine HUIs in a static database [12, 16, 18, 19]. Some HUIM algorithms have been proposed with transaction insertion [6, 13, 14]. The original database is still, however, required to be rescanned for maintaining and updating the HUIs in some cases. The problem of combination explosion based on level-wise approach is also a critical issue to be solved.

In this paper, a memory-based incremental approach is proposed with transaction insertion to efficiently discover HUIs. The proposed algorithm inherits the HUI-Miner algorithm [15] to build the utility-list structures for mining HUIs with transaction insertion. An Estimated Utility Co-Occurrence Structure (EUCS) [10] is also applied in the proposed algorithm to speed up the performance of the proposed approach. Based on the designed algorithm, it outperforms the two-phase algorithm [16] and the state-of-the-art FHM algorithm [10] in batch mode and other previous algorithms for incremental mining [13, 14].

2   Related Work

Utility mining [7, 20] is concerned as an extension of the frequent itemsets by considering both the quantities and profits of items to discover the valuable itemsets than the frequent ones. An itemset is concerned as a HUI if its utility is larger than or equal to the minimum utility count. Chan et al. first proposed the top-k objective-directed data mining to mine the top-k closed utility patterns based on business objective [7]. Yao et al. proposed the utility model to firstly consider both quantities and profits of the items to mine the HUIs [20]. Liu et al. proposed the two-phase model [16] to mine HUIs based on the developed transaction-weighted downward closure (TWDC) property. Many algorithms been proposed to mine HUIs based on two-phase model. Lin et al. designed a high utility pattern tree (HUP-tree) algorithm [12] to compress the original database into a tree structure. Tseng et al. then proposed the UP-tree structure with UP-growth and UP-growth+ mining algorithms to efficiently mine HUIs [18]. Liu et al. then proposed a HUI-Miner algorithm [15] to compress the database into the utility-list structures. Each entry in the utility-list structure stores transaction IDs (TIDs), the utility of itemset X in the transaction (Iutility), and the rest utilities of itemsets except X in the transaction (Rutility). Based on the HUI-Miner algorithm and the designed pruning strategy of the enumeration tree, the HUIs can be easily discovered. Fournier-Viger et al. then modified the

HUI-Miner algorithm and designed an Estimated Utility Co-Occurrence Structure (EUCS) to keep the relationships between 2-itemsets, thus speeding up the computations compared to the HUI-Miner algorithm [10].

Most algorithms process the static database to mine HUIs. Ahmed et al. proposed an IHUP algorithm with three tree structures for mining HUIs with transaction insertion [6]. Lin et al. proposed an incremental algorithm (FUP-HUI-INS) [13] for updating the discovered HUIs based on the FUP concept [9] and two-phase model [16] with transaction insertion. Lin et al. then also proposed an improved pre-large concept for mining high-utility itemsets with transaction insertion (PRE-HUI-INS) [14]. Since FUP-HUI-INS and PRE-HUI-INS algorithms are processed by two-phase model, an additional database rescan is still necessary performed to find the actually HUIs. Besides, both of them required computations to find the HTWUIs based on the pattern-growth approach.

3   Preliminaries and Problem Statement

Definition 1. An itemset X is a set of k distinct items $\{i_1, i_2, ..., i_k\}$ in a quantitative database $D = \{T_1, T_2, ..., T_n\}$, in which k is the length of an itemset. An itemset X is contained in a transaction T_n if $X \subseteq T_n$.

Definition 2. The utility of an item i_j in T_q is defined as $u(i_j, T_q) = q(i_j, T_q) \times p(i_j)$, in which $q(i_j, T_q)$ is the quantity of an item i_j in T_q, $p(i_j)$ is the profit value of an item i_j.

Definition 3. The utility of an itemset X in transaction T_q is denoted as $u(X, T_q)$, which can be defined as:

$$u(X, T_q) = \sum_{i_j \in X \wedge X \subseteq T_q} u(i_j, T_q).$$

Definition 4. The utility of an itemset X in D is denoted as $u(X)$, which can be defined as:

$$u(X) = \sum_{X \subseteq T_q \wedge T_q \in D} u(X, T_q).$$

Definition 5. The transaction utility of transaction T_q is denoted as $tu(T_q)$, in which m is the number of items in T_q. Thus $tu(T_q)$ can be defined as:

$$tu(T_q) = \sum_{j=1}^{m} u(i_j, T_q).$$

Definition 6. Total utility of D is denoted as TU^D, which can be defined as:

$$TU^D = \sum_{T_q \in D} tu(T_q).$$

Definition 7. Given a database D and the minimum high utility threshold is set as ε, a high-utility itemset X in D is denoted as $HUI^D(X)$, which can be defined as:

$$HUI^D(X) = \sum_{X \subseteq T_q \wedge T_q \in D} u(X, T_q) \geq \varepsilon \times TU^D.$$

Problem Statement. Given a transactional database D, its total utility is defined as TU^D from D, a minimum utility threshold is set at $0 < \varepsilon \leq 1$, the HUIM is to find the completely k-itemsets whose utilities larger than or equal to minimum utility count as $(\varepsilon \times TU^D)$.

Since the downward-closure property of ARM is not kept in HUIM, the transaction-weighted downward closure property (TWDC) was thus proposed by two-phase model [16].

Definition 8. The transaction-weighted utility of an itemset X is the sum of all transaction utilities $tu(T_q)$ containing an itemset X, which is defined as:

$$TWU(X) = \sum\nolimits_{X \subseteq T_q, T_q \in D} tu(T_q).$$

Definition 9. An itemset X is defined as a high transaction-weighted utilization itemset (HTWUI) if $TWU^D(X) \geq \varepsilon \times TU^D$.

Property 1. The transaction-weighted downward closure (TWDC) property of two-phase model is that if an itemset X is not as a HTWUI, the subsets of X could not be HUIs.

4   Proposed Incremental Algorithm

In this paper, the HUI-Miner algorithm [15] is adopted to design the incremental algorithm for HUIM. Before transactions are inserted into the original database, the utility-list structures are built in advance to keep not only the HTWUIs but also those itemsets which are not the high transaction-weighted utilization itemsets from the original database to avoid the database rescan with transaction insertion.

4.1   Utility-List Structures

Definition 10. An entry of X in the utility-list structure consisted of the set TIDs for X in T_q ($X \subseteq T_q \in D$), the set of utility for X in T_q (*Iutility*), and the set of remaining utility for X in T_q (*Rutility*), in which *Rutility* is defined as:

$$X.Rutility(T_q) = \sum\nolimits_{i_j \in T_q \wedge i_j \notin X} u(i_j, T_q).$$

The construction algorithm can be found from HUI-Miner approach [15]. In the construction process, the itemsets are sorted in ascending order of their transaction-weighted utility (TWU). For the *Rutility* of an itemset X in a transaction, it keeps the rest utilities in the transaction except the processed itemset X. Since the TWU values of the itemsets are changed with transaction insertion, the sorted order of the utility-list structures and the *Rutility* value should also be changed. The number of inserted transactions is, however, very small compared to the original database. In the proposed algorithm, the sorted order of the itemsets in the inserted transactions follows the initially order of itemsets in the original database.

Definition 11. The $X.Iutility.sum$ is to sum the utilities of an itemset X in database D as:

$$X.Iutility.sum = \sum_{X \subseteq T_q \wedge T_q \in D} X.Iutility(T_q).$$

Definition 12. The $X.Rutility.sum$ is to sum the rest utilities except an itemset X in database D as:

$$X.Rutility.sum = \sum_{X \subseteq T_q \wedge T_q \in D} X.Rutility(T_q).$$

For more k-itemsets, the utility-list structures are recursively constructed until no candidates are generated for determination.

Definition 13. Any extension of an itemset X is a combination of X with the itemset(s) after an itemset X, which is denoted as X'.

For example, assume the itemset B, then the extension of B is BC, BD and so on, both BC and BD can denoted as B'.

Based on the HUI-Miner [15], a pruning strategy can also be adopted to compress the border for determination than the TWDC property.

Property 2. Given the utility-list structure of an itemset X, if the summation of $Iutility$ and $Rutility$ of an itemset X in D is less than the minimum utility count, any extension X' of X is not a HUI.

In addition, the Estimated Utility Co-occurrence Pruning (EUCP) strategy [10] is also adopted in the proposed algorithm to further keep the relationship of 2-itemsets, thus eliminating the extension itemsets with lower utility without re-constructing the utility-list structures.

4.2   Proposed Algorithm

Based on the above properties inheriting from HUI-Miner and EUCS structures, the proposed incremental algorithm is described in Fig. 1.

Algorithm 1: HUI-list-INS algorithm
INPUT: D, the original database; d, the incremental database; TU^D, the total utility in D; TU^d, the total utility in d; *ptable*, the profit table; ε, the minimum utility threshold; $EUCS$, the Estimated Utility Co-Occurrence Structure; $DB.UL$, the utility list of D; $db.UL$, the utility list of d; $U.UL$, the utility list of U; $X.UL$, the utility list of itemset X.
OUTPUT: High-utility itemsets.
BEGIN Procedure
1. $DB.UL=NULL$, $db.UL=NULL$, $U.UL=NULL$, $X.UL=NULL$;
2.   **FOR** each T_q in D **DO**
3.   **FOR** each X in T_q **DO**
4. $X.UL \leftarrow \{T_q, Iutility, Rutility\}$.
5. $EUCS \leftarrow \{X, X'\}$.   //$X'$, the extension of X.
6.   **END FOR**
7.   **END FOR**
8. $DB.UL \leftarrow \cup X.UL$.

Fig. 1. Pseudo code of the proposed HUI-list-INS algorithm

```
9.    FOR each T_q in d DO
10.     FOR each X in T_q DO
11.         X.UL ← {T_q, Iutility, Rutility}.
12.         update the TWU(X) in EUCS.
13.     END FOR
14.   END FOR
15.   db.UL ← ∪X.UL.
16.   call merge-list(DB.UL, db.UL, U.UL).
17.   FOR each X in U.UL DO
18.     IF X.Iutility.sum ≥ (TU^D+TU^d) × ε
19.         HUIs ← X.
20.     END IF
21.     IF X.Iutility.sum + X.Rutility.sum ≥ (TU^D+TU^d) × ε THEN
22.         extULs ← null.      // extULs, the set of utility list of all X's 1-extensions;
23.         FOR each Y after X in U.UL DO
24.             IF ∃(X,Y,Z)∈ EUCS and Z ≥ (TU^D+TU^d) × ε THEN
25.                 extULs ← extULs + Construct(X.UL, Y, Z).
26.             END IF
27.         END FOR
28.         call HUI-list-INS(X, extULs, ε).
29.     END IF
30.   END FOR
END Procedure
```

Fig. 1. (*continued*)

For the designed incremental algorithm with transaction insertion, the original database is firstly scanned to construct the utility-list structures for all 1-itemsets and the EUCS structure for each item (Line 2-8). Similarly, the inserted transactions are processed in the same way. Each related TWU values of items in the built EUCS are also updated by the inserted transactions (Line 9-15). The designed merge-list algorithm is used to combine the utility-list structures from the original database and inserted transactions into the updated utility-list structures (Line 16). After that, the 1-extensions of an itemset X is recursively processed (Line 17-28) by using a depth-first procedure. Each itemset X is then determined by the designed condition to check whether it is a HUI (Line 18-20). The extensions of the processed itemset are then determined by the designed condition (Line 21) for depth-first search. The updated EUCS structure is also used to prune the unpromising itemset, thus reducing the search space for mining high-utility itemsets (Line 24-26). The construction of utility-list structure algorithm is then performed to construct the *extULs* of X. The proposed HUI-list-INS algorithm is then recursively performed to mine HUIs (Line 21-29). The algorithm is then terminated until no itemsets are generated. The construction procedure of utility-list structures are recursively processed for k-itemsets if it is necessary to process the depth-first search in the search space. The construction algorithm is similar as the algorithm shown in [15]. The proposed merge-list algorithm to combine original database and the incremental one is described in Fig. 2.

Algorithm 2: merge-list algorithm

INPUT: *DB.UL* is the utility-list of *D*; *db.UL* is the utility-list of *d*; *U.UL* is the
        utility-list of *U*.

OUTPUT: *U.UL*.

BEGIN Procedure

1.   *U.UL = null*, *X.UL = null*.
2.   *FOR* each itemset *X* and *X.UL∈ DB.UL DO*
3.   *IF X.UL ≠ null THEN*
4.   search itemset *X∈ DB.UL* in *db.UL*
5.   *IF ∃(X∈ DB.UL* and *X∈db.UL) THEN*
                /* *E_i* is the element *of X.UL* */
6.   *FOR* each element *E_i∈ X.UL* and *X.UL∈ db.UL DO*
7.   *X.Iutility.sum ← X.Iutility.sum + E_i.Iutility;*
8.   *X.Rutility.sum ← X.Rutility.sum + E_i.Rutility;*
9.   *X.UL ←E_i*.
10.   *END FOR*
11.   *END IF*
12.   *U.UL ← X.UL*.
13.   *END IF*
14. *END FOR*
15. *RETURN U.UL*.

END Procedure

Fig. 2. Pseudo code of the proposed merge-list algorithm

5   Experimental Evaluation

Several experiments in terms of execution time, and memory consumption are
conducted to show the performance of the proposed algorithm in four database
including three real-life [1] and a synthetic databases [5]. The two-phase algorithm
[16], the state-of-the-art FHM algorithm [10], two incremental FUP-HUI-INS [13]
and PRE-HUI-INS [14] algorithms are used to compare the proposed algorithm. The
values of quantities and profits were assigned to the purchased items in all databases
by the two-phase model [16] except foodmart database. Parameters and characteristics
for four databases are respectively described in Tables 1 and 2.

Table 1. Parameters descriptions

#|D|	Total number of transactions
#|I|	Number of distinct items
AvgLen	Average length of transactions
MaxLen	Maximal length of transactions

Table 2. Characteristics of used databases

Databases	#|D|	#|I|	AvgLen	MaxLen
foodmart	21,556	1,559	4	11
retail	88,162	16,470	10.3	76
chess	3,196	75	37	37
T10I4D100K	100,000	870	10.1	29

5.1 Runtime

Experiments were made to show the runtime of the proposed algorithm compared to the two-phase and FHM algorithms in batch mode and the other two incremental algorithms. The runtime includes the construction and mining phases. Experiments are then conducted to show the comparisons under various minimum utility thresholds (MUs) with a fixed insertion ratio (IR). The results are shown in Fig. 3.

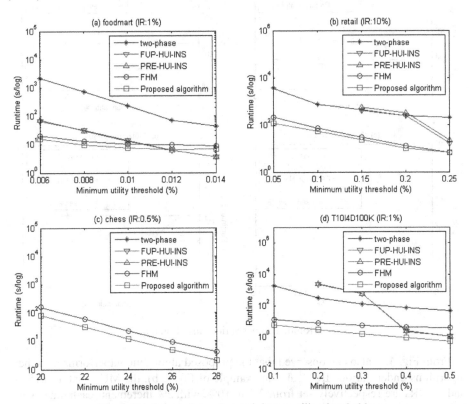

Fig. 3. Runtime under various minimum utility thresholds

From Fig. 3, it can be observed that the proposed algorithm has better performance than the two-phase and FHM algorithms in batch mode and the incremental FUP-HUI-INS and PRE-HUI-IN algorithms. The runtime is decreasing along with the increasing of MU. The observation is reasonable since less candidates of HUIs are generated when MU is set higher. When MU is set lower, the gap between the proposed algorithm and other three algorithms becomes large except the FHM algorithm, which indicates that the other three algorithms required more runtime than the proposed algorithm. Since the FHM algorithm uses the similar pruning strategies as the proposed approach, there is no great difference between them. The FHM is, however, performed in batch mode, thus requiring database rescan each time when the transactions are inserted into the original database. Experiments are then conducted to show the comparisons under different IRs with a fixed MU. The results are shown in Fig. 4.

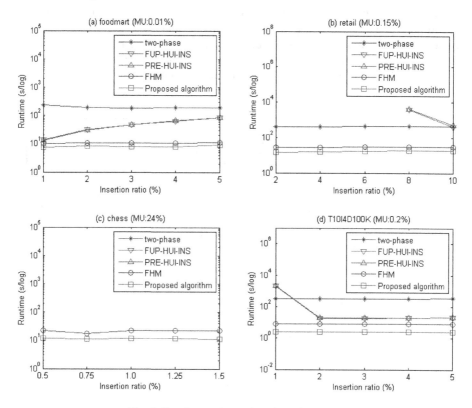

Fig. 4. Runtime under various insertion ratios

From Fig. 4, it also can observed that the proposed algorithm outperforms the other algorithms under various IRs. Take an example of Fig. 4(b), the MU is set at 0.15%, and the IRs are respectively set from 2% to 10%, with 2% increments each time. Two incremental FUP-HUI-INS and PRE-HUI-INS algorithms have worse performance than the other algorithms. When the IR is set lower than 8%, the average runtime of two-phase algorithm is 420 seconds, the FHM is 28 seconds, and the proposed algorithm is 16 seconds. The runtime of FUP-HUI-INS and PRE-HUI-INS algorithms exceed 10^4 seconds. The reason is that FUP-HUI-INS and PRE-HUI-INS algorithms could have "combination explosion" problem when MU or IR is set lower. This situation may frequently occur depending on the database characteristics. From the above results, the other algorithms have worse performance in chess database except the FHM and the proposed algorithm, which can be easily observed from Fig. 3(c) and Fig. 4(c). Since the chess belongs to dense database with long patterns in the transactions, a great amount of HTWUIs are generated by the two-phase, FUP-HUI-INS and PRE-HUI-INS algorithms.

5.2   Memory Consumption

Memory consumption of the propose algorithm compared to the other algorithms is then evaluated. Experiments are then conducted to show the comparisons under various MUs with a fixed IR. The results are shown in Fig. 5.

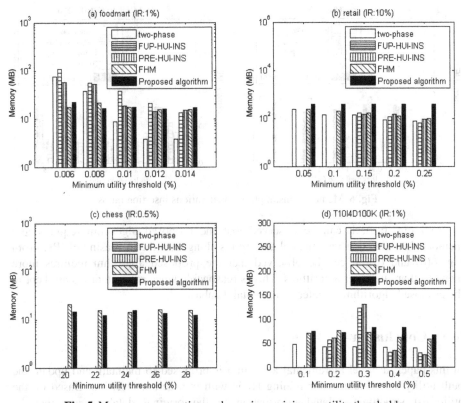

Fig. 5. Memory consumption under various minimum utility thresholds

From Fig. 5, it can be observed that the FHM and the proposed algorithms requires steady memory along with the increasing of MUs compared to the other algorithms. This is because that the FHM and the proposed algorithms are necessary to build the utility-list structures for keep the itemsets. When MU is set lower, the proposed algorithm requires fewer memory than the other algorithms, which can be observed from Fig. 5(a). Experiments are then conducted to show the comparisons under various IRs with a fixed MU. The results are shown in Fig. 6.

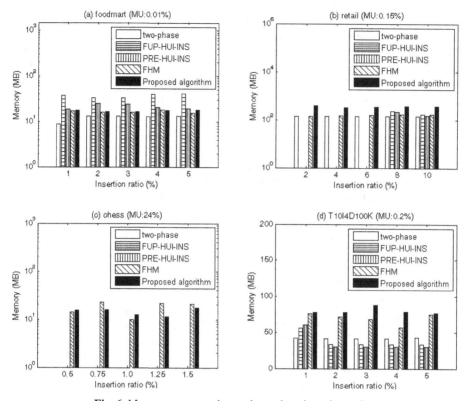

Fig. 6. Memory consumption under various insertion ratios

From Fig. 6(a), it can be observed that the proposed algorithm requires less memory than the other incremental algorithms along with the increasing of IRs. From Fig. 6(b) and (d), it can be observed that the proposed algorithm requires more memory than the other algorithms. This is reasonable since more itemsets are kept in the proposed algorithm for later incremental database.

6   Conclusion

In this paper, a novel incremental algorithm is proposed to maintain and update the built utility-list structures for mining HUIs with transaction insertion. Based on the utility-list structures, related information in the original database can thus be compressed. The proposed algorithm also applies the Estimated Utility Co-Occurrence Structure (EUCS) to keep the information between 2-itemsets, thus speeding up the computations. Without the level-wise approach for generate-and-test candidates, HUIs can be easily discovered based on the designed algorithm for the incremental database. Experimental results show that the performance of the proposed algorithm outperforms than other algorithms.

Acknowledgement. This research was partially supported by the Shenzhen Peacock Project, China, under grant KQC201109020055A, by the Natural Scientific Research Innovation Foundation in Harbin Institute of Technology under grant HIT.NSRIF.2014100, and by the Shenzhen Strategic Emerging Industries Program under grant ZDSY20120613125016389.

References

1. Frequent itemset mining dataset repository (2012), http://fimi.ua.ac.be/data/
2. Abdullah, Z., Herawan, T., Deris, M.: Mining Significant Least Association Rules using Fast SLP-growth Algorithm. In: Kim, T.-h., Adeli, H. (eds.) AST/UCMA/ISA/ACN 2010. LNCS, vol. 6059, pp. 324–336. Springer, Heidelberg (2010)
3. Agrawal, R., Imielinski, T., Swami, A.: Database Mining: A Performance Perspective. IEEE Transactions on Knowledge and Data Engineering 5, 914–925 (1993)
4. Agrawal, R., Srikant, R.: Fast Algorithms for Mining Association Rules in Large Databases. In: The International Conference on Very Large Data Bases, pp. 487–499 (1994)
5. Agrawal, R., Srikant, R.: Quest synthetic data generator (1994), http://www.Almaden.ibm.com/cs/quest/syndata.html
6. Ahmed, C.F., Tanbeer, S.K., Jeong, B.S., Lee, Y.K.: Efficient Tree Structures for High Utility Pattern Mining in Incremental Databases. IEEE Transactions on Knowledge and Data Engineering 21, 1708–1721 (2009)
7. Chan, R., Yang, Q., Shen, Y.D.: Mining high utility itemsets. In: IEEE International Conference on Data Mining, pp. 19–26 (2003)
8. Chen, M.S., Han, J., Yu, P.S.: Data Mining: An Overview from a Database Perspective. IEEE Transactions on Knowledge and Data Engineering 8, 866–883 (1996)
9. Cheung, D.W., Han, J., Ng, V., Wong, C.Y.: Maintenance of Discovered Association Rules in Large Databases: An Incremental Updating Technique. In: International Conference on Data Engineering, pp. 106–114 (1996)
10. Fournier-Viger, P., Wu, C.W., Zida, S., Tseng, V.S.: FHM: Faster High-Utility Itemset Mining Using Estimated Utility Co-Occurrence Pruning. Foundations of Intelligent Systems 8502, 83–92 (2014)
11. Hong, T.P., Lin, C.W., Wu, Y.L.: Incrementally Fast Updated Frequent Pattern Trees. Expert Systems with Applications 34, 2424–2435 (2008)
12. Lin, C.W., Hong, T.P., Lu, W.H.: An Effective Tree Structure for Mining High Utility Itemsets. Expert Systems with Applications 38, 7419–7424 (2011)
13. Lin, C.W., Lan, G.C., Hong, T.P.: An Incremental Mining Algorithm for High Utility Itemsets. Expert Systems with Applications 39, 7173–7180 (2012)
14. Lin, C.W., Hong, T.P., Lan, G.C., Wong, J.W., Lin, W.Y.: Incrementally Mining High Utility Patterns based on Pre-large Concept. Applied Intelligence 40, 343–357 (2014)
15. Liu, M., Qu, J.: Mining High Utility Iitemsets without Candidate Generation. In: ACM International Conference on Information and Knowledge Management, pp. 55–64 (2012)
16. Liu, Y., Liao, W.K., Choudhary, A.: A Two-Phase Algorithm for Fast Discovery of High Utility Itemsets. In: Ho, T.-B., Cheung, D., Liu, H. (eds.) PAKDD 2005. LNCS (LNAI), vol. 3518, pp. 689–695. Springer, Heidelberg (2005)
17. Nath, B., Bhattacharyya, D.K., Ghosh, A.: Incremental Association Rule Mining: A Survey. WIREs Data Mining Knowledge Discovery 3, 157–169 (2013)

18. Tseng, V.S., Shie, B.E., Wu, C.W., Yu, P.S.: Efficient Algorithms for Mining High Utility Itemsets from Transactional Databases. IEEE Transactions on Knowledge and Data Engineering 25, 1772–1786 (2013)

19. Wu, C.W., Shie, B.E., Yu, P.S., Tseng, V.S.: Mining Top-k High Utility Itemsets. In: ACM International Conference on Knowledge Discovery and Data Mining, pp. 12–16 (2012)

20. Yao, H., Hamilton, H.J.: Mining Itemset Utilities from Transaction Databases. Data & Knowledge Engineering 59, 603–626 (2006)

Mining Frequent Closed Sequential Patterns with Non-user-defined Gap Constraints*

Wentao Wang[1], Lei Duan[1,2,3,**], Jyrki Nummenmaa[4], Song Deng[5],
Zhongqi Li[1], Hao Yang[1], and Changjie Tang[1]

[1] School of Computer Science, Sichuan University, China
[2] West China School of Public Health, Sichuan University, China
[3] State Key Laboratory of Software Engineering, Wuhan University, China
[4] School of Information Sciences, University of Tampere, Finland
[5] Institute of Advanced Technology, Nanjing University of Posts and
Telecommunications, China
wangwentao91@hotmail.com, {leiduan,cjtang}@scu.edu.cn,
jyrki.nummenmaa@uta.fi, ds16090311@163.com, zqli.scu@gmail.com,
hyang.cn@outlook.com

Abstract. Frequent closed sequential pattern mining plays an important role in sequence data mining and has a wide range of applications in real life, such as protein sequence analysis, financial data investigation, and user behavior prediction. In previous studies, a user predefined gap constraint is considered in frequent closed sequential pattern mining as a parameter. However, it is difficult for users, who are lacking sufficient priori knowledge, to set suitable gap constraints. Furthermore, different gap constraints may lead to different results, and some useful patterns may be missed if the gap constraint is chosen inappropriately. To deal with this, we present a novel problem of mining frequent closed sequential patterns with non-user-defined gap constraints. In addition, we propose an efficient algorithm to find the frequent closed sequential patterns with the most suitable gap constraints. Our empirical study on protein data sets demonstrates that our algorithm is effective and efficient.

Keywords: frequent closed sequential pattern, gap constraint, sequence data mining.

1 Introduction

Ever since sequential pattern mining problem was introduced by Agrawal and Srikant [1], sequential pattern mining has become a significant task in data mining and has received wide attention. Different types of sequential patterns have been proposed, such as frequent sequential pattern [2], distinguishing sequential pattern [3], closed sequential pattern [4], and periodic sequential pattern [5].

---

* This work was supported in part by NSFC 61103042, SRFDP 20100181120029, SKLSE2012-09-32, and China Postdoctoral Science Foundation 2014M552371.
** Corresponding author.

X. Luo, J.X. Yu, and Z. Li (Eds.): ADMA 2014, LNAI 8933, pp. 57–70, 2014.

Moreover, sequential pattern mining has been applied in various fields. For instance, in biological field [6,7], mining protein and nucleotide sequences helps researchers to identify distinguished protein families in DNA. In commercial field, analyzing customers' behavior sequences can help managers get more useful information about customers' personalized tastes.

Gap constraint, which can make the expression of a pattern more flexible and general, has been widely used in sequential pattern mining. Specifically, a gap constraint defines the minimum and maximum numbers of wildcards[1] between two elements in a sequence. For example, given a sequence $S = A\$C\$\$F$, a valid gap constraint can be $[1,2]$.

Recent research [8] indicated that mining closed sequential patterns instead of all frequent sequential patterns could get a more compact yet complete result set, and the efficiency of mining is improved greatly. In previous studies on frequent closed sequential pattern mining, such as [9,10], a predefined gap constraint is necessary as a parameter. However it is difficult for users to set suitable gap constraints. Without enough a priori knowledge, users have to pick up a gap constraint randomly or do a large number of experiments to find a suitable one. Please note that different gap constraints lead to different results. Example 1 demonstrates this situation.

Example 1. Given a set $\mathcal{D}$ of sequences in Table 1, we mine subsequences that appear in all sequences. If the gap constraint is $[0,2]$, the result is $\{BA, CC\}$. If the gap constraint is $[0,3]$, the result is $\{BA, CC, CA\}$.

Table 1. An example of sequential data set

Id	Sequence
S1	ABDAFCAC
S2	CACDFABBA
S3	BCAECAFC
S4	CCBAEDFB
S5	ECBDFACCE

To break the limitation of predefining gap constraints, we propose an algorithm, named FOUNTAIN, for mining frequent closed sequential patterns with non-user-defined gap constraints. To our best knowledge, there is no previous work considering mining frequent closed sequential patterns without user predefined gap constraints. The main contributions of this work include: (1) analyzing the impact of gap constraint selection to the mining result; (2) proposing the problem of mining frequent closed sequential patterns with non-user-defined gap constraints; (3) designing an efficient algorithm to solve this problem, and conducting experiments on real data sets to verify it's effectiveness and efficiency.

---

[1] In this paper, we use '$\$$' to represent a wildcard.

The rest of the paper is organized as follows. Section 2 summarizes the related work. Section 3 provides the detailed definitions, as well as the necessary notations and terms. Section 4 presents our FOUNTAIN algorithm. Section 5 reports an experimental study on the data sets concerning protein families. Finally, Section 6 discusses the concluding remarks and future work.

2 Related Work

Frequent closed sequential pattern mining has been extensively studied in recent years, because of it's more compact yet complete result set and better efficiency. The CloSpan algorithm [4] was proposed for mining closed sequential patterns. It needs to maintain a candidate set to prune the search space and check the closure property. As maintenance of candidate set consumes much memory, Wang et al. [8] presented BIDE algorithm to mining frequent closed sequential pattern without candidate maintenance. The new pruning strategies and closure checking scheme make BIDE more efficient than previous algorithms.

Introducing gap constraints into sequence data mining can make the sequential patterns more flexible and general. For instance in biology analysis, gap constraints are widely considered in DNA or protein sequence mining [6,11,12,13].

Gap constraints have been used in mining a variety of types of sequential patterns. Antunes et al. [14] proposed an algorithm to handle the sequence pattern mining problem with gap constraints based on PrefixSpan [15]. Xie et al. [12] studied the frequent pattern mining under the condition of predefined minimum support and gap constraints. Zhang et al. [5] mined frequent periodic patterns satisfying a gap constraint. Shah et al. [16] introduced contrast patterns with gap constraint to peptide folding prediction. Ji et al. [3] designed ConSGapMiner to find minimal contrast subsequence patterns with gap constraints, which are frequent in one class and infrequent in the other. Wang et al. [17] proposed the concept of density in the mining task. He et al. [10] put forward a method of mining closed sequential patterns with gap constraints to analyze the growing log database. The Gap-BIDE algorithm, an expansion of BIDE by Li et al. [9], discovers frequent closed sequential patterns with gap constraints.

To the best of our knowledge, [9] is the work most related to ours. However, Gap-BIDE requires a predefined (fixed) gap constraint, which is hard for users to set, if they lack sufficient a priori knowledge. In contrast, our method doesn't require a predefined gap constraint. Our method finds the most suitable gap constraint for each frequent closed sequential pattern. As a result, our method is more user-friendly and practical.

3 Problem Definition and Analysis

In this section, we give the formal definition for frequent closed sequential patterns with non-user-defined gap constraints. First of all, we present some notations and terms used in this paper.

Let Σ be the alphabet, which is a finite set of distinct symbols. A sequence S over Σ is an ordered list with the form $S = s_1 s_2 ... s_n$, where $s_i \in \Sigma$ $(1 \leq i \leq n)$. The *length* of S is the number of elements it contains, denoted by $|S|$.

We use $S_{[i]}$ to denote the i-th element in S $(1 \leq i \leq |S|)$. For two elements $S_{[i]}$ and $S_{[j]}$ in S satisfying $1 \leq i < j \leq |S|$, the *gap* between $S_{[i]}$ and $S_{[j]}$, denoted by $Gap(S, i, j)$, is the number of elements between $S_{[i]}$ and $S_{[j]}$ in S. That is, $Gap(S, i, j) = j - i - 1$.

For two sequences S and S' satisfying $|S| > |S'|$, if there exist integers $1 \leq k_1 < k_2 < ... < k_{|S'|} \leq |S|$, such that $S' = S_{[k_1]} S_{[k_2]} ... S_{[k_{|S'|}]}$, we say that sequence S' is a sub-sequence of S and S is a super-sequence of S'.

Example 2. Given an alphabet $\Sigma = \{A, C, G, T\}$ and a sequence $S = ACGTCAT$, we have $|S| = 7$, $S_{[1]} = A$, $S_{[3]} = G$, $S_{[5]} = C$, $S_{[7]} = T$. Moreover, the gap between the first 'A' and the last 'C', i.e. $Gap(S, 1, 5)$, is 3.

A *gap constraint* γ is an interval that consists of two non-negative integers, $\gamma.min$ (lower bound) and $\gamma.max$ (upper bound), satisfying $\gamma.min \leq \gamma.max$. That is, $\gamma = [\gamma.min, \gamma.max]$. The *width* of γ, denoted by $||\gamma||$, is $\gamma.max - \gamma.min + 1$. Take $\gamma = [2, 5]$ as an example, $||\gamma|| = 5 - 2 + 1 = 4$.

For two sequences S and S' satisfying $S' = S_{[k_1]} S_{[k_2]} ... S_{[k_{|S'|}]}$ $(1 \leq k_1 < k_{|S'|} \leq |S|)$, given a gap constraint γ, if for any k_i and k_{i+1} $(1 \leq i < |S'|)$, $\gamma.min \leq Gap(S, k_i, k_{i+1}) \leq \gamma.max$, we say sequence S' is contained in sequence S with gap constraint γ, denoted by $S' \sqsubseteq_\gamma S$. We denote an *instance* of S' in S by $< k_1, k_2, ..., k_{|S'|} >$.

Example 3. Let $S = ACGTCAT$, $S' = ACT$, and $\gamma = [0, 3]$. Then $S' \sqsubseteq_\gamma S$, and both $< 1, 2, 4 >$ and $< 1, 5, 7 >$ are instances of S' in S.

Given a set $\mathcal{D}$ of sequences, the *support* of sequence P with gap constraint γ, denoted by $Sup(P, \gamma)$, is the fraction of sequences in $\mathcal{D}$ that contain P with γ. Specifically,

$$Sup(P, \gamma) = \frac{|\{S \in \mathcal{D} \mid P \sqsubseteq_\gamma S\}|}{|\mathcal{D}|} \quad (1)$$

Given a support threshold α, P is *frequent* in $\mathcal{D}$ with gap constraint γ, if $Sup(P, \gamma) \geq \alpha$.

Intuitively, closed patterns, instead of all patterns, result in a more compact yet complete presentation and better mining efficiency. However, as analyzed in [9], the downward closure property no longer holds in the gap-constrained sequential pattern. For example, given $S = ACGTCAT$, $P = ATA$, $P' = AA$, and $\gamma = [1, 2]$, we can see that $P \sqsubseteq_\gamma S$, while $P' \not\sqsubseteq_\gamma S$.

To break this limitation, a concept called *contiguous sub-sequences* is introduced in the definition of closed sequential pattern [9]. We also adopt this concept in this work.

For two sequences S and S' satisfying $|S| > |S'|$, if there exists $|S'|$ integers, $\overset{|S'|-1}{\underset{i=0}{\cup}} k + i$ $(k \geq 1)$, such that $\forall 0 \leq i \leq |S'| - 1$: $S'_{[i+1]} = S_{[k+i]}$, we say sequence

S' is a contiguous sub-sequence of S and S is a contiguous super-sequence of S'. For example, for $S = ACGT$, $S_1 = ACG$ and $S_2 = CGT$ are contiguous sub-sequences of it, yet $S_3 = ACT$ is not.

Lemma 1. *Let sequence P' be a contiguous sub-sequence of P. Given a gap constraint γ, then $Sup(P', \gamma) \geq Sup(P, \gamma)$.*

Proof. Since P' is a contiguous sub-sequence of P and $P \sqsubseteq_\gamma S$, then we have $P' \sqsubseteq_\gamma S$. Thus,

$$|\{S \in \mathcal{D} \mid P' \sqsubseteq_\gamma S\}| \geq |\{S \in \mathcal{D} \mid P \sqsubseteq_\gamma S\}|$$

By Equation 1, $Sup(P', \gamma) \geq Sup(P, \gamma)$.

Corollary 1. *Suppose sequence P is frequent with gap constraint γ, then all contiguous sub-sequences of P are frequent with γ.*

Intuitively, a gap constraint with smaller width is more accurate and useful. In this work, gap constraints with smaller width are preferred. In the case of same width, the gap constraint with smaller lower bound is preferred. More often than not, users do not want to see gap constraints with large width as answers, since large width is hard to understand and may be impractical. Thus, we assume a *maximum gap constraint width $\omega > 0$*, and consider only gap constraints whose width are not greater than ω.

Definition 1 (Frequent Closed Sequential Pattern with the Most Compact Gap Constraint). *Given a set $\mathcal{D}$ of sequences, a support threshold α, and a maximum gap constraint width ω, the tuple T consisting of sequence P and gap constraint γ ($\|\gamma\| \leq \omega$), denoted by $T = <P, \gamma>$, is a frequent closed sequential pattern with the most compact gap constraint, if conditions (1–3) are true:*

1. *(frequency condition) $Sup(P, \gamma) \geq \alpha$;*
2. *(closure condition) for any contiguous super-sequence P' of P, $Sup(P, \gamma) > Sup(P', \gamma)$;*
3. *(the most compact gap constraint) there is no gap constraint γ' satisfying either $\|\gamma'\| < \|\gamma\|$, or $\|\gamma'\| = \|\gamma\|$ and $\gamma'.min < \gamma.min$, such that conditions 1 and 2 hold.*

Given a set of sequences, a support threshold and a maximum gap constraints width, the *problem of mining frequent closed sequential patterns with non-user-defined gap constraints* is to discover all frequent closed sequential patterns with the most compact gap constraints as defined in Definition 1.

Example 4. Observing the data set in Table 1, suppose $\alpha = 0.8$. For the sequence $P = CA$, the tuple $< P, \gamma >$ always satisfies the frequency and closure conditions, when $\gamma = [0, 1]$, $\gamma = [0, 2]$, or $\gamma = [2, 3]$. From these three candidates, we select $< P, [0, 1] >$ as output because the width of gap constraint $[0, 1]$ is shorter than $[0, 2]$ and the lower bound of $[0, 1]$ is smaller than $[2, 3]$. Furthermore, all frequent closed sequential patterns generated by FOUNTAIN include: $\{< CF, [2, 4] >, < CA, [0, 1] >, < CC, [0, 1] >, < CAC, [0, 3] >, < BF, [1, 5] >, < BA, [0, 1] >, < AF, [2, 3] >, < AC, [0, 1] >, < DF, [0, 1] >\}$.

4   Mining Algorithm

In this section, we present our algorithm FOUNTAIN for discovering frequent closed sequential patterns with non-user-defined gap constraints. In general, FOUNTAIN consists of three main steps: candidate generation, frequency condition checking, and closure condition checking.

4.1   The Framework of FOUNTAIN

As a frequent closed sequential patterns with the most compact gap constraint (Definition 1) is a tuple consisting of a sequence and a gap constraint, FOUNTAIN generates candidate sequences and for each candidate sequence, FOUNTAIN checks whether there exists a gap constraint such that the frequency and closure conditions are satisfied.

In order to find all frequent closed patterns, FOUNTAIN enumerates each candidate sequence by the approach of set-enumeration tree [18], which takes a total order on the set, the symbols in alphabet in the context of our problem, and then enumerates all possible combinations systematically. FOUNTAIN traverses the set-enumeration tree in a depth-first search manner. Figure 1 illustrates an example of a set-enumeration tree with $\Sigma = \{A, B, C\}$. Algorithm 1 presents the framework of FOUNTAIN.

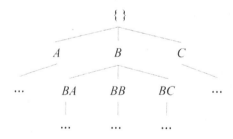

Fig. 1. An example of a set-enumeration tree

4.2   Frequency Checking

For a candidate sequence P, FOUNTAIN first checks whether there exists a gap constraint γ, such that $< P, \gamma >$ is frequent. According to Equation 2, we can get the upper bound of $< P, \gamma >$'s support.

Proposition 1. *For a sequence P, given a gap constraint γ, the smallest upper bound of $Sup(P, \gamma)$, denoted by $Sup_{max}(P)$, is*

$$Sup(P, \gamma) \leq \frac{|\{S \in \mathcal{D} \mid P \text{ is a sub-sequence of } S\}|}{|\mathcal{D}|} = Sup_{max}(P) \qquad (2)$$

Algorithm 1. The Framework of FOUNTAIN

Input: $\mathcal{D}$: set of sequences, α: support threshold, ω: maximum gap constraint width
Output: R: set of frequent closed sequential patterns with the most compact gap
    constraints
 1: traverse the sequence set-enumeration tree in a depth-first search manner;
 2: **for** each sequence P **do**
 3:    $\Gamma \leftarrow \{\gamma \mid Sup(P, \gamma) \geq \alpha\}$; // Section 4.2
 4:    **if** $\Gamma \neg \emptyset$ **then**
 5:       $\gamma \leftarrow$ the most compact gap constraint in Γ;
 6:       **if** $< P, \gamma >$ is closed **then**
 7:          $R \leftarrow R \cup \{< P, \gamma >\}$; // Section 4.3
 8:       **end if**
 9:    **end if**
10: **end for**
11: **return** R;

By Proposition 1, if $Sup_{max}(P)$ is less than the support threshold α, then P cannot be frequent with any gap constraint. We get the following pruning rule.

Pruning Rule 1. *For sequence P, if $Sup_{max}(P)$ is less than the support threshold α, P and its child nodes in the set-enumeration tree are pruned.*

Let ℓ be the maximum length of a sequence in data set $\mathcal{D}$. All gap constraints are contained in interval $[0, \ell - 2]$. Therefore, we can get a set of candidate gap constraints $\Gamma = \{\gamma \mid 0 \leq \gamma.min \leq \gamma.max \leq (\ell - 2)\}$. Algorithm 2 presents the pseudo-code of generating candidate gap constraints with time complexity $O(\ell^2)$. Note that the result of Algorithm 2 may include non-compact gap constraints.

Algorithm 2. A Straightforward Way to Generate Candidate Gap Constraints

---

Input: ℓ: maximum sequence length in $\mathcal{D}$, ω: maximum gap constraint width
Output: Γ: set of candidate gap constraints
 1: $\Gamma \leftarrow \varnothing$;
 2: **for** $i \leftarrow 0$ to $\ell - 2$ **do**
 3:    **for** $j \leftarrow i$ to $i + \omega - 1$ **do**
 4:       **if** $j \leq \ell - 2$ **then**
 5:          $\Gamma \leftarrow \Gamma \cup \{[i, j]\}$;
 6:       **end if**
 7:    **end for**
 8: **end for**
 9: **return** Γ;

Example 5. In Table 1, the length of the longest sequence in $\mathcal{D}$ is 9. So for every candidate gap constraint γ, we have $0 \leq \gamma.min \leq \gamma.max \leq 7$. Let sequence $P = DC$. Then there are four instances, which correspond to three gap constraints: [2,2], [3,3], [4,4]. Namely, the minimal gap between 'D' and 'C' is 2. Thus, gap constraints $[0, 0]$ and $[1, 1]$ are redundant candidates for P.

For the sake of efficiency, inspired by merge sort [19], we design a new method to generate candidate gap constraints, and select the most compact gap constraint one by one. For any two gap constraints γ_1, γ_2, a new gap constraint γ, merged from γ_1 and γ_2 (denoted by $\gamma_1 \cup \gamma_2$), is

$$\gamma.min = MIN\{\gamma_1.min, \gamma_2.min\}, \gamma.max = MAX\{\gamma_1.max, \gamma_2.max\}$$

For example, if $\gamma_1 = [1, 4]$, $\gamma_2 = [2, 5]$, then $\gamma_1 \cup \gamma_2 = [1, 5]$.

Definition 2 (Minimal Single Instance Gap Constraint). *Given two sequences P and S satisfying $P = S_{[k_1]}S_{[k_2]}...S_{[k_{|P|}]}$ ($1 \leq k_1 < k_{|P|} \leq |S|$), then the gap constraint $\gamma = [MIN\{Gap(S, k_i, k_{i+1})\}, MAX\{Gap(S, k_i, k_{i+1})\}]$ ($1 \leq i < |P|$) is a minimal single instance gap constraint.*

Example 6. For the data in Table 1, let sequence $P = CAC$. Then P have five different minimal single instance gap constraints: [0,0], [0,1], [1,3], [0,3] and [0,4].

Let Γ_s be the set of minimal single instance gap constraints. We use B_l to denote the lower bounds of gap constraints in Γ_s. That is, $B_l = \{\gamma.min \mid \gamma \in \Gamma_s\}$. For each $i \in B_l$, all candidate upper bounds, denoted by $U(i)$, are

$$U(i) = \{\gamma_2.max \mid \exists \gamma_1, \gamma_2 \in \Gamma_s, i = \gamma_1.min, i \leq \gamma_2.min, \gamma_1.max \leq \gamma_2.max\} \quad (3)$$

FOUNTAIN sorts $U(i)$ in ascending order. Algorithm 3 gives the pseudo-code for finding the most compact gap constraint for a candidate sequence.

Algorithm 3. Frequency Checking with the Most Compact Gap Constraint

---

Input: Γ_s: set of single instance gap constraints, P: candidate sequence, α: support threshold

Output: γ: the most compact gap constraint satisfying $Sup(P, \gamma) \geq \alpha$

1: $B_l \leftarrow \{\gamma'.min \mid \gamma' \in \Gamma_s\}$;
2: **for** each $i \in B_l$ **do**
3:    generate $U(i)$ by Equation 3;
4:    sort $U(i)$ by ascending order;
5: **end for**
6: **repeat**
7:    $R \leftarrow \{[i, j] \mid i \in B_l, j \text{ is the first element in } U(i)\}$;
8:    $\gamma \leftarrow$ the most compact gap constraint in R;
9:    $U(\gamma.min) \leftarrow U(\gamma.min) \setminus \{\gamma.max\}$;
10:   **if** $Sup(P, \gamma) \geq \alpha$ **then**
11:       **return** γ;
12:   **end if**
13: **until** $R = \emptyset$;
14: **return** null;

Example 7. Let $\Gamma_s = \{[0,2], [0,4], [1,1], [1,3], [2,3], [2,6]\}$. Then $B_l = \{0,1,2\}$ and $U(0) = \{2,3,4,6\}$, $U(1) = \{1,3,6\}$, $U(2) = \{3,6\}$. The selection process goes as follows. First, FOUNTAIN selects the gap constraint $\gamma = [1,1]$ (the width is minimum), and then checks whether $< P, \gamma >$ satisfies the closure condition for sequence P. If it does, $< P, \gamma >$ is output. Otherwise, 1 is removed from $U(1)$, and FOUNTAIN continues to select the most compact gap constraint from $U(0), U(1),$ and $U(2)$. Figure 2 demonstrates this process.

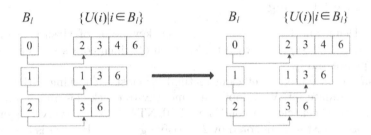

Fig. 2. Removal of a gap when selecting the most compact gap constraint

Lemma 2. *Given a set $\mathcal{D}$ of sequences, let P be a contiguous sub-sequence of Q. Then $MAX\{Sup(P,\gamma) \mid \gamma \in \Gamma\} \geq MAX\{Sup(Q,\gamma) \mid \gamma \in \Gamma\}$.*

Proof. Assume, for contradiction, that there exists a gap constraint $\gamma' \in \Gamma$ making $MAX\{Sup(P,\gamma) \mid \gamma \in \Gamma\} < Sup(Q,\gamma')$ is true. Then $Sup(P,\gamma') \leq MAX\{Sup(P,\gamma) \mid \gamma \in \Gamma\} < Sup(Q,\gamma')$.

From Equation 1, we get

$$Sup(P,\gamma') = \frac{|\{S \in \mathcal{D} \mid P \sqsubseteq_{\gamma'} S\}|}{|\mathcal{D}|} < Sup(Q,\gamma') = \frac{|\{S \in \mathcal{D} \mid Q \sqsubseteq_{\gamma'} S\}|}{|\mathcal{D}|} \quad (4)$$

However, P is a contiguous sub-sequence of Q. Therefore according to Lemma 1, $Sup(P,\gamma') \geq Sup(Q,\gamma')$. This contradicts with Equation 4, completing the proof.

According to Lemma 2, we get Pruning Rule 2.

Pruning Rule 2. *If $Sup\{(P,\gamma) \mid \forall \gamma \in \Gamma\} < \alpha$, then we can prune all children nodes of P in set-enumeration tree.*

After getting the most compact gap constraint γ and $Sup(P,\gamma) \geq \alpha$, considering the correlation between sub-sequence and super-sequence, we get Pruning Rule 3.

Pruning Rule 3. *Let P be a contiguous sub-sequence of Q and γ the most compact gap constraint of P. When searching for the most compact gap constraint for Q, if $Sup(P,\gamma) \geq \alpha$, then we can prune all gap constraints before γ in the sorted order, if they correspond to Q and are selected from the set Γ.*

For example, suppose $Sup(P, [2, 5]) \geq \alpha$, and $[2,5]$ is the most compact gap constraint for P. Then for any super-sequence of P, we can ignore all gap constraints whose width is less then 4, and select the most compact one starting from gap constraint $[2, 5]$.

When we get the most compact gap constraint γ, the next step is checking whether $< P, \gamma >$ satisfies frequency and closure conditions. In the next subsection, we illustrate the implementation of closure checking.

4.3   Closure Checking

Recalling Definition 1, given $< P, \gamma >$, the key point of closure checking is the comparison of $Sup(P, \gamma)$ with tuples, each of which consists P's contiguous super-sequence and γ.

Li *et al.* [9] presented an efficient method for closure checking without maintaining a candidate set. We adopt the same way to check closure condition.

In order to check closure condition, FOUNTAIN scans every sequence, in which P has an instance, denoted by I, satisfying $\gamma_I \subseteq \gamma$. Here γ_I is the minimal single instance gap constraint of I. Suppose $S_{[f]}$ and $S_{[e]}$ are the first and last elements of I in S.

Let $\Phi(S) = \{S_{[k]} \mid f - \gamma.max - 1 \leq k \leq f - \gamma.min - 1\}$ and $\Psi(S) = \{S_{[k]} \mid e + \gamma.min + 1 \leq k \leq e + \gamma.max + 1\}$. After FOUNTAIN has finished the scan process, if

$$\bigcap_{P \sqsubseteq_\gamma S} \Phi(S) = \emptyset \text{ and } \bigcap_{P \sqsubseteq_\gamma S} \Psi(S) = \emptyset$$

then $< P, \gamma >$ is a closed sequential pattern, i.e. $< P, \gamma >$ is a frequent closed sequential pattern with the most compact gap constraint.

Example 8. For the data in Table 1, suppose the current candidate pattern is $< CF, [2, 4] >$. We can obtain $\Psi\{S2\} = \{A, B\}$, $\Psi\{S5\} = \{C, E\}$ through scanning all instances of $P = CF$. Now $\Psi\{S2\} \cap \Psi\{S5\} = \emptyset$, in other words there is no contiguous super-sequence of P that has the same support with P when gap constraint is $[2, 4]$. So, $< CF, [2, 4] >$ is a frequent closed sequential pattern with gap constraint $[2, 4]$.

5   Empirical Evaluation

In this section, we report a systematic empirical study using real data sets to verify the effectiveness and efficiency of our method. All experiments were conducted on a PC computer with an Intel Core i7-4770 3.40 GHz CPU, and 8 GB main memory, running Windows 7 operating system. All algorithms were implemented in Java and compiled by JDK 7.

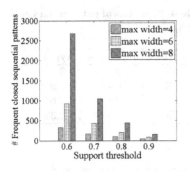
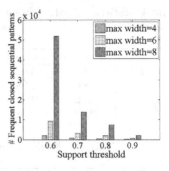

Fig. 3. Distribution of results on CbiX   **Fig. 4.** Distribution of results on DUF1694

5.1 Effectiveness

To have realistic and interesting mining results, we select two protein sequential data sets CbiX and DUF1694 from Pfam[2]. Statistics of these two data sets are listed in Table 2.

Table 2. Statistics of data sets

Data set	# sequences	# symbols	Average length of sequences (Avg. len.)
CbiX	76	40	106
DUF1694	16	22	123

In order to verify the effectiveness of our method, FOUNTAIN, we apply it to mining frequent closed sequential patterns from CbiX and DUF1694, respectively. Figure 3 (Figure 4) illustrates the number of patterns discovered from CbiX (DUF1694) using different support thresholds (α) and values of maximum gap constraint width (ω). We can see that the number of patterns decreases as α increases. In addition, as α increases, the numbers of patterns discovered using different ω are closer to each other. The reason is that more sequences are pruned by Pruning Rules 1 and 2 as the increase of α, no matter what the value of ω is.

Recall that in this paper we don't predefine a gap constraint when mining frequent closed sequential patterns. Instead, we find the most compact gap constraints in the process of mining. Table 3 and Table 4 show the statistics of patterns on CbiX when $\alpha = 0.9$ and $\omega = 8$, respectively.

Table 5 lists the mining results of FOUNTAIN in CbiX when $\alpha = 0.9$ and $\omega = 2$. We can see that patterns with different gap constraints are discovered by FOUNTAIN. Thus, our method is more flexible compared with previous closed sequential pattern mining methods which need user to predefine a gap constraint.

---

[2] Protein Family Database. http://www.sanger.ac.uk/

Table 3. Statistics of patterns on CbiX ($\alpha = 0.9$)

ω	# patterns	Avg. len.
4	44	2.023
6	84	2.083
8	156	2.321
10	282	2.585

Table 4. Statistics of patterns on CbiX ($\omega = 8$)

$\alpha * |CbiX|$	# patterns	Avg. len.
46	2686	3.371
53	1049	2.991
61	447	2.678
75	156	2.321

Table 5. Mining results in CbiX ($\alpha = 0.9$ and $\omega = 2$)

$< EL, [0,1] >$	$< LA, [0,1] >$	$< LE, [0,1] >$	$< PL, [0,1] >$
$< LL, [1,2] >$	$< VL, [2,3] >$	$< LV, [7,8] >$	$< GL, [8,9] >$
$< AL, [11,12] >$	$< GA, [13,14] >$		

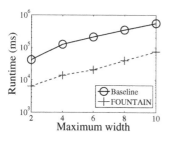

Fig. 5. Runtime in CbiX when $\alpha = 0.9$ **Fig. 6.** Runtime in DUF1694 when $\alpha = 0.9$

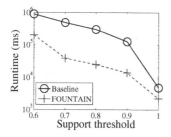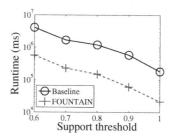

Fig. 7. Runtime in CbiX when $\omega = 4$ **Fig. 8.** Runtime in DUF1694 when $\omega = 4$

5.2 Efficiency

To the best of our knowledge, there was no existing method similar to our method. So we adopt Algorithm 2 as a baseline to generate candidate gap constraints and compare it with FOUNTAIN in running time. We applied FOUNTAIN as well as baseline on the protein families to mining frequent closed sequential patterns.

Figures 5 and 6 show the efficiency test on CbiX and DUF1694 with respect to ω when $\alpha = 0.9$, respectively. We can see that FOUNTAIN runs faster than the baseline, because the number of candidate gap constraints generated by FOUNTAIN is smaller than that of the baseline.

Figures 7 and 8 compare the runtime of FOUNTAIN with the baseline on CbiX and DUF1694, respectively. When α is larger, the number of candidate patterns is smaller, so the decrease in runtime is significant. FOUNTAIN is more efficient than the baseline because Pruning Rule 1 speeds up the search process.

6   Conclusions

The previous frequent closed sequential pattern mining problem requires users to predefine a gap constraint as a parameter. But setting a suitable gap constraint is difficult for users who lack enough background knowledge, so many interesting patterns may be missed. To deal with this problem, we present an algorithm, named FOUNTAIN, to find frequent closed sequential patterns with the most suitable gap constraints. The experiments on protein data sets verify the effectiveness and efficiency of FOUNTAIN.

As future work, we attempt to apply our method to more medical data analysis and study the situation of multi-items instead of one item on mining frequent closed sequential patterns with non-user-defined gap constraints.

References

1. Agrawal, R., Srikant, R.: Mining sequential patterns. In: Proc. of the 11th Int'l Conf. on Data Engineering, Taipei, Taiwan, pp. 3–14 (1995)
2. Zaki, M.J.: SPADE: An efficient algorithm for mining frequent sequences. Mach. Learn. 42(1-2), 31–60 (2001)
3. Ji, X., Bailey, J., Dong, G.: Mining minimal distinguishing subsequence patterns with gap constraints. Knowl. Inf. Syst. 11(3), 259–286 (2007)
4. Yan, X., Han, J., Afshar, R.: CloSpan: Mining closed sequential patterns in large databases. In: Proc. of the 3rd SIAM Int'l Conf. on Data Mining, San Francisco, CA, USA, pp. 166–177 (2003)
5. Zhang, M., Kao, B., Cheung, D.W., Yip, K.Y.: Mining periodic patterns with gap requirement from sequences. ACM Trans. Knowl. Discov. Data 1(2) (August 2007)
6. Ferreira, P.G., Azevedo, P.J.: Protein sequence pattern mining with constraints. In: Jorge, A.M., Torgo, L., Brazdil, P.B., Camacho, R., Gama, J. (eds.) PKDD 2005. LNCS (LNAI), vol. 3721, pp. 96–107. Springer, Heidelberg (2005)
7. She, R., Chen, F., Wang, K., Ester, M., Gardy, J.L., Brinkman, F.S.L.: Frequent-subsequence-based prediction of outer membrane proteins. In: Proc. of the 9th ACM SIGKDD Int'l Conf. on Knowl. Discov. and Data Mining, pp. 436–445. ACM, New York (2003)
8. Wang, J., Han, J., Li, C.: Frequent closed sequence mining without candidate maintenance. IEEE Trans. on Knowl. and Data Engineering 19(8), 1042–1056 (2007)
9. Li, C., Yang, Q., Wang, J., Li, M.: Efficient mining of gap-constrained subsequences and its various applications. ACM Trans. Knowl. Discov. Data 6(1), 2:1–2:39 (2012)
10. He, H., Wang, D., Chen, G., Zhang, W.: An alert correlation analysis oriented incremental mining algorithm of closed sequential patterns with gap constraints. Appl. Math 8(1L), 41–46 (2014)

11. Wu, X., Zhu, X., He, Y., Arslan, A.N.: PMBC: Pattern mining from biological sequences with wildcard constraints. Comput. Biol. Med. 43(5), 481–492 (2013)
12. Xie, F., Wu, X., Hu, X., Gao, J., Guo, D., Fei, Y., Hua, E.: MAIL: Mining sequential patterns with wildcards. Int. J. Data Min. Bioinformatics 8(1), 1–23 (2013)
13. Gusfield, D.: Algorithms on Strings, Trees, and Sequences: Computer Science and Computational Biology. Cambridge University Press, New York (1997)
14. Antunes, C., Oliveira, A.L.: Generalization of pattern-growth methods for sequential pattern mining with gap constraints. In: Perner, P., Rosenfeld, A. (eds.) MLDM 2003. LNCS, vol. 2734, pp. 239–251. Springer, Heidelberg (2003)
15. Pei, J., Han, J., Mortazavi-asl, B., Pinto, H., Chen, Q., Dayal, U., Hsu, M.-c.: PrefixSpan: Mining sequential patterns efficiently by prefix-projected pattern growth. In: Proc. of the 17th Int'l Conf. on Data Engineering, ICDE 2001, pp. 215–224. IEEE Computer Society, Washington, DC (2001)
16. Shah, C.C., Zhu, X., Khoshgoftaar, T.M., Beyer, J.: Contrast pattern mining with gap constraints for peptide folding prediction. In: Proc. of the 21st Int'l FLAIRS Conf., Coconut Grove, Florida, USA, pp. 95–100 (2008)
17. Wang, X., Duan, L., Dong, G., Yu, Z., Tang, C.: Efficient mining of density-aware distinguishing sequential patterns with gap constraints. In: Bhowmick, S.S., Dyreson, C.E., Jensen, C.S., Lee, M.L., Muliantara, A., Thalheim, B. (eds.) DASFAA 2014, Part I. LNCS, vol. 8421, pp. 372–387. Springer, Heidelberg (2014)
18. Rymon, R.: Search through systematic set enumeration. In: Proc. of the 3rd Int'l Conf. on Principles of Knowl. Representation and Reasoning, pp. 539–550. Cambridge (1992)
19. Cormen, T.H., Leiserson, C.E., Rivest, R.L., Stein, C.: Introduction to Algorithms, 2nd edn. MIT Press, Cambridge (2001)

High Utility Episode Mining Made Practical and Fast

Guangming Guo[1,2], Lei Zhang[3], Qi Liu[1],
Enhong Chen[1], Feida Zhu[2], and Chu Guan[1]

[1] School of Computer Science and Technology
University of Science and Technology of China, Hefei 230027, China
[2] School of Information Systems, Singapore Management University
[3] School of Computer Science and Technology, Anhui University
guogg@mail.ustc.edu.cn

Abstract. This paper focuses on the problem of mining high utility episodes from complex event sequences. Episode mining, one of the fundamental problems of sequential pattern mining, has been continuously drawing attention over the past decade. Meanwhile, there is also tremendous interest in the problem of high utility mining. Recently, the problem of high utility episode mining comes into view from the interface of these two research areas. Although prior work [11] has proposed algorithm UP-Span to tackle this problem, their method suffers from several performance drawbacks. To that end, firstly, we explicitly interpret the high utility episode mining problem as a complete traversal of the lexicographic prefix tree. Secondly, under the framework of lexicographic prefix tree, we examine the original UP-Span algorithm and present several improvements on it. In addition, we propose several clever strategies from a practical perspective and obtain much tighter utility upper bounds of a given node. Based on these optimizations, an efficient algorithm named TSpan is presented for fast high utility episode mining using tighter upper bounds, which reduces huge search space over the prefix tree. Extensive experiments on both synthetic and real-life datasets demonstrate that TSpan outperforms the state-of-the-art in terms of both search space and running time significantly.

1 Introduction

Sequential pattern mining [6], one of the most important data mining problems, has been continuously drawing attention from researchers. And when the sequential data becomes event sequence, the task of *frequent episodes mining* (FEM) [7] is introduced. FEM reveals a lot of useful information hidden in the event sequence with a wide range of applications [2,4,7,8,10]. However, the discovered frequent episode is still too simple and primitive. In some cases, FEM may lose some important information. It only takes the presence and absence of events into account and neglects the semantic information of different events. However, in reality, events in the same sequence can be significantly different from each other. For example, different web pages/items are of different importance/cost/profit for the decision makers (both producers and consumers). If we only apply FEM to event sequences, some truly interesting episodes, such as high-profit ones, may be filtered out due to their low frequency.

X. Luo, J.X. Yu, and Z. Li (Eds.): ADMA 2014, LNAI 8933, pp. 71–84, 2014.

To tackle the above challenge, the concept of utility is introduced as an alternative measure aside from frequency [3,12]. In *high utility mining*, unlike the traditional frequency based pattern mining, the *downward closure property* does not keep. In other words, when appending a new item/event to an itemset/episode one by one, the frequency of an itemset/episode decreases monotonously or remains unchanged, but the utility varies irregularly. As a result, previous optimization methods in frequent pattern mining become invalid in high utility mining. Without efficient pruning strategies, high utility mining becomes prohibitive due to the curse of dimensionality. Fortunately, exploiting *transaction/sequence weighted utility* (*TWU/SWU*) [5,13], we can first generate candidates efficiently using *TWU/SWU*'s downward closure property, and then identify high utility ones from far smaller number of candidates.

However, previous high utility mining works mainly focus on the transactional databases, seldom consider other types of databases, like sequence database. As far as we know, Wu et al. [11] presented the first attempt to solve the problem of *high utility episode mining*[1] in complex event sequence. But the proposed algorithm UP-Span suffers from low efficiency in both running time and memory consumption. More importantly, the proposed utility upper bound named *Episode Weighted Utility* (*EWU*) is only a loose and basic utility bound for episodes. In practice, we may need to check a large number of candidates when using *EWU*, which becomes a bottleneck for high utility episode mining in large databases. To that end, in this paper we present several improvements over UP-Span, and an efficient algorithm named TSpan, which has two much tighter upper bounds than *EWU* for high utility episode mining.

Contribution. In this paper, we first explicitly tackle high utility episode mining under the framework of a traversal of lexicographic prefix tree. Under this framework, we discuss how to implement the UP-Span algorithm in a much more efficient manner, which can save a lot of search space and running time. Moreover, leveraging the lexicographic prefix tree, we propose algorithm named TSpan (**T**ighter upper bound when **Span**ning prefixes), which consists of two tighter upper bounds for operations related to spanning episodes. These upper bounds are able to maintain the preferable downward closure property, which effectively reduces search space over the tree. Last but not least, we conduct experiments with both synthetic and real-life datasets and clearly show that TSpan can effectively reduce both search space and running time.

2   Preliminaries and Lexicographic Prefix Tree

Fig. 1 shows a simple complex event sequence, where each event in the sequence is associated with an event type and occurrence time. It is called complex event sequence because at each occurrence time there can be events occurring simultaneously and the frequency of events at a given time is different, as shown in Fig. 1. Following [11], events occur at the same time point are called simultaneous events SE. For high utility episode mining, each event is associate with an external utility value and Table 1 shows an example utility table for the example event sequence in Fig. 1.

---

[1] Following [11], when talking about the word episode, we specifically refer to serial episode. Other type of episodes are good candidates for future research.

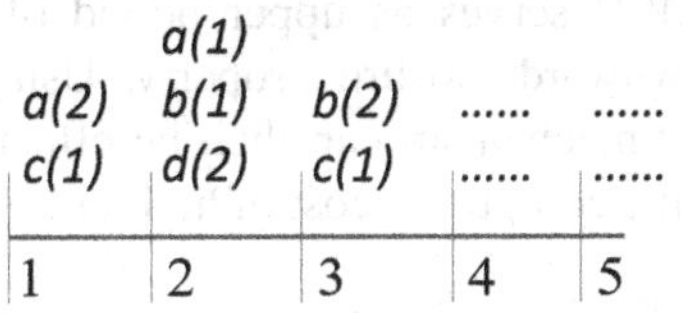

```
        a(1)
a(2)   b(1)   b(2)   ......   ......
c(1)   d(2)   c(1)   ......   ......

1      2      3      4      5
```

Fig. 1. A simple complex event sequence

Table 1. External utilities for events in Fig. 1

Event	a	b	c	d
Utility	1	1	1	1.5

Introduced in [7], the number of minimal occurrences is a popular measure for frequent episode mining. Formal definition of minimal occurrence is as follows: an occurrence of an episode α, $[T_s, T_e]$, is minimal iff α does not occur at any subinterval $[T_s', T_e'] \subset [T_s, T_e]$. For simplicity, here we denote a minimal occurrence of episode α as $mo(\alpha)$. The set of all minimal occurrences of α is denoted as $moSet(\alpha)$. For instance, in Fig. 1, the time interval $[1,2]$ is one of the minimal occurrences of episode $\langle ab \rangle$ and $moSet(\langle a, b \rangle) = \{[1,2],[2,3]\}$.

Based on minimal occurrence, we can define the utility of an episode. Given an episode $\alpha = \langle SE_1, SE_2, ..., SE_k \rangle$, where each simultaneous events SE_i is associated with a time point T_i, the utility of episode α w.r.t. minimal occurrence $mo(\alpha) = [T_s, T_e]$ is $u(\alpha, mo(\alpha)) = \sum_{i=1}^{k} u(SE_i, T_i)$. For example, $u(\langle ab \rangle, [1,2]) = u(a, T_1) + u(b, T_2) = 2 + 1 = 3$. As one episode may have multiple minimal occurrences, the utility of an episode can be naturally defined as the sum of episode's utility w.r.t. each minimal occurrence, i.e., $u(\alpha) = \sum_{mo(\alpha) \in moSet(\alpha)} u(\alpha, mo(\alpha))/u(CES)^2$. For episode $\langle ab \rangle$ in Fig. 1, $u(\langle ab \rangle) = (u(\langle ab \rangle, [1,2]) + u(\langle ab \rangle, [2,3]))/11 = (3+3)/11 = 0.55$.

In real world cases, the events of an episode usually happen together within a reasonable time period. Following [11], we name maximum time duration of an episode as MTD. For any minimal occurrence $mo(\alpha) = [T_s, T_e]$ of α, it must satisfy the condition that $T_e - T_s + 1 \leq MTD$. With MTD, we can then formally introduce the concept *EWU* presented in [11] as follows.

Definition 1 (EWU(α)). *Given episode* $\alpha = \langle (SE_1), (SE_2), ..., (SE_n) \rangle$ *and* $mo(\alpha) = [T_s, T_e]$ *is one of its minimal occurrence, EWU (Episode[3] Weighted Utility) value of* α *w.r.t.* $mo(\alpha)$, *i.e.,* $EWU(\alpha, mo(\alpha)) = \sum_{i=1}^{n} u(SE_i, T_i) + \sum_{i=e}^{s+MTD-1} u(CES_i, T_i)$. *Then given* $moSet(\alpha) = \{mo(\alpha)_1, mo(\alpha)_2, ..., mo(\alpha)_k\}$, $EWU(\alpha) = \sum_{i=1}^{k} EWU(\alpha, mo(\alpha)_i)$ $/u(CES)$.

---

[2] For simplicity of comparison, episode's utility is defined a ratio between its utility value and complex event sequence's total utility value.

[3] Under the current context, the term "episode" specifically denotes the largest episode when expanding α given a user specified MTD.

Similar to *TWU/SWU*, *EWU* serves as upper bound of an episode's utility and maintains the favorable downward closure property. Using this upper bound, the candidate-generation-and-test mechanism can then be effectively used in high utility episode mining. In this way, the computing cost of high utility episode mining becomes acceptable.

Different from [11], we view high utility episode mining from complex event sequences as a complete traversal over the lexicographic prefix tree. Under this framework, a more clear understanding of the process can be obtained, and further development of optimization becomes easier. Before formally introducing lexicographic prefix tree, we would like to first define two operations over the prefix tree, i.e., I-Concatenation and S-Concatenation, which are closely related to this concept.

Definition 2 (I-Concatenation and S-Concatenation). *Assume that we have an l-episode[4] α, appending an event to the end of α will lead to an expanded episode, which is an $(l+1)$-episode. We call this operation as* concatenation. *Specifically, when the time duration of α does not change, we denote this operation as* I-Concatenation. *However, when the time duration of α is increased by 1, we denote this operation as* S-Concatenation.

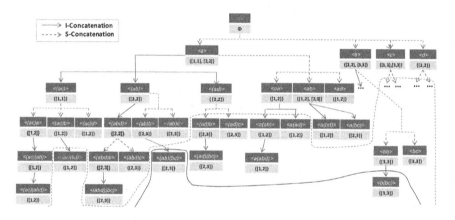

Fig. 2. Lexicographic prefix tree for event sequence in Fig. 1 (MTD=2, min_util=0.7)

Definition 3 (Lexicographic prefix tree). *In lexicographic prefix tree, a) the root node[5] of the prefix tree is empty, b) all the child nodes are generated from the I-Concatenation or S-Concatenation of a parent node, c) all the child nodes are listed in an incremental and lexicographic order.*

Fig. 2 shows an example lexicographic prefix tree for the example event sequence $\langle(ac)(abd)(bc)\rangle$ in Fig. 1. In Fig. 2, we also present the minimal occurrences associated with each episode. As it can be seen, even for such short an event sequence with MTD=2 and min_util=0.7, large number of nodes need checking if there is no pruning strategy employed. In Fig. 2, the black solid line shows the search space boundary over the

---

[4] An l-episode means that the episode's length is l.

[5] Without ambiguity, the terms episode and node will be used interchangeably in the following.

prefix tree when only using $EWU(\alpha)$ as the upper bound estimation of episode's utility for pruning. In the following, we will also take Fig. 2 as example for explaining our improvements over UP-Span and proposed pruning strategies. For easy reference, we summarize the notations defined above in Table 2.

Table 2. Summary of notations described above

CES	The total complex event sequence
SE_k	Simultaneous events at time T_k
$mo(\alpha)$	The minimal occurrence of episode α
$moSet(\alpha)$	The set of all minimal occurrences of episode α
$u(\alpha)$	The utility value of α
MTD	Maximum time duration of episodes
$EWU(\alpha)$	Episode Weighted Utility of α
$I\text{-}Concatenation$	Concatenation operation that increases episode length by 1, but episode's time duration size keeps unchanged
$S\text{-}Concatenation$	Concatenation operation that increases episode length by 1, and also increases episode's time duration size by 1

3   Efficient High Utility Episode Mining

In this section, we will first discuss the improvements over the original UP-Span algorithm we have made. Secondly, we will present our proposed pruning strategies, which are more tighter and efficient than EWU. Our strategies effectively improve the algorithm's efficiency by further reducing the search space. In the example lexicographic prefix tree (Fig. 2) for event sequence $\langle (ac)(abd)(bc) \rangle$, the search space boundary using our pruning strategies is shown by the dashed cut line (red). Compared to EWU, this example clearly shows that our proposed upper bound based pruning strategies are much more effective and efficient.

3.1   Efficient Implementation of the UP-Span Algorithm

The algorithm UP-Span (*high Utility ePisode mining by Spanning prefixes*) presented in [11], adopts idea similar to that of [13,9], all following the same *prefix-growth* (also known as *pattern-growth*) paradigm. When spanning prefixes in the complex event sequences, there are two kinds of candidate events, i.e., the simultaneous events and serial events, to extend with. Thus, we need to define two different procedures for spanning episode prefixes in high utility episode mining. The UP-Span algorithm uses the term *miningSimultHUE* and *miningSerialHUE* to denote these two procedures. Readers can refer to [11] for the details of the UP-Span algorithm. However, we'd like to adopt *I-Concatenation* and *S-Concatenation* [13] to denote these two different procedures, since these two terms explicitly indicating that these two operations are on the lexicographic prefix tree. The details of our implementation of *I-Concatenation* and *S-Concatenation* are shown in Algorithm 1 and 2. The procedure named **Prefix-Growth** is omitted since it is simply a successive call of these two operations.

Algorithm 1. I-Concatenation

Input : (1) α: episode (2) $moSet(\alpha)$: all minimal occurrences of α (3) MTD:
maximum time duration (4) $min\_utility$: minimum utility threshold

Output: The set of high utility simultaneous episodes w.r.t prefix α

1   **for** *each* $mo(\alpha) = [T_s, T_e] \in moSet(\alpha)$ **do**

2     $IES = \{e \mid$ event e occurs at T_e *and* e *is after the last event in* α $\}$;

3   **for** *each event* $e \in IES$ **do**

4       $\beta = I\text{-}concatenate(\alpha, e)$;

5       $mo(\beta) = mo(\alpha)$;

6       $betaSet = betaSet \cup \beta$;

7       $moSet(\beta) = moSet(\beta) \cup mo(\beta)$;

8   **for** *each* $\beta \in betaSet$ **do**

9   **if** $EWU(\beta) \geq min\_utility$ **then**

10   **if** $u(\beta) \geq min\_utility$ **then**

11         $HUE\_Set = HUE\_Set \cup \beta$;

12   **Prefix-Growth**$(\beta, moSet(\beta), MTD, min\_utility)$;

Algorithm 2. S-Concatenation

Input : (1) α: episode (2) $moSet(\alpha)$: all minimal occurrences of α (3) MTD:
maximum time duration (4) $min\_utility$: minimum utility threshold

Output: The set of high utility simultaneous episodes w.r.t prefix α

1   **for** *each* $mo(\alpha) = [T_s, T_e] \in moSet(\alpha)$ **do**

2   **for** *each time point* t *in* $[T_e + 1, T_s + MTD + 1]$ **do**

3       $SES_t = \{e \mid$ event e occurs at $t\}$;

4   **for** *each event* $e \in SES_t$ **do**

5         $\beta = S\text{-}concatenate(\alpha, e)$;

6         $mo(\beta) = [T_s, t]$;

7         $M = \{mo \mid mo \in moSet(\beta)$ *and* $mo \subseteq mo(\beta)\}$;

8   **if** $M = \emptyset$ **then**

9           $N = \{mo \mid mo \in moSet(\beta)$ *and* $mo(\beta) \subset mo\}$;

10   **if** $N \neq \emptyset$ **then**

11             $moSet(\beta) = moSet(\beta) - N$;

12             $moSet(\beta) = moSet(\beta) \cup mo(\beta)$;

13   **else**

14           $betaSet = betaSet \cup \beta$;

15           $moSet(\beta) = moSet(\beta) \cup mo(\beta)$;

16   **for** *each* $\beta \in betaSet$ **do**

17   **if** $EWU(\beta) \geq min\_utility$ **then**

18   **if** $u(\beta) \geq min\_utility$ **then**

19         $HUE\_Set = HUE\_Set \cup \beta$;

20   **Prefix-Growth**$(\beta, moSet(\beta), MTD, min\_utility)$;

Both *I-Concatenation* and *S-Concatenation* follow the same framework: firstly, generate all the valid candidates *betaSet* and their corresponding minimal occurrence set $moSet(\beta)$ by means of prefix growth from episode α (the first **for** loop); then each candidate is checked to see if its *EWU* value is above the threshold $min\_utility$ to decide whether or not to recursively call the procedure ***prefix-growth*** (the second **for** loop). If the candidate β's exact utility is above $min\_utility$, it will be added to *HUE_Set* (Line 11 and 19 respectively). After this recursive procedure *prefix-growth* is executed on all the 1-episodes in lexicographic order, it exactly finishes the complete search on the lexicographic prefix tree and all the high utility episodes will be collected into the set *HUE_Set*. It is worth noting that the sub-procedure *I-Concatenate/S-Concatenate* performs simultaneous/serial concatenation operation on α and $e \in IES/SES$ to form candidate episode β (Line 4 and 5 respectively), and the sub-procedure *EWU* and $u(\alpha)$ computes the *EWU* and *utility* value of an episode respectively.

Discussion. Although the main procedure of the above presented algorithms is the same as the UP-Span algorithm, we make a couple of improvements compared to the original UP-Span algorithm.

First of all, we conduct the search process explicitly under the framework of lexicographical prefix tree. Before the search process begins, the simultaneous events at each time point are sorted in the lexicographic order as it is required in lexicographical tree. Consequently, we have the following nice property: Using the I-Concatenation procedure described in Algorithm 1, 1) all the candidate simultaneous episodes *betaSet* with prefix α will be generated, and 2) each $\beta \in betaSet$ has exactly the complete and correct minimal occurrence set $moSet(\beta)$[6].

Secondly, when spanning prefixes using *S-Concatenation* (Algorithm 2), we do not simply first collect all the serial candidate episode sets *betaSet* and their corresponding minimal occurrence set $moSet(\beta)$, and then filter out the illegal minimal occurrences of each β. Instead, whenever adding a new minimal occurrence of β, we check if there exists a minimal occurrence mo in $moSet(\beta)$ such that $mo \subseteq mo(\beta)$ (this set of mo is denoted as M in Line 7) *or* $mo \supset mo(\beta)$ (this set of mo is denoted as N in Line 9). If M is not $\emptyset$, then $mo(\beta)$ is not a valid minimal occurrence for episode β, and no further steps are taken. On the other hand, if M is $\emptyset$ (Line 8), and N is not $\emptyset$, the set N will be deleted from the $moSet(\beta)$ (N is not minimal occurrence because of the new $mo(\beta)$); otherwise, $mo(\beta)$ will be directly added to $moSet(\beta)$ (Line 10-15).

Furthermore, as we can see, different from the UP-Span algorithm, the projected database is not utilized in these algorithms, which is a common trick in sequential pattern mining to facilitate the prefix-growth process. It is because that not only the episode α itself, but also its minimal occurrence set $moSet(\alpha)$ are stored during the mining process. Together with the original event sequence and MTD, $moSet(\alpha)$ already provides a sparse representation of projected database w.r.t. α, which can be used to compute the *EWU* value and *utility* of α efficiently.

The final optimization of our implementation is that we propose to compute the exact utility of candidate episodes incrementally since the prefix-growth paradigm generates patterns recursively. As every minimal occurrence $mo(\beta)$ of the candidate episode β

---

[6] The proof is omitted due to space limit.

will be generated, $u(\beta, mo(\beta))$ can be directly obtained from the sum of the corresponding $mo(\alpha)$'s utility and the newly appended event e's utility. Thereafter, the utility of β can be obtained from the sum of $u(\beta, mo(\beta))$. In this way, we can avoid the frequent scans of the original database to compute the value of $u(\beta)$.

3.2   Efficient Pruning Strategies

Although the pruning strategies proposed using *EWU* is effective as demonstrated in [11], we argue here that the performance of this algorithm can be further improved. Our proposed upper bounds are able to further prune the search space over the lexicographic prefix tree and reduce the running time of algorithm significantly.

Table 3. Notations used in the following

$E_{last}(\alpha)$	The last event of α
$IES_i(\alpha)$	Event Set for *I-Concatenation* w.r.t. $mo_i(\alpha)$
$SES_i(\alpha)$	Event Set for *S-Concatenation* w.r.t. $mo_i(\alpha)$
$IES(\alpha)$	Event Set for *I-Concatenation* w.r.t. α
$SES(\alpha)$	Event Set for *S-Concatenation* w.r.t. α

Improved *EWU* for I-Concatenation As described in **Definition** 1, the computation of *EWU*(α) w.r.t. one minimal occurrence $mo(\alpha) = [T_s, T_e]$ is the sum of two parts, episode α's own utility w.r.t. $mo(\alpha)$, and the partial utility of CES starting from time point T_e to $T_s + MTD - 1$. Taking all the minimal occurrences of α into account, we can then get *EWU*(α) by summing all the *EWU*($\alpha, mo_i(\alpha)$). However, the estimated utility upper bound is still very loose. In fact, the first and second part have overlaps with each other during computation. That is, if $\alpha = < SE_s, SE_{s+1}, ..., SE_e >$, the utility of event set SE_e is counted twice during computation.

Under the framework of lexicographic prefix tree, events in each simultaneous event set are sorted in the lexicographic order. Hopefully, the repetitive calculation of utility of the events $e \in SE_e$ can be avoided. Specifically, we propose to compute an upper bound in this way: the first part keeps unchanged; the second part becomes the sum of utility of events after E_{last} in CES_e, i.e. $u(IES(\alpha))$, and utility of events in subsequent CES_i after CES_e, i.e. $u(SES(\alpha))$. As there often exist events in the set SE_e, this method can be seen as an improved upper bound of $u(\alpha)$ before I-Concatenation. We denote this new upper bound as *IEIC* (Improved *EWU* for *I-Concatenation*).

$$IEIC(\alpha, mo_i(\alpha)) = \sum_{j=1}^{k} u(SE_j, T_j) + u(IES_i(\alpha)) + u(SES_i(\alpha)), \quad (1)$$

where $\alpha = < (SE_1), (SE_2), ..., < SE_j >, ..., (SE_k) >$, $mo_i(\alpha) = [T_{si}, T_{ei}]$, and the meaning of $IES_i(\alpha)$ and $SES_i(\alpha)$ is illustrated in Table 3.

Taking all the minimal occurrences $mo(\alpha)$ of episode α into consideration, we can then obtain the value of $IEIC(\alpha)$:

$$IEIC(\alpha) = (u(\alpha) + u(IES(\alpha)) + u(SES(\alpha)))/u(CES). \quad (2)$$

This newly proposed upper bound $IEIC(\alpha)$, can be directly applied to replace the function of EWU used in Algorithm 1, and 2. In this way, we get a more efficient upper bound for pruning the search space.

Improved EWU for S-Concatenation. When it comes to *S-Concatenation* in the *prefix-growth* procedure, an improved upper bound of $u(\alpha)$ can be obtained similar to *IEIC*. Any *S-concatenation* episode β of α w.r.t. $mo(\alpha) = [T_s, T_e]$ only considers the event $e \in \{CES_i \mid e+1 \leq i \leq s + MTD - 1\}$, i.e., $SES(\alpha)$. Therefore, there is no need to take the utility of event set $IES(\alpha)$ into account when estimating the upper bound of $u(\alpha)$ for *S-Concatenation*. Similar to *IEIC*, we denote this new upper bound as *IESC* (Improved estimation of EWU for S-Concatenation).

$$IESC(\alpha) = (u(\alpha) + u(SES(\alpha)))/u(CES). \tag{3}$$

This efficient estimation of utility upper bound can only be applied before the sub-procedure of *S-Concatenation*. Thus, the *prefix-growth* procedure for high utility episode mining can be improved by employing this upper bound. We call this updated *prefix-growth* procedure as **TSpan**, which is illustrated as follows.

Algorithm 3. TSpan

    Input : (1) α: episode (2) $moSet(\alpha)$: all minimal occurrences of α (3) MTD:
                maximum time duration (4) $min\_utility$: minimum utility threshold
    Output: The set of high utility episodes with α as the prefix
1 *I-Concatenation*$(\alpha, moSet(\alpha), MTD, min\_utility)$;
2 **if** $IESC(\alpha) \geq min\_utility$ **then**
3      $\lfloor$ *S-Concatenation*$(\alpha, moSet(\alpha), MTD, min\_utility)$;

Discussion. Replacing function $EWU(\alpha)$ with $IEIC(\alpha)$ and procedure *Prefix-Growth* with TSpan (Algorithm 3) in Algorithm 1, 2, we finally obtain our improved algorithm TSpan over UP-Span. As TSpan adopts strategies including tighter upper bound estimation of $u(\alpha)$ and tighter upper bound estimation for *S-Concatenation*, much more search space is pruned compared to UP-Span. The efficiency of TSpan can be shown clearly in the example lexicographic prefix tree in Fig. 2. The black solid line shows the search space boundary of the UP-Span algorithm, while the red dashed line shows the search space boundary of TSpan.

To be specific, the total utility of $u(\langle(ac)(abd)(bc)\rangle) = 11$ under the above settings. Therefore, the utility of a given node α, i.e., $u(\alpha)$ must be greater than 7.7(11 * 0.7) to be recognized as a high utility episode. Using UP-Span, only nodes $\langle(ab)(bc)\rangle$ and $\langle(b)(bc)\rangle$ are pruned, while TSpan is much more efficient at reducing search space in this example. For instance, the child nodes of $\langle(ad)\rangle$ are not pruned by UP-Span since $EWU(\langle(ad)\rangle) = (u(\langle(ad)\rangle) + u(\langle(abd)(bc)\rangle))/11 = 1.1$, greater than 0.7. But $IEIC(\langle(ad)\rangle) = (u(\langle(ad)\rangle) + u(\langle(bc)\rangle))/11 = 7/11 < 0.7$, thus the child nodes of $\langle(ad)\rangle$ is not searched using our proposed algorithm. Before performing *S-Concatenation* operation on node $\langle(b)\rangle$, the original upper bound $EWU(\langle(b)\rangle)$ is sure to be larger than 0.7 due to two minimal occurrences of $\langle(b)\rangle$. But $IESC(\langle(b)\rangle) =$

$(u(\langle\langle b\rangle\rangle, [2,2]) + u(\langle\langle bc\rangle\rangle) + u(\langle\langle b\rangle\rangle, [3,3]))/11 = 6/11 < 0.7$. As a result, all the serial concatenation episodes of $\langle\langle b\rangle\rangle$ are pruned from the search space. Similar results can also be demonstrated for other pruned nodes.

4 Performance Evaluation

In this section, we will conduct experiments on various algorithms to evaluate the performance of our proposed strategies.The main characteristics of datasets used in experiments are presented in Table 4. The synthetic dataset is generated with the IBM synthetic data generator for transactional database [1]. The main parameters to generate these dataset include T: the average size of transactions; I: the average size of maximal potential pattern; D: the total number of transactions; N: the number of distinct items. Aside from synthetic dataset, three real-life datasets are also employed during evaluation. The first one is Foodmart2000 database[7], a well known example database for business intelligence. The second one is Retail dataset, obtained from the FIMI dataset repository[8]. The last one is Chainstore dataset, a large dataset downloaded from NU-MineBench repository[9].

Table 4. Statistical information of different datasets

DataSet	#Trans	#Items	Avg. Length
T20I12N1KD10K	10,000	1,000	20
FoodMart2000	5,581	1,559	15.6
Retail	88,162	8,600	11.2
ChainStoreSmall	10,000	46,086	14.3

It should be noted that both the synthetic and real-life datasets can be seen as transactional databases. But they can also be viewed as a single long complex event sequence by considering each item as an event and each transaction as a simultaneous event set at a time point. In the experiments, we only use the first 10k transactions of dataset Retail and ChainStore as this size is enough for comparison and bigger size of the dataset will cost too much time to run. Only the FoodMart2000 and Chainstore dataset have unit profits (i.e., external utility) and product sales (i.e., internal utility). Therefore, we generate unit profits and product sales as follows: the unit profits of each item are generated using a log-normal distribution ranging from 1 to 1000 and sales of each items are generated randomly between 1 and 5.

To evaluate the proposed algorithm for high utility episodes mining, we compare TSpan with three baseline algorithms, that is, the original UP-Span algorithm (UP-Span), improved algorithm with strategy IEIC (IEIC Only), improved algorithm with strategy IESC (IESC Only) and our proposed algorithm TSpan with

---

[7] http://msdn.microsoft.com/en-us/library/
    aa217032(v=sql.80).aspx
[8] http://fimi.ua.ac.be/data/
[9] http://cucis.ece.northwestern.edu/projects/DMS/MineBench.html

both strategies (TSpan). As the optimization over the implementation of operations *I-Concatenation* and *S-Concatenation* in Section 3.1 is in very fine granularity and is sure to save running time and memory consumption, we focus on the performance comparison between TSpan and UP-Span. It is worth noting that both the baselines and TSpan here adopt the implementation improvement in Section 3.1.

4.1 Evaluation on Synthetic Dataset

As shown in Fig. 3, the tendency between search space and running time are very consistent. That is, the more the searched nodes, the longer the running time. Both the proposed strategies IEIC and IESC take effects, reducing the running time and search space as much as 25% when $min\_utility = 0.05$. And the combination of both proposed pruning strategies leads to further improved efficiency. As it can be seen, the difference between the four algorithms' performance becomes larger when the $min\_utility$ decreases, and we can forecast that the gap between algorithms will be larger as $min\_utility$ continues to decrease. Since experiments are conducted under the setting that MTD = 8 and utilities of events follow the lognormal distribution, high utility episodes can only be generated when $min\_utility$ is very small. All these facts lead to the conclusion that the proposed strategies are practically useful.

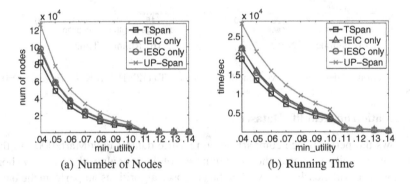

(a) Number of Nodes   (b) Running Time

Fig. 3. Evaluation under varied $min\_utility$ on dataset T20I12N1KD10K (MTD=8)

We also tested the scalability of the four algorithms when the size of the transactions (i.e., the size of the complex event sequence) increases. The results are reported in Fig. 4. As we generate the dataset using the same algorithm and all other settings are the same, the number of searched nodes keeps the same. But the running time of these algorithms grows linearly, indicating that the algorithms' scalability is very good. In Fig. 5, we vary the average length of transactions, i.e., the average length of simultaneous event set, to compare the performance difference between the algorithms. TSpan grows smoothly in both number of searched nodes and running time, while the baselines grow much faster, demonstrating that TSpan is more efficient.

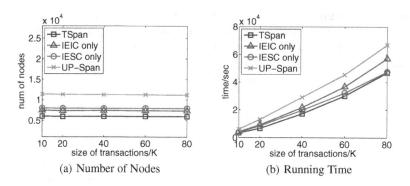

Fig. 4. Evaluation under varied #Trans on dataset T20I12N1KDxK ($min\_utility$=0.1, MTD=8)

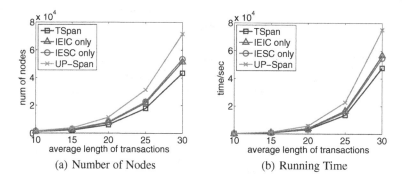

Fig. 5. Evaluation under varied Avg.Length on dataset TxI12N1KD10K ($min\_utility$=0.1, MTD=8)

4.2 Evaluation on Real-life Dataset

Fig. 6 shows the performance comparison on real-life dataset FoodMart2000. As this dataset is quite small and sparse, the execution time of the four algorithms is very short. However, we can still clearly observe that the proposed algorithms outperform the baseline by a large margin. For example, when $min\_utility = 0.08$, TSpan finishes in only one fifth of the time needed by UP-Span. For dataset Retail, due to its inherent characteristics and our settings on external utilities, the running time of these algorithms becomes too long to observe when $MTD > 4$. So we only report the performance comparison when $MTD = 4$, which is still capable of demonstrating the performance advantage of our proposed strategies.We can predict from the curve's trend of Fig. 7 (a) that the gap between the TSpan and the baselines will become even larger when $min\_utility$ is smaller than 0.1. In Fig. 7 (b), we can conclude that TSpan is faster than the UP-Span algorithm by more than 2000 seconds in most cases. Fig. 8 shows that TSpan is always the winner compared with others, similar to the other two real-life datasets. And strategy IEIC and IESC can compensate each other well to further improve the algorithm's performance.

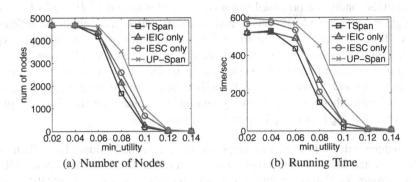

(a) Number of Nodes

(b) Running Time

Fig. 6. Evaluation under varied minimum utility thresholds on dataset FoodMart2000 (MTD=8)

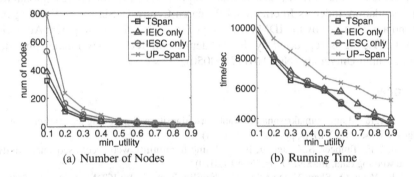

(a) Number of Nodes

(b) Running Time

Fig. 7. Evaluation under varied minimum utility thresholds on dataset Retail (MTD=4)

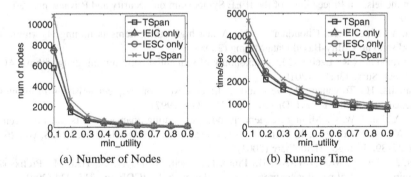

(a) Number of Nodes

(b) Running Time

Fig. 8. Evaluation under varied minimum utility thresholds on dataset ChainStoreSmall (MTD=6)

5 Conclusion and Future Work

In this paper, we focus on proposing practical and fast high utility episode mining algorithms. The most important thing we did is that we tackle the problem under the framework of a complete traversal of the lexicographic prefix tree. Using this framework, we first presented four efficient improvements over the original implementation of UP-Span. Secondly, we proposed two novel strategies named $IEIC$ and $IESC$ to obtain tighter upper bounds of a given node's utility, which bring about the efficient algorithm TSpan. Finally, we demonstrated the effectiveness of these strategies systematically on both synthetic and real life datasets. Experimental results show that the proposed strategies can improve the performance significantly by reducing the number of searched nodes and the running time. We believe that these proposed strategies can be incorporated with other similar high utility mining tasks, and more effective strategies can be proposed using lexicographic prefix tree in the future.

Acknowledgement. This work was supported in part by the National High Technology Research and Development Program of China under Grant No. 2014AA015203, the Anhui Provincial Natural Science Foundation under Grant No. 1408085QF110 and the Fundamental Research Funds for the Central Universities of China under Grant No. WK0110000042. This work was also partially supported by the Singapore National Research Foundation under its International Research Centre @ Singapore Funding Initiative and administered by the IDM Programme Office, Media Development Authority (MDA). Lei Zhang is supported by the Academic and Technology Leader Imported Project of Anhui University (NO. J10117700050).

References

1. Rakesh Agrawal and Ramakrishnan Srikant. Fast algorithms for mining association rules in large databases. In: VLDB, pp. 487–499 (1994)
2. Ahmed, C.F., Tanbeer, S.K., Jeong, B.S.: Mining high utility web access sequences in dynamic web log data. In: SNPD, pp. 76–81 (2010)
3. Chan, R., Yang, Q., Shen, Y.-D.: Mining high utility itemsets. In: ICDM, pp. 19–26 (2003)
4. Lee, W., Stolfo, S.J., Mok, K.W.: A data mining framework for building intrusion detection models. In: Proceedings of the IEEE Symposium on Security and Privacy, pp. 120–132 (1999)
5. Liu, Y., Liao, W.K., Choudhary, A.N.: A fast high utility itemsets mining algorithm. In: Workshop on Utility-Based Data Mining (2005)
6. Mabroukeh, N.R., Ezeife, C.I.: A taxonomy of sequential pattern mining algorithms. ACM Comput. Surv. 43(1), 3 (2010)
7. Mannila, H., Toivonen, H., Verkamo, A.I.: Discovery of frequent episodes in event sequences. Data Min. Knowl. Discov. 1(3), 259–289 (1997)
8. Ng, A., Fu, A.W.-C.: Mining frequent episodes for relating financial events and stock trends. In: Whang, K.-Y., Jeon, J., Shim, K., Srivastava, J. (eds.) PAKDD 2003. LNCS, vol. 2637, pp. 27–39. Springer, Heidelberg (2003)
9. Pei, J., Han, J., Mortazavi-Asl, B., Pinto, H., Chen, Q., Dayal, U., Hsu, M.: Prefixspan: Mining sequential patterns by prefix-projected growth. In: ICDE, pp. 215–224 (2001)
10. Saxena, K., Shukla, R.: Significant interval and frequent pattern discovery in web log data. CoRR, abs/1002.1185 (2010)
11. Wu, C.-W., Lin, Y.-F., Tseng, V.S., Yu, P.S.: Mining high utility episodes in complex event sequences. In: KDD (2013)
12. Yao, H., Hamilton, H.J., Butz, C.J.: A foundational approach to mining itemset utilities from databases. In: The 4th SIAM International Conference on Data Mining, pp. 482–486 (2004)
13. Yin, J., Zheng, Z., Cao, L.: Uspan: an efficient algorithm for mining high utility sequential patterns. In: KDD, pp. 660–668 (2012)

Community Dynamics: Event and Role Analysis in Social Network Analysis

Justin Fagnan, Reihaneh Rabbany, Mansoureh Takaffoli,
Eric Verbeek, and Osmar R. Zaïane

Department of Computing Science,
University of Alberta,
Edmonton, Alberta, Canada
{fagnan,rabbanyk,takaffol,everbeek,zaiane}@ualberta.ca

Abstract. Social networks are analyzed and mined to find communities, or groupings of interrelated entities. Community mining provides this higher level of structure and offers greater understanding, but networks change over time. Their constituent communities change, and the elements of those communities change over time as well. By performing event analysis, the evolutions of communities are abstracted in order to see structure in the dynamic change over time. This higher level of analysis has a counterpart that deals with the fine grain changes in community members with relation to their communities or the global network. We discuss here an approach to analyzing community evolution events and entity role changes to uncover critical information in dynamic networks.

1 Introduction

Many complex information networks and social networks can be modelled by graphs of interconnected nodes to represent the interaction of individuals or entities with one another. For instance, the graph of co-authorship relationships between researchers, the interaction between posters on an on-line forum, the graph of web pages inter-connected through hyperlinks, and Protein-Protein Interaction (PPI) networks are examples of complex networks. In these networks, understanding the underlying structure and determining the structural properties of the network facilitates the global understanding of the system and benefits applications such as targeted marketing and advertising, influential individuals identification, information diffusion modelling, and much more. One way to gain information about the network is the identification of communities, which are sets of densely connected individuals that are loosely connected to others [18]. There has been a considerable amount of work done to detect communities in static graphs, such as modularity methods [18,8,13], stochastic methods [2,9], and heterogeneous clustering methods [24,3]. For a comparative surveys, see, for example [15,19].

Social networks usually model systems that are evolving over time as the entities change their activities and interactions (authors publish papers with new

X. Luo, J.X. Yu, and Z. Li (Eds.): ADMA 2014, LNAI 8933, pp. 85–97, 2014.

co-authors, old pages are deleted while new ones are added to the web, etc.). Furthermore, the communities in these dynamic networks usually have fluctuating members and these communities grow and shrink over time. However, studying these dynamic networks as static graphs discards the temporal information associated with the interaction. In order to explicitly address the dynamic nature of the interactions, the dynamic social network can be modelled by a series of static snapshots. In these models, each snapshot corresponds to a discrete time interval, constituted of the interactions during that specific interval. In some scenarios the size of such a time interval is determined, while for many cases the interval size is arbitrary. However, the size of a time interval has a great impact on the observation found by the dynamic network analysis. Recently, Caceres et al. [6] propose an algorithm to determine the appropriate time interval by finding a balance between minimizing the noise and loss of temporal information. After modeming the dynamic social network with the appropriate time interval, the temporal evolution of the network can be studied. Leskovec et al. [16] study the patterns of growth for large social networks based on the properties of large networks, such as the degree of distribution. The problem of mining patterns of link formations and link predictions in a time evolving graph is proposed in [21]. All the aforementioned studies considered the macroscopic properties on the graph level and overlooked the mesoscopic properties on the level of communities.

Tantipathananandh and Berger [27] formulate the detection of dynamic communities as a graph colouring problem and prove that their algorithm is a small constant factor approximation. Falkowski et al. [10] discover the evolution of communities by applying clustering on a graph formed by all detected communities at different time points. A number of researchers are working on identifying critical events that characterize the evolution of communities in dynamic social networks. Palla et al. [20] identify events by applying Clique Percolation Method (CPM) community mining on a graph formed by the communities discovered at two consecutive snapshots. Then, based on the results of the community mining algorithm, events pertaining to the communities are specified. Asur et al. [4] define critical events between detected communities at two consecutive snapshots which are implement in the form of bit operations. However, these events do not cover all of the transitions that may occur for a particular community. Greene et al. [12] describe a weighted bipartite matching to map communities and then characterized each community by a series of events.

In all of the above work, the communities in each snapshot are mined independently without considering the temporal information and their relationship to communities at previous snapshots. This independent community mining approach is suitable for the social networks with highly dynamic community structures. Another approach is to use incremental community mining, where the community mining at a particular time snapshot is influenced by the communities detected in previous timeframes. Thus, the incremental community mining approach finds a sequence of communities with temporal similarity and hence, is suitable for networks with community structures that are more stable over time. Incremental community mining approaches, which consider both current and historic information in the mining process, are proposed in [17,28,5,25].

In our previous work, we provide an event-based framework, MODEC (Modelling and Detecting the Evolutions of Communities), to capture the events and transitions of communities and individuals over the entire observation time [26]. However, in the MODEC framework, the reason why a community or an individual experiences a specific event is not addressed. The changes in the role of individuals in a community can have a high influence on the development of the community and can act as triggers to evoke community changes. For example, if the leader of a community leaves, it might cause the remaining community members to become less active or disperse to other communities.

In this paper, we illustrate event analysis and change of individual's roles. Our main contribution is to propose a framework to reveal the relation between the structure of communities and the behavior of individuals in a dynamic scenario. In the following section, we present how the evolutions of communities are abstracted in order to see structure in the dynamic change over time. Then we describe different roles that an individual can play in the whole network and in their communities and also how these roles change after events. Finally through the visualizations in our last section, we demonstrate that analyzing community evolution events and entity role events uncovers critical information in dynamic networks. We implemented the proposed visualizations in our tool Meerkat, a social network analysis system that encompasses our MODEC framework [7].

2   Community Dynamics Modelling

In order to analyze dynamic social networks and study the evolution of their communities and individuals, we propose a two-stage framework, called MODEC, that analyzes the dynamic evolution of communities [26]. Our framework assumes that the communities are independently extracted in each snapshot by an arbitrary community mining algorithm. In the first stage of the framework we employ a one-to-one matching algorithm to match the communities extracted in different snapshots. A meta community, which is a series of similar communities detected by the matching algorithm in different timeframes, is then constructed. In the second stage, we identify a series of significant events and transitions which are used to explain how the communities and individuals of a meta community evolve over time. In this section, we review the MODEC framework and the event definitions proposed to track communities or individuals over time.

We model the dynamic social network as a sequence of graphs $\{G_1, G_2, ..., G_n\}$, where $G_i = (V_i, E_i)$ denotes a graph containing the set of individuals and their interactions at a particular snapshot i. The set $C_i = \{C_i^1, C_i^2, ..., C_i^{n_i}\}$ denoted the n_i communities detected at the ith snapshot, where community $C_i^p \in C_i$ is also a graph represented by (V_i^p, E_i^p). Here, we distinguish between a community and a meta community. A community contains individuals that are densely connected to each other at a particular snapshot, whereas a meta community is a series of similar communities from different snapshots which represents the evolution of its constituent communities over time.

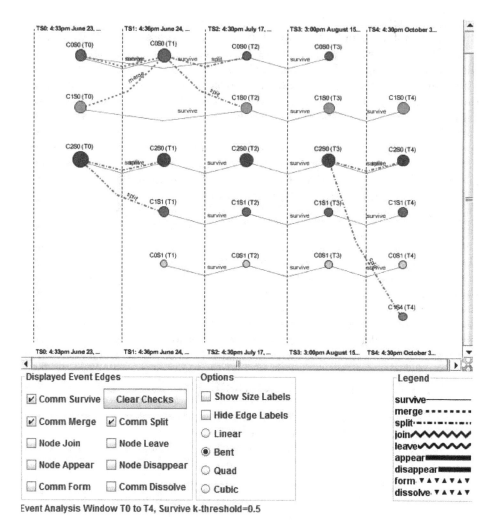

Fig. 1. Visualization of captured MODEC defined events along the time dimension (Visualization by Meerkat http://meerkat.aicml.ca/)

In order to capture the changes that are likely to occur for a community, we consider five events (form, dissolve, survive, split, and merge) and define four transitions (size, compactness, leader, and persistence transition). A community may *split* at a later snapshot if it fractures into multiple communities. It can *survive* if there exists a similar community in a future snapshot. In the case where there is no similar community at a later snapshot, the community *dissolves*. A set of communities may also *merge* together at a later snapshot. Finally, at any snapshot there may be newly *formed* communities which are defined as communities that have no similar community in any previous snapshots. The meta community is then a sequence of *survival* communities ordered by time, from the timeframe where it first appears to the timeframe where it is last observed. Furthermore, a community may undergo different transitions at the same time. The size transition occurs when the number of nodes of a community increases (i.e. expand) or decreases (i.e. shrink) over time. Moreover, a community compacts or diffuses at a later snapshot if its normalized number of edges increases or decreases respectively. For the case when the number of nodes and edges of a community remains the same, the community persists. Finally, when the most central member of a community shifts from one node to the other, the community experiences leader shift. Figure 1 presents an example of such events and changes in communities while the complete definition of our proposed events and transitions can be found in [26].

The key concept for the detection of the events, meta community, and also transitions is the notion of similarity between communities from different snapshots. The similarity between different communities can be determined using the similarity measures such as Jaccard, correlation-based, and more. However, in this paper, we consider two communities discovered at different snapshots as similar if the percentage of their mutual members exceed a given threshold $k \in [0, 1]$. After selecting the similarity measure, the set of communities extracted by a community mining algorithm at a given snapshot have to be matched to the communities at previous snapshots based on their similarity. This matching is non-trivial, because a community may be similar to several communities at the same time. We use greedy matching, and a weight bipartite matching [26] to match communities at different snapshots.

In order to analyze the behaviour of individuals in communities, we define four events involving individuals (appear, disappear, join, leave) [26]. A node appears at a snapshot when it exists in that snapshot but was not present in the previous snapshots. It may disappear from one snapshot if it exists in that snapshot but will not occur in the next snapshots. A node joins to a community if it exists in that community but did not belong to a community with the same meta community in the previous snapshots. Finally, it leaves a community if it exists in that community but will not belong to a community with the same meta community in the next snapshots.

To capture the behavioural characteristics of the individuals, we define two metrics [26]: the stability metric calculates the tendency of an individual to interact with the same nodes over the observation time; the influence metric determines how one individual influences others to join or leave a community. In the next section we present a more comprehensive list of possible roles that an individual can play both within its community and in the whole network and how these roles are related to the temporal events.

3   Role Analysis and Event Triggers

Walton [29] states that the changing nature of an individual community and its leadership are of central importance to the explanation of community action. Here, we explore different roles associated with individuals in a social networks, and in the later section we will provide examples of how the changes in these roles affect the events detected for communities.

In our discussion of roles within dynamic community mining, we explore some domain agnostic, generic network roles. These roles should be interpretable in any network, although the names may not apply all of their connotations. The presence of these roles in a given analysis is dependent upon the specific dataset, the community mining algorithm used, the event detection framework applied, and any thresholds applied for the roles.

We define these generic roles across two possible role scopes. The role scopes are either global to the network, or limited to the community the individual is in. Some of the defined roles have both global and community bound versions, while some roles cannot have a community bound version. There is also some notion of connectivity incorporated in the definition of roles; which could be the betweenness metric, node degree, or another appropriate metric. The choice of metric will have an effect on the results of the analysis, but the role definitions do not rely on the selected metric.

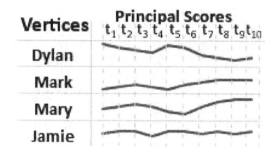

Fig. 2. An token example showing a change of role of individuals vis-à-vis principal role. While Dylan was authoritative in t1, he lost his principal role by t10. Mary on the other hand became authoritative in t10 when she had a low principal score in t5.

We will start with the more well-studied role, the **"outlier role"**. The outlier role reflects individuals who are not a part of any community. Gao et al. [11] and Aggarwal et al. [1] introduced the concept of community outliers and then proposed algorithms to find these community outliers. However, not all mining algorithms allow individuals to be excluded from all communities, so its presence is mining algorithm dependent. This role does not have an associated threshold, nor is there a community role equivalent.

The **"principal role"** (or authority role) indicates an individual who is very central due to high connectivity. *Community leaders* are evaluated within the community, the induced subnetwork, they belong to, whereas *global principals* have high centrality within the entire network. Note that communities with flat intra-community degree distribution for their member nodes will be without a principal, which is acceptable. In the case the distribution is nearly flat, with a few extremes with only slightly higher degree, we see a need to scale the threshold relative to the community density to prevent assignment of a principal by blind faith. When looking at the principal role in a social domain, we might expect some followers to move with the principal if a principal changes communities (see Figure 2).The concept of principle role in general and more specifically community leader is also used to detect communities in the social network [22,23].

The **"peripheral role"** indicates individuals with the least connectivity. These individuals have the lowest levels of centrality within their community, or centrality within the network for the global version. To identify these individuals that are not well integrated into a community, we apply a threshold to select the individuals with the least connectivity, relative to the network or community density, in the same way as for the principal role. If the threshold is not relative to edge density, sparsely and evenly connected communities could consist entirely of members who are both peripheral and principal. One could argue that this would be acceptable, but our perspective is that such communities are best described as having no member in either of the principal or peripheral roles.

The **"mediator role"** indicates an individual which has high centrality but does not belong solely to one specific community. This can include individuals who happen to belong to multiple communities when the mining algorithm supports overlap and hub detection, or those that are excluded from community membership but are still highly connected, but insufficiently to be a community member.

The community based equivalent of the mediator role is the **"extrovert role"** whom must belong to some community. Unlike the rest of the roles defined so far, the extrovert is detected by comparing their inter-community connections to their intra-community connections. The threshold can be set somewhere slightly above 1.0, since community members tend to have many of their connections within their community. Having slightly more reaching out indicates an extrovert. Depending on the community mining algorithm used, extroverts could have significantly more inter-community edges. If this is the case, the threshold can be made much higher than 1.0, particularly when the mining algorithm detects many communities where conservative algorithms detect very few.

Note that both the mediator and extrovert roles differ from hubs as defined by the HITS algorithm [14], because a hub is defined as connecting multiple authorities, whereas these two roles do not have to connect identified authorities but any set of individuals from more than one community. Likewise, the principal role is not defined as resulting from connections from multiple hubs.

It is desirable to scale the roles' thresholds by the community density as they are applied to each community. This is defined as the number of edges present divided by the number of possible edges density. When this is not done, very sparse or very dense communities may have a preponderance or lack of members filling the roles, or members filling the roles when our intuitive sense of the roles would expect none.

Events usually indicate structural change in a network, excepting the survive event, which can occur with no accompanying changes such joining and leaving

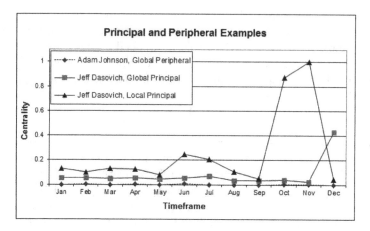

Fig. 3. An example showing how one can monitor changes in roles over time

members (or it can occur despite accompanying changes). For the remaining community events (merge, split, form, dissolve), significant structural changes are almost always indicated, and individual roles are dependent upon network and community structures. When individual events occur (join, leave, appear, disappear), the likelihood of an impact upon individual roles is less likely, or at least dependent upon there being a large number of individual events, or individuals with more central roles coming into competition.

Some clear examples of the effects of events upon role changes may be described. A community authority (or having a principal role) may leave one community, and afterwards they cannot be an authority for the community they just left. If they join another community, they might or might not have a principal role in the new community. They may also have been a global authority, but the changes that led to them leaving their original community might also have caused them to lose connectivity to the greater network, and they may no longer be globally authoritative. The same sort of considerations apply to merge and split events, where the community context has changed an old principal now has competition from others, or has lost connections vital to their authoritative role.

For any given role change, we can identify the community events involved. Our current method of role change attribution consists of associating each significant role change with all events involving that individual directly, as well as events involving the communities that the individual was involved with before and after the role change. This will not capture domain dependent attributions, that is, attributions that require domain knowledge. For example, a change in authority may be associated with a merge of the authority's community with two other communities, resulting in a loss of authority role for that individual. There might also have been splits and joins elsewhere in the network, not involving this individual or these three networks. If the domain contained causal connections between domain specific events and the change in role, and if those domain specific events had no clean relation to the generic network events, the

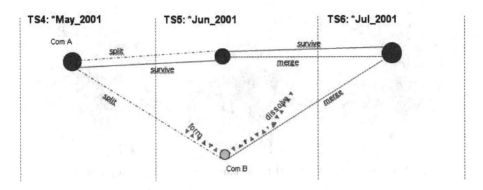

Fig. 4. An example of event visualization

causal connection would go undetected in the network analysis. This is a domain
dependent risk, and it is not clear that a generic analysis can ever deal with all
such possibilities.

Other than these community-event-individual associations, it might be possible to give deeper attributions of events upon role changes. This would require a
much more in depth analysis framework, and would involve questions of ambiguous causality at the generic network level. If tentative causal links are identified,
they could then be associated with real domain events that are not represented in
SNA models. For example, if an individual changes from being a community authority to being an authority in another community following a leave/join event,
there might be a discrete real-world event corresponding with this change, such
as the individual being promoted from being director of one department to being
a director of another department. It is our position that to understand network
dynamics at the domain level, it is necessary to have the generic, quantitative
backing from dynamic SNA, including both community event analysis as well
as individual role analysis. The generic network methods are intended to offer
evidence and act as a modelling lens for domain specific hypotheses and descriptions. These two sources of knowledge may be compared and contrasted with
the combination of theoretical analysis and statistical modelling found in experimental scientific disciplines. Without network analytic evidence to support
domain hypotheses or measure domain events, full understanding cannot occur.

4   Visualization and Analysis

In this section, we illustrate the practical application of the proposed role analysis on the Enron email dataset. The Enron email dataset contains the emails
exchanged between employees of the Enron Corporation. The entire dataset includes a period of 15 years and the corresponding graph for the entire data has
over 80,000 nodes and several hundred thousand edges, where nodes are employees and edges are emails between them. We study the year 2001 and consider a

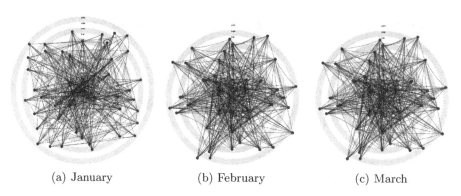

(a) January   (b) February   (c) March

Fig. 5. Changing the role of an individual, Paul Kaufman, from January till March, changing from a peripheral role to principal role (Visualization by Meerkat http://meerkat.aicml.ca/).

total of 285 nodes and 23559 edges, with each month being one snapshot. For each of the 12 snapshots, one graph is constructed with the extracted employees as the nodes and email exchanged between them as the edges. Due to computational efficiency, we apply the weighted local community mining algorithm [8] to produce sets of disjoint communities for each snapshot.

Figure 3 illustrated an example of the changes in roles over time. Here we focus on two individuals. One of them, Adam, is globally peripheral for all timeframes. The other, Jeff, is relatively central both globally and locally to his communities, but displays changes in his principal role value over time. In May, Jeff is a member of the blue community. However, in June, he is a member of the cyan community. Due to his move, the blue community itself splits to the blue and green communities. Then in July when Jeff comes back to the blue community, the communities that had split, merge back into one. So it seems that Jeff acts as a glue between the two, which shows an example of the involvement of a role in the community evolutions.

Community events can be visualized as a second order network. Vertices correspond to communities, with names and colours identifying those that correspond across multiple timeframes. Events themselves are portrayed as edges that connect the communities. We see in Figure 4 a subset of communities from the Enron data. This diagram depicts community events occurring across timeframes May, June, and July only for communities, blue, cyan, and green. The split that occurs here, where blue community splits and cyan community is formed, corresponds with the move of Jeff from community blue to green, and other members following him there. The subsequent merge also relates to Jeff's move.

These event diagrams facilitate the abstraction of communities over time into communities that evolve via community events. Furthermore, it allows users to explore and pick out interesting and critical changes occurring in their data over time. The depicted diagram is a filtered subset for purposes of illustration.

A more refined level of detail could be obtained by a series of concentric circles that visually show the role metric scores for all vertices in a selected community, as depicted in Figure5. Those vertices closer to the centre of the circles have high metric scores and those on the outskirts have low metric scores. Here, we only provide the circles for blue community in the first three months. Using these concentric circles we can visually observe that for instance an individual name Paul Kaufman (marked by red circles), at the beginning is among peripheral nodes, but as times goes he develop his relationship and become more central within his community.

5 Conclusion

Social networks are dynamic. The change affects the network as a whole, the individual communities, as well as the particular entities in the network. We depicted in this paper the MODEC defined events and transitions that pertain to the evolution of communities and the events associated to individuals moving between these communities. We also described the detectable roles individuals may take globally or within their respective community (principal, peripheral, outlier, mediator, etc.). We explained the relationship between the evolution of communities, the movement of individuals between these communities and changes in the role of those individuals. We further illustrated that changes in the role of individuals in a community have a direct relationship with the development of the community. The role change can act as triggers to evoke community changes. Role modification can affect the dynamics of communities and the events in the communities can alter the role of individuals. Through our visualizations, we demonstrated that analyzing community evolution events and entity role events gives us valuable insights on the dynamics of networks.

References

1. Aggarwal, C.C., Zhao, Y., Yu, P.S.: Outlier detection in graph streams. In: Proceedings of the 2011 IEEE 27th International Conference on Data Engineering, ICDE 2011 (2011)
2. Airoldi, E.M., Blei, D.M., Fienberg, S.E., Xing, E.P.: Mixed membership stochastic blockmodels. Journal of Machine Learning Research 9, 1981–2014 (2008)
3. Alqadah, F., Bhatnagar, R.: A game theoretic framework for heterogenous information network clustering. In: Proceedings of the 17th ACM SIGKDD International Conference on Knowledge Discovery and Data Mining (2011)
4. Asur, S., Parthasarathy, S., Ucar, D.: An event-based framework for characterizing the evolutionary behavior of interaction graphs. In: Proceedings of the 13th ACM SIGKDD International Conference on Knowledge Discovery and Data Mining (2007)
5. Aynaud, T., Guillaume, J.-L.: Multi-step community detection and hierarchical time segmentation in evolving networks. In: Proceedings of the Fifth International Workshop on Social Network Mining and Analysis (2011)
6. Caceres, R.S., Berger-Wolf, T., Grossman, R.: Temporal scale of processes in dynamic networks. In: Proceedings of the IEEE ICDM 2011 Workshop on Data Mining in Networks, DaMNet (2011)

7. Chen, J., Fagnan, J., Goebel, R., Rabbany, R., Sangi, F., Takaffoli, M., Verbeek, E., Zaiane, O.R.: Meerkat: Community mining with dynamic social networks. In: IEEE International Conference on Data Mining (ICDM), Sydney, Australia (December 2010)

8. Chen, L., Roy, A.: Event detection from flickr data through wavelet-based spatial analysis. In: Proceedings of the ACM CIKM International Conference on Information and Knowledge Management (2009)

9. Choi, D.S., Wolfe, P.J., Airoldi, E.M.: Stochastic blockmodels with growing number of classes. CoRR, abs/1011.4644 (2010)

10. Falkowski, T., Bartelheimer, J.: Applying social network analysis methods to explore community dynamics. In: Serdult, U., Taube, V. (eds.) Applications of Social Network Analysis 2005, pp. 189–212. Wissenschaftlicher, Berlin (2008)

11. Gao, J., Liang, F., Fan, W., Wang, C., Sun, Y., Han, J.: On community outliers and their efficient detection in information networks. In: Proceedings of the 16th ACM SIGKDD International Conference on Knowledge Discovery and Data Mining, KDD 2010 (2010)

12. Greene, D., Doyle, D., Cunningham, P.: Tracking the evolution of communities in dynamic social networks. In: Proceeding of the International Conference on Advances in Social Networks Analysis and Mining (2010)

13. Kenley, E.C., Cho, Y.-R.: Entropy-based graph clustering: Application to biological and social networks. In: Proceedings of the 11th IEEE International Conference on Data Mining (2011)

14. Kleinberg, J.M.: Hubs, authorities, and communities. ACM Comput. Surv (December 31, 1999)

15. Lancichinetti, A., Fortunato, S.: Community detection algorithms: a comparative analysis (September 2010)

16. Leskovec, J., Kleinberg, J., Faloutsos, C.: Graphs over time: densification laws, shrinking diameters and possible explanations. In: Proceedings of the eleventh ACM SIGKDD International Conference on Knowledge Discovery in Data Mining (2005)

17. Lin, Y.-R., Chi, Y., Zhu, S., Sundaram, H., Tseng, B.L.: Facetnet: a framework for analyzing communities and their evolutions in dynamic networks. In: Proceeding of the 17th International Conference on World Wide Web (2008)

18. Newman, M.E.J.: Finding community structure in networks using the eigenvectors of matrices. Physical Review E 74(3), 036104 (2006)

19. Orman, G.K., Labatut, V., Cherifi, H.: Qualitative comparison of community detection algorithms. In: Cherifi, H., Zain, J.M., El-Qawasmeh, E. (eds.) DICTAP 2011 Part II. CCIS, vol. 167, pp. 265–279. Springer, Heidelberg (2011)

20. Palla, G., Barabasi, A.-L., Vicsek, T.: Quantifying social group evolution. Nature 446(7136), 664–667 (2007)

21. Prakash, A.K.J., Comar, M., Tan, P.-N.: Linkboost: A novel cost-sensitive boosting framework for community-level network link prediction. In: Proceedings of the 11th IEEE International Conference on Data Mining (2011)

22. Rabbany, R., Chen, J., Zaïane, O.R.: Top leaders community detection approach in information networks. In: The Fifth ACM workshop on Social Network Mining and Analysis, SNA-KDD (2010)

23. Shah, D., Zaman, T.: Community detection in networks: The leader-follower algorithm. In: Workshop on Networks Across Disciplines: Theory and Applications (2010)

24. Sun, Y., Yu, Y., Han, J.: Ranking-based clustering of heterogeneous information networks with star network schema. In: Proceedings of the 15th ACM SIGKDD International Conference on Knowledge Discovery and Data Mining (2009)
25. Takaffoli, M., Rabbany, R., Zaïane, O.R.: Incremental local community identification in dynamic social networks. In: International Conference on Advances in Social Networks Analysis and Mining, ASONAM 2013 (2013)
26. Takaffoli, M., Sangi, F., Fagnan, J., Zaïane, O.R.: Tracking changes in dynamic information networks. In: The International Conference on Computational Aspects of Social Networks (2011)
27. Tantipathananandh, C., Berger-Wolf, T.: Constant-factor approximation algorithms for identifying dynamic communities. In: Proceedings of the 15th ACM SIGKDD International Conference on Knowledge Discovery and Data Mining (2009)
28. Tantipathananandh, C., Berger-Wolf, T.Y.: Finding communities in dynamic social networks. In: Proceedings of the 11th IEEE International Conference on Data Mining (2011)
29. Walton, J.: Differential patterns of community power structure: An explanation based on interdependence. The Sociological Quarterly 9, 3–18 (1968)

Local Community Extraction
for Non-overlapping and Overlapping
Community Detection

Zhan Bu[1,*], Guangliang Gao[2], Zhiang Wu[1], Jie Cao[1], and Xiao Zheng[3]

[1] Jiangsu Provincial Key Laboratory of E-Business,
Nanjing University of Finance and Economics, Nanjing, China
[2] School of Computer Science and Engineering,
Nanjing University of Science and Technology, Nanjing, China
[3] School of computer science and technology,
Anhui University of Technology, Maanshan, China
buzhan@nuaa.edu.cn

Abstract. The scale of current networked system is becoming increasingly large, which exerts significant challenges to acquire the knowledge of the entire graph structure, and most global community detection methods often suffer from the computational inefficiency. Local community detection aims at finding a community structure starting from a seed vertex without global information. In this article, we propose a Local Community Extraction algorithm (LCE) to find the local community from a seed vertex. First, a local search model is carefully designed to determine candidate vertices to be preserved or discarded, which only relies on the local/incomplete knowledge rather than the global view of the network. Second, we expand LCE for the global non-overlapping community detection, in which the labels of detected local communities are seen as vertices' attributive tags. Finally, we adopt the results of LCE to calculate a membership matrix, which can been used to detect the global overlapping community of a graph. Experimental results on four real-life networks demonstrate the advantage of LCE over the existing degree-based and similarity-based local community detection methods by either effectiveness or efficiency validity.

Keywords: Complex Network, Incomplete Knowledge, Local Community Extraction, Non-overlapping Community, Overlapping community.

1 Introduction

A complex network is composed by a large number of highly interconnected dynamical nodes. Social , biological, and computer science networks are only a few examples of complex networks, and they often display a common topological feature-community structure. Discovering the latent communities therein is

---

* Corresponding author.

X. Luo, J.X. Yu, and Z. Li (Eds.): ADMA 2014, LNAI 8933, pp. 98–111, 2014.

a useful way to infer some important functions. e.g., in social network, communities can be defined as subgroups whose members are "friends" to each other. A common formulation of the problem of community detection is to find a partitioning, $\mathcal{P} = \{\mathcal{C}_1, \cdots, \mathcal{C}_K\}$, of disjoint or joint subsets of vertices of the graph representing the network, in a meaningful manner.

In general, a community should be thought of a set of nodes that has more and/or better-connected edges between its members than between its members and the remainder of the network. The existing community definitions in the literature can be roughly divided into three categories, one is global-based [3,4,15], the other based on the vertex-similarity [8,17], and the third is local-based [13,16]. 1) The global-based definitions consider the graph as a whole, they follow the assumption that a graph has community structure if it is different from a random graph, e.g., the null model of Newman and Girvan [15]. 2) The vertex-similarity-based definitions are based the assumption that communities are groups of vertices similar to each other. e.g., the graph vertices can be embed in an n-dimensional Euclidean space, then the similarity of any vertex pair can be measured by their Euclidean distance. 3) The local-based definitions compare the internal and external cohesion of a sub-graph. The first recipe of this kind is LS-set [13], which stems from social network analysis. Some other local-based definitions can be found in recent literature.

Existing global-based and the vertex-similarity-based community detection approaches require clear pictures of the entire graph structure. And they are often described as global community detection(in short as GCD henceforth). As the scale of current networked system is becoming increasingly large, which exerts significant challenges to acquire the knowledge of the entire graph structure, and most GCD methods often suffer from the computational inefficiency. In spite of these limitations, local community detection(in short as LCD henceforth) would be very useful. Several LCD methods have been proposed to find the community containing a particular starting vertex for decades. According to the ways of how to evaluate the quality of a local community, the existing approaches to LCD can be classified into two main categories, namely, degree-based methods and similarity-based methods. 1)Degree-based methods [2,6,1,7,14] evaluate the local community quality by investigating vertices' degrees. 2)Similarity-based methods utilize similarities between vertices to help evaluate the local community quality [9,12,5]. Although LCD methods based degree and similarity are extensively studied, further study is still needed on finding a nice balance between the high efficiency of local search models and the high accuracy of detected communities. And how to ingeniously use the results of LCD to discovery the global non-overlapping and overlapping community structures is worth an in-depth study and concern.

In this work, we attempt to design a novel similarity-based method(LCE) for extracting local communities from large-scaled networks. In particular, LCE picks the neighbor vertex with the largest structure similarity as the candidate vertex and calculate the modularity gain to determine whether it should be added to the local community or not. Our method is naturally a heuristic, since it does not examine all of vertices in the network, and the structural similarity of each pair of vertices in LCE is calculated only once by using a dynamical priority queue. So the execution of LCE is accelerated and the accuracy remains high. We further expand LCE for the global non-overlapping community detection, in which the labels of detected local communities are used as vertices' attributive tags. Finally, we adopt the results of LCE to calculate a membership matrix, which can been used to detect the global overlapping community of a graph. Experimental results on four real-life networks demonstrate the superiority of LCE over the classic degree-based and similarity-based LCD methods by either effectiveness or efficiency validity.

2 Problem Definition and Preliminaries

Let $\mathcal{G} = (V, E, w)$ be a given weighted undirected graph, where V is the set of nodes ($|V| = n$), E is the set of edges ($|E| = m$) that connect the nodes in V, and w is the weight of every edge in E. LCD is formulated as finding a subset of graph $\mathcal{G}' = (V', E')$ from a seed $v_s \in V$. Note that in LCD, the entire network structure is unknown at the beginning. Besides the detected local community, only partial information, i.e., the local community's neighbors and their linkage information, are available after each detection process. To be specific, we divide the explored graph into three regions: the local community $\mathcal{C}$, the boundary area $\mathcal{B}$ and a larger unknown area $\mathcal{U}$. Initially, we add the seed v_s to $\mathcal{C}$. Then, all of neighbors of nodes in $\mathcal{C}$ (e.g., v_s) are added to $\mathcal{B}$. In such local search model, $\mathcal{C}$ can be locally expanded from v_s with a predefined criterion.

Generally, a community is measured by a specific property of the vertices within it. For this task, different community measurements have been proposed [16,1,7,14] in recent years. In this paper, we adopt a structural similarity measure from the cosine similarity function [9] to effectively denotes the local connectivity density of any two adjacent vertices in a weighted network. Here, we first formalize some notions of the local community.

Definition 1 (Structural Similarity). *The structural similarity between two adjacent vertices v_i and v_j is defined as $s_{i,j} = \frac{\sum_{v_k \in \Gamma(v_i) \cap \Gamma(v_j)} w_{i,k} w_{j,k}}{\sqrt{\sum_{v_k \in \Gamma(v_i)} w_{i,k}^2} \sqrt{\sum_{v_k \in \Gamma(v_j)} w_{j,k}^2}}$, where $\Gamma(v_i) = \{v_j \in V | \{v_i, v_j\} \in E\}$.*

The criterion we use to extract the local community containing the seed v_s is derived from [18], which finds a community with a large number of edges within itself and a small number of edges to the rest of the network.

Definition 2 (Local Modularity). *The local modularity of a community $\mathcal{C}$, denoted as $W(\mathcal{C})$, is given as $W(\mathcal{C}) = \frac{I(\mathcal{C})}{|\mathcal{C}|^2} - \frac{O(\mathcal{C})}{|\mathcal{C}||\mathcal{C}^c|}$, where $\mathcal{C}^c$ is the complement of $\mathcal{C}$, $I(\mathcal{C}) = \sum_{v_i, v_j \in \mathcal{C}} A_{ij}$, $O(\mathcal{C}) = \sum_{v_i \in \mathcal{C}, v_j \in \mathcal{C}^c} A_{ij}$, $A = [A_{ij}]$ is an $n \times n$ adjacency matrix of the graph $\mathcal{G}$.*

Based on the definition of local modularity, we have the following theorem.

Theorem 1. *The local modularity value of a community C will increase when C has high intra-cluster density and low inter-cluster density.*

PROOF:The term $I(C)$ is twice the number of the edges within C, and $O(C)$ represents the number of edges between C and the rest of the network. Each term is normalized by the total number of possible edges in each case. Note that we normalize the first term by $|C|^2$ rather than $|C|(|C|-1)$ in order to conveniently derive the modularity gain discussed below, but in practice this makes little difference. Subject to this small difference, the local modularity can be described as the intra-cluster density minus the inter-cluster density. $\square$

In Definition 2, we make an adjustment in the spirit of the ratio cut as $\hat{W}(C) = |C||C^c|(\frac{I(C)}{|C|^2} - \frac{O(C)}{|C||C^c|})$, where the factor $|C||C^c|$ penalizes very small and very large communities and produces more balanced solutions.

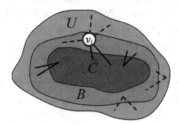

Fig. 1. The $\hat{W}$ variant when a vertex v_i joins C

Suppose a community C is detected from a certain vertex v_s. We explore the adjacent vertices in the boundary area B of C, as shown in Fig. 1. We distinguish three types of links: those internal to the community $C(L)$, between C and the vertex $v_i(L_{in})$, between C and others vertices in $B(L_{out})$. To simplify the calculations, we express the number of external links in terms of L and k_i(the degree of vertex v_i), so $L_{in} = a_1 L = a_2 k_i$, $L_{out} = b_1 L$, with $b_1 \geq 0$, $a_1 \geq \frac{1}{L}$, $a_2 \geq \frac{1}{k_i}$(since any v_i in B at least has one neighbor in C). So, the value of $\hat{W}$ for the current community can be written as $\hat{W}(C) = \frac{n-|C|}{|C|}2L - (a_1 + b_1)L$.

Definition 3 (Modularity Gain). *The modularity gain for the community C adopting a neighbor vertex v_i can be denoted as:*

$$\triangle \hat{W}_C(v_i) = (\frac{n-|C|-1}{|C|+1}2L(1+a_1) - (b_1 L + k_i - a_2 k_i)) - (\frac{n-|C|}{|C|}2L - (a_1+b_1)L)$$

$$= 2n\frac{a_2 k_i |C| - L}{|C|(|C|+1)} - k_i. \tag{1}$$

$\triangle \hat{W}_C(v_i)$ can be utilized as a criterion to determine whether the candidate vertex v_i should be included in the community C or not.

3   Local Community Extraction(LCE)

In this section, we propose a Local Community Extraction algorithm (in short as LCE henceforth). First, we introduce the basic idea of LCE and then present algorithmic details including the complexity analysis for LCE. Second, we introduce how to use LCE to detect the global non-overlapping and overlapping community structures.

To find the densely connected local community containing vertex v_s, LCE works with two iterative steps: update step and join step. First, the starting vertex v_s is added in $\mathcal{C}$. In the update step, LCE refreshes the the boundary area $\mathcal{B}$, and calculate the structural similarities between vertices in the community $\mathcal{C}$ and their neighbor vertices in $\mathcal{B}$. In the joining step, LCE tries to absorb a vertex in $\mathcal{B}$ having highest structural similarity with vertices in $\mathcal{C}$. If $\triangle \hat{W}_\mathcal{C}(v_i) > 0$, then the vertex v_i will be inserted into $\mathcal{C}$. Otherwise, it will be removed from $\mathcal{B}$. The two procedures above will be repeated in turn until set $\mathcal{B}$ is empty. Then, the whole community $\mathcal{C}(v_s)$ is discovered. We further select the vertex with maximum degree in $\mathcal{C}(v_s)$ as the core vertex, which can be also seen as the label of detected community. The pseudo-code of LCE is given in the following.

Algorithm 1. LCE(v_s)

Require: v_s
Ensure: $\mathcal{C}(v_s)$, $l(\mathcal{C}(v_s))$
 1: $\mathcal{C} \leftarrow \{v_s\}$
 2: $\mathcal{B} \leftarrow \{v_i | v_i \in \Gamma(v_s)\}$
 3: **while** $\mathcal{B} \neq \emptyset$ **do**
 4:     $v_i^* = \arg max_{v_i \in \mathcal{B}} \sum_{v_j \in \mathcal{C}} s_{i,j}$
 5:     **if** $\triangle \hat{W}_\mathcal{C}(v_i^*) > 0$ **then**
 6:         $\mathcal{C} \leftarrow \mathcal{C} \cup \{v_i^*\}$
 7:         $\mathcal{B} \leftarrow \mathcal{B} \cup \{v_j' | v_j' \in \Gamma(v_i^*), v_j' \notin \mathcal{C}\}$
 8:     **else**
 9:         $\mathcal{B} \leftarrow \mathcal{B} - \{v_i^*\}$
10:     **end if**
11: **end while**
12: $\mathcal{C}(v_s) \leftarrow \mathcal{C}$
13: $l(\mathcal{C}(v_s)) \leftarrow \arg max_{v_j \in \mathcal{C}} k_j$

Remark. Unlike existing methods [1,7,14], which calculate the quantitative metrics for each vertex in $\mathcal{B}$ and select the vertex who produces the greatest increment of the metric to join $\mathcal{B}$, LCE picks the neighbor vertex with the largest structure similarity as the candidate vertex v_i^* and calculate $\triangle \hat{W}_\mathcal{C}(v_i^*)$ to determine whether it should be added to $\mathcal{C}$ or not. The structural similarity reflects the local connectivity density of the graph. The larger the similarity between a vertex inside $\mathcal{C}$ and a vertex outside it, the more common neighbors the two vertices share, and the more probability they are at the same community. Furthermore, the structural similarity of each pair of vertices in LCE is calculated only once by using a dynamical priority queue. So the execution of LCE is accelerated and the accuracy remains high.

Complexity Analysis. The running time of LCE is mainly consumed in line 4 of Algorithm 1. We can implement it using a binary Fibonacci heap H [9], which takes two steps: 1) Extract Step (extract the maximum element from H). As each Extract operation of H takes $O(logn')$ time and the body of the while loop is executed n' times, the total time for all Extract Steps is $O(n'logn')$, where n' is the number of vertices inferred(vertices in $\mathcal{C} \cup \mathcal{B}$). 2) Update Step (for each vertex in current $\mathcal{B}$, we update its sum of structure similarities with vertices in $\mathcal{C}$). First, the sum of structure similarities with vertices in $\mathcal{C}$ for each vertex $v_i \in \mathcal{B}$ should be computed, which can be completed in $O(k')$ time, where k' is the mean degree of inferred vertices. For vertices which are not in H, we insert them to H in $O(1)$ time; otherwise, it takes $O(1)$ time to make an Increase-Key operation. As the above steps are executed $O(m')$ times, where m' is the number of edges in $\mathcal{C} \cup \mathcal{B}$. Therefore, the total time of the Update Step is $O(m'k')$. Adding all together, the total time complexity is $O(m'k' + n'logn')$ for LCE.

Non-overlapping Community Detection. Non-overlapping community detection aims to find a good K-way partition $\mathcal{P} = \{\mathcal{C}_1, \cdots, \mathcal{C}_K\}$, where $\mathcal{C}_k$ is the k-th community, and $\mathcal{C}_1 \cup \cdots \cup \mathcal{C}_K \subseteq V$, $\mathcal{C}_k \cap \mathcal{C}_{k'} = \emptyset \; \forall \; k \neq k'$. Our assumption is that LCE inputted with similar adjacent vertices will return analogous community structures, in which the core vertices are almost unanimous. Therefore, if LCE returns the the same community label, the input vertices are likely to be in the same community. The process of LCE expansion algorithm for non-overlapping (in short as LCEnO henceforth) is given in Algorithm 2. Note that, the line 2 can be paralleled executed. Therefore, LCEnO could be completed in $O(m^*k^* + n^*logn^*)$ time, where n^*, m^* are the number of vertices and edges in the largest $\mathcal{C} \cup \mathcal{B}$, and k^* is the mean degree of inferred vertices in it.

Algorithm 2. LCEnO($\mathcal{G}$)

Require: $\mathcal{G} = (V, E, w)$
Ensure: $L = [l_s], s = 1, \cdots, n$
1: **for** $s = 1; s <= n; s + +$ **do**
2:    $[\mathcal{C}(v_s), l_s] \leftarrow$ LCE(v_s)// Parallel Computing
3: **end for**

Overlapping Community Detection. For an overlapping partition, overlapping communities can be represented as a membership matrix $\mathbf{U} = [u_{i,k}], i = 1, \cdots, n, k = 1, \cdots, K$, where $0 \leq u_{i,k} \leq 1$ denotes the ratio of membership that node i belongs to C_k. If node i belongs to only one community, $u_{i,k} = 1$, and it clearly follows that $\sum_{k=1}^{K} u_{i,k} = 1$ for all $1 \leq i \leq n$. With the detected communities of LCE, $u_{i,k}$ can be calculated as follows:

$$u_{i,k} = \frac{\sum_{s=1,\cdots,n \wedge l_s = k} \chi(v_i, \mathcal{C}(v_s))}{\sum_{s=1,\cdots,n} \chi(v_i, \mathcal{C}(v_s))}, \quad \chi(v_i, \mathcal{C}(v_s)) = \begin{cases} 1 & if v_i \in \mathcal{C}(v_s) \\ 0 & otherwise. \end{cases} \quad (2)$$

The process of LCE expansion algorithm for overlapping (in short as LCEO henceforth) is given as follows. The running time of LCEO is mainly consumed in lines 4-7 of Algorithm 3, which is calculating the membership that node i belongs to C_k. The total time of those steps is $O(nK)$. Adding the local community extraction steps, the total time complexity is $O(m^*k^*+n^*logn^*+nK)$ for LCEO.

Algorithm 3. LCEO($\mathcal{G}$)

Require: $\mathcal{G} = (V, E, w)$
Ensure: $\mathbf{U} = [u_{i,k}], i = 1, \cdots, n, k = 1, \cdots, K$
 1: **for** $s = 1$; $s <= n$; $s + +$ **do**
 2:    $[\mathcal{C}(v_s), l_s] \leftarrow$ LCE(v_s)// Parallel Computing
 3: **end for**
 4: **for** $i = 1$; $i <= n$; $i + +$ **do**
 5:    **for** $k = 1$; $i <= K$; $k + +$ **do**
 6:        $u_{i,k} \leftarrow \dfrac{\sum_{s=1,\cdots,n \,\wedge\, l_s=k} \chi(v_i, \mathcal{C}(v_s))}{\sum_{s=1,\cdots,n} \chi(v_i, \mathcal{C}(v_s))}$
 7:    **end for**
 8: **end for**

We introduce the main framework of our approach by an example as shown in Fig. 2. The original graph includes 12 vertices and 20 edges. First, a state-of-the-art GCD algorithm known as FUC [3] is applied to identify its communities, the community structure is shown below the original graph in Fig. 2. Second, we employ LCE starting from all the vertices to detect their local communities. e.g., LCE starting from v_1 detects a local community including vertices v_1, v_5 and v_9, in which v_5 has the maximum degree. Therefore, the label of this community is marked as 5, which is also the attributive tag of vertex v_1. When we acquire all the attributive tags of 12 vertices, the global non-overlapping community structure of the graph has been detected. Finally, all local communities are assembled to be a membership matrix $\mathbf{U} = [u_{i,k}]$, which promulgates the global overlapping community structure.

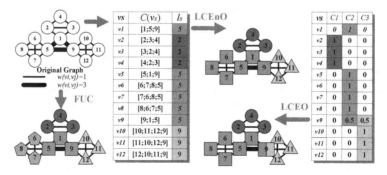

Fig. 2. An illustrative example

4 Experimental Results

Four real-world undirected networks: Karate, NCAA, Facebook and PGP are used for experiments. Some characteristics of these networks are shown in Table 1, where $|V|$ and $|E|$ indicate the numbers of nodes and edges respectively in the network, and $< k >$ indicates the average degree. Karate is a well known social network that describes the friendship relations between members of a karate club. NCAA is a representation of the schedule of American Division I college football games. Vertices in the network represent teams, which are divided into eleven communities(or conferences) and five independent teams. Edges represent regular season games between the two teams they connect. Facebook has been anonymized by replacing the Facebook-internal ids for each user with a new value. Each edge tells whether two users have the same political affiliations. PGP is a large scale social network, where each node represents a peer and each tie points out that one peer trusts the other.

Table 1. Real-world networks

Network	$|V|$	$|E|$	$< k >$
Karate	34	78	4.59
NCAA	115	616	10.71
Facebook	4,039	88,234	43.69
PGP	10,680	24,340	4.56

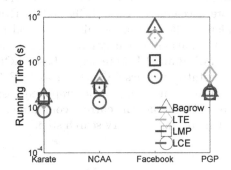

Fig. 3. Comparison on efficiency

4.1 Performance of LCE

The Effectiveness. To test the effectiveness of LCE, the results are compared with the ground truth communities of each network. To be special, let $\mathcal{T}(v_s)$ be the ground truth community including the vertex v_i, we can compare $\mathcal{T}(v_s)$ and $\mathcal{C}(v_s)$ in the framework of Precision, Recall and F-measure (PRF) to assess our results. A higher value of precision (P) indicates fewer wrong classifications, while a higher value of recall (R) indicates less false negatives. It is common to use the harmonic mean of both measurements, called F-measure, which weighs precision and recall equally important. They are calculated as follows:

$$P(v_s) = \frac{|\mathcal{C}(v_s) \cap \mathcal{T}(v_s)|}{|\mathcal{C}(v_s)|}, \quad R(v_s) = \frac{|\mathcal{C}(v_s) \cap \mathcal{T}(v_s)|}{|\mathcal{T}(v_s)|}, \quad F1(v_s) = \frac{2P(v_s)R(v_s)}{P(v_s) + R(v_s)}. \quad (3)$$

Since the last two networks (Facebook and PGP) have no ground truth, we apply FUC [3] to identify communities of them, and utilize its detection results as the ground truth for the LCD algorithms. This is based on the intuition that a LCD method is acceptable if it can achieve an approximate result as a GCD approach does, because LCD methods usually perform faster than GCD approaches. As the global community quality metrics such as the well-known Modularity metric [15] are not suitable to evaluate the quality of the detected local community, we use each vertex in a community as a seed and report algorithms' average precision, recall and F1-measure. We compare LCE with classical LCD algorithms, such as LWP [14], ELC [1], LTE [9]. The comparison results are presented in Table 2, from which we can observe that: 1)the recall values for all methods are overall worse than precision values, this is because LCD methods are based on the greedy search, which will trend to find a local optimal solution; 2)LCE almost achieves the high precision for all datasets, which demonstrate the superiority of its local search model over the other methods; 3)LCE usually outperforms LMR and ELC, and have a slight advantage over LTE, even though the later has been proven by extensive experiments to be one of the most accurate algorithms among previous LCD methods in[9].

The Efficiency. Fig. 3 shows the average running time of LDC methods starting from each vertex in the four test graphs. Apparently, the execution of LCE is more accelerated. Both LCE and LTE are similarity-based algorithms, their difference lies at the definition of local modularity. Compared with LCE, the calculation of modularity gain in LTE is more complex, which will consume extra time. LMR and ELC are degree-based LCD algorithms, which need calculate the quantitative metrics for each vertex in $\mathcal{B}$. The metric calculations are somewhat duplicate, which can not be simplified. Especially, the stopping criteria for ELC is to jude whether the current community is a "p-strong community", which will cost more time in every search step.

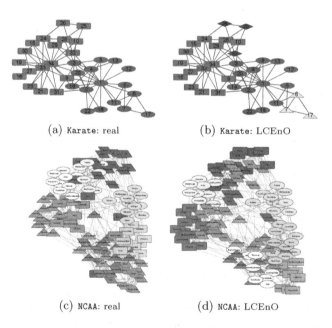

(a) Karate: real   (b) Karate: LCEnO

(c) NCAA: real   (d) NCAA: LCEnO

Fig. 4. LCEnO on small social networks

Table 2. Accuracy comparison of LCD on real-world networks

Network Comm.	size	LCE			LWP			ELC			LTE		
		P	R	F1	P	R	F1	P	R	F1	P	R	F1
Karate-A	16	1.00	0.58	**0.73**	0.94	0.49	0.64	0.93	0.49	0.64	1.00	0.49	0.66
Karate-B	18	0.97	0.47	0.63	0.97	0.44	0.61	0.89	0.48	0.63	1.00	0.57	**0.73**
NCAA-AC	9	1.00	1.00	**1.00**	0.70	0.48	0.57	0.68	0.56	0.61	1.00	1.00	**1.00**
NCAA-BE	8	1.00	1.00	**1.00**	0.48	0.47	0.48	0.51	0.67	0.58	0.80	1.00	0.89
NCAA-Ten	12	1.00	1.00	**1.00**	0.33	0.26	0.29	0.17	0.21	0.19	1.00	1.00	**1.00**
NCAA-SE	12	1.00	1.00	**1.00**	0.81	0.55	0.65	0.83	0.85	0.84	1.00	1.00	**1.00**
NCAA-PT	10	0.91	0.82	**0.86**	0.68	0.58	0.62	0.68	0.73	0.70	0.91	0.82	**0.86**
NCAA-Others	5	0.12	0.24	0.16	0.21	0.40	**0.27**	0.14	0.52	0.22	0.19	0.32	0.24
NCAA-MA	13	1.00	0.50	0.67	0.78	0.48	0.60	0.81	0.78	**0.79**	0.86	0.50	0.64
NCAA-MV	8	1.00	1.00	**1.00**	0.76	0.70	0.73	0.67	0.70	0.69	1.00	1.00	**1.00**
NCAA-WA	10	1.00	1.00	**1.00**	0.65	0.45	0.53	0.67	0.60	0.63	1.00	1.00	**1.00**
NCAA-Twelve	12	1.00	1.00	**1.00**	0.67	0.40	0.52	0.61	0.56	0.35	1.00	1.00	**1.00**
NCAA-SB	7	0.64	0.51	**0.57**	0.49	0.61	0.54	0.23	0.69	0.35	0.64	0.51	0.56
NCAA-USA	10	0.74	0.66	**0.70**	0.41	0.32	0.36	0.25	0.23	0.24	0.74	0.66	**0.70**
Facebook-1	341	1.00	0.16	0.28	0.99	0.05	0.10	0.88	**0.40**	0.55	1.00	0.15	0.26
Facebook-2	66	0.88	0.48	0.61	0.42	0.14	0.21	0.16	0.96	0.27	0.94	0.57	**0.71**
Facebook-3	308	0.94	0.26	0.41	0.92	0.07	0.13	0.41	0.15	0.22	0.97	0.18	**0.30**
Facebook-4	25	0.96	1.00	0.98	1.00	0.36	0.53	0.97	0.59	0.74	1.00	1.00	**1.00**
Facebook-5	206	1.00	0.33	**0.50**	1.00	0.09	0.17	0.97	0.31	0.47	1.00	0.33	0.49
Facebook-6	62	0.94	0.42	0.58	0.90	0.19	0.31	0.56	0.32	0.41	0.99	0.44	**0.61**
Facebook-7	408	0.94	0.58	0.71	0.38	0.04	0.07	0.21	0.17	0.18	0.96	0.60	**0.74**
Facebook-8	483	0.94	0.19	**0.31**	0.81	0.05	0.09	0.16	0.13	0.14	0.97	0.16	0.27
Facebook-9	442	0.98	0.30	**0.45**	0.97	0.07	0.14	0.97	0.20	0.33	1.00	0.24	0.38
Facebook-10	73	0.94	0.92	0.93	0.53	0.19	0.28	0.06	0.12	0.08	1.00	1.00	**1.00**
Facebook-11	237	0.99	0.87	**0.92**	0.26	0.07	0.10	0.15	0.03	0.04	1.00	0.82	0.90
Facebook-12	226	0.98	0.68	**0.80**	0.96	0.13	0.23	0.10	0.21	0.14	0.99	0.46	0.63
Facebook-13	554	0.98	0.18	0.31	0.96	0.06	0.10	0.63	0.37	**0.46**	0.99	0.18	0.21
Facebook-14	548	1.00	0.11	0.20	0.99	0.03	0.07	0.98	0.24	**0.39**	1.00	0.08	0.12
Facebook-15	60	1.00	0.33	**0.50**	0.98	0.13	0.23	0.99	0.33	**0.50**	0.98	0.14	0.24
PGP-1	395	0.95	0.16	**0.28**	0.86	0.16	0.27	0.82	0.12	0.21	0.96	0.12	0.21
PGP-2	303	0.93	0.22	**0.36**	0.92	0.22	**0.36**	0.73	0.19	0.30	0.93	0.19	0.32
PGP-3	974	0.94	0.13	0.24	0.74	0.13	0.22	0.84	0.17	**0.29**	0.94	0.18	0.31
PGP-4	379	0.99	0.12	0.21	0.78	0.12	0.20	0.90	0.17	**0.29**	0.99	0.13	0.21
PGP-5	1457	0.93	0.11	**0.20**	0.88	0.11	**0.20**	0.76	0.08	0.15	0.93	0.09	0.17
PGP-6	798	0.98	0.06	0.11	0.94	0.06	0.11	1.00	0.08	**0.16**	0.98	0.08	0.15
PGP-7	1289	0.96	0.14	0.24	0.80	0.14	0.23	0.76	0.13	0.22	0.96	0.17	**0.29**
PGP-8	513	0.97	0.17	**0.29**	0.87	0.17	0.28	0.87	0.11	0.20	0.97	0.11	0.20
PGP-9	417	0.93	0.17	0.28	0.93	0.17	0.28	0.92	0.27	**0.41**	0.92	0.13	0.22
PGP-10	1091	0.93	0.22	0.36	0.86	0.22	0.35	0.83	0.19	0.31	0.93	0.31	**0.47**

4.2   Performance of LCEnO

Here, we first apply LCEnO to the two small social networks with ground truth: Karate and NCAA. The purpose is to gain a direct understanding of non-overlapping community detection by network visualization. Then, we further compare LCEnO with classical GCD methods, such as FNM [15], FUC [3], Metis [10], and Cluto [11].

Karate is split into two parties following a disagreement between an instructor (node 1) and an administrator (node 34), which serves as the ground truth about the communities in Fig. 4(a). We employ LCEnO to extract non-overlapping communities from the network. The result is shown in Fig. 4(b), which supplements the division of the club with more information. More interestingly, LCEnO actually tends to partition this network into four rather than two communities, as indicated by the nodes in four colors/shapes in Fig. 4(b). This implies that there exits a latent sub-party (including vertices 6, 7, 11) inside the party led by node 1, and a latent sub-party (including vertices 25, 26, 32) inside the party led by node 34.

The ground truth of NCAA labels vertices with their actual conferences, corresponding twelve different colors/shapes in Fig. 4(c). As shown in Fig. 4(d), LCEnO generally well captures the "sharp-cut" teams in conferences "AC", "BE", "Ten", "SE", "MV", "WA", and "Twelve" respectively, although there yet exists some teams assigned mistakenly. Note that nearly all the "Orangered rectangle" in Fig. 4(c) are totally detected mistakenly by LCEnO. This is indeed reasonable since those vertices have very few internal connections, actually, they represent five independent teams (Utah State, Navy, Notre Dame, Connecticut and Central Florida) in NCAA.

We compare LCEnO with GCD methods, such as FNM [15], FUC [3], Metis [10], and Cluto [11] on the effectiveness. For each method/network, Table 3 displays the modularity that is achieved and the running time. The modularity obtained by LCEnO are slightly lower than FUC's, but it outperforms nearly all the other methods. In terms of running time, Metis has a great advantage due to its parallel processing modules. However, it perform poor on graphs with obscure community structure, e.g., Karate and NCAA. While LCEnO keeps a nice balance between high modularity and short running time, which can be applied large scale network community detection.

Table 3. Modularity and running time comparison

Network	LCEnO	FNM	FUC	Metis	Cluto
Karate	0.38/0.03s	0.38/0.05s	**0.42**/0.03s	0.24/**0.01s**	0.36/0.02s
NCAA	0.58/0.20s	0.57/0.20s	**0.60**/0.06s	0.08/**0.01s**	0.60/0.03s
Facebook	0.73/2.68s	0.78/8.45m	**0.84**/6.29s	0.79/**0.53s**	0.82/4.24s
PGP	0.67/**0.44s**	0.85/179.42m	**0.88**/22.50s	0.83/1.76s	0.72/11.90s

4.3 Performance of LCEO

To evaluate the performance of LCEO, we also employ the PRF framework. Let $\hat{C}_k$ be the k-th overlapping community, which obeys $\hat{C}_1 \cup \cdots \cup \hat{C}_K \subseteq V$. In the following, we introduce a membership threshold α, $0 < \alpha \leq 1$, to control the scale at which we want to observe the overlapping communities in a network.

Definition 4 (α-Overlapping Community). *The k-th α-overlapping community, denoted by $\hat{C}_k(\alpha)$, is defined as $\hat{C}_k(\alpha) = \{v_i | u_{i,k} \geq \alpha\}$.*

We can use each vertex in a overlapping community as a seed and report LCEO's average PRF. Fig. 5 shows the accuracy in the function of α for the four test graphs, from which we can observe that: 1)the recall values for LECHO have a significant improvement in all scales, compared with previous LCD algorithms; 2)the values of α in the range $[0.6, 0.8]$ are optimal, in the sense that overlapping communities extracted by LCEO in this region have a high F1-measure; 3)LECHO performs better in dense networks rather than in sparse networks.

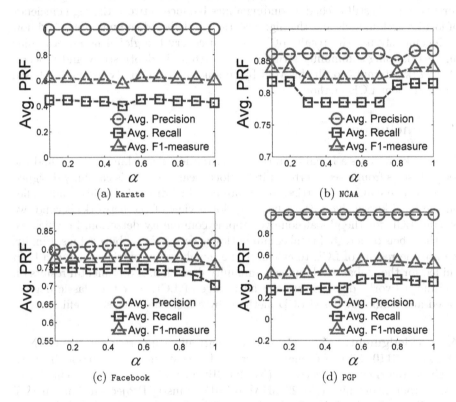

Fig. 5. The accuracy for different α on the four test networks

5 Related Work

Existing approaches to LCD can be classified into two main categories, namely, degree-based methods and similarity-based methods. 1)Degree-based methods evaluate the local community quality by investigating verticesdegrees. Some naive solutions, such as l-shell search algorithm [2], discovery-then-examination approach [6], and outwardness-based method [1], only consider the number of edges inside and outside a local community. Clauset [7] defines local modularity by considering the boundary points of a sub-graph, and proposes a greedy algorithm on optimizing this measure. Similarly, Luo et al. [14] present another measurement as the ratio of the internal degree and external degree of a subgraph. Both methods can achieve high recall but suffer from low precision due to including many outliers [6]. 2)Similarity-based methods utilize similarities between vertices to help evaluate the local community quality. LTE algorithm [9] is a representative of this kind, using a well-designed metric for local community quality known as Tightness. There are a few alternative similarity-based metrics such as VSP [12] and RSS [5] that can also help evaluate the local community quality, although they are not originally designed for LCD.

Although LCD methods based degree and similarity are extensively studied, further study is still needed on finding a nice balance between the high efficiency of local search models and the high accuracy of detected communities. And how to ingeniously use the results of LCD to discovery the global non-overlapping and overlapping community structures is worth an in-depth study and concern. Our work attempts to fill this void by conducting community extraction based on an efficient LCE method.

6   Conclusion

This work proposes a Local Community Extraction algorithm (LCE) to find the local clusters from a seed vertex. First, a local search model is carefully designed to determine candidate vertices to be preserved or discarded, which only relies on the local knowledge rather than the global view of the network. Second, we expand LCE for the global non-overlapping community detection, in which we use the labels of detected local communities as vertices' attributive tags. Finally, we use the results of LCE to calculate the membership matrix, which can been guided for the global overlapping community detection. Experimental results on real-life networks demonstrate the advantage of LCE over the classic degree-based and similarity-based LCD methods by either effectiveness or efficiency.

Acknowledgment. This research was partially supported by NSFC (Nos. 71372188, 61103229), National Center for International Joint Research on E-Business Information Processing (No. 2013B01035), National Key Technologies R&D Program of China (No. 2013BAH16F01), Industry Projects in Jiangsu S&T Pillar Program (No. BE2012185), the Priority Academic Program Development of Jiangsu Higher Education Institutions (PAPD),and Key/Surface Project of Natural Science Research in Jiangsu Provincial Colleges and Universities (Nos. 12KJA520001, 14KJA520001, 14KJB520015).

References

1. Bagrow, J.P.: Evaluating local community methods in networks. Journal of Statistical Mechanics: Theory and Experiment 2008(05), 5001 (2008)
2. Bagrow, J.P., Bollt, E.M.: Local method for detecting communities. Physical Review E 72(4), 46108 (2005)

3. Blondel, V.D., Guillaume, J.-L., Lambiotte, R., Lefebvre, E.: Fast unfolding of communities in large networks. Journal of Statistical Mechanics: Theory and Experiment 2008(10), 10008 (2008)
4. Bu, Z., Zhang, C., Xia, Z., Wang, J.: A fast parallel modularity optimization algorithm (fpmqa) for community detection in online social network. Knowledge-Based Systems 50, 246–259 (2013)
5. Chen, H.-H., Gou, L., Zhang, X.L., Giles, C.L.: Discovering missing links in networks using vertex similarity measures. In: Proceedings of the 27th Annual ACM Symposium on Applied Computing, pp. 138–143. ACM (2012)
6. Chen, J., Zaïane, O., Goebel, R.: Local community identification in social networks. In: International Conference on Advances Social Network Analysis and Mining, ASONAM 2009, pp. 237–242. IEEE (2009)
7. Clauset, A.: Finding local community structure in networks. Physical Review E 72(2), 26132 (2005)
8. Hlaoui, A., Wang, S.: A direct approach to graph clustering. In: Neural Networks and Computational Intelligence, pp. 158–163 (2004)
9. Huang, J., Sun, H., Liu, Y., Song, Q., Weninger, T.: Towards online multiresolution community detection in large-scale networks. PloS one 6(8), e23829 (2011)
10. Karypis, G.: Multi-constraint mesh partitioning for contact/impact computations. In: Proceedings of the 2003 ACM/IEEE conference on Supercomputing, p. 56. ACM (2003)
11. Karypis, G., Han, E.H., Kumar, V.: Chameleon: Hierarchical clustering using dynamic modeling. Computer 32(8), 68–75 (1999)
12. Li, K., Pang, Y.: A vertex similarity probability model for finding network community structure. In: Tan, P.-N., Chawla, S., Ho, C.K., Bailey, J. (eds.) PAKDD 2012, Part I. LNCS, vol. 7301, pp. 456–467. Springer, Heidelberg (2012)
13. Luccio, F., Sami, M.: On the decomposition of networks in minimally interconnected subnetworks. IEEE Transactions on Circuit Theory 16(2), 184–188 (1969)
14. Luo, F., Wang, J.Z., Promislow, E.: Exploring local community structures in large networks. Web Intelligence and Agent Systems 6(4), 387–400 (2008)
15. Newman, M.E., Girvan, M.: Finding and evaluating community structure in networks. Physical Review E 69(2), 26113 (2004)
16. Radicchi, F., Castellano, C., Cecconi, F., Loreto, V., Parisi, D.: Defining and identifying communities in networks. Proceedings of the National Academy of Sciences of the United States of America 101(9), 2658–2663 (2004)
17. Von Luxburg, U.: A tutorial on spectral clustering. Statistics and Computing 17(4), 395–416 (2007)
18. Zhao, Y., Levina, E., Zhu, J.: Community extraction for social networks. Proceedings of the National Academy of Sciences 108(18), 7321–7326 (2011)

Discovery of Tampered Image with Robust Hashing

Zhenjun Tang[1,2], Junwei Yu[2], Xianquan Zhang[1,2], and Shichao Zhang[1,2,3,*]

[1] Guangxi Key Lab of Multi-source Information Mining & Security,
Guangxi Normal University, Guilin 541004, P.R. China
[2] Department of Computer Science, Guangxi Normal University, Guilin 541004, P.R. China
{tangzj230,zxq6622}@163.com,
zhangsc@gxnu.edu.cn, 444139840@qq.com
[3] Faculty of Information Technology, University of Technology, Sydney, NSW 2007, Australia

Abstract. Tampered image discovery from similar images is a challenging problem of multimedia security. Aiming at this issue, we propose a robust image hashing with invariant moments. Specifically, the proposed hashing firstly converts the input image into a normalized image by interpolation, filtering and color space conversion. Then it divides the normalized image into overlapping blocks and extracts invariant moments of blocks to form a feature matrix. Finally, the feature matrix is compressed to make a short hash. Hash similarity is determined by measuring similarity between hash segments with correlation coefficient. Experimental results indicate that our hashing is robust against normal digital operations and can efficiently distinguish tampered images from similar images. Comparisons show that our hashing is better than some notable hashing algorithms in classification performances between robustness and content sensitivity.

Keywords: Robust hashing, image hashing, invariant moment, tampering discovery.

1 Introduction

Nowadays, powerful tools make image editing much easier than ever before. Therefore, people can easily modify images to suit specific applications. But this also brings some problems. For example, people may compress a given image for transmission, but malicious attackers may tamper content of the given image, such as inserting an object or deleting an object. Fig. 1 (a) is an original image, (b) is a compressed image with JPEG quality factor 20, and (c) is a tampered image by deleting a vehicle. Clearly, Figure 1 (a) and (b) are visually similar, but (c) is significantly different from (a) and (b). Therefore, sophisticated techniques are in demand for distinguishing tampered images from similar images (including the original image and the compressed images). In this work, we study an emerging technology called image hashing. It not only can identify similar images, but also discovers tampered images.

Image hashing is a novel multimedia technology [1]. It derives a content-based compact representation, called image hash, from an input image, and has been widely

---

* Corresponding author.

X. Luo, J.X. Yu, and Z. Li (Eds.): ADMA 2014, LNAI 8933, pp. 112–122, 2014.

used in image authentication [2], image indexing [3], copy detection, digital water-marking [4], tampering discovery [5], image quality assessment [6], image forensics [7], image search[8, 9], and so on. In general, it produces the same or very similar hashes for those perceptually identical images. This is the hashing property of percep-tual robustness [10]. For different images, it generates different image hashes. This is called discrimination. In addition, it should be sensitive to content change when it is applied to authentication or forensics. In other words, it should create significantly different hashes for discovering tampered images. Actually, it is a challenging task [11] to develop image hashing algorithms reaching good performances between ro-bustness and content sensitivity.

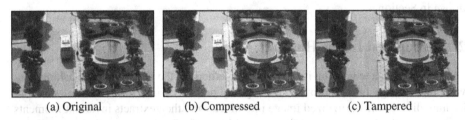

(a) Original   (b) Compressed   (c) Tampered

Fig. 1. Original image and its compressed and tampered versions

In the past, many researchers have contributed to developing image hashing algo-rithms. For example, Venkatesan et al. [12] used statistics of coefficients in discrete wavelet transform (DWT) domain to construct image hashes. This method is robust against JPEG compression and median filtering, but fragile to gamma correction and contrast adjustment. Li and Chang [13] designed a hashing method based on invariant relation between DCT coefficients at the same position in separate blocks. Their me-thod can distinguish JPEG compression from malicious attacks. Swaminathan et al. [14] proposed to calculate image hashes based on coefficients of Fourier-Mellin trans-form. This algorithm is robust against moderate geometric transforms and filtering, but its discrimination is not good. Monga and Mihcak [15] were the first to use non-negative matrix factorization (NMF) to derive image hashing. This method is robust against many popular digital operations, but sensitive to watermark embedding. Ou and Rhee [16] used Radon transform (RT) combining with discrete cosine transform (DCT) to design image hashing. The RT-DCT hashing is insensitive to content change and its discrimination should be improved. Kang et al. [17] introduced a com-pressive sensing-based image hashing. This method is sensitive to image rotation. In [18], Li et al. extracted image hashes by random Gabor filtering (GF) and dithered lattice vector quantization (LVQ). The GF-LVQ hashing has better performances than the well-known algorithms [14], [17], but its discrimination is also not desirable enough. In another study [19], Tang et al. convert RGB color image into YCbCr and HSI color spaces, extract invariant moments (IM) from each color component, and use them to form image hashes. The IM-based hashing is resilient to rotation, but has limitation in measuring local content change. Recently, Tang et al. [20] exploit color vector angle and DWT to generate image hashes. This method can resist JPEG com-pression and image rotation within 5°.

Although various hashing algorithms have been reported, there are still some problems. For example, more efforts are still needed to develop efficient algorithms reaching a desirable balance between perceptual robustness and content sensitivity. In this work, we exploit invariant moments in a new way to design a robust image hashing, which not only can achieve a good robustness, but also has an excellent capability of tampering discovery. We exploit many images to illustrate efficiency of our algorithm. Comparisons with existing notable algorithms are also done. The results show that our algorithm is better than the compared algorithms in classification performances between robustness and content sensitivity.

The rest of this paper is organized as follows. Section 2 introduces the proposed image hashing. Section 3 presents experimental results and conclusions are finally drawn in Section 4.

2   Proposed Image Hashing

Our proposed image hashing is a three-step method. In the first step, our hashing converts input image into a normalized image by preprocessing. In the second step, our hashing divides the normalized image into blocks and then extracts invariant moments from image blocks to construct feature matrix. In the final step, the feature matrix is compressed to make a short image hash. Section 2.1 briefly reviews invariant moment, Section 2.2 describes detailed steps of our hashing and Section 2.3 presents calculation of hash similarity.

2.1   Review of Invariant Moments

The well-known invariant moments are firstly introduced by Hu [21]. Since invariant moments have indicated excellent property (such as geometric invariant), they have been widely used in image classification [22], image matching [23], character recognition, and so on. Let $f(x, y)$ be gray value of a pixel in a digital image sized $m \times n$, where $1 \leq x \leq m$ and $1 \leq y \leq n$. Thus, seven invariant moments [21] are defined as follows:

$$\phi_1 = \eta_{20} + \eta_{02} \tag{1}$$

$$\phi_2 = (\eta_{20} - \eta_{02})^2 + 4\eta_{11}^2 \tag{2}$$

$$\phi_3 = (\eta_{30} - 3\eta_{12})^2 + (3\eta_{21} - \eta_{03})^2 \tag{3}$$

$$\phi_4 = (\eta_{30} + \eta_{12})^2 + (\eta_{21} + \eta_{03})^2 \tag{4}$$

$$\phi_5 = (\eta_{30} - 3\eta_{12})(\eta_{30} + \eta_{12})[(\eta_{30} + \eta_{12})^2 - 3(\eta_{21} + \eta_{03})^2]$$
$$+ (3\eta_{21} - \eta_{03})(\eta_{21} + \eta_{03})[3(\eta_{30} + \eta_{12})^2 - (\eta_{21} + \eta_{03})^2] \tag{5}$$

$$\phi_6 = (\eta_{20} - \eta_{02})[(\eta_{30} - \eta_{12})^2 - (\eta_{21} - \eta_{03})^2]$$
$$+ 4\eta_{11}(\eta_{30} + \eta_{12})(\eta_{21} + \eta_{03}) \tag{6}$$

$$\phi_7 = (3\eta_{21} - \eta_{03})(\eta_{30} + \eta_{12})[(\eta_{30} + \eta_{12})^2 - 3(\eta_{21} + \eta_{03})^2]$$
$$- (\eta_{30} - 3\eta_{12})(\eta_{21} + \eta_{03})[3(\eta_{30} + \eta_{12})^2 - (\eta_{21} + \eta_{03})^2] \tag{7}$$

where η_{pq} $(p, q = 0,1,2,\ldots)$ is the normalized central moment defined as:

$$\eta_{pq} = \frac{\mu_{pq}}{\mu_{00}^{\gamma}} \tag{8}$$

in which γ is determined by:

$$\gamma = \frac{p+q}{2} + 1 \qquad p+q = 2,3,\cdots \tag{9}$$

and μ_{pq} is the central moment calculated by:

$$\mu_{pq} = \sum_{x=1}^{m} \sum_{y=1}^{n} (x-\overline{x})^{p} (y-\overline{y})^{q} f(x,y) \tag{10}$$

where

$$\overline{x} = \frac{M_{10}}{M_{00}}, \quad \overline{y} = \frac{M_{01}}{M_{00}} \tag{11}$$

and M_{pq} is the $(p+q)$-th order moment:

$$M_{pq} = \sum_{x=1}^{m} \sum_{y=1}^{n} x^{p} y^{q} f(x,y) \tag{12}$$

In this study, we choose local invariant moments as image features, which can make our image hash discriminative and sensitive to content change. Note that the seventh invariant moment ϕ_7 is generally used to distinguish mirror images [21]. Since mirror operation in local image blocks is not considered here, the seventh invariant moment is not selected in our hashing. In other words, only the first six invariant moments are taken as features in our work.

2.2   Detailed Steps of Our Hashing

The details of our image hashing algorithm are presented as follows.

(1) *Preprocessing.* To make a normalized image for invariant moment extraction, some digital operations are applied to the input image. Firstly, bilinear interpolation is used to resize the input image to a standard size $M \times M$. This is to make our method robust against image rescaling. Secondly, the resized image is passed through a Gaussian low-pass filter. This can reduce influence of losing those insignificant high-frequent details caused by JPEG compression. Finally, for RGB color image, the normalized image is then converted into HSV color space and the **V** component is taken for representing the input image.

(2) *Local invariant moment extraction.* To reflect local content of the input image, the **V** component is divided into $m_1 \times m_1$ overlapping blocks, where overlapping size between each two neighbor blocks is $m_1/2$. For simplicity, let M be integral multiple of m_1. Thus, the total number of image blocks is $K = N_1^2$, where $N_1 = (2M/m_1 - 1)$. For the i-th block, the vector $\mathbf{u}_i$ forming by its first six invariant moments is selected for representation. Therefore, a feature matrix **U** sized $6 \times K$ is obtained as follows.

$$\mathbf{U}=[\mathbf{u}_1, \mathbf{u}_2, ..., \mathbf{u}_K] \tag{13}$$

Next, data normalization [24] is applied to each row of $\mathbf{U}$. And the result of $\mathbf{U}$ is denoted as $\mathbf{U}^{(1)}$.

(3) *Feature compression.* To conduct feature compression, we generate a reference vector $\mathbf{p}_0=[p_0(1), p_0(2), ..., p_0(6)]^T$, whose elements are randomly picked from each row of $\mathbf{U}^{(1)}$. Let $\mathbf{p}_i=[p_i(1), p_i(2), ..., p_i(6)]^T$ be the i-th column of $\mathbf{U}^{(1)}$ $(i=1,2,..., K)$. Thus, Euclidean distance between $\mathbf{p}_i$ and $\mathbf{p}_0$ is calculated as follows.

$$d_i = \sqrt{\sum_{j=1}^{6}[p_i(j) - p_0(j)]^2} \tag{14}$$

As K is still a big number, we merge neighbor elements to make a short hash. To do so, the sequence $[d_0, d_1, ..., d_K]$ is firstly reshaped to a square matrix $\mathbf{B}$ sized $N_1 \times N_1$. For each row, the mean of each two adjacent elements is calculated. Then, a small matrix sized $N_1 \times N_1/2$ is obtained. Similarly, for each column, the mean of each two adjacent elements is calculated. And finally, a feature matrix of size $N_1/2 \times N_1/2$ is available. Therefore, a sequence $\mathbf{s}=[s(1), s(2), ..., s(L)]$ is obtained by concatenating elements of the small feature matrix, where $L=N_1/2 \times N_1/2$. To reduce storage, each element $s(l)$ $(l=1, 2, ..., L)$ is quantized to an integer as follows.

$$h(l)=\text{Round}\,[s(l)\times1000 + 0.5] \tag{15}$$

Consequently, our image hash $\mathbf{h}$ is available as follows.

$$\mathbf{h}=[h(1), h(2), ..., h(L)] \tag{16}$$

2.3   Hash Similarity Evaluation

Let $\mathbf{h}_1=[h_1(1), h_1(2), ..., h_1(L)]$ and $\mathbf{h}_2=[h_2(1), h_2(2), ..., h_2(L)]$ be a pair of hashes of two input images. We divide hash sequences into overlapping segments sized $1 \times k$, where the overlapping size is $k/2$. Let the i-th segments of $\mathbf{h}_1$ and $\mathbf{h}_2$ be $\mathbf{h}_1^{(i)}$ and $\mathbf{h}_2^{(i)}$, respectively. Thus, their similarity is evaluated by the well-known correlation coefficient.

$$S_i = \frac{\sum_j \left(h_1^{(i)}(j)-\mu_1^{(i)}\right)\left(h_2^{(i)}(j)-\mu_2^{(i)}\right)}{\sqrt{\sum_j \left(h_1^{(i)}(j)-\mu_1^{(i)}\right)^2 \times \sum_j \left(h_2^{(i)}(j)-\mu_2^{(i)}\right)^2}+\varepsilon} \tag{17}$$

where ε is a small constant to avoid zero denominator, and $\mu_1^{(i)}$ and $\mu_2^{(i)}$ are the means of $\mathbf{h}_1^{(i)}$ and $\mathbf{h}_2^{(i)}$, respectively. Thus, the similarity between $\mathbf{h}_1$ and $\mathbf{h}_2$ can be determined by:

$$S =\min\{S_1, S_2, ..., S_J\} \tag{18}$$

where $\min\{...\}$ means the minimum value and J is the total segments. The range of S is $[-1, 1]$. The bigger the S, the more similar the evaluated hashes and then the more similar the corresponding images. If S is bigger than a threshold, the images of the input hashes are considered as visually similar images. Otherwise, they are the different images or one is a tampered version of the other.

3   Experimental Results

In the following experiments, our parameter settings are as follows. In the preprocessing, the input image is resized to 512×512, a 3×3 Gaussian low-pass filter with zero mean and a unit standard deviation is taken, the block size is 64×64 and the segment size is 7. Therefore, our hash length is 64 integers. Section 3.1 presents our classification between robustness and discrimination. Section 3.2 validates our capability of discovering tampered images, and Section 3.3 show performance comparisons with some existing algorithms.

3.1   Classification between Robustness and Discrimination

Various images are exploited to test robustness of our hashing and the results validate our efficiency. For space limitation, typical results are presented here. Fig. 2 is eight standard test images. We exploit Photoshop, MATLAB and StirMark [25] to generate visually similar versions of these test images. The used digital operations (parameter settings) include brightness adjustment (−20, −10, 10, 20), contrast adjustment (−20, −10, 10, 20), gamma correction (0.75, 0.9, 1.1, 1.25), 3×3 Gaussian low-pass filtering (0.3, 0.4, ..., 1.0), JPEG compression (30, 40, ..., 100), watermark embedding (10, 20, ..., 100), image scaling (0.5, 0.75, 0.9, 1.1, 1.5, 2.0), the operation of rotation (0.5, 0.75, 0.9, 1.1, 1.5, 2.0), cropping and rescaling and image noise (1, 2, 3, 4). Thus, each test image has 56 similar versions and 8×56=448 pairs of visually similar images are obtained.

(a) Airplane   (b) Baboon   (c) Girl   (d) House

(e) Peppers   (f) Sailboat   (g) Splash   (h) Lena

Fig. 2. Standard test images for robustness validation

We calculate hash similarity between each pair of similar images. Table 1 presents statistics of S under different digital operations. It is observed that, the means of all digital operations are bigger than 0.84, and their standard deviations are small. In addition, the minimum S of the test operations are all bigger than 0.6, except contrast adjustment, gamma correction, and the operation of rotation, cropping and rescaling. Therefore, if we choose 0.6 as the threshold, 98.44% visually similar images are correctly detected.

Table 1. Statistics of S under different digital operations

Operation	Maximum	Minimum	Mean	Standard deviation
Brightness adjustment	0.9998	0.6778	0.9557	0.0711
Contrast adjustment	0.9994	0.5464	0.9399	0.1034
Gamma correction	0.9985	0.5825	0.9290	0.0932
3×3 Gaussian low-pass filtering	1.0000	0.9972	0.9997	0.0006
JPEG compression	0.9988	0.8965	0.9680	0.2930
Watermarking embedding	1.0000	0.7544	0.9686	0.0438
Image scaling	0.9989	0.9157	0.9730	0.0249
Rotation, cropping and rescaling	0.9966	-0.0130	0.8438	0.2046
Image noise	0.9994	0.9328	0.9865	0.0187

To test discrimination of our hashing, we collect a large database with 200 different color images via the Internet, digital camera, and the Ground Truth Database [26], whose image sizes range from 256×256 to 2048×1536. We apply our hashing to the image database, extract 200 image hashes, calculate similarity S between each pair of hashes, and then obtain 200×(200−1)/2=19900 results. It is observed that, the minimum and the maximum S are −0.9992 and 0.5141, respectively. And the mean and standard deviation are −0.5918 and 0.1981, respectively. Clearly, if we choose 0.6 as the threshold, no image will be falsely considered as similar image. If the threshold is 0.50, only 0.0089% different images will be mistakenly detected as visually identical images. From the above analysis, we conclude that the threshold 0.6 can reach a good balance between robustness and discrimination.

3.2   Capability of Discovering Tampered Images

To test our capability of discovering tampered images, we take the 200 color images used for discrimination as the source images and convert them to a standard size 512×512. We create tampered versions of these normalized images by randomly selecting a 64×64 block from a 502×651 template image, as shown in Fig. 3, and then using the block to randomly replace the content of the normalized images. Therefore, 200 tampered images are available. Fig. 4 indicates some typical results, where the first row is the original images and the second row is their tampered versions. We extract hashes of the original images and their tampered versions, calculate their similarities and then obtain 200 results. It is found that, our hashing can correctly detect 80.5% tampered images when the threshold is 0.6. If the threshold increases, the correct detection rate will improve. For example, the correct detection rates will be 88.50% and 92.50% when the thresholds are 0.8 and 0.9.

Fig. 3. A template image

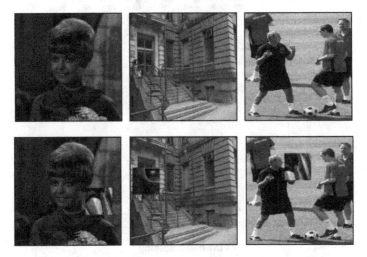

Fig. 4. Original images and their tampered versions

3.3   Performance Comparisons

To show advantages of our hashing, we compare it with two popular hashing algorithms: RD-DCT hashing [16] and IM-based hashing [19]. The same images used in Sections 3.1 and 3.2 are adopted to validate robustness and content sensitivity of the compared algorithms. To make quantitative analysis, correction detection rates of similar images and tampered images are both calculated. As different thresholds will lead to different detection rates, we choose many thresholds for each algorithm, and then obtain a set of detection rates. Fig. 5 presents detection rate comparisons among different algorithms. It is observed that as the correct detection rate of similar images increases, the correct detection rate of tampered images will inevitably decrease. In general, intersection of the curves of the two detection rates reaches a desirable balance between robustness and content sensitivity. The bigger the correct detection rate of the intersection is, the better the whole performance of the algorithm has. Clearly, the correct detection rate of our intersection is much bigger than those of the compared algorithms. This means that our hashing is better than the compared algorithms in classification performances between robustness and content sensitivity.

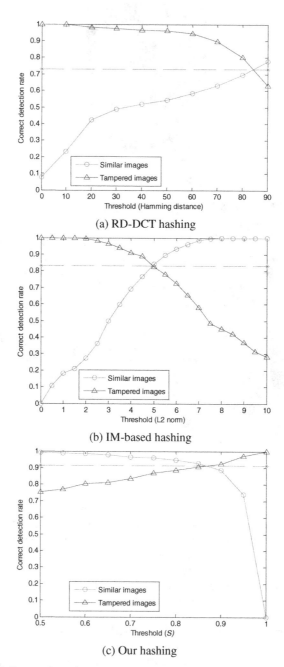

(a) RD-DCT hashing

(b) IM-based hashing

(c) Our hashing

Fig. 5. Correct detection rate comparisons among different algorithms

4   Conclusions

In this work, we have proposed a robust image hashing for discovering tampered images. The proposed hashing exploits invariant moments of overlapped blocks to represent image features, uses a novel similarity metric to measure local change in the hash sequence, and thus efficiently distinguishes tampered images from similar images. Experiments have illustrated that our hashing is robust against many normal digital operations, such as JPEG compression, brightness/contrast adjustment and image scaling. Comparisons with the RT-DCT hashing and the IM-based hashing have been also conducted and the results have showed that our hashing is better than compared algorithms in classification performances between robustness and content sensitivity.

Acknowledgements. This work is supported in part by the Australian Research Council (ARC) under large grant DP0985456; the China 863 Program under grant 2012AA011005; the China 973 Program under grant 2013CB329404; the Natural Science Foundation of China under grant 61300109, 61363034, 61170131; the Guangxi Natural Science Foundation under grants 2012GXNSFBA053166, 2012GXNSFGA060004; Guangxi "Bagui Scholar" Teams for Innovation and Research; the Scientific and Technological Research Projects of Guangxi Education Administration under grants YB2014048, ZL2014005; the Project of the Guangxi Key Lab of Multi-source Information Mining & Security under grant 14-A-02-02; and the Project of Outstanding Young Teachers' Training in Higher Education Institutions of Guangxi.

References

1. Zhu, X., Zhang, L., Huang, Z.: A sparse embedding and least variance encoding approach to hashing. IEEE Transactions on Image Processing 23, 3737–3750 (2014)
2. Ahmed, F., Siyal, M.Y., Abbas, V.U.: A secure and robust hash-based scheme for image authentication. Signal Processing 90, 1456–1470 (2010)
3. Winter, C., Steinebach, M., Yannikos, Y.: Fast indexing strategies for robust image hashes. Digital Investigation 11, S27–S35 (2014)
4. Fridrich, J., Goljan, M.: Robust hash functions for digital watermarking. In: IEEE International Conference on Information Technology: Coding and Computing, pp. 178–183. IEEE Press, New York (2000)
5. Tang, Z., Wang, S., Zhang, X., Wei, W., Su, S.: Robust image hashing for tamper detection using non-negative matrix factorization. Journal of Ubiquitous Convergence and Technology 2, 18–26 (2008)
6. Lv, X., Wang, Z.J.: Reduced-reference image quality assessment based on perceptual image hashing. In: IEEE International Conference on Image Processing, pp. 4361–4364. IEEE Press, New York (2009)
7. Lu, W., Wu, M.: Multimedia forensic hash based on visual words. In: IEEE International Conference on Image Processing, pp. 989–992. IEEE Press, New York (2010)

8. Zhu, X., Huang, Z., Cheng, H., Cui, J., Shen, H.: Sparse hashing for fast multimedia search. ACM Transactions on Information Systems 31, 9 (2013)
9. Zhu, X., Huang, Z., Cheng, H., Shen, H., Zhao, X.: Linear cross-modal hashing for efficient multimedia search. In: the 21st ACM International Conference on Multimedia, pp. 143–152. ACM, New York (2013)
10. Tang, Z., Dai, Y., Zhang, X., Zhang, S.: Perceptual image hashing with histogram of color vector angles. In: Huang, R., Ghorbani, A.A., Pasi, G., Yamaguchi, T., Yen, N.Y., Jin, B. (eds.) AMT 2012. LNCS, vol. 7669, pp. 237–246. Springer, Heidelberg (2012)
11. Tang, Z., Wang, S., Zhang, X., Wei, W.: Structural feature-based image hashing and similarity metric for tampering detection. Fundamenta Informaticae 106, 75–91 (2011)
12. Venkatesan, R., Koon, S.-M., Jakubowski, M.H., Moulin, P.: Robust image hashing.. In: IEEE International Conference on Image Processing, pp. 664–666. IEEE Press, New York (2000)
13. Lin, C.Y., Chang, S.F.: A robust image authentication system distinguishing JPEG compression from malicious manipulation. IEEE Transactions on Circuits System and Video Technology 11, 153–168 (2001)
14. Swaminathan, A., Mao, Y., Wu, M.: Robust and secure image hashing. IEEE Transactions on Information Forensics and Security 1, 215–230 (2006)
15. Monga, V., Mihcak, M.K.: Robust and secure image hashing via non-negative matrix factorizations. IEEE Transactions on Information Forensics and Security 2, 376–390 (2007)
16. Ou, Y., Rhee, K.H.: A key-dependent secure image hashing scheme by using Radon transform. In: IEEE International Symposium on Intelligent Signal Processing and Communication Systems, pp. 595–598. IEEE Press, New York (2009)
17. Kang, L., Lu, C., Hsu, C.: Compressive sensing-based image hashing. In: IEEE International Conference on Image Processing, pp. 1285–1288. IEEE Press, New York (2009)
18. Li, Y., Lu, Z., Zhu, C., Niu, X.: Robust image hashing based on random Gabor filtering and dithered lattice vector quantization. IEEE Transactions on Image Processing 21, 1963–1980 (2012)
19. Tang, Z., Dai, Y., Zhang, X.: Perceptual hashing for color images using invariant moments. Applied Mathematics & Information Sciences 6, 643S–650S (2012)
20. Tang, Z., Dai, Y., Zhang, X., Huang, L., Yang, F.: Robust image hashing via colour vector angles and discrete wavelet transform. IET Image Processing 8, 142–149 (2014)
21. Hu, M.K.: Visual pattern recognition by moment invariants. IRE Transaction on Information Theory 8, 179–187 (1962)
22. Hsia, T.C.: A note on invariant moments in image processing. IEEE Transactions on Systems, Man, and Cybernetics 11, 831–834 (1981)
23. Goshtasby, A.: Template matching in rotated images. IEEE Transactions on Pattern Analysis and Machine Intelligence 7, 338–344 (1985)
24. Tang, Z., Zhang, X., Dai, X., Yang, J., Wu, T.: Robust image hash function using local color features. AEÜ-International Journal of Electronics and Communications 67, 717–722 (2013)
25. Petitcolas, F.A.P.: Watermarking schemes evaluation. IEEE Signal Processing Magazine 17, 58–64 (2000)
26. Ground Truth Database,
    http://www.cs.washington.edu/reseach/imagedatabase/groundtruth/

TweeProfiles: Detection of Spatio-temporal Patterns on Twitter*

Tiago Cunha, Carlos Soares, and Eduarda Mendes Rodrigues

Faculdade de Engenharia da Universidade do Porto
Rua Dr. Roberto Frias, s/n
4200-465 Porto Portugal
{tiagodscunha,csoares,eduarda}@fe.up.pt

Abstract. Online social networks present themselves as valuable information sources about their users and their respective behaviours and interests. Many researchers in data mining have analysed these types of data, aiming to find interesting patterns. This paper addresses the problem of identifying and displaying tweet profiles by analysing multiple types of data: spatial, temporal, social and content. The data mining process that extracts the patterns is composed by the manipulation of the dissimilarity matrices for each type of data, which are fed to a clustering algorithm to obtain the desired patterns. This paper studies appropriate distance functions for the different types of data, the normalization and combination methods available for different dimensions and the existing clustering algorithms. The visualization platform is designed for a dynamic and intuitive usage, aimed at revealing the extracted profiles in an understandable and interactive manner. In order to accomplish this, various visualization patterns were studied and widgets were chosen to better represent the information. The use of the project is illustrated with data from the Portuguese twittosphere.

Keywords: Data Mining, Clustering, Spatio-temporal patterns, Visualization.

1   Introduction

In recent years, social media services have adapted the paradigm of social networks and achieved a huge importance in social life and also in business strategies for companies, as they are regarded as a timely and cost-effective source of spatio-temporal and behavioural information [1]. The massive adhesion and the number of platforms that provide social interaction lead to a growth in the data stored within these services and its usage by researchers.

Twitter has proven to be a popular data source within social media due to the large number of active users and the easy access to their public API. As such, it has fuelled several studies [2–4]. Twitter data can probably be organized into

---

* An earlier version of this work was presented at the Encontro Nacional de Inteligência Artificial e Computacional (Brazilian AI meeting - ENIAC).

X. Luo, J.X. Yu, and Z. Li (Eds.): ADMA 2014, LNAI 8933, pp. 123–136, 2014.
© Springer International Publishing Switzerland 2014

subgroups that represent profiles of tweets and, thus, of the users. These profiles can be useful for many tasks (marketing, political science, government, product development, etc.). However, given the amount of data as well as its complex nature (space, time, content and social), these patterns are not easily extracted using classical data mining strategies. Moreover, the ability to assign different importance to each data dimension may prove useful to control the data mining process and further reveal hidden patterns.

Therefore, our goal is the design and development of an automatic tool to extract and display the patterns mined. The contributions of this project are: the development of a Data Mining process that enables a weighted combination of multiple clusterings in various dimensions, and a web application that facilitates a flexible and interactive exploration of those patterns. This is indeed the greatest contribution of this paper: a weighted combination of multiple clusterings, each one obtained in various heterogeneous dimensions (i.e. time, space, content and social relations). The approach is applied to data from the Portuguese twittosphere to illustrate the type of patterns that can be obtained.

This paper is organized as follows: Section 2 contains the state of the art for the scientific fields of clustering algorithms and distance measures. In Section 3, the concepts and decisions for the data mining process are explained in detail, namely the distance measures used and the combination process. Section 4 presents the visualization tool developed to represent and analyse the patterns extracted with our methodology. In Section 5, we present some results obtained in the case study considered in this project. Finally, in Section 6 we list our conclusions and tasks for future work.

2   Related Work

Clustering is formally defined as the process of grouping a set of data objects into multiple groups, or clusters, so that objects within a cluster have high similarity, but are very dissimilar to objects in other clusters [5]. Similarity assessment is calculated through distance functions. Clustering is the logical choice for our project: clustering algorithms are able to extract patterns from unlabelled data, such as Twitter posts; and clustering provides groupings of similar objects, which can be regarded as representing profiles of tweets.

Clustering algorithms can be sorted into 4 different types: Partitioning, Hierarchical, Density-based and Grid-based [5]. Partitioning algorithms organize the objects to create partitions accordingly to a particular criterion. They are known for generating clusters of spherical shape by using distance-based techniques to group the objects. They generally use mean or medoid to represent cluster centres and have proven effective up to medium size sets [5]. Within this set of algorithms, the most well known are k-Means and k-medoids [5].

A Hierarchical clustering method works by grouping data objects into a hierarchy or a tree of clusters [5]. This method can either be agglomerative (if it starts with small clusters and recursively merge them to find a single final cluster) or divisive (all objects are in a single cluster and iteratively are divided until each one has only one object or the objects are very similar). Usually, the results of hierarchical algorithms are represented by a dendrogram (i.e. tree diagram). The most representative algorithms within this class are BIRCH [6] and Chameleon [7].

Density-based clustering algorithms follow the strategy of modelling clusters as dense regions in the data space, separated by sparse regions [5]. The most well known algorithms are DBSCAN [8] and DENCLUE (DENsity-based CLUstEring) [9].

DBSCAN finds core objects (i.e. points with dense neighbourhood) and iteratively connects them to the neighbours if these are in the core object's ε--neighbourhood. The ε -neighbourhood is defined through a user defined parameter: the radius ε and states that a point is in the core object's ε--neighbourhood if it is within the pre-defined radius. Therefore, for two points p and q, we can say that p is directly density reachable from q if it is in the ε--neighbourhood of q. Another user input is *MinPts* that determines if a point is a core object. If within the ε-neighbourhood there are at least *MinPts* points, then we are in the presence of a core object. The algorithm takes in account the two previous concepts and iteratively connects core objects to its ε-neighbourhood until all objects are processed.

Grid-based algorithms use a space-driven approach instead of a data-driven approach as in the previous algorithms [5]. They partition the space into cells of a multi-resolution grid data structure. This ensures a fast processing time independent from the size of the data set, although it is affected by the resolution of the grid. One notable examples of this clustering algorithms is STING (STatistical INformation Grid) [10].

Clustering algorithms need distance functions in order to calculate dissimilarities between objects and to group these objects by similarity. The objective function aims for high intra-cluster similarity and low inter-cluster similarity [5]. The distance functions chosen for clustering depend on the data types and the representation spaces. Therefore, one must divide the different distance functions accordingly to the dimensional types we use. In this paper we consider the following dimensions: spatial, temporal, content and social. In the spatial dimension, data is defined by latitude and longitude, which are numeric values extracted from the tweets. Therefore, similarity functions between numeric values must be explored. The 4 most important distances of this type in a euclidean space are the Euclidean Distance, the Manhattan Distance, the Minkowski Distance, the Mahalanobis Distance and the Chebychev Distance [5]. For the specific case of latitude and longitude coordinates there is a better suited distance measure that considers the earth's shape: the haversine distance.

As far as the temporal dimension goes, contrary to the spatial dimension, which is mapped in R^2, time is represented in R, which facilitates the difference calculation. For each pair of tweets, the timestamp values are used to compute the temporal distance. However, any of the previous distance functions for euclidean space is applicable to the temporal dimension. We note that this distance function is limited to explore the time difference between posts, whereas other patterns could be mined, such as for instance seasonality of events. In this way, clusters could reflect events in the same weekday, for instance, and therefore extract events than happen regularly.

Considering the connections between users, it is possible to assume the existence of a social graph in Twitter. Therefore, the social distance can be simplified to a distance between nodes within a graph. Two distance measures for graphs are the Geodesic Distance and SimRank [5]. Network Similarity [11] and a pseudo-distance measure [12] can also be used.

In order to calculate the similarity between two texts, one must explore Text Mining distance functions. The Cosine similarity distance [5] (as the most commonly used) and a variation denominated Tanimono distance. Lastly, a variation of Jaccard similarity complemented with Dice's coefficient [13] can also be used. In order to apply a distance function on text, document representations must be specified in a previous stage. The main idea is to create a document-term matrix and extract the vectors to compute their dissimilarity. The main document representation techniques are TF [14] and TF-IDF [15].

3   Clustering on Multiple Dimensions

We present in this section our methodology to tackle this problem, including the distance functions used, the clustering algorithm chosen and the strategy proposed for combining dimensions.

Although many other distance functions than those presented here provide a more elaborate and possibly more powerful approach, they tend to be more difficult to interpret. This decision is transversal to all the distance functions chosen, since our approach must provide a visual platform to navigate through the patterns. It is therefore essential to use intuitive metrics that are easily represented and interpreted.

We consider that each tweet is formally defined as t_i, where i is the index identifier on the tweet data collection. The distance functions between two tweets t_i and t_j are defined as $dist^X(t_i, t_j)$, where X is the dimension on which the function maps the values. X can take the values Sp, T, C, So which are related respectively to the spatial, temporal, content and social dimensions.

In order to calculate the spatial distance between two points, these must be mapped in space. The data received from Twitter for the spatial dimension consists of the latitude and longitude of each tweet. For each pair of tweets t_i and t_j, the distance function uses the latitudes ϕ_{t_i} and ϕ_{t_j} and longitudes λ_{t_i} and λ_{t_j} to determine the distance. The value R is the earth's radius:

$$dist^{Sp}(t_i, t_j) = 2R sin^{-1}\left(\left[sin^2\left(\frac{\phi_{t_i} - \phi_{t_j}}{2}\right) + cos\phi_{t_i} cos\phi_{t_j} sin^2\left(\frac{\lambda_{t_i} - \lambda_{t_j}}{2}\right)\right]^{0.5}\right)$$

The temporal distance was simply calculated as the difference of timestamps, in seconds. The conversion to seconds is necessary due to the use of a generic clustering algorithm, which is, thus, unable to process timestamped differences.

For each pair of tweets t_i and t_j, the timestamp values Δ_i and Δ_j are used to compute the distance:

$$dist^T(t_i, t_j) = |\Delta_i - \Delta_j| \tag{1}$$

To calculate text similarity, the choice fell on TFIDF for vector representation and cosine similarity to determine the dissimilarity between two texts. TFIDF was chosen to reduce the importance of frequently used words and therefore give preference to discriminative ones. On the other hand, the cosine similarity is a traditional yet powerful metric also used in many papers [13, 16].

For a tweet t_i we define the tweet text as α_i and calculate its TFIDF representation in a document matrix D. The following equation elucidates on this transformation, where TF represents a term frequency and IDF represents Inverse Document Frequency:

$$TFIDF(\alpha_i, D) = TF(\alpha_i, D) * IDF(\alpha_i) \tag{2}$$

The cosine function takes as input two TFIDF representations, β_i and β_j, for two tweets, t_i and t_j, and returns a similarity value between 0 and 1, which we adopt for the distance between texts:

$$dist^C(t_i, t_j) = 1 - \frac{\beta_i.\beta_j}{||\beta_i||\beta_j||} \tag{3}$$

Finally for social distance, the geodesic distance between the users who make the posts in the graph defined by the follow relations was chosen. This distance measure introduces a problem that was not present in the previous measures: the possibility of infinite distances. These occur when two nodes are not connected (neither directly nor through other nodes). This case must be treated before combining distances for all dimensions. We replace the infinite value for the total number of nodes plus one. This ensures that the distance between two users that are not connected is always greater than the distance between any two connected users.

All these distance measures were applied to all pairs of tweets and stored in dissimilarity matrices for further processing. The dissimilarity matrices are formally defined as D^X, where X is the dimension. X can take the values Sp, T, C, So which are related respectively to the spatial, temporal, content and social dimensions. Within each dissimilarity matrix D^X, the distance value in row i and column j is defined as $d^X_{(i,j)}$. These matrices are hereby defined as single dimensional distance matrices.

The combination of distances refers to the weight system applied over the all four dimensions to obtain a new multi-dimensional dissimilarity matrix. The weights determine the importance of the dimensions, given as percentage values, while the overall result shows the importance that should be assigned to each dimension. The multi-dimensional dissimilarity matrix will contain the sum of all dissimilarities multiplied by the weight value for the corresponding dimension, always ensuring that this sum is equal to 100%. More formally, for each weight value w_{Sp}, w_T, w_C, w_{So}, we have that $w_{Sp}, w_T, w_C, w_{So} \in \{0, 0.25, 0.5, 0.75, 1\}$ | $w_{Sp} + w_T + w_C + w_{So} = 1$.

However, in order to obtain relevant results, one must normalize the single dimensional distance matrices, since the scales for each dimension are different. The purpose is to use a similar scale for all dimensions to ensure that the importance of each dimension is determined by the weights and not its scale. Therefore, a min-max normalization [5] was applied to all single dimensional distance dissimilarity matrices D^X with the goal of switching from a scale $[min_{D^X}, max_{D^X}]$ into a new scale $[newmin_{D^X}, newmax_{D^X}]$ which is similar in all dimensions. The normalizing equation is applied to all distance values $d^X_{(i,j)}$ in order to obtain the normalized values $d'^X_{(i,j)}$.

In our problem, the minimum value must always be zero since this is the lowest possible value for distance. Therefore, the normalization formula can be simplified, since both min_{D^X} and $newmin_{D^X}$ equals zero:

$$d'^X_{(i,j)} = \frac{d^X_{(i,j)}}{max_{D^X}} * newmax_{D^X} \tag{4}$$

The application of this formula results in scales that lie between 0 and $newmax_{D^X}$. This is also important for defining the clustering algorithm parameters which directly depend on the scale of each dissimilarity matrix.

The multi-dimensional dissimilarity matrix of all four dimensions D^{4D} is the sum of each single dimensional distance matrices $(D^{Sp}, D^T, D^C, D^{So})$ multiplied by the respective weight value $(w_{Sp}, w_T, w_C, w_{So})$:

$$D^{4D} = w_{Sp}D^{Sp} + w_T D^T + w_C D^C + w_{So} D^{So} \tag{5}$$

The clustering algorithm was then applied to the multi-dimensional dissimilarity matrix, in order to obtain the clusters to display in the visualization tool. The clustering algorithm chosen was DBSCAN due to various reasons:

- Density based algorithms are able to detect arbitrarily shaped clusters [17].
- It is not necessary to specify the number of clusters to calculate. This is important in our project because one of the goals is to find patterns although there is no guarantee that the data contain clusters. The enforcement of the number of clusters is unsuitable for some real world applications [17].
- Microblog messages contain noise and using a density based approach, this noise is considered as an outlier and filtered from the results [17].

4   TweeProfiles Tool

The visualization tool is responsible for showing the information mined in the most intuitive and simple way possible, while enabling interaction. To achieve such goals, we represent the 4 dimensions as:

- Spatial: map. The clusters are represented as red circles while the tweets as dots. Each dot is colored depending on the cluster it belongs. Click events are associated to both tweets and clusters to display further information;

- Temporal: timeline. It containing bars with the start and end date for each cluster and with a click event associated to each cluster;
- Content: wordcloud. It displays the words in a cluster with sizes proportional to their frequencies;
- Social: minimum spanning tree. It contains the most important users in a cluster (There is a limitation of 10 users to be displayed in the tool) and a click event to view the user's profile in Twitter;

For the map, the Google Maps JavaScript API[1] was the choice, due to the completeness of information displayed, intuitive interactivity, ability to integrate data from various sources and JavaScript libraries and the typically fast response time to load and refresh the webpage.

D3[2] was used to display the tweets, plotted on top of the map. This representation was overlapped on the map using a Google Maps class named Overlay. An Overlay containing the tweets is anchored to specific latitudes and longitudes so when the map is changed, the points change accordingly. Each point is also associated with an on-click event listener to show a tooltip with the related data to the point specified. If the points are not clustered, they are not presented in the visualization tool to avoid displaying unnecessary information.

The last objects to be displayed on the map are the cluster circles. These circles were generated using Google Maps Polygons and are also inserted in an Overlay class. The event associated with clicking on this circle is the display of a new division in the webpage. This division is used to present a summary of cluster information and also a D3-based interactive social graph representing the most relevant users in the cluster and their connections.

Besides the map representation, the timeline is also used to present all the clusters. The JavaScript library used was the Timeglider JQuery plugin.[3] This widget requires only the input of a JSON file with a specific syntax and it does all the mapping of clusters in horizontal bars. It guarantees that the bars displayed do not overlap and therefore all clusters are accessible through mouse events.

When a single bar is clicked on, it shows a tooltip with a summary of information for the specific cluster. On the other hand, by simply hovering the mouse on the bar, the cluster summary division appears, in the same manner as when the clusters circles in the map are clicked on.

The final widget we implemented is the social graph through a D3 interactive graph example. This implementation was chosen due to the ability to drag a node in the graph, which triggers a transformation that re-arranges all the nodes and edges. Additionally, the names of the users represented by the nodes and the edges weight are available in tooltips.

The previous widgets are responsible for representing the information in various dimensions. Interaction with them enables the analysis of different patterns. To navigate through the subsets and to change the weights of the dimensions, we

---

[1] https://developers.google.com/maps/documentation/javascript/reference
[2] http://d3js.org/
[3] http://timeglider.com/widget/index.php

used slidebars. We used the JavaScript implementation of DHTMLX slidebar.[4] This widget enables the definition of minimum, maximum and step values. For each dimension, since they are represented as weight percentages, the minimum value of the sidebar is set to 0 and maximum to 100. The step value is 25, since this was the step value used for the computation of the combined dissimilarities.

5 Results

The data extraction was done using a tool that retrieves data from online social networks, namely SocialBus.[5] By using this platform and filtering the data for this project's purposes, 119,558 tweets were retrieved with spatial, temporal and content attributes. This dataset contains data from May 2012 to February 2013, with tweets written in several languages (mostly Portuguese and English) and published from various countries (although the majority of tweets are from Portugal and Brazil).

To evaluate the social dimension, it was necessary to extract more data from Twitter in order to build a social graph of users, since the only information available at this point were the usernames of each tweet author. This data was retrieved from the Twitter RESTful API and the graph was built, defining nodes as the users and links as the following relationship. We retrieved only 9,794 edges for our 9,362 users. This result is justified by the fact that most users are not directly connected to any user in our database and are, therefore, excluded from the final graph. This lead to the existence of only 932 connected nodes in the graph.

At this point, we applied the methodology explained in Section 3. However, due to the size of the data, we had to split the original dataset into smaller ones. This decision, together with the weighting methodology led to several hundred clusterings. As we cannot display all results in this paper, we will just present results of some illustrative combinations. The first result we discuss illustrate the effect of setting different values for the dimensions weights on the results. Figure 1 shows a clustering with 100% importance set to the spatial dimension. It creates 3 clusters: Europe (blue points), America (orange points) and Africa (yellow points). It is visible in the timeline that all clusters occupy the entire time span (from June until October 2012).

When we set 100 % importance to the temporal dimensions for the same data (Figure 2), we obtain 2 clusters: one with tweets between June 3 and June 6 2012 (blue points) and another cluster between August 16 and October 13 2012 (orange points). In the spatial dimension, it is visible that there is no clear separation among clusters and that the clusters are, in fact, overlapped in this dimension, as could be expected given the weights selected.

However, when we assign 50% importance to both the spatial and temporal dimensions (Figure 3), the results are a combination from both the previously presented clusterings. Now, we have 4 clusters: Europe between June 4 and

---

[4] http://dhtmlx.com/docs/products/dhtmlxSlider/
[5] http://reaction.fe.up.pt/socialbus/

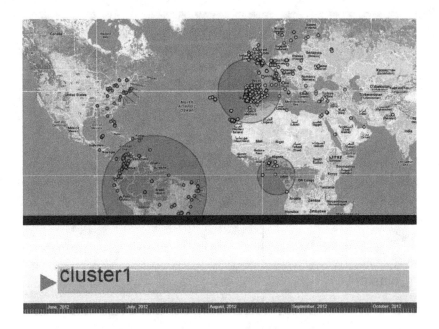

Fig. 1. Example clustering with 100% importance assigned to the spatial dimension

June 6 (blue points), America in the same time period (orange points), Europe between August 16 and October 13 (yellow points) and America in the same time period (green points).

If we were to assign weights to the other dimensions, the outcome would reflect also the same behaviour as in these examples. The difference is that it is harder to evaluate them in the platform, due to the large amount of information displayed and the fact that the content and social dimensions are harder to represent and interpret in clustering.

The following results represent a couple of interesting patterns. Only two examples are presented due to lack of space. Figures 4 and 5 exemplify some patterns retrieved using our Data Mining approach and represented on our visualization tool. Figure 4 shows different users that practise cycling and used the same sports application to publish their performance on Twitter. The positions of each user are visible, as well as the dates of the event. Figure 5 shows a cluster of posts by users who attended the same sports event, namely a football match between Portugal and Northern Ireland on the 16 October 2012.

The methodology proposed here assumes static data. Additionally, it has some scalability problems due to the high computational costs in calculating and manipulating the matrices. One solution to these issues is the use of stream clustering methods [18].

This work has several shortcomings, such as computational costs, scalability, granularity of control of the weights and orthogonal treatment of different dimensions. However, it represents a simple solution to a complex problem. Most importantly, some of these shortcomings present interesting opportunities for

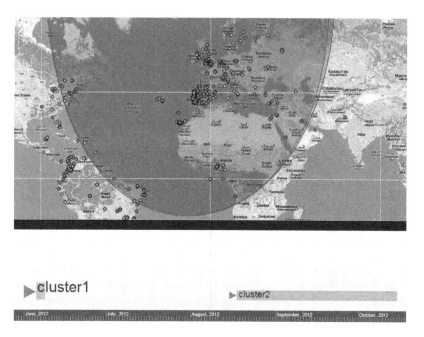

Fig. 2. Example clustering with 100% importance assigned to the temporal dimension

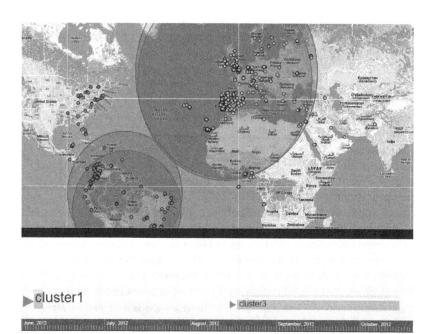

Fig. 3. Example clustering with 50% importance assigned to both the spatial and temporal dimension

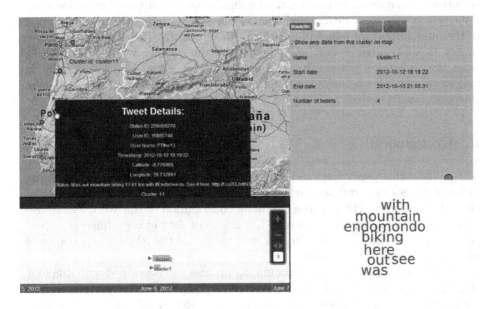

Fig. 4. Pattern presenting users using the same smartphone application for sports practising

Fig. 5. Pattern presenting users in a sports event: football match between Portugal and Northern Ireland at Dragão Stadium

scientific development. In particular, the simultaneous analysis of heterogeneous data, which was dealt with in a simple way in this project, is recognized as one of the most important issues in the Big Data community [19]. Furthermore, the ability to control the weight of each of those dimensions, which is important from the end user's perspective and that was also solved in a simple way in this project, is also a challenge from the computational, algorithmic and user interaction perspectives.

6   Conclusion

The goals of this paper were to develop a data mining approach for combining different types of information and also to apply our approach to Twitter data in several dimensions. The purpose and main contribution of this work was to use clustering to identify patterns across dimensions of data of very different nature (e.g., space and content), enabling the user to control the relative importance of each one of them in the process.

To accomplish these goals, a data mining process was developed with different stages: data preparation, dissimilarity matrices computation, normalization and combination and lastly, clustering using those matrices. A different distance function was chosen for each dimension (Haversine for space, Time Interval for time, cosine similarity on a TFIDF representation of the content and geodesic distance for social). A min-max normalization function was applied to all matrices. Given the computational cost of the process, the matrices were combined using a set of pre-defined weights, representing different levels of importance of each dimension. Clusterings were obtained by running DBSCAN on these matrices.

The visualization tool represents those patterns. Namely, a map, a timeline and a graph were implemented using various JavaScript libraries to create interactive widgets. These widgets enabled the simultaneous representation of the same information in different dimensions and to interact with them for a deeper and more flexible exploration of the results presented.

The first task for future work is to study better ways for represent the social and content dimensions, in order to ease the interpretation of results. Another task is to adapt this methodology for stream clustering while enabling a direct connection to Twitter's Streaming API. We can also change the dimensions and research whether this is indeed a generic methodology that accomplishes multidimensional clustering. Finally, since the proposed method for clustering in several dimensions is generic, one must also evaluate clustering with other dimensions, as for instance images.

Acknowledgments. This work was partially supported by projects REACTION (Retrieval, Extraction and Aggregation Computing Technology for Integrating and Organizing News - UTA-Est/MAI/0006/2009) and POPSTAR (Public Opinion and Sentiment Tracking, Analysis, and Research - PTDC/CPJ-CPO/116888/2010); "NORTE-07-0124-FEDER-000059" and "NORTE-07-0124-FEDER-000057" *funded by the North Portugal Regional Operational

Programme (ON.2 – O Novo Norte), under the National Strategic Reference Framework (NSRF), through the European Regional Development Fund (ERDF) and by national funds, through the Portuguese funding agency, Fundação para a Ciência e a Tecnologia (FCT); and a research grant assigned by the Doctoral Program in Informatics Engineering at the Faculdade de Engenharia da Universidade do Porto; and Sapo Labs / UP from Portugal Telecom.

References

1. Lee, C.-H., Yang, H.-C., Chien, T.-F., Wen, W.-S.: A Novel Approach for Event Detection by Mining Spatio-temporal Information on Microblogs. In: 2011 International Conference on Advances in Social Networks Analysis and Mining, pp. 254–259 (2011)
2. Bosnjak, M., Oliveira, E.: TwitterEcho: a distributed focused crawler to support open research with twitter data. In: Proc. of the Intl. Workshop on Social Media Applications in News and Entertainment (SMANE 2012), ACM 2012 International World Wide Web Conference (2012)
3. Golder, S.: Tweet, Tweet, Retweet: Conversational Aspects of Retweeting on Twitter. HICSS 2010 Proceedings of the 2010 43rd Hawaii International Conference on System Sciences, 1–10 (2010)
4. Abel, F., Gao, Q., Houben, G.J., Tao, K.: Analyzing Temporal Dynamics in Twitter Profiles for Personalized Recommendations in the Social Web. In: Proceedings of ACM WebSci 2011, 3rd International Conference on Web Science (2011)
5. Han, J., Kamber, M., Pei, J.: Data Mining: Concepts and Techniques, 3rd edn. Elsevier Science & Technology, Massachussets (2006)
6. Ramakrishnan, R., T. Zhang, M.L.: BIRCH: an efficient data clustering method for very large databases. In: Procedings SIGMOD 1996 Proceedings of the 1996 ACM SIGMOD International Conference on Management of Data, vol. 1, pp. 103–114 (1996)
7. Karypis, G., Han, E.H., Kumar, V.: Chameleon: Hierarchical clustering using dynamic modeling. Computer 32(8), 68–75 (1999)
8. Ester, M., Kriegel, H.P., Sander, J., Xu, X.: A density-based algorithm for discovering clusters in large spatial databases with noise. In: Proceedings of the 2nd International Conference on Knowledge Discovery and Data Mining, pp. 226–231 (1996)
9. Hinneburg, A., Keim, D.A.: An efficient approach to clustering in large multimedia databases with noise. In: Proceedings of 4th International Conference in Knowledge Discovery and Data Mining (KDD 1998), pp. 58–65 (1998)
10. Wang, W., Yang, J., Muntz, R.: STING: A statistical information grid approach to spatial data mining. In: Proceedings of the 23rd International Conference on Very Large Data Bases, pp. 186–195 (1997)
11. Akcora, C.G., Carminati, B., Ferrari, E.: Network and profile based measures for user similarities on social networks. In: 2011 IEEE International Conference on Information Reuse & Integration, pp. 292–298 (2011)
12. Dekker, A.: Conceptual Distance in Social Network Analysis. Journal of Social Structure 6 (2005)
13. Ryu, H., Lease, M., Woodward, N.: Finding and exploring memes in social media. In: Proceedings of the 23rd ACM conference on Hypertext and social media - HT 2012, p. 295 (2012)

14. Manning, C.D., Raghavan, P., Schtze, H.: Introduction to Information Retrieval. Cambridge University Press, New York (2008)
15. Lopes, A.A., Pinho, R., Paulovich, F.V., Minghim, R.: Visual text mining using association rules. Computers & Graphics 31(3), 316–326 (2007)
16. Rangrej, A., Kulkarni, S., Tendulkar, A.V.: Comparative study of clustering techniques for short text documents. In: Proceedings of the 20th International Conference Companion on World Wide Web - WWW 2011 (2011)
17. Lee, C.-H.: Mining spatio-temporal information on microblogging streams using a density-based online clustering method. Expert Systems with Applications 39(10), 9623–9641 (2012)
18. Mahdiraji, A.: Clustering data stream: A survey of algorithms. International Journal of Knowledge-Based and Intelligent Engineering Systems 13(2), 39–44 (2009)
19. Provost, F., Fawcett, T.: Data Science and its Relationship to Big Data and Data-Driven Decision Making. Big Data 1(1), 51–59 (2013)

Real-Time Event Detection Based on Geo Extraction and Temporal Analysis

Xiao Feng[1], Shuwu Zhang[1], Wei Liang[1], and Zhe Tu[2]

[1] Institue of Automation, Chinese Academy of Sciences, Beijing, China
{xiao.feng,shuwu.zhang,wei.liang}@ia.ac.cn
[2] Beijing University of Chemical Technology, Beijing, China
waytosea214@gmail.com

Abstract. Microblogging is an important source of information about what is happening in the real world. In this work, we propose a novel approach for real-time event detection targeting accident and disaster events (ADEs) using microblogs from Sina Weibo. Our aim is to detect out every microblog which reports a real-world occurrence of a target event from the microblog stream. We formulate the event detection problem as a classification problem using microblog-based features, linguistic features, content features, and event features. We propose a street-level location extraction method based on the textual content to cooperate geo-information extraction. In order to deliver fresh events, we use a temporal analysis method to filter away past events. We compare our method with two state-of-the-art baselines on event detection, and achieve improvements in both precision and recall.

Keywords: event detection, microblogs, geo-information extraction, temporal analysis.

1 Introduction

Microblogging services, such as Twitter and Sina Weibo, allow people to report and share short messages (limited to 140 characters) about what is happening. Yet Twitter users publish more than 200 million tweets daily, and Sina Weibo is leading the microblogging market in China since Twitter is unavailable[1]. Microblog users present the most up-to-date information and buzz about current events at any time. Nowadays, microblogs have become an important complementary source of information on current events. Clearly, we can benefit from real-time event detection from individual microblogs[2].

The Topic Detection and Tracking (TDT) project defines an event as something that happens at some specific time and place, and the unavoidable consequences[3]. Under the TDT definition, specific accidents, crimes and natural disasters are examples of events[4]. Intuitively, we regard an event as something that actually happens, and a topic as something that people discuss. We take the TDT definition as the definition of events in this paper, and target Accident and Disaster related Events(ADEs), such as car accidents, fire disaster, or earthquake, as they are very important types of events, and the real-time detection of ADEs is highly significant.

X. Luo, J.X. Yu, and Z. Li (Eds.): ADMA 2014, LNAI 8933, pp. 137–150, 2014.

Previous work in event detection from microblogs mainly relied on clustering algorithms[5,6,7] and topic models[2], [8]. Generally, both clustering algorithms and topic models have relatively high computational complexity due to the large scale of microblogs, and thus the process of event detection is inevitably time consuming. To address the problem of time delay, some researchers applied hashing algorithm to accelerate the similarity computation[9,10,11]. Through investigation, we find that almost all events detected by these methods have already attracted attention of many users, which means microblogs about those events have already been frequently published or widely forwarded. However, events which are quite valuable for particular application (e.g. local accident reporting) but yet have not attracted enough attention always cannot be detected. Besides, the real-time nature of microblogging has not been well studied in previous work. Studies on event detection system that monitors microblog stream and delivers target event reports relevant to user needs are still rare.

In this work, we propose a scheme for real-time ADE detection using Chinese microblogs from Sina Weibo. We monitor the microblog stream and attempt to detect out every microblog which reports a real-world occurrence of a target event timely. However, microblog has several characteristics which present unique challenges for this task.

Firstly, because microblog users can talk about whatever they choose, it is often difficult to identify whether they are truly describing a real-time occurrence or just making a story, or merely expressing personal feelings. Moreover, most of the time, microblog users mention mundane events in their daily lives (such as what they ate for lunch), and the distinction between these mundane events and ADEs is not easy.

Further, real-time ADE detection requires examining whether a detected event is newly occurred and filtering away reports of past events, but temporal analysis of an event is non-trivial, especially when dealing with microblogs in Chinese, because the Chinese language is quite flexible in the use of tense and the expression of time.

Finally, though there are number of researches on event detection from English microblogs [12,13,14], the proposed methods may not fit into applications dealing with microblogs in Chinese, because of differences in expression style and cultural background.

To detect target events fast and precisely, we refer to the approach presented in [15] and apply a support vector machine (SVM) [18] to classifying a microblog as either belonging to a positive or negative class, which corresponds to the detection of a target event. Moreover, as a microblog often contains rich information, such as the text content of the message, its posting time, the GPS tag, and the hashtag etc., we can extract the geolocation and analyze the temporal pattern by utilizing these information, and further extend the features used in [15] for classification. Experiment results show that we achieve improvements in both precision and recall.

Our main contributions include: (i) a real-time ADE detection scheme using Sina Weibo microblogs, (ii) a geo-Information extraction approach based on POS-tags and Markov chain model, (iii) a temporal analysis method for detecting newly-occurred events.

The remainder of this paper is organized as follows. Section 2 introduces some related work. Section 3 introduces the scheme of real-time event detection and the proposed methods. We evaluate the performance of proposed methods in Section 4 and we finally conclude our work in Section 5.

2   Related Work

2.1   Domain-Specific Event Detection

Our work focuses on real-time event detection and targets accident and disaster events. The detection of specific types of real-world events needs to extract useful microblogs out of the huge amount of available data, because in this case only a small fraction of the available microblogs is relevant. Hence, the performance of domain-specific event detection mainly relies on the filtering approaches.

The works closest to ours are [13] and [15]. [13] proposed a Twitter-based Event Detection and Analysis System(TEDAS), to detect newly-occurred Crime and Disaster related Events(CDE), such as shooting, car accidents, or tornado, and analyze the spatial and temporal pattern of a detected event, and identify the significance of events. The system utilizes two kinds of features, Twitter-specific features and CDE-specific features, to train a classifier to determine whether a tweet is related to a CDE. [15] propose an event notification system monitoring Japanese tweets to detect an earthquake before an earthquake actually arrives. A SVM is applied to classify a tweet into positive and negative classes, which corresponds to the detection of a target event. Features for the classification include the keywords in a tweet, the number of words, the context of event words, etc. We take the proposed methods in these two works as baselines in experiments to evaluate the performance of our approach.

2.2   Geo-information Extraction

There are three ways of acquiring geo-information from microblogs: GPS-tagging through Local Based Service(LBS) or IP address, location field and time zone from the user profile, and location extraction from the textual content[16]. The first two methods are easy to implement, but they will fail when the user is not willing to share the location where the message is sent, or the location where the target event happens is totally different from the user-registered location. Therefore, we use the third way to extract geo-information. [12] utilized a multinomial naive Bayes classifier to predict user-level geolocation for each event-related tweet. [16] proposed a method to automatically identify location keywords and further estimating a Twitter user's city-level location based purely on the textural contents. Both these two approaches focus on geo-information estimation from English texts and their geolocation datasets only include geo-names of the USA. In this work, we attempt to extract street-level locations of an event from the content written in Chinese. According to current knowledge, our work is the first try to address this problem.

2.3   Temporal Analysis

We perform temporal analysis to ensure that the detected target events are newly-occurred and filter away past events. [15] proposed a temporal model to estimate the probability of a natural disaster(e.g. an earthquake) occurrence at time t, based on the exponential distribution. However, it is not suitable for the temporal analysis of a specific local accident (e.g. a car accident), because the number of microblogs that report it can be very limited. To deliver the freshest relevant information to people, [17] introduce a temporal evaluation of each keyword's usage based on the assumption that a term can be regarded as emerging if it frequently occurs in the specified time interval and it was relatively rare in the past. This work shares the same goal with our work, and accordingly we can filter away some past events by detecting keywords which frequently appears in previous time intervals.

3   Real-Time Target Event Detection

3.1   Scheme Overview

In this paper, we target Accident and Disaster Events, such as car accidents, fire hazard, or earthquake. Accordingly, an event that we would like to detect is a target event. Figure 1 shows the overview of the real-time event detection scheme. The overall flow is the following:

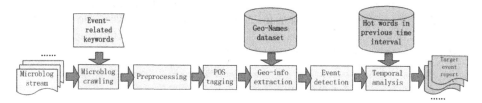

Fig. 1. The overview of the real-time target event detection scheme

1. Crawl microblogs which include keywords related to a target event type. The keywords can be either set by the user or selected from specific domain dictionary.
2. Remove all "@" symbols together with usernames, extract out all external links, and accomplish Chinese word segmentation as well as stop-word elimination by using ICTCLAS[1] in the preprocessing stage.
3. Use ICTCLAS as POS tagger to extract named entities and annotate POS tag for each word in the microblog.
4. Apply our proposed method and the geo-names dataset to extracting street-level geolocations, since we can identify all country-level, province-level and city-level geo-entities of China by using ICTCLAS. Moreover, we will take the use of geo-information as an event feature for event detection.

---

[1] http://ictclas.org/

5. Construct the feature vector and classify the microblog into a positive or negative class, corresponding to the detection of a target event.
6. Make temporal analysis of the microblog classified into positive class in 5, and deliver the newly-occurred target events.

3.2   Street-Level Geolocation Extraction

According to our investigation, we find that more than 80% microblogs reporting target events contain geo-information. The geo-information can be in the form of either text description in the content or GPS-tag at the end of the microblog. Moreover, the actual observations show that the usage of text description is much more than the usage of GPS-tag. It is probably because that users are not willing to disclose the private information (their specific geolocation) when making event reports.

Based on a large number of observations, we find that the smaller the geographic scope is described, the more likely that a target event is detected. Therefore, our goal is to extract more specific geo-information based purely on the textual content of a microblog.

With the help of ICTCLAS, all country-level, province-level and city-level locations can be identified and tagged as "ns" after POS-tagging. But, ICTCLAS fails to identify street-level locations in most circumstances. For example, here is a microblog, such as "I see a car accident happens near Nanchang University Commercial Street.". ICTCLAS can identify the city-level location "Nanchang", but it fails to identify the street-level location "Nanchang University Commercial Street".

The task of street-level geolocation extraction is to extract out each phrase which describes a street-level location ("S-L loc", for short) from the textual content of a microblog.

First, we manually annotate 383 phrases of street-level locations from 370 event-related microblogs as the training set.

Then, we extract 65 words which are commonly used as the last words of phrases in the training set, such as "Road", "Bridge", "Avenue", "Square", "Street", and so on. We designate these words as the symbol words which can indicate street-level locations. For each word of the symbol words, we locate it in the textual content, which means that we find the end position of a S-L loc phrase.

Next, we search the start position of a S-L loc phrase based on the probability of a sequence of words being a phrase of S-L loc. Figure 2 shows the S-L loc phrase in the microblog "I see a car accident happens near Nanchang University Commercial Street".

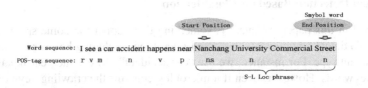

Fig. 2. An example of the S-L loc phrase in a microblog

Because there are so many variations of word sequences and so much priori knowledge needs to be prepared, we use the corresponding sequence of POS-tags to approximately replace the sequence of words when calculating the probability. The probability of a sequence of words being a phrase of S-L loc is computed as follows:

$$P(S-L\,loc\,|\,w_{i-k},\cdots,w_{i-1},w_i) \propto P(S-L\,loc\,|\,t_{i-k},\cdots,t_{i-1},t_i),\quad i=1,2,3,\cdots\cdots \text{ and } k=1,2,\cdots i-1 \quad (1)$$

where w_i is the symbol word which indicates the S-L loc, and meanwhile w_i is the i-th word of a microblog, and t_i is the corresponding POS-tag of w_i.

We apply Bayes' theorem and the First-order Markov chain model to (1), and obtain results as follows:

$$P(S-L\,loc\,|\,t_{i-k},\cdots,t_{i-1},t_i) \propto P(t_{i-k},\cdots,t_{i-1},t_i\,|\,S-L\,loc)$$

$$= P(t_i\,|\,S-L\,loc)\prod_{k=1}^{i-1} P(t_{i-k+1}\,|\,t_{i-k},S-L\,loc)$$

$$= \frac{\#(t_i \text{ is the POS tag of a symbol word})}{\#(t_i,S-L\,loc)}\prod_{k=1}^{i-1} \frac{\#(t_{i-k}t_{i-k+1},S-L\,loc)}{\#(t_{i-k},S-L\,loc)} \tag{2}$$

where $\#(t_i,S-L\,loc)$ is the frequence of t_i appearing in phrases of S-L locs in the training set, and $\#(t_{i-k}t_{i-k+1},S-L\,loc)$ is the frequence of the sequence $t_{i-k}t_{i-k+1}$ appearing in phrases of S-L locs in the training set.

For each $k=1,2,\cdots\cdots i-1$, we compute the probability $P(S-L\,loc\,|\,t_{i-k},\cdots;t_{i-1},t_i)$ using (2), and give a threshold δ to decide the start position of a S-L loc phrase. If $P(S-L\,loc\,|\,t_{i-k},\cdots;t_{i-1},t_i)\leq\delta$ and $P(S-L\,loc\,|\,t_{i-k+1},\cdots;t_{i-1},t_i)>\delta$, we take $i-k+1$ as the word index of the start position. Moreover, if $P(S-L\,loc\,|\,t_{i-k+1},\cdots;t_{i-1},t_i)>\delta$ for all $k=1,2,\cdots i-1$, we take the beginning of the microblog as the start position. However, if we obtain the start position when $k=1$ or fail to find any symbol word in the textual content, we believe that there is not any phrase describing S-L loc in the textual content.

Finally, we use the extracted phrase as keywords to search the most similar geo-name in the our Geo-Names dataset, and we take the most similar search result as the street-level geo-information extracted from the microblog. If there are not any similar geo-names in the dataset, we submit the phrase for manually processing.

3.3   Event Detection Based on Classification

As described in this paper, an event is something that happens at some specific time and place. To detect target events, we crawl microblogs including keywords related to the target event type. For instance, we use "car accident", "fire", and "earthquake" as crawling keywords. However, even if a microblog contains the crawling keywords, it might not be appropriate as an event report[15]. For instance, microblogs such as "Recently I always encounter car accidents, can driving people be more careful?" or

"Fire is relentless, and the prevention is important!". These microblogs are truly related to the target events, but they are not reports of real-world occurrences.

Therefore, we formulate the event detection problem as a classification problem. Given a microblog m, the task is to classify m into a positive or negative class. A microblog which is truly referring to an actual target event occurrence is denoted as a positive class. We manually annotate positive and negative examples as a training set to train a SVM to classify microblogs automatically into positive and negative categories.

Table 1. Group of Features for Event Detection

Group	Feature	Definition
Microblog-based Features	num-of-words	the number of words
	num-of-links	the number of web links
	num-of-hst	the number of hashtags
Linguistic Features	pent-of-vrb	percentage of words that are verbs.
	num-of-NE	The number of named entities identified by ICTCLAS
Content Features	TFIDF-of-feature-word	TFIDF values of feature words selected from training set based on IG
Event Features	num-of-time-words	the number of time (or date) words identified by ICTCLAS
	loc-of-wide-geo-scope	the number of locations identified by ICTCLAS
	loc-of-small- geo-scope	the number of street-level locations extracted by using the proposed method
	GPS-tag	whether the microblog contains a GPS-tag

Table 1 contains 4 groups of features extracted from each microblog for event detection, organized as follows:

- Microblog-based Features: Generally, a microblog which reports a real-time ADE contain less number of words and less number of external links to web pages. A hash tag always indicates a topic, which may refer to a significant event happened before.
- Linguistic Features: Verbs are used a lot when people describe a specific event, and the corresponding subject of a specific event is often a named entity.
- Content Features: We select a certain number feature words from the vocabulary of the training set using feature selection method based on Information Gain (IG)[19]. These feature words are either closely related to the positive class or closely related to the negative class. The TFIDF value of a feature word indicates the extent that the word can represent the microblog.
- Event Features: Microblogs reporting specific events often contain time or location information (either text description in the content or GPS-tag). Moreover, a large number of observations show that the smaller the geographic scope is described, the more likely that a target event is detected.

We compare the usefulness of our features and that of features proposed in [13] and [15] in Section 4. Using the trained model, we can identify whether a microblog refer to an actual target event occurrence.

3.4 Temporal Analysis

The real-time event detection requires that the detected target events are newly-occurred, and thus we need to make temporal analysis of a target event report to filter away past events. First of all, we give the definition of a newly-occurred event and a past event as follows:

Definition 1. An event can be defined as newly-occurred if it happens in the currently-considered time interval.

Definition 2. A past event can be defined as an event that happened in a previous time interval.

The duration of the intervals is set by the user. If we set the duration of the intervals to be one calendar day, the examples of a newly-occurred event and two past events are as follows:

An Example of a Newly-occurred Event: "A car accident happens on the Yangtze River Bridge. A motorcycle hit a truck, people cannot move, full of blood!"

An Example of a Past Event: "A car accident happened in Linhe yesterday. A Cadillac hit a tricycle."

In this example, it is clear that the word "yesterday" indicating a past event.

Another Example of a Past Event: "A fire burned 7 grain barns of SINOGRAIN."

The above example is posted on June 2, 2013, but the event happened on May 31, 2013, which is obviously in a previous time interval. Moreover, the name entity "SINOGRAIN" was a very hot word in Sina Weibo from May 31 to June 2, 2013.

We formulate the problem of filtering away past events as a Naive Bayes classification problem. Given an event report microblog m and a currently-considered time interval I_c, the task is to decide whether the detected event in m happens in I_c. If the detected event happens in I_c, we denote m as C; if not, we denote m as $\overline{C}$. We will classify m into C if $P(C \mid m) > P(\overline{C} \mid m)$. $P(C \mid m)$ and $P(\overline{C} \mid m)$ are computed as follows:

$$P(C \mid m) \propto P(m \mid C) = \prod_{i=1}^{n} P(feature_i \mid C) = \prod_{i=1}^{n} \frac{\#(feature_i, C)}{\#(C)} \qquad (3)$$

$$P(\overline{C} \mid m) \propto P(m \mid \overline{C}) = \prod_{i=1}^{n} P(feature_i \mid \overline{C}) = \prod_{i=1}^{n} \frac{\#(feature_i, \overline{C})}{\#(\overline{C})} \qquad (4)$$

We set the duration of the intervals to be one calendar day and prepare examples of C and $\bar{C}$ as the training set, we can train a model to classify microblogs automatically into C and $\bar{C}$ categories.

Table 2. Features for Temporal Analysis

Feature	Definition
use-of-links	whether m contains any web links
use-of-hst	whether m contains any hashtags
is-forwarded	whether m is forwarded from others
nearest-time-word	the nearest time (or date) word identified by ICTCLAS of the crawling keyword
word-related-to-C	whether m contains the word w which is closely related to C based on $\chi^2(w, C)$
word-related-to-$\bar{C}$	whether m contains the word w which is closely related to $\bar{C}$ based on $\chi^2(w, \bar{C})$
previous hot NE	whether m contains the named entity e which appear frequently in previous time intervals

Table 2 shows our features extracted from each microblog for the classification. We propose these features based on empirical assumptions and a large number of observations. We find that most microblogs that contain web links report past events, and hash tags often refer to significant events which happened in the past. Moreover, if the microblog is forwarded from others, its content tends not to be fresh. Through investigation, we find that the nearest word denoting time or date of the crawling keyword always plays an important role in deciding whether the event happened in the past. As we cannot infer the tense of a sentence based on the form of verbs in Chinese, we must rely on the words denoting time and date. In addition, we select a certain number of feature words using the statistical method χ^2 [19], to improve the model's capability of distinguishing. Based on the assumption in [17], we believe that if a microblog contains the name entity which appeared frequently in previous time intervals, it probably refers to a hot event which happened in the previous time.

4   Experiments

In this section, we describe the experiment results and make evaluation of proposed methods.

4.1   Data Set

The datasets used in this paper are crawled with the methods proposed in [15] from Sina Weibo. We use the subset of microblogs between June 1, 2013 and June 3, 2013 to simulate a live tweet stream. To ensure that there is not any overlapping part between the training set and the test set, we remove all sample duplications in the dataset. We manually picked a certain number of examples including positive examples and negative examples as the training sets and the test sets for car accident, fire disaster, and earthquake events. Some statistics about the data sets are presented in Table 3.

Table 3. Statistics of Data Set from Sina Weibo

Data set	Num of microblogs including crawling keywords		
	"car accident"	"fire"	"earthquake"
Before de-duplication	13546	34753	164472
After de-duplication	5577	6276	23956
Traning set	1140 (positive) 1260 (negative)	4594 (positive) 406 (negative)	2208 (positive) 192 (negative)
Test set	254 (positive) 346 (negative)	1184 (positive) 92 (negative)	552 (positive) 48 (negative)

From Table 3, we can find that the negative examples including the crawling keyword of car accident are more than the positive examples. In contrast, the positive examples of fire disaster and earthquake are much more than the negative examples. This shows that by using the appropriate crawling keywords, we can obtain more than 80% precision of event detection of fire disaster and earthquake.

4.2   Evaluation of Street-Level Geo-Information Extraction

As the natural disaster events, such as earthquake, usually cover a wider geographic scope, we rarely can extract street-level locations from microblog reporting a natural disaster. Therefore, we manually annotate 383 street-level locations from 370 microblogs referring to car accident events and fire events as the training set, and another 272 street-level locations from 312 microblogs referring to car accident events as the test set.

As described in this paper, the task of street-level geo-information extraction is to extract out the phrases describing street-level locations from the text content of a microblog. To evaluate the performance of proposed method, we compare the phrase extracted by our method and the phrase annotated manually. If the difference is no more than one word, then we consider the phrase extracted by our method is precise, because the difference of one word will not affect the performance of event detection, and can be easily corrected by using geo-names dataset in practical application. Moreover, we use the recall to measure whether our proposed method can extract each phrase of a street-level location from the text content. Table 4 shows the Precision, Recall and F-value with the different δ (see Section 3).

Table 4. Evaluation of Geo-information Extraction

δ	Precision	Recall	F-value
0	**83.6%**	**80.2%**	**81.89%**
1×10^{-4}	85.4%	77.3%	81.22%
5×10^{-4}	87.2%	62.1%	72.55%
1×10^{-3}	88.3%	43.6%	58.42%

The experiment results shows that the highest F-value is achieved on the set test when $\delta = 0$, and the recall falls rapidly along with the increasing of δ. Therefore, we set $\delta = 0$ in subsequent experiments to ensure the high recall.

4.3   Evaluation of Event Detection Based on Microblog Classification

The task of event detection is to classify a microblog m including the crawling keyword into a positive or negative class. The microblog which is truly referring to an actual target event occurrence is denoted as a positive class.

In this experiment, we use the dataset described in Table 3 to compare the usefulness of our features (described in Table 1) and that of features proposed in [13], named baseline 1, and [15] , named baseline 2. We use the linear kernel SVM implemented in Weka[2] as the classifier. The highest classification performance of different features is presented in Table 5.

Table 5. Evaluation of Event Detection

Evaluation	Microblog Features	Linguistic Features	Content Features	Event Features	Baseline 1	Baseline 2	All proposed features
Car accident events							
Precision	64.3%	53.2%	82.1%	63.1%	82.3%	76.8%	**88.7%**
Recall	82.5%	69.7%	47.5%	96.4%	89.1%	62.5%	**94.5%**
F-value	72.3%	60.3%	60.2%	76.3%	85.6%	68.9%	**91.5%**
Fire events							
Precision	86.1%	74.8%	91.6%	78.4%	89.1%	87.3%	**92.7%**
Recall	89.5%	76.7%	63.2%	95.6%	92.5%	84.5%	**93.1%**
F-value	87.8%	75.7%	74.8%	86.1%	90.8%	85.9%	**92.9%**
Earthquake events							
Precision	85.6%	75.4%	92.1%	73.5%	94.7%	92.5%	**95.4%**
Recall	87.3%	80.0%	68.3%	93.8%	93.2%	83.2%	**97.2%**
F-value	86.4%	77.6%	78.4%	82.4%	93.9%	87.6%	**96.3%**

We obtain the highest F-value when using all proposed features. Microblog-based Features, Linguistic Features and Event Features do not contribute much to the precision. Although Content Features can achieve higher precision, the recall is always low, because of their good capability of distinguishing. Event features often produce

---

[2] http://www.cs.waikato.ac.nz/ml/weka/

much higher recall than other features, because they can capture the natural character-istics of events. Our method achieves better performance than two baselines, especial-ly in event detection of car accident, because people tend to describe more specific geolocations when reporting a local accident.

4.4   Evaluation of Temporal Analysis

The task of temporal analysis is to decide whether an event reported by microblog m happens in the currently-considered time interval I_c, by classifying m into C or $\bar{C}$. If the event happens in I_c, m is classified into C.

In this experiment, we set the duration of the time intervals to be one calendar day, and only focus on hot name entities which frequently appear on the previous day of the currently-considered interval. We manually annotate 582 microblogs posted on June 3, 2013 reporting actual ADE occurrences, and among them there are 251 microblogs reporting events truly happened on June 3, 2013. These 251 microblogs are examples of C, while others are examples of $\bar{C}$.

We take 66% of the annotated examples as the training set, and the remaining as the test set to compare the usefulness of proposed features (described in Table 2). We use the Naive Bayes classifier implemented in Weka. The previous hot NEs are name entities which appeared frequently on June 2, 2013. We divide the features into 4 groups, and put all proposed features into the fifth group. The highest classification performance of each group of features is presented in Table 6.

Table 6. Evaluation of Temporal Analysis

Group	Features	Precision	Recall	F-value
1	use-of-links use-of-hst is-forwarded	71.1%	86.8%	78.2%
2	nearest-time-word	58.3%	85.2%	69.2%
3	word-related-to-C word-related-to-$\bar{C}$	89.6%	74.4%	81.3%
4	previous hot NE	66.2%	88.4%	75.7%
5	All proposed features	**87.1%**	**85.6%**	**86.3%**

From Table 6, we can see that the group containing all proposed features achieves the highest F-value. Group 3 achieve the highest precision but the recall is much low-er, while Group 1, Group 2 and Group 4 achieve higher recall but low precision. The experiment results show that features of Group 3 have great capability of distinguish-ing, and features of Group1, Group2, and Group4 can capture the natural characteris-tics of new events.

4.5   Practical Case Study

In this case study, we use microblogs posted on June 3, to simulate a live tweet stream, and apply our approach to detect newly-occurred car accident, fire disaster and earthquake events. The duration of time interval is one-calendar day. Figure 3 shows the number of microblogs reporting newly-occurred target events in every hour on June 3, 2013, detected by our system.

We can find that there are two peaks of reporting car accident events, which are from 8:00 to 9:00 and from 17:00 to 18:00. The peaks indicate that car accidents happen a lot during the morning and evening rush hours in China. A large number of microblogs reporting fire events were posted from 9:00 to 12:00, because a serious fire disaster happened around 6:00 in Jilin province of China on June 3, 2013. Two peaks of earthquake reports show that people felt the quake more strongly between 00:00 and 1:00, and between 23:00 and 24:00.

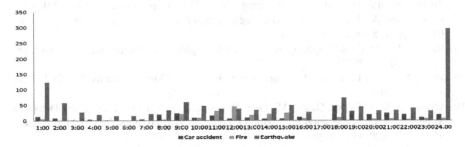

Fig. 3. The number of microblogs reporting newly-occured events on June 3, 2013

5   Conclusion

This paper proposes a scheme for real-time detection targeting accident and disaster events from Sina Weibo. Due to the geo-information is an important characteristic of an event, we propose a content-based method to extract street-level locations. Further, we propose useful features to improve the performance of event detection based on microblog classification. To deliver fresh events, we propose a temporal analysis method to automatically filter away past events. Experiment results show improvements achieved by our method, when compared to the state-of-the-art baselines. In the future work, we will focus on tracing and summarizing the target events which have been detected.

Acknowledgements. The wok is supported by the National Key Technology R&D Program of China under Grant No. 2012BAH04F02, 2012BAH88F02, 2013BAH61F01 and 2013BAH63F01, and the International S&T Cooperation Program of China under Grant No. 2013DFG12980.

References

1. Gao, Q., Abel, F., Houben, G.-J., Yu, Y.: A Comparative Study of Users' Microblogging Behavior on Sina Weibo and Twitter. In: Masthoff, J., Mobasher, B., Desmarais, M.C., Nkambou, R. (eds.) UMAP 2012. LNCS, vol. 7379, pp. 88–101. Springer, Heidelberg (2012)
2. Ritter, A., Mausam, Etzioni, O.: Open Domain Event Extraction from Twitter. In: KDD, pp. 1104–1112. ACM (2012)
3. TDT 2004: Annotation manual,
   http://www.ldc.upenn.edu/Projects/TDT2004
4. McMinn, A., Moshfeghi, Y., Jose, J.: Building a Large-scale Corpus for Evaluating Event Detection on Twitter. In: CIKM, pp. 409–418. ACM (2013)
5. Pohl, D., Bouchachia, A., Hellwagner, H.: Automatic Sub-Event Detection in Emergency Management Using Social Media. In: WWW, pp. 683–686. ACM (2012)
6. Lee, C., Yang, H., Chien, T., Wen, W.: A Novel Approach for Event Detection by Mining Spatio-temporal Information on Microblogs. In: ASONAM, pp. 254–259. IEEE (2011)
7. Li, C., Sun, A., Datta, A.: Twevent: Segment-based Event Detection from Tweets. In: CIKM, pp. 155–164. ACM (2012)
8. Rao, Y., Li, Q.: Term Weighting Schemes for Emerging Event Detection. In: WI-IAT, pp. 105–112. IEEE (2012)
9. Zhou, X., Chen, L.: Event Detection over Twitter Social Media Streams. VLDB J 23(3), 381–400 (2014)
10. Watanabe, K., Ochi, M., Okabe, M., Onai, R.: Jasmine: a Real-time Local-event Detection System based on Geolocation Information Propagated to Microblogs. In: CIKM, pp. 2541–2544. ACM (2011)
11. Petrović, S., Osborne, M., Lavrenko, V.: Streaming First Story Detection with Application to Twitter. In: NACL, pp. 181–189. ACL (2010)
12. Baldwin, T., Cook, P., Han, B., Harwood, A., Karunasekera, S., Moshtaghi, M.: A Support Platform for Event Detection Using Social Intelligence. In: EACL, pp. 69–72. ACL (2012)
13. Li, R., Lei, K., Khadiwala, R., Chang, K.: TEDAS: a Twitter-based Event Detection and Analysis System. In: ICDE, pp. 1273–1276. IEEE (2012)
14. Popescu, A., Pennacchiotti, M.: Detecting Controversial Events from Twitter. In: CIKM, pp. 1873–1876. ACM (2010)
15. Sakaki, T., Okazaki, M., Matsuo, Y.: Earthquake Shakes Twitter Users: Real-time Event Detection by Social Sensors. In: WWW, pp. 851–860. ACM (2010)
16. Cheng, Z., Caverlee, J., Lee, K.: You Are Where You Tweet: a Content-based Approach to Geo-locating Twitter Users. In: CIKM, pp. 759–768. ACM (2010)
17. Cataldi, M., Di Caro, L., Schifanella, C.: Emerging Topic Detection on Twitter based on Temporal and Social Terms Evaluation. In: MDMKDD. ACM (2010)
18. Joachims, T.: Text Categorization with Support Vector Machines. In: Nédellec, C., Rouveirol, C. (eds.) ECML 1998. LNCS, vol. 1398, pp. 137–142. Springer, Heidelberg (1998)
19. Yang, Y., Pedersen, J.O.: A Comparative Study on Feature Selection in Text Categorization. In: ICML, vol. 97, pp. 412–420 (1997)

An Automatic Unsupervised Method Based on Context-Sensitive Spectral Angle Mapper for Change Detection of Remote Sensing Images

Tauqir Ahmed Moughal[1,2] and Fusheng Yu[1]

[1] Laboratory of Complex Systems and Intelligent Control,
School of Mathematical Sciences,
Beijing Normal University, Beijing 100875, P.R. China
[2] Department of Mathematics and Statistics,
Allama Iqbal Open University,
Islamabad 44000, Pakistan
tauqir_mughal@yahoo.com, yufusheng@263.net

Abstract. This paper proposes an automatic unsupervised method for change detection at pixel level of Landsat-5 TM images based on spectral angle mapper (SAM). In most existing studies, conventional use of SAM does not take into account contextual information of a pixel. The proposed method incorporates spatio-contextual information both at feature and decision level for improved change detection accuracy. First, a similarity image is created using context-sensitive spectral angle mapper, and then it is segmented into two segments changed and unchanged using k-means algorithm to create a change map. The quantitative as well as qualitative comparison of the experiment results shows that the proposed method gives better results than the other existing method.

Keywords: Remote sensing, change detection, spectral angle mapper, spatio-contextual information.

1 Introduction

Remote sensing images have become major source of information for land use and land cover change detection. This is particularly important in understanding natural and artificial changes effecting ecological system at regional and global level. It is assumed that change may have taken place between two (or more) time instances which needs to be detected using remote sensing images. Change detection is the process of identifying differences in the state of an object or phenomenon by observing it at different times [1].

Change can only be detected in supervised paradigm if some prior information about the nature of change is available before analysis. This prior information is required in the form of training set for each class which is used for the learning process of classifier. However, availability of training set poses a critical restriction on the scope of supervised classification methods because in many real life change detec-

X. Luo, J.X. Yu, and Z. Li (Eds.): ADMA 2014, LNAI 8933, pp. 151–162, 2014.

tion problems prior information is seldom available. On the other hand, unsupervised classification methods do not require any prior information, which is more realistic and frequent situation encountered in real life where prior information (training set) is not known in advance. Thus the usefulness of unsupervised technique is more than a supervised one for a change detection problem.

Based on nature of change, change detection techniques can be broadly categorized as binary change (i.e. change and no change) and "from to" change detection (e.g. bare soil to vegetation). Former technique involves only detection of changed area and provide no information about nature of change, while later technique provide detailed information about the type of land cover change (e.g., from forest to agriculture) for every pixel under examination [2,3].

In the simplest form of change detection process only two images (bitemporal) are used for binary change detection (change and no change). Several techniques are proposed for this purpose including image thresholding techniques [4,5,6,7], decision and feature level fusion [8,9], principal component analysis [10,11], neighborhood correlation analysis [12], hypothesis testing [13,14,15] and machine learning algorithms [3]. More exhaustive and general discussion about change detection is found in review papers [1,16,17,18,19,20].

Most change detection techniques assume that pixels are independently distributed in an image which is not a realistic assumption in real life change detection problems. This paper proposes an unsupervised context-sensitive technique for a binary change detection problem. The proposed methodology is based on spectral angle mapper (SAM) which incorporates spatio-contxtual information of pixels at feature and decision level. Unlike other context sensitive techniques such as markov random field (MRF) [21] and EM algorithm [22] which requires fitting a probability distribution such as Gaussian or mixture of Gaussians for modeling of changed and unchanged class, the proposed technique is distribution free which implies that it neither require any estimation of statistical terms nor the modeling of statistical distribution of classes. The proposed technique is also unsupervised in nature and do not require any prior information (in terms of training set) about changed and unchanged class. Therefore, the major advantage of proposed technique is that it can be even applied to those situations where least amount of information is available for a change detection problem.

SAM has been applied to broad range of applications including earth sciences [23], urban studies [24], vegetation research [25] and planetary studies [26]. SAM is able to directly compare the similarities of pixels between two images. This method treats both pixels as vectors and calculates the spectral angle between them. The major motivation for using SAM for change detection is that it is independent of magnitude of reflectance value hence making it robust against albedo. Another advantage of SAM is that it does not require any feature extraction and can be applied to any number of bands efficiently. Fig. 1 illustrates the effectiveness of SAM over change vector analysis (CVA), which is another commonly used distance-based change detection technique based on difference image.

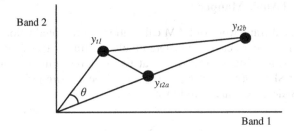

Fig. 1. Two-dimensional space having change vectors with the same angle (θ) but different distance

In Fig. 1, point y_{t1} and y_{t2a} represents pixel value at time 1 and time 2, respectively. Point y_{t2b} is the noisy version of y_{t2a} which reflects the increase in magnitude because of illumination. It is clear from this figure that angle between y_{t1} and y_{t2a} is the same as angle between y_{t1} and y_{t2b}. Therefore on the basis of angle, y_{t2a} and y_{t2b} will be assigned to same class of y_{t1} because they have same angle with y_{t1}. However distance of y_{t1} and y_{t2a} is not same as distance of y_{t1} and y_{t2b}. Therefore on the basis of distance, y_{t2a} will be assigned to same class of y_{t1} but y_{t2b}, which actually should belongs to same class of y_{t2a}, will be wrongly assigned to different class because of noise which results in larger distance from y_{t1}. It indicates that presence of noise affects the change detection capability badly. In many situations noise free images are hardly available. This suggests that distance is not a right choice for detecting change in presence of noise because it will result in more false alarms as compare to SAM. SAM appears to be more robust against noise and exhibit its scale invariant property, hence becomes more suitable choice for addressing problem of change area overestimation.

Two experiments are performed to evaluate the performance of proposed technique. Results are compared with maximum likelihood classifier (MLC) which is state-of-art supervised learning algorithm. The proposed technique found to be superior with respect to accuracy.

This paper comprises of four sections. Section 2 briefly describes the proposed change detection technique. The data sets used in the experiments and the results obtained are described in section 3. Finally, conclusions are drawn in section 4.

2   Methodology

2.1   Preprocessing

Two major requirements for bitemporal change detection are radiometric and atmospheric corrections and precise image coregistration. However, unsupervised change detection does not require any radiometric and atmospheric calibration [27]. Therefore only image coregistration is required in this study. The root mean square error (RMSE) is a measure of precision and used to determine how accurately two images are coregistered. Bitemporal images were coregistered precisely such that RMSE < 0.5 pixel which is satisfactory for a change detection problem. A running mean filter was also used to reduce the noise affecting reflectance value of pixels.

2.2   Spectral Angle Mapper

The mathematical framework of SAM calculation is presented below. Let us consider two radiometrically corrected coregistered multispectral images I_1 and I_2 of size $l \times m$ acquired over same geographical region at two different time instances t_1 and t_2, respectively. Let SI $= \{\theta(r,s),\ 1 \leq r \leq l,\ 1 \leq s \leq m\}$ be the similarity image obtained by applying the SAM technique as follows:

$$\theta_{SI_{(r,s)}} = \cos^{-1}\left[\frac{\sum\limits_{i=1}^{nb} \rho I_{1_{(r,s)i}} \rho I_{2_{(r,s)i}}}{\left(\sum\limits_{i=1}^{nb} \rho I_{1_{(r,s)i}}^{2}\right)^{1/2}\left(\sum\limits_{i=1}^{nb} \rho I_{2_{(r,s)i}}^{2}\right)^{1/2}}\right] \qquad (1)$$

where $\rho I_{j_{(r,s)}}$ $(j \in \{1,2\})$ is the reflectance value of $(r,s)^{\text{th}}$ pixel in j^{th} image and $\theta_{SI_{(r,s)}}$ is the angle of $(r,s)^{\text{th}}$ pixel in the similarity image calculated from corresponding pixels of the images I_1 and I_2 having nb bands. A small angle between two vectors represents that two pixels are similar to each other while a large angle shows large dissimilarity.

2.3   Integrating Spatio-contextual Information at Feature Level

In order to incorporate the spatio-contextual information, similarity image (SI) is calculated for each pixel (r,s) in I_1 and I_2 image in its spatial neighborhood. For a given pixel at position (r,s) its spatial neighborhood is denoted by $N_d(r,s)$. Every pixel, except for edges, has four vertical and horizontal neighbors and four diagonal neighbors which collectively constitute 8 neighbors. These eight neighbors denoted by $N_8(r,s)$ are also referred to as a 3x3 spatial window. Fig. 2 depicts the layout of horizontal and vertical, diagonal and 3x3 window.

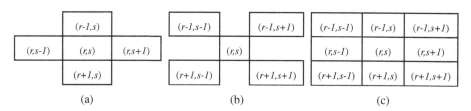

(a)   (b)   (c)

Fig. 2. Neighbors of a pixel at position (r,s). (a) Horizontal and vertical neighbors, (b) Diagonal neighbors and (c) 3x3 window

Bigger spatial window such as 5x5 and 8x8 are also used if change is expected to spread over a large area. Choices for size of spatial window are application based and heavily rely on analyst experience. Selecting a too small spatial neighborhood might be helpful in detecting smaller change but at the same time may introduce salt-and-pepper noise. On the other hand, a bigger spatial neighborhood is robust against salt-and- pepper noise but might not able to detect smaller changes. Therefore, a manual trial and error method was adopted. Three choices for spatial window size 3x3, 5x5 and 8x8 were used and optimal spatial window size was determined as the one which gives maximum over all accuracy by comparing the resultant image with the reference image. In this study, 3x3 window is used as it gives better result when compared with 5x5 and 8x8 windows.

2.4   K-Means Clustering

Detection of change from no change area can be think of as clustering problem in which goal is to partition SI into two clusters such that F: SI→{0,1}, where F is k-means algorithm. K-means algorithm basically minimizes the following objective function

$$Q(X;\gamma) = \sum_{i=1}^{n}\sum_{j=1}^{m} u_{ij} \| x_i - \gamma_j \|^2 \qquad (2)$$

where $u_{ij} \in \{0,1\}$ is entity of the partition matrix, x_i is the ith feature and γ_j is j^{th} cluster center. For binary change detection, k-means algorithm produces two clusters with centers γ_1 and γ_2. Now these clusters are labeled as changed and unchanged by comparing it with reference image. Let W_c and W_u be changed and unchanged clusters with cluster centers γ_c and γ_u, respectively. Each pixel of SI is assigned to one of the two clusters based on its distance from γ_c and γ_u. The pixels are assigned to the cluster having minimum distance. Final change map is obtained by

$$F_{(r,s)} = \begin{cases} 1, \ \| \theta_{SI_{(r,s)}} - \gamma_c \| \leq \| \theta_{SI_{(r,s)}} - \gamma_u \| \\ 0, \ otherwise \end{cases} \qquad (3)$$

where ‖.‖ is Euclidean distance. Change map F consists of zeros (white) and ones (black) indicating unchanged and changed areas, respectively.

2.5   Integrating Spatio-contextual Information at Decision Level

In order to make the proposed technique more robust against change area overestimation, spatio-contextual information is also augmented at decision level. This goal is achieved by integrating decision level spatial information in cluster labels by applying a running neighborhood voting procedure. Thus a pixel at $(r,s)^{th}$ location will be kept as unchanged unless majority of its neighborhood pixels are changed. The pixel of interest is reassigned the value of the most frequently occurring label within its neighborhood. Once again, like feature level spatio-contextual information, the size of neighborhood is decided by manual trial and error method by considering it as a function of accuracy. Several window sizes were tested and 8x8 appears to be better than 3x3 and 5x5 window for integration of decision level contextual information. This step is particularly important to remove any small, discontinuous, or noisy clusters.

3   Experimental Studies

3.1   Description of Datasets

In order to show the effectiveness of proposed technique, we considered two bitemporal data sets corresponding to geographical area of the Island of Elba, Italy and California, USA. Spatial resolution for each of dataset is 30m. Detailed description of each dataset is given below.

Dataset Related to Island of Elba, Italy. The first of the two datasets used in our experiments consist of two multispectral images acquired by Landsat-5 Thematic Mapper (TM) in the Island of Elba, Italy on August 1994 and September 1994. A subset of 414x326 pixels has been selected as test data from entire available Landsat scene. Fig. 3 (a) and (b) show channel 4 of the August and September images, respectively. It is apparent that between the two aforementioned dates, a wildfire destroyed significant portion of vegetation. A reference map (Fig. 4) corresponding to the location of wildfire was manually defined by detailed visual analysis of available bitemporal images and their similarity image. Reference map contains 1943 changed and 133021 unchanged pixels which indicate that changed area is relatively small. Experiments were performed, in an automatic manner, to detect this small change as accurately as possible.

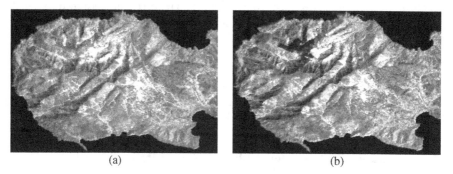

(a)   (b)

Fig. 3. Images of Island of Elba, Italy (a) channel 4 of Landsat image acquired in August 1994 (b) channel 4 of Landsat image acquired in September 1994

Fig. 4. Reference image of Island of Elba, Italy

Dataset Related to Island of California, USA. The second dataset used in our experiments composed of two multispectral images acquired by Landsat-5 Thematic Mapper (TM) in California, USA on July 05, 2009 and October 25, 2009. From entire available Landsat scene, a section of 2100x1212 pixels has been selected as test data. Fig. 5 (a) and (b) show false color composite of channel 7, 4 and 2 of the July and October images, respectively. A wild fire took place between the aforementioned dates and destroyed very large area. Similar to case of Italy dataset, in this case a reference map (Fig. 6) was also developed by visual inspection of both the available images and their similarity image. In reference map, 657179 changed and 1888021 unchanged pixels are identified. Unlike Italy dataset, change is spread over a significantly large area.

    (a)   (b)

Fig. 5. Images of California, USA (a) false color composite of channel 7, 4 and 2 of Landsat image acquired on July 5, 2009 (b) false color composite of channel 7, 4 and 2 of Landsat image acquired on October 25, 2009

Fig. 6. Reference image of California, USA

3.2 Results and Discussion

In order to prove the effectiveness of proposed technique, both qualitative and quantitative comparisons were performed.

Qualitative Analysis. For both Italy and California datasets, SI is generated by using channel 4 and 7 as these are reported to be very effective for burned area detection.

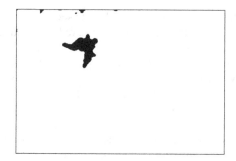

Fig. 7. Change detection map for dataset related to Island of Elba, Italy by using proposed technique

Fig. 8. Change detection map for dataset related to California, USA by using proposed technique

Fig. 7 and Fig. 8 shows the binary change map obtained by proposed technique for Italy and California datasets, respectively. The resulting change map for both datasets can easily be compared with their respective reference map. This eyeballing gives coarse idea about the quality of produced change maps. For Italy dataset, resulting change map (Fig. 7) looks very similar to the reference map (Fig. 4). However, careful inspection reveals that there are few false alarms at top-left of the scene. For California dataset, again some false alarms can be seen in the change map (Fig. 8), mostly at the bottom-left of the scene, when compared with reference map (Fig. 6). This visual inspection is a starting point and performance of proposed technique needs to be further investigated in terms of accuracy. Therefore, in the following section quantitative comparison is provided.

Quantitative Analysis. Quantitative assessment is performed in terms of overall accuracy (OA), kappa coefficient (K), false alarms (i.e. unchanged pixels which are labeled as changed pixels – FA) and missed alarms (i.e. changed pixels which are identified as unchanged pixels – MA). OA is simply the percentage of correctly detected pixels. K is considered to be a robust indicator of accuracy as compared to OA because it takes into account the possibility of agreements occurring by chance in random detection. Table 1 provides above mentioned performance assessment indicators for Italy and California dataset.

Table 1. Overall accuracy, kappa coefficient, missed alarms and false alarms obtained by proposed technique

Dataset	OA (%)	K	MA (%)	FA (%)
Italy	99.65	0.88	8.13	0.24
California	97.15	0.93	5.49	1.93

From these results it is obvious that proposed technique performed well in detecting change and provides 99.65% and 97.15% OA for Italy and California dataset, respectively. Problem of overestimation of changed area is also well handled as FA is far less than MA for both the datasets. It is also worthy to note that smaller change (Italy dataset) as well as larger change (California dataset), in both cases proposed method is able to produce high OA and K.

As mentioned earlier, we also compared our results with one supervised change detection technique known as direct multidate comparison [1,20]. In this method, bitemporal images are first stacked and then classified at one time in supervised mode. We used maximum likelihood classifier (MLC) for the purpose of classification. In order to have a comparative analysis between proposed technique and MLC, Table 2 and Table 3 show the results for Italy and California dataset, respectively. It is clear from these results that proposed technique performed better (OA=99.65 for Italy dataset and OA=97.15 for California dataset) than MLC (OA=99.54 for Italy dataset and OA=96.59 for California dataset). MLC do not take care of contextual information and also require training set. On the other hand, proposed technique utilizes spatio-contextual information and do not require any prior information in form of training set and still produce better results than MLC.

Table 2. Overall accuracy, kappa coefficient, missed alarms and false alarms for Italy dataset

Method	OA (%)	K	MA (%)	FA (%)
Proposed Technique	99.65	0.88	8.13	0.24
MLC	99.54	0.84	0.26	13.43

Table 3. Overall accuracy, kappa coefficient, missed alarms and false alarms for California dataset

Method	OA (%)	K	MA (%)	FA (%)
Proposed Technique	97.15	0.93	5.49	1.93
MLC	96.59	0.91	3.16	3.48

In terms of K, the proposed technique is 4% better in case of Italy dataset and 2% better in case of California dataset, which show significant improvement when compared with MLC. Unlike K, OA is slightly better than MLC for both datasets (see Table 2 and Table 3). Therefore, in order to evaluate whether the increase in OA of the proposed technique is statistically significant or not, a statistical test of significance, known as two-proportion z-test, was performed. Table 4 gives the results for this test.

Table 4. Statistical significance of the difference between OA of proposed technique and MLC

Dataset	z-value	p-value
Italy	4.16	0.000
California	35.55	0.000

Under the assumption of normal distribution, if p-value is less than significance level ($\alpha=0.05$), then hypothesis of equal proportion is rejected. It is clear from Table 4 that hypothesis of equality of OA of proposed technique and MLC is rejected for both datasets on basis of p-value. Therefore, the proposed technique is significantly better than MLC in terms of OA for both datasets.

As already mentioned, the reason behind improved results by proposed technique is that it uses contextual information and robust against changed area overestimation, which results in less FA and improved OA and K.

4   Conclusion

An effective unsupervised method based on SAM is proposed and tested in this paper. Feature and decision level spatio-contextual information is incorporated and its performance is evaluated in two different experiments. To this aim, two multitemporal Landsat-5 Thematic Mapper (TM) images are used as experimental data. Experimental results show that proposed method obtains higher accuracy than other comparison method. The proposed method is also more practical in nature because of the facts that it is distribution free and unsupervised and can be applied in those situations where least amount of information is available. In the future, further research will focus on utilizing spatio-contextual information for "from to" change.

Acknowledgement. This work is supported by Beijing Natural Science Foundation (No.4112031: Clustering and forecasting of large scale temporal data based on knowledge-guidance and optimal granulation of information) and the Fundamental Research Funds for the Central Universities.

References

1. Singh, A.: Digital Change Detection Techniques using Remotely Sensed Data. Int. J. Remote Sens. 10, 989–1003 (1989)
2. Im, J., Jensen, J.R.: A Change Detection Model Based on Neighborhood Correlation Image Analysis and Decision Tree Classification. Remote Sens. Environ. 99, 326–340 (2005)
3. Chan, J.C.W., Chan, K.P., Yeh, A.G.O.: Detecting the Nature of Change in an Urban Environment: A Comparison of Machine Learning Algorithms. Photogramm. Eng. Rem. S. 67, 213–225 (2001)
4. Sezgin, M., Sankur, B.: A Survey over Image Thresholding Techniques and Quantitative Performance Evaluation. J. Electron. Image. 13, 146–165 (2004)
5. Xue, J.H., Zhang, Y.J.: Ridler and Calvard's, Kittler and Illingworth's and Otsu's Methods for image Thresholding. Pattern Recognition Letters 33, 793–797 (2012)
6. Rosin, P.L., Ioannidis, E.: Evaluation of Global Image Thresholding for Change Detection. Pattern Recognition Letters 24, 2345–2356 (2003)
7. Orlando, J.T., Rui, S.: Image Segmentation by Histogram Thresholding Using Fuzzy Sets. IEEE Trans. Image Processing. 11, 1457–1465 (2002)
8. Du, P., Liu, S., Gamba, P., Tan, K., Xia, J.: Fusion of Difference Images for Change Detection over Urban Areas. IEEE J. Sel. Top. Appl. Earth Obs. Remote Sens. 5, 1076–1086 (2012)
9. Du, P., Liu, S., Xie, J., Zhao, Y.: Information Fusion Techniques for Change Detection from Multi-temporal Remote Sensing Images. Information Fusion 14, 19–27 (2013)
10. Celik, T.: Unsupervised Change Detection in Satellite Images Using Principal Component Analysis and k-Means Clustering. IEEE Geoscience and Remote Sensing Letters 6, 772–776 (2009)
11. Deng, J.S., Wang, K., Deng, Y.H., Qi, G.J.: PCA based Land use Change Detection and Analysis using Multitemporal and Multisensor Satellite Data. Int. J. Remote Sens. 29, 4823–4838 (2008)
12. Im, J.: Neighborhood Correlation Image Analysis for Change Detection Using Different Spatial Resolution Imagery. Korean Journal of Remote Sensing 22, 337–350 (2006)
13. Teng, S.P., Chen, Y.K., Cheng, K.S., Lo, H.C.: Hypothesis-test-based Landcover Change Detection using Multi-temporal Satellite Images – A Comparative Study. Adv. Space Res. 41, 1744–1754 (2008)
14. Meola, J., Moses, R.L.: Detecting Changes in Hyperspectral Imagery Using a Model-Based Approach. IEEE Trans. Geosci. Remote Sens. 49, 2647–2661 (2011)
15. Krylov, V.A., Moser, G., Voisin, A., Serpico, S.B., Zerubia, J.: Change Detection with Synthetic Aperture Radar Images by Wilcoxon Statistic Likelihood Ratio Test. In: 19th IEEE International Conference on Image Processing, pp. 2093–2096 (2012)
16. Lu, D., Mausel, P., Brondízio, E., Moran, E.: Change Detection Techniques. Int. J. Remote Sens. 25, 2365–2407 (2004)
17. Coppin, P., Jonckheere, I., Nackaerts, K., Muys, B.: Digital Change Detection Methods in Ecosystem Monitoring: A Review. Int. J. Remote Sens. 25, 1565–1596 (2004)
18. Radke, R.J., Andra, S., Al-Kofahi, O., Roysam, B.: Image Change Detection Algorithms: A Systematic Survey. IEEE Trans. Image Process 14, 294–307 (2005)
19. Bhagat, V.S.: Use of Remote Sensing Techniques for Robust Digital Change Detection of Land: A Review. Recent Patents on Space Technology 2, 123–144 (2012)
20. Hussain, M., Chen, D., Cheng, A., Wei, H., Stanley, D.: Change Detection from Remotely Sensed Images: From Pixel-based to Object-based Approaches. ISPRS J. Photogramm. 80, 91–106 (2013)

21. Kasetkasem, T., Varshney, P.K.: An Image Change Detection Algorithm based on Markov Random Field Models. IEEE Trans. Geosci. Remote Sens. 40, 181–1823 (2002)
22. Dempster, A.P., Laird, N.M., Rubin, D.B.: Maximum Likelihood from Incomplete Data via the EM Algorithm. J. Roy. Stat. Soc. 39, 1–38 (1977)
23. Meer, F.V.D.: The Effectiveness of Spectral Similarity Measures for the Analysis of Hyperspectral Imagery. Int. J. Appl. Earth Obs. 8, 3–17 (2006)
24. Schiefer, S., Hostert, P., Damm, A.: Correcting Brightness Gradients in Hyperspectral Data from Urban Areas. Remote Sens. Environ. 101, 25–37 (2006)
25. Leeuw, H.D., Jia, H., Yang, L., Liu, X., Schmidt, K., Skidmore, A.K.: Comparing Accuracy Assessments to Infer Superiority of Image Classification Methods. Int. J. Remote Sens. 27, 223–232 (2006)
26. Schmidt, F., Doute, S., Schmitt, B.: WAVANGLET: An Efficient Supervised Classifier for Hyperspectral Images. IEEE Trans. Geosci. Remote Sens. 45, 1374–1385 (2007)
27. Healey, S.P., Cohen, W.B., Zhiqiang, Y., Krankina, O.N.: Comparison of Tasseled Cap-based Landsat Data Structures for use in Forest Disturbance Detection. Remote Sens. Environ. 97, 301–310 (2005)

A Personalized Travel System
Based on Crowdsourcing Model

Yi Zhuang[1], Fei Zhuge[1], Dickson K.W. Chiu[2], Chunhua Ju[1], and Bo Jiang[1]

[1] College of Computer and Information Engineering, Zhejiang Gongshang University
[2] Faculty of Education, The University of Hong Kong
{zhuang,jch,nancyjiang}@mail.zjgsu.edu.cn,
dicksonchiu@ieee.org

Abstract. With the proliferation of the online tourism markets, and the rapid change of tourists demands, existing online travel platforms cannot satisfy tourists to some extent, since their tourism demands tend to be more personalized and dynamic. Based on the above motivations, we design and develop a personalized tourism system based on a novel cooperation crowdsourcing model through the Internet. More importantly, data quality control based on the crowdsourcing model is a key problem which affects the accuracy and effectiveness of tourist recommendation. To address this problem, we propose three data quality control schemes for personalized tours based on the crowdsourcing model. Extensive experiments validate the effectiveness of our proposed approach.

Keywords: independent travel, crowdsourcing, quality control.

1 Introduction

With the improvement of living standards, tourism has become a popular activity for fun. Meanwhile, the rapid development of technology and sufficient social public facilities make independent tourism [10] become one of the new tourism modes. Now more and more people take the way that traveling by themselves as the main leisure. Nowadays there are many excellent online travel service platforms, such as Ctrip, TripAdvisor, etc. Most of these platforms, however, employ "search-select" mode based on reservation or coupon model, or share tourism experience as its feature. Occasionally, some platforms provide personalized services. The problem is that customized services are done by some travel agencies, but often this requires much manpower, and often the custom recommendation quality cannot really meet personalization requirements. Fortunately, the emergence of the crowd- sourcing paradigm can help us solve these problems.

Crowdsourcing model [2] is based on personal choice, collecting the public's knowledge, skills, information and techniques to solve the complex and diverse problems. It reflects a public participatory culture. Sufficiently converging wisdom from the crowd is the remarkable feature. Recently, platforms based on crowdsourcing have drawn attention of the public [11,12], and many excellent cases have succeeded in the application of this model, such as Wikipedia, Amazon Mechanical Turk [1] and so on. In addition, it is convenient to acquire data information for people nowadays who

X. Luo, J.X. Yu, and Z. Li (Eds.): ADMA 2014, LNAI 8933, pp. 163–174, 2014.

have access to both practical experience and network resources [9]. Therefore, it is not only a theory of innovation, but also a viable practice to the personalized travel system we designed.

Lease [8] regarded the advent of crowdsourcing as a benefit as well as a detrimental. We must also objectively recognize the key to applying crowdsourcing model into a personalized travel system includes maximizing the power of social networks of all groups, getting rid of the negative factors that exist in the traditional travel services, and improving the travel experience of tourists. However, the ultimate problem of crowdsourcing is *data quality control.*

This paper introduces our new personalized tourism system based on the crowdsourcing model. It is not exactly a travel service platform that is similar to the traditional ticket reservation, hotel reservations, or contemporary popular tourism platforms offering the relevant products, but an online self-service tourism platform based on crowdsourcing. We also adopt the new travel services model based on customization to achieve the goal of personalized tourism.

Many researchers pay attentions to data quality control in the field of crowdsourcing and propose a variety of strategies which are mainly divided into the following three categories:

- *Assess the quality of the crowdsourcing results;*
- *Propose the processes of staged quality control;* and
- *Other quality control strategies;*

The quality of data, in a certain extent, can be influenced by these strategies more or less, but these strategies also have their limitations. So, we apply a crowdsourcing model-based data quality control strategy to personalized tourism. The data quality control is enforced according to analyzing the content and the behavior. The contributions of the paper are as follows:

1. We present a framework of combining online tourism services with the crowdsourcing model.

2. We propose three schemes of data quality control for the crowdsourcing model based on content and behavior analysis.

3. We design and develop a prototype system and perform extensive experiments to testify the effectiveness of our data quality control schemes.

The rest of the paper is organized as follows. We present related work in Section 2. In Section 3, detailed analysis of our novel data quality control strategies are described. Specific algorithms of the strategies are given in Section 4. In Section 5, we analyze our experiment result. The conclusion and future work are summarized in the last section.

2   Related Work

There are many well-known online travel service platforms. For example, for those mainly depend on the services of ticket booking and hotel reservation, Ctrip is the representative. The platform, 17u.com, is used to introduce the information of tourism destinations. Qunar.com is the first travel search engine in China comparing domestic flights and hotel prices online. The website, called mafengwo, provides services to share travel experience. TripAdvisor allows tourists to publish travel review.

The well-known platform named qyer.com advocates independent travel with the concept of 'pinch pennies' and provides travel information, etc. Many of such platforms do not aware of upcoming personalized tourist requirements. Only a few realize the change trend of requirements, and corresponding services are provided in time. However, tourist experience for personalization service is not satisfactory because of the actual needs constrained by resources, and the goal of adequate personalization has not been achieved. Taking these cases into account, we can observe that personalized travel service is often supported by a big team with tourism expertise. No matter how big the team is, it is not bigger than the scale of Internet users. Nevertheless, we should identify a reality when access to the mass wisdom of Internet users. Generally, due to the difference of the crowd's quality (including profession and morality), the quality of tasks completed in the way of crowdsourcing is often poor. Thus, an effective data quality control strategy for personalized travel system is crucial.

Recently, the state-of-the-art research has focused on the quality control in the field of crowdsourcing. In this paper, strategies are classified into three categories roughly. First, results of crowdsourcing are evaluated directly. Second is to put forward the staged quality control processes. Third, there are other quality control strategies.

Result quality assessment methods based on the crowd- sourcing model can be divided into two kinds. One is supervised method [3], which picks works with known correct answers from gold standard datasets to estimate the ability of each worker. For instance, all of workers are required to be qualified before beginning the assignment by passing several tasks chosen from the gold standard dataset. This approach has its limitation, because it spends much effort on preparing the gold standard datasets and it is difficult to obtain the unique result for tasks. The other is unsupervised approach [3] that employs redundancy to control quality. It distributes a small task to many crowd workers, integrating information by voting or other sophisticated statistical techniques. This kind of approach would consider the characteristics of each worker or task, such as worker ability or task difficulty. But, most existing methods are assumed in structured content and are not appropriate for the unstructured.

In the respect of the staged quality control process, existing strategies can be roughly classified into two types. One is dynamic quality control in multi-stage [4]. The method sets testing point in each stage in the process of completing the whole task, to assess the quality of task completed in the previous stage. If the quality of results submitted is poor, the worker will be stopped from participating in the next phase and the results submitted at this stage will be removed. Meanwhile, new workers are chosen to continue to finish the rest of tasks. It needs to spend extra time for testing and replacement, and has a long period before completing the tasks, because of setting a testing point, and the replacement strategy is not necessarily the most appropriate. Another approach is two-stage workflow composed of creation stage and review stage [5]. Supervised and unsupervised mode are involved as well, so extensive domain knowledge including the features of information and the gold standard data sets[3] are needed, and heavy workload is also an issue.

There are three other quality control strategies. First, think about how to design a good crowdsourcing task [4] to reach the goal of obtaining high quality results. We can witness such a phenomenon that there are strong dependencies between rewards credits of mission and the quality of results. When a task is released at a low salary, very few people will be interested in it, leading to the low speed of task acceptance; while it will be easier to attract people with low efficiency when a task is issued at high rewards credits. Second, the management of workers' quality is the guarantee of high data quality. This strategy manages the quality of workers mainly by studying the error-masking technology and bad workers detection technology [6]. It is on the basis of assumption that qualified workers provide high quality data and unqualified workers give garbage. But, it is flawed to judge the data quality with the consideration about the quality of workers separately. Third, a series of avoiding strategies are proposed in view of the analysis of low quality categories [7]. The method is given priority to prompt message and aims to reduce the probability of committing low quality information by workers. It is associated with the quality of workers themselves but is not able to solve the problem.

Compared with the state-of-the-art methods, in this paper, we propose three innovative data quality control strategies based on the crowdsourcing. As the tourist information is the primary raw data, we control the quality of data based on the pattern of crowdsourcing through the analysis of the contents and behaviors. Specifically, we will propose three data quality control strategies, namely the *content analysis*(CA)-based scheme, the *behavior analysis*(BA)-based one, and *the hybrid one*.

3   System Overview

First of all, we introduce an overall framework of the data quality control strategy processing based on crowdsourcing model shown in Fig.1.

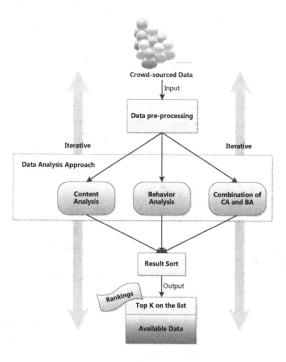

Fig. 1. The flow chart of data quality control strategy based on the crowdsourcing model

In Fig.1, the crowdsourcing data from the personalized travel platform should be preprocessed to filter the obvious noise and correct some available data (i.e., descriptive text) expressed in a non-uniform way. Then, a data analysis processing is performed which includes three schemes: *a content analysis-based method, behavior analysis-based method* and *the combination of the two ones are presented above*. These three schemes are compared in Section 4, and the most effective one will be concluded from the experimental results. Data are sorted decreasingly after the analysis of the previous phase by using a high efficient sorting algorithm, called quick sort. The reason why quick sort algorithm is chosen in this paper is that it shows superior performance in chaotic situation than other sorting algorithms, considering the time and space complexity. Not any more will we analyze the details about sorting algorithms because it is not the focus in this paper. After sorting, the top K (K is a variant) on the ranking are taken as candidates of valid data. The process will be repeatedly iterative and constantly improve the accuracy of valid data set with the update of crowdsourcing data.

4   Three Data Quality Control Schemes

The paper mainly focuses on the study of data quality control scheme based on crowdsourcing model. Before introducing the three schemes, the list of the symbols to be used throughout the paper is first summarized in Table 1.

Table 1. Meaning of the symbols used

Symbol	Meaning
K	a candidate scenic spot set of valid attraction
S	an existing set of scenic spot
M	a cluster of scenic spots in a city recommended by N guiders
k	a valid spot set
m	a scenic spot set filtered from M
α, β, μ	credibility value
$\theta 1, \theta 2, \theta 3$	proportionality coefficient
p	positive ratio
n	negative ratio

In Fig. 1, the data preprocessing before data analysis makes contributions to improving the overall efficiency of data analysis, and is considered as the function of preliminary de-noising which is beneficial to improve the accuracy of valid data obtained at the last stage. Thus, the process is essential. For example, the following two cases on the basis of personalized travel service are taken into account. One of the cases is that the data like those ones which are obviously not related to tourist attractions, such as a hotel or restaurant advertisement, should be filtered out when a guider recommends some tourist attractions of a certain city to visitors. In the other kind of case, the data should be retained. Take Hangzhou into consideration. A guider uses abbreviated names of scenic spots to do some recommendation for visitors.

For example, the sight spot named '*Lingyin Temple*' may be called '*Lingyin*', and the same thing happens to '*Taiziwan*', which is abbreviation of the '*Taiziwan Park*'. Finally, three kinds of data quality control schemes are described in detail.

4.1   CA-Based Data Quality Control Scheme

In this subsection, the content analysis(CA)-based data quality control strategy assumes that the data preprocessing has completed. The process of data analysis employs the word frequency statistics and probability analysis method to obtain the frequencies of scenic spots recommended by all local guiders from a certain city. The process of data quality control is shown in Fig. 2. The system can judge whether the attractions given by tour guiders are valid spots. The system platform uses statistical method to analyze all tourist attractions of a certain city recommended by N guiders synchronously, and uses quick sort algorithm to order the result of statistical analysis in reversed sequence. Then, the top K sites are selected as the candidates of valid spots from the sorted list. The process above will be repeated iteratively to revise valid attractions of cities constantly.

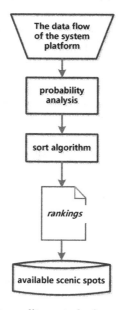

Fig. 2. The process of data quality control scheme based on content analysis

As for example, suppose a tourist attraction set of a city as S. A city's scenic spots set M recommended by N guiders are gathered through our system platform, and the set may have the duplicate data. The set M is selected by using the method of word frequency statistics. Simultaneously, non-repetitive scenic spot set m is collected. The top $|K|$ spots (size of set K is smaller than the size of set m) selected from ranking which is ordered by the value of data in the set m are regarded as valid attractions after an iteration. Iteration repeats afterwards. Finally, it is the time for comparison of result set $|K|$ and attraction set $|S|$. Valid spot set k is proper subset of set K and set S. The performance of the schemes is evaluated by using two metrics (i.e., recall and precision).

4.2   BA-Based Data Quality Control Scheme

Similarly, the data preprocessing has been finished before the data quality control strategy based on behavior analysis (BA), and the data that are irrelevant to tourist attractions are filtered. The comments, offered by visitors after traveling, are treated as a factor that determines whether a certain spot is a valid attraction of a city. First of all, the results of evaluation for scenic spots are classified according to Table 2.

Table 2. Statistical table of evaluations for scenic spots

SCENIC SPOT	POSITIVE	NEGATIVE	FREQUENCY
spot 1	0	0	0
spot 2	n1	0	n1
spot 3	0	n2	n2
spot4	n3	n4	n3+n4
......	....	....	......

Then, the results are analyzed according to Table 2, in order to sort the degree of good reputation of non-repeated spots. We argue that a higher degree means a higher probability of identifying a spot that is a valid attraction. The data in the table are analyzed and each spot is assigned a value which indicates the credibility being a valid spot. We illustrate the assignment depending on the data shown in Table 2. The credibility value of spot which does not receive any evaluation (e.g., spot 1) is α and those have been evaluated (e.g., spot 2, spot 3, spot 4) is β. Considering the spots have been evaluated as valid attractions, obviously, is inappropriate. We discuss the three categories according to the evaluations. Eq. (1) shows the assignment of the credibility being a valid attraction for the spot that has comment. μ is the value of credibility on the condition of having the credibility β. $\theta1, \theta2, \theta3$ are the proportionality coefficients allocated in three different cases, and $\theta1+\theta2+\theta3=1$. γ is also the value of credibility when a spot has negative feedback only. We argue that negative review can be divided into two categories.

The first is that the city does have the spot, but visitors think it is not worth to visit, and the second is that the tourist attraction is not present in the city. Here, we assign a value of 0.5 to γ. Besides, p, n denotes positive ratio and negative ratio, respectively. Take the evaluation of spot 4 in Table 2 for example. Positive ratio $p=n3/(n3+n4)$, and negative ratio $n= n4/(n3+n4)$.

$$\mu = \begin{cases} \theta1 & negative = 0, frequency \neq 0 \\ \theta2 \cdot \gamma & positive = 0, frequency \neq 0 \\ \theta3 \cdot \frac{|p-r \cdot n|}{|p+r \cdot n|} & positive \neq 0, negative \neq 0 \end{cases} \quad (1)$$

The same sorting algorithm described in the previous sub-section is employed to rank the credibility value of various non-repeated spots in a certain city, and the top $|K|$

data are selected to be the candidates of valid spots. Similarly, a valid spot set k is obtained by comparing the candidate set K with set S. Finally, values are put into Eq. (2), calculating the recall and precision of the strategy respectively.

$$recall = \frac{|rel \cap ret|}{|ret|} \quad , \quad precision = \frac{|rel \cap ret|}{|rel|} \tag{2}$$

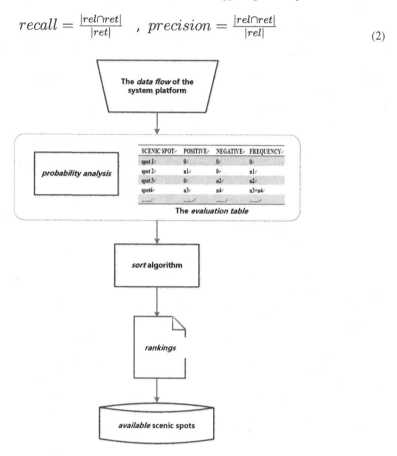

Fig. 3. The process of data quality control based on content analysis and behavior analysis

4.3   Hybrid Data Quality Control Scheme

This method combines the content and behavior analysis-based strategies. In view of the evaluations for the tourists attractions and the sorting results by using probability statistics for content analysis, the spots that are not in the list of top $|K|$ have chance to be packed into the list of valid spots, for they have positive reviews more frequently. Also, the spots in the list of top $|K|$ are probably pushed out, since they frequently have negative valuations. Fig. 3 is the process of data quality control based on content analysis and behavior analysis. We define R_c as the credibility being valid spot, the value is given by the method of probabilistic analysis, in the process of content analysis, and R_b denotes the credibility being valid spot assigned in the process of behavior analysis. R measures the credibility for a spot comprehensively through multiplying the credibility value of two analytical methods, i.e., $R=R_c \cdot R_b$. Later, the scenic spots will be sorted in terms of credibility. We take the top $|K|$ scenic spots as the valid spots, and compare it with the attraction set S to obtain the tourist attraction set k which is the proper subset of attraction set S. Likewise, all the value are plugged into Eq. (2) to calculate the recall rate and precision of this strategy. Finally, the data are provided to compare the effectiveness of three schemes.

5   Experiments

In this section, we perform extensive simulation experiments to compare the validity and accuracy of the three data quality control schemes. We collected the recommended data of three cities (e.g., *Beijing*, *Hangzhou* and *Nanjing*) provided by 50 tourism guiders with different backgrounds. The attraction set S mentioned above is collected in advance through the Internet in which the 40 most popular tourist attractions are recommended in the three cities, respectively to verify the correctness of results in the following experiments.

5.1   A Prototype System

The personalized online travel platform based on crowd- sourcing model is developed and used to testify the effectiveness of the data quality control schemes. The three data quality control schemes are adopted in the prototype system respectively. Fig. 4 shows the main interface of the system platform. The interface of the recommendation form filled by the local guiders is illustrated by Fig. 5. Fig. 6 is the interface of the results filtered from the attractions recommended by using the scheme provided in this paper.

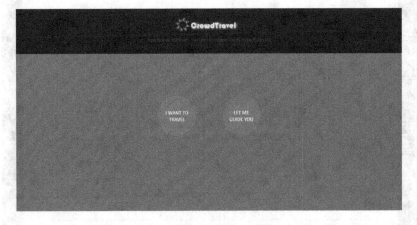

Fig. 4. The entry of the system

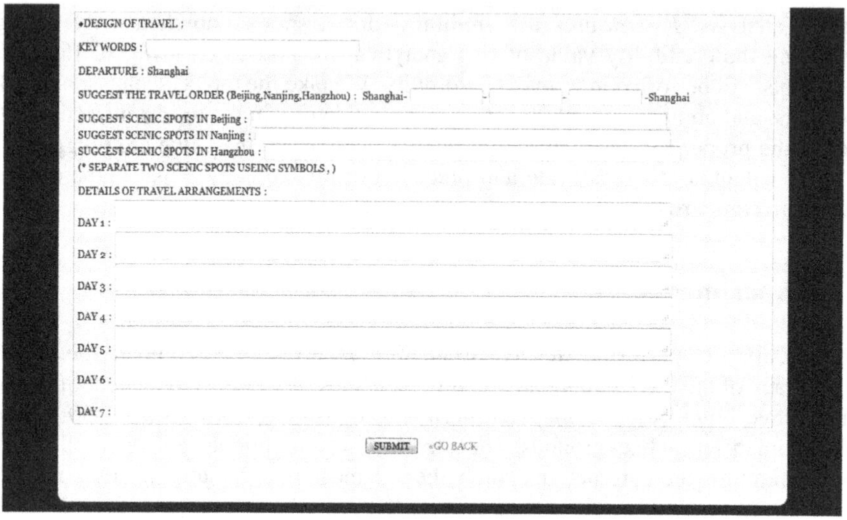

Fig. 5. The interface of the tourist recommendation

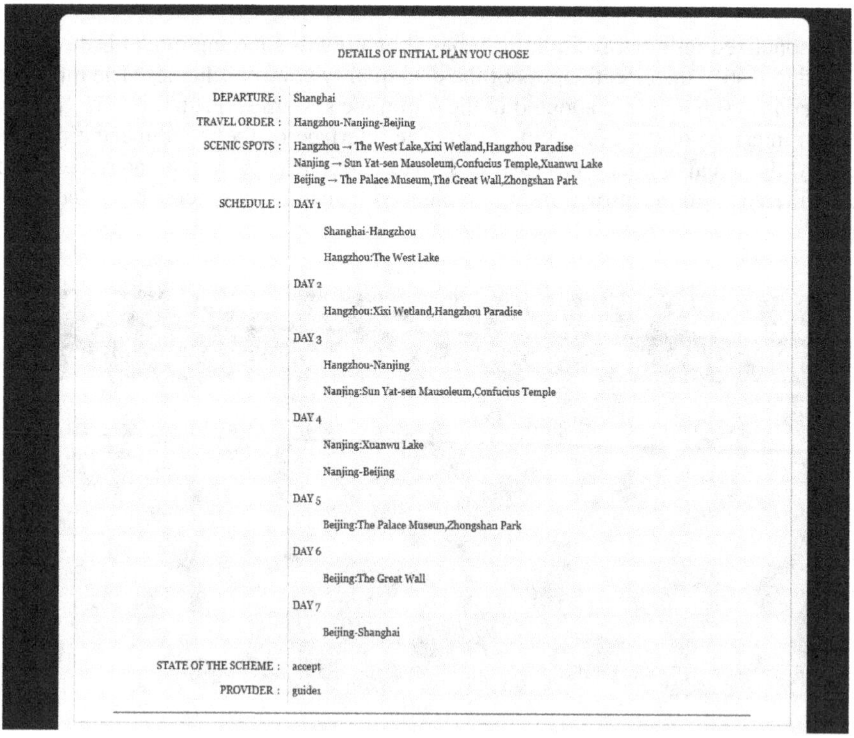

Fig. 6. The results of recommendation

5.2   Effect of K

In this experiment, we testify the effectiveness of K by using the datasets mentioned above. Three schemes are compared is the effectiveness of the data quality control. Denoting the set of ground-truth as *rel*, and the set of results returned by a data quality control method as *ret*, the recall and precision achieved by this quality control method are defined in Eq.(2).

Fig. 7 illustrates *recall-precision* curves for the performance comparisons of the three strategies proposed. In particular, it compares the data quality control result of scenic spots given by the 50 local guiders from a certain city (e.g., *Beijing, Hangzhou,* and *Nanjing*) that are randomly chosen from the database. Fig. 7 shows that the hybrid scheme is better than other two. The results conform to our expectation. This is because the hybrid scheme can benefit from the both advantages through sorting twice. From the other view, the recall ratios of the hybrid one increase as the percentage of k grows (Fig.7(a)). The lower the percentage value is, the fewer the numbers of the scenic spots and the valid spots are. The growth rates, however, remain level. Whereas, the precision ratios of the hybrid one fluctuate as the percentage of k grows (Fig.7(b)). On the whole, yet, the precision ratios of the hybrid scheme and content analysis scheme peak at beginning. Lower the percentage of k is, larger the proportion of valid spots in scenic spots is. In contrast, the precision ratio of behavior analysis scheme at the end of curve is slightly larger than that of it at beginning.

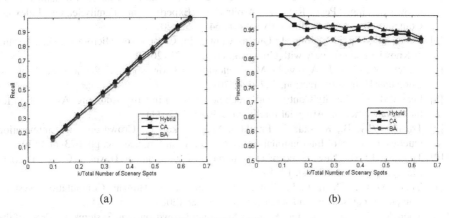

(a)   (b)

Fig. 7. Effect of K

6   Conclusions and Future Work

In this paper, we have developed a personalized online travel system based on crowdsourcing model. For the sake of taking full advantages of crowdsourcing in the system, and improving the users' experiences, three schemes of the data quality control are proposed based on information content and user behavior analysis. Extensive experiments are conducted to demonstrate the effectiveness of the strategies.

In our future work, we will further study the following issues. In the terms of content analysis, better methods can be used to obtain the credibility value of being valid spots.

Moreover, in the behavior analysis, more effective taxonomy for positive and negative evaluations can be applied to improve the accuracy. In addition, more comprehensive experiments can be conducted.

Acknowledgment. This paper is partially supported by the Program of National Natural Science Foundation of China under Grant Nos. 61003074 and 61272188; the Program of Natural Science Foundation of Zhejiang Province under Grant Nos. LY13F020008, and LY13F020010; the Ministry of Education of Humanities and Social Sciences Project under Grant No. 14YJCZH235. The Science & Technology Innovative Team of Zhejiang Province under Grant No. 2012R10041-06.

References

[1] Amazon's Mechanical Turk. (2012), http://www.mturk.com
[2] Lan, X., Changchun, G.: Crowdsourcing Changes Enterprise's Innovation Model. Shanghai Journal of Economics (3), 35–41 (2010)
[3] Baba, Y., Kashima, H.: Statisitcal Quality Estimation for General Crowdsourcing Tasks. Knowledge Discovery in Database (8), 554–562 (2013)
[4] Zhang, Z.-Q., Pang, J.-S., Xie, X.-Q.: Research on Crowdsourcing Quality Control Strategies and Evaluation Algorithm. Chinese Journal of Computer (8), 1636–1649 (2013)
[5] Hansen, D.L., Schone, P., Corey, D., et al.: Quality Control Mechanisms for Crowdsourcing: Peer Review, Arbitration,&Expertise at FamilySearch Indexing. Computer Supported Collaborative Work, 649–659 (2013)
[6] Venetis, P., Garcia-Molina, H.: Quality Control for Comparison Microtasks. Data Mining & Knowledge Discovery with Crowdsourcing, 15–21 (2012)
[7] Willett, W., Heer, J., Agrawala, M.: Strategies for Crowdsourcing Social Data Analysis. Computer Human Interaction, 227–236 (2012)
[8] Lease, M.: On Quality Control and Machine Learning in Crowdsourcing. Association for the Advancement of Artificial Intelligence, 97–102 (2011)
[9] Trushkowsky, B., Kraska, T., Franklin, M.J., Sarkar, P.: Crowdsourcing Enumeration Queries. In: Proc. of International Conference on Data Engineering, pp. 673–684 (2013)
[10] Tingting, J.: The New Trend of Development of Independent Travel in China. Tourism Management Research (4), 12–13 (2013)
[11] Yuen, M.-C., Chen, L.-J., King, I.: A Survey of Human Computation System. Computational Science and Engineering, 723–728 (2009)
[12] Yuen, M.-C., King, I., Leung, K.-S.: A Survey of Crowdsourcing Systems. In: Proc. of the Third IEEE International Conference on Social Computing, pp. 766–773 (2011)

Difference Factor' KNN Collaborative Filtering Recommendation Algorithm

Wenzhong Liang[1], Guangquan Lu[1], Xiaoyu Ji[1], Jian Li[1], and Dingrong Yuan[2]

[1] Wuzhou University, No.82 Fu Min 3 Road, Wuzhou, Guangxi, China, 543002
[2] Guangxi Normal University, No.15 Yu Cai Road, Guilin, Guangxi, China, 541004
wzxylwz@163.com, guangquan185@gmail.com,
{kippy2001,lijian727}@sina.com,
dryuan@mailbox.gxnu.edu.cn

Abstract. With the development of electronic commerce, Collaborative Filtering Recommendation system emerge, which uses machine learning algorithms for people provide a set of N items that will be of interest. In many user-based collaborative filtering applications based on KNN(K nearest neighbor algorithm), they only use similarity information(cosine similarity) between users, in some case, they have not use the difference information, so the precision and recall is not well. To address these problem, we propose a Difference Factor' K-NN collaborative filtering method, called DF-KNN. DF-KNN is an instance-based learning method and the key step in algorithms is how to use the difference factor and how to compute, the second step is mix similarity together. Our experimental evaluation on the MovieLens datasets show that the proposed DF-KNN and NDF-KNN(Normal Different Factor's KNN) are much efficient than the traditional user-neighborhood based KNN and provide recommendations whose quality is up to 13% better.

Keywords: collaborative filtering, recommendation system, K nearest neighbor.

1 Introduction

The amount of information in the world is increasing far more quickly than our ability to process it. Before recommendation system is born, based on the user-behavior data applications is very popular in the sites. For example, the hot lists and the trend charts. Also it is very easy to statistics but it can winning many customers in the process. So collaborative filtering is very important for e-commerce. Using personalized recommendations algorithms is to effectively promote sales for company. For example, Greg Linden, developer of Amazon's recommendation engine[1], reported that in 2002 over 20% of Amazon's sales resulted from personalized recommendations. Recommendation algorithm based on user behavior analysis is an important algorithm for personalized recommendation system, also rename as collaborative filtering algorithm. In real world applications, the recommendation system bring benefits to companies and the clients, however, it is difficulty to help client to find their interest.

This paper proposes a new collaborative filtering algorithm to handle recommendation for clients, called DF-KNN(Difference Factor' KNN), which is an instance-based learning method in machine learning. It is also efficient for calculating the

X. Luo, J.X. Yu, and Z. Li (Eds.): ADMA 2014, LNAI 8933, pp. 175–184, 2014.

nearest neighbor measure. Specifically, this approach uses a difference factor and similarity measures to mix together during the process of searching for the nearest neighbor under the assumption which require the instance i to have a similarity feature as the instance j when calculating the cosine measures. We evaluate the performance of our method using MovieLens datasets[2].The experimental results show that our approach is superior to k-NN using the cosine measure and whose quality is up to 13% better.

The rest of this paper is organized as follows. In section 2 we briefly review collaborative filtering recommendation system's related work. Section 3, we design our DF-KNN(Different Factor K Nearest Neighbor) algorithm and NDF-KNN(Normal Different Factor K Nearest Neighbor) algorithm. Section 4 describes our experiments on MovieLens and discussion of the results. The final section provides some conclude remarks and directions for future research.

2   Related Work

In this section we briefly present some of the research literature related to collaborative filtering, recommender systems, data mining and personalization. Collaborative filtering is to predict the utility of items to a certain user based on a database of user interest behavior from a sample or population of other users.

User behavior data exist on the site that is in the form of log. Recommendation systems depending on the user behavior data improve the clients' personal experience. Several ratings-based automated recommendation systems were developed, like the GroupLens research system[3,4] provides a pseudonymous collaborative filtering solution for Usenet news and movies. Ringo[5] and Video Recommender[6] are email and web-based systems that generate recommendations on music and movies, respectively.

Yehuda.[10] proposes a collaborative filtering recommendation framework which is based on viewing user feedback on products as ordinal, rather than the more common numerical view. Yi Cai [11] borrow ideas of object typicality from cognitive psychology and propose a novel typicality-based collaborative filtering recommendation method. Zhang,Yu[12] use the collaborative filtering to select features in social network relationship which is used to build a vector space model, and propose a distributed algorithm based on factorization machines and genetic algorithm.

Other technologies can be applied to recommender systems for improving the effective, including Bayesian networks, clustering, association rules[16,18,19], classification[20], missing data imputation[17,22,23], Dimension-Reduction method[21]. Bayesian networks create a model based on a training set with a decision tree at each node and edges representing user information. The resulting model is very small, very fast, and essentially as accurate as nearest neighbor method[7]. Bayesian networks may prove practical for environments in which knowledge of user preferences changes slowly with respect to the time needed to build the model but are not suitable for environments in which user preference models must be updated rapidly or frequently. Yue Shi[13] summarize and analyze recommendation scenarios involving information sources and the collaborative filtering algorithms that have been recently developed, and also provide a comprehensive introduction to a large body of research.

The accuracy of the recommendation is an important indicator whether it is an effective in recommendation system. We use the precision accuracy and the recall rate to evaluate the algorithms. In application, we also consider the algorithm spends on time and space complexity. Our DF-KNN collaborative filtering algorithm is different from the KNN algorithm, it can improve the precision accuracy and the recall rate by adding the different factor.

3   The DF-KNN Algorithm

3.1   User-Based Top-N Collaborative Filtering Algorithms

User-based Collaborative filtering[8,9] is the most successful technology for building recommendation systems to date, and is extensively used in many commercial recommendation systems. These approach rely on the fact that the person which have interest together also have some similarly-behaving. User-based top-N collaborative filtering algorithm analyze the user-item matrix to identify user group between the different interest, and then use the user interest behaving to compute the list of top-N collaborative filtering. The key motivation behind these approach is that the same group will have the similarity behaving, for example to purchase similarity goods, to browse the same movies or news. So these schemes need to identify the neighborhood of similar users when a recommendation is requested. Many of different strategy have been proposed to compute the relations between the different users based on either probabilistic approaches or more traditional user-to-user correlations.

In this paper we study user-based top-N collaborative filtering algorithms that use user-to-user similarity to compute the relations between the users. Building the model, for each user i, we find the k most similar user-behaving data $\{i_1, i_2, \ldots \ldots, i_k\}$, then compute their corresponding similarities $\{S_{i_1}, S_{i_2}, \ldots \ldots S_{i_k}\}$. this similarities is used to compute the top-N users' interest. Last, we recommend the items which they have not understand but they should like them.

3.2   The Similarity Computation

Similarity computation[14] between users or items is an important step in collaborative filtering algorithms. For a user-based collaborative algorithm, we calculate the similarity, $\text{sim}(u, v)$, between the user u and the user v have the same interest items.

3.2.1   Correlation-Based Similarity

In this case, the users u and v, we calculate the two items i and j, is measured by computing the Pearson correlation or other correlation-based similarities. Pearson correlation measures the extent to which two variables linearly relate with each other. For the user-based algorithm, the Pearson correlation between users u and v is given by:

$$\text{sim}(u, v) = \frac{\sum_{u \in U}(R_{u,i} - \overline{R_I})(R_{u,j} - \overline{R_J})}{\sqrt{\sum_{u \in U}(R_{u,i} - \overline{R_I})^2}\sqrt{\sum_{u \in U}(R_{u,j} - \overline{R_J})^2}} \tag{1}$$

where $R_{u,i}$ is the rating of user u on item i, $\overline{R_I}$ is the average rating of the ith item by those users.

3.2.2   Vector Cosine-Based Similarity

In this section, two users behaving are thought of as two vectors in the m dimensional user-space. The similarity between them is measured by computing the cosine of the angle between these two vectors[15]. We can adopt this form for collaborative filtering, where users take the role of documents, titles take the role of words, and votes take the role of word frequencies. We use that under the KNN algorithm, formally, if R is the m × n user-item matrix, the similarity between two items u and v, is denoted by $sim(u, v)$ is given by

$$sim(u, v) = \cos(\vec{u}, \vec{v}) = \frac{\vec{u} \cdot \vec{v}}{||\vec{u}||_2 ||\vec{v}||_2} \tag{2}$$

where "." denotes the dot-product of the two vectors. In the top-N collaborative filtering approach often employ K nearest neighbor(KNN) find the top-N items. Our DF-KNN algorithm is based on cosine similarity and add the different factor.

3.3   DF-KNN Algorithm

We have introduced Pearson correlation and cosine similarity here, but actually there are many more ways to measure similarity between two sets of data. The best one to use will depend on application, and it is worth trying Euclidean distance. So we use different factor to change and promote similarity computing base on cosine similarity. Give two users u and v, let $N(u), N(v)$ denote the user u and the user v had positive feedback item set. We can compute the cosine similarity as:

$$W_{uv} = \frac{|N(u) \cap N(v)|}{\sqrt{|N(u)||N(v)|}}, \tag{3}$$

For fig. 1, user-based collaborative filtering compute the two users interest similarity. The user A have behavior on item $\{a, b, d\}$, The user B have behavior on item $\{a, c\}$, use the cosine similarity to compute the user A and the user B interest similarity as:

$$W_{AB} = \frac{|\{a, b, d\} \cap \{a, c\}|}{\sqrt{|\{a, b, d\}||\{a, c\}|}} = \frac{1}{\sqrt{6}}$$

In the same way, we can compute the user A and the user C, the user D interest similarity as:

$$W_{AC} = \frac{|\{a, b, d\} \cap \{b, e\}|}{\sqrt{|\{a, b, d\}||\{b, e\}|}} = \frac{1}{\sqrt{6}}$$

$$W_{AD} = \frac{|\{a, b, d\} \cap \{c, d, e\}|}{\sqrt{|\{a, b, d\}||\{c, d, e\}|}} = \frac{1}{3}$$

Careful you find the cosine similarity only use users' common interest information, and not use their different information. Base on this weakness, we propose a different factor to improve the users interest similarity computing as:

$$W_f = \frac{1}{1 + (|N(u)| - |N(v)|) + (|N(v)| - |N(u)|)} \tag{4}$$

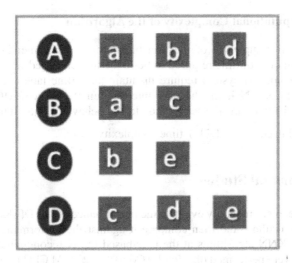

Fig. 1. users behavior records

$$DFW_{uv} = \frac{|N(u) \cap N(v)|}{\sqrt{|N(u)||N(v)|}} \times W_f \tag{5}$$

Adjust cosine similarity computing and employ the KNN approach we called our strategy DF-KNN(Different Factor K nearest neighbor), and it is described below in Fig.2:

The DF-KNN algorithm is presented as follows:

---

1   For the user, Compute DFW_{uv} base on Equation (5).

2   Get k Nearest Instances, find the interest similarity users set.

3   In the interest similarity users set, find the items that target user likes, and recommend the items to target user.

Fig. 2. The DF-KNN algorithm

another lifting scheme we normalize the different factor, are computed as:

$$W_f' = \frac{W_{fmax} - W_{fmin}}{(W_f - W_{fmin}) + \varepsilon} \tag{6}$$

$$NDFW_{uv} = \frac{|N(u) \cap N(v)|}{\sqrt{|N(u)||N(v)|}} \times W_f' \tag{7}$$

We use the equation (7) instead of equation (5), and it is called NDF-KNN(Normal Different Factor K nearest neighbor) approach.

3.4   The Computational Complexity of the Algorithm

The computational complexity of the user-based top-N recommendation algorithm depends on the amount of time required to build the model. Based KNN collaborative filtering recommendation system require maintain an offline table. Assumption that there are M users and N items(the user interest items), we need $O(M * M)$ space complexity, also if there are K user records to the behavior of the items, user-based KNN or DFKNN need $O(N * \left(\frac{K}{N}\right)^2)$ time complexity.

4   Experimental Studies

In this section we experimentally evaluate the performance of our DF-KNN algorithms or NDF-KNN algorithms and then compare it against the performance of the using cosine similarity KNN algorithms at the user-based top-N recommendation systems. All experiments were performed on a Intel® Core™ i3-2330M CPU based workstation running at 2.20GHz, 4GBytes of memory, and windows7-based operating system.

4.1   Data Sets

The MovieLens dataset[2] is applied to compare the performances of the three algorithms. The dataset is used by the GroupLens research group at the university of Minnesota to create and maintain. In this paper, we choose the public 100K dataset, there are 943 users give 1682 films. the scores record. We split the dataset to train data and test data, in our experiments($M = 10$), In order to evaluate the quality of the top-N collaborative filtering algorithms we split each of the datasets into a training and test set, by randomly selecting one of the non-zero entries of each row to be part of the test set, and used the remaining entries for training. For example, we use $M - $ fold cross validation method, and choose one for testing, the other $M - 1$ data are used for training. Last, we create user interest model in train dataset, and forecast user behavior. To ensure that results of evaluation is not an over fitting, we need do M experiments, and put the average for the final evaluation.

4.2   Experimental Evaluation on Prediction Accuracy and Recall rate

First, the DF-KNN approach is evaluated on MovieLens dataset in order to demonstrate the approach's effectiveness. As we had no prior information about the optimal k for a specified application, the optimal value of k will be obtained by experimental tests in our algorithm, i.e. k, varied from 1 to 100. The metric was measured using the precision accuracy, the recall rate and the F measure..

Let $R(u), T(u)$ denote recommendation system for the user u recommend N items set and the items set that user u like in the test. we compute three metric, i.e. Precision(P), Recall(R) and F measure as:

$$P = \frac{\sum_u |R(u) \cap T(u)|}{\sum_u |R(u)|}, \tag{8}$$

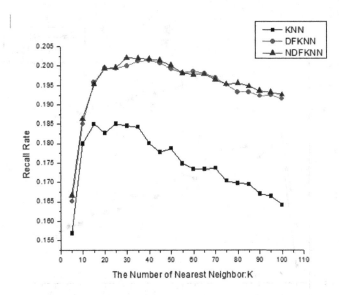

Fig. 3. Experimental result on the MovieLens for three algorithms recall rate

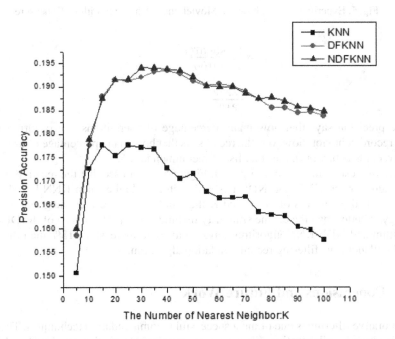

Fig. 4. Experimental result on the MovieLens for three algorithms precision accuracy

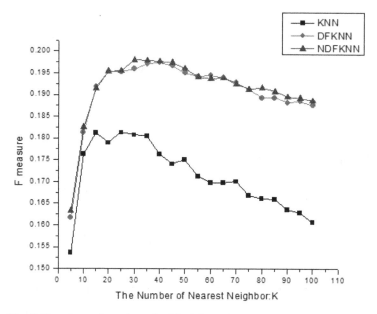

Fig. 5. Experimental result on the MovieLens for three algorithms F-measure

$$R = \frac{\sum_u |R(u) \cap T(u)|}{\sum_u |T(u)|}, \qquad (9)$$

$$F = \frac{2 \times P \times R}{P + R} \qquad (10)$$

The precision says that how many percentage of users-items that occurred in the score record behavior, however, the recall says that how many percentage of user-items score records behavior in the final list of recommendation.

From the results in Figure 3, Figure 4, Figure 5 we can see that the recall, precision, F-measure for DF-KNN and NDF-KNN algorithm is higher than KNN based cosine similarity and regardless of the varied K(the number of the nearest neighbors), so our strategy is better than the cosine similarity methods. The performance of the DF-KNN algorithm and NDF-KNN algorithm have an average increase of 13% than the KNN top-N collaborative filtering recommendation algorithm.

5   Conclusions and Future Work

Collaborative filtering is one of most successful recommendation techniques. There are memory-based collaborative filtering techniques such as the neighborhood-based collaborative algorithm; model-based collaborative filtering techniques such as Bayesian belief nets collaborative filtering algorithms, clustering collaborative algorithm etc. In our paper, we discuss the DF-KNN, NDF-KNN and KNN memory-based collaborative filtering techniques. We know memory-based collaborative filtering algorithms are easy to implement and have good performances for dense datasets[14].

As we have seen, DF-KNN and the NDF-KNN are the top-N collaborative filtering recommendation algorithm. Be different from the cosine similarity computing, we add the different factor to mix the similarity to compute. By the experiments, we see that it obtain a better effect than KNN based cosine similarity computing in precision and the recall, F-measure, and our provide recommendations whose quality is up to 13% better.

We believe that the DF-KNN and NDF-KNN collaborative filtering algorithms presented in this paper can be improved by item-based approaches also. The future researches will focus on the different factor and model-based collaborative filtering, such as SVD, Matrix Factorization. It will be measured and computed the user-user' interest behavior or the item-item similarity for much more applications. Because artificial data often not reliable due to the characteristics of collaborative filtering tasks, we need more real-world datasets from experiments to recommendation system research.

Acknowledgments. Funding for this research was provided in part by the Guangxi Key Lab of Multi-source Information Mining & Security under grants MIMS13-02, We like to thank anonymous reviewers for their valuable comments.

References

1. Linden, G., Smith, B., York, J.: Amazon.com recommendations: Item-to-item collaborative filtering. IEEE Internet Computing 7(1), 76–80 (2003)
2. http://www.grouplens.org/node/73
3. Resnick, P., Iacovou, N., Suchak, M., Bergstrom, P., Riedl, J.: GroupLens: An Open Architecture for Collaborative Filtering of Netnews. In: Proceedings of CSCW 1994, Chapel Hill, NC (1994)
4. Konstan, J., Miller, B., Maltz, D., Herlocker, J., Gordon, L., Riedl, J.: GroupLens:Applying Collaborative Filtering to Usenet News. Communications of the ACM 40(3), 77–87 (1997)
5. Shardanand, U., Maes, P.: Social Information Filtering: Algorithms for Automating 'Word of Mouth'. In: Proceedings of CHI 1995, Denver,CO (1995)
6. Hill, W., Stead, L., Rosenstein, M., Furnas, G.: Recommending and Evaluating Choices in a Virtual Community of Use. In: Proceedings of CHI 1995 (1995)
7. Breese, J.S., Heckerman, D., Kadie, C.: Empirical Analysis of Predictive Algorithms for Collaborative Filtering. In: Proceedings of the 14th Conference on Uncertainty in Articial Intelligence, pp. 43–52 (1998)
8. Konstan, J., Miller, B., Maltz, D., Herlocker, J., Gordon, L., Riedl, J.: GroupLens: Applying collaborative filtering to Usenet news. Communications of the ACM 40(3), 77–87 (1997)
9. Resnick, P., Iacovou, N., Suchak, M., Bergstrom, P., Riedl, J.: GroupLens: An open architecture for collaborative filtering of net news. In: Proceedings of CSCW (1994)
10. Koren, Y., Sill, J.: Collaborative filtering on ordinal user feedback. In: Proceeding IJCAI 2013 Proceedings of the Twenty-Third International Joint Conference on Artificial Intelligence, pp. 3022–3026. AAAI Press (2013)
11. Cai, Y., Leung, H.-F., Li, Q.: Typicality-Based Collaborative Filtering Recommendation. IEEE Transactions on Knowledge and Data Engineering 26 (2014)

12. Yu, Z., Xiaomin, Z., Qiwei, S.: A recommendation model based on collaborative filtering and factorization machines for social networks. In: 2013 5th IEEE International Conference on Date of Conference Broadband Network & Multimedia Technology (IC-BNMT), November 17-19 (2013)

13. Shi, Y., Larson, M., Hanjalic, A.: Collaborative Filtering beyond the User-Item matrix: A Survey of the State of the Art and Future Challenges. Journal ACM Computing Surveys(CSUR) 47(1) (June 2014)

14. Su, X., Khoshgoftaar, T.M.: A Survey of Collaborative Filtering Techniques. Hindawi Publishing Corporation Advances in Artificial Intelligence, Article ID 421425, 19 pages (2009)

15. Salton, G., McGill, M.: Introduction to Modern Information Retrieval. McGraw-Hill, New York (1983)

16. Zhang, S., Zhang, C., Yan, X.: Post-mining: maintenance of association rules by weighting. Inf. Syst. 28(7), 691–707 (2003)

17. Zhang, S., Qin, Z., Ling, C.X., Sheng, S.: Missing Is Useful': Missing Values in Cost-Sensitive Decision Trees. IEEE Trans. Knowl. Data Eng. 17(12), 1689–1693 (2005)

18. Wu, X., Zhang, S.: Synthesizing High-Frequency Rules from Different Data Sources. IEEE Trans. Knowl. Data Eng. 15(2), 353–367 (2003)

19. Wu, X., Zhang, C., Zhang, S.: Efficient mining of both positive and negative association rules. ACM Trans. Inf. Syst. 22(3), 381–405 (2004)

20. Wu, X., Zhang, C., Zhang, S.: Database classification for multi-database mining. Inf. Syst. 30(1), 71–88 (2005)

21. Zhao, Y., Zhang, S.: Generalized Dimension-Reduction Framework for Recent-Biased Time Series Analysis. IEEE Trans. Knowl. Data Eng. 18(2), 231–244 (2006)

22. Qin, Y., Zhang, S., Zhu, X., Zhang, J., Zhang, C.: Semi-parametric optimization for missing data imputation. Appl. Intell. 27(1), 79–88 (2007)

23. Zhu, X., Zhang, S., Jin, Z., Zhang, Z., Xu, Z.: Missing Value Estimation for Mixed-Attribute Data Sets. IEEE Trans. Knowl. Data Eng. 23(1), 110–121 (2011)

A Reputation-Enhanced Recommender System

Ahmad Abdel-Hafez, Xiaoyu Tang, Nan Tian, and Yue Xu

Queensland University of Technology, Brisbane, Australia
{a.abdelhafez,n.tian,yue.xu}@qut.edu.au,
xiaoyu.tang@connect.qut.edu.au

Abstract. Reputation systems are employed to provide users with advice on the quality of items on the Web, based on the aggregated value of user-based ratings. Recommender systems are used online to suggest items to users according to the users, expressed preferences. Yet, recommender systems will endorse an item regardless of its reputation value. In this paper, we report the incorporation of reputation models into recommender systems to enhance the accuracy of recommendations. The proposed method separates the implementation of recommender and reputation systems for generality. Our experiment showed that the proposed method could enhance the accuracy of existing recommender systems.

Keywords: Recommender System, Reputation System, Personalization, User profile, Enrichment, Merging Ranked Lists.

1 Introduction

Today, recommender systems are an essential part of many Web 2.0 sites. Therefore, enhancing the accuracy of current recommender systems can significantly improve services provided by these websites and positively affect customer satisfaction [1]. Recommender systems suggest a list of items that are personalized based on the opinions of similar members in a target user's local community, while reputation systems provide the opinions of the whole community. The systems are similar in that they both collect user item data [2]. However, to our knowledge, only modest efforts have been made to incorporate item reputations in the recommendation process [2]. We suggest that combining item reputations with recommendations can enhance the accuracy of recommender systems.

Recommender systems use two main filtering methods to generate lists. These are collaborative filtering and content-based filtering. The collaborative filtering (CF) method exploits user ratings to identify other users with similar tastes to the target user, and then predicts items the target user might like based on the similar-user preferences. An item-to-item correlation system is applied in content-based filtering (CBF). Thus, the system recommends an item to the target user if the item content is similar to the content of an item the target user has previously liked or viewed. Recently, a third, hybrid system which combines both methods has emerged. In this paper, we made use of the user-based CF recommendation method for evaluation. However, the proposed method was designed to be general and can be combined with other recommendation methods.

X. Luo, J.X. Yu, and Z. Li (Eds.): ADMA 2014, LNAI 8933, pp. 185–198, 2014.
© Springer International Publishing Switzerland 2014

User-based CF recommender systems assume that people have similar tastes and will respond similarly to various items. Therefore, data from similar users is employed to generate recommendations for the target user. Item-based CF is a different approach that uses item similarities. This method detects similar items, rather than similar users. Similar items are those the system expects groups of users to prefer. In general, the CF method depends on the accuracy of the similarity functions to find the most similar users or items. A lack of sufficient data about users or items (e.g., in the case of cold start situations or sparse datasets) can negatively affect the accuracy of the recommendation. In these cases, the predicted items generated by CF may not reflect the relevance of the predicted items to the target user. This means that an item with no relevance to the target user may still earn high prediction value.

An item's reputation is calculated by a specific aggregation method based on ratings given by many users. The final aggregated value reflects the opinions of the whole community toward a specific item. High item-reputation scores can indeed reflect the quality of an item in the view of the whole community. Consequently, these scores can predict whether more (interested) users will like the item. However, if applied alone, reputation scores do not predict whether an individual user will like an item with high accuracy. This is because the reputation score does not consider the individual's specific preference; therefore, reputation scores are not personalized. This means that the individual user may not like a highly reputed item.

In this paper, we introduce a method to combine the two separate systems and enhance the accuracy of the top-N recommendations generated by a CF recommender system. We conducted experiments to evaluate our method using a real dataset with different sparsity levels. The resulting accuracy of the proposed system was consistently better than the system that used only the CF method. The generality is one of the advantages of the proposed method, as any recommendation or reputation method can be used in conjunction. We employed a user-based CF method [4] and the Dirichlet reputation model [8]. Previous work in recommender and reputation systems is discussed in section 2. The detailed method is introduced in section 3. Section 4 describes the experiment and presents a discussion of the results.

2   Related Work

Recommender systems represent an essential component of many websites. Resnick and Varian suggested that recommender systems work similarly to word-of-mouth recommendations [9]. Resnick et al. introduced GroupLens, a system for the CF of net-news, in 1994 [10]. They defined the CF system as the one that helps people make choices based on the opinions of others. It worked, they said, by detecting users with similar tastes (neighbors) and then offering recommendations to the target user based on this neighbor data.

The CF approaches are classified into model-based, memory-based, and hybrid approaches. Memory-based algorithms depend on user profiles to predict ratings or to generate the top-N recommended items. The memory-based CF approaches can be classified into user-based and item-based approaches. The user-based approach generates a neighborhood of like-minded users (K-Nearest Neighbor [KNN]) based on profile similarity measures. Common similarity measures include the Pearson correlation coefficient (PCC) and the cosine similarity. These measures calculate predictions using weighted averages of the ratings given by other users in the neighborhood, where the weight is proportional to the similarity value between the target user and the neighborhood users. The same method can be applied for the item-based approach [9][11].

Model-based CF algorithms apply the user's earlier ratings to develop a model, which is then used to predict ratings for unrated items. The approaches used with the model-based CF include k-means clustering [12], the multiple multiplicative factor model [13], the Markov decision process [14], the restricted Boltzmann machine model [15], and the latent factor models based on the matrix factorization technique (i.e., singular value decomposition [SVD]) [16].

Reputation models use different methods to generate aggregated values that represent reputation scores; the Naïve model uses the average of the ratings of an item to measure the item's reputation, while many other models use the weighted average method as an aggregator to calculate item reputations based on item ratings. The weight can represent the user's reputation score, the time when the rating was given, or the distance between the current reputation score and the rating received [6,7]. Abdel-Hafez et al. [20,21] used the normal-distribution to generate weighted average reputation model which explicitly reflects the distribution of ratings of items.

The reputation model we used in our research was introduced by Jøsang and Haller and based on the Dirichlet probability distribution [8]. The authors used a cumulative vector $\vec{R}_y$ to represent the aggregated ratings for agent y. $\vec{R}_y = (R_y(i) \mid i = 1, ..., k)$ and $R_y(i)$ is the number of ratings of the level i. They added a decay factor to calculate the aggregate ratings, assuming that human agents change their behavior over time. They then calculated a single reputation score based on the multinomial probabilities derived from the aggregated ratings, which is defined in equation (1). $S_y(i)$ is the probability of rating i that other agents give to agent y. The overall reputation is calculated by equation (2), which is the weighted sum of the rating probabilities with weights $v(i)$ evenly distributed in the range [0,1].

$$\vec{S}_y = \left(S_y(i) = \frac{R_y(i) + Ca(i)}{C + \sum_{j=1}^{k} R_y(j)} \mid i = 1 \dots k \right) \tag{1}$$

$$\sigma = \sum_{i=1}^{k} v(i) \times S_y(i) , v(i) = \frac{i-1}{k-1} \tag{2}$$

where σ represents the overall reputation value, $\vec{S}_y$ represents the score vector of each rating level, C is a constant value, and $a(i)$ is the base rate, which is equal to $1/k$.

Recently, research has focused on improving the accuracy of recommender systems by combining the traditional recommendation methods with reputation systems [2]. Ku and Tai [17] proposed an exploratory framework to investigate the effects of recommendation and reputation systems on user purchase intentions toward recommended products. Their results showed that the opinions of other consumers influenced consumer attitudes about purchasing the recommended product through normative social influence. This revealed the effectiveness of recommendation systems that considered online reviews to influence consumers. Jøsang et al. [2] suggested that combining reputation scores with recommendation scores would provide more accurate recommendations. They used the same belief model they had introduced in a previous work [18] to calculate reputation scores. The authors mentioned different methods for combining resulted scores, but they adopted the Cascading Minimum Common Belief Fusion (CasMin) method. This method ensured that the values from the recommender and reputation systems would need to be both high to produce a high value in the CasMin fusion method.

3   A Reputation-Enhanced Recommender System

Our goal was to introduce a new reputation-aware recommender system that could enhance the accuracy of recommendations by filtering low-quality items based on reputation. The proposed method uses two ranked lists of items; the first list is generated by a recommender system, such as the user-based CF recommender system [4], and the second list is generated based on item reputations calculated using a reputation model, such as the Dirichlet reputation model [8]. The two ranked lists are then combined to enhance the accuracy of the recommendations. The proposed method is general, as it separates the implementation of the recommender system, the reputation system, and the merging process. In other words, we can apply any other recommendation method to generate the first list of items, and any other reputation model to generate the second list.

3.1   Definitions

The input of the proposed item reputation-aware recommender system is user ratings. To make this model generalizable and applicable for any website, we intentionally did not use any other content information. The reputation and recommendation scores are generated from the available ratings and are considered input data. The following definitions for the input data are used throughout the paper.

- Users: $U = \{u_1, u_2, \ldots, u_{|U|}\}$ is a set of users who have rated at least one item.
- Items: $P = \{p_1, p_2, \ldots, p_{|P|}\}$ is a set of items that are rated at least one time by a user in U.
- Users-Ratings: This is a user-rating matrix defined as a mapping $ur: U \times P \to [0, r]$. If the user u_i has rated the item p_j with rating a, then $ur(u_i, p_j) = a$; otherwise, $ur(u_i, p_j) = 0$ such that $0 < a <= r$, and r is the maximum rating.
- Item-Reputation Score: $S = \{s_1, s_2, \ldots, s_{|P|}\}$, where s_i is the reputation score for item p_i.
- Item Recommendation Score: $T = \{t_1, t_2, \ldots, t_{|P|}\}$ where t_i is the recommendation score for item p_i. This value is used to generate the candidate list of top-M recommendation using equation (3).

$$TopM_{u_i} = \underset{1 \to M}{argmax} \ T_{u_i}, u_i \in U \qquad (3)$$

3.2 Generating Recommendations by Merging the Two Ranked Lists

We propose two methods, the re-sorting and the weighted Borda-count methods, to combine the recommendation and reputation scores in order to generate the final top-N recommendations. Before discussing the merging methods, we want to emphasize the differences between the two lists, as this was the justification behind the selection of the two methods. The recommender-generated lists represent personalized item recommendations for users. The reputation lists reflect the community opinion about items and are not related to individual user preferences. Therefore, we assumed that recommendation lists would be more accurate than would be using only impersonalized reputation lists. Thus, we prioritized the use of the recommender-generated lists over the use of the reputation-based lists and chose recommendation lists as the primary candidate recommendations.

Re-sorting Method. In this method, we used the top-M recommendation list as the primary candidate recommendations for the target user. In the next step, we sorted the candidate list of items according to their reputation scores. In this case, we guaranteed that all the recommended items were personalized and that all the candidate recommendations were related to the user. We wanted to recommend the items with the best quality, measured by the reputation model, assuming that a higher-quality item would have a greater influence on consumer behaviors. Finally, we recommended the Top-N in the final list, $N < M$.

Fig. 1 shows an example of the re-sorting method. It reveals that any item in the top-M recommendation list is a candidate for recommendation, and the final list is selected based on the reputation scores. This method has the advantages of both recommender and reputation systems for two primary reasons. First, all the candidate items are personalized and related to the user tastes, since they have been generated by a recommender system. Afterward, sorting items based on reputation elevates the highly reputed items. In other words, the final recommended items will be more highly reputed and more closely related to the user preferences.

The value of M in this method has a great impact on the accuracy of the resulted recommendation. In this paper, we consistently use $M = 3 \times N$, this value is selected based on the experiment. The value of M will be tested further in section 5.4.

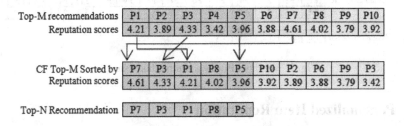

Fig. 1. Re-sorting method example

Weighted Borda-Count Method. The Borda-count (BC) [19] method is a popular voting method that uses points to represent the multiple selections of a candidate; that is, if the list contains N items, the top ranked item is given the score N and the next one is $N - 1$, and so on. Every item that is outside the Top-N list will receive a score of zero. This score is the BC. Two ranked lists are merged by summing up the two BCs of the same item in the two lists. The final ranked list is sorted based on the BC sums of items. For an item $p \in P$, the sum of the BCs for this item is denoted $SBC(p)$. The items with the highest SBC will appear at the top of the list. We adopted the BC method to merge a recommendation list and a reputation list. For a user u and an item $p \in P$, let $BC_{rec}(p)$ be the BC of p in the recommendation list and $BC_{rep}(p)$ the BC of p in the reputation list. Then, the sum BC was $SBC(p) = BC_{rec}(p) + BC_{rep}(p)$. The Top-N recommendation for the user u is defined in equation (4).

$$TopN_u^{BC} = argmax_{p \in P}^N SBC(p) \tag{4}$$

As mentioned, the recommendation list had a higher priority than the reputation list because the recommendation list was personalized. To distinguish the difference between the two lists and to emphasize the importance of the recommendation list, we proposed a weighted Borda-count (WBC) method by introducing a weight in the BC method. The weighted sum of BC $WSBC$ and the top-N recommendations are defined below, where $0 < \alpha < 1$:

$$WSBC(p) = \alpha \times BC_{rec}(p) + (1 - \alpha) \times BC_{rep}(p) \tag{5}$$

$$TopN_u^{WBC} = argmax_{p \in P}^N WSBC(p) \tag{6}$$

Based on the experiment, we set $\alpha = 0.7$. This value will give higher weight for the recommender system generated list. The example provided in Fig. 2 shows how this method works.

Weighted Borda Count For Recommendation List	P1	P2	P3	P4	P5	P6	P7	P8	P9	P10					
	7	6.3	5.6	4.9	4.2	3.5	2.8	2.1	1.4	0.7					

Weighted Borda Count For Reputation List	P13	P7	P3	P14	P1	P16	P21	P8	P11	P5					
	3	2.7	2.4	2.1	1.8	1.5	1.2	0.9	0.6	0.3					

Combined WBC	P1	P2	P3	P4	P5	P6	P7	P8	P9	P10	P13	P14	P16	P21	P11
	8.8	6.3	8.0	4.9	4.5	3.5	5.5	3.0	1.4	0.7	3.0	2.1	1.5	1.2	0.6

Top-N Recommendation	P1	P3	P2	P7	P4

Fig. 2. Weighted Borda-count method example

4   Personalized Item Reputation

An item's reputation is the global community opinion about it. At a specific time, the ranking of items based on item reputation is the same for all users. This means that the top ranked items on the reputation-based list are not necessarily the items that a particular user likes. If the item recommendation is determined only based on item reputation, then the same items with the highest reputations will be recommended for all users. Similarly, when this list is combined with the recommender-generated list, the items at the top of the reputation list will dictate the recommendation list and will always have advantages over all other items for all users.

The other major problem with using the reputation-ranked list in recommendation systems is that items with high reputations can appear in the recommendation list despite that they are outside the scope of the individual user's preferences. This causes a drop in system accuracy. Therefore, we propose a personalized reputation for the items to tackle this problem. The idea was to build a user-preference profile based on previous user ratings, and then to use this profile to filter the items that were outside the preference scope.

4.1   Implicit Item Category

To produce the personalized reputation-based item list, we needed to cluster items based on user ratings. Items that were rated by similar users are grouped in the same cluster. Each item cluster reflected certain common features shared by users with similar interests, and each was called an "implicit item category". In many application domains, the ontologies or taxonomies of the item/product categories are available; in such cases, we could use the provided ontology directly instead of undertaking a clustering method.

In the experiment, we assumed that each implicit item category reflected a certain user preference for items. We could build an individual user's preferences by collecting the categories of items the user had rated. We used only the positive ratings, as the items with negative ratings were not preferred. The implicit item category and user item preference are defined below:

- Implicit Item Categories $C = \{C_1, C_2, ..., C_{|C|}\}$ is the set of categories wherein items in P belong to $C_i = \{p | p \in P\}$ and $C_i \cap C_j = \emptyset$.
- User Item Preference $P_u = \left\{p | p \in P, ur(u, p) \geq \frac{(r+1)}{2}\right\}$, r is the maximum rating and contains all the user's preferred items.

A user item preference P_u is a set of items that the user has rated positively. Ratings that are larger than or equal to $\frac{(r+1)}{2}$ were considered positive ratings, where r was the maximum rating. Based on user item preferences, we defined user category preference as described below:

- User Category Preference $F_u = \{C_i | C_i \in C, (C_i \cap P_u) \neq \emptyset\}$ contains item categories in which the user's preferred or positively rated items belong. A user category preference F_u is a set of categories that are preferred by the user u.

The personalized reputation was defined as the degrading process for all the items in the reputation-ranked list that did not belong to the user preference. To apply the personalization to the reputation model, we degraded the reputation of all the items that belong to those categories which are not in the user preference. This step ensured that the items which are outside user's interest scope will not be recommended. The purpose of using reputation systems remained, as we did not change the reputation values of the other items, but kept the global community opinion. We only preserved or degraded the items based on the user's individual preferences. The derived resulting list is called personalized item reputation (PIR). The use of PIR guaranteed that the reputation-based ranked list was different for each group of users, which meant that a greater variety of items would be considered compared to the number of items considered using reputation without personalization. Equation (7) shows PIR_p calculation where S_p is the reputation for the item p.

$$PIR_p = \begin{cases} S_p, & p \in C_i, C_i \in F_u \\ 0, & Otherwise \end{cases} \quad (7)$$

4.2   User Preferences Enrichment

Using the PIR method raised a new concern regarding sparse datasets. Specifically, this was because it is common for a user to rate only a very small number of items. In this case, the number of categories in the user profile is low and, consequently, every item that belongs to other categories is degraded. We solved this problem by "enriching" the user preferences for those users whose profiles have less number of categories than the predefined minimum number. The minimum number of categories should be related to the average of ratings for a user. We enriched the profile with other categories that appeared in neighbors' profiles until the threshold number is reached. Then, we began to add categories according to the number of times they appeared in the neighbor' profiles. The result was an *enriched personalized item reputation (EPIR)* which was calculated exactly as the PIR but after performing the enrichment process described in algorithm 1. The user neighborhood is defined below.

- User Neighborhood $N_{u_i} = \left\{ u_j | u_j \in maxK\{sim(u_i, u_j)\} \right\}$ $u_j \in U$ is the set of nearest neighbors of user $u_i \in U$:, where maxK{} is required to obtain the top K large values.

5   Experiment

We conducted the top-N recommender system experiment. We aimed to demonstrate that combining item reputation with user-based CF could enhance the accuracy of the top-N recommendations.

Algorithm 1. Enrichment Process

---

1. for all users $u_k \in N_{u_i}$
2.   for all categories $C_j \in C$
3.   if $C_j \notin F_{u_i}$
4. $frequency[j] = frequency[j] + 1$
5. while $|F_{u_i}| < min$ //min is the minimum number of categories per user profile
6.   find C_j with $Max(frequency[j])$
7.   add C_j to F_{u_i}
8. $frequency[j] = 0$

5.1 Dataset

We used the MovieLens movie ratings dataset extracted from Grouplens.org. The dataset contained around 100,000 ratings on 1,682 movies provided by 943 users. We used this dataset in three different ways: 1) using all 2) using only 10%, and 3) using only 5% of the ratings. The purpose of the three tests was to observe the effects of this method on recommendation accuracy over dense and sparse datasets. The numbers of users and movies did not change in the three datasets; the only factor that changed was the number of ratings. Table 1 presents some of the statistics for each dataset.

For each of the generated datasets, the ratings were selected randomly per user. However, we defined the minimum number of ratings selected for any user at 10 for the ML10 dataset and five for the ML5 dataset. This was because, when we split the dataset into training and testing sets, we wanted to ensure that there was at least two items in testing for the ML10 dataset and 1 item for the ML5 dataset. For both da-tasets (ML10 and ML5), we generated 10 randomly selected additional subsets using the same method to perform a 10-fold experiment. We split each dataset into training and testing sets by randomly selecting 80% of each user's ratings into a training da-taset and the rest into a testing dataset. For the MLC dataset, we performed a 5-fold experiment, where each time a different 20% of the dataset was selected for testing. We calculated the average of the results at the end. The sparsity for the datasets was calculated using equation (8).

$$Sparsity = 1 - \frac{\# \, of \, Ratings}{\# \, of \, Users \times \# \, of \, Items} \tag{8}$$

5.2 Evaluation Metrics

We evaluated the top-N recommendation experiment with the globally used precision and recall metrics. The recommended item was considered a hit if it appeared in the user-testing dataset and the user has granted the item a rating $>= 3$. We used the value of 3 because any rating < 3 in a 5-star scale system employed by this system indicates that the user did not like the item. Finally, we used the F1-score metric to represent the results of both precision and recall.

Table 1. Datasets statistics

	MovieLens 5% (ML5)	MovieLens 10% (ML10)	MovieLens Complete (MLC)
Number of ratings	6,515	13,077	100,000
Sparsity	0.99589	0.99175	0.93695
Min ratings per user	5	10	20
Max ratings per user	36	73	737
Average ratings per user	6.849	13.867	106.044
Min ratings per movie	59	114	583
Max ratings per movie	0	0	1
Average ratings per movie	3.840	7.774	59.453

The three metrics were calculated during the experiment for each user. At the end, we used the average to provide one score for the recommender system. The higher the metrics result, the better the top-N recommendations.

5.3   Experiment Settings

We conducted the experiment in three runs for each dataset using the values of the recommendation list $top\text{-}N = 20$, the candidate list $top\text{-}M = 60$, and the nearest neighbors $K = 20$. The experiment comprised three parts: 1) the user-based CF, 2) the Dirichlet reputation model, and 3) the ranked lists proposed merging methods.

User-Based CF. We implemented the user-based recommender system introduced in [4] based on the best choices mentioned in the work. We first calculated the similarities between users [4] using the PCC method. After we obtained the similarity data, we generated the neighborhood of size k for each user by simply selecting the k users with the highest similarity values.

In addition, we noticed that adding a threshold value for the minimum number of common items between any two users could dramatically enhance the accuracy. Hence, we punished the user similarities between users who shared fewer than the predetermined value. We set threshold to 30, 3 and 1 for the MLC dataset, ML10 and ML5, respectively. The selection of these values is based upon the average ratings per user, which are presented in Table 1.

Next, we generated the item predictions to select the top-N items. According to [4], the best results for the top-N recommendations were achieved using the most frequent items in the neighborhood. If items had similar frequencies, we sorted them using the prediction value. At the end of this stage, we had developed a ranked list of recommended items.

Personalized Item's Reputation. The second part of the experiment comprised generating a ranked list of items using the personalized items reputation. We implemented and tested the Dirichlet reputation model, which was calculated using equations (1, 2) [8]. We chose this model because it added uncertainty to the reputation score, which can provide better results when the number of ratings per item is low. It is worth mentioning that the implemented reputation model affected the final result at this stage.

We used the movie categories provided with the MovieLens dataset to generate user category preferences. Then, the ranked list was generated after enriching the user preferences and personalizing the item reputation-ranked list. The personalization and enrichment processes are explained in details in section 4.

Combining Two Ranked Lists. We implemented the two proposed merging-ranked-lists methods. Each one of these methods was used with four reputation-generated ranked lists. The reputation methods tested were:

1. DIR: the Dirichlet reputation model
2. PIR: the personalized item reputation; we used the (DIR) method as the basic reputation method
3. EPIR: an enriched version of the PIR. We first checked the number of categories rated by the user, and if the number was less than the determined number, we proceeded to the enrichment process. Based on the experiment we used (6,12,24) as minimum numbers of categories for the ML5, ML10, and MLC datasets, respectively.

5.4   Results and Discussion

Table 2 shows the precision, recall, and F1-scores for each of the implemented methods over the three tested datasets. It also includes the results from the CasMin method proposed by Jøsang et al. in [2]. First, we will discuss the effects of the merging method adopted on the CF accuracy. Afterwards, we will examine the effects of the different reputation methods used.

Discussion of Merging Methods Results. The first thing we noticed from the results was that the re-sorting method produced the best results among all the merging methods when the personalized reputation scores were combined. It is because the re-sorting incorporated the CF candidate list's top-M as the basic list, all the candidate movies were personalized for the user. Surprisingly, when we sorted them according to reputation, the final recommended items were more relevant if the reputation was personalized; otherwise, the recommended items were less relevant. The only explanation was that the items in the top-M list generated by the CF did not belong to the set of categories that the user preferred. However, the personalized reputation system was able to filter those items so that better results could be obtained. From this observation, we can say that items' reputation can have a positive impact on recommendation accuracy if they were personalized.

In contrast, the WBC method obtained best performance with the non-personalized reputation scores, although its results were still not good as the re-sorting method. The WBC method often incorporated items with high reputation scores even if they did not appear in the recommendation-candidate list's top-M. Thus, when the reputation was personalized, the WBC results were better than the CF method results. This was because even the high reputation items populated outside the CF list remained within the categories of items the user preferred. However, the re-sorting method still performed better than the WBC.

We noticed that most of the implemented methods had lower F1-scores than the CF method; this proved that reputation itself was not an important factor which is associated with the recommendation accuracy. In contrast, using personalized versions of reputation lists could significantly enhance the reputation accuracy. We found that the two methods (re-sorting and WBC) enhanced CF accuracy when the PIR and EPIR reputations were used.

Discussion of Reputation Methods Used. It is now clear that the proposed personalized methods of reputation model generated better results than did the original reputation lists. We had two versions of this kind of reputation: the PIR and the EPIR. Using the ML5 and ML10 datasets, the EPIR produced slightly better results than did the PIR method. This meant that the neighbor categories could be used to enrich the user categories by increasing the diversity of recommendations, while still producing more accurate results.

When we used the MLC dataset, both methods produced exactly the same results. This was because no enrichment was required for these dense datasets. Moreover, the Dirichlet reputation model produced results different from those of the Naïve method, which indicated that the reputation method should be carefully selected to enhance results.

Table 2. Results of top-N recommendation accuracy using three datasets

Method	Merging Method	ML5			ML10			MLC		
		Precision	Recall	F1-score	Precision	Recall	F1-score	Precision	Recall	F1-score
CF	N/A	0.0061	0.0684	0.0112	0.0079	0.0723	0.0142	0.0283	0.0229	0.0253
CF - DIR	CasMin [2]	0.0004	0.0032	0.0007	0.0004	0.0034	0.0007	0.0063	0.0087	0.0073
CF - DIR	Re-sorting	0.0077	0.0611	0.0137	0.0089	0.0785	0.0160	0.0472	0.0431	0.0451
CF - PIR		0.0182	0.1661	0.0328	0.0259	0.0903	0.0402	0.0598	0.0602	0.0600
CF - EPIR		0.0201	0.1812	0.0362	0.0259	0.0920	0.0404	0.0598	0.0602	0.0600
CF - DIR	Weighted Borda Count	0.0075	0.0665	0.0134	0.0079	0.0729	0.0143	0.0336	0.0283	0.0308
CF - PIR		0.0131	0.1249	0.0237	0.0146	0.0858	0.0249	0.0465	0.0448	0.0456
CF - EPIR		0.0136	0.1301	0.0246	0.0149	0.0885	0.0255	0.0465	0.0448	0.0456

Impact of Varying Top-M Value. The size of the candidate recommendation list's top-M had a huge effect on the accuracy of the results using the re-sorting merging method. In this test, we varied the values of M to compare the accuracy of the results and to choose the best value of M. Fig. 3 displays the results, starting with the size of $M = N$, which behaved exactly as the CF method.

We noticed that the system accuracy was better when we increased the size of the candidate list to a certain level. After that, the curve began declining until the results were worse than those of the CF method. These results made sense, as when $M \gg N$ the effect of the reputation system became stronger than the CF system, the accuracy was low. Selecting the optimal size for M was important to obtain the best results; in our experiment, $M = 3 \times N$ is chosen for the ML10 dataset.

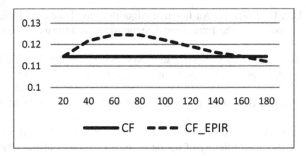

Fig. 3. Re-sorting method F1-scores with different top-M values using ML10 dataset

6   Conclusions

In this paper, we presented a new method for enhancing the accuracy of top-N recommendations using reputation systems. We introduced a personalized reputation method to render the utility of using reputation to improve the performance of recommender systems. Based upon the evaluations, we have important findings to share. First, reputation models do not necessarily produce better results when they are incorporated with recommender systems. On the contrary, reputation models without personalization can reduce the accuracy of the recommendations. The second significant finding is that personalized reputation scores can be very helpful for improving the accuracy of recommender systems.

References

1. Zhou, X., Xu, Y., Li, Y., Josang, A., Cox, C.: The state-of-the-art in personalized recommender systems for social networking. Artificial Intelligence Review 37(2), 119–132 (2012)
2. Jøsang, A., Guo, G., Pini, M.S., Santini, F., Xu, Y.: Combining recommender and reputation systems to produce better online advice. In: Torra, V., Narukawa, Y., Navarro-Arribas, G., Megías, D. (eds.) MDAI 2013. LNCS, vol. 8234, pp. 126–138. Springer, Heidelberg (2013)
3. Papagelis, M., Plexousakis, D., Kutsuras, T.: Alleviating the sparsity problem of collaborative filtering using trust inferences. In: Herrmann, P., Issarny, V., Shiu, S.C.K. (eds.) iTrust 2005. LNCS, vol. 3477, pp. 224–239. Springer, Heidelberg (2005)
4. Sarwar, B., Karypis, G., Konstan, J., Riedl, J.: Analysis of recommendation algorithms for e-commerce. In: Proceedings of the 2nd ACM Conference on Electronic Commerce, pp. 158–167 (2000)
5. Liang, H., Xu, Y., Li, Y., Nayak, R., Tao, X.: Connecting users and items with weighted tags for personalized item recommendations. In: Proceedings of the 21st ACM Conference on Hypertext and Hypermedia, pp. 51–60 (2010)
6. Riggs, T., Wilensky, R.: An Algorithm For Automated Rating Of Reviewers. In: Proceedings of the First ACM/IEEE-CS Joint Conference on Digital Libraries, pp. 381–387 (2001)

7. Ayday, E., Lee, H., Fekri, F.: An Iterative Algorithm For Trust And Reputation Management. In: Proceedings of the International Symposium on Information Theory, pp. 2051–2055 (2009)

8. Jøsang, A., Haller, J.: Dirichlet Reputation Systems. Proceedings of the Second International. In: Proceedings of Conference on Availability, Reliability and Security, pp. 112–119 (2007)

9. Resnick, P., Varian, H.R.: Recommender systems. Community of ACM 40(1997), 56–58 (1997)

10. Resnick, P., Iacovou, N., Suchak, M., Bergstrom, P., Riedl, J.: GroupLens: an open architecture for collaborative filtering of net-news. In: Proceedings of the 1994 ACM Conference on Computer Supported Cooperative Work, pp. 175–186 (1994)

11. Abdullah, N.: Integrating collaborative filtering and matching-based search for product recommendation, Doctor of Philosophy, Faculty of Science and Technology, Queensland University of Technology, QUT ePrints (2012)

12. Shepitsen, A., Gemmell, J., Mobasher, B., Burke, R.: Personalized recommendation in social tagging systems using hierarchical clustering. In: Proceedings of the 2008 ACM Conference on Recommender Systems, pp. 259–266 (2008)

13. Marlin, B., Zemel, R.S.: The multiple multiplicative factor model for collaborative filtering. In: Proceedings of the Twenty-First International Conference on Machine Learning, pp. 73–81 (2004)

14. Shani, G., Heckerman, D., Brafman, R.I.: An MDP-based recommender system. Journal of Machine Learning Research 6(2006), 453–460 (2006)

15. Salakhutdinov, R., Mnih, A., Hinton, G.: Restricted Boltzmann machines for collaborative filtering. In: Proceedings of the 24th International Conference on Machine Learning, pp. 791–798 (2007)

16. Koren, Y.: Factorization meets the neighborhood: a multifaceted collaborative filtering model. In: Proceedings of the 14th ACM SIGKDD International Conference on Knowledge Discovery and Data Mining, pp. 426–434 (2008)

17. Ku, Y.-C., Tai, Y.-M.: What Happens When Recommendation System Meets Reputation System? The Impact of Recommendation Information on Purchase Intention. In: International Conference on System Sciences (HICSS), pp. 1376–1383 (2013)

18. Jøsang, A.: A logic for uncertain probabilities. International Journal of Uncertainty. Fuzziness and Knowledge-Based Systems 9(2001), 279–311 (2001)

19. Dummett, M.: The Borda count and agenda manipulation. Social Choice and Welfare 15(2), 289–296 (1998)

20. Abdel-Hafez, A., Xu, Y., Jøsang, A.: A normal-distribution based reputation model. In: Eckert, C., Katsikas, S.K., Pernul, G. (eds.) TrustBus 2014. LNCS, vol. 8647, pp. 144–155. Springer, Heidelberg (2014)

21. Abdel-Hafez, A., Xu, Y., Jøsang, A.: A rating aggregation method for generating product reputations. In: Proceedings of the 25th ACM Conference on Hypertext and Social Media, pp. 291–293 (2014)

An Empirical Methodology to Analyze the Behavior of Bagging

Fábio Pinto, Carlos Soares, and João Mendes-Moreira

INESC TEC/Faculdade de Engenharia, Universidade do Porto
Rua Dr. Roberto Frias, s/n
Porto, Portugal 4200-465
fhpinto@inescporto.pt csoares@fe.up.pt jmoreira@fe.up.pt

Abstract. In this paper we propose and apply a methodology to study the relationship between the performance of bagging and the characteristics of the bootstrap samples. The methodology consists of 1) an extensive set of experiments to estimate the empirical distribution of performance of the population of all possible ensembles that can be created with those bootstraps and 2) a metalearning approach to analyze that distribution based on characteristics of the bootstrap samples and their relationship with the complete training set. Given the large size of the population of all ensembles, we empirically show that it is possible to apply the methodology to a sample. We applied the methodology to 53 classification datasets for ensembles of 20 and 100 models. Our results show that diversity is crucial for an important bootstrap and we show evidence of a metric that can measure diversity without any learning process involved. We also found evidence that the best bootstraps have a predictive power very similar to the one presented by the training set using naive models.

Keywords: Ensemble Learning, Bagging, Diversity, Metalearning.

1 Introduction

Bagging is an ensemble learning technique that allows to generate multiple predictive models and aggregate their output to provide a final prediction [1]. Typically, the aggregation function is the mean (if the outcome is a quantitative variable) or the mode (if the outcome is a qualitative variable). The models are built by applying a learning algorithm to bootstrap replicates of the learning set. Empirical studies show that bagging is able to reduce the error in comparison with single models and is very competitive with other ensemble learning techniques [2].

In this paper, we propose and apply a methodology to study the performance of the bagging algorithm. We investigate the reasons that affect the influence of a bootstrap (and corresponding model) in the space of sub-ensembles. For that, we compute specific bootstrap characteristics. These measures are then compared

X. Luo, J.X. Yu, and Z. Li (Eds.): ADMA 2014, LNAI 8933, pp. 199–212, 2014.

with the importance of a bootstrap[1] on the predictive performance of ensembles that include the model generated by applying a learning algorithm to it.

Our study is based on Metalearning (MtL) techniques. MtL is the study of principled methods that exploit metaknowledge to obtain efficient models and solutions by adapting machine learning and data mining processes [3]. We aim to gain knowledge about the performance and intrinsic behavior of the bagging algorithm. So, we use MtL in a descriptive approach instead of the more typical predictive framework. For that, we adapted several metafeatures already proposed in the literature [4,5] and we also introduce some new ones that are very specific of our problem domain.

We tested our proposed methodology empirically by executing experiments with 53 classification datasets collected from the UCI repository [6]. For each dataset, we generated bagged ensembles of decision trees with 20 and 100 models. We were able to generate and test all possible combinations of the ensembles with 20 models. However, for computational reasons, we were forced to sample the number of combinations tested for ensembles with 100 models. We present results that indicate the validity of this sampling procedure. All the insights collected from the metadata describing ensembles with 100 models are compared with the 20 models case. This allowed a validation of our sampling procedure.

Given the descriptive aim of our work, we used standard exploratory data analysis procedures to extract knowledge from the metadata that we generated. The main contributions of this paper are: 1) a methodology based on an extensive experimental procedure and on MtL for empirically studying the performance of bagging; 2) new metafeatures that characterize the relationship between bootstrap samples and the complete training data; 3) an exploratory MtL approach using visualization and a statistical method applied to UCI datasets, yielding interesting observations concerning the relationship between the characteristics of the bootstrap sample and the performance of the bagging ensemble.

This paper is organized as follows. Section 2 describes the related work in the field of ensemble learning particularly focused on the bagging algorithm. Section 3 presents the empirical methodology for studying ensembles and a study of the representativeness of the results obtained by sampling from all the possible ensembles with 100 models. Section 4 describes the MtL approach used in this work as well as the metafeatures. In Section 5, we present the descriptive study on the characteristics of a bootstrap and its importance on the predictive performance of an ensemble. Finally, Section 6 concludes the paper with some final remarks and future work.

2   Related Work

Several papers propose theoretical frameworks that provide important insights on the effectiveness and reasons behind the success of bagging. Breiman [1] argued that aggregating can transform good predictors into nearly optimal ones,

---

[1] We define an important bootstrap as a bootstrap which its correspondent model belongs to the best combinations of tested ensembles in terms of performance.

highlighting however the importance of using unstable learners (small variations in the training set must generate very distinct models [7]).

Friedman [8] related bagging with the bias and variance decomposition of the error. Shortly, the error is split into two components: bias, associated with the intrinsic error of the learner generalization ability; and variance, associated with the error assigned to the variation in the model from one bootstrap to another. In the context of bagging, Friedman claimed that the variance component is reduced (because of the bootstrapping procedure) without changing the bias.

Domingos [9] presented two alternative hypotheses for the success of bagging: although rejecting the possibility of approximation to the optimal procedure of Bayesian model averaging with an appropriate implicit prior probability distribution, he proved that bagging works effectively because it shifts the prior to a more appropriate region of model space. However, Domingos recognized one important fact: none of the above frameworks relate the success of bagging with the domain characteristics.

Friedman and Hall [10] confirmed Breiman's claim by showing that bagging is most successful when used with highly nonlinear estimators such as decision trees and neural networks. In this study they also found evidence that sub-sampling is virtually equivalent to traditional bootstrap sampling. Bühlmann and Yu [11] provided theoretical explanations of the same claim.

Grandvalet [12] provided an interesting study in which he found that bagging equalizes the influence of examples in a predictor. Bootstrapping a dataset implies that fewer examples have a small influence, while the highly influential ones are down-weighted. The author claims that bagging is useless when all examples have the same influence on the original estimate, is harmful when high impact examples improve accuracy, and is otherwise beneficial.

For the ensemble learning literature, it is important to gain understanding of ensembles performance. One way to understand the behavior of learning processes is MtL. Some papers use MtL in a more descriptive manner with the intention of extracting interesting and useful knowledge of a specific domain. Kalousis et al. [13] used MtL for a meta-descriptive symmetrical study in which they found similarities among classification algorithms and datasets. In another domain, Wang et al. [14] published a paper that focuses on rule induction for forecasting method selection by understanding the nature of historical forecasting data. They provide useful rules that rely on metafeatures for suggesting a specific method. Our application of MtL in this paper resembles more these two papers.

3   Empirical Methodology to Characterize Bagging Performance

Formally, an ensemble F gathers a set of predictors of a function f denoted as $\hat{f}_i$. Therefore, $F = \{\hat{f}_i, i = 1, ..., k\}$ where the ensemble predictor is defined as $\hat{f}_f$.

We propose a methodology to empirically analyze the behavior of bagging. Given a set of k bootstrap samples (also referred to in this paper as bootstraps, for simplicity), we estimate the empirical distribution of performance of the bagging ensembles that can be generated from all elements of its power set. In other words, we estimate the empirical distribution of performance of all possible ensembles of size 2, 3, ... k that can be generated from those k bootstraps.

This distribution can be used to study the role of a given bootstrap (and respective predictive model $\hat{f}_i$) in the performance of $2^k - 1$ possible ensembles, as done in this paper. Additionally, the distribution can be used to analyze the joint relationship between the bootstrap samples in each ensemble and its performance.

It is easy to understand that is impossible to execute the complete set of experiments for ensembles with a realistically large size, such as k=100, given that the number of combinations to test is $2^k - 1$. Therefore, the only possibility is to estimate the distribution of the performance of all ensembles that can be generated with the set of k bootstraps by sampling from its power set. To investigate the validity of this approach, we carried out the following study.

3.1   Estimating the Distribution of Performance by Sampling from the Power Set of Bootstraps

To validate our methodology based on sampling, we executed the full methodology with k=20 and then we studied the impact of sampling. Based on these results, we extrapolate our findings for k=100. We used the Kullback-Leibler Divergence [15] (KLD) to measure the difference between the probability distributions P and Q, defined as $D_{KL}(P||Q) = \sum_n P_n log_2 \left(\frac{P_n}{Q_n} \right)$ where P is the results obtained by testing $2^k - 1$ combinations of k models and Q a sample of those results. Since the KLD measure is not symmetric, we averaged the divergences, then $D_{KL} = \frac{D_{KL}(P||Q) + D_{KL}(Q||P)}{2}$. Given that this experiment implies a large component of randomness, we executed each sampling procedure 100 times and we averaged the values obtained.

In the first experiment, for each dataset, we progressively increased the sampling proportion and systematically computed the KLD between the sample and the population with k=20. Figure 1 shows, as expected, that as the sampling proportion increases, the divergence between the samples and respective population decreases. One can see that for most of the datasets the fall of the curve is rather fast. Figure 2 shows the same result but the values for the 53 datasets are averaged for each sampling proportion. Again, as expected, the standard deviation and the mean KLD decreases as the proportion of sampling increases.

To assess the hypothesis that increasing the number of models in an ensemble changes the sampling results, we repeated the experiment for ensembles with different k values, from 10 to 19. Figure 3 shows a slight increase in the divergence between the samples of equal proportion and respective populations as k increases. This result is expected given that the introduction of a new model can possibly change the inter-relations between the models and therefore affect the performance of some subsets of models. However, all the curves[2] present a very

---

[2] Estimated using a Local Polynomial Regression (LOESS).

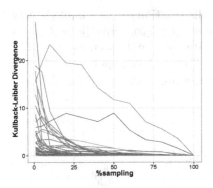

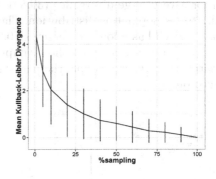

Fig. 1. KLD between % of sample and population. Each line represents a different dataset.

Fig. 2. Mean KLD (and standard deviation, through vertical lines) between % of sample and population

similar pattern. This is indicative that a similar curve could be assumed for an ensemble with $k=100$.

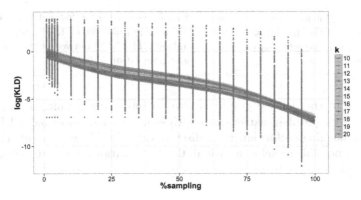

Fig. 3. Sampling and Kullback-Leibler Divergence, averaged for all datasets

3.2 Discussion

Although the evidence showed previously gives us confidence in the sampling variant of our methodology, we still lack sensitivity on the KLD measure to be able to interpret the values of this experiment more reliably. It is difficult by just looking to the graphs if we are actually losing significative information by sampling.

Figures 4 and 5 show two density graphs for a 10% sample and the corresponding complete population. The first concerns the *dis* dataset. One can see that even for a very large divergence (30.89), the distribution of the sample is very similar to the population distribution. The second graph concerns the *acetylation* dataset, which has a lower divergence (0.23). Most of the datasets show similar values of divergence between their samples and respective populations. This is indicative that we can sample the performance of an ensemble with $k=100$ and proceed our study.

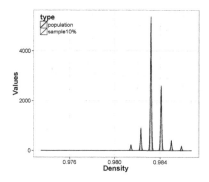
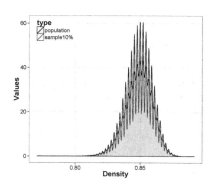

Fig. 4. Density plot for a 10 % sample and population of the *dis* dataset. The KLD between this sample and population is 30.89.

Fig. 5. Density plot for a 10 % sample and population of the *acetylation* dataset. The KLD between this sample and population is 0.23.

The shape of the density graphs is also an interesting result. Both graphs presented a very peculiar pattern of multiple peaks. This is explained by the fact that the bagging performance is a discrete variable. The number of accuracy values that is possible to achieve with all the combinations of a finite set models is limited.

4　Metalearning to Understand Bagging

The methodology presented in Section 3 can be used to provide insights on the types of bootstraps, in terms of how they contribute to the performance of the ensemble. Additionally, it can be combined with a MtL approach to analyze the relationship between the characteristics of the bootstrap sample and the performance of the ensemble.

　　The main issue in MtL is defining the metafeatures. The most used ones are simple, statistical and information-theoretic metafeatures [3]. In this group we can find the *number of examples* of the dataset, *correlation between numeric attributes* or *class entropy*, to name a few. The use of these kinds of metafeatures provides not only informative data characteristics but also interpretable knowledge about the problems. Other kinds of metafeature are model-based [16]. These capture some characteristic of a model generated by applying a learning algorithm to a dataset, *i.e.*, the number of leaf nodes of a decision tree. Finally, a metafeature can also be a landmarker [5]. These are generated by making a quick performance estimate of a learning algorithm on a particular dataset.

For this work, we relied on simple, statistical, information-theoretic and landmarker metafeatures. For the first group, we selected several metafeatures already present in the literature which were first used for MtL in the METAL and Statlog projects [3]. We also introduce a new metafeature based on the Jensen-Shannon distance [17] between a bootstrap and the training set. This metafeature aims to measure how different is the bootstrap from the original dataset. It can also be seen as a diversity measure that focuses directly on the bootstrap sample and not on the predictions made by the generated model.

We used two landmarkers: a decision stump and a Naive Bayes classifier. Given the different bias of the algorithms, it is expected that the metafeatures can help capture different patterns. We also used two diversity measures proposed in the ensemble learning literature: the Q-Statistic [18] and Classifier Output Difference [19] (COD) measures. Kuncheva et al. [18] state that the Q-Statistic is the diversity measure with greater potential for providing useful information about ensemble performance. We adapted the Q-Statistic to the specificities of our problem. Kuncheva et al. present it as a metric to measure the diversity of an ensemble. We use it to measure the diversity between the predictions of two models: one generated by applying a learning algorithm to a bootstrap (b) and the other to the original dataset (d). Using such a measure in this study gives a different perspective on its usefulness. Formally, our adapted Q-Statistic is $Q_{b,d} = \frac{N^{bb}N^{dd} - N^{db}N^{bd}}{N^{bb}N^{dd} + N^{db}N^{bd}}$ where each element is formed as in Table 1.

Table 1. Relationship between a pair of classifiers

	$f_b$correct	$f_d$correct
$f_b$correct	N^{bb}	N^{bd}
$f_d$correct	N^{db}	N^{dd}

The COD metric has been proposed as a measure to estimate the potential of combining classifiers

$$C\hat{O}D_T(\hat{f}_b, \hat{f}_d) = \frac{\sum_{x \in Ts} \begin{cases} 1, & \text{if } \hat{f}_b(x) = \hat{f}_d(x) \\ 0, & \text{otherwise} \end{cases}}{|Ts|}$$

in which Ts is test or validation set.

Lee and Giraud-Carrier [20] published a paper on unsupervised MtL in which they study the application of several diversity measures for ensemble learning as a distance function for clustering learning algorithms. In their experiments, only one measure, COD, presents results that indicate that it can be a good measure for this kind of task. This is indicative that the metric can also be useful in our problem.

In summary, the metafeatures used for this work are: *number of examples* of a bootstrap, *number of attributes, proportion of symbolic attributes, proportion of missing values, proportion of numeric attributes with outliers, class entropy, average entropy between symbolic attributes, average mutual information between symbolic attributes and the class, average mutual information between pairs of symbolic attributes, average absolute correlation between numeric attributes, average absolute skewness between numeric attributes, average kurtosis between numeric attributes, canonical correlation* of the most discriminating single linear combination of numeric attributes and the class distribution, *Jensen-Shannon distance* between the dataset and bootstrap, decision stump *landmarker*, Naive Bayes *landmarker*, Q-Statistic and COD.

The experiments that we carried with the UCI datasets allowed to collect results from the performance of the bagging algorithm in very distinct learning problems. Given that our goal is to understand the importance of each model (and respective bootstrap) in the ensemble space, we need to aggregate the results obtained for each one of them and compute an estimate of importance.

We adapted the measure NDCG [21] (Normalized Discounted Cumulative Gain) to form our metatarget. We consider the performance of the ensembles (in decreasing order) to which the bootstrap k belongs, for each dataset, as $acc_{1,d}$, $acc_{1,d}$, ..., $acc_{n,d}$ where n represents an ensemble and d a dataset. Therefore, for each bootstrap k of the dataset d, we calculate the respective DCG

$$DCG_{k,d} = \sum_{n=1}^{100} acc_{n,d} + \sum_{101}^{n} \frac{acc_{n,d}}{log_{100}n}$$

and we normalize it by an ideal ranking ($IDCG_d$) in which the best ensembles (testing all bootstraps) for each dataset are selected. Then,

$$NDCG_{k,d} = \frac{DCG_{k,d}}{IDCG_d}$$

In order to allow a more concise exploratory analysis of the metadata, we discretized the metatarget. This process is done using the Fisher-Jenks [22] algorithm. The method was chosen since it is well suited to find the optimal partition into different classes of a continuous variable.

5   What Makes a Good Bootstrap?

Most of the metafeatures described characterize the bootstrap in isolation. For instance, the *class entropy* metafeature focuses on the bootstrap and does not relate it with the original dataset. One exception is the diversity measure that characterizes the difference between a set of predictions from a model learned on a bootstrap and another model learned in the original training set. Furthermore, some metafeatures computed for bootstraps of the same dataset show very similar values. For instance, it is not expected that the class entropy varies significantly across bootstrap samples of the same training set. Additionally, the range of values of a metafeature for different datasets is expected to be quite different. However, we need metafeatures with values in comparable ranges across datasets to be able to extract useful insights with our MtL approach. In summary, we need to transform the metafeatures in order for them to 1) discriminate between bootstrap samples from the same dataset and 2) be comparable across datasets.

So, we applied one of two simple transformations to each meta-variable: **1**) proportional difference of the metafeature computed for the bootstrap in relation to the metafeature computed for the original training set ($\frac{metafeature_d - metafeature_b}{metafeature_d}$) **2**) proportional difference of the metafeature computed for the bootstrap in relation to the maximum value computed for all the bootstraps of the dataset. Then, it is rescaled in order to keep the natural interpretation of the variables by subtracting ($1 - \frac{Max(metafeature_b) - metafeature_b}{Max(metafeature_b)}$). The first transformation was applied to all the metafeatures except the Jensen-Shannon distance, Q-Statistic and COD. To these metafeatures, since we could not compute them in original training set, we applied the second transformation.

The results of the discretization of the metatarget can be verified in Figures 6 and 7. One can see that the discretized values are grouped in a descending order of the value of the metatarget, as it is desirable. Through the analysis of the results we will mention the concept of importance. We consider that bootstraps of class A are more important than bootstraps of class B or C, therefore, we are interested in understanding the characteristics of important bootstraps.

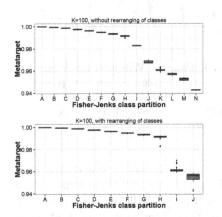

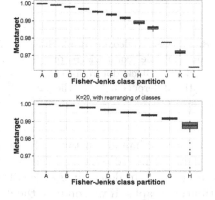

Fig. 6. Boxplot of numeric metatarget (k=100) vs classes found by Fisher-Jenks algorithm

Fig. 7. Boxplot of numeric metatarget (k=20) vs classes found by Fisher-Jenks algorithm

However, some classes group very few observations. It can become problematic to analyze those groups. We decided to merge these classes and reduce the sparsity of the discretization. The graphs at the bottom of Figures 6 and 7 show the boxplots of the metatarget variable after that rearrangement.

5.1 Exploratory Analysis

To assist our analysis, we used Kruskal-Wallis one-way analysis of variance with Wilcoxon pairwise rank sum test as post hoc procedure (0.95 confidence interval) with Holm adjustment method. This analysis was carried to check for significant different medians of the metafeatures among the classes of the metatarget. Figures 8 and 9 show the results of Wilcoxon test for the metafeatures that the Kruskal-Wallis test showed a *p-value* below 0.05. One can see that the metafeatures *avg.symb.pair.mutual.information*, *nb.landmarker* and *q.statistic* are the most discriminative ones. We will focus on metafeatures that are more interesting for the ensemble learning literature and withdraw the analysis of the remaining metafeatures due to space limitations.

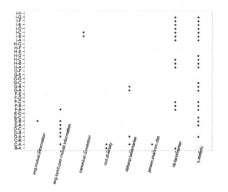 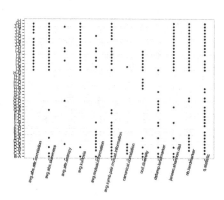

Fig. 8. Pairwise Wilcoxon Rank Sum test for multiple comparison procedures (k=20). Black dot represents a significative difference between the pair of classes.

Fig. 9. Pairwise Wilcoxon Rank Sum test for multiple comparison procedures (k=100). Black dot represents a significative difference between the pair of classes.

The Jensen-Shannon distance shows a very interesting pattern that can be verified in Figure 10. One can see that the gradient of the colors associated with each class (in descending order of importance) is reflected in the density distribution graphs. If we compare the distribution of the most important bootstraps (classes A, B, C...) with the less important ones it is clear that, as the Jensen-Shannon distance decreases, the importance of the bootstraps associated with that value also decreases. In other words, bootstraps that are very similar with the original training dataset do not generate a useful model for a bagging ensemble. This is not new for the ensemble learning literature, however, here we measure diversity without any learning process involved. However, this result can not be verified in Figure 11 which represents the metadata with k=20. This can be explained by the fact that since the k=20 experiment generates fewer bootstraps it is harder to find bootstraps with low importance (we can see in

Figure 7 that the range of the metatarget in this experiment is smaller than in the $k=100$ experiment). However, this remains to be confirmed, which could be done by repeating these experiments for other values of k.

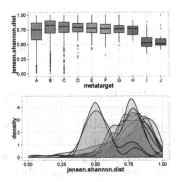

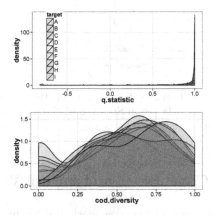

Fig. 10. Boxplot and density distribution of the Jensen-Shannon distance with $k=100$

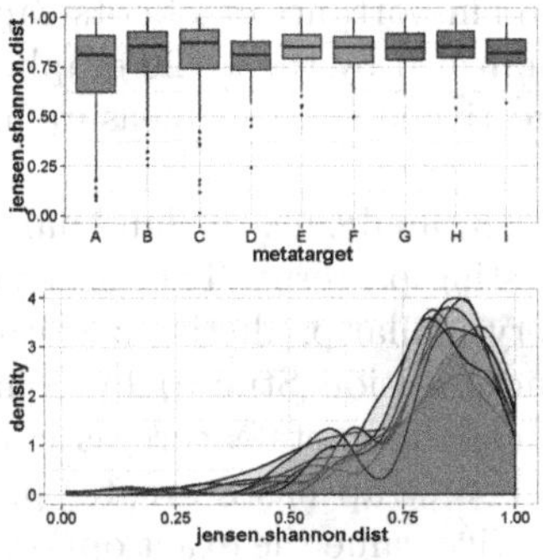

Fig. 11. Boxplot and density distribution of the Jensen-Shannon distance with $k=20$

Figures 12 and 13 show the density distribution of the diversity measures along the classes of the metatarget. Concerning the Q-Statistic (the bigger, the lesser is the diversity), the results are highly unclear. Although the Wilcoxon test shows that this metafeature has discriminating power, that is not visible graphically. The values of all classes are extremely biased to 1. This may seem contradictory to existing knowledge in the ensemble learning literature, where the Q-Statistic is known to be a good diversity indicator [18].

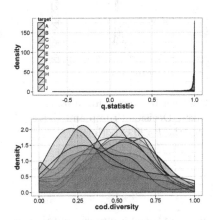

Fig. 12. Density distribution of the metafeatures Q-Statistic and COD for the $k=100$ experiment

Fig. 13. Density distribution of the metafeatures Q-Statistic and COD for the $k=20$ experiment

However, we must note that the Q-Statistic is usually computed between models of bootstrap samples while in our case, it is between models of a bootstrap sample and the training set. On the other hand, the COD metric (higher the value, higher is the diversity) shows a very clear direct relationship between diversity and importance of a bootstrap in the $k=100$ experiment. Again, the result is not confirmed by the $k=20$ graph. However, we consider this result indicative of the effectiveness of this measure in estimating the potential of combining two classifiers.

Finally, by analyzing the landmarker metafeatures in Figures 14 and 15 we can see interesting patterns. The most prominent one is that important bootstraps have a very similar predictive performance using naive algorithms (such as Naive Bayes and Decision Stump) by comparison against the training set: since we transformed this metafeature as explained previously, a negative value means that the bootstrap has a greater predictive performance than the training set and a positive value the exact opposite. Moreover, we can also see a protuberant peak of the classes that gather the worst bootstraps in the density curves at the left side of the graphs. This indicates that *bad* bootstraps have a superior predictive performance than the training sets.

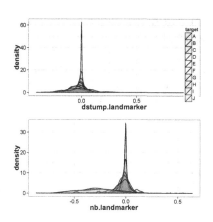 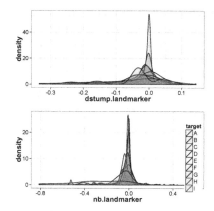

Fig. 14. Density distribution of the landmakers Decision Stump and Naive Bayes for the $k=100$ experiment

Fig. 15. Density distribution of the landmakers Decision Stump and Naive Bayes for the $k=20$ experiment

6   Conclusions and Future Work

This paper proposes a methodology based on an extensive experimental procedure and on MtL for empirically studying the performance of an ensemble learning algorithm, more particularly, bagging. We executed experiments with 53 UCI classification datasets using ensembles of decision trees. Initially, we generated 20 models for each dataset and we tested all possible combinations of those models in the sub-ensemble space. We also executed experiments in which we generated 100 models for each dataset but, due to computational reasons, we were forced to sample the number of combinations tested of the individual models. The results obtained gives us confidence about the effectiveness of the sampling procedure, meaning that it is possible to investigate the distribution of performance of all bagging ensembles obtained with an algorithm by sampling the results. It would be interesting to repeat the experiments with another base learner but we leave that for future work.

To relate the distribution of performance with the characteristics of the bootstrap samples, we adopted an MtL approach. We used several metafeatures proposed in the literature and we introduce three new ones that are very specific of our domain. From our point of view, ensembles are a very promising application of MtL concepts and techniques both to gain a better understanding of their behavior as well as to develop new ensemble methods.

We focused on understanding the characteristics of bootstraps that generate models that are important for the bagging ensemble. We used exploratory data analysis techniques for that goal. Results show interesting patterns that are discriminative of a bootstrap predictive power 1) the bootstrapping procedure should result in a bootstrap sample that is significantly different from the training set, according to the analysis of the Jensen-Shannon distance; 2) the predictions of a model learned from of a bootstrap should be different from the predictions of a model learned from the training, as is known in the ensemble learning literature. However, this observed with the COD metric but not with the Q-Statistic metafeature; 3) the predictive power of a good bootstrap is very similar to the one presented by the training set using naive models.

We plan to extend the work presented in this paper for a predictive MtL approach. The knowledge obtained here can be used to prune a set of bootstraps that can be transformed into an pruned ensemble. It would also be interesting to extend and adapt the methodology proposed in this paper to other ensemble learning algorithms like boosting or random forests. This would bring challenges in the development of the metafeatures in order to deal with probabilistic and random processes.

Acknowledgements. This work is partially funded by FCT/MEC through PIDDAC and ERDF/ON2 within project NORTE-07-0124-FEDER-000059, a project financed by the North Portugal Regional Operational Programme (ON.2 O Novo Norte), under the National Strategic Reference Framework (NSRF), through the European Regional Development Fund (ERDF), and by national funds, through the Portuguese funding agency, Fundação para a Ciência e a Tecnologia (FCT).

References

1. Breiman, L.: Bagging predictors. Machine Learning 24(2), 123–140 (1996)
2. Dieterich, T.G.: An experimental comparison of three methods for constructing ensembles of decision trees: Bagging, boosting, and randomization. Machine Learning 40(2), 139–157 (2000)

3. Brazdil, P., Carrier, C.G., Soares, C., Vilalta, R.: Metalearning: Applications to data mining. Springer (2008)

4. Brazdil, P.B., Soares, C., Da Costa, J.P.: Ranking learning algorithms: Using ibl and meta-learning on accuracy and time results. Machine Learning 50(3), 251–277 (2003)

5. Pfahringer, B., Bensusan, H., Giraud-Carrier, C.: Tell me who can learn you and i can tell you who you are: Landmarking various learning algorithms. In: Proceedings of the 17th ICML, pp. 743–750 (2000)

6. Blake, C., Merz, C.J.: {UCI} repository of machine learning databases (1998)

7. Breiman, L., et al.: Heuristics of instability and stabilization in model selection. The Annals of Statistics 24(6), 2350–2383 (1996)

8. Friedman, J.H.: On bias, variance, 0/1loss, and the curse-of-dimensionality. Data Mining and Knowledge Discovery 1(1), 55–77 (1997)

9. Domingos, P.: Why does bagging work? a bayesian account and its implications. In: KDD, pp. 155–158. Citeseer (1997)

10. Friedman, J.H., Hall, P.: On bagging and nonlinear estimation. Journal of Statistical Planning and Inference 137(3), 669–683 (2007)

11. Büchlmann, P., Yu, B.: Analyzing bagging. Annals of Statistics, 927–961 (2002)

12. Grandvalet, Y.: Bagging equalizes influence. Machine Learning 55(3), 251–270 (2004)

13. Kalousis, A., Gama, J., Hilario, M.: On data and algorithms: Understanding inductive performance. Machine Learning 54(3), 275–312 (2004)

14. Wang, X., Smith-Miles, K., Hyndman, R.: Rule induction for forecasting method selection: Meta-learning the characteristics of univariate time series. Neurocomputing 72(10), 2581–2594 (2009)

15. Kullback, S., Leibler, R.A.: On information and sufficiency. The Annals of Mathematical Statistics, 79–86 (1951)

16. Peng, Y.H., Flach, P.A., Soares, C., Brazdil, P.B.: Improved dataset characterisation for meta-learning. In: Lange, S., Satoh, K., Smith, C.H. (eds.) DS 2002. LNCS, vol. 2534, pp. 141–152. Springer, Heidelberg (2002)

17. Lin, J.: Divergence measures based on the shannon entropy. IEEE Transactions on Information Theory 37(1), 145–151 (1991)

18. Kuncheva, L.I., Whitaker, C.J.: Measures of diversity in classifier ensembles and their relationship with the ensemble accuracy. Machine Learning 51(2), 181–207 (2003)

19. Peterson, A.H., Martinez, T.: Estimating the potential for combining learning models. In: Proceedings of the ICML Workshop on Meta-learning, pp. 68–75 (2005)

20. Lee, J.W., Giraud-Carrier, C.: A metric for unsupervised metalearning. Intelligent Data Analysis 15(6), 827–841 (2011)

21. Järvelin, K., Kekäläinen, J.: Cumulated gain-based evaluation of ir techniques. ACM Transactions on Information Systems 20(4), 422–446 (2002)

22. Fisher, W.D.: On grouping for maximum homogeneity. Journal of the American Statistical Association 53(284), 789–798 (1958)

Context-Aware Recommendation Using GPU Based Parallel Tensor Decomposition

Benyou Zou[1,2], Mengwei Lan[1,2], Cuiping Li[1,2,*], Liwen Tan[1,2], and Hong Chen[1,2]

[1] Key Lab of Data Engineering and Knowledge Engineering of MOE,
Renmin University of China, Beijing, China, 100872
[2] School of Information, Renmin University of China, Beijing, China, 100872
`{zoubenyou,qianjie,licuiping,tan1204,chong}@ruc.edu.cn`

Abstract. Recommender system plays an important role in many practical applications that help users to deal with information overload and provide personalized recommendations to them. The context in which a choice is made has been recognized as an important factor for recommendation systems. Recently, researchers extend the classical matrix factorization and allows for a generic integration of contextual information by modeling the data as a tensor. However, current tensor factorization methods suffer from the limitation that the computing cost can be very high in practice. In this paper, we propose GALS, a GPU based parallel tensor factorization algorithm, to accelerate the tensor factorization on large data sets to support the efficient context-aware recommendation. Experiments show that the proposed method can achieve 10 times faster than the current tensor factorization methods.

Keywords: recommendation algorithm, collaborative filtering, tensor, GPU.

1 Introduction

With the development of the internet, the amount of network information increase rapidly. How to make recommendation for users from the massive information quickly and efficiently is a major problem in recommender system field currently. Taking electronic commerce as an example, the electronic commerce websites, such as Taobao, Jingdong, Dangdang, Amazon, eBay, contain various kinds and large quantities of goods. For example, the number of existing products in Tmall has already exceeded 4 million according to incomplete statistics. In this platform, it is necessary to have an excellent recommendation system. On the other hand, the data contained in these sites is not the only user and commodities, but also includes a lot of other information, such as users' age, address, and the manufacturer of the product, date, etc. This information is also greatly affects a user's preferences for products. In addition, some dynamic information, such as weather, mood, etc., also affect people's decisions to some extent. However, traditional recommendation algorithms based on matrix decomposition cannot handle similar information. This is certainly a loss in terms of the recommendation algorithm. To this end, researchers proposed to incorporate context information into the

---

* Corresponding author.

X. Luo, J.X. Yu, and Z. Li (Eds.): ADMA 2014, LNAI 8933, pp. 213–226, 2014.

recommender systems. They extend the classical matrix factorization and allows for a generic integration of contextual information by modeling the data as a tensor. Currently, there are some works on the use of tensors for context-aware recommendation. For example, Rendle et al. [15] suggested to use tensor to model contextual information and perform the rating prediction, Karatzoglou et al.[10] proposed the multiverse recommendation method based on the Tucker tensor factorization. The experimental results show that context-aware recommendation improves upon non-contextual matrix factorization method. However, all these methods suffer from the limitation that the computation cost can be very high in practice. One essential operation of tensor factorization is to compute a sequence of consecutive n-mode tensor matrix products. The size of real data can cause computation to take very long time to complete. This delay is unacceptable in most real environments, as it severely limits productivity. Ideally, we would like to develop new hardware-accelerated solutions that can offer improved processing power to tackle the expensive tensor operations which are inherently involved in the computation. In this paper, we propose GALS, a GPU based parallel tensor factorization algorithm, to accelerate the tensor factorization on large data sets to support the efficient context-aware recommendation. The experiment shows that the GALS has about 10 times faster than the traditional methods.

The rest of this paper is organized as follows. Section 2 gives the background information of our study. Section 3 introduces tensor factorization method . Section 4 introduce the GPU based parallel tensor factorization method we proposed in this paper. A performance analysis of our methods is presented in Section 5. We conclude the study in Section 6.

2   Related Works

2.1   Tensor Factorization

As a generalization of matrix factorization, tensor factorization has been studied from an algebraic perspective and recently witnessed a renewed interest, especially in relation to data analysis. For instance, [9] applied the HOSVD to identifying handwritten digits, [11]applied Tucker decompositions to separate conversations in online chatrooms, [1]utilized tensor to perform social network analysis, [4] employed tensor to extend the Latent Semantic indexing to cross-language information retrieval, [13] used Tucker decomposition for analyzing clickthrough data.

2.2   GPU Applications

The GPUs have no longer been confined to processing graphics recently. Due to GPUs' efficient parallelism and flexible programmability, an increasing number of researchers and commercial organizations have achieved success in general-purpose computation on GPU, such as database operations, geometric computing, grid computing, scientific computing and so on. Methods have been proposed to enhance relational joins on database [8], information retrieval [5]and k nearest neighbor query [6]. The tensor-matrix multiplication, which is the vital operation of tensor factorization, is suitable to exploit parallel programming. The challenges and advances in the implementation of sparse matrix multiplication under parallel architecture are discussed in[2].

Table 1. Symbols

Symbol	Definition and Description
$\mathbf{A},\mathbf{B}$	Matrices (bold upper case)
$\mathbf{A}(i,j)$	An element of $\mathbf{A}$
$\mathbf{A}(i,:)$ or $\mathbf{A}(:,i)$	i-th row or column of $\mathbf{A}$
$\mathcal{A},\mathcal{B}$	Tensors (calligraphic letters)
$\mathcal{A}(i_1,\ldots,i_N)$	An element of tensor $\mathcal{A}$
$\mathcal{A}(\ldots,i_{n-1},:,i_{n+1},\ldots)$	the n-mode vectors of $\mathcal{A}$
N	The order of a tensor
$\mathrm{uf}(\mathcal{A},n)$ or $\mathbf{A}_{(n)}$	The n-mode unfolding matrix of $\mathcal{A}$
$\mathbf{U}^{(n)}$	n-mode basis matrix of $\mathbf{A}_{(n)}$
$\times_n$	the n-mode product
$\mathcal{A}_{[s_1,\ldots,s_N]}$	A block of tensor $\mathcal{A}$
$r_{u,i}$	rating of item i from user u
$\bar{r}_u$	Average rating score from user u
$\bar{r}_i$	Average rating score of item i
$\hat{r}_{u,i}$	Predict score of item i from user u

3   Tensor Factorization

The symbols and meanings used in this paper are shown in table 1.

3.1   Notations and Assumptions

Definition 1 (Tensor). *Tensors are higher-order generalizations of vectors and matrices. High-order tensors are denoted as: $\mathcal{A} \in \Re^{I_1 \times I_2 \times \cdots \times I_N}$, where the order of $\mathcal{A}$ is N ($N > 2$) while a vector a tensor of order 1 and a matrix is order 2 tensor, respectively.*

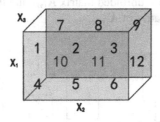

Fig. 1. An example of 3-order Tensor $\mathcal{A}$

Definition 2 (Mode). *Each dimension of a tensor is called a mode. For example, $\mathcal{A} \in \Re^{I_1 \times I_2 \times \cdots \times I_N}$ is a tensor with N modes (called N^{th} order tensor). An element of $\mathcal{A}$ is denoted as $\mathcal{A}(i_1,\ldots,i_n,\ldots,i_N)$, where $1 \leqslant i_n \leqslant I_n$. The n-mode vectors of $\mathcal{A}$ are defined as the I_n-dimensional vectors obtained from $\mathcal{A}$ by varying its index in while keeping all the other indices fixed. In this paper we denote it as $\mathcal{A}(\ldots,i_{n-1},:,i_{n+1},\ldots)$.*

Definition 3 (Unfolding). *Unfolding a tensor $\mathcal{A} \in \Re^{I_1 \times \cdots I_n \cdots I_N}$ along the nth mode is denoted as $uf(\mathcal{A},n)$, which results in a matrix, $\mathbf{A}_{(n)} \in \Re^{I_n \times (I_1 I_2 \cdots I_{n-1} I_{n+1} \cdots I_N)}$, where the column vectors of $\mathbf{A}_{(n)}$ are the n-mode vectors of $\mathcal{A}$.*

For example in figure 1, the matrices of unfolding tensor $\mathcal{A}$ is as follows:

$$A(1) = \begin{bmatrix} 1 & 2 & 3 & 7 & 8 & 9 \\ 4 & 5 & 6 & 10 & 11 & 12 \end{bmatrix}$$

$$A(2) = \begin{bmatrix} 1 & 7 & 4 & 10 \\ 2 & 8 & 5 & 11 \\ 3 & 9 & 6 & 12 \end{bmatrix}$$

$$A(3) = \begin{bmatrix} 1 & 4 & 2 & 5 & 3 & 6 \\ 7 & 10 & 8 & 11 & 9 & 12 \end{bmatrix}$$

Definition 4 (Mode Product). *The n-mode product of a tensor $\mathcal{A} \in \Re^{I_1 \times \cdots I_n \cdots I_N}$ and a matrix $U \in \Re^{J_n \times I_n}$, denoted as $\mathcal{A} \times_n U$, is a tensor in $\Re^{I_1 \times \cdots J_n \cdots I_N}$ with entries:*

$$(\mathcal{A} \times_n U)(i_1, ..., j_n, ..., i_N) = \sum_{i_n} \mathcal{A}(i_1, ..., i_N) U(j_n, i_n).$$

For example, if tensor $\mathcal{A}$ is in the shape of $3 \times 4 \times 5$ while $\mathbf{U}$ is in the shape of 2×3. The result of $\mathcal{A} \times_1 \mathbf{U}$ is a tensor of shape $2 \times 4 \times 5$. More details about the multilinear algebra, please refer to [14].

Tensor Decomposition. At present, many tensor factorization models have been put forward, such as Tucker models [16] , PARAFAC models[7] and CANDECOMP models [3]. In [14], Tucker3 with orthogonality constraints on the components, has been named as n-mode SVD (or HOSVD). We follow the HOSVD formulation here. A N-order tensor $\mathcal{A}$ is factorized into N basis matrices $\mathbf{U}^{(1)} \in \Re^{I_1 \times R_1}$, $\mathbf{U}^{(2)} \in \Re^{I_2 \times R_2}$, ..., $\mathbf{U}^{(N)} \in \Re^{I_N \times R_N}$, and one core tensor $\mathcal{C} \in \Re^{R_1 \times R_2 \times ... \times R_N}$. In this case:

$$\mathcal{A} \approx \mathcal{C} \times_1 \mathbf{U}^{(1)} \times_2 \mathbf{U}^{(2)} ... \times_N \mathbf{U}^{(N)}$$

where $\mathbf{U}^{(n)} \in \Re^{I_n \times R_n}$ has orthonormal columns for $1 \leqslant i \leqslant N$. It is obtained by first performing regular SVD on unfolded matrix $\mathbf{A}_{(n)}$ and then extracting the first R_n columns of the left singular matrix of $\mathbf{A}_{(n)}$.

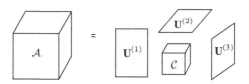

Fig. 2. An example of 3-mode SVD

Figure 2 depicts an example of 3-mode SVD. A 3-mode tensor is factorized into 3 basis matrices $\mathbf{U}^{(1)}, \mathbf{U}^{(2)}, \mathbf{U}^{(3)}$, and one core tensor $\mathcal{C}$. The HOSVD factorization allows an interplay between a factor and any factor in other modes. Furthermore, it allows for full control over the dimensionality by performing truncations on the factors, which is very important for large real data sets. Now the aim is to find a decent approximation $\tilde{\mathcal{A}}$ for $\mathcal{A}$ such that the loss function $L(\tilde{\mathcal{A}}, \mathcal{A})$ between observed and estimated values is minimized.

Definition 5 (Rank-$(R_1, R_2, ..., R_N)$ Approximation). *Given a Nth order tensor $\mathcal{A} \in \Re^{I_1 \times I_2 \times ... \times I_N}$, tensor $\tilde{\mathcal{A}} \in \Re^{I_1 \times I_2 \times ... \times I_N}$ is a rank-$(R_1, R_2, ..., R_N)$ approximation of $\mathcal{A}$ if $Rank_n(\tilde{\mathcal{A}}) = R_n \leqslant Rank_n(\mathcal{A})$ and for all n, the following least-squares cost function is minimized:*

$$\tilde{\mathcal{A}} = arg \min_{\tilde{\mathcal{A}}} \|\mathcal{A} - \tilde{\mathcal{A}}\|^2.$$

When R_n is much smaller than I_n for all n, the best rank approximation $\tilde{\mathcal{A}}$ gives a compact approximation of the original tensor $\mathcal{A}$, resulting in good data compression for large real data sets.

The basis matrices $\mathbf{U}^{(i)}$ is orthogonal, so $U^{(i)} U^{(i)^T} = U^{(i)^T} U^{(i)} = E$. Given basis matrices $\mathbf{U}^{(1)}$, $\mathbf{U}^{(2)}$, ..., $\mathbf{U}^{(N)}$, The core tensor $\mathcal{C}$ can be readily computed as $\mathcal{C} \approx \mathcal{A} \times_1 \mathbf{U}^{(1)^T} \times_2 \mathbf{U}^{(2)^T} ... \times_N \mathbf{U}^{(N)^T}$. Thus, the optimization problem focuses on the computation of the basis matrices only. In this paper, we seek efficient computation of the best rank-$(R_1, R_2, ..., R_N)$ approximation of $\mathcal{A}$.

Given basis matrices $\mathbf{U}^{(1)}$, $\mathbf{U}^{(2)}$, ..., $\mathbf{U}^{(N)}$, the core tensor $\mathcal{C}$ can be readily computed as $\mathcal{C} = \mathcal{A} \times_1 \mathbf{U}^{(1)^T} \times_2 \mathbf{U}^{(2)^T} ... \times_N \mathbf{U}^{(N)^T}$. Thus, the optimization problem focuses on the computation of the basis matrices only. In this paper, we seek efficient computation of the best rank-$(R_1, R_2, ..., R_N)$ approximation of $\mathcal{A}$.

3.2   Our Method

In this section, we will discuss how to find the best rank-$(R_1, R_2, \ldots, R_N)$ approximation of $\mathcal{A}$ efficiently. We want to minimize:

$$\|\mathcal{A} - \tilde{\mathcal{A}}\|^2 = \|\mathcal{A}\|^2 - 2\langle \mathcal{A}, \tilde{\mathcal{A}} \rangle + \|\tilde{\mathcal{A}}\|^2.$$

Based on the definition of scalar product, we can get:

$$\langle \mathcal{A}, \tilde{\mathcal{A}} \rangle = \langle \mathcal{A}, \mathcal{C} \times_1 \mathbf{U}^{(1)} \times_2 \mathbf{U}^{(2)} \times ... \times_N \mathbf{U}^{(N)} \rangle$$
$$= \langle \mathcal{A} \times_1 \mathbf{U}^{(1)^T} \times_2 \mathbf{U}^{(2)^T} \times ... \times_N \mathbf{U}^{(N)^T}, \mathcal{C} \rangle \cdot$$
$$= \|\mathcal{C}\|^2$$

Since $\mathbf{U}^{(n)}(1 \leqslant n \leqslant N)$ has orthonormal columns, they don't affect the Frobenius norm, so we have: $\|\tilde{\mathcal{A}}\|^2 = \|\mathcal{C}\|^2$. Thus, we have: $\|\mathcal{A} - \tilde{\mathcal{A}}\|^2 = \|\mathcal{A}\|^2 - \|\mathcal{C}\|^2$. Now, The best rank-$(R_1, R_2, \ldots, R_N)$ tensor approximation can be computed by maximizing $\|\mathcal{C}\| = \|\mathcal{A} \times_1 \mathbf{U}^{(1)^T} \times_2 \mathbf{U}^{(2)^T} ... \times_N \mathbf{U}^{(N)^T}\|$.

The most popular algorithm for computing the best rank-$(R_1, R_2, \ldots, R_N)$ tensor approximation is the Alternating Least Squares (ALS) [14]. The ALS algorithm is an iterative algorithm. In each iterative step, it optimizes only one of the basis matrices, while keeping other $N - 1$ basis matrices fixed.

For instance, in the $l+1$ step, with $\mathbf{U}^{(1)}_{l+1}, \ldots, \mathbf{U}^{(n-1)}_{l+1}, \mathbf{U}^{(n+1)}_{l+1}, \ldots, \mathbf{U}^{(N)}_l$ fixed, tensor $\mathcal{A}$ is projected onto the $(R_1, \ldots, R_{n-1}, R_{n+1}, \ldots, R_N)$-dimensional space as follows:

$$\mathcal{A}' = \mathcal{A} \times_1 \mathbf{U}^{(1)^T}_{l+1} \times_2 \mathbf{U}^{(2)^T}_{l+1} ... \times_{(n-1)} \mathbf{U}^{(n-1)^T}_{l+1}$$
$$\times_{(n+1)} \mathbf{U}^{(n+1)^T}_l ... \times_N \mathbf{U}^{(N)^T}_l . \tag{1}$$

Then the $U_{l+1}^{(n)}$ can be get from the SVD decomposition of $uf(\mathcal{A}', n)$, which is the n-mode unfolding of tensor $\mathcal{A}'$.

One bottleneck of the Equation 1 is in the computation of mode product where a sequence of consecutive n-mode tensor matrix products needs to be finished. The computation cost can be very high in practice. The size of real data can cause computation to take very long time to complete. This delay is unacceptable in most real environments, as it severely limits productivity.

4 Parallel Tensor Factorization

In this section, we develop new hardware-accelerated solutions that can offer improved processing power to tackle the expensive tensor operations. Based on ALS, we propose GALS, a parallel tensor factorization algorithm, to accelerate the tensor factorization on large data sets to support the efficient context-aware recommendation. The basic idea is to exploit the inherent parallelism and high memory bandwidth of GPU to perform tensor related operations in parallel.

4.1 Blockwise N-Mode Product

In contrast with the main memory, GPU memory is relatively small; it is hard to hold the whole tensor $\mathcal{A}$. A natural way to factorize the tensor in parallel based on GPU is to partition $\mathcal{A}$ into blocks and perform the n-mode product block by block. We will introduce two strategies to solve this problem for sparse tensor and dense tensor.

For sparse tensor, we use Compressed Row Storage (CRS)[1] model to represent the unfold matrix of tensor, this model can save a lot of storage space, and we can easily access the column data of the matrix. So, we block the tensor by rows, every row of a tensor is a block. For example, in Figure 3, tenor 2-mode product with matrix, $\mathcal{A} \times_2 \mathbf{U}^{(2)^T}$, $\mathcal{A}$ is $6 \times 5 \times 2$, $\mathbf{U}^{(2)^T}$ is 5×3, we divided tensor $\mathcal{A}$ to 6 parts, the is divided to 3 parts $\mathbf{U}^{(2)^T}$. If we use 3 threads to do this job, the first 3 parts of tensor $\mathcal{A}$ and the matrix $\mathbf{U}^{(2)^T}$ to do mode product and then the remaining 3 parts of tensor $\mathcal{A}$, for example, T1-B1, T2-B2 and T3-B3.

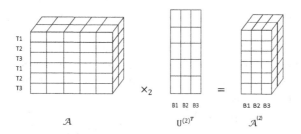

Fig. 3. Blockwise 2-mode product for sparse tensor

---

[1] http://netlib.org/linalg/htmltemplates/node91.html

For dense tensor, given a tensor $\mathcal{A} \in \Re^{I_1 \times I_2 \times \dots \times I_N}$, following [17], we require each block remains as an Nth-order sub-tensor after partitioning. Assuming each dimension I_n is evenly partitioned into S_n segments (zeros are filled for the case that I_n is not dividable by S_n), each segment has a size of $\lceil I_n/S_n \rceil$. Then each block itself is a tensor, denoted as: $\mathcal{A}_{[s_1, s_2, \dots, s_N]} \in \Re^{\lceil I_1/S_1 \rceil \times \lceil I_2/S_2 \rceil \times \dots \times \lceil I_N/S_N \rceil}$

Correspondingly, the basis matrix $\mathbf{U}^{(n)} \in \Re^{J_n \times I_n}$ for $1 \leqslant i \leqslant N$ can be partitioned in the same way. An example of partitioning on tensors and basis matrices is shown in Figure 4. The first dimension J_n and the second dimension I_n both should be partitioned into S_n segments. Each block of $\mathbf{U}^{(n)}$ is denoted as:$\mathbf{U}^{(n)}_{[t_n, s_n]}$.In order to ensure that the blocks can still fit the GPU memory and would not increase their size after n-mode products, $\lceil J_n/T_n \rceil$ should not larger than $\lceil I_n/S_n \rceil$.

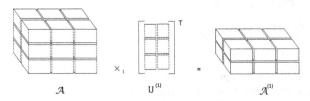

Fig. 4. Blockwise 1-mode product for dense tensor

For instance, in Figure 4, tensor $\mathcal{A}$ is partitioned into $3 \times 3 \times 2$ blocks while $\mathbf{U}^{(1)}$ is partitioned into 2×3 blocks. The outcome of the 1-mode product of $\mathcal{A}$ and $\mathbf{U}^{(1)^T}$ is still a tensor $\mathcal{A}^{(1)}$, but with $2 \times 3 \times 2$ blocks.

4.2 The GPU Based Parallel Tensor Factorization

The pseudo-code of tensor ALS approximation is given in algorithm 1 as follows ($GMP()$ means GPU based mode product):

Algorithm 1 provides a natural parallel approach for blockwise tensor factorization. However, there is one drawback of Algorithm 1 in the SVD calculation of $uf(\mathcal{A}', n)$(line 12). And another bottleneck is that the product of $\mathcal{A}$ and a series of dense basis matrices $\mathbf{U}^{(n)}$ ($1 \leqslant n \leqslant N$) will be very large and dense although the tensor $\mathcal{A}$ is sparse; therefore, the intermediate products $\mathcal{A}'$ (line 7 and 10) are dense and may be too large to fit GPU memory even if the final tensor does fit. This problem is called "intermediate data explosion". To compensate for these two drawbacks, we propose two optimization strategies in the following.

Column Factor Optimization: The purpose of this optimization strategy is reducing the intermediate data size. According to [12], we can avoid the "intermediate data explosion" problem without compromising calculating speed by handling the computation in a piecemeal fashion. We will introduce how this strategy can be used to optimize the GPUTENSOR algorithm we proposed as well.

Given basis matrices $\mathbf{U}^{(n)} \in \Re^{J_n \times I_n}$ ($1 \leqslant n \leqslant N$)and a tensor $\mathcal{A} \in \Re^{I_1 \times I_2 \times \dots \times I_N}$, Equation 1 can be computed one column with less memory at a time:

Algorithm 1. The GALS Algorithm

Input: A tensor $\mathcal{A}$, and $R_1, R_2, \ldots, R_N$;
Output: Core tensor $\mathcal{C}$, and basis matrix $\mathbf{U}^{(n)}$ $(1 \leqslant n \leqslant N)$
1: Initialize $\mathbf{U}_0^{(n)}$ $(1 \leqslant n \leqslant N)$;
2: $\mathcal{C}_0 = \mathcal{A} \times_1 \mathbf{U}_0^{(1)^T} \times_2 \mathbf{U}_0^{(2)^T} \ldots \times_N \mathbf{U}_0^{(N)^T}$;
3: Let $l=0$;
4: For each $n \in [1, N]$ Do
5:    $\mathcal{A}' = \mathcal{A}$;
6:    For each $m \in [1, n-1]$ and $m \neq n$ Do
7:        $\mathcal{A}' = GMP(\mathcal{A}', \mathbf{U}_{l+1}^{(m)^T}, m)$;
8:    Endfor
9:    For each $m \in [n+1, N]$ Do
10:       $\mathcal{A}' = GMP(\mathcal{A}', \mathbf{U}_l^{(m)^T}, m)$;
11:   Endfor
12:   $(\mathbf{U}_{l+1}^{(n)}, \Sigma_{l+1}^{(n)}, \mathbf{V}_{l+1}^{(n)}) = \text{SVD}(uf(\mathcal{A}', n), R_n)$;
13: Endfor
14: $\mathcal{C}_{l+1} = \mathcal{A} \times_1 \mathbf{U}_{l+1}^{(1)^T} \times_2 \mathbf{U}_{l+1}^{(2)^T} \ldots \times_N \mathbf{U}_{l+1}^{(N)^T}$;
15: If $\|\mathcal{C}_{l+1}\|^2 - \|\mathcal{C}_l\|^2 < \varepsilon$ for a given tolerance ε, QUIT;
16: Otherwise, continue from Line 4 with l increased by 1.

$$\mathcal{A}'(j_1, \ldots, j_{n-1}, :, j_{n+1}, \ldots, j_N)$$
$$= \mathcal{A} \times_1 \mathbf{U}_{l+1}^{(1)^T}(:, j_1) \times_2 \ldots \times_{(n-1)} \mathbf{U}_{l+1}^{(n-1)^T}(:, j_{n-1})$$
$$\times_{(n+1)} \mathbf{U}_l^{(n+1)^T}(:, j_{n+1}) \times_{(n+2)} \ldots \times_N \mathbf{U}_l^{(N)^T}(:, j_N).$$

where $j_n = 1, \ldots, J_n$, $\mathbf{U}_{l+1}^{(1)^T}(:, j_1)$ is the j_1-th column of $\mathbf{U}_{l+1}^{(1)^T}$. Here, the largest intermediate result is a vector of size I_n (instead of $J_1 \times J_2 \ldots \times I_n \times \ldots \times J_N$ in the standard calculation of Equation 1). Equation 1 can be calculated column by column using this strategy, which not only greatly reduced the intermediate data size, but also dramatically lowered the transfer cost between main memory and GPU device memory.

SVD Acceleration Optimization: SVD acceleration optimization strategy is aimed to improve the efficiency. In line 12 of Algorithm 1, we have noticed that unfolding matrix of $\mathcal{A}'$ required one SVD operation. Because the result of $uf(\mathcal{A}', n)$ is a wide matrix $\mathbf{A}'_{(n)} \in \Re^{I_n \times (R_1 R_2 \ldots R_{n-1} R_{n+1} \ldots R_N)}$, it will spend quite a long time to compute SVD for it. A simple strategy to improve the efficiency is to perform the SVD on the small matrix $\mathbf{A}'_{(n)} \mathbf{A}'^T_{(n)} \in \Re^{I_n \times I_n}$. Next, we will introduce how it works.

Assuming $\text{SVD}(\mathbf{A}'_n) = \mathbf{U}\Sigma\mathbf{V}$, we have:

$$SVD(\mathbf{A}'_{(n)}\mathbf{A}'^T_{(n)}) = \mathbf{U}\Sigma^2\mathbf{U}^T \tag{2}$$

From Equation 2 we can get left singular matrix $\mathbf{U}$ and the core matrix Σ. For the right singular matrix $\mathbf{V}$, we just need a mathematical transform as following:

$$\mathbf{V} = \Sigma^{-1}\mathbf{U}^T\mathbf{U}\Sigma\mathbf{V} = \Sigma^{-1}\mathbf{U}^T\mathbf{A}'_{(n)}$$

Since the size of $\mathbf{A}'_{(n)}\mathbf{A}'^{T}_{(n)} \in \Re^{I_n \times I_n}$ is small in most real cases, the SVD of $\mathbf{A}'_{(n)}\mathbf{A}'^{T}_{(n)}$ can be calculated highly effective. However, as mentioned in [17], the matrix product $\mathbf{A}'_{(n)}\mathbf{A}'^{T}_{(n)}$ needs to be computed blockwise before SVD since $\mathbf{A}'_{(n)}$ is large. Algorithm 2 is the pseudo-code of GPU Mode Product (GMP).

Algorithm 2. $GMP(\mathcal{A}, \mathbf{U}, n)$

Input: Tensor $\mathcal{A} \in \Re^{I_1 \times I_2 \times \cdots \times I_N}$, matrix $\mathbf{U}$, mode n

Output: Result tensor $\mathcal{B}$

  1: Allocate space in GPU memory for matrix $\mathbf{U}$;
  2: Transfer data of $\mathbf{U}$ into GPU memory;
  3: For $i = 1$ to I_n ;
  4:     Allocate space for slice of tensor $S_n = \mathcal{A}_{i_1,\ldots,i_{n-1},:,:,i_{n+2},\ldots,i_N}$ in GPU ;
  5:     Transfer data of S_n into GPU memory;
  6:     Allocate space for result R in GPU memory;
  7:     If S_n is sparse;
  8:         Allocate column($\mathbf{U}$) process blocksevery process block has thd threads;
  9:         $SGMPKernel(S_n, U, R)$;
 10:     Otherwise;
 11:         Allocate $\lceil i_n/thd \rceil \times \lceil i_{n+1}/thd \rceil$ process blocksevery process block has $thd \times thd$ threads;
 12:         $DGMPKernel(S_n, U, R)$;
 13:     Transfer data of R back to main memory;
 14:     Free GPU memory for S and R;
 15: Endfor;
 16: Free GPU memory for $\mathbf{U}, S_n$ and R;
 17: Construct $\mathcal{B}$ using returned matrices ;

---

5 Experiments

All of our following experiments are using C++ language. The type of GPU is NVidia GeForce GTX 580, with 1536MB memory, 512 streaming multiprocessor.

5.1 Data Sets

The data we used in our experiments are summarized in Table 2, the details are as follows:

Table 2. Details of the Data Sets

Data	I_1	I_2	I_3	Nonzeros
Sythetic	1k ~ 8K	1k ~ 8K	100	3M ~ 24M
Movielens-1	943	1682	5	100K
Movielens-2	6040	3952	5	1M
Epinions	8K	8K	2	1M

Algorithm 3. $SGMPKernel(S, \mathbf{U}, R)$

Input: Matrices S and $\mathbf{U}$
Output: The result matrix R
1: Get the block index bid and thread index tid
2: Allocate the shared memory to $data[]$
3: For $i=tid$ to RowOf($\mathbf{U}$) step tid
4:    $data[i]=U[i][bid]$
5:    waiting for every thread finish
6: For $i=0$ to RowOf(S) step tid
7:    $sum \leftarrow 0$
8:    Find elements $S_{i,*}$(i, :) in the i-th row of S
9:    For each element $S_{i,j}$ in $S_{i,*}$
10:       Find the element $data[j]$
11:       $sum+=data[j] * S_{i,j}$
12:    $R[i][bid] \leftarrow sum$
13: Endfor

Synthetic Data: Random tensor of size . The size of and vary from 1k to 8K, and the number of nonzeros varies from 3M to 24M.

Movielens Data: this data set is available at GroupLens [2] . It allows users to rate various movies they have watched, giving the preference score range from 1 to 5. We call the small one as Movielens-1 and the large one as Movielens-2. To construct the third dimension, following [10], we discrete the user age into 5 groups.

Epinions Data: this data set is derived from Epinions [3] by filtering entries whose values are below a threshold. It allows users to review items (such as cars, books, movies, software, etc.) and also assign them numeric ratings in the range of 1 to 5. We filter the data by users who has ratings more than 20. The similar method is used to construct the third dimension. The Epinions has the trust network for users, we use the the trust network in recommendation.

5.2   Compare Method

We first present a comprehensive study using the synthetic datasets, which shows high efficiencies of our algorithms. We then evaluate the effectiveness and efficiencies of our algorithms on two real data sets, the MovieLens and the Epinions data. The runtime reported in all experiments includes the I/O time. We compare the performance of the following three algorithms:

   CALS: the ALS algorithm for HOSVD, we implement it using core dual core cpu without parallel methods.

   GALS: the parallel ALS algorithm for HOSVD we proposed using GPU

---

[2] http://www.grouplens.org/node/73
[3] http://www.epinions.com

Algorithm 4. $DGMPKernel(S, \mathbf{U}, R)$

Input: Matrices S and $\mathbf{U}$
Output: The result matrix R
 1:    shared $A[thd][thd]$
 2:    shared $B[thd][thd]$
 3:    $tidr \leftarrow$ threadIndex.x; $tidc \leftarrow$ threadIndex.y;
 4:    $vidr \leftarrow$ blockIndex.x; $tidc \leftarrow$ blockIndex.y;
 5:    $sum \leftarrow 0$
 6:    for $i = 0$ to Column(S) step thd
 7:        $A[tidr][tidc] \leftarrow S[tidr + bidr][tidc + i]$
 8:        $B[tidr][tidc] \leftarrow U[tidr + i][tidc + bidc]$
 9:        waiting for every thread finish
10:        for $j = 0$ to thd
11:            $A[tidr][j] * B[j][tidc]$
12:        waiting for every thread finish
13:    $R[tidr + bidr][tidc + bidc] \leftarrow sum$

GALSN: to evaluate the accuracy of tenor based recommendation method, we use GALS without context information (GALSN). We only consider the user-item matrix, the GALSN in the same to SVD method.

5.3 Efficiencies Evaluation

This set of experiments is conducted to evaluate the efficiencies of our parallel method.

Experiments on Synthetic Data Sets. In this section, we evaluate the average time of every iteration step of ALS(line 4-12 in Algorithm 1). Figure 5(a) shows the time spent by both algorithms on different size of data sets. We varies the size of the first two dimensions of the synthetic data set from 1k to 8k. We can see from the figure that GALS achieves substantial saving in time than CALS. For a tensor data, we enjoy a 86% saving in factorization time. The results clearly indicate that our parallel algorithm for tensor factorization is efficient. The savings in time mainly come from the inherent parallel ability of GPU.

The core tensor dimension size also affects the efficiency of these algorithms. In this experiment, we fix the tensor size at 5k×5k×100 and the data density at 0.3%, $R_1 = R_2 = R_3 = k$ where k varies from 10 to 100. Fig.5(b) shows the run time of both algorithms with respect to varying the core tensor dimension size k. although both algorithms have linear scalability, the run time of GALS scales much better than CALS. Please note that there is one cross point in Fig. 5(b) the reason is that GALS needs time to transfer the original data from CPU memory to GPU memory. In contrast, CALS doesn't have such transfer cost. When the core tensor dimension size k is small, the transfer time takes the larger part of the total cost compared to the computation time, so GALS is slower in this case than CALS. When k is large, the transfer cost can be nearly ignored, GALS outperforms CALS.

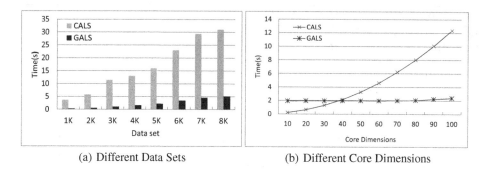

(a) Different Data Sets   (b) Different Core Dimensions

Fig. 5. Performance on Different Hardware Platforms

Experiments on Real Data Sets. This set of experiments is used to evaluated the performance of our algorithm on three real data sets: Movielens-1, Movielens-2, and Epinions. Table 3 shows the whole process time of Tensor ALS method, we can see that the GALS is about 10 times faster than CALS.

Table 3. Efficiency on Different Real Data Sets(Time in seconds)

Algorithm	MovieLens-1	MovieLens-2	Epinions
CALS	252.469	701.071	983.680
GALS	23.234	61.497	100.375

5.4   Effectiveness Evaluation

In this experiment, we conduct experiments to validate the effectiveness of our proposed algorithms on different synthetic data sets. We adopted the widely-used measure, Root Mean Squared Error (RMSE), to evaluate the accuracies of predicted ratings. Given a tensor $\mathcal{A} \in \Re^{I_1 \times I_2 \times \dots \times I_N}$, the RMSE between the predicted and actual ratings is given by:

$$RMSE = \sqrt{\frac{1}{TS} \sum (\tilde{\mathcal{A}}(i_1, \dots, i_N) - \mathcal{A}(i_1, \dots, i_N))^2}.$$

where TS represents the tensor size of tensor $\mathcal{A}$.

To evaluate the performance of tensor based context-aware recommendation, we use GALSN without context information to compare the performance of GALS. In our experiments, 80% data is used as the training data and the remaining 20% is used as the testing set. Table 4 shows the RMSE results of GALSN and GALS on different data sets. Clearly, one can find that our GALS methods achieve higher accuracy compared to the no context method GALSN.

Table 4. RMSE of the Real Data Set

Algorithm	MovieLens-1	MovieLens-2	Epinions
GALSN	0.9054	0.9401	0.9029
GALS	0.8713	0.8830	0.8801

6 Conclusion

This paper addresses the issues of parallel computation of tensor factorization for context-aware recommendation. We have proposed GALS, a parallel tensor factorization algorithm, to accelerate the tensor factorization. The experimental results verified the feasibility and accuracy of the algorithm we proposed, but there are still many places where need to be improved. For example, the SVD method used in calculation of RANK must load the entire matrix in the decomposition process. Therefore, due to the limitations of memory size, our approach cannot meet the demand of big data yet. How to design an asynchronous decomposition method is undoubtedly our future direction. In addition, the parallel algorithm we implemented is a sub-block-level parallelism granularity, or row level. In future work, we will consider extending the parallel granularity into fragmentation level.

Acknowledgements. This work was supported by National Basic Research Program of China (973Program) (No.2014CB340402, No. 2012CB316205), National High Technology Research and Development Program of China (863 Program) (No.2014AA015204), NSFC under the grant No.61272137, 61033010, 61202114, 61165004 and NSSFC (No: 12&ZD220). It was partially done when the authors worked in SA Center for Big Data Research in RUC. This Center is funded by a Chinese National 111 Project Attracting.

References

1. Agrawal, R., Imieliski, T., Swami, A.N.: Mining association rules between sets of items in large databases. Sigmod Record 22, 207–216 (1993)
2. Buluç, A., Gilbert, J.R.: Challenges and advances in parallel sparse matrix-matrix multiplication. In: ICPP 2008: Proceedings of the 2008 37th International Conference on Parallel Processing, pp. 503–510. IEEE Computer Society, Washington, DC (2008)
3. Carroll, J.D., Chang, J.-J.: Analysis of individual differences in multidimensional scaling via an n-way generalization of eckart-young decomposition. Psychometrika 35(3), 283–319 (1970)
4. Chew, P.A., Bader, B.W., Kolda, T.G., Abdelali, A.: Cross-language information retrieval using PARAFAC2. In: Knowledge Discovery and Data Mining, pp. 143–152 (2007)
5. Ding, S., He, J., Yan, H., Suel, T.: Using graphics processors for high performance ir query processing. In: WWW 2009: Proceedings of the 18th International Conference on World Wide Web, pp. 421–430. ACM, New York (2009)
6. Garcia, V., Debreuve, E., Barlaud, M.: Fast k nearest neighbor search using gpu. In: 2008 IEEE Computer Society Conference on Computer Vision and Pattern Recognition Workshops, pp. 1–6 (2008)

7. Harshman, R.A.: Foundations of the PARAFAC procedure: Model and conditions for an "explanatory" multi-mode factor analysis. UCLA Working Papers in Phonetics 16, 1–84 (1970)

8. He, B., Yang, K., Fang, R., Lu, M., Govindaraju, N.K., Luo, Q., Sander, P.V.: Relational joins on graphics processors. In: SIGMOD 2008: Proceedings of the 2008 ACM SIGMOD International Conference on Management of Data, pp. 511–524. ACM, New York (2008)

9. Kapteyn, A., Neudecker, H., Wansbeek, T.: An approach to n -mode components analysis. Psychometrika 51, 269–275 (1986)

10. Karatzoglou, A., Amatriain, X., Baltrunas, L., Oliver, N.: Multiverse recommendation: n-dimensional tensor factorization for context-aware collaborative filtering. In: Proceedings of the Fourth ACM Conference on Recommender Systems, RecSys 2010, pp. 79–86. ACM, New York (2010)

11. Kleinberg, J.M.: Authoritative sources in a hyperlinked environment. Journal of The ACM 46, 604–632 (1999)

12. Kolda, T., Sun, J.: Scalable tensor decompositions for multi-aspect data mining. In: Eighth IEEE International Conference on Data Mining, ICDM 2008, pp. 363–372 (2008)

13. Kolda, T.G., Bader, B.W., Kenny, J.P.: Higher-Order Web Link Analysis Using Multilinear Algebra. In: IEEE International Conference on Data Mining, pp. 242–249 (2005)

14. Lathauwer, J.V.L.D., Moor, B.D.: A multilinear singualr value decomposition. SIAM Journal of Matrix Analysis and Applications 21, 1253–1278 (2000)

15. Rendle, S., Balby Marinho, L., Nanopoulos, A., Schmidt-Thieme, L.: Learning optimal ranking with tensor factorization for tag recommendation. In: KDD 2009: Proceeding of the 15th ACM SIGKDD International Conference on Knowledge Discovery and Data Mining, New York, NY, USA, pp. 31:279–31:311 (2009)

16. Tucker, L.: Some mathematical notes on three-mode factor analysis. Psychometrika 31(3), 279–311 (1966)

17. Wang, H., Wu, Q., Shi, L., Yu, Y., Ahuja, N.: Out-of-Core Tensor Approximation of Multi-Dimensional Matrices of Visual Data. In: SIGGRAPH, vol. 24, pp. 527–535 (2005)

Location-Based Recommendation Using Incremental Tensor Factorization Model

Benyou Zou[1,2], Cuiping Li[1,2,*], Liwen Tan[1,2], and Hong Chen[1,2]

[1] Key Lab of Data Engineering and Knowledge Engineering of MOE,
Renmin University of China, Beijing, China, 100872
[2] School of Information, Renmin University of China, Beijing, China, 100872
{zoubenyou,licuiping,tan1204,chong}@ruc.edu.cn

Abstract. Newly emerging location-based online social services, such as Meetup and Douban, have experienced increased popularity and rapid growth. The classical *Matrix Factorization* methods usually only consider the user-item matrix. Recently, Researchers have extended the matrix adding location context as a tensor and used the *Tensor Factorization* methods for this scenario. However, in real scenario, the users and events are changing over time, the classical *Tensor Factorization* methods suffers the limitation that it can only be applied for static settings. In this paper, we propose a general *Incremental Tensor Factorization* model, which models the appearance changes of a tensor by adaptively updating its previous factorized components rather than recomputing them on the whole data every time the data changed. Experiments show that the proposed methods can offer more effective recommendations than baselines, and significantly improve the efficiency of providing location recommendations.

Keywords: Location-Based Social Networks, Location-Based Services, Recommendation Systems, Incremental Tensor Factorization.

1 Introduction

The advances in location-acquisition and wireless communication technologies enable people to add a location dimension to traditional social networks, a bunch of location-based social networking services (LBSNs) [16] , where users can easily share experiences in the physical world via mobile devices. e.g., Foursquare[1], GeoLife [17], Meetup[2] and DoubanEvent[3], have provided convenient online platforms for people to create, distribute and organize social events[8]. Many of these networks have attracted a huge number of users and have experienced rapid business growth. For example, Meetup has 9.5 million active users, creating 280,000 social events every month.

While the information about users' location and relationships is important to accurately model their behaviors and improve their experience, researchers proposed to incorporate location information into the recommender systems. Wei Zhang *et al.* [14]

---

* Corresponding author.
[1] https://foursquare.com/
[2] www.meetup.com
[3] www.douban.com/events/

X. Luo, J.X. Yu, and Z. Li (Eds.): ADMA 2014, LNAI 8933, pp. 227–238, 2014.

combines latent factor model with location features for group recommendation, which considers location features, social features, and implicit patterns simultaneously in a unified model. As tensor model offers a natural representation for multiple facets, Nan Zheng *et al.* [15] propose a tensor decomposition-based group recommendation model to combine semantic tags with social relations to suggest groups to users. However, all these methods suffer from the limitation:

1. The nature of real-world events indicates location information should be considered in the proposed methods as well. Yet, previous research methods in group and event recommendation do not exploit location features of users and groups.
2. It only works for static settings. In many real-world scenarios, such as Meetup's event recommendation, Foursquare's group recommendation, and Netflix's movie recommendation, data is changing over time. In this case, the existing methods have to rebuild the model on updated data, that means, all previous factorized components have to be recomputed. This is typically an expensive task. Our aim is to develop an incremental method that can continuously update the model to reflect its present nature.

In this paper, we adopt the tensor factorization model to the location-based recommendation. we observe that in real-world applications, the content universe changes rapidly, thus redoing the factorization on updated data is often infeasible. To address this limitation, we propose an incremental tensor factorization algorithm, an effective incremental method which models the appearance changes of a tensor by adaptively updating its previous factorized components rather than recomputing them on the whole dataset every time data changes.

The rest of this paper is organized as follows. Section 2 gives the background information of our study. Section 3 introduces the incremental tensor factorization method for dynamic data sets. A performance analysis of our methods is presented in Section 4. We conclude the study in Section 5.

2   Preliminaries

2.1   Location-Based Recommendation

Recommendations in location based social network has been studied in recent years[2,10,15]. Figure 1 show the overview of location-based recommendations. Given a query city, the system recommender the events based on the user's location history. Zheng[15] proposed an proposed tensor-based location recommendation. While in real LBSN, the users and events number are increasing rapidly, so we proposed an incremental tensor based method to apply the location-based recommendation.

2.2   Tensor Factorization

Notations and Assumptions. Here, we briefly recall the necessary preliminaries on tensors. Table 1 lists the main symbols we use throughout the paper. Without loss of generality, in this paper, matrices are denoted by bold upper case letters, such as **A**, **B**,

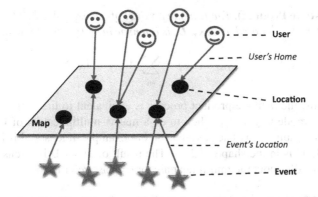

Fig. 1. Overview of Location-Based Events Social Networks

Table 1. Symbols

Symbol	Definition and Description
$\mathbf{A},\mathbf{B}$	Matrices (bold upper case)
$\mathbf{A}(i,j)$	An element of $\mathbf{A}$
$\mathbf{A}(i,:)$ or $\mathbf{A}(:,i)$	i-th row or column of $\mathbf{A}$
$\mathcal{A},\mathcal{B}$	Tensors (calligraphic letters)
$\mathcal{A}(i_1,\ldots,i_N)$	An element of tensor $\mathcal{A}$
$\mathcal{A}(\ldots,i_{n-1},:,i_{n+1},\ldots)$	the n-mode vectors of $\mathcal{A}$
N	The order of a tensor
$uf(\mathcal{A},n)$ or $\mathbf{A}_{(n)}$	The n-mode unfolding matrix of $\mathcal{A}$
$\mathbf{U}^{(n)}$	n-mode basis matrix of $\mathbf{A}_{(n)}$
$\times_n$	the n-mode product
$\mathcal{A}_{[s_1,\ldots,s_N]}$	A block of tensor $\mathcal{A}$

and tensors by calligraphic letters, such as $\mathcal{A}$, $\mathcal{B}$. Tensors are higher-order generalizations of vectors and matrices. High-order tensors are denoted as: $\mathcal{A} \in \Re^{I_1 \times I_2 \times \ldots \times I_N}$, where the order of $\mathcal{A}$ is N ($N > 2$) while a vector and a matrix is a tensor of order 1 and 2, respectively.

Each dimension of a tensor is called a *mode*. For example, $\mathcal{A} \in \Re^{I_1 \times I_2 \times \ldots \times I_N}$ is a tensor with N modes (called N^{th} order tensor). An element of $\mathcal{A}$ is denoted as $\mathcal{A}(i_1, \ldots, i_n, \ldots, i_N)$. The n-mode vectors of $\mathcal{A}$ are defined as the I_n-dimensional vectors obtained from $\mathcal{A}$ by varying its index in while keeping all the other indices fixed. In this paper we denote it as $\mathcal{A}(\ldots, i_{n-1}, :, i_{n+1}, \ldots)$.

Definition 1 (Matrix Unfolding). *Unfolding a tensor* $\mathcal{A} \in \Re^{I_1 \times \ldots I_n \ldots I_N}$ *along the* n^{th} *mode is denoted as* $uf(\mathcal{A}, n)$, *which results in a matrix,* $\mathbf{A}_{(n)} \in \Re^{I_n \times (I_1 I_2 \ldots I_{n-1} I_{n+1} \ldots I_N)}$, *where the column vectors of* $\mathbf{A}_{(n)}$ *are the* n-*mode vectors of* $\mathcal{A}$.

For example, Figure 4 shows a 4-order tensor of shape $2 \times 2 \times 2 \times 3$ and its four unfolding matrices along different modes. The color difference in the figure is used for the description of later incremental update and can be ignored now.

Definition 2 (Mode Product). *The n-mode product of a tensor* $\mathcal{A} \in \Re^{I_1 \times \cdots I_n \cdots I_N}$ *and a matrix* $U \in \Re^{J_n \times I_n}$, *denoted as* $\mathcal{A} \times_n U$, *is a tensor in* $\Re^{I_1 \times \cdots J_n \cdots I_N}$ *with entries:*

$$(\mathcal{A} \times_n U)(i_1, ..., j_n, ..., i_N) = \sum_{i_n} \mathcal{A}(i_1, ..., i_N)U(j_n, i_n).$$

In other words, the n-mode product process is equivalent to first unfold the tensor $\mathcal{A}$ along the nth mode to get $\mathcal{A}_{(1)}$, then to do a matrix multiplication of U and $\mathcal{A}_{(1)}$, finally to fold the result matrix back as a tensor. For example, if tensor $\mathcal{A}$ is in the shape of $3 \times 4 \times 5$ while U is in the shape of 2×3. The result of $\mathcal{A} \times_1 U$ is a tensor of shape $2 \times 4 \times 5$. More details about the multilinear algebra, please refer to [6].

HOSVD-Decomposition. In literature, a great many tensor factorization models have been proposed, including Tucker models [12] and PARAFAC models [3] or CANDE-COMP models [5]. In [6], Tucker3 with orthogonality constraints on the components, has been named as n-mode SVD (also called HOSVD). In this paper, we follow the HOSVD formulation in which a N-order tensor $\mathcal{A}$ is factorized into N basis matrices $U^{(1)} \in \Re^{I_1 \times R_1}$, $U^{(2)} \in \Re^{I_2 \times R_2}$, ..., $U^{(N)} \in \Re^{I_N \times R_N}$, and one core tensor $\mathcal{C} \in \Re^{R_1 \times R_2 \times \cdots \times R_N}$. In this case:

$$\mathcal{A} \approx \mathcal{C} \times_1 U^{(1)} \times_2 U^{(2)} \ldots \times_N U^{(N)}$$

where $U^{(n)} \in \Re^{I_n \times R_n}$ has orthonormal columns for $1 \leqslant i \leqslant N$. It is obtained by first performing regular SVD on unfolded matrix $\mathbf{A}_{(n)}$ and then extracting the first R_n columns of the left singular matrix of $\mathbf{A}_{(n)}$.

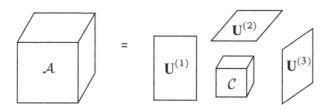

Fig. 2. An example of 3-mode SVD

Figure 2 shows an example of 3-mode SVD in which a 3-mode tensor is factorized into 3 basis matrices $\mathbf{U}^{(1)}, \mathbf{U}^{(2)}, \mathbf{U}^{(3)}$, and one core tensor $\mathcal{C}$. The HOSVD factorization allows an interaction between a factor with any factor in the other modes. Furthermore, it allows for full control over the dimensionality by performing truncations on the factors, which is very important for large real data sets. The aim now is to find a good approximation $\widetilde{\mathcal{A}}$ for $\mathcal{A}$ such that the loss function $L(\widetilde{\mathcal{A}}, \mathcal{A})$ between observed and estimated values is minimized.

Definition 3 (Rank-$(R_1, R_2, ..., R_N)$ Approximation). *Given a Nth order tensor $\mathcal{A} \in \Re^{I_1 \times I_2 \times \cdots \times I_N}$, rank-$(R_1, R_2, ..., R_N)$ approximation of $\mathcal{A}$ is formulated as finding a* *lower-rank tensor* $\tilde{\mathcal{A}} \in \mathfrak{R}^{I_1 \times I_2 \times \ldots \times I_N}$, *with* $Rank_n(\tilde{\mathcal{A}}) = R_n \leqslant Rank_n(\mathcal{A})$, *for all* n, *such that the following least-squares cost function is minimized:*

$$\tilde{\mathcal{A}} = arg \ \min_{\tilde{\mathcal{A}}} \|\mathcal{A} - \tilde{\mathcal{A}}\|^2.$$

When R_n is much smaller than I_n for all n, the best rank approximation $\tilde{\mathcal{A}}$ gives a compact approximation of the original tensor $\mathcal{A}$, resulting in good data compression for large real data sets.

Given basis matrices $\mathbf{U}^{(1)}, \mathbf{U}^{(2)}, \ldots, \mathbf{U}^{(N)}$, the core tensor $\mathcal{C}$ can be readily computed as $\mathcal{C} = \mathcal{A} \times_1 \mathbf{U}^{(1)^T} \times_2 \mathbf{U}^{(2)^T} \ldots \times_N \mathbf{U}^{(N)^T}$. Thus, the optimization problem focuses on the computation of the basis matrices only. In this paper, we seek efficient computation of the best rank-$(R_1, R_2, \ldots, R_N)$ approximation of $\mathcal{A}$ as well as its incremental update.

3 Incremental Tensor Decomposition Based Events Recommendation Model

Before we present the proposed online incremental tensor decomposition method, we first give a brief review of the related background as well as the introduction to the notations and symbols we use.

3.1 Introduction to SVD-Updating

The SVD-Updating method [9], Given an $m \times n$ matrix A,the rank-k approximation of A is denoted as:

$$A_k = U_k \times \Sigma_k \times V_k^T \qquad (1)$$

Let D denote the new appended matrix to process, then D is a $m \times p$ matrix, D is appended to the columns of the rank-k approximation of the $m \times n$ matrix A, so that the k-largest singular values and corresponding singular vectors of

$$B = (A_k | D) \qquad (2)$$

are computed. The process was illustrated in Figure 3. Let B from Equation(1) and define SVD(B) $= U_B \Sigma_B V_V^T$. Then

$$U^T B \begin{pmatrix} V_k \\ & I_d \end{pmatrix} = (\Sigma_K | U_K^T D),$$

since $A_k = U_K \Sigma_k V_K^T$. If $F = (\Sigma_k | U_k^T D)$,and SVD(F)$=U_F \Sigma_F V_F^T$, then it follows that:

$$U_B = U_K U_F \qquad (3)$$

$$V_B = \begin{pmatrix} V_k \\ & I_d \end{pmatrix} V_F, \qquad (4)$$

since $(U_F U_K)^T B \begin{pmatrix} V_k \\ & I_d \end{pmatrix} = \Sigma_F = \Sigma_B$. Hence U_B and V_B are $m \times k$ and $(n+d) \times (k+d)$ dense matrices, respectively.

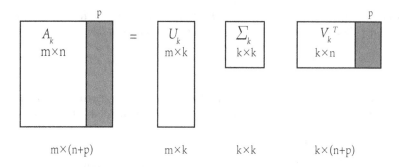

Fig. 3. Mathematical representation of row SVD updating

3.2   Incremental Rank-$(R_1, R_2, \ldots, R_N)$ High-Order Tensor Approximation

Based on HOSVD and inspired by [7], we proposed an Incremental Rank-$(R_1, R_2, \ldots, R_N)$ Approximation method for N-order tensor(IRHOTA), our method presented below efficiently identifies the dominant projection subspaces of N-order tensor, and is capable of incrementally updating these subspaces when new data arrive. Given the SVD of the mode-k unfolding matrix $A_{(k)}(1 \leq k \leq N)$ for a N-order tensor $\mathcal{A} \in \Re^{I_1 \times I_2 \times \ldots \times I_N}$, IRHOTA is able to efficiently compute the SVD of the mode-i unfolding matrix $A_i^*(1 \leq k \leq N)$ for $\mathcal{A} = (\mathcal{A}^*|\mathcal{F}) \in \Re^{I_1 \times I_2 \times \ldots \times I_n^* \ldots \times I_N}$ where $\mathcal{F} \in \Re^{I_1 \times I_2 \times \ldots \times I_n' \ldots \times I_N}$ in a new N-order subtensor and $I_n^* = I_n + I_n'$.

In this article, we take the increased dimension of tensor as the last order, we denote the new tensor as

$$\mathcal{A} = (\mathcal{A}^*|\mathcal{F}) \in \Re^{I_1 \times I_2 \times \ldots \times I_N^*} \tag{5}$$

and newly added subtensor

$$\mathcal{F} \in \Re^{I_1 \times I_2 \times \ldots \times I_N'} \tag{6}$$

where $I_N^* = I_N + I_N'$, the mode-i unfolding matrix

$$A_{(i)}^* \in \begin{cases} \Re^{I_1 \times (I_2 \times \ldots \times I_N^*)} & \text{if } i = 1 \\ \Re^{I_N^* \times (I_1 \times \ldots I_{N-1})} & \text{if } i = N \\ \Re^{I_i \times (I_{i+1} \times \ldots \times I_N^* \times I_1 \times \ldots I_{i-1})} & \text{otherwise} \end{cases} \tag{7}$$

To facilitate the description, we take a 4-order tensor as example, Figuer 1 is used for illustration. In the left half of Figure 1, four identical tensors are unfolded in four different modes. For each tensor, the white regions represent the original subtensor while the dark regions denote the newly added subtensor, The four unfolding matrices corresponding to the new data subtensors, the column spaces of $A_{(1)}^*$, $A_{(2)}^*$ and $A_{(3)}^*$ are extended at the same time when the row space of $A_{(4)}^*$ is extended. Consequently, IRHOTA needs to track the changes these three unfolding spaces, and needs to identify the dominant projection subspaces for a compact representation of the tensor. It is note that $A_{(2)}^*$ and $A_{(3)}^*$ can be decomposed as: $A_{(2)}^* = (A_{(2)}|F_{(2)}) \cdot P_2 = B_2 \cdot P_2$, $A_{(3)}^* = (A_{(3)}|F_{(3)}) \cdot P_3 = B_3 \cdot P_3$ where $B_2 = (A_{(2)}|F_{(2)})$, $B_3 = (A_{(3)}|F_{(3)})$ and P_2, P_3 is

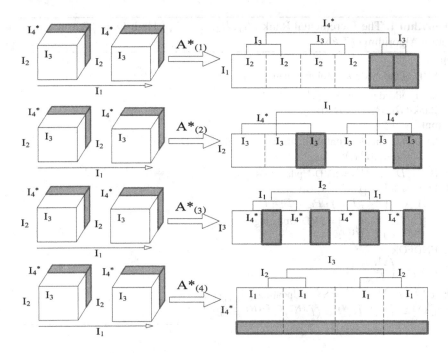

Fig. 4. Illustration of unfolding a (4 - order) tensor

orthonormal matrix obtained by column exchange and transpose operations on identity matrix G. Let

$$G_i = (E_1|Q_1|E_2|Q_2|...|E_k|Q_k), k = I_1 \times I_{i-1}$$

$$E_i, Q_i \in \Re^{(I_1 \times ...I_{k-1}I_{k+1} \times ...I_N^*) \times (I_{k+1} \times ...I_N)}$$

which is generated by partitioning G_i into $2I_1 \times I_{i-1}$ blocks in the column dimension. Consequently, the orthonormal matrix P_i is formulated as:

$$P_i = (E_1|E_2|...|E_{I_1 \times I_{i-1}}|Q_1|Q_2|...|Q_{I_1 \times I_{i-1}})^T. \tag{8}$$

In this way, $A^*_{(2)}$'s and $A^*_{(3)}$'s SVD is efficiently computed on the basis of P and B's SVD. Further more, $A^*_{(1)}$'s SVD is efficiently obtained by performing SVD on the matrix $(A_{(1)}|F(1))$. Similarly, $A^*_{(3)}$'s SVD is efficiently obtained by performing SVD on the matrix $\left(\dfrac{A_{(3)}}{F_{(3)}}\right)^T$. The specific procedure of IRHOTA is list in Algorithm 1.

4   Experiments

In this section, we first describe the settings of experiments including the data sets, comparative approaches, and the evaluation method. Then we report the experimental results on both the recommendation effectiveness and efficiency of our recommendation model.

Algorithm 1. The Incremental Rank-$(R_1, R_2, \ldots, R_N)$ High-Order Tensor Approximation Algorithm (ITF)

Input:

SVD of the mode-k unfolding matrix $A_{(k)}$,i.e. $U^{(k)} D^{(k)} V^{(k)^T}$;

newly added tensor $\mathcal{F} \in \Re^{I_1 \times I_2 \times \ldots \times I'_N}$;

Rank-$(R_1, R_2, \ldots, R_N)$;

Output:

SVD of the mode-k unfolding matrix $A^*_{(k)}$,i.e. $\widehat{U}^{(k)} \widehat{D}^{(k)} \widehat{V}^{(k)^T}$;

1: $A^*_{(1)} = (A_{(1)} | F_{(1)})$;

2: $[\widehat{U}^{(1)}, \widehat{D}^{(1)}, \widehat{V}^{(1)}] = $ SVD-Updating$(A^*_{(1)})$;

3: FOR each $k \in [1, N-1]$ Do

4:     $A^*_{(k)} = (A_{(k)} | F_{(k)}) \cdot P_k = B_k \cdot P_K$;

5:     $[\widehat{U}^{(k)}, \widehat{D}^{(k)}, \widetilde{V}^{(k)}] = $ SVD-Updating(B_k);

6:     $\widehat{V}^{(k)} = P_k^t \cdot \widetilde{V}^{(k)}$;

7: ENDFOR

8: $A^*_{(N)} = \left(\dfrac{A_{(N)}}{F_{(N)}} \right)$;

9: $[\widetilde{U}^{(N)}, \widetilde{D}^{(N)}, \widetilde{V}^{(N)}] = $ SVD-Updating$((A^*_{(N)})^T)$;

10: $\widehat{U}^{(N)} = \widetilde{V}^{(N)}, \widehat{D}^{(N)} = (\widetilde{D}^{(N)})^T, \widehat{V}^{(N)} = \widetilde{U}^{(N)}$.

4.1 Data Sets

The DoubanEvent dataset we use is provided by Hongzhi Yin [4]. DoubanEvent is China's largest event-based social networking site, users can publish and participate in social events. This dataset consists 100,000 users, 300,000 events and 3,500,000 positive definite RSVPs. The following information is recorded when collecting the data: 1) user information, including user-id, user-name and user-home city; 2) event information, consisting of event-id, eventname, event-latitude, event-longitude, event-summary and its category.

As discussed before, our incremental tensor factorization model can handle the user-event-location tensor without redo factorization when the tensor changed. To evaluate the performance of our incremental tensor factorization model and compare the time cost of the different algorithms, we separate the training data in 6 parts: we take 50% training data as base training data, the remaining 50% training data was divided to 5 parts, as incremental training data, every part was 10% of training data. We take 5 steps to add the incremental training data to base training data.

4.2 Comparative Approaches

We compare our proposed method with the flowing competitor methods:

1. **Matrix factorization(MF):** It is a fundamental type of latent factor model for recommendation[13][4]. We consider the user-event matrix and use MF to recommender the event to users.

---

[4] http://net.pku.edu.cn/daim/yinhongzhi/index.html

2. **Tensor factorization(TF)**: The classical tensor factorization model we introduced in 2, used by [15] [11]. The method considers the user's location as one dimension of the user-event-location tensor.
3. **Incremental Tensor factorization(ITF)**: The method we proposed which described in Section 3. The method aims to reduce the computation time when the user-event-location tensor changed, eg. the number of users increased or new events were organized.

4.3 Evaluation Measures

We evaluate both the effectiveness of the suggested recommendations and the efficiency for generating online recommendations with the baseline solutions.

Recommendation Effectiveness. To make the effectiveness evaluation, we select the location history generated in a querying city as a test set and use the rest of the user's location history as a training set for us to learn the user's preference. We regard the venues that a user has visited in the querying city as the ground truths and match the recommended events against these venues. Based on the given city and user's location history, some events will be recommended by our system. To evaluate the recommender models, we adopt the testing methodology and the measurement precision and recall applied in[1].

$$precision = \frac{number\ of\ recovered\ ground\ truths}{total\ number\ of\ recommendations} \qquad (9)$$

$$recall = \frac{number\ of\ recovered\ ground\ truths}{total\ number\ of\ ground\ truths} \qquad (10)$$

Recommendation Efficiency. The Incremental Location Recommendation Model we proposed is to updating the previous factorized components, while the classical Tensor Factorization method will redo the factorization when the data changed. We test the efficiency of our proposed method using compare the time costs of different recommendation.

4.4 Experimental Results

In this subsection, we first report the effectiveness and then compare the efficiency of different recommendation algorithms.

Effectiveness. Figure 5 and Figure 6 reports the performance of the recommendation algorithms on DoubanEvent dataset. We show only the performance where k is in the range [3...30], because a greater value of k is usually ignored for a typical top-k recommendation task. As shown in Figure 5(a) , our proposed model outperforms the MF

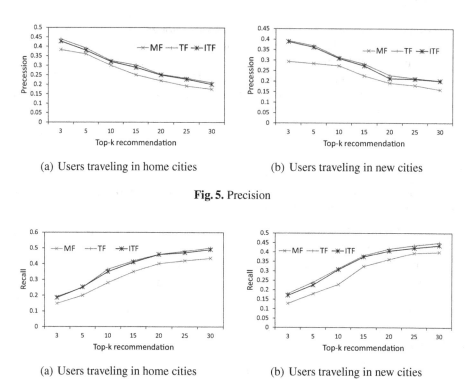

(a) Users traveling in home cities   (b) Users traveling in new cities

Fig. 5. Precision

(a) Users traveling in home cities   (b) Users traveling in new cities

Fig. 6. Recall

method, and nearly the same as the classical TF method, because our incremental tensor factorization model is an incremental way to do the tensor factorization, it can significantly speed up the factorization without redo the factorization operation when the tenor changed.

In Figure 6(a) and Figure 6(b), we report the recall of all recommendation algorithms where query occur when users are travel in home cites and user's travel to another cities. From the figure we can see that the trend of comparison result is similar to that presented in Figure 5.

From the figures, we observe that the tensor-based method can efficiently aware the location infirmations, it can give the better recommend events to users.

Efficiency. In the efficiency study of the models, we aims to compare the time consumption of these methods when the tensor changed. For the MF method and TF method, they will be redo the factorization operations when tensor changed, it take a lot of time, while the ITF method we proposed can adaptively updating its previous factorized components saves a lot of time. Figure 7 shows that our method is significantly improve the time consumption, when the tensor and matrix scale up, the superiority is more significant.

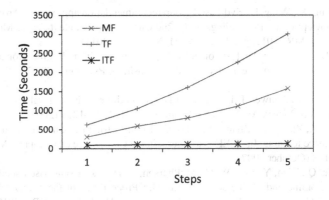

Fig. 7. Time cost of every step

5 Conclusion

This paper proposed a tensor-based location-aware recommender system, which provides a user with spatial item recommendations with the querying city based on the location preferences of users. Based on the observation that the context information changes dynamically and frequently, we present an incremental method models can adaptively updating its previous factorized components rather than recomputing them on the whole data. We evaluated our system using extensive experiments on real data set. According to the experimental results, our approach significantly outperforms the MF method and can significantly save computing times.

Acknowledgements. This work was supported by National Basic Research Program of China (973Program) (No.2014CB340402, No. 2012CB316205), National High Technology Research and Development Program of China (863 Program) (No.2014AA015204), NSFC under the grant No.61272137, 61033010, 61202114, 61165004 and NSSFC (No: 12&ZD220). It was partially done when the authors worked in SA Center for Big Data Research in RUC. This Center is funded by a Chinese National 111 Project Attracting.

References

1. Bao, J., Zheng, Y., Mokbel, M.F.: Location-based and preference-aware recommendation using sparse geo-social networking data. In: Proceedings of the 20th International Conference on Advances in Geographic Information Systems, SIGSPATIAL 2012, pp. 199–208. ACM, New York (2012)
2. Cremonesi, P., Koren, Y., Turrin, R.: Performance of recommender algorithms on top-n recommendation tasks. In: Proceedings of the Fourth ACM Conference on Recommender Systems, RecSys 2010, pp. 39–46. ACM, New York (2010)
3. Harshman, R.: Foundations of the Parafac Procedure: Models and Conditions for an "explanatory" Multimodal Factor Analysis. In: UCLC Working Papers in Phonetics, University of California at Los Angeles (1970)

4. Hu, X., Meng, X., Wang, L.: Svd-based group recommendation approaches: An experimental study of moviepilot. In: Proceedings of the 2nd Challenge on Context-Aware Movie Recommendation, CAMRa 2011, pp. 23–28. ACM, New York (2011)

5. Carroll, J.D., Chang, J.: Analysis of individual differences in multidimensional scaling via an n-way generalization of eckart-young decomposition. Psychometrika 35, 35:283–35:319 (1970)

6. Vandewalle, J., Lathauwer, L.D., Moor, B.D.: A multilinear singualr value decomposition. SIAM Journal of Matrix Analysis and Applications 21, 1253–1278 (2000)

7. Li, X., Hu, W., Zhang, Z., Zhang, X., Luo, G.: Robust visual tracking based on incremental tensor subspace learning. In: IEEE 11th International Conference on Computer Vision, ICCV 2007, pp. 1–8 (October 2007)

8. Liu, X., He, Q., Tian, Y., Lee, W.-C., McPherson, J., Han, J.: Event-based social networks: Linking the online and offline social worlds. In: Proceedings of the 18th ACM SIGKDD International Conference on Knowledge Discovery and Data Mining, KDD 2012, pp. 1032–1040. ACM, New York (2012)

9. O'Brien, G.W.: Information management tools for updating an svd-encoded indexing scheme (1994)

10. Scellato, S., Noulas, A., Lambiotte, R., Mascolo, C.: Socio-spatial properties of online location-based social networks. In: Adamic, L.A., Baeza-Yates, R.A., Counts, S. (eds.) ICWSM. The AAAI Press (2011)

11. Sun, J., Zeng, H., Liu, H., Lu, Y., Chen, Z.: Cubesvd: A novel approach to personalized web search. In: Proceedings of the 14th International Conference on World Wide Web, WWW 2005, pp. 382–390. ACM, New York (2005)

12. Tucker, L.: Some mathematical notes on three-mode factor analysis. Psychometrika 31(3), 279–311 (1966)

13. Vozalis, M.G., Margaritis, K.G.: Applying svd on item-based filtering. In: Proceedings of the 5th International Conference on Intelligent Systems Design and Applications, ISDA 2005, pp. 464–469. IEEE Computer Society, Washington, DC (2005)

14. Zhang, W., Wang, J., Feng, W.: Combining latent factor model with location features for event-based group recommendation. In: Proceedings of the 19th ACM SIGKDD International Conference on Knowledge Discovery and Data Mining, pp. 910–918. ACM (2013)

15. Zheng, N., Li, Q., Liao, S., Zhang, L.: Flickr group recommendation based on tensor decomposition. In: Proceedings of the 33rd International ACM SIGIR Conference on Research and Development in Information Retrieval, SIGIR 2010, pp. 737–738. ACM, New York (2010)

16. Zheng, Y.: Location-based social networks: Users. In: Zheng, Y., Zhou, X. (eds.) Computing with Spatial Trajectories, pp. 243–276. Springer, New York (2011)

17. Zheng, Y., Xie, X., Ma, W.-Y.: Geolife: A collaborative social networking service among user, location and trajectory. IEEE Data Eng. Bull. 33(2), 32–39 (2010)

Mining Continuous Activity Patterns from Animal Trajectory Data

Yuwei Wang[1,2], Ze Luo[1], Baoping Yan[1], John Takekawa[3],
Diann Prosser[4], and Scott Newman[5]

[1] Computer Network Information Center,
Chinese Academy of Sciences, Beijing, China
[2] University of Chinese Academy of Sciences, Beijing, China
[3] US Geological Survey, Western Ecological Research Center, California, USA
[4] US Geological Survey, Patuxent Wildlife Research Center, Maryland, USA
[5] EMPRES Wildlife Health and Ecology Unit, Animal Production and Health
Division, Food and Agriculture Organization of the United Nations, Rome, Italy

Abstract. The increasing availability of animal tracking data brings us
opportunities and challenges to intuitively understand the mechanisms
of animal activities. In this paper, we aim to discover animal movement
patterns from animal trajectory data. In particular, we propose a notion
of continuous activity pattern as the concise representation of underlying similar spatio-temporal movements, and develop an extension and
refinement framework to discover the patterns. We first preprocess the
trajectories into significant semantic locations with time property. Then,
we apply a projection-based approach to generate candidate patterns and
refine them to generate true patterns. A sequence graph structure and
a simple and effective processing strategy is further developed to reduce
the computational overhead. The proposed approaches are extensively
validated on both real GPS datasets and large synthetic datasets.

Keywords: movement patterns, continuous activity patterns.

1 Introduction

The increasing available movement data brings us an opportunity to understand
mobility-related phenomena. Many movement patterns have been revealed from
the trajectory data, like relationship patterns [1,2], moving sequential patterns
[3,4,6] and periodic patterns [5]. Some of these techniques have been applied
to animal trajectory data. Studies on animal activity patterns [7,8] could help
reveal intrinsic links between animal movements and external stimuli such as
change of environmental conditions or spread of disease.

In this paper, we define a notion of continuous activity patterns to represent
similar activity sequences, which is an extension of the spatio-temporal sequential pattern. Traditional approaches [4,9,3] are not effective enough to discover
these patterns. Firstly, these approaches usually transform raw trajectories into
sequences of dense regions (i.e. the regions given beforehand or frequently visited

X. Luo, J.X. Yu, and Z. Li (Eds.): ADMA 2014, LNAI 8933, pp. 239–252, 2014.

by the moving objects) before pattern mining. The generated patterns are not precise because false or very large regions could be formed owing to numerous stochastic locations [10]. The final pattern instances may be far apart. Secondly, dense regions are constructed from locations in the spatial dimension, ignoring the time dimension [11]. This also leads to imprecision when we seek spatio-temporal patterns. Thirdly, existing approaches either have no constraint on the continuity of the activities in a pattern or require strict continuity [3,4,9]. In the first case, the generated pattern may be not significant because the activities in the pattern are not closely linked. In the second case, interesting patterns involving non-strict continuous activities may be overlooked.

We propose an extension and refinement framework to discover continuous activity patterns from trajectory data. First, individual potential behaviors with temporal and spatial attributes are extracted from raw trajectories. Then, a refinement procedure is dynamically intertwined with an extension procedure to derive all precise patterns, i.e. similar activity sequences. A certain degree of continuity of the activities is guaranteed while some small individual divergences are tolerated. The main contributions presented in this paper are: 1) An introduction of the notion of continuous activity patterns, including a framework to discover these patterns; 2) development of a sequence graph structure and use of a simple and efficient strategy to improve the performance of the framework; 3) validation of our proposed approaches using a real GPS dataset and large synthetic datasets.

The rest of the paper is organized as follows. Section 2 discusses related work. Section 3 formally defines our problem and Section 4 presents the extension and refinement framework. Section 5 shows our experimental study and Section 6 summarizes this paper.

2   Related Work

As an extension of frequent sequential pattern mining, the spatio-temporal sequential pattern mining has attracted much attention in recent years. For moving to the continuous spatio-temporal context, a basic challenge is that the spatial coordinates do not repeat themselves exactly [4]. A popular approach is to transform the data into sequences of regions by discretizing the geographical space and to consider the movement in terms of regions that moving objects traverse [11]. The regions of interest either are given beforehand or are identified automatically as the frequently visited regions. Morzy [12] divides the space into cells and then mines frequent cell sequences based on the well-known Apriori and PrefixSpan algorithms [13]. Cao et al. [4] first decompose the raw trajectory into line segments, and combine frequent similar line segments into popular regions. They then mine frequent region sequences using a substring tree structure.

Some studies further capture the temporal information, and propose novel patterns. Giannotti et al. [3] introduce the temporally annotated sequential pattern, namely T-Pattern, in which the movement between two regions is annotated with a typical transition time. Instead of the static spatial decomposition, dense regions are incrementally identified locally on the trajectory projections. Hence, the latter part of the pattern becomes more precise. Tseng et al. discover the temporal movement patterns in sensor networks [14]. The transition times in the movements complying with a pattern must be equal. Lee et al. [9] partition the space into cells and propose a graph-based algorithm to mine movement patterns. Li et al. [15] aim to discover similar movements with similar transition times from users' trajectories, and uses them to calculate the user similarity.

The approaches outlined above need improvement as imprecise regions may be formed due to the presence of many irregular locations in the trajectories which do not actually comply with the mobility pattern. In addition, these approaches either have no adjacency constraint regarding the spatial and temporal gaps between consecutive pattern elements [3,12,15] or enforce an immediate adjacency constraint [4,9], and therefore lose the flexibility to discover real-life similar movements with a certain degree of divergence.

3   Problem Definition

In this section, we formulate the problem of continuous activity pattern mining. The main notations used in this paper are listed in Table 1.

Table 1. List of Notations

Notation	Explanation	Notation	Explanation
$traj, traj_i$	trajectories	C, C_i	activity clusters
$dist(p,q)$	the spatial distance between p and q	P, P_i	continuous activity patterns
sp, sp_i	activity spots	$sp_{l,i}$	i-th activity spot in C_l
g, d	gap and distance thresholds of the activity spot	EPS	spatio-temporal distance threshold
$center(sp)$	center of sp	min_sup	minimum support
$sp.s, sp.e, sp.d$	starting, ending and duration times of sp	ext	max extendible time of the continuous activity pattern
$traj(sp)$	trajectory that sp belongs to	$proj(P)$	projection of the spot sequence set with regard to P

Definition 1 (Trajectory). *A trajectory $traj = \langle p_1, p_2, \ldots, p_n \rangle$ is a series of points generated by a moving object, where $p_i (1 \leq i \leq n)$ contains latitude lat, longitude lon and timestamp t, and $p_i.t < p_{i+1}.t (0 < i < n)$.*

Where the activities occur and how long the activities last are important clues to understand underlying activities. For example, according to [8], for migratory birds, a stay of a month in the summer indicates a possible breeding activity, and a stay of more than two months may mean moult, while a short stay means that the bird stops over there. We propose the concept of the activity spot to abstract potential activities from raw trajectories.

Definition 2 (Activity Spot). *Given a trajectory $traj = \langle p_1, p_2, \ldots, p_n \rangle$, a distance threshold d, a gap threshold g. An activity spot sp represents a maximal (possibly nonconsecutive) subsequence $traj = \langle p_{k_1}, p_{k_2}, \ldots, p_{k_m} \rangle$ of $traj$, where $0 < k_i - k_{i-1} \leq g$ ($\forall 1 < i \leq m$) and $dist(p_{k_i}, center(sp)) < d$ ($\forall 1 \leq i \leq m$).*

The activity spot means that the object remains in a spatial range. The points in an activity spot could be relaxed continuous in the raw trajectory because of inherent noise in trajectories. We use the distance to the center because animals often have a pendulum-like movement relative to central locations such as a nest. For simplicity, we use the word "spot" to refer to "activity spot".

Meaningful underlying activities are believed to be shared by multiple objects. These can be identified by clustering spots according to spatio-temporal similarities. We extend the notion of density-based clustering in [16] to the spatio-temporal context (the generated clusters are called **activity clusters**). Due to space limitations, we only give the concepts of the spatio-temporal neighborhood and the core, which are different from those in [16].

Definition 3 (Spatio-temporal Neighborhood). *Given a set of spots $S = \{sp_1, sp_2, \ldots, sp_n\}$, and a variable EPS in which $EPS.d$ and $EPS.t$ denote the distance thresholds in space and time separately. The spatio-temporal neighborhood of a spot sp with regard to EPS in S is the set $NER_{EPS}^{S}(sp) = \{sp_i \in S \mid |sp.d - sp_i.d| \leq EPS.t, dist(sp, sp_i) \leq EPS.d\}$.*

Definition 4 (Core Spot). *Given N trajectories, a set of spots S, the spatio-temporal distance threshold EPS, a minimum support min_sup, a spot sp is a core spot if $|\{traj(sp_i) \mid sp_i \in NER_{EPS}^{S}(sp)\}| \geq N * min\,\mathrm{sup}$.*

An activity cluster stands for an imprecise form of a common underlying activity. Along the pattern mining process, an activity cluster would evolve into one or several true underlying activities after eliminating the interference of some occasional spots. In the following we define the relaxed continuity among the spots, and then give the definition of the continuous activity pattern.

Definition 5 (Successor and Continuous Activity Sequence). *Let ext be the extendible time threshold, $SP = \langle sp_1, sp_2, \ldots, sp_n \rangle$ be the spot sequence of an object, $sp_i, sp_j \in SP$ and $sp_i \neq sp_j$. If $0 \leq sp_i.s - sp_j.e \leq ext$, then sp_i is a successor of sp_j. Let $SP' = \langle sp_{k_1}, sp_{k_2}, \ldots, sp_{k_m} \rangle$ be a (possibly nonconsecutive) subsequence of SP. SP' is called a continuous activity sequence if $sp_{k_{l+1}}$ is a successor of sp_{k_l} for $0 < l < m$.*

Definition 6 (Continuous Activity Pattern (CAP) and Pattern Instance). *Given N objects, EPS, min_sup and ext, a sequence $P = \langle C_1, C_2, \dots,$ $C_L \rangle$ is a Continuous Activity Pattern (CAP) of length L iff: 1) Similarity: $\forall 1 \le l \le L$, the spots in C_l constitute an activity cluster; 2) Continuity: $\forall sp_i \in C_l(1 \le l \le L)$, there exists a continuous activity sequence $inst =$ $\langle sp_{k_1}, sp_{k_2}, \dots, sp_{k_L} \rangle$ $(sp_{k_j} \in C_j, 1 \le j \le L)$ such that $sp_i = sp_{k_i}$; 3) Maximality: for any continuous activity sequence $inst = \langle sp_{k_1}, sp_{k_2}, \dots, sp_{k_L} \rangle$, if $\forall 1 \le l \le L$, $C_l \cup sp_{k_l}$ is an activity cluster, then $sp_{k_l} \in C_l$. We say that each continuous activity sequence $inst = \langle sp_{k_1}, sp_{k_2}, \dots, sp_{k_L} \rangle$ from P is a pattern instance of P if $sp_{k_l} \in C_l$ for $\forall 1 \le l \le L$.*

4   Continuous Activity Pattern Discovery

4.1   The Extension and Refinement Framework

We propose the Extension and Refinement Framework (**ERF**) for discovering CAPs from spot sequences. First, we extract the spots from raw trajectories. Each trajectory is scanned once and is transformed into a spot sequence. The process of spot extraction is straightforward and therefore omitted due to space limitations. Then, ERF follows the pattern-growth strategy in the well-known PrefixSpan algorithm [13], as shown in Algorithm 1. It mainly consists of two intertwined procedures: extension and refinement. For a CAP P, we try to extend it based on its projection and generate a set of candidate CAPs (Line 6). To ensure similarity constraint in Definition 6, each candidate is tested, updated and divided into several patterns in the refinement procedure (Line 9).

In the extension procedure (Algorithm 2), we first collect the subsequent spots which are in ext time distance of the last elements of current instances (Line 1), and then cluster these spots (Line 2). For each cluster, we append its elements to the corresponding pattern instances of P, and get a candidate CAP P' (Line 4). Then, we remove the instances which can not be extended (Line 6) and generate the projection of P' based on $proj(P)$ (Line 7). After the extension, P' needs to be verified, because the cluster structure may be different from that in P after pruning. Here, to avoid the unnecessary verification for any unchanged cluster $C_l(1 \le l \le L + 1)$, we record the previous size of C_l in an array $P'.sizes$ (Line 5). If the current size of C_l after pruning remains unchanged, the test on C_l in the refinement can be skipped.

In Example 1, we illuminate the projecting process of one sequence in D (Please refer to [13,3] for more information). For readability, we use the same symbol (e.g. a) instead of absolute spatio-temporal representation (i.e. the coordinates and time) to represent those spots, which currently are considered similar (in the same cluster currently).

Example 1. Given a spot sequence $SP = \langle (a, 1, 3), (b, 4, 5), (a, 6, 7), (c, 8, 9),$ $(d, 10, 12) \rangle$ in which the three units stand for a spot with its starting time and ending time. For the pattern $P = \langle a \rangle$, the projection of SP on P is $\{\langle (b, 4, 5),$ $(a, 6, 7), (c, 8, 9), (d, 10, 12) \rangle, \langle (c, 8, 9), (d, 10, 12) \rangle\}$. When extending P to $P' =$

Algorithm 1. The Extension and Refinement Framework(ERF)

Input: A set of spot sequences $D = \{SP_1, SP_2, \ldots, SP_N\}$, EPS, min_sup, ext
Output: A set of CAPs R
1 $P = \langle \rangle$, $proj(P) = D$; // initialize a CAP and its projection database
2 $EnQueue(Q, \langle P, proj(P) \rangle)$; // add the CAP and projection pair into Q
3 **while** $Q \neq \emptyset$ **do**
4 $\quad$ $T = Q.DeQueue()$; // $T = \langle P, proj(P) \rangle$
5 $\quad$ $R.add(T.P)$;
6 $\quad$ $S' = Extension(T, EPS, min\_sup, ext, N)$;
7 $\quad$ **for** $\langle P', proj(P') \rangle \in S'$ **do**
8 $\quad\quad$ $SUB = \{i \mid 1 \leq i \leq L\}$; // the subscripts of the clusters which need to be refined in P'
9 $\quad\quad$ $S = Refinement(\langle P', proj(P') \rangle, EPS, min\_sup, ext, N, SUB)$;
10 $\quad\quad$ $Q.EnQueueAll(S)$; // add all pairs into Q

11 **return** R;

Algorithm 2. Extension

Input: A CAP and projection pair $T = \langle P, proj(P) \rangle$, EPS, min_sup, ext, N
Output: A set of candidate CAP and projection pairs S'
1 $SP_{ext} = CollectSuccessors(T, ext)$;
2 $CS = ClusterSP(SP_{ext}, EPS, min\_sup, N)$;
3 **for** $C \in CS$ **do**
4 $\quad$ $P' = ExtendInstances(P, C)$;
5 $\quad$ $P'.sizes = Append(P.sizes, |C|)$;
6 $\quad$ $PruneInvalidInstances(P')$;
7 $\quad$ $proj(P') = ReProject(P', proj(P))$; // set $proj(P')$ from $proj(P)$
8 $\quad$ $S'.Add(\langle P', proj(P') \rangle)$;

9 **return** S';

$\langle a, b \rangle$, the projection on P' becomes $\{\langle (a, 6, 7), (c, 8, 9), (d, 10, 12) \rangle\}$, which is generated from the projection on P incrementally.

In the refinement procedure (Algorithm 3), we check whether all spots in the l-th ($l \in SUB$) cluster still constitute a cluster (Lines 4~5); If they do (Line 6), we continue to check the next until SUB becomes empty and returns the CAP (Line 17). Otherwise, we update P' with regard to the generated cluster set CS, split P' into new candidates (Line 8), and refine each candidate (Lines 9~15).

4.2 The Extension and Refinement Framework with the Sequence Graph Structure

In the naive ERF, a pattern instance is stored in a spot list. Owing to the use of the relaxed continuity, more than one new sequence could be generated from a current instance of length L in the extension.

Algorithm 3. Refinement

Input: $T' = \langle P', proj(P') \rangle$, EPS, min_sup, ext, N, SUB

Output: A set of CAP and projection pairs S

```
1  while SUB ≠ ∅ do
2  │   l = SUB.next() ; // get a subscript and remove it from SUB
3  │   if P'.sizes[l] = |Cl| then continue ; // Cl has not been changed
4  │   CS = ClusterSP(Cl, EPS, min_sup, N);
5  │   if Compare(CS, Cl) = true then
6  │   │   P'.sizes[l] = |Cl|;
7  │   else
8  │   │   S' = Split(T', CS) ; // split P' in T' into |CS| new candidates
9  │   │   for j = 1 to |CS| do
10 │   │   │   T'' = S'[j] ; // T'' = ⟨P'', proj(P'')⟩
11 │   │   │   CopyTo(P'.sizes, P''.sizes) ; // copy P'.sizes to P''.sizes
12 │   │   │   P''.size[l] = |CS[j]| ; // the new size of l-th cluster in P''
13 │   │   │   SUB'' = {k|1 ≤ k ≤ L, k ≠ l} ;
14 │   │   │   S'' = Refinement(T''', EPS, min_sup, ext, N, SUB'');
15 │   │   └   S.addAll(S'');
16 │   └   return S;
17 return {T'};
```

Example 2. Given a spot sequence $\langle (a, 1, 3), (b, 4, 5), (a, 6, 7), (b, 8, 8), (b, 9, 10),$ $(d, 11, 11), (b, 12, 13), (d, 14, 15), (f, 16, 17), (g, 18, 19), (f, 20, 23), (h, 24, 27) \rangle$ with the same notation as in Example 1. Assume that ext is 6 and $EPS.t$ is 3, and the current considered pattern is $\langle a, b, d, f \rangle$.

Considering Example 2, we can get 15 instances of length 4. This costs 60 memory units. Actually, these instances have some shared fragments. In order to reuse such fragments, we develop a graph structure (called **sequence graph**) to record all instances of a (possibly candidate) pattern. The sequence graph is a directed graph whose vertices are the spots and whose edges are successor pointers from the spots to their immediate successor spots in the instances. In a sequence graph, the spots in each level l (i.e. in C_l) are stored in the same array, and the pointer to the successor spots in level $l + 1$ to represent the sequential relationship of the activities. Fig. 1(a) shows a part of the sequence graph corresponding to the sequence in Example 2.

We call the revised ERF using the sequence Graph structure **ERF_G**. The main revision lies in the pruning step in the extension and in the splitting step in the refinement. When extending P of length L, in the pruning step we consider the spots in C_L instead of the whole instances. For $sp_{L,i}$, we first check whether it has no successors. If that is true, we do not directly remove all spots which belong to the same instances with $sp_{L,i}$. Instead, we first remove $sp_{L,i}$. Then, we examine $sp_{L-1,j}$ in Level $L - 1$. If the set of successors of $sp_{L-1,j}$ becomes empty, $sp_{L-1,j}$ should also be removed and the operation will influence its ancestors iteratively. Otherwise, $sp_{L-1,j}$ survives and the examination process will

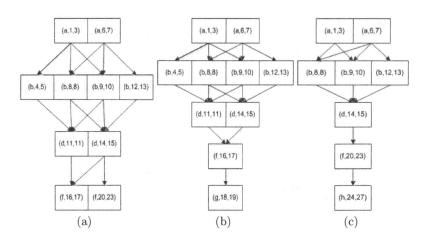

Fig. 1. Illustration of the sequence graph. (a) The corresponding part of the graph on the sequence in Example 2; (b)The change when $(f, 20, 23)$ can not be extended; (c) The change when $(f, 16, 17)$ can not be extended.

be terminated. Consider Example 2. As shown in Fig. 1, if the spot $(f, 20, 23)$ could not be extended, it will be removed. Other elements remain unchanged. If $(f, 16, 17)$ could not be extended, Levels 2~4 will change. In the splitting step of the refinement, we also need to prune invalid instances for each new candidate CAP. The pruning is similar to that in the extension except that the process should be conducted both forward and backwards.

Within each level we note the changes in the sequence graph (i.e. the number of removed spots). We can accelerate the refinement by using a simple strategy, i.e. these levels are processed in descending order of the degree of change. Larger change of a level may lead to a greater shrinkage of the cluster, and therefore the candidate pattern can be updated more quickly. We call the ERF using the sequence Graph and the Sorted processing strategy **ERF_GS**.

5   Experiments

We systematically evaluate the effectiveness and efficiency of our approaches based on a real GPS telemetry dataset and synthetic datasets. All the algorithms are implemented in Java and executed on a machine with Intel(R) Core 2 Quad CPU Q8400@ 2.66 GHz and 4 GB memory.

5.1   Experimental Data

The real dataset is collected from 157 wild birds. The tracking was carried out in four countries (China, India, Mongolia and Bangladesh) from 2007 to 2011. The dataset is preprocessed by retaining at most one sample daily. The average length of generated trajectories is 211 days. These data are accessible on the USGS website.

We design a trajectory generator based on the real dataset which is controlled by these parameters: N, α, γ, ζ, ε, σ, β, λ and TL. To generate a trajectory, we randomly select a real spot as the initial reference and use the corresponding sequence as the reference sequence. We perturb the spot with the spatial variance ζ and the temporal variance ε separately. We then randomly generate raw points relative to the perturbed spot with variance σ. Then, the object moves to the next spot along the reference sequence. On the way, we generate in-between points in random directions under the control of λ. λ denotes the maximum deviation between the actual direction and the excepted direction. An object can change its reference sequence with possibility β. In the change, we randomly select a nearby spot within distance γ. We randomly generate $\alpha * N$ trajectories as noise. The average length of the trajectories is TL. The default parameters of the generator are: $TL = 600$, $N = 1000$, $\alpha = 0.2$, $\gamma = 100$ km, $\zeta = 50$ km, $\varepsilon = 5$ days, $\sigma = 10$ km, $\beta = 0.1$, $\lambda = 45°$.

5.2   Effectiveness

We evaluate the effectiveness using the real GPS telemetry dataset of wild bird movements. Here, we only present the results of ERF because the effects of ERF, ERF_S and ERF_GS are consistent. We compare ERF with the following two baselines.

Baseline 1. We revise the T-Pattern mining algorithm [3] by incorporating the continuity constraint. The regions are incrementally identified using the projections of the patterns found currently. We use the duration time that the object stays in the region instead of the transition time in original T-Pattern algorithm. We call this Continuous T-Pattern Mining algorithm ($CTPM$).

Baseline 2. We implement an algorithm by revising CTPM to incorporate the Spot extraction process (called $CTPM\_SP$). The patterns are mined based on the spot sequences. The regions are incrementally identified based on the sequence projections.

The parameter settings depend on specific tracking scenes. According to the ecological study on wild birds [8], we set d as 30 km, g as 3, min_sup as 0.02, ext as 20 days, $EPS.d$ as 80 km and $EPS.t$ as 15 days. We use the same settings of min_sup and ext in CTPM and CTPM_SP, and set their time tolerance threshold $\tau = EPS.t = 15$ days and the cell width $\omega = EPS.d = 80$ km (Please refer to [3] for more information).

Fig. 2(a) shows the three longest patterns and a large region generated by CTPM. The three patterns are $P_2 = \langle 1, 2, 3 \rangle$, $P_3 = \langle 1, 4, 5 \rangle$ and $P_1 = \langle 1, 6, 7 \rangle$ where the numbers represent the region IDs. Using CTPM, only 31 short patterns are discovered, and the maximum length is 3. One important reason for this is that movements in a large region (eg. Region 0)are considered as a whole. Another reason is the use of the casual grid-based space-partitioning scheme. One may argue that if we use a smaller ω or a higher min_sup, some new patterns can be generated. However, it is obvious that the current interesting patterns involving Region 1 will disappear.

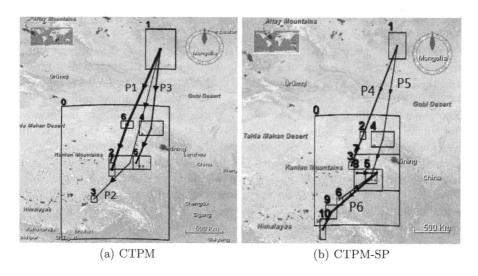

(a) CTPM   (b) CTPM-SP

Fig. 2. Some patterns discovered by CTPM and CTPM_SP. Each rectangle is a discovered dense region, and the number above its upper left corner is the region ID. Each directed path represents a movement pattern along the corresponding regions.

Fig. 2(b) shows 4 of 74 patterns found by CTPM_SP. Compared with CTPM, CTPM_SP can find longer and more detailed patterns. The reason is potential activities are represented by the spots and subsequent mining is based on sequence projections. From Fig. 2, we can see that the patterns discovered by both CTPM and CTPM_SP are not precise, especially for the earlier parts (e.g. Region 0 in the pattern $P_6 = \langle 0, 7, 8, 9, 10 \rangle$). This will lead to ambiguous understanding of animal behavior patterns.

We present 5 of 224 CAPs discovered by ERF in Fig. 3 as examples. Compared with the patterns discovered by CTPM and CTPM_SP, these patterns are more accurate and easier to interpret. For example, the longest CAP is $P_7 = \langle 0\,(51 \sim 60), 1\,(1 \sim 3), 2\,(2 \sim 23), 3\,(1 \sim 13), 4\,(7 \sim 20), 5\,(1 \sim 17), 6\,(1 \sim 8), 7\,(1 \sim 9) \rangle$ in Fig. 3(a). Here, the intervals given in brackets are the activity duration times. P_7 means that the birds breed in Region 0, and then stopover in Regions 1~7 (from Mongolia to China) successively. P_8 in Fig. 3(a) suggests a successive stopover sequence from Huangheyuan Wetland, China to Ganges, India. $P_{10} = \langle 13\,(13 \sim 42), 14\,(1 \sim 4), 15\,(6 \sim 19), 15\,(126 \sim 138), 17\,(1 \sim 17), 18\,(1 \sim 20) \rangle$ in Fig. 3(b) denotes that birds stay in Zhaling Lake, pass Zhamucuo Wetland and Selincuo Nature Reserve, winter in Lhasa Valley (Region 16) for more than 126 days, and then migrate back to Namucuo Lake (Region 17) and Zhamucuo Wetland (Region 18). In Fig. 3(b) the CAPs $P_9 = \langle 19\,(1 \sim 13), 20\,(8 \sim 20), 21\,(1 \sim 9), 22\,(1 \sim 13), 23\,(1 \sim 9) \rangle$ and $P_{11} = \langle 24\,(145 \sim 170), 25\,(1 \sim 6), 26\,(2 \sim 4), 27\,(25 \sim 46), 28\,(6 \sim 12) \rangle$ are generated from the same candidate CAP through the refinement procedure. However, they represent two significantly different behaviors. The former is a stopover sequence, while the latter reflects the birds' breeding and moulting period in Qinghai Lake,China, followed

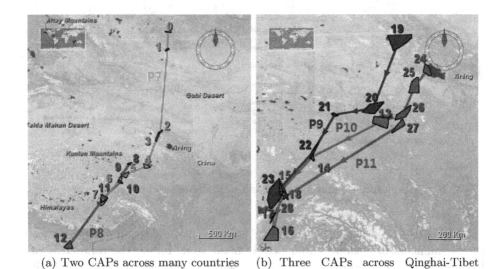

(a) Two CAPs across many countries   (b) Three CAPs across Qinghai-Tibet Plateau

Fig. 3. Some examples of Continuous Activity Patterns (CAP) discovered by ERF. Each polygon with a number is an activity cluster with its ID. Each directed path connecting the clusters (denoted as the polygons) is a CAP.

by a southward migration. This illustrates that the refinement procedure is important to identifying animal behavior. It is worth noting that the birds involved each CAP in Fig. 3 are from a distinct GPS tracking project. This reveals the superior discriminating power of CAP on different migratory behaviors.

We also try to set *ext* as 1 which means that successive activities in the pattern must be strictly continuous in the original trajectories. In this case only a few short patterns emerge, and many interesting patterns such as those in Fig. 3 disappear. Hence,can tolerate some noise and minor divergences using the relaxed continuity.

5.3   Efficiency

We study the efficiency of ERF, ERF_G and ERF_GS on synthetic datasets with regard to different parameter settings. Here, we only present some results for N, min_sup and ext. The default setting used here are: $d = 30$ km, $g = 3$, min_sup, $ext = 12$ days, $EPS.d = 80$ km and $EPS.t = 12$ days.

In Fig. 4, we report on performance of ERF, ERF_G and ERF_GS with regard to N. We present the consuming times of the extension and the refinement separately. As seen in Fig. 4(a), the runtime cost is sensitive to N, since the main cost lies in clustering steps and the time complexity of the clustering algorithm [16] is $O(n^2)$ for n spots without a spatial index. The cost of refinement dominates for all approaches. This is because the refinement needs to be executed multiple times after the extension. We also see that the extension runtime costs

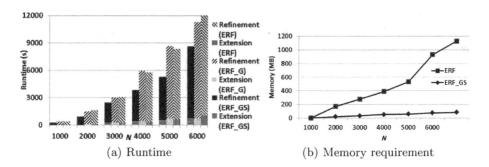

(a) Runtime   (b) Memory requirement

Fig. 4. Algorithm performance related to the number of trajectories N

of ERF, ERF_G and ERF_GS are similar because their extension processes have similar numbers of clustering and based on the same reason the total costs of ERF and ERF_G are similar. It is easy to see that ERF_GS outperforms ERF and ERF_G by about 30%. For brevity, we only show total runtime costs of ERF and ERF_GS for runtime evaluation in the following. We show the memory requirements for ERF and ERF_GS due to the memory allocation consistency of ERF_G and ERF_GS. As shown in Fig. 4(b), ERF_GS outperforms ERF by about one order of magnitude on the memory requirement. This is because many fragments of instances of candidate patterns could be shared under the adoption of the sequence graph structure.

The performances of ERF and ERF_GS with regard to min_sup and ext is shown in Figs. 5 and 6. As expected, ERF_GS outperforms ERF significantly. As min_sup decreases or ext increases, the runtime of ERF increases more quickly than that of ERF_GS. Interestingly, the change of runtime is smoother when varying ext, and there is a concave effect around $ext = 12$ days. This is because that when ext has reached some value, it is sufficient to tolerate small divergences in similar movements. With the further increase of ext, the runtime of ERF_GS decreases slightly. The reason is that more spots can be extended and thus the clusters in newly generated candidate patterns are more easily formed, and therefore the refinement is terminated more quickly.

Fig. 5 shows that the performance of ERF regarding memory requirements significantly deteriorates with the decrease of min_sup. The change is almost quadratic. This is because too many spots could be retained during the mining, and resulting in a large number of candidate instances. However, the memory requirement of ERF_GS is low, and its change is slight. This indicates that a large number of shared fragments exist in the candidate instances, and the sequence graph structure can reuse them to save a large amount of memory.

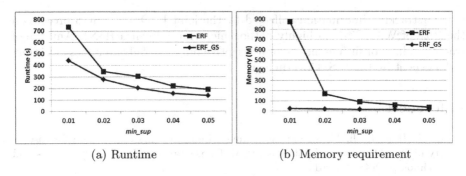

(a) Runtime   (b) Memory requirement

Fig. 5. Algorithm performance in relation to the minimum support min_sup

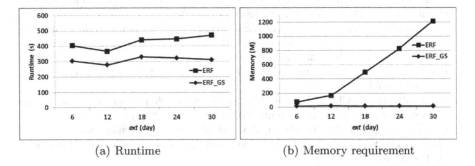

(a) Runtime   (b) Memory requirement

Fig. 6. Algorithm performance in relation to the extendible time threshold ext.

6 Conclusion

In this paper, we investigate the problem of discovering continuous activity patterns from animal trajectories. Unique from earlier proposed movement patterns, the continuous activity pattern is able to model more accurate and more significant collective activity sequences. We propose an extension and refinement framework to discover the patterns, and further propose a sequence graph structure and a processing strategy to improve the performance. Our approach benefits biologists in understanding collective animal activities in an ecologically insightful way. This work also provides a potential approach towards studying the relationships among animals from their trajectories collected during different periods.

Acknowledgments. Funding was provided by the Natural Science Foundation of China (Nos. 61361126011 and 90912006); the Special Project of Informatization of Chinese Academy of Sciences in "the Twelfth Five-Year Plan" (No. XXH12504-1-06); the National R&D Infrastructure and Facility Development Program of China (No.BSDN2009-18; United States Geological Survey; the Food & Agriculture Organization of The United Nations (FAO). The authors also thank Movebank (https://www.movebank.org/) that serves as a data repository for tracking information collected by the joint FAO and USGS project on the Role of Wild Birds in highly pathogenic avian influenza H5N1. Any use of trade, product, or firm names is for descriptive purposes only and does not imply endorsement by the U.S. Government.

References

1. Li, Z., Han, J., Ji, M., et al.: MoveMine: Mining moving object data for discovery of animal movement patterns. ACM Transactions on Intelligent Systems and Technology 2(4), 37 (2011)
2. Jeung, H., Yiu, M.L., Zhou, X., et al.: Discovery of convoys in trajectory databases. In: Proceedings of the VLDB Endowment, pp. 1068–1080 (2008)
3. Giannotti, F., Nanni, M., Pinelli, F., et al.: Trajectory pattern mining. In: Proceedings of the 13th SIGKDD, pp. 330–339 (2007)
4. Cao, H., Mamoulis, N., Cheung, D.W.: Mining frequent spatio-temporal sequential patterns. In: Proceedings of the 5th ICDM (2005)
5. Cao, H., Mamoulis, N., Cheung, D.W.: Discovery of periodic patterns in spatiotemporal sequences. IEEE Transactions on Knowledge and Data Engineering 19(4), 453–467 (2007)
6. Ye, Y., Zheng, Y., Chen, Y., et al.: Mining individual life pattern based on location history. In: Proceedings of the tenth MDM, pp. 1–10 (2009)
7. Prosser, D.J., Cui, P., Takekawa, J.Y., et al.: Wild bird migration across the Qinghai-Tibetan plateau: A transmission route for highly pathogenic H5N1. PloS ONE 6(3), e17622 (2011)
8. Cui, P., Hou, Y., Tang, M., et al.: Movement patterns of Bar-headed Geese Anser indicus during breeding and post-breeding periods at Qinghai Lake, China. Journal of Ornithology 152(1), 83–92 (2011)
9. Lee, A.J., Chen, Y.-A., Ip, W.-C.: Mining frequent trajectory patterns in spatial–temporal databases. Information Sciences 179(13), 2218–2231 (2009)
10. Huang, G., Zhang, Y., He, J., Ding, Z.: Efficiently retrieving longest common route patterns of moving objects by summarizing turning regions. In: Huang, J.Z., Cao, L., Srivastava, J. (eds.) PAKDD 2011, Part I. LNCS (LNAI), vol. 6634, pp. 375–386. Springer, Heidelberg (2011)
11. Parent, C., Spaccapietra, S., et al.: Semantic trajectories modeling and analysis. ACM Computing Surveys 45(4), article 42 (2013)
12. Morzy, M.: Mining frequent trajectories of moving objects for location prediction. In: Perner, P. (ed.) MLDM 2007. LNCS (LNAI), vol. 4571, pp. 667–680. Springer, Heidelberg (2007)
13. Pei, J., Pinto, H., et al.: Prefixspan: Mining sequential patterns efficiently by prefix-projected pattern growth. In: Proceedings of the 29th ICDE, pp. 215–224 (2001)
14. Tseng, V.S., Lin, K.W.: Energy efficient strategies for object tracking in sensor networks: A data mining approach. Journal of Systems and Software 80(10), 1678–1698 (2007)
15. Li, Q., Zheng, Y., Xie, X., et al.: Mining user similarity based on location history. In: Proceedings of the 16th ACM SIGSPATIAL, pp. 298–307 (2008)
16. Ester, M., Kriegel, H.P., et al.: A density-based algorithm for discovering clusters in large spatial databases with noise. In: Proceedings of the 2th KDD, pp. 226–231 (1996)

Personalized Privacy Protection
for Transactional Data

Li-e. Wang[1,2] and Xianxian Li[1,2,⋆]

[1] Guangxi Key Lab of Multi-source Information Mining & Security,
Guangxi Normal University, Guilin, China
[2] College of Computer Science and Information Technology,
Guangxi Normal University, Guilin, China
{wanglie,lixx}@gxnu.edu.cn

Abstract. Privacy protection in publication of transactional data is an important problem. However,the bulk of existing methods focus on a universal approach that exerts the same amount of preservation for all users and all sensitive values, without considering different requirements of users and different sensitivity of concrete attribute values. Motivated by this, we introduce a framework which provides personalized privacy protection based on the form of bipartite graphs via partition and generalization. Our approach can preserve privacy of sensitive associations between entities and retain the largest amount of nonsensitive associations to provide better data utility. Experiments have been performed on real-life data sets to measure the accuracy of answering aggregate queries. Experimental results show that our approach offer strong trade-offs between privacy and utility.

Keywords: Personalized, Privacy-preserving, Bipartite Graph.

1   Introduction

Transactional data are increasingly used in applications, such as business decision, personalized recommendation or research purposes, since such data has immense economic and social value for revealing relationships, dependencies, and performing predictions of outcomes and behaviors[1–3].Such data can be naturally modeled as a bipartite graph where nodes represent entities and edges indicate the associations between them. Private data often comes in the form of associations between entities. Purchase records are typical examples of transactional data. Let's take purchase records for example. Maybe the set of products being sold in a store and their attributes is public knowledge. However, a sensitive product such as AdultToy bought by a particular individual is considered sensitive, since it is indicative of his/her secret preference. So these associations should not be revealed in anonymized data. Many proposals have been proposed to protect the privacy of published transactional data including

---

⋆ Corresponding author.

X. Luo, J.X. Yu, and Z. Li (Eds.): ADMA 2014, LNAI 8933, pp. 253–266, 2014.

item editing methods[4, 5],generalizing-based partitioning [6–9]and bucketing approaches[10, 11].However, all the previous works have overlooked a very important fact, that is, different individuals may have different privacy preference and there is a big difference in sensitivity to different values of sensitive attribute. For example, some users are shy of publishing their purchase records, and others do not worry about it. Different products may also have the same problem. For instance, we all know that AdultToy is more sensitive than Printer. That is, different sensitive values require different levels of privacy protection. Therefore, providing an identical privacy protection to all users and all sensitive values may not be fair, even may cause excessive information loss which would make data useless.

Contributions. To overcome the problems mentioned above, we present a new framework which provides personalized privacy preserving services based on the form of bipartite graphs via partition and generalization. We define the associations in different forms such as sensitive and nonsensitive for the first time. In this paper, we only define two forms of sensitivity, sensitive and nonsensitive. We also can enlarge to more forms if need be. Whether an association is sensitive or not depends on the relevant customer and product. The association is sensitive while its relevant customer or product is sensitive. Otherwise, the association is nonsensitive. The sensitive associations should be protected before published and the nonsensitive associations can publish directly.

Combined with the properties of transactional data, we devise a partition approach based on bipartite graphs which can handle high dimensional data well. In our approach, the pair of vertexes associated with sensitive associations should be partitioned into different subareas for protecting privacy. And the pair of vertexes associated with nonsensitive associations should coexist on the same subarea as more as possible, since nonsensitive associations can publish directly. Meanwhile, we introduce a sensitive ratio μ for trying to keep a balance between customers and products in a subarea since the size of subareas is adaptive. Moreover, we guarantee the privacy of sensitive associations to certain individuals via generalization. The associations between subareas should be generalized into together represented as weighted edges in partitioned bipartite graphs.

We devise an effective algorithm for generating safe subareas which satisfy k-anonymous and l-diversity based on personalized requirements of customers and different sensitivity of sensitive values. We evaluate experimentally our approach with real datasets, and experimental results show that the approach preserves data utility for aggregate analysis to a degree not achieved by previous methods.

Organization. The rest of the paper is organized as follows: Section 2 surveys related work on anonymization. Section 3 introduces some notions of graph model, privacy model and utility metrics. Section 4 describes our approach and algorithm description in detail. Section 5 demonstrates our approach by experimental study comparing with prior works. Section 6 concludes the paper with directions for future work.

2   Related Work

The problem of how to anonymize and publish data for others to analyze and study has attracted lots of studies recently, such as k-anonymization[12, 13] , l-diversity[14], t-closeness[15] and anatomy[16]. However, these models are initially designed for relational database having predetermined and fixed quasi-identifier and sensitive attributes, which cannot be well used for transactional data publishing because transactional data have variable length and high dimensionality[6].

So approaches to anonymizing transactional data have been proposed recently. Most of them employ generalization[6–8, 17, 18]or suppression[5] or generalization and suppression[9, 19] , a few of them adopt permutation[10, 11].

Terrovitis et al.[6, 7]proposed a new model called k^m-anonymity based on k-anonymity model for relational data. The basic idea in k^m-anonymity is that for any transaction in the data set and any subset of m items in that itemset, there are at least k-1 other transactions with the same m items. He et al.[8]proposed a k-anonymity based principle by adopting a top-down, local generalization approach for preserving better data utility. Loukides et al.[17] suggested a constraint-based k-anonymity model to prevent identity disclosure through specific itemsets. Recently, Loukides et al.[18] proposed a rule-based privacy model to prevent both identity and sensitive information disclosure by allowing data publishers to express fine-grained protection requirements. Xu et al.[5] modeled the power of attackers by the maximum size of public itemsets that may be acquired as prior knowledge, and proposed a novel privacy notion called (h,k,p)-coherence for transactional databases. J. Liu et al.[9] and Cao et al.[19] integrated generalization with suppression to anonymize set-valued data. With[19], both sensitive and nonsensitive items may be used to associate an individual with a sensitive item. Ghinita et al.[10] proposed privacy degree, an adoption of l-diversity[14], and integrated the band matrix technique with bucketization[10, 11] for anonymization.

The relevant works to this paper is the research of k-anonymous by Cormode et al.[20] and Zhou et al.[21]. And Cormode et al.[20] proposed a safe (k,l)-groupings approach for masking the mapping between nodes and entities via grouping the nodes of the graph, which could preserve the structural information of the graph perfectly. But the approach can not effectively prevent from background knowledge attacks. To resist against such attacks, Zhou et al.[21]adopted a generalized approach via generalizing the graphs to super-nodes and super-edges for anonymizing bipartite graphs data based on [20] . However, the generalization approach incurred excessive information loss and reduced the utility of data for aggregate analysis. Besides, the problem of personalized anonymization is not considered in those works, which is non-trivial for improving data utility.

3   Preliminarties

3.1   Graph Model

Throughout, we focus on problems of anonymizing transactional data and we use bipartite graphs to model such data as $G = (V, W, E, Sv, Sw)$. Here the bipartite graph G consists of $n = |V|$ nodes of one type, $m = |W|$ nodes of other type, and a set of $|E|$ edges $E \subseteq V \times W$. Sv and Sw is the sensitivity labels set of node V, W respectively.Note that here, the graph is relatively sparse: each customer typically buys only a small fraction of products, and each product is bought by only a few customers. We show an illustrative example of a bipartite graph, which consists of a set of customer nodes V (such as $c1,c2,c3,\ldots$) and a set of product nodes W (such as $p1,p2,p3,\ldots$ in Fig.1(a)). Each node in V and W has an identity (such as id) and a label of sensitivity (s indicates sensitive; n indicates nonsensitive) as shown in Fig.1 (a). An edge ($v \in V, w \in W$) in E indicates that the customer represented by node v has bought the product represented by node w. The products can be bought by each customer in various combinations.

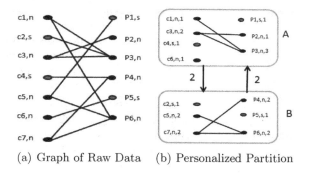

(a) Graph of Raw Data   (b) Personalized Partition

Fig. 1. Examples of Personalized Partition Approach

A partitioned bipartite graph is shown in Fig. 1(b). A subarea is represented by a box and each node in the box belongs to the subarea. The additional information on labels is the degree of each node. Each edge from subarea A to subarea B indicates that one customer node in subarea A has bought a product in subarea B. The weight of edges is presented the number of customer nodes in subarea A who have bought the products in subarea B, and vice versa.

3.2   Privacy Model

Given that G is a bipartite graph over two types of nodes V, W, edges E and labels Sv, Sw. Our objective is to construct a published version G', which should preserve nonsensitive properties of G as many as possible, while masking sensitive associations between the entities. This should hold even under attacks from an adversary who knows some limited information about the original data. That is, the anonymized version G' should not provide any information for the adversary with partial knowledge of G to infer additional sensitive associations between entities. For convenience, we define the relative notions of k-anonymous for bipartite graphs listed below.

Definition 1. Outdegree. Given a subarea T, V and W are the node set of customers and products of T respectively. The outdegree of subarea T is the number of outgoing edges which are from T to others, that is, the customer nodes in V have bought the product nodes which are not in W.

Definition 2. Indegree. Given a subarea T, V and W are the node set of customers and product of T respectively. The indegree of subarea T is the number of incoming edges which are from other subareas to T, that is, the customer nodes which are not in V have bought the product nodes in W.

Definition 3. K-anonymous subarea. Given a subarea T, the subarea T is k-anonymous if and only if the outdegree and indegree of T are both more than k.

Definition 4. K-anonymous graph. Given a bipartite graph G, the graph G is k-anonymous if and only if all subareas of G are both k-anonymous.

According to above definitions, we know that the size of each subarea is uncertain, but the outdegree and indegree of the subareas must be larger than k (k is anonymous parameters) in a k-anonymous graph. For example, there is a 2-anonymous graph (k=2) in Fig.1 (b). The outdegree and indegree of subarea A are 2 and 3 respectively. So the probability of deducing an association which is from customer nodes in subarea A to product nodes in subareas B is $1/2$ and the probability of deducing an association which is from customer nodes in subarea B to product nodes in subareas A is $1/3$. So the bipartite graph satisfies 2-anonymous.

Definition 5. Sensitive association. An association is sensitive association if any of its relevant entities is sensitive.

Definition 6. Nonsensitive association. Nonsensitive association which can publish directly is between nonsensitive entities.

Generally speaking, whether a customer is sensitive or not depends on himself. According to customers'needs, the option of anonymous is optional when e-shopping. If checked, the customer is sensitive. Otherwise, the customer is nonsensitive. And the levels of sensitive values are specified by data owner or domain experts before anonymity. For example, private goods such as AdultToy can be specified as sensitive product and office supplies such as Printer can be specified as nonsensitive product. According to Definition 5, an association is sensitive while its corresponding customer nodes or product nodes is sensitive. In contrast, the associations between nonsensitive customers and nonsensitive products which can be revealed are considered as nonsensitive associations.

Our objective is to protect sensitive associations and preserve nonsensitive associations for improving the balance performance between privacy protection and beneficial uses of data. Information such as $c6$ bought $p5$ is considered to be sensitive since $p5$ is a sensitive product. All relevant associations of $c4$ are sensitive associations because $c4$ is a sensitive customer. In contrast, information such as $c1$ bought $p6$ allows to be deduced because $c1$ and $p6$ are both nonsensitive. Thus, the final published data should guarantee that the sensitive associations can not be revealed.

3.3   Utility Metrics

In this paper, utility is judged based on the quality with various queries which can be answered on the anonymized graph. In [20], the author listed three typical types of queries with increasing complexity. The queries of type-0 is related with graph structure only, type-1 is related with attribute predicate on one side only, and type-2 is related with attribute predicate on both sides. Observe that here, queries of type-0 and type-1 can be answered accurately since we keep the degrees of entities. Queries of type-2 cannot guarantee perfect accuracy, because we can not determine exactly which edges is selected. Just as be proved in [20], finding the best upper and lower bounds for answering the aggregate query of type-2 is NP-Hard. And we use the parameter of expected error $|\mu - Q|/Q$ to precisely evaluate utility by providing bounds and expected values, where the correct answer on original data is Q and the expected answer on anonymized data is μ for each query. Smaller values of the expected error indicate better utility. We give an example for further illustrating the estimation.

Example 1. *The Expected Error of Answering Aggregate Query of Type-2*
    Let us take answering the query *"Count the total number of sensitive products bought by nonsensitive customers "*in Fig.1 for example. We compute a lower bound estimation L, an upper bound estimation U, and an expected value μ. If the correct answer to the query is Q, we compute the expected error as $|\mu - Q|/Q$. We analyze each subarea in turn. First, we can see that there are both 1 nonsensitive customer which is not full degree in subarea A and B, and there are both 1 sensitive product in subarea A and B, and there are both 2 edges between subarea A and B. So we can find the upper bound $U_1 = min(1, 2, 1) = 1$, which mean that a maximum of two nonsensitive customers may have bought sensitive products. Similarly, we can find $U_2 = min(1, 2, 1) = 1$.We also can count the lower bound $L_1 = max(0, 2 - 1 - 1) = 0$ for subarea A, and $L_2 = max(0, 2 - 1 - 1) = 0$ for subarea B. Then the expected answer can count as $(U_1 + L_1)/2 + (U_2 + L_2)/2 = 1$. It is easy to find that the correct answer Q is 2 from the original data in Fig.1. So the expected error is $|1 - 2|/2 = 0.5$.

4   Partition and Generalization Anonymization

4.1   Partition and Generalization Approach

To solve the problem above, we propose a personalized privacy-preserving approach which employs partition based on random walk and generalization on edges. Our approach can be divided into two steps as follows. First, we partition nodes into subareas by correlative degree which is used to measure the concurrence of customers and products for keeping more nonsensitive associations. Second, we generalize sensitive edges to super-edges for concealing sensitive associations. Meanwhile, we preserve node degrees by additional labels in anonymized bipartite graph and represent super-edges with weight for keeping more useful information for aggregate analysis. In detail, all nodes are districted according to the associations between them. The pair of nodes can place in the same subarea if and only if the association between them is nonsensitive. Otherwise, the pair of nodes must divide into two different subareas. For convenience, we give a definition about degree information of nodes below.

Definition 7. Degree . Given a node v in subarea A, node v is full degree if all relevant associations of v are nonsensitive and both exist in subarea A. Otherwise, node v is under degree.

In our approach, the size of subareas is adaptive.One may argue that there will incur a new privacy problem if there is only one node under degree in a subarea, which is called homogeneity attack in prior works. So we devise a safe condition for addressing the problem. We introduce a parameter l for guaranteeing l-diversity. It limits the rate which is the number of nodes under degree divided by total number of its subarea. Meanwhile, we devise a parameter α to control the number of sensitive nodes for keeping privacy.

Definition 8. Safe condition .Given a k-anonymous subarea T which contains n nodes with full degree, x nodes under degree, and h sensitive nodes, the subarea T is safe if and only if it satisfies safe condition as follows: $x/(x + n) > l$ and $h/(x + n) < \alpha$.

In detail, the method of partition can be divided into three steps as follows. First, we partition nonsensitve nodes into subarea by correlative degree. The output of the first step is a subarea sequence without sensitive nodes. Second, we add sensitive nodes and the isolated nodes into subarea under safe condition. Third, we add the degree of nodes on the labels for reducing information loss. Let us take a concrete example as an illustration.

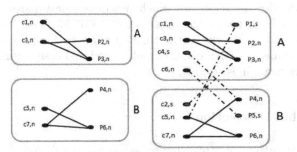

(a) Partitioning Nonsensi- (b) Adding Sensitive Node
tive Node

Fig. 2. Examples of Partition

Example 2. *Partitioning via Random Walking*

Fig.2 illustrates step by step how the anonymized routine works via partitioning nodes into 2-anonymous subareas. The original database contains sex transactions, as shown in Fig.1 (a). We create a new subarea A and select $c1$ as a start point which has big degree, and then walk to one of its neighbors such as $p3$. The Anonymize routine is called as customer walk to product. So $c1$ and $p3$ is placed in subarea A. We will check the indegree and outdegree of subarea A after each walk and go on walking until k-anonymous is satisfied. $P3$ is considered as a start point in the next walk. And the next walk is from $p3$ to $c3$ and from $c3$ to $p2$, until there is unable to go on. Then, we can select another nonsensitive customer node as a new start point as shown in Fig.2 (a). In the process, we recursively invoke Anonymize routine on each subarea by correlative degree until all nonsensitive nodes have been processed.

Next, we will tackle with sensitive nodes by adding the pair vertex of sensitive associations in two different subareas. As shown in Fig.2 (b), we add $c4$ in subarea A because $p4$ is in subarea B and there is a sensitive association between $c4$ and $p4$. Then we add $c2$ in subarea B since $p3$ is in subarea A and there is a sensitive association between $c2$ and $p3$.The reason is same for placing $c4$ in subarea A for $p4$ and placing $p1$ into subarea A for $c5$. $C6$ is an isolate node since it has only bought a sensitive product. So $c6$ can join in subarea A or subarea B. But placing $p5$ in subarea A may raise the problem of homogeneity attack. So we place $c6$ in subarea A while adding $p5$ in subarea B.

In the end, we generalize the sensitive associations to super-edges called as indegree or outdegree of each subarea for guaranteeing privacy. The super-edges are represented by weight edges in partitioned bipartite graphs. Meanwhile we add degree information in the labels of nodes for preserving more useful information for aggregate analysis. The result of the whole process is shown in Fig.1 (b). We can see that each subarea can contain k or more nodes, but the outdegree and indegree of a subarea must larger than k for guaranteeing privacy, that is, the probability of a customer in subarea A who buy products in subarea B is not less than $1/k$.

We devise an effective algorithm which generates safe subareas based on the proposed approach. In detail, the process can be divides into three parts: the k-anonymous partition algorithm, the algorithm for checking l-diversity and the algorithm for adding sensitive nodes or isolate nodes. These detailed algorithms descriptions are given in section B.

4.2   Algorithm Description

The partition algorithm of generating a k-anonymous subarea of V is shown in Algorithm 1. The inputs of the algorithm are the original nodes list of V, W and the corresponding labels Sv and Sw, and the anonymous parameter k. Line 1 adds all nonsensitive customer nodes into $Vnonsensitive$ from V by Sv, and adds all nonsensitive product nodes into $Wnonsensitive$ from W by Sw. Initially, the cluster sequence VG is set to empty.

Algorithm 1. *The k-anonymous partition algorithm*

Input: V, W, Sv, Sw, k

Output: *A k-anonymous subarea sequence VG*

1: Add all nonsensitive customers into $Vnonsensitive$ from V by Sv, add all nonsensitive products into $Wnonsensitive$ from W by Sw , $VG \leftarrow \phi$;

2: **while** ($Vnonsensitive$ is not empty) **do**

3:    Select a customer v from $Vnonsensitive$;

4:    create a new subarea T, $T = \{v\}$, $Vnonsensitive = Vnonsensitive - \{v\}$;

5:    **while** ($indegree(T) < k$) **do**

6:       Select a nonsensitive product w from Neighbor (v);

7:       $T = T \cup \{w\}$, $Wnonsensitive = Wnonsensitive - \{w\}$;

8:       **while** ($outdegree(T) < k$) **do**

9:          Select a nonsensitive customer v' from Neighbor (w);

10:          $T = T \cup \{v'\}$; $Vnonsensitive = Vnonsensitive - \{v'\}$;

11:       **end while**

12:    **end while**

13:    $VG = VG \cup \{T\}$;

14: **end while**

The loop of lines 2-14 tries to find a desired subarea which satisfies k-anonymous by greedy algorithm. In detail, the greedy procedure works as follows. Line 2 checks whether the list of $Vnonsensitive$ is empty or not. If the list of $Vnonsensitive$ is not empty, the algorithm will select a customer node v as a starting point. Line 4 creates a new subarea which contains v alone, and deletes v from $Vnonsensitive$. Loop of lines 5-12 tries to generate a subarea via random walk and tackles with the problem of k-anonymous for the subarea T. Line 5 checks whether the indegree of subarea T is more than k or not. If the indegree of subarea T is less than k, the algorithm will select a nonsensitive product w from its neighbor set randomly in line 6. Then add w into a subarea T and delete w from $Wnonsensitive$. The process terminates while the indegree of subarea T is more than k. Similarly, line 8 checks whether the outdegree of subarea T is more than k or not. Otherwise, the algorithm selects a nonsensitive customer node v' from w's neighbor set randomly, then add v' into a subarea T and delete v' from $Vnonsensitive$ until the outdegree of subarea T is more than k. In the end, line 13 adds the subarea T into VG. The whole process terminates while $Vnonsensitive$ is empty. In other words, all nonsensitive nodes have been processed.

The algorithm for guaranteeing l-diversity is shown in Algorithm 2. The inputs of the algorithm are the k-anonymous subarea sequence VG from Algorithm 1 and diversity parameter l. The algorithm will check each subarea from VG in turn. Line 2 counts the number of customer nodes under degree according to Definition 7. The loop of lines 3-9 tries to tackle with the l-diversity problem of outdegree. In detail, line 3 checks whether the rate of the number of nodes under degree divided by total number of its subarea is more than l or not. If it does not satisfy l-diversity principle, the algorithm will try to find another subarea T' which does not satisfy l-diversity principle either. Then the algorithm tries to

Algorithm 2. *Checking l-diversity algorithm*

Input: *VG, l*
Output: *A k-anonymous and l-diversity subarea VG'*
1: **for** each *T* in *VG* **do**
2:    *n*=count the number of the customer nodes under degree ,*m*= count the total number of customer nodes ;
3:    **while** (*n/m < l*) **do**
4:        *T'*= next subarea in *VG*;
5:        **if** $(l - diversity(T') < l)$ **then**
6:            Customer-SWAP (*T, T'*);
7:        **end if**
8:        update *n(T,T')*;
9:    **end while**
10:    *n*=count the number of the product nodes under degree ,*m*= count the total number of product nodes ;
11:    **while** (*n/m < l*) **do**
12:        *T'*= next subarea in *VG*;
13:        **if** $(l - diversity(T') < l)$ **then**
14:            Product-SWAP (*T, T'*);
15:        **end if**
16:        update *n(T,T')*;
17:    **end while**
18: **end for**

---

swap customer node in subarea *T* for another subarea's for ensuring *l*-diversity in line 6. Line 8 updates the number of customer nodes under degree and moves on to the next round. Similarly, line 10 counts the number of product nodes under degree according to Definition 7. The loop of lines 11-17 tries to tackle with the *l*-diversity problem of indegree. The process terminates while all subarea satisfy *k*-anonymous and *l*-diversity.

Algorithm 1-2 show the operation of the algorithm and the algorithm for adding sensitive nodes or isolate nodes is not listed due to the paper space limit. We partition nodes based on personalized requirements and different levels of sensitivity for keeping nonsensitive associations to the most degree and generalize sensitive associations for guaranteeing privacy. This is crucial difference between our approach and prior works.

5   Experiments

5.1   Experimental Framework

In this section, we evaluate the performance and utility of our privacy-preserving approach. We implement all algorithms in C++ and apply them to DBLP, BMS-WebView-2(BMS2) and BMS-POS (POS), whose characteristics are listed in Table 1. *Sv* and *Sw* are synthetic data following the example of ADULT dataset since these datasets do not include attribute informationand. We set the proportion of the *sensitive: nonsensitive* to be 1:2 arbitrarily.

We first give an overview of how our approach called Personalized partition performs compared with Cormode et al. [20] called Safe (k,l)-grouping on DBLP dataset. And we compare our personalized partition with Zhou et al.[21] called Generalization on the first 10000 transactions of BMS2. We then vary different parameters to see how performance changes for our approach on BMS2. And we compare the quality of query on three different datasets on the performance of our approach. For the parameters in the privacy model, unless specified otherwise, we fix k in k-anonymity at 10 and l in l-diversity at 0.5 and α in α-sensitive constraint at 0.4.

Table 1. Characteristics of the three datasets

dataset	#Trans	#Distinct items	# Max.trans.size	# Avg.trans.size
DBLP	216753	170371	71	5.4
BMS-WebView-2	77370	3336	50	4.5
BMS-POS	306983	1177	5	2.65

As in prior work [20], we evaluated data quality using the expected error $|\mu - Q|/Q$ for query, where Q is the correct answer on original data and μ is the expected answer on anonymized data for each query. To clearly show the trends, we repeat each experiment over ten random choices of predicates and show the mean expected error. Smaller values of the expected error indicate better utility. We study the accuracy of three queries with different properties for evaluating the utility of the anonymized data. The three queries are:

- Query A: Find the total number of customers buying only a single product satisfying Pa. This is a type-1 query with both attribute predicates and structural predicates. We vary the selectivity of Pa from 0.1 to 0.9.
- Query B: Find the average number of customers buying sensitive products satisfying Pa. This is a type-2 query which is related with attribute predicate on both sides. The selectivity of Pa is varied as above.
- Query C: Find the total number of products satisfying Pn being bought by who satisfy Pa. We vary the selectivity of both Pn and Pa.

5.2   Experimental Results

Fig. 3 and Fig. 4 plot the expected error vs. selectivity of Pa under above approaches, respectively on DBLP dataset and BMS2 dataset. Fig. 3 shows the results of Personalized partition compared with Safe (k,l)-grouping running on DBLP dataset for different kinds of queries. We find that Personalized partition performs much better than Safe (k,l)-grouping over those kinds of queries. This is expected, because Safe (k,l)-grouping assumes that all entities are with identical sensitivity, while our approach introduces the notion of personalization and tries to keep the greatest part of nonsensitive associations. So our approach can reduce

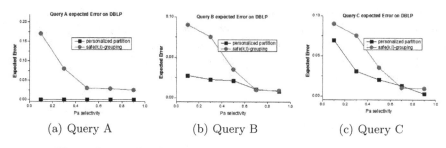

Fig. 3. Personalized partition compared with Safe (k,l)-grouping

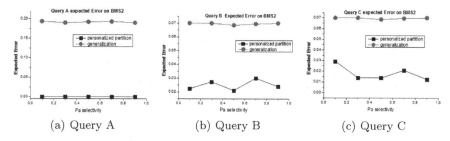

Fig. 4. Personalized partition compared with Generalization

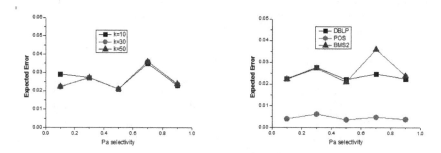

Fig. 5. Effect of k on data quality over BMS2   **Fig. 6.** Data quality of query B on three datasets

information loss significantly for aggregate analysis. Fig. 4 shows a comparison of Personalized partition and Generalization running on BMS2 dataset for different kinds of queries. The quality of queries results shows that Personalized partition proposed in our paper also consistently outperforms Generalization on all three kinds of queries, using the expected error metric. Note that generalization would incur excessive information loss and even make data useless, especially for sparse high-dimensional data. However, as stated above, our personalized privacy protection can reduce information loss to the most degree. Combined with the experimental results of Fig. 3 and Fig. 4, we can conclude that our personalized approach can provide more accurate results for aggregate analysis than those works without considering personalized privacy protection.

Fig. 5 illustrates the effect of parameter k on the performance of Personalized partition running on BMS2 dataset. Observe that here, the results are not raised obviously as k increases. This is because Personalized partition won't incur too much information distortion or information loss with the increases of k. Besides, each subarea can preserve more nonsensitive associations while k increases, especially in case the number of sensitive associations is small enough. So our approach is more stable than above approaches. Fig. 6 shows the results of Personalized partition running on three different datasets. We can find that the expected error on POS dataset is more stable than other datasets, using the same parameter setting. We note that POS dataset is sparser than others. That is, Personalized partition performs better on sparser dataset. Thus, we can conclude that our approach is suitable for anonymizing transactional data which is sparse and high-dimensional.

6 Conclusion

In this paper, we introduce a novel personalized privacy-preserving approach based on the form of bipartite graphs, which can handle high dimensionality well. The personalized privacy-preserving approach focuses on both different requirements of customers and different sensitivity of sensitive values via partition based on random walking and generalization on edges. The paper devises a personalized safe-partitioning algorithm based on sensitive associations and nonsensitive associations. In our experiment, we compare our approach to prior works proposed in [20, 21], and measure the quality of data by the expected error of answering aggregates queries. These results demonstrate that our approach preserves data utility much better for aggregate analysis than previous methods.

Our future work could try to apply a suitable anonymization to high-dimensional dataset with multiple sensitive attributes which can be modeled as a multipartite graph. Further, we plan to investigate how graph anonymization techniques can be applied to other types of data, e.g. spatial data, stream data.

Acknowledgments. The research is supported by the National Key Basic Research Program of China (973 Program, No. 2012CB326403), National Science Foundation of China (No. 61272535), Guangxi Bagui Scholar Teams for Innovation and Research Project, Guangxi Collaborative Innovation Center of Multisource Information Integration and Intelligent Processing,Guangxi Natural Science Foundation (No. 2013GXNSFBA019263), and Science and Technology Research Projects of Guangxi Higher Education (No. 2013YB029).

References

1. Tene, O., Polonetsky, J.: Big data for all: Privacy and user control in the age of analytics (2013)
2. Shichao, Z., Zhenxing, Q., Charles, X.L., Shengli, S.: missing is useful: Missing values in cost-sensitive decision trees. IEEE Trans. Knowl. Data Eng. 17(12), 1689–1693 (2005)

3. Xindong Wu, S.Z.: Synthesizing high-frequency rules from different data sources. IEEE Trans. Knowl. Data Eng. 15(2), 353–367 (2003)
4. Atzori, M., Bonchi, F., Giannotti, F., Pedreschi, D.: Anonymity preserving pattern discovery. The VLDB Journal the International Journal on Very Large Data Bases 17(4), 703–727 (2008)
5. Xu, Y., Wang, K., Fu, A.W.C., Yu, P.S.: Anonymizing transaction databases for publication. In: Proceedings of the 14th ACM SIGKDD International Conference on Knowledge Discovery and Data Mining, pp. 767–775. ACM (2008)
6. Terrovitis, M., Mamoulis, N., Kalnis, P.: Privacy-preserving anonymization of set-valued data. Proceedings of the VLDB Endowment 1(1), 115–125 (2008)
7. Terrovitis, M., Mamoulis, N., Kalnis, P.: Local and global recoding methods for anonymizing set-valued data. The VLDB Journal the International Journal on Very Large Data Bases 20(1), 83–106 (2011)
8. He, Y., Naughton, J.F.: Anonymization of set-valued data via top-down, local generalization. Proceedings of the VLDB Endowment 2(1), 934–945 (2009)
9. Liu, J., Wang, K.: Anonymizing transaction data by integrating suppression and generalization. In: Zaki, M.J., Yu, J.X., Ravindran, B., Pudi, V. (eds.) PAKDD 2010, Part I. LNCS (LNAI), vol. 6118, pp. 171–180. Springer, Heidelberg (2010)
10. Ghinita, G., Tao, Y., Kalnis, P.: On the anonymization of sparse high-dimensional data. In: IEEE 24th International Conference on Data Engineering, ICDE 2008, pp. 715–724. IEEE (2008)
11. Ghinita, G., Kalnis, P., Tao, Y.: Anonymous publication of sensitive transactional data. IEEE Transactions on Knowledge and Data Engineering 23(2), 161–174 (2011)
12. Samarati, P.: Protecting respondents identities in microdata release. IEEE Transactions on Knowledge and Data Engineering 13(6), 1010–1027 (2001)
13. Sweeney, L.: k-anonymity: A model for protecting privacy. International Journal of Uncertainty, Fuzziness and Knowledge-Based Systems 10(05), 557–570 (2002)
14. Machanavajjhala, A., Kifer, D., Gehrke, J., Venkitasubramaniam, M.: l-diversity: Privacy beyond k-anonymity. ACM Transactions on Knowledge Discovery from Data (TKDD) 1(1), 1–52 (2007)
15. Li, N., Li, T., Venkatasubramanian, S.: t-closeness: Privacy beyond k-anonymity and l-diversity. In: ICDE, vol. 7, pp. 106–115 (2007)
16. Xiao, X., Tao, Y.: Anatomy: Simple and effective privacy preservation. In: Proceedings of the 32nd International Conference on Very Large Data Bases, VLDB Endowment, pp. 139–150 (2006)
17. Loukides, G., Gkoulalas-Divanis, A., Shao, J.: Anonymizing transaction data to eliminate sensitive inferences. In: Bringas, P.G., Hameurlain, A., Quirchmayr, G. (eds.) DEXA 2010, Part I. LNCS, vol. 6261, pp. 400–415. Springer, Heidelberg (2010)
18. Loukides, G., Gkoulalas-Divanis, A., Shao, J.: Efficient and flexible anonymization of transaction data. Knowledge and information systems 36(1), 153–210 (2013)
19. Cao, J., Karras, P., Raïssi, C., Tan, K.L.: ρ-uncertainty: inference-proof transaction anonymization. Proceedings of the VLDB Endowment 3(1-2), 1033–1044 (2010)
20. Cormode, G., Srivastava, D., Yu, T., Zhang, Q.: Anonymizing bipartite graph data using safe groupings. Proceedings of the VLDB Endowment 1(1), 833–844 (2008)
21. Zhou, J., Jing, J., Xiang, J., Wang, L.: Privacy preserving social network publication on bipartite graphs. In: Askoxylakis, I., Pöhls, H.C., Posegga, J. (eds.) WISTP 2012. LNCS, vol. 7322, pp. 58–70. Springer, Heidelberg (2012)

A Hybrid Algorithm for Privacy Preserving Social Network Publication

Peng Liu[1], Lei Cui[1], and Xianxian Li[2,*]

[1] School of Computer Science and Engineering, Beihang University, Beijing, China
[2] College of Computer Science and Information Technology,
Guangxi Normal University, Guilin, China
{liupeng,lixx}@gxnu.edu.cn, cuilei@act.buaa.edu.cn

Abstract. With the rapid growth of social networks, privacy issues have been raised for publishing data to third parties. Simply removing the identifying attributes before publishing the social network data is considered to be an ill-advised practice, because the structural characteristic may reveal the users privacy. We discuss the current techniques for publishing social network data and define a privacy preserving social network data publishing model with confidence p. Then we devise a hybrid privacy preserving algorithm satisfying the defined model for publishing social network data. Combining the features of k-anonymity with randomization, the algorithm uses the k-anonymous concept to hide the sensitive information into the natural groups of social network data and employs random approach to process the residual data. We conduct the algorithm on several real-world datasets, the experimental results show that our algorithm is practical and efficient. Compared with the related k-anonymity and random methods, our algorithm is stable and modifies the original data less than the existing algorithms.

Keywords: privacy, social network, k-anonymity, randomization.

1 Introduction

While the popularity of social network services consistently rises, people have a more convenient way to interact digitally with each other. According to recent report [1], people continue to spend more time on social networks than on any other category of sites; total time spent on social media in the U.S. across PCs and mobile devices increased by 37 percent up to 121 billion minutes in July 2012, compared to 88 billion in July 2011.

User profiles and user-generated contents are central features of social network service. A profile is generated from answers to questions age, location, interests, etc. Some sites allow users to upload pictures, add multimedia content, and modify the look of the profile [2]. The user-generated contents contain blogs, photos, videos comments and rating to other users. The data generated by social network services is often referred to as the social network data. The

---

* Corresponding author.

X. Luo, J.X. Yu, and Z. Li (Eds.): ADMA 2014, LNAI 8933, pp. 267–278, 2014.
© Springer International Publishing Switzerland 2014

social network data is useful in many application domains such as government applications, business applications, scientific applications, and educational[3,4]. For example, the government can use the data to get the opinion of the public quickly and adjust the policy properly; the company can use the data to make better targeted advertising than any other method can currently provide. The social network data needs to be shared or published to third parties for the utilizing purpose. But social network data usually contains sensitive information about the users. Simply publishing the data in its original form will violate the individual privacy. Thus, the identifiable attributes such as security number and name are removed before publishing the data. Backstorm et al. show this naive protecting technique is not enough to ensure privacy because the structural characteristic of target users may disclose the privacy [5]. To strengthen the privacy, we need to publish another version of data with more modification; however, this will sacrifice the data utility.

As the utility and privacy of the social networks both are not neglected, it is necessary for a sanitizing algorithm to balance between privacy protection and utility loss during modification. Thus, how to protect individuals privacy while preserving the utility of social network data becomes a critical problem in recent years.

1.1 Related Work

Many researchers have realized that the privacy issue is significantly important when using the social network data. Several important models have been proposed, such as k-anonymity [6,7,8,9], l-diversity[10,11,12], and random approach[13,14,15], to preserve the privacy.

According to the hiding information strategy, existing privacy-preserving mechanisms can be classified into two categories: k-anonymity-based approach and random-based approach.

The k-anonymity-based approach uses the concept of k-anonymity proposed by Sweeney for the privacy preserving of micro data [16]. It hides the individual in a group of similar attributes and the target cannot be uniquely identified from at least k-1 individuals in the published data for the purpose of privacy preserving.

Liu and Terzi [6] studied the problem of identity re-identification based on degree structural characteristic. They assumed that the adversary only has prior knowledge of degrees of certain vertices, and use the information to re-identify certain nodes in the published graph data. They defined the graph anonymization problem and proposed a two-step algorithm to achieve k-anonymity.

Zhou and Pei [7] assumed the adversary has the background knowledge about how the neighboring vertices of a targeted vertex are connected themselves. The victim may be re-identified from a social network, even if the victims identity is preserved using the conventional anonymization techniques, by the unique neighborhood structure. They proposed k-neighborhood anonymous model: for every node, there exist at least other k-1 nodes sharing isomorphic neighborhoods, and a novel technique called Neighborhood Component Coding to solve the graph isomorphism tests problem. In paper [10], they extends their work by include the l-diversity technique to protect the sensitive vertex label.

Zou et al. [17] proposed a k-Automorphism model based on the assumption that the adversary can learn and know the sub graph around the target entity. A graph is k-Automorphism if and only if for every node there exist at least k-1 other nodes do not have any structure difference with it. Similar to k-Automorphism model, Wu et al. [8] proposed a k-Symmetry model protecting privacy against re-identification by sub graph information.

The random-based approach uses the perturbation to protect the data privacy by increasing the adversarys uncertainty about the true identity of individuals. It adds random noise to the original social network data by modifying the vertices and edges, and protects the data against re-identification in a probabilistic manner.

Hay et al. [13] proposed a random protection model which modifies the edges by randomly adding m false edges followed by deleting m true edges. Ying and Wu [14,15] extent this approach to protect link relations.

However, the random modification process may not change some vertices according probability. The random model may not provide sufficient protection for every vertex in the social network data graph.

1.2   Contribution and Organization

The privacy preservation in publishing social network data is a new challenging problem that cannot be solved by one shot. In this paper, we propose a novel privacy preserving method combining the k-anonymity with random concept for publishing social network data and balance the issues between privacy preservation and data utility. We introduce some measurements of information loss for anonymizing graph, and conduct the experiments on some real-world datasets.

The rest of the paper is organized as follows: We identify and define the privacy problem in Section 2. A practical solution is described in Section 3, and examines our solution by real-world datasets in Section 4. We discuss future works and conclude the paper in Section 5.

2   Problem Description

The social network data is usually viewed or represented as a graph that contains vertices and connections between them. Formally, we model a social network as an undirected, symmetric and simple graph $G(V, E, L, \varphi)$, where V is a finite set of vertices denoting social network users, $E \subseteq V \times V$ is a set of edges denoting social relationships between vertices, L is a set of identifiers, and a function $\varphi : V \to L$ maps the vertices to the labels.

The social network data contains valuable information as well as sensitive information of users. The data must undergo a sanitized procedure before being published to third parties. The adversary usually relies on background knowledge to breach the privacy and then learns the sensitive information from the sanitized social network data. In this paper, we assume that an adversary only has the background knowledge of the structural attributions of some target individuals. We define the privacy breach of a social network as follows:

Definition 1. (Privacy breach) Given the published social network data, the privacy breach occurs when an adversary can successfully map the target real-world entity to a vertex in the published data.

This privacy breach is also named by identity disclosure. A simply naive way to protect the individual from being re-identified is removing the identifiable attributes such as name or social security number. Fig.1b shows an example of naive anonymization of original social network data, illustrated as Fig.1a, by replacing the identifications of each user with an integer. According to the notion in paper [5], we give a formal definition of social network data naive anonymization.

Definition 2. (Naive anonymization) The naive anonymization of a social network data $G(V, E, L, \varphi)$ is an isomorphic graph $G_n(E_n, V_n)$ where $E_n = E$ and $V_n = V$.

Contrast to the intuition, the naive anonymization is not enough to preserve the privacy. For example, if an adversary knows Tim has four friends, he can uniquely re-identify that Tim is the vertex with label 0 in the naive anonymized graph shown in Fig.1b.

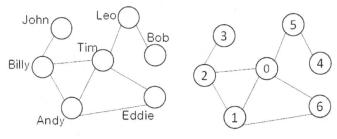

(a) Original social network data
graph G

(b) Naive anonymized data
G_n

Fig. 1. Social network data graphs

The aim of this paper is to devise a sanitizing method that can generate another version of the original social network data and preserve the privacy while maintaining data utilization. Motivated by the k-anonymity and random notion, we devise a hybrid privacy-preserving method for social network data publishing. The algorithm applies the k-anonymous concept to hide the sensitive information of individual into the natural groups of social network data and uses random approach to process the residual data. The hybrid privacy-preserving method generate a graph $G_h(E_h, V_h)$ from the original graph $G(V, E, L, \varphi)$, where $V_h = V$, by edges modification.

We use two measure functions to gauge the privacy disclosure risk: the prior disclosing risk $r(x)$ and posterior disclosing risk $r(x|G_h)$ [9,11].

The risk function $r(x)$ is the adversary's prior belief on the event of successfully re-identifying a target individual x with the background knowledge of the degree information, and the risk function $r(x|G_h)$ is the adversary's posterior belief after getting the released graph data G'.

We assume that the adversary only has the background knowledge of the target's degree and uses the information to find the vertices with the same degree in the released data. For each individual in the social network data, the prior disclosing risk

$$r(x) = max(1 - E[L(x, \alpha)]) = maxP(X = \alpha) = 1/n, \tag{1}$$

where $E[L(x, \alpha))]$ is the expectation of the loss function

$$L(x, \alpha) = \begin{cases} 1, x = \alpha \\ 0, x \neq \alpha \end{cases}, \tag{2}$$

The function $L(x, \alpha)$ is a penalizing counter for the adversary guessing the wrong mapping function φ^{-1} of α.

If the adversary has the released data, we define the posterior disclosed risk of each individual as

$$r(x|G_h) = max(1 - E[L(x, \alpha)]|G') = max(P(X = \alpha)|G') \tag{3}$$

Our privacy preserving goal is to prevent attackers from re-identifying any user and acquiring the privacy information with published social network data. The posterior disclosing risk $rw(x|G_h)$ of the whole graph,

$$rw(x|G_h) = max_{x \in V}(P(X = x|G_h)), \tag{4}$$

is the posterior disclosed risk of the most vulnerable individual in the graph data G_h.

In this paper, we restrict the sanitizing operations to edges modification. That is, the graph G_h is generated from G only by adding and deleting a set of edges. This modification will protect the data privacy but sacrifice some data quality. So, we choose the difference between the original data and the sanitized data to measure the data utility lost, and call it information loss. The information loss is defined as

$$IL(G, G_h) = (|E \cup E_h|) - (|E \cap E_h|), \tag{5}$$

where $IL(G, G_h)$ is the information loss between G and G_h, E and E_h are the edge sets of G and G_h.

Finally, we define the privacy preserving data publishing problem as follows:

Definition 3. (Privacy preserving data publishing with confidence p). A sanitized graph $G_n(V_n, E_n)$ is generated from the a social network data graph $G(V, E, L, \varphi)$ by edges modification .If the posterior disclosing risk $rw(x|G_h) \leq (1 - p)$, the $G_n(V_n, E_n)$ is a privacy preserving data publishing with confidence p.

3   Hybrid Privacy-Preserving Data Publishing Algorithm

In this section, we analyze the relative privacy preserving methods of k-anonymity and random approach. We combine the property of hiding in group of k-anonymity with the perturbation of random approach and devise a hybrid privacy-preserving data publishing method. Then, we prove that our method satisfies the previous definition 3 for the data publishing.

3.1 k-Anonymity and Random Approaches

The k-anonymity approach hides the individual in a group of similar attributes such that each individual cannot be uniquely distinguished from at least k-1 individuals in the published data. To achieve this, the original data is modified by suppressing or generalizing according to the hiding attribution. Because of the complexity and dependency among vertices of social network data, a small modification of an edge or vertex can spread across the whole network. Aggarwal G.[9] showed that the k-anonymity problem is NP-hard by generalization even when the attribute values are ternary .

The random-based approach uses perturbation to protect the data privacy. It adds random noise to the original social network data by randomly adding or deleting edges and protects the data against the re-identification risk in a probabilistic manner. However, existing random modification approaches may not change some vertices due to the uncertainty of probability. The random model may not provide sufficient protection for every vertex in the social network data. The random model does not use the innerly grouping attribute of the data, and will lead to more information loss.

Motivated by the preceding approaches, we propose a novel privacy preserving method which combining the k-anonymity with random concept for publishing social network data. We use k-anonymity concept to protect the individuals who can be grouped together into equivalent class, and protect the individuals who do not satisfy k-anonymity by random approach.

3.2   The Hybrid Privacy-Preserving Data Publishing Algorithm

The hybrid privacy-preserving data publishing algorithm contians five steps.

1. We use the naive anonymization method to remove the identifiable attributes of original social network data $G(V, E, L, \varphi)$ and get the naive anonymization graph $G_n(V_n, E_n)$ with reallocated integer indexes.

2. Divide the vertices in set V_n, into two sets: the anonymized set V_k consists of vertices satisfying k-anonymity and the unanonymized set V_r contains otherwise vertices.

3. For the vertices in the set V_r, we randomly add m false edges between them if they are not connected already, followed by deleting m edges if they have edge between them.

4. Checking each vertex in set V_r, we partition the set V_r into V_{ra} containing the vertices that the degree changed in the third step, and V_{rb} containing the vertices otherwise.

Table 1. Table of Notation

Symbol	Meaning
G, G_h	The original graph, published graph
$d_i, d_i{}'$	The degree of node i of G and G_h
n, l, N	The number of nodes and links of G, $N = n(n-1)/2$
$r(x), r(x\|G_h)$	The prior and posterior privacy disclosure risk
m	The number of deleting and adding edges in the algorithm
d_{id}	The deleting edge number of step 3
d_{ia}	The adding edge number of step 3

5. For each vertex in set V_{rb} , we randomly select another vertex in V_{rb}; if there is an edge between them, delete the edge, otherwise add an edge between them.

To be convenient, we summarize the commonly used symbols of this paper in table 1.

This algorithm will ensure that the output dada satisfies the definition 3, privacy preserving data publishing with confidence p. We prove it as follows.

We assume that the adversary uses the following attacking approach: first, he gets the degree information of the target victim, and then finds the vertices having same degree in the published graph G_h. If the searching result contains more than one vertex, one vertex will be randomly selected as the victim. We can calculate the posterior privacy disclosing risk $rw(X|G_h)$.

$$rw(X|G_h) = max_{x \in V}(P(X = x|G_h))$$
$$\leq max(max_{x \in V_k}(P(X = x|G_h)), max_{x \in V_r}(P(X = x|G_h))) \qquad (6)$$
$$= max(\tfrac{1}{k}, max_{x \in V_r}(P(X = x|G_h))$$

We analyze the proceeding step to calculate the disclosing risk of random protecting part. After step 3, we can calculate the degree distribution of $P(d_i{}' = x|d_i), d_i \in V_r$. For each edge in the original graph, the probability of being deleted is $p_d = m/l$, and each non-existing link can be added with a probability $p_a = m/(N - l)$, where $N = n(n-1)/2$.

Let a random variable X be the deleting number of edges of vertex i after the deleting process in the step 3. The range of X is $\{x|0 \leq x < min(m, d_i)\}$. Let $d_{id} = min(m, d_i)$, the distribution of X given d_i is specified by the condition probabilities:

$$P(X = x|d_i) = B(x, d_{id}, p_d) = \binom{d_{id}}{x}p_d{}^x(1 - p_d)^x. \qquad (7)$$

Let a random variable Y be the adding number of edges of vertex i by the adding procedure in the step 3. The range of Y is $\{y|0 \leq y < min(m, n-1-d_i)\}$. Let $d_{ia} = min(m, n - 1 - d_i)$, the distribution of Y given d_i is specified by the condition probabilities :

$$P(Y = y|d_i) = B(y, d_{ia}, p_a) = \binom{d_{ia}}{y}p_a{}^y(1 - p_a)^y. \qquad (8)$$

Let a random variable Z be the number of edges of vertex i after the step 3. We have $Z = d_i + Y - X$; Y and X are independent. The range of Z is $\{z | 0 \leq z < d_i + d_{ia}\}$. The distribution of Z given d_i is specified by the condition probabilities:

$$
\begin{aligned}
&P(Z = z | d_i) \\
&= P(d_i + Y - X = z | d_i) \\
&= \sum_{(x,y):d_i+Y-X=z} P(x, y | d_i) \\
&= \sum_{all\ x} P(x, z + x - d_i | d_i) \\
&= \sum_{x=0}^{d_{id}} B(x, d_{id}, p_d) B(z + x - d_i, d_{ia}, p_a).
\end{aligned}
\tag{9}
$$

If $d_i + x - z$ does not satisfy the condition $0 \leq d_i + x - z \leq d_{ia}$, we can get $B(d_i + x - z, d_{ia}, p_a) = 0$.

After step 5, the degrees of every vertices in set V_r are deferent with their original degrees. Thus the adversary can only randomly select a vertex from the G_h as the target individual. The posterior privacy disclosing risk

$$
rw(X | G_h) = \frac{1}{k}
\tag{10}
$$

Note that the step 3 is necessary. If we omit step 3, there is a subtle risk that the adversary may exploit the weakness of step 5 and search the possible targets within ± 1.

4   Experiments

In this section, we present an empirical study to evaluate the performance and utility of our hybrid privacy-preserving method and compare it with the existing state-of-the-art techniques, such as k-anonymity [6] and random approach [14,15]. As mentioned before, the purpose of releasing social networks data is to use them in safe and secure manner. The anonymization process should not change the data property significantly. To evaluate the effect of our algorithm, we use a set of evaluation measures listed below.

Information Loss (IL): The metric is defined in section 2. The IL is the difference between the original graph and the sanitized graph.

Clustering Coefficient (CC): It measures the degree to which nodes in a graph tend to be clustered together. In this paper, we use the clustering coefficient of the whole graph

$$
C' = \frac{1}{n} \sum_{i=1}^{n} C_i
\tag{11}
$$

where C_i is the local clustering coefficient of a vertex introduced by Watts and Strogatz [18], C' is the average of the local clustering coefficients of all the vertices.

We then use three diverse real-world datasets to examine the efficiency and practice of our algorithm. Our datasets are described as follows:

Table 2. Structural Properties of the Datasets

Dataset	#Vertices	#Edges	CC
Facebook	534	4813	0.5437
Citation	2555	6101	0.1623
Movies	27312	122514	0.1180

Table 3. The average-degree changing probability

Dataset	Changing Fraction				
	10%	30%	50%	70%	90%
Facebook	0.091	0.530	0.743	0.820	0.865
Citation	0.006	0.130	0.279	0.475	0.713
Movies	0.009	0.183	0.211	0.368	0.512

Facebook dataset: It is a 'circles' (or 'friends lists') from Facebook. It is collected from survey participants using this Facebook app and available at http://snap.stanford.edu/data/ egonets-Facebook.html [19]. It contains 534 vertices and 4813 edges.

Citation dataset: It is a citation graph dataset, consists of 2555 papers and 6101 citation relationship [20]. The dataset is available at http://www.datatang.com/data/17310.

Movies dataset: A data set consists of movies, actors, directors, writers, and various relationships between them crawled from http://en.wikipedia.org/wiki/Category:English-language-films. It contains 27312 vertices and 122514 relations, and is available at http://www.datatang.com/data/17311.

All datasets have undirected edges, with self-loops removed. We eliminated a small percentage of disconnected vertices in each dataset. Their tructural properties are summarized in Table 2.

The experiments were conducted on a workstation running the Linux operating system, Debian 6.0.2, with a Intel Xeon Processor(2630) and 16GB memory. The program was implemented in C++ and compiled by g++ with -03 parameter.

The goal of our first experiment is to evaluate the impact of edges modification in step 3 on the data utility. We test three datasets with different modifying edges number m from $\%10|V_r|$ to $\%100|V_r|$. The X-axis is the confidence p, range from 50% to 95%, and the Y-axis is the IL. Figure 2 shows the results of the three real datasets. We can observe that the IL increase dramatically with the confidence p and m. With the same p, adding and deleting more edges will bring more IL. From the equation 11, we can calculate the probability of the vertices changing to their degrees, $P(Z \neq d_i|d_i) = 1 - P(Z = d_i|d_i)$. The average-degree changing probability is showed in table 3. We conclude that $30\%|V_r|$ is a suitable parameter for our algorithm.

To illustrate the effectiveness and superiority of our algorithm, we compare our work with the related k-anonymity [6] and random approach [14,15].

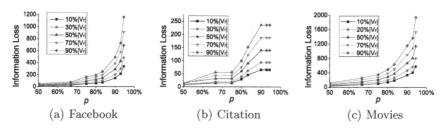

Fig. 2. The Information Loss of different random size

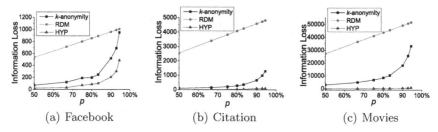

Fig. 3. Information Loss results for different p

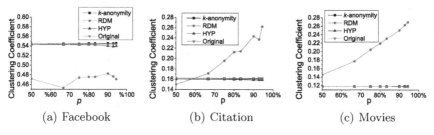

Fig. 4. Clustering Coefficient results for different p

Fig. 3a,3b,3c demonstrate the Information Loss results of the three data sets when p values from 50% to 95%. The HYP algorithm performs better than the other algorithms; it only increases a little with the confidence p. The random approach has approximated liner increasing rate. The IL of k-anonymity lies between the two algorithms, but it increases dramatically when p is more than 80%.

Finally in Fig. 4a, 4b, and 4c, we illustrate the values of the Clustering Coefficient of the three data sets when p values from 50% to 95%. The k-anonymity and HYP affect the CC slightly, but the random approach affects the CC dramatically. Especially for social network data with higher CC, the random approach will destroy the group attribution and decrease the CC dramatically.

5 Conclusion

In this paper we track the novel and important problem of preserving privacy in social networks data publishing, and propose an efficient algorithm to solve the degree attacking problem. We devise a hybrid privacy-preserving method for publishing social network data. The hybrid algorithm combines the techniques of k-anonymity and randomization and it can efficiently protect the data privacy with less utility loss. We compare our algorithm with the related k-anonymity and random approach on several real-world data sets. The experimental results show that our algorithm is stable and modifies the original data less than the existing algorithms.

Social networks data is much more complicated than tabular data, since a single change of an individual data can affect all the neighbor individual data. Moreover, any topological structure combined other especial properties of the social networks data can be potentially used to breach private information. The future works in this field of social networks privacy preserving are still plentiful.

One of our future work is trying to develop combining privacy model for social networks data and investigating how well different strategies protect privacy (identity, link privacy, and attribute privacy) when adversaries exploit various background knowledge in their attacks.

Acknowledgments. The research is supported by the National Key Basic Research Program of China (973 Program, No. 2012CB326403), National Science Foundation of China (Nos. 61272535 and 61165009), Guangxi Natural Science Foundation (Nos. 2012GXNSFAA053219 and 2013GXNSFBA019263), Guangxi "Bagui Scholar" Teams for Innovation and Research Project, and Science and Technology Research Projects of Guangxi Higher Education (No.2013YB029).

References

1. Company, T.N.: State of the media: The social media report 2012, pp. 1–12 (2012)
2. Ellison, N.B., et al.: Social network sites: Definition, history, and scholarship. Journal of Computer-Mediated Communication 13(1), 210–230 (2007)
3. Zhang, S., Qin, Z., Ling, C.X., Sheng, S.: "Missing is usefu": Missing values in cost-sensitive decision trees. IEEE Trans. Knowl. Data Eng. 17(12), 1689–1693 (2005)
4. Zhang, C., Zhang, S.: Association Rule Mining. LNCS (LNAI), vol. 2307. Springer, Heidelberg (2002)
5. Backstrom, L., Dwork, C., Kleinberg, J.: Wherefore art thou r3579x?: Anonymized social networks, hidden patterns, and structural steganography. In: Proceedings of the 16th International Conference on World Wide Web, pp. 181–190. ACM (2007)
6. Liu, K., Terzi, E.: Towards identity anonymization on graphs. In: Proceedings of the 2008 ACM SIGMOD International Conference on Management of Data, pp. 93–106. ACM (2008)
7. Zhou, B., Pei, J.: Preserving privacy in social networks against neighborhood attacks. In: IEEE 24th International Conference on Data Engineering, ICDE 2008, pp. 506–515. IEEE (2008)

8. Wu, W., Xiao, Y., Wang, W., He, Z., Wang, Z.: k-symmetry model for identity anonymization in social networks. In: Proceedings of the 13th International Conference on Extending Database Technology, pp. 111–122. ACM (2010)

9. Aggarwal, G., Feder, T., Kenthapadi, K., Motwani, R., Panigrahy, R., Thomas, D., Zhu, A.: Approximation algorithms for k-anonymity. Journal of Privacy Technology (JOPT) (2005)

10. Zhou, B., Pei, J.: The k-anonymity and l-diversity approaches for privacy preservation in social networks against neighborhood attacks. Knowledge and Information Systems 28(1), 47–77 (2011)

11. Machanavajjhala, A., Kifer, D., Gehrke, J., Venkitasubramaniam, M.: l-diversity: Privacy beyond k-anonymity. ACM Transactions on Knowledge Discovery from Data (TKDD) 1(1), 3 (2007)

12. Yuan, M., Chen, L., Yu, P.S.: Personalized privacy protection in social networks. Proceedings of the VLDB Endowment 4(2), 141–150 (2010)

13. Hay, M., Miklau, G., Jensen, D., Weis, P., Srivastava, S.: Anonymizing social networks. Computer Science Department Faculty Publication Series, 180 (2007)

14. Ying, X., Pan, K., Wu, X., Guo, L.: Comparisons of randomization and k-degree anonymization schemes for privacy preserving social network publishing. In: Proceedings of the 3rd Workshop on Social Network Mining and Analysis, p. 10. ACM (2009)

15. Ying, X., Wu, X.: On link privacy in randomizing social networks. Knowledge and Information Systems 28(3), 645–663 (2011)

16. Sweeney, L.: k-anonymity: A model for protecting privacy. International Journal of Uncertainty, Fuzziness and Knowledge-Based Systems 10(05), 557–570 (2002)

17. Zou, L., Chen, L., Özsu, M.T.: K-automorphism: A general framework for privacy preserving network publication. Proceedings of the VLDB Endowment 2(1), 946–957 (2009)

18. Watts, D.J., Strogatz, S.H.: Collective dynamics of small-world networks. Nature 393(6684), 440–442 (1998)

19. McAuley, J.J., Leskovec, J.: Learning to discover social circles in ego networks. In: NIPS, vol. 272, pp. 548–556 (2012)

20. Tang, J., Sun, J., Wang, C., Yang, Z.: Social influence analysis in large-scale networks. In: Proceedings of the 15th ACM SIGKDD International Conference on Knowledge Discovery and Data Mining, pp. 807–816. ACM (2009)

Achieving Absolute Privacy Preservation in Continuous Query Road Network Services

Yankson Herbert Gustav[1], Xiao Wu[2], Yan Ren[2,*],
Yong Wang[1], and Fengli Zhang[1]

[1] School of Computer Science and Technology, University of Electronic Science and
Technology of China, Chengdu, China
[2] National Computer Network Emergency Response Technical Team Coordination
Center of China, Beijing, China
hgustav.yankson@gmail.com, cla@uestc.edu.cn, ry@cert.org.cn

Abstract. Research have shown that location semantics have lead to privacy leakages especially when two or more users in a cloaked region depict similar semantic locations. This implies that, to achieve absolute privacy(query privacy, location privacy and semantic location privacy) protection for a client on road network, it is important that cloaked users have their locations distinctly diverse with diverse semantics, and making diverse service request thus satisfying the k-anonymity and l-diversity conditions for privacy. Unfortunately, the determination of semantic location of a mobile user online is a challenge which makes the achievement of absolute privacy protection more challenging. In this paper, we developed a privacy preserving algorithm that protects a client's absolute privacy for continuous query road network services. We employed an offline trajectory clustering algorithm and semantic location graph to aid the selection of cloaked users that will effectively protect the absolute privacy of a client. We evaluated the effectiveness of our algorithm on a real world map with two defined metrics, and it exhibited an excellent anonymization success rate in a very good query processing time for the entire period of continuously querying road network services.

Keywords: Location-based Services (LBS), Privacy Preservation Algorithm, Trajectory Clustering Algorithm, Semantic Location Privacy.

1 Introduction

Location based queries have become very common mobile applications because of the convenience that it provide its users. Despite the convenience in its usage, it threatens the privacy of users. The threat of privacy results from the fact that, the disclosed user location for service can be combined with other kinds of data to allow an adversary make unwanted inferences on the activity of the user at some given time in the past [1].

---

* Corresponding author.

X. Luo, J.X. Yu, and Z. Li (Eds.): ADMA 2014, LNAI 8933, pp. 279–292, 2014.

There are two types of privacy issues namely location privacy and query privacy. Location privacy is related to the disclosure of exact locations that a user has visited, whereas query privacy is related to the disclosure of sensitive information in the query itself and its association to the user. An adversary having knowledge of these two, can reveal places of frequent visit and personal preferences of the user, hence there is the need to protect it [2]. In recent times, location semantics have also shown to leak privacy especially when two or more users in a cloaked region depict similarity in semantic locations [3]. Therefore, to achieve absolute privacy protection(query privacy, location privacy and semantic location privacy) for a client on a road network, it is important that cloaked users must have their locations distinctly diverse with diverse semantics, and making diverse service requests thus satisfying the k-anonymity and l-diversity conditions for privacy. However, determining the semantic location of a mobile user online is a challenge, which makes the achievement of absolute privacy protection more challenging.

The random segment sampling and network expansion cloaking methods are the two known cloaking methods in literature for road networks. The network expansion cloaking method blurs a user's location into a cloaked set of k-users from connected road segments S such that S satisfies client's privacy. Whereas the random segment sampling model blurs a user's location into a cloaked set of k-users from randomly selected road segments [4]. Employing the network expansion cloaking methods to obtain absolute privacy protection may not be achievable, as all cloaked users may assume similar semantics because of the short distances amongst cloaked users. For example, cloaking a client with k-1 other users in a big university campus with such a cloaking technique will have all k cloaked users' location depicting the university as its semantics and hence will leak their privacy. Employing the random segment sampling method seems a better option because of its random segment cloaking, however, the inability to determine the semantics of users online makes it a challenge. We believe a mechanism where the query, location and its semantic of an online mobile user could be determined offline would make the random segment sampling model appropriate. In that way, the characteristics of users that protect the absolute privacy of clients could be determined offline before going ahead to cloak it online.

In this paper, we develop a privacy preserving algorithm that protects the absolute privacy of clients for continuous query road network services. Our contributions in this work are as follows:

1. Develop an algorithm that cluster trajectories of mobile users offline and use the derived movement trends online to aid the selection of cloaked users that will enhance the absolute privacy protection of clients.

2. Develop an algorithm to generate a sematic location graph to aid the determination of users' semantic locations online.

3. Develop a privacy preservation algorithm to protect the absolute privacy of clients continuously querying road network services, using the derived

movement trends from the offline trajectory clustering algorithm and the semantic location graph.

4. We introduce two metrics namely Anonymisation Success Rate and Query Processing Time to test the efficiency of our algorithm on a real world map.

The rest of the paper is organized as follows; Section 2 discusses related work on privacy preservation in LBS. Section 3 discusses the preliminaries including designing goals, system architecture, and the road network model. Section 4 discusses the development of the trajectory clustering algorithm and the semantic location graph. Section 5 discusses the privacy preserving algorithm and its security. Sections 6 presents the experiment, and conclude in section 7.

2   Related Work

In this section, we will categories the related work into privacy preservation in euclidean space and those constrained by the underlying road network.

2.1   Privacy Preservation in Euclidean Space

To protect the semantic location of clients, Damiani et al [5] employed the semantic location cloaking method which allows users to define a personalised privacy profile stating specified sensitive place types and the desired degree of privacy for each type. Unfortunately, semantic location cloaking method have been designed to work only in unconstrained space in which users can move without restrictions, but in real world setting, movement is confined to road network and may therefore lead to privacy leakages. B.Lee et al [3] proposed location privacy protection technique, which protects the location semantics from an adversary. They employed a trusted anonymizing server that uses the location semantic information for cloaking users with semantically heterogeneous locations. Similarly, their work considered euclidean space that will lead to privacy leakages under road network restrictions. C. Chow et al [6] proposed a spatial cloaking technique for snapshot and continuous location-based queries that clearly distinguishes between location privacy and query privacy using k-sharing region and memorization properties. They adopted a minimum area A_{min} within which a client wants to be anonymised. However, their work employed a cloaking technique that did not consider semantic location of its cloaked users hence are likely to include users with same semantics leading to privacy leakages.

2.2   Privacy Preservation on Road Networks

To protect privacy on road networks, Li et al [7] proposed a personality privacy-preserving cloaking framework for the protection of sensitive positions on road network environment. In their client-server architecture scheme, a user expressed his privacy requirements by specifying some types of sensitive semantics and used popularity ratio of those places to measure the degree of semantic diversity.

Yigitoglu et al [8] presented an extension of the semantic location cloaking model for location sharing under road-network constraints, that relies on the trusted anonymizer. Our work is different because we did not consider some places as sensitive but rather all locations of clients as sensitive.

The network expansion method is known to be less attack resilient [4], and therefore not surprising that [4], [9], and [10] employed a modified version of network expansion cloaking method to make it attack resilient. However, their works may lead to privacy leakages when considering semantic locations due to their closeness of segments which will likely make all cloaked users depict same semantics. We intend to use a more attack resilient random segment sampling model employing the semantic location graph and trajectory clustering algorithm to aid the selection of cloaked users that protect the absolute privacy of clients.

In other related research work, Binh Han et al [11] proposed a frame work called NEAT, a road network aware algorithm for fast and effective clustering of spatial trajectories of mobile objects travelling on road networks which takes into account the physical constraints of the road network, network proximity and the traffic flows among consecutive road segments.

3   Preliminaries

3.1   Assumptions and Architectural Systems

We adopt a trusted third party architecture consisting of a mobile client(MC), Anonymous Server(AS) and location based server(LBS) [12]. We assume that the anonymous server has been supplied with the initial trajectory and service request (query content) database of users by the cellular service provider. The location and service request database may be built from clients' regular phone call and query of LBS. If such an initial database does not exist, we assume a location and service request sample collection phase by the anonymous server. The sampling location and service request collection phase should last for some few days. More location data will be obtained from mobile users during their requests for LBS. We also assume that MC is allowed a privacy profile k, being the number of users he would want to be anonymized with.

The core of the system is the AS which consist of the Cloaked Repository (CR) and an offline Trajectory Clustering Engine (TCE). Their functions can be defined as follow:

1. The cloaked repository keeps some previously cloaked results and use them to generate new cloaked regions.
2. The Trajectory clustering Engine performs the clustering of a database of mobile users' trajectories

The architecture is as shown in Fig. 1.

3.2   Road Network Model

A road network is represented by a single directed graph G = (V, E), composed of the junction nodes V = $(n_0, n_1 \ldots n_n)$ and directed edges E = $(s_{id},$

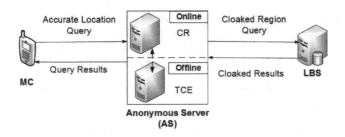

Fig. 1. The System Architecture

$n_i n_j)|n_i, n_j \in V)$. An edge e = $(s_{id}, w_0, w_1, con, n_i n_j) \in$ E representing a road segment connecting two junctions n_i and n_j in the real road network with attributes such as the segment identifier s_{id}, the segment classification w_0, the traffic density w_1, and types of service request con $(s_{r1}, s_{r2}...s_{rn} \in con$ where each s_r is a service request). The order $n_i n_j$ indicates the direction from n_i to n_j of the road segment. For a bi-directional road segment s_{id}, we use edge e = $(s_{id}, n_i n_j)$ and e' = $(s_{id}, n_j n_i)$ to denote the bi-directional lanes with road segment identifier s_{id}. The length of a road segment e = $(s_{id}, n_i n_j$) is denoted by length $\mid n_i n_j \mid$. We classify the segments according to their speed limits. The segments are classified primary (speed limit <40km/hr), auxiliary (40km/hr$\leq$ speed limit <70km/hr), highway (70km/hr$\leq$ speed limit <100km/hr), express way (speed limit$\geq$100km/hr) denoted by p, a, h, ex respectively. Therefore segment classification w_0 could be represented by p, a, h, or ex. For example $w_o =$ ex represents express road classification. We denote the position of a user on a road segment s_{id} with coordinates (x, y) at a time stamp t by l=(s_{id}, x, y, t).

Definition1 Trajectory: A trajectory denoted by TR = $\{t_{id}, l_0, l_1....l_n\}$, is a time-ordered sequence of locations $l_0, l_1, ..., l_n$ of a user on the road network over time and uniquely identified by a trajectory identifier t_{id}. For a mobile user, his/her trips with beginning location and destination location forms a trajectory.

Definition2 Trajectory-fragment: A trajectory-fragment of TR, denoted by tf = $\{t_{id}, s_{id}, l_k l_{k+m}\}$, represents a sub-trajectory $l_k, l_{k+1}...l_{k+m}$ consisting of m + 1 consecutive points extracted from TR which lie on the same road segment s_{id}.

Definition3. Base cluster: A base cluster b with respect to a segment is a group of distinct trajectory-fragments with similar characteristics. Each of these trajectory-fragments belongs to a distinct trajectory TR and is associated with a segment s_{id}. A group of base clusters in a segment s_{id} is called a segment cluster $c_{s_{id}}$.

Definition4. Class cluster: For a given set of trajectories T = $\{TR_1... TR_n\}$ on a road network, a class cluster c_{w_0} is a set of all segment clusters $c_{s_{id}}$ in all segments with similar segment classification w_0. Therefore, class clusters are c_p, c_a, c_h, c_{ex} where each class cluster could be represented generally as c_{w_0}.

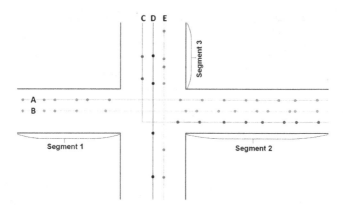

Fig. 2. A section of road network depicting trajectories A,B,C,D,E broken into trajectory fragments along segments

Definition5. Cluster : For a given set of trajectories T = $\{TR_1... TR_n\}$ on a road network, cluster C is the set of all class clusters ie C = $\{c_p, c_a, c_h, c_{ex}\}$. C therefore represents all class clusters with segment classifications w_0.

4   Developing Trajectory Clustering Algorithm and Semantic Location Graph

4.1   Trajectory Clustering Algorithm

One technique that is widely known to group objects of a database into a set of meaningful subclasses is the clustering algorithm [11]. We intend to use this tool to cluster trajectories of users into subclasses according to their similar movement characteristics on same road segment offline. We believe users within a segment can be considered close in terms of network proximity and will therefore display a group of subclass characteristics in their movements which will reflect that of any online mobile object on that segment. We therefore can estimate from the offline characteristics if its inclusion will protect privacy of the client before going ahead to cloak a user with such characteristics online. We intend to cluster trajectories of mobile users offline into subclasses along a road segment according to segment id, service request, direction of flow, speed, temporal details, length of the segment and it's speed limits. We use these properties because we believe they influence movement trends of a mobile user on a road segment.

With this prior information, we intend to develop a trajectory clustering algorithm on users' trajectories. Consider a set of trajectories denoted by T = $\{TR_1, TR_2... TR_k\}$ in the TCE, we examine a single trajectory $TR_i = \{t_{id}, l_0, l_1...l_n\}$ from the first location l_0 to the last location l_n. We take every two consecutive points in the trajectory, say l_i and l_{i+1}, and check to obtain the road junction node that intersects two road segments using the map-matching approach [13].

Next, we insert the obtained junction node(s) as new points in between l_i and l_{i+1} in the trajectory being examined. After examining every point in a given trajectory TR_i, the sequence of junction nodes added to TR_i will serve as the trajectory splitting points used to partition the trajectory into trajectory fragments (tf) along segments. For example, in Fig.2 trajectory C has a two trajectory fragments broken along segment 2 and segment 3. The period of time travelled by the individual trajectory fragments are analyzed and categorized into different time periods of the day(t_b). This procedure is repeated for all set of trajectories in T.

We group the trajectory-fragments by their road segment id, direction of flow, speed (v), segment length, time period of travel(t_b) and speed limits v_l on that segment to form a collection of base clusters. We then compute the resulting traffic density w_1 on each base cluster in a segment as the number of trajectory fragments. The density gives the likely number of users on that segment with a particular characteristics at a given time period of the day. For example in Fig. 2, if all trajectory fragments have similar characteristics then segment2 will have a traffic density of 3, whiles segment1 has a density of 2. A group of base clusters under the same segment id forms a segment cluster $c_{s_{id}}$. The algorithm output a base cluster according to segment id as b= $(s_{id}, w_0, s_r, v, w_1, t_b, n_i n_j)$, being the group of users characteristics on the segment at the time t_b. Finally, we abstract all segments clusters under similar road classifications w_0 into class clusters denoted as c_p, c_a, c_h, $c_{ex} \in C$. We abstract according to segment classification because we believe it will aid cloaking of users with diverse location semantics. A group of class clusters form a cluster C. The detailed description is as shown in Algorithm 1.

Algorithm 1 Trajectory Clustering Algorithm

Input: $< GraphG >$, $<T = TR_1, TR_2... TR_n>$,$<TR_i = t_{id}, l_0\ l_1...l_n>$
Output: $< clusterC = c_p, c_a, c_h, c_{ex} >$,
1: Let G composed of the junction nodes V$(n_i, n_j \in V)$ and directed edges E (e $\in$ E);
2: **for** TR_i,i=1...n **do**
3:     examine every l_i and l_{i+1} to obtain the sequence of road junction nodes n_i
4:     insert the obtained junction node(s) as new points in between l_i and l_{i+1}
5:     use new junction node to break TR into tf
6:     Assign identities s_{id} for each e
7:     categorise t_b of all tf
8:     group all tf along e according to s_{id}, v, v_l, t_b, $n_i n_j$ and ∣ $n_i n_j$ ∣ into b
9: **end for**
10: Evaluate w_1 in each b
11: group all b in each e as $c_{s_{id}}$;
12: group all $c_{s_{id}}$ with same w_0 as c_{w_0} ;
13: output all c_{w_0} as C;

4.2   Semantic Location Graph

In this subsection, we discuss an algorithm to build a semantic location graph to aid the selection of users for cloaking in order to protect semantic location privacy. We define semantic location privacy as the disclosure of the semantics associated with a location visited by a user. The essence of protecting semantic location privacy is to avoid a user's activities at that location from being inferred by an adversary. For example, if a location visited by a user is a cancer hospital, it may be inferred from its semantics that a user may be a cancer patient. Therefore, to cloak a location into a region such that it becomes anonymous, it's important that the semantic locations of cloaked users are made l-diverse to avoid privacy leakages. To aid the achievement of l-diversity in semantic locations, we need to build a semantic location graph.

To build such a graph, we need to determine what constitutes a semantic location. We define semantic location as a location where a kind of service is provided and therefore visited by many people who stay for a period of time. People visit locations mostly with a reason. We go to restaurants to have food, schools to attend classes, and hospitals to see a doctor. Since we have reasons for a visit, we stay for a while at a location for those reasons. Moreover, we spend different amount of time based on these reasons from which an adversary is able to link a location to a user's activity. Using the duration of stay and the service provided at the location as factors, different point of interest (POI) representing different semantic locations can therefore labelled [3]. Unfortunately labelling is not the focus of this research, so we assume point of interest (POI) collections and their location coordinates which are open source information. [14].

Let's consider a set of POI locations SL=$(L_1, L_2, L_3....L_n)$ where each L represents a semantic location. We categorise the semantic locations according to their similarity of service provided at that location denoted $SL_u=(L_1,... L_m)$, where u is a category of service provided at $L_1,.. L_m$. For example, we group clinic, health post, hospitals, dental clinics, etc under the category health, where health is category of service provided. We do the categorisation to help us avoid cloaking user locations with similar service to enhance l-diversity in their semantic locations. Using the set of semantic locations SL, we employ the map matching approach to locate their exact positions on various segments on our road network model. We then generate a semantic location graph G' depicting various semantic locations and their services provided. The algorithm is as shown below;

Algorithm 2 Semantic location Graph Algorithm

Input: $< GraphG >$, $<$SL $= (L_1, L_2, L_3.....L_n)>$
Output: $< SemanticLocationGrapgG' >$ and $< SL_u >$
1: Let G composed of the junction nodes V($n_i,n_j \in$ V) and directed edges E (e $\in$ E);
2: **for** L_i,i=1...n **do**
3:     label each L_i according to service provided
4:     group all L_i with same u
5:     output each group as SL_u
6: **end for**
7: **for** L_i,i=1...n **do**
8:     insert in G
9:     output G'
10: **end for**

---

5   Privacy Preserving Algorithm and Security Analysis

5.1   Privacy Preserving Algorithm

In this subsection, we develop the privacy preserving algorithm using trajectory clustering algorithm and the semantic location graph. The mobile client(MC) sends a new query q in the form q=$(q_{id}, l, t_i, t_f, k, s_r$), where l is the location coordinates, t_i is the query initiation time and t_f is the expiration time. The service request is s_r, privacy profile k, and q_{id} is the client id. On receiving a new query q, AS determines the time of q, and use the location to find the segment from which it was issued, classification of the segment, semantic location L associated with q from G' and the category of service provided SL_u to which it belongs.

We define a traffic density threshold σ, below which it will not be appropriate for online cloaking to be executed. $w_1 \geq \sigma$ implies we are sure to find enough mobile users on that segment at the time period t_b. Traffic density threshold σ is determined by AS based on history.

To find users online to anonymise MC, TCE searches through all class clusters c_{w_0} other than that containing MC, to find the k-1 other segments at random and select a base cluster each with time t_b that satisfies the time of q from the selected segments. The selected base clusters must satisfy the cloaking conditions designed to aid the achievement of absolute privacy protection employing the random segment sampling method.

Cloaking Conditions:- All selected base clusters must satisfy;

1. k-1 in number,
2. The time period $t_{b_{k-1}}$ of the $(k-1)^{th}$ selected base cluster b_{k-1} must satisfy the time t of the query q and the time t_{b_1} of first selected base cluster b_1, ie $t \in t_{b_1}$; $t \in t_{b_{k-1}}$.
3. The traffic density of any selected base cluster $w_1(b) \geq \sigma$ ie $w_1(b_1) \geq \sigma$, $w_1(b_{k-1}) \geq \sigma$.
4. The service request $s_r(b_{k-1})$ of the $(k-1)^{th}$ selected base cluster b_{k-1} must not be the same as that of q and that of the first selected base cluster b_1, ie $s_r(b_1) \neq s_r(b_{k-1}) \neq s_r(q)$.
5. The category of service provided $SL_u(b_{k-1})$ of the semantic location L on the selected segment of the $(k-1)^{th}$ selected base cluster should not be in the same category as q and that of the first selected base cluster b_1 , ie $SL_u(b_1) \neq SL_u(b_{k-1}) \neq SL_u(q)$,
6. The segment classification $c_{w_0}(b_{k-1})$ of $(k-1)^{th}$ selected base cluster should not be the same as that of q and the first selected base cluster b_1, ie $c_{w_0}(b_{k-1}) \neq c_{w_0}(b_1) \neq c_{w_0}(q)$.

The cloaking conditions ensures that we are likely to find k-1 mobile users online with k-1 different locations associated with k-1 different semantics from k-1 different segments requesting k-1 different service request at the time of the query. When the cloaking conditions are met, AS cloaks k-1 other online users q' with the characteristic of the k-1 selected base clusters from their respective segments into a cloaked region with q. All queries not meeting these conditions are suppressed. Client's id is removed and replaced with quasi-id and put into a cloaked region R_i where i represent the ith snapshot with cloaked region identity R_{id}. The cloaked region R_i containing k users is then forwarded to the LBS. The LBS provides the result for all k users and forwards it to AS. AS knowing the exact location of MC and his request submits the appropriate result to MC.

For continuous query LBS, a query will continuously be issued periodically by AS within the period $(t_f - t_i)$. AS cloaks users requesting continuous service for the first snapshot if client is requesting continuous service so as to maintain same cloaked users at t_i throughout the query period, whiles ensuring cloaking conditions at all times. Further, we keep a repository of already cloaked users request to use at a later time when its related to the same segment. The number of cloaked sets that meets cloaking conditions is denoted by n. The privacy preserving algorithm is as described in algorithm 3.

Algorithm 3 Privacy Preserving Algorithm

Input: $<queryq = q_{id}, l, k, t_i, t_f, s_r>, < classclusterc_{w_0} = (c_p, c_a, c_h, c_{ex}) >, < G' >, < SL_u >,$
Output: $< cloakedregion R >$
1: for q issued at t=t_i =i;
2: determine t_i of q
3: identify e, c_{w_o}, L and SL_u of q
4: randomly output a base cluster (b_1) in e of $c_{w_0}(b_1) \neq c_{w_0}(q)$ and $t_i \in t_{b_1}$
5: ensure $w_1(b_1) \geq \sigma$, $SL_u(b_1) \neq SL_u(q)$ and $s_r(b_1) \neq s_r(q)$
6: goto line 11
7: for $k > 2$ do
8:   ensure $w_1(b_{k-1}) \geq \sigma$; $c_{w_0}(b_1)$ $\neq c_{w_0}(q) \neq c_{w_0}(b_{k-1})$; $s_r(b_1)$ $\neq s_r(b_{k-1}) \neq s_r(q)$;
     $SL_u(b_1) \neq SL_u(b_{k-1}) \neq SL_u(q)$
9: **end for**
10: **if** line 8 is satisfied **then**
11:     select q' online with characteristics of b_1 to b_{k-1} at their respective e
12:     cloak q with k-1 other q' into cloaked region R_i
13:     replace q_{id} with quasi id
14:     assign region identity R_{id}
15: **else**
16:     suppress query
17: **end if**
18: forward R_i to LBS
19: for t>t_i=i do
20:     Repeat 2-18
21: **end for**

5.2   Security Analysis

An adversary with the exact position of users, sample of cloak set of some snapshot on road segments, knowledge of some sample query contents and cloaking algorithm may be able to launch query tracking attack [6], query homogeneity attacks [15], query semantic homogeneity attacks and location similarity attack [3]. We employed the randomly selected segment cloaking technique which is attack resilient to enhance the security of our algorithm.

Let us consider a scenario, in which all users from a cloaked region are requesting for the same type of service such as the location of a special club.

In this case, even if an adversary cannot link an individual query to a specific user, it is still known to the adversary that all the users in the cloaked region including the client are interested in that special club. This kind of attack is referred to as query homogeneity attack. To avoid query homogeneity attack, we avoided cloaking together any two users requesting the same service such that the probability of linking a request to a client is $\frac{1}{k}$.

Assuming cloaked users in a region show a primary school, high school, and a university as their semantic locations, an adversary may be able to infer from the semantics that the users are likely students and so is the query client. This attack is called query semantic homogeneity attack. If the cloaked users locations depict a big university campus, an adversary may be able to infer from their locations that they may be students of that university from their location similarity. This attack is termed location similarity attacks. To overcome query semantic homogeneity attacks, we avoided cloaking users with similar semantic locations and with similar services provided at that location whiles for location similarity attacks we cloaked users at diverse location from different road segment with different classifications such that linking a client to its semantics or location is $\frac{1}{k}$.

Query tracking attacks are attacks in which an adversary aggregates different snapshots of continuous query and takes an intersection of all snapshots such that the user that appears across all regions is identified as the query client. Alternatively, an adversary could model the mobility of any user from a cloaked region at t_i to a cloaked region at t_f as a Markov chain [16] where each state of the Markov chain represents a cloaked region. The m-step transition probability of a user can be defined as the probability that a user at an initial cloaked region t_i will be at a final region t_f across m cloaked regions, which can be expressed as;

$$P^m(t_i, t_f) = P(X_m = t_f \mid X_0 = t_i) \tag{1}$$

The adversary evaluates the transition probability for all users in the cloaked regions, and the user that appear across all regions will have $P^m(t_i, t_f) = 1$ indicating its the query client otherwise its not.

To overcome query tracking attacks, we ensure that all users in a cloaked region for the first snapshot are querying continuously. This is realistic since our trajectory clustering algorithm gives us a fair idea of the segment in which to find such a user. However, different users query for different time periods, so where the time for the query client is greater than the other cloaked users, we reuse repository query for the same segment where necessary to protect the privacy of the query client at all times. Where the time for the query client is less than the other cloaked users, the query client will then be fully protected across its query period. In this way, aggregating all snapshots and taking intersection will lead to the same number of users appearance in all cloaked regions thus making the probability of identifying the query client $\frac{1}{k}$. An adversary using equation (1) will also have $P^m(t_i, t_f) = 1$ for all cloaked users since they will all appear at all cloaked regions hence making it difficult to identify the query client. An adversary having the algorithm is not anticipated because AS is trusted.

6   Experiment and Evaluation

6.1   Evaluation Criteria and Metrics

Success Rate. Anonymisation Success Rate (S) measures the ability of our algorithm to avoid suppression of a query. We evaluate this metric as the ratio of the number of successfully cloaked snapshots n to the total number of successful and unsuccessful cloaked regions C_{total} within an active query period.

$$SuccessRate(S) = \frac{n}{C_{total}} \times 100 \qquad (2)$$

Query Processing Time. Query Processing Time is the time required by the algorithm to find the k-1 other users. The average cloaking time T_{avg} for continuous query that has just elapse its active period consisting of n snapshots can be evaluated as;

$$T_{avg} = \frac{\sum_{i=1}^{n} T_{R_i}}{n} \qquad (3)$$

where T_{R_i} is the cloaking time of the query with region R_i.

6.2   Experimental Setup

Using the Thomas Brinkhoff Network-based Generator of Moving Object [17], we generated 10000 mobile users moving along the map of shanghai with varied speeds and assigned varied service requests. All road segments were assigned speed limits according to our classifications and with segment id. We employed 20148 POI GPS data set consisting of 50 different categories of service provided in Shanghai city obtained from GPS Data Team [18] as our semantic locations. We recorded 100 snapshots at an interval of 5seconds. With the simulated data, we implemented our algorithms using a laptop with 6GB memory and a Core (TM) i3-2330M 2.20 GHz Intel processor.

6.3   Discussions

We studied the effects of our defined metrics with some snapshot queries for a cloaked region of k-users and l-diverse segment $(k - l)$ values. From Fig.3, there was a high processing time at the first snapshot which is due to lots of time required for processing to meet the initial cloaking conditions. The querying process time decreased sharply from the initial snapshot until the first ten snapshots. The decrease may be the results of the cloak users requesting continuous service at the initial query hence there was less time required because it involved same users. There was a slight increase in the processing time from the 10th to 20th snapshot which might be due to users changing segments and therefore some processing was required. Thereafter, there was a steady decrease in processing time. This trend might be due to the introduction of repository queries hence a reduction in the processing time. Generally, the query processing time decreased with increase in snapshots even when k-l values was increased.

From Fig.4, the anonymization success rate remained almost constant for the first ten snapshots which is because users cloaked at t_i were querying continuously hence most of snapshots met the cloaking principles. There was a sharp decrease in success rate thereafter until the 20th snapshot which may be due to mobile objects changing segments with different classification and different semantic locations hence most of the snapshots could not meet the cloaking conditions. There was a steady increase in success rate thereafter which is due to the gradual introduction of repository queries hence most queries met cloaking conditions. When k-l was increased, similar trends was exhibited. Generally, our algorithm exhibited an average success rate of about 87.8 percent per snapshot within the 100 snapshots evaluated.

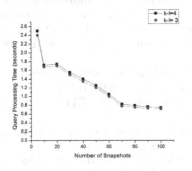 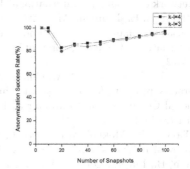

Fig. 3. Graph showing a measure of Query Processing Time

Fig. 4. Graph showing a measure of Anonymization Success Rate

7   Conclusion

In this paper, we developed a privacy preserving algorithm that protects a client's absolute privacy for continuous query road network services. We employed an offline trajectory clustering algorithm and semantic location graph to aid the selection of cloaked users that will effectively protect the absolute privacy of a client. We evaluated the effectiveness of our algorithm on a real world map with two defined metrics, and it exhibited an excellent anonymization success rate in a very good query processing time for the entire period of continuously querying road network services.

References

1. Silvestri, C., Yigitoglu, E., Damiani, M.L., Abul, O.: Sawlnet: Sensitivity aware location cloaking on road-networks. In: 2012 IEEE 13th International Conference on Mobile Data Management (MDM), pp. 336–339. IEEE (2012)

2. Dewri, R., Ray, I., Whitley, D.: Query m-invariance: Preventing query disclosures in continuous location-based services. In: 2010 Eleventh International Conference on Mobile Data Management (MDM), pp. 95–104. IEEE (2010)

3. Lee, B., Oh, J., Yu, H., Kim, J.: Protecting location privacy using location semantics. In: Proceedings of the 17th ACM SIGKDD International Conference on Knowledge Discovery and Data Mining, pp. 1289–1297. ACM (2011)

4. Wang, T., Liu, L.: Privacy-aware mobile services over road networks. Proceedings of the VLDB Endowment 2, 1042–1053 (2009)

5. Damiani, M.L., Silvestri, C., Bertino, E.: Fine-grained cloaking of sensitive positions in location-sharing applications. IEEE Pervasive Computing 10, 64–72 (2011)

6. Chow, C.-Y., Mokbel, M.F.: Enabling private continuous queries for revealed user locations. In: Papadias, D., Zhang, D., Kollios, G. (eds.) SSTD 2007. LNCS, vol. 4605, pp. 258–275. Springer, Heidelberg (2007)

7. Li, M., Qin, Z., Wang, C.: Sensitive semantics-aware personality cloaking on road-network environment. International Journal of Security & Its Applications 8 (2014)

8. Yigitoglu, E., Damiani, M.L., Abul, O., Silvestri, C.: Privacy-preserving sharing of sensitive semantic locations under road-network constraints. In: 2012 IEEE 13th International Conference on Mobile Data Management (MDM), pp. 186–195. IEEE (2012)

9. Hossain, A., Hossain, A.A., Chang, J.W.: Spatial cloaking method based on reciprocity property for users' privacy in road networks. In: 2011 IEEE 11th International Conference on Computer and Information Technology (CIT), pp. 487–490. IEEE (2011)

10. Chow, C.Y., Mokbel, M.F., Bao, J., Liu, X.: Query-aware location anonymization for road networks. GeoInformatica 15, 571–607 (2011)

11. Han, B., Liu, L., Omiecinski, E.: Neat: Road network aware trajectory clustering. In: 2012 IEEE 32nd International Conference on Distributed Computing Systems (ICDCS), pp. 142–151. IEEE (2012)

12. Wang, Y., He, L.P., Peng, J., Zhang, T.T., Li, H.Z.: Privacy preserving for continuous query in location based services. In: Proceedings of the 2012 IEEE 18th International Conference on Parallel and Distributed Systems, pp. 213–220. IEEE Computer Society (2012)

13. Weber, M., Liu, L., Jones, K., Covington, M.J., Nachman, L., Pesti, P.: On map matching of wireless positioning data: a selective look-ahead approach. In: Proceedings of the 18th SIGSPATIAL International Conference on Advances in Geographic Information Systems, pp. 290–299. ACM (2010)

14. Haklay, M., Weber, P.: Openstreetmap: User-generated street maps. IEEE Pervasive Computing 7, 12–18 (2008)

15. Liu, F., Hua, K.A., Cai, Y.: Query l-diversity in location-based services. In: Tenth International Conference on Mobile Data Management: Systems, Services and Middleware MDM 2009, pp. 436–442. IEEE (2009)

16. Papoulis, A., Pillai, S.U.: Probability, random variables, and stochastic processes. Tata McGraw-Hill Education (2002)

17. Brinkhoff, T.: A framework for generating network-based moving objects (2008), http://iapg.jade-hs.de/personen/brinkhoff/generator/

18. Gps data team website, http://www.gps-data-team.com/

A C-DBSCAN Algorithm for Determining Bus-Stop Locations Based on Taxi GPS Data

Wu Wang[1], Li Tao[1], Chao Gao[1], Binfeng Wang[1],
Hui Yang[2], and Zili Zhang[1,3,*]

[1] School of Computer and Information Science
Southwest University, Chongqing, China
[2] Department of Information System, School of Business
City University of Hong Kong, China
[3] School of Information Technology
Deakin University, VIC 3217, Australia
{vincentwang,tli,cgao,wangbf,zhangzl}@swu.edu.cn, huiyang@gmail.com

Abstract. Determining suitable bus-stop locations is critical in improving the quality of bus services. Previous studies on selecting bus stop locations mainly consider environmental factors such as population density and traffic conditions, seldom of them consider the travel patterns of people, which is a key factor for determining bus-stop locations. In order to draw people's travel patterns, this paper improves the density-based spatial clustering of applications with noise (DBSCAN) algorithm to find hot pick-up and drop-off locations based on taxi GPS data. The discovered density-based hot locations could be regarded as the candidate for bus-stop locations. This paper further utilizes the improved DBSCAN algorithm, namely as C-DBSCAN in this paper, to discover candidate bus-stop locations to Capital International Airport in Beijing based on taxi GPS data in November 2012. Finally, this paper discusses the effects of key parameters in C-DBSCAN algorithm on the clustering results.

Keywords: Bus-stop locations, Public transport service, Taxi GPS data, Centralize density-based spatial clustering of applications with noise.

1 Introduction

The design and optimization of bus stops is one of public transportation issues which deserve the long-term attention, as reasonable locations of bus stops will strengthen the social operation efficiency, and improve citizens' satisfaction of public transportation service. However, previous studies on selecting bus-stop locations mainly consider the factors of population density, the regional environment and traffic conditions rather than travel patterns of people, which may lead to the deviation of bus-stop locations and actual demands. Therefore, it is still a critical and valuable problem for conducting scientific research on how to select a reasonable public transportation site. In the era of big data, the taxi trajectory data based on the Global Positioning System (GPS) provides the

---

* Corresponding author.

X. Luo, J.X. Yu, and Z. Li (Eds.): ADMA 2014, LNAI 8933, pp. 293–304, 2014.

possibility for analyzing the bus-stop locations. The GPS data of taxis contains the real-time information of license plate number, current time, longitude and latitude positions and so on. Through clustering pick-up locations by using the GPS data, hot pick-up locations which reflect the large amount of passengers demands could be considered as bus-stop locations.

Spatial clustering methods based on the GPS data can support the analysis of hot pick-up locations. As a typical density-based clustering analysis algorithm, density-based spatial clustering of application with noise (DBSCAN) algorithm can discover clusters in arbitrary shapes and shield effectively the interference of noise data with easier implementation and better clustering effects [1]. Nevertheless, the classic DBSCAN algorithm has its deficiency in determination of bus-stop locations for it is unable to determine cluster centers.

This paper aims to determine the locations of candidate bus stops through analyzing the big GPS data of taxis. The main work in this paper includes two folds. On the one hand, this paper improves the DBSCAN algorithm to mine candidate bus stops based on the GPS data. The promoted centralized density-based spatial clustering of applications with noise (C-DBSCAN) algorithm introduces spherical distance to measure the distance between two data points (i.e., pick-up or drop-off locations) and calculates a cluster center by averaging the latitudes and longitudes of all the data points in a cluster. On the other hand, based on the taxi GPS data in Beijing in November 2012, this paper utilizes the C-DBSCAN algorithm to find some candidate bus-stop locations to Capital International Airport in Beijing through the following three steps: (1) utilizing C-DBSCAN algorithm to cluster the taxi GPS data to find a certain number of hot pick-up locations; (2) aggregating the discovered hot pick-up locations at the first step by reapplying C-DBSCAN algorithm to generate a few refined hot pick-up locations, the number of which may not be too much to be considered as candidate bus-stop locations; (3) ranking the refined hot locations based the number of pick-up passengers to find the candidate bus-stop locations that should be considered for when planning or improving current bus routes and bus stops.

The reminder of this paper is organized as follows. Section 2 reviews the related work about present methods for determining bus-stop locations, the research achievements in the area of taxi data analysis and the GPS data analysis algorithm, and also introduces the DBSCAN clustering algorithm. Section 3 shows the improved C-DBSCAN algorithm in detail. Section 4 presents the clustering results and discusses how different parameters affect the clustering results. Section 5 concludes this paper.

2   Related Work

2.1   Determining Bus-Stop Locations

In the existing research of bus-stop locations, Wang et al. [2] have studied the planning method of city bus terminals and proposed a spacing optimization model to minimize the travel time of all passengers and they have discussed the method for bus-stops' capacity calculation when setting many bus-stops in one hot area. Koshy and Arasan [19] have developed a HETERO-SIM simulation model to examine the influence of bus stops on traffic flows by taking into account bus stops composition. Fernandez et al. [20] have built a user behavior model to analyze the interactions between traffic and bus stops that are located close to a signal controlled junction. Li and Chen [3] have conducted the regression analysis for the bus stops' survey data in the Nanjing city, and then calibrated parameters and established a regression model to calculate the length of the bus stop. However, previous studies on selecting bus-stop locations mainly consider environmental factors such as population density and traffic conditions, seldom of them consider the travel patterns of people, which is a key factor for determining bus-stop locations. As the behaviors of taxis pick-up/drop-off passengers may reflect peoples travel patterns to some extent, mining the taxi GPS data, which records taxis behaviors during the cruise, provides a potential effective way for determining which locations should be considered as bus stops.

2.2 GPS Data Based Studies for Transportation Related Problems

There are several previous studies attempting to address the public transportation related issues based on taxi GPS data. In order to promote the taxi service and profitability, some studies designed models and implemented algorithms to get the best parking place to wait for the potential passengers [5][6], to provide taxi drivers with personalized and optimal driving paths [7], and to ensure the quality of taxi services [8]. In order to strengthen the convenience of passengers, some existing studies recommended places for passengers to wait for the empty taxis and predict the average waiting time [4]. At the same time, the model for predicting the time cost for every trips have been presented in order to achieve the optimal allocation of social resources [9]. And the integrated recommendation system could provide convenience for drivers and passengers [10]. Meanwhile carpool recommendation is provided to maximize the effectiveness of the driver and passengers [11]. As the taxi GPS data reflect travel patterns of people, the analysis of the real taxi GPS data could improve the quality of transportation service.

2.3 Clustering Algorithms Using GPS Data

Some previous studies utilize various clustering algorithms to identify dense and sparse traffic areas, which is beneficial to mining the potential common quality in a global view so as to make effective decisions. Li et al. [12] have utilized the Parallel K-Means (PKM) clustering algorithm to realize more scientific and effective division of traffic control time in a day based on real-time traffic data. In order to solve the problem in identification of traffic accident locations, Sun and Wang [13] have introduced the density-based clustering (DENCLUE) algorithm for identifying accurate points with high occurrence rate. Cai and Yang [14] have improved the accuracy of DBSCAN clustering algorithm by narrowing the search radius and applied it to the identification of bus-stop locations. The cluster algorithm mentioned in previous study have their deficiencies in determining bus-stop locations, and DBSCAN algorithm could be a suitable method to solve this problem.

2.4 DBSCAN Algorithm

DBSCAN is a density-based clustering algorithm which is designed to discover clusters in arbitrary shapes. The key feature of DBSCAN is that it confines the minimum number of objects in a given radius in every cluster. The DBSCAN not only discovers clusters in arbitrary shapes but also shields the interference of noise data effectively. Here are some key definitions of the DBSCAN algorithm [15]:

Definition 1 (Density): Refers to the number of points in a circle which set a certain point as the center and set Eps as the radius.

Definition 2 (Neighborhood): Refers to the collection of points in a circle which set a certain point as the center and set Eps as the radius, set as $N_{Eps}(p)=\{q \in D \mid dist(p,q) \leq Eps\}$.

Definition 3 (Directly Density-reachable): A point p is directly density-reachable from a point q wrt. *Eps*, *MinPts* if p belongs to $N_{Eps}(q)$ and core object p satisfied $N_{Eps}(q) \geq MinPts$.

Definition 4 (Density-reachable): A point p is density-reachable from a point q wrt. Eps, MinPts if there is a chain of points $p_1, \cdots, p_n, p_1 = q, p_n = p$ such that p_{i+1} is directly density-reachable from p_i

Definition 5 (Density-connected): A point p is density-reachable to a point q wrt. *Eps*, *MinPts* if there is a point o such that both p and q are density-reachable from o wrt. *Eps* and *MinPts*.

DBSCAN algorithm is described as follows:

1. After inputting the original data and choosing a data point X at random, the algorithm would check Eps-neighborhood of the point X.

2. If X is a core object and not belong to any certain class, all point which are density-reachable from point X would form a cluster.

3. If X is not a core object, X would be set as a noise point.

4. After finishing the procession of point X, the algorithm would go back to the first step and repeat the algorithm until all the points in the original data is processed.

However, DBSCAN algorithm meets challenges in discovering candidate bus-stop locations:

1. DBSCAN can not cluster data sets well in large different densities, since the sole *MinPts-Eps* combination cannot then be chosen appropriately for all clusters.

2. The algorithm can not obtain the cluster center. The cluster center is the key point in some practical problem sometime, in order to solve this problem, the improved DBSCAN would be put forward to obtain the cluster center.

3   C-DBSCAN: A Centralize-DBSCAN Algorithm

The improvement of C-DBSCAN clustering algorithm mainly contains two aspects: (1) introducing spherical distance to measure the distance between two data points (i.e., pick-up or drop-off locations) (2) calculating a cluster center by averaging the latitudes and longitudes of all the data points in a cluster.

3.1   The Calculation of Spatial Distance between Data Points

Many formulas such as the Euclidean distance, absolute value distance, Chebyshev distance, Markov distance formula are the main method in calculating the distance between entities. In processing latitudes and longitudes of the GPS data, because of the sparsity and space phenomenon in high dimensional space [16], those above formula can not eliminate the influence of the "Curse of Dimensionality", which leads to the inaccuracy of results. So the existing deviation of data dimension should be paid more attention when choosing the distance measurement.

The spherical distance formula calculates the distance between two points [17] as follows:

$$S = \Delta\sigma \times R \tag{1}$$

$$\Delta\sigma = 2\arcsin\left(\sqrt{\sin^2(\Delta\varphi/2) + cos\varphi_A cos\varphi_B \sin^2(\Delta\gamma/2)}\right) \tag{2}$$

where S means actual spherical distance between two points, R represents the average radius of the earth; $\Delta\sigma$ which is calculated in formula (2) represents the central angel of two lines which link two points with the center of sphere respectively; φ and γ are radians of each points latitude and longitude $\Delta\varphi$ and $\Delta\gamma$ denote the difference values between two radian corresponding to the longitude and latitude, respectively.

3.2   The Determination of the Cluster Center

To determine cluster centers, we utilize formulas (3) and (4) to attain cluster centers.

$$lat_{center} = \left(\sum_{i=1}^{n} latA\right)\Big/n \tag{3}$$

$$lon_{center} = \left(\sum_{i=1}^{n} lonA\right)\Big/n \tag{4}$$

where n is the number of elements in the cluster; $latA$ and $lonA$ mean the latitude and longitude of the point A, respectively; lat_{center} and lon_{center} represent the latitude and longitude of the cluster center, respectively.

Algorithm 1. C-DBSCAN Clustering Algorithm

Input:

Data Collection pointsList, Radius Eps and Density Threshold MinPts

Output:

Clustering C

1: C = 0;

2: for each unvisited point P in dataset pointsList

3:   mark P as visited;

4:   return all points within P's eps-neighborhood;

5:   **if** sizeof(NeighborPts) ¡MinPts, mark P as NOISE;

6:   **else** set C as a next cluster and add P to cluster C;

7:   **for** each point Q in NeighborPts

8:   **if** Q unvisited, mark Q as visited, return all points within Q's eps-neighborhood;

9:   **if** sizeof(NeighborPts) ≥ MinPts,NeighborPts = NeighborPts joined with NeighborPts;

10:   **end if**

11:   **end if**

12:   **if** Q is not yet member of any cluster, add Q to cluster C;

13:   **end if**

14:   **end for**

15:   **end if**

16: set the average of the cluster as the center point of cluster

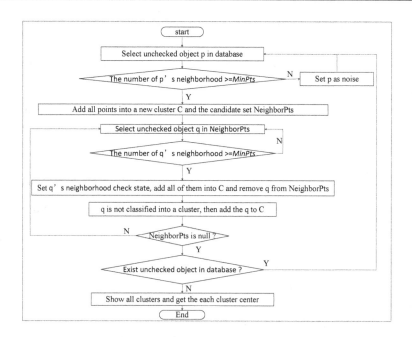

Fig. 1. The description of the C-DBSCAN clustering algorithm

3.3   The Description of the C-DBSCAN

The C-DBSCAN clustering algorithm is described in Fig.1. As shown in Fig.1, firstly, through selecting unchecked object p and comparing the number of its' neighborhood with the threshold value MinPts, object p is set as a noise or all object p's neighborhood is added into a new cluster and the candidate set NeighborPts. Secondly, the algorithm would check the object q in NeighborPts in the same way until the candidate set NeighborPts is null. Thirdly, the algorithm keep checking other unchecked objects in database the same as object p. Finally, the algorithm is set to get all clusters and calculate cluster centers.

4   Experiments

The experimental analysis of this study mainly includes two aspects: realize the recommendation of candidate bus-stop locations within three steps by using the C-DBSCAN clustering algorithm and discuss the effects of key parameters in C-DBSCAN algorithm on the clustering results. This section first introduces the data set and data processing. Then, this section utilizes the C-DBSCAN algorithm to find candidate bus-stop locations to Capital International Airport in Beijing through three steps. Finally, this section discusses the effects of key parameters in C-DBSCAN algorithm on the clustering results, so that put up some constructive suggestion for bus-stop locations.

4.1   Data Set and Data Preprocessing

In this work, we use the GPS trajectories of taxicab to Capital International Airport in Beijing from datatang[1]. This data set includes 12000 GPS trajectories of taxi generated in Nov. 2012. The data is shown as ASCII text with a comma separator (e.g., data item and order: car_number, car_trigger, car_state, car_time, longitude, latitude, speed, direction, car_state, and the corresponding records: 123456, 0, 1, 20110414160613, 116.4078674, 40.2220650, 212, 174, 1). Based on the taxi GPS data, we extract the latitudes and longitudes for locations that people get on and off taxis for determining bus-stops locations.

The preprocessing of the taxi GPS data includes the following steps:

1. Getting the pick-up and drop-off data of the 12000 taxis in Beijing within a month in original data set (i.e., value of attribute car_trigger is 0 or 1);

2. Grouping the data set according to the car_number and get all pick-up and drop-off data of each taxis within a month, and sort the data set in reverse chronological order of the attribute car_time;

3. Querying the latitude and longitude range of the Capital International Airport by using the Google earth software (longitude:116.57351110000002 to 116.627945- 59999998, latitude: 40.0461603 to 40.1079176), and get the experimental data.

---

[1] http://www.datatang.com/data/44502

4.2 Recommendation of Bus Stop Locations

Based on the taxi GPS data in Beijing in November 2012, this section utilizes the C-DBSCAN algorithm to find some candidate bus-stop locations to Capital International Airport in Beijing through the following three steps: (1) In order to get a certain number of hot pick-up locations, we utilize C-DBSCAN algorithm to cluster the processed taxi GPS data and get cluster centers; (2) Through aggregating the discovered hot pick-up locations at the first step by reapplying C-DBSCAN algorithm, this step groups these clusters and generates a few refined hot pick-up locations, the number of which may not be too much to be considered as candidate bus-stop locations; (3) The refined hot locations would be ranked based on the number of pick-up passengers so as to find the candidate bus-stop locations that should be considered for when planning or improving current bus routes and bus stops.

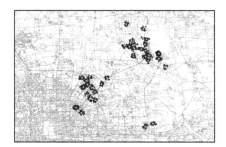 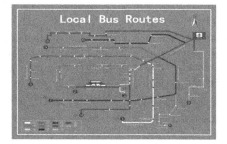

Fig. 2. The results of C-DBSCAN algorithm in first step (Eps=0.5, MinPts =50)

Fig. 3. The map of airport bus route

In the first step, through using C-DBSCAN algorithm, this step divides the processed GPS data into different clusters and gets cluster centers. In the C-DBSCAN algorithm, the Eps which means maximus radius of the neighbourhood and the MinPts which means the Minimum number of points in an Eps-neighbourhood of that point are the two global parameters of C-DBSCAN algorithm. The value of these two parameters affect the accuracy of clustering results. The urban road traffic planning and design specification (GB50220-95) in our country recommended that set the reasonable stop spacing in the urban bus and trolley (normal line) as 500 m-800 m, as well as set the suburb as 800 m-1000 m [20]. Because the experiment was conducted within the urban public transportation on bus-stop locations, in this experiment the proper distance between the stops is 500 meters, the value of parameter Eps is 0.5 km. As the value of the parameter MinPts, 50 is a proper value after considering the actual problems and results which have compared the different values for many times.

The C-DBSCAN cluster result which sets Eps as 0.5 km and sets MinPts as 50 by using the data in Nov. 2012 is shown in Fig.2. Fig.2 shows the visualization of the clustering results in ArcMap platform, which sets the Beijing road map as background and marks the adjacent clusters with different colors as well as determine the center of each cluster by using the yellow point.

Fig.3 shows the existing bus route map of Capital International Airport in Beijing, which is obtained from the website of Capital International Airport in Beijing co., LTD. In the bus route map, the airport buses link the airport with 11 districts such as Fangzhuang, Xidan, Beijing station, Gongzhufen, Zhongguancun, the Olympic Village, West Railway station, Shangdi, E-town, and Tongzhou. Compared with stops of 11 bus routes, the clustering result we gain from the taxi GPS data indicates strong demand of the transportation in different areas. So the improvement of bus-stop locations would be beneficial to improve the quality of public transport service.

In order to provide more reasonable recommendations for bus-stop locations, the second step divides cluster centers obtained in the first step into different traffic modules by using C-DBSCAN algorithm. Considering with the scale of different traffic modules and the first step result, the C-DBSCAN clustering sets Eps as 2 km and sets MinPts as 2. As shown in Fig.4, the isolate cluster center is set as one traffic module.

$$IM = M * AD(i)/(\sum\nolimits_{i=1}^{n} AD(i)) + N * PD(i)/(\sum\nolimits_{i=1}^{n} PD(i)) \qquad (5)$$

Table 1. The Calculation of IM (traffic module A as an example)

Point	Longitude	Latitude	AD	PD	IM
Point1	116.5424673	40.0666117	1.50275	223	0.295399904
Point2	116.5570673	40.0751933	1.40250	64	0.157945563
Point3	116.5528637	40.0650462	1.26108	153	0.220004292
Point4	116.5622956	40.0619868	1.25144	120	0.192316835
Point5	116.5631632	40.0683761	1.21921	52	0.134333406

In the third step, we rank the refined hot locations based on the number of pick-up passengers to find the candidate bus-stop locations. The importance measurement which is used to rank the hot locations is described in formula(5). In this formula, we introduce parameters of pick-up degree (PD) and average distance (AD) from one area to any other areas in each traffic module. And IM represents the area Importance Measurement which evaluates the importance of each cluster center. We set M=N=0.5 and get the area importance measurement in each traffic module. As shown in Table 1, we set traffic module A as an example and find that the IM of Point 1 is the biggest value in this traffic module, so that area around Point 1 can be considered as a candidate bus-stop location.

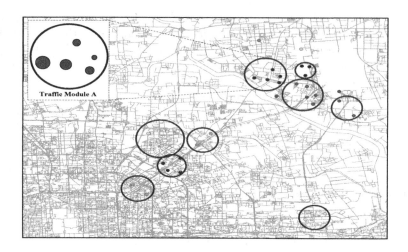

Fig. 4. The result of C-DBSCAN in second step (Eps=2, MinPts=2))

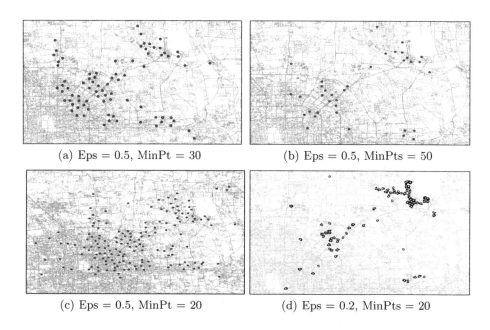

(a) Eps = 0.5, MinPt = 30

(b) Eps = 0.5, MinPts = 50

(c) Eps = 0.5, MinPt = 20

(d) Eps = 0.2, MinPts = 20

Fig. 5. Clustering results under different parameter settings

4.3 Discussions

In the experiments, the values of two global parameters (Eps and MinPts) in C-DBSCAN affect the accuracy of clustering results. We then discuss the effects of the two parameters as follows:

The Fig.5(a) and Fig.5(b) describe the effect of parameter MinPt. As shown in Fig.5(a) and Fig.5(b), by using the data in the period of 21:00-5:00 in November

2012, we can get the result of 81 clusters (Eps = 0.5, MinPt = 30) and 36 clusters (Eps = 0.5, MinPts = 50). These two figures show that the bigger MinPts value is, the less number of cluster would be. So after determining appropriate value of Eps, the bigger value of MinPts (about 50) which indicates the traffic information in a macroscopic view is a fine choice if carrying out the whole transportation planning. Also while optimizing the stops in a small area, the smaller value of MinPts which indicates the detailed and accurate traffic information is beneficial to the optimization of key points.

The Fig.5(c) and Fig.5(d) describe the effect of parameter Eps. By using the data in first half of the November in 2012, Fig.5(c) and Fig.5(d) show the result of 162 clusters (Eps = 0.5, MinPt = 20) and 79 clusters (Eps = 0.2, MinPts = 20). These two figures show that after setting appropriate value of MinPts, the proper value of Eps (the statutory public transportation site spacing: about 0.5 km) is a fine choice if carrying out the transportation planning in the area with proper density. As to a dense area, the smaller value of Eps would be a fine choice.

5   Conclusion

This paper firstly presented the improved DBSCAN algorithm for determining bus stop locations by using the real GPS data of taxis in Beijing. Then this paper further utilized the C-DBSCAN algorithm to discover candidate bus-stop locations to Capital International Airport in Beijing based on taxi GPS data in November 2012. Finally, with the discussion of effects of two parameters, this study would provide the constructive suggestion and powerful support for determining the bus-stop locations.

Future research will consider the following issues. On the one hand, we will improve the existing C-DBSCAN clustering algorithm and optimize the selection mechanism of parameters. On the other hand, as the discovered bus-stop locations in this paper are a little bit roughly, we will fine-tune the discovered bus-stop locations so as to provide constructive suggestions for designing and improving bus-stop locations in Beijing.

Acknowledgments. This work was supported by the National Science and Technology Support Program (No. 2012BAD35B08), National High Technology Research and Development Program of China (863 Program) (No. 2013AA013801), National Natural Science Foundation of China (Nos. 61402379, 61403315), the Specialized Research Fund for the Doctoral Program of Higher Education (No. 20120182120016), Natural Science Foundation of Chongqing (Nos. cstc2012jjA40013, cstc2013jcyjA400-22), and the Fundamental Research Funds for the Central Universities (Nos. XDJK2012B016, XDJK2012C018, XDJK2014C142).

References

1. Xi, J.K., Tan, H.Q.: Spatial clustering analysis and its evaluation. Computer Engineering and Design 7, 1712–1715 (2009)
2. Wang, W., Yang, X.M., Chen, X.W.: The planning and management of the public transportation system, pp. 138–153. Science Press (2002)
3. Li, N., Chen, X.W.: The study of the parking capability and design length of parking space of bus stop. China Civil Engineering Journal 36(7), 72–77 (2003)
4. Lee, J., Shin, I., Park, G.: Analysis of the passenger pick-up pattern for taxi location recommendation. In: The Fourth International Conference on Networked Computing and Advanced Information Management 2008, pp. 199–204 (2008)
5. Zheng, X., Liang, X., Xu, K.: Where to wait for a taxi. In: Proceedings of the ACM SIGKDD International Workshop on Urban Computing, pp. 149–156 (2012)
6. Qi, G., Pan, G., Li, S.: How long a passenger waits for a vacant taxi–large-scale taxi trace mining for smart cities. In: 2013 IEEE and Internet of Things (iThings / CPSCom), IEEE International Conference on and IEEE Cyber, Physical and Social Computing Green Computing and Communications (Green Com), pp. 1029–1036 (2013)
7. Yuan, J., Zheng, Y., Xie, X.: Driving with knowledge from the physical world. In: Proceedings of the 17th ACM SIGKDD International Conference on Knowledge Discovery and Data Mining, pp. 316–324 (2011)
8. Zhang, D., Li, N., Zhou, Z., Chen, C., Sun, L., Li, S.: Ibat: detecting anomalous taxi trajectories from GPS traces. In: Proceedings of the 13th International Conference on Ubiquitous Computing, pp. 99–108 (2011)
9. Gao, M., Zhu, T., Wan, X.: Analysis of travel time patterns in urban using taxi GPS data. In: Internet of Things (iThings / CPSCom), IEEE International Conference on and IEEE Cyber, Physical and Social Computing Green Computing and Communications (Green Com), pp. 512–517 (2013)
10. Ma, S., Zheng, Y., Wolfson, O.: T-share: a large-scale dynamic taxi ridesharing service. In: The 29th International Conference on Data Engineering, pp. 410–421 (2013)
11. Calvo, R.W., Luigi, F.D., Haastrup, P., Maniezzo, V.: A distributed geographic information system for the daily carpooling problem. Computer Operation Research. Elsevier Science Ltd (2004)
12. Li, Y., Li, W., Wang, H.C.: An application of cluster analysis algorithm in traffic control. Systems Engineering 22(2), 66–68 (2004)
13. Wang, H.Y., Sun, L., You, K.: Accident-prone location identification method based on DENCLUE clustering algorithm. Journal of Transportation Engineering and Information 11(2), 5–10 (2013)
14. Cai, Y.W., Yang, B.R.: Improved DBSCAN algorithm for public bus stop cluster. Computer Engineering 34(10), 56–67 (2008)
15. Han, J.W., Kamber, M.: Data mining: concepts and techniques. Morgan Kaufmann Publishers, San Francisco (2000)
16. Beyer, K., Goldstein, J., Ramakrishnan, R., Shaft, U.: When is nearest neighbor meaningful? In: Beeri, C., Bruneman, P. (eds.) ICDT 1999. LNCS, vol. 1540, pp. 217–235. Springer, Heidelberg (1998)
17. Sinnott, R.W.: Virtues of the Haversine. Sky and Telescope 68(2), 159–170 (1984)
18. GB50220-95, Code for transport planning on urban road
19. Koshy, R.Z., Arasan, V.T.: Influence of bus stops on flow characteristics of mixed traffic. Journal of Transportation Engineering 131(8), 640–643 (2005)
20. Corts, C., Fernndez, R., Burgos, V.: Modeling passengers, buses and stops in traffic microsimulators: review and extensions. Journal of Advanced Transportation 44(2), 72–88 (2010)

A Secure and Efficient Privacy-Preserving Attribute Matchmaking Protocol in Proximity-Based Mobile Social Networks

Solomon Sarpong and Chunxiang Xu

Department of Computer Science
University of Electronic Science and Technology of China
Chengdu, China
sarpong.uestc@gmail.com, chxxu@uestc.edu.cn

Abstract. Since the advent of matchmaking protocols, many such protocols have been proposed. The bane of most of these protocols has been with the preservation of users' privacy and reduce or remove completely the leaking of users' private information. Most of the existing matchmaking protocols simply match-pair persons without checking if they have enough common attributes to be an appropriate pair. Also, in most matchmaking protocols, since the inputs are private, malicious participants may choose their sets arbitrarily and use this flexibility to affect the results or learn more about the input of an honest individual. As an improvement, we propose a novel hybrid matchmaking protocol. In this proposed protocol, the initiator sets a threshold number of common attributes that a-would-be pair should have to qualify as a match-pair. With the use of certification of attributes to ensure that the inputs are not arbitrary and a preset threshold number of common attributes defined, an initiator can adequately find pair(s) without leaking any private information. In addition to helping find the most appropriate pair, our proposed protocol also has the ability to resist semi-honest and malicious attacks.

Keywords: Hybrid, nonspoofability, multiparty, proximity-based, threshold.

1 Introduction

In recent years, many researchers have dedicated substantial effort in designing efficient mobile social network (MSN) protocols. This is to facilitate the sharing of sensitive and private information. The sharing of information should be such that there is no leakage and unauthorized persons should not know anything about the individuals' information being shared. In lieu of this, the research community has foreseen the need for mechanisms to enable limited (privacy-preserving) sharing of sensitive information and a number of effective solutions have been proposed. Among them, Private Set Intersection (PSI) techniques are particularly appealing for scenarios where two parties wish to compute an

X. Luo, J.X. Yu, and Z. Li (Eds.): ADMA 2014, LNAI 8933, pp. 305–318, 2014.

intersection of their respective sets of items without revealing to each other any other information [1].

The advent of online social network has received a lot of attention as it provides the online community of users an information sharing platform. These MSN applications have been very successful because they attract millions of users, some of whom have little or no prior experience using such applications. With the pace at which the internet is growing and it being used for all sort of activities, privacy issues have become a major concern [4]. Increasing dependence on anytime-anywhere availability of data and the commensurately increasing fear of losing privacy motivate the need for privacy-preserving techniques. One interesting and common problem occurs when two parties need to privately compute an intersection of their respective sets of data. In doing so, one or both parties must obtain the intersection (if one exists), while neither should learn anything about other's data set. In most cases, since each party is not willing to disclose the content of its list, ordinary private set intersection protocols will not be appropriate. Privacy consideration of individual information is important in private set intersection however, authentication of users' attributes is also an important issue. In light of this, authorisation of the input sets of the persons in the protocol is more appropriate as it prevents potent attack such as semi-honest attack.

Hence, our research seeks to formulate a hybrid matchmaking protocol that is very efficient and secure against semi-honest and malicious attacks. The novelty of this proposed protocol is that, the initiator sets a threshold number of common attributes that individual(s) should have to qualify as a pair. Hence among the many would-be possible matching-pairs, the initiator will be able to find a pair(s) that has at least the threshold number of common attributes. When such individual(s) is found, apart from the matching-pair(s) that is privy to the number of common attributes, no one else does. After the matching-pair(s) have been found, they can then compute to know their actual attributes.

The rest of this paper is organized as follows: in section 2 we take a brief look at private set intersection; also in section 3, we present related works. Our protocol together with the algorithms are presented in section 4. Also, the security and simulation of the algorithm are in section 5. Finally, we conclude this paper in section 6.

2   Private Set Intersection

Generally speaking, Private Set Intersection (PSI) is a cryptographic protocol that involves two individuals, say Alice and Bob, each with a private set of attributes. Their goal is to compute the intersection of their respective attributes, such that minimal information is revealed in the process. In other words, Alice and Bob should learn the attributes (if any) common to both sets and nothing else.

The problem of private set intersection is, if Alice and Bob hold attributes S_A and S_B, respectively and want to compute the intersection of their attributes, their wish is to jointly compute the intersection in such a way that reveals as little as possible about S_A to Bob and S_B to Alice. In other words, both Alice and Bob should learn only $S_A \cap S_B$ but nothing more [10]. While this task could be completed with general secure multiparty techniques, it is far more efficient to have a dedicated protocol. A problem common to all these protocols is that the input attributes S_A and S_B can be chosen arbitrarily by Alice and Bob [1]. Hence, a dishonest person can therefore insert fabricated attributes in the set that s/he suspects the other individual might have. The intersection will reveal if the other person indeed has those attributes in his/her set.

In order to address this issue, Authorized Private Set Intersection (APSI) and its variants [8], [24] and [25] ensure that each party can only use attributes certified by a certification authority in the intersection protocol. In [25], two parties are able to privately share their information while enforcing complex policies after each party's set has been authorised. In particular, we consider the scenario where two parties each hold a set of attributes and wish to find the intersection of their attributes without revealing other attributes that are not in the intersection. In such applications, it is important to ensure that attributes being exchanged are properly authenticated or authorized by a trusted authority in the intersection protocol [3]. When authorization is done, it thwarts dishonest behaviour. Hence, unless some form of authentication is required, a malicious individual can claim possession of fictitious attributes, in an attempt to find out whether the other individual possesses them.

The problem of authentication of mutually suspicious parties is becoming more and more important in distributed systems. A user in a distributed system may not only need to verify the identity of the system, but may require that the system, or another user or node in the system, verifies itself to him/her. Moreover, both sides may require some degree of authentication before they release any information about themselves [4]. A trusted third party is used in such cases, however this may not be practical in a highly distributed system. But for well established reasons, this cannot be appropriate. The use of cryptographic matchmaking protocols allows users with various attributes to verify whether or not they have some attributes in common without revealing the attributes to each other or a third party.

Certifying attributes restricts their input and hence reduces the strength of attacks from malicious individuals. Bob though malicious, will follow the protocol but as he wishes to learn as much as possible about the private set of Alice, can strategically populate his private set with all of his best guesses for Alice's private set and hence, his private set will be as large as possible. This maximizes the amount of information Bob learns about Alice's private set. In the extreme case, Bob may claim his private set contains all possible attributes, which will always reveal the private set of Alice. Bob may also vary his private set over multiple runs of the protocol, in order to learn more information over time. These attacks are even more powerful when the protocol can be executed anonymously. It must be noted that all these behaviours are permitted in any protocol which allows the individuals to choose their private inputs arbitrarily [12].

3   Related Work

In this section we will take a look at some previous works on matchmaking for mobile social networking, and then focus on reviewing cryptographic protocols for matchmaking.

3.1   Mobile Social Networking Applications

In private set intersection problem, two individuals each having a set of attributes compute the intersection of their attributes such that no individual learns information other than the intersecting attributes. Matchmaking protocol is a private set intersection problem where a match-pair is made by computing the intersection of their individual attributes. In some matchmaking protocols, there is the use of a trusted server. Such protocol applications can be found in [6], [15] and [16]. Another way is the fully distributed technique, which requires no trusted server in the whole matchmaking process. Matchmaking protocols in [5], [21] and [23] use such techniques. The attributes of the initiator and the candidates are shared among multi-parties using Shamir Secret Sharing Scheme, the computing of common attributes set are conducted among multi-parties as well [11]. The third technique in use is a hybrid, where a trusted centralized server is needed only for the purpose of management and verification, and it does not participate in the matchmaking operations. This mechanism can provide efficient matchmaking services with relatively high scalability. In [7], [11], [13], [19] and [26], are the protocols based on hybrid mechanisms designed to support privacy preserving matchmaking functions for MSN.

In [9], with two persons involved in matchmaking, at the end of the protocol only one of them, say Alice, computes the intersection set. This may lead to information asymmetry as Alice may decide not to continue with the matchmaking after knowing the attributes of the other user. Furthermore as the proposed protocol is one way, several malicious attacks can be launched by the individuals in the protocols. This protocol was improved upon in [7] by removing the likelihood of malicious attack by persons involved in the protocol. Also, both persons in the protocol perform the intersection set. The proposed protocol does not take into consideration if the would-be pair has enough common attributes to qualify to be paired. In Wang et. al. [11], the proposed protocol lets a user (called initiator) find the best match among multi-parties (called candidates). In their protocol, the best match means the user that has the maximum intersection set size with the initiator. Also, Li et. al. [38] defined two increasing levels of privacy with decreasing amounts of revealed profile information. Using a set of privacy-preserving profile matching schemes for proximity-based mobile social networks, an initiating user can find from a group of other users the one whose profile best matches with him/her.

Apart from [11] and [38], all the existing matchmaking protocols do not consider if the initiator and a user have enough common attributes before pairing them. Even these two papers that tried to match-pair users based on the number of attributes they have in common, simply assumed that the best match is the user (among other candidates) that has the maximum intersection set size with the initiator. It can be noted that these protocols are not very adequate as the best match does not necessarily mean the pair has enough common attributes to make a good pair. However in [26], privacy-preserving scalar computation was used to find a user that has at least a threshold number of common attributes with the initiator. If the number of common attributes is at least the threshold set by the initiator, they then become a match-pair.

3.2   Private Matchmaking Protocols

Oblivious transfer (OT) is a protocol which allows a server to transfer one of two items to client, such that the client can choose which item s/he wants, keep his/her choice hidden from server, and learn nothing about the other items. OT can be used to construct private set intersection protocols [18], [24] and [27]. These were improved upon by Kissner and Song [2]. Using threshold cryptosystem, they proposed efficient protocols to solve privacy preserving set intersection and privacy-preserving set matching problems. Thus, private set intersection (PSI) protocols enable two parties each holding a set of input to jointly compute the intersection of their inputs without leaking any other information. However these are less efficient than specialized protocols, and efficiency decreases when elements are chosen from larger domains hence, cannot be used in a distributed environment. Variants of this protocol can be found in [1] and [37]. In [36], there was the use of Shamir Secret Sharing to guarantee the privacy of the intersection set and prevent malicious attacks.

Freedman et. al. [14] introduced the oblivious pseudorandom function based protocols, which implemented private set intersection and private cardinality of set intersection protocol based on pseudorandom function. Hazay and Lindell [17] using efficient secure protocols for set intersection and pattern matching, securely computed the set intersection functionality based on secure pseudo-random function evaluations, in contrast to previous protocols that are based on polynomials. In addition, utilizing specific properties of the Naor-Reingold pseudorandom function, a secure pseudorandom function evaluation in order to achieve secure pattern matching. In [18], there is an improvement in oblivious pseudorandom function.

In [12], when individuals want to query the common items in their database, they compute the hash values of their individual items. They then exchange the hash values of their individual items. In this way, they are able to find the common items in their intersection without revealing any other items that are not in the intersection. Other researchers use commutative encryption to achieve private set intersection and private cardinality of set intersection. Agrawal et. al. [9], suggested the power function, $f_e(x) = x^e \bmod p$, as an example of a commutative encryption function.

4   Our Protocol

The quest of this paper is to propose a mutual matchmaking protocol that will help an initiator find the most appropriate pair(s) among other match-pair seekers. The initiator, Alice, sets a threshold number of common attributes, $A_{Threshold}$, that a person(s) should possess in order to qualify as a matching-pair. After the initiator and the other person(s) have calculated the number of attributes they have in common, if the number of common attributes is at least $A_{Threshold}$, then they become a match-pair. At this point the initiator and the other person(s) only know the number of attributes they have in common. In this protocol, some privacy levels are worth considering;

Privacy level 1: At the end of the algorithm, Alice and the person(s) in the protocol mutually learn the size of their intersection set. An adversary should learn nothing.

Privacy level 2: At the end of the protocol, Alice and the person(s) with at least $A_{Threshold}$ number of attributes will mutually learn the actual attributes they have in common. An adversary should learn nothing.

Table 1. Explanation to some notations used

Notation	Explanation
$A_{Threshold}$	Threshold number of attributes set by Alice
S_i	Certification of Alice's by CA; $S_i = H(ID_A \parallel a_i)^d \bmod N$
σ_{jk}	Certification of Candidates' by CA; $\sigma_{jk} = H(ID_j \parallel b_{jk})^d \bmod N$
R_A	Random number chosen by Alice; $R_A \longleftarrow_r Z_{N/2}$
R_j	Random number chosen by each candidate; $R_j \longleftarrow_r Z_{N/2}$
$\mid I_{Aj} \mid$	Number of attributes in the intersection between Alice and a Candidate
$\mid I_{jA} \mid$	Number of attributes in the intersection between a Candidate and Alice

4.1   Initial Phase

Our system consists of Y users (persons) denoted as $P_1, ..., P_Y$, each possessing a portable device. Each device of a person in the protocol communicate through wireless interfaces such as Bluetooth or WIFI. For simplicity, we assume every participating device is in the communication range of each other. Alice launches the matchmaking process to find an individual(s) that has at least the preset threshold number of common attributes, $A_{Threshold}$, among the other persons. Alice has a set of attributes $\{a_i\}, i = 1, \ldots, n$ whilst each of the other j persons' profile consists of a set of attributes $\{b_{jk}\}, j = 1, \ldots, m; k = 1, \ldots, p$. Note that, we assume the system adopts some standard way to describe every attribute, so that two attributes are exactly the same if they are the same semantically. Table 1 explains some of the notations used in this paper.

Creation of Secret Keys for Communication: The CA generates an RSA key pair, (e, d), and $N = pq$, where $p = 2 \cdot p' + 1$ and $q = 2 \cdot q' + 1$; p, q, p' and q' are large prime numbers. The CA makes N and e public. The CA also outputs a collision resistant cryptographic hash function H. Each person looking for a pair also creates RSA key pair (e_Y, d_Y) and makes e_Y public. Each person further chooses a username and an ID. The ID is the hash of his/her RSA private key.

Certification of Attributes: Let the attributes of Alice and the j individuals looking for a match-pair be $a = \{a_1, a_2, \ldots, a_n\}$ and $b_j = \{b_{j1}, b_{j2}, \ldots, b_{jp}\}$ respectively. Each person in the protocol then exponentiates the personal attributes using the public key of the CA. Alice's attributes hence becomes, $a^e = \{a_1^e, a_2^e, \ldots, a_n^e\}$, whist the attributes of the individuals become $b_j^e = \{b_{j1}^e, b_{j2}^e, \ldots, b_{jp}^e\}$, $1 \leq j \leq m$. Alice, and the individuals, then encrypt their attributes, ID, username, and the public key pair of his/her RSA key using the public key of the CA and send it to the CA. Thus, Alice sends $E_e\{a^e \parallel ID_A \parallel username \parallel RSApublickey, e_A\}$ to the CA. The individuals also send $E_e\{b_j^e \parallel ID_j \parallel username \parallel RSApublickey, e_j\}$ to the CA.

When the CA receives the attributes from the initiator and the individuals, the CA certifies them and sends them back to Alice and the individuals. Thus, the CA computes and sends $A = \{(a_1, S_1), (a_2, S_2), \ldots, (a_n, S_n)\}$ to Alice; where $S_i = H(ID_A \parallel a_i)^d \bmod N$. Also, the CA computes and sends $B_j = \{(b_{j1}, \sigma_{j1}), (b_{j2}, \sigma_{j2}), \ldots, (b_{jp}, \sigma_{jp})\}$ to each of the individuals; where $\sigma_{jk} = H(ID_j \parallel b_{jk})^d \bmod N$. This process is done just once for each person in the protocol. In the event that a person wants to update the attributes, s/he goes through the same process again.

4.2   Matchmaking Phase

Alice has private inputs $A = \{(a_1, S_1), (a_2, S_2), \ldots, (a_n, S_n)\}$ and for all $i = 1, \ldots, n$, she chooses a random number $R_{A:i} \longleftarrow_r Z_{N/2}$ and calculates $M_{A:i} = S_i \cdot g^{R_{A:i}} \bmod N$. Alice then sends $M_{A:i}, i = 1, \ldots, n$ to each individual. Thus, Alice sends $M_{A:1} \parallel M_{A:2} \parallel \ldots \parallel M_{A:n}$ to each individual. Each individual with private input $B_j = \{(b_{j1}, \sigma_{j1}), (b_{j2}, \sigma_{j2}), \ldots, (b_{jp}, \sigma_{jp})\}$ also chooses a random number $R_{B_j:k} \longleftarrow_r Z_{N/2}$ and calculates $M_{B_j:k} = \sigma_{jk} \cdot g^{R_{B_j:k}} \bmod N$. The individuals then send $M_{B_j:k}, j = 1, \ldots, m; k = 1, \ldots, p$ to Alice. Thus, each individual sends $M_{B_j:1} \parallel M_{B_j:2} \parallel \ldots \parallel M_{B_j:p}$ to Alice.

Alice further chooses and stores another random number, $R_A \longleftarrow_r Z_{N/2}$ and after receiving $M_{B_j:k}, j = 1, \ldots, m; k = 1, \ldots, p$, Alice computes $Z_A = g^{eR_A} \bmod N$. Furthermore, for all $j = 1, \ldots, m$ and $k = 1, \ldots, p$, Alice calculates $M'_{B_j:k} = (M_{B_j:k})^{eR_A} \bmod N$. Also, for all $i = 1, \ldots, n$, Alice computes $\{a_i\}^{R_A}$, thus Alice computes the ordered set $\{a_1^{R_A}, a_2^{R_A}, \ldots, a_n^{R_A}\}$. Using the random permutation ξ_A, Alice computes $\xi_A\{a_1^{R_A}, a_2^{R_A}, \ldots, a_n^{R_A}\}$ which reorders the set such that, it is not possible for anyone to know the actual order of the set in polynomial time. Alice then sends $Z_A \parallel M'_{B_j:k}, j = 1, \ldots, m; k = 1, \ldots, p \parallel \xi_A\{a_1^{R_A}, a_2^{R_A}, \ldots, a_n^{R_A}\}$ to each individual. Each individual also chooses and stores a random number $R_j \longleftarrow_r Z_{N/2}$ and calculates $Z_j = g^{eR_j} \bmod N$. For all $i = 1, \ldots, n$, each individual then calculates $M'_{A:i} = (M_{A:i})^{eR_j} \bmod N$. For all $k = 1, \ldots, p$, each individual computes the ordered set $\{b_{j1}^{R_j}, b_{j2}^{R_j}, \ldots, b_{jp}^{R_j}\}$. Also, using the random permutation ξ_j each individual computes $\xi_j\{b_{j1}^{R_j}, b_{j2}^{R_j}, \ldots, b_{jp}^{R_j}\}$ which reorders the set such that, it is not possible for anyone to know the actual order of the set in polynomial time. Each individual then sends $Z_j \parallel M'_{A:i}, i = 1, \ldots, n \parallel \xi_j\{b_{j1}^{R_j}, b_{j2}^{R_j}, \ldots, b_{jp}^{R_j}\}$ to Alice.

Alice then encrypts and sends her random number, R_A, to each individual using their public keys. Thus, Alice sends $E_{e_j}(R_A)$ to each individual. Also, each individual encrypts and sends his/her random number, R_j, to Alice using her public key. Thus, each individual sends $E_{e_A}(R_j)$ to Alice. Alice then computes, $\xi_A\{a_1^{R_A R_j}, a_2^{R_A R_j}, \ldots, a_n^{R_A R_j}\}$ and each individual also computes $\xi_j\{b_{j1}^{R_j R_A}, b_{j2}^{R_j R_A}, \ldots, b_{jp}^{R_j R_A}\}$.

Alice computes $D_{A:i} = M'_{A:i} \cdot Z_j^{-R_{A:i}} \bmod N \; \forall i = 1, \ldots, n$ and $j = 1, \ldots, m$. After Alice has output $D_{A:i} = \xi_j\{b_{j1}^{R_j R_A}, b_{j2}^{R_j R_A}, \ldots, b_m^{R_j R_A}\}$, Alice finds the number of attributes in the intersection between $\xi_A\{a_1^{R_A R_j}, a_2^{R_A R_j}, \ldots, a_n^{R_A R_j}\}$ and $\xi_j\{b_{j1}^{R_j R_A}, b_{j2}^{R_j R_A}, \ldots, b_{jp}^{R_j R_A}\}$. Denoting the intersection between Alice and an individual by I_{Aj}, Alice can only know the number of attributes, $|I_{Aj}|$ because of the random permutation ξ_j used by each of the individuals. Each individual computes $D_{B_j:k} = M'_{B_j:k} \cdot Z_A^{-R_j} \bmod N \; \forall j = 1, \ldots, m$ and $k = 1, \ldots, p$. Also, after each individual has output $D_{B_j:k} = \xi_A\{a_1^{R_A R_j}, a_2^{R_A R_j}, \ldots, a_n^{R_A R_j}\}$, each individual finds the number of attributes in the intersection between $\xi_j\{b_{j1}^{R_j R_A}, b_{j2}^{R_j R_A}, \ldots, b_{jp}^{R_j R_A}\}$ and $\xi_A\{a_1^{R_A R_j}, a_2^{R_A R_j}, \ldots, a_n^{R_A R_j}\}$. Denoting the intersection between an individual and Alice by I_{jA}, each individual can only know the number of attributes, $|I_{jA}|$ because of the random permutation ξ_A used by Alice.

At the end of the algorithm, Alice outputs the number of attributes in the intersection, $|I_{Aj}|$, with each individual; likewise, each individual also outputs the number of attributes in the intersection, $|I_{jA}|$, with Alice. Alice then checks which individuals' intersection set is at least the threshold number of common attributes. Hence, the individual(s) with $|I_{Aj}| \geq A_{Threshold}$ then becomes the match-pair of Alice. At this point, Alice and the individual(s) in the protocol know only the number of attributes they have in common.

For simplicity, let us assume Bob was the only individual in the protocol who has the number of attributes that is at least $A_{Threshold}$. Alice and Bob at this point know only the number of attributes they have in common. Assume Bob's random permutation and random number are ξ_B and R_B respectively. Also, the intersection set computed by Alice and Bob are I_{AB} and I_{BA} respectively. In order for them to know the actual attributes they have in common, they exchange their random permutations. Alice sends her random permutation $E_{e_B}(\xi_A)$ to Bob. With the knowledge of ξ_A, Bob can compute ξ_A^{-1} and hence recover $\{a_1^{R_A R_B}, a_2^{R_A R_B}, \ldots, a_n^{R_A R_B}\}$ from $\xi_A\{a_1^{R_A R_B}, a_2^{R_A R_B}, \ldots, a_n^{R_A R_B}\}$. With the knowledge of R_A given to Bob by Alice in step 7, Bob is able to compute and know the actual attributes he has in common with Alice. Also, Bob sends his random permutation $E_{e_A}(\xi_B)$ to Alice. With the knowledge of ξ_B, Alice can compute ξ_B^{-1} and hence recover $\{b_1^{R_B R_A}, b_2^{R_B R_A}, \ldots, b_m^{R_B R_A}\}$ from $\xi_B\{b_1^{R_B R_A}, b_2^{R_B R_A}, \ldots,$

$b_m^{R_B R_A}\}$. With the knowledge of R_B given to Alice by Bob in step 7, Alice is able to compute and know the actual attributes she has in common with Bob.

Algorithm 1. Computing the Number of Common Attributes

Require: Let $\{N, e, g, H\}$ be inputs common to Alice and the individual(s) from the CA.

1: Private attributes of Alice, $(a_1, a_2, \ldots, a_n)$; After certification, her private set becomes $A = \{(a_1, S_1), (a_2, S_2), \ldots, (a_n, S_n)\}$, where $S_i = H(ID_A \parallel a_i)^d \bmod N$.

2: The j individuals with k attributes have private input, $(b_{j1}, b_{j2}, \ldots, b_{jp})$; After certification, their private input set become $B_j = \{(b_{j1}, \sigma_{j1}), (b_{j2}, \sigma_{j2}), \ldots, (b_{jp}, \sigma_{jp})\}$, where $\sigma_{jk} = H(ID_j \parallel b_{jk})^d \bmod N$.

3: For all $i = 1, \ldots, n$, Alice chooses a random number $R_{A:i} \longleftarrow_r Z_{N/2}$ and calculates $M_{A:i} = S_i \cdot g^{R_{A:i}} \bmod N$. Alice then sends $M_{A:i}, i = 1, \ldots, n$ to each individual.

4: For all $j = 1, \ldots, m$ and $k = 1, \ldots, p$, each individual chooses a random number $R_{B_j:k} \longleftarrow_r Z_{N/2}$ and calculates $M_{B_j:k} = \sigma_{jk} \cdot g^{R_{B_j:k}} \bmod N$. Each individual then sends $M_{B_j:k}, j = 1, \ldots, m, k = 1, \ldots, p$ to Alice.

5: Alice further chooses and stores another random number, $R_A \longleftarrow_r Z_{N/2}$, computes $Z_{A:i} = g^{eR_A} \bmod N$. Alice then computes $M'_{B_j:k} = (M_{B_j:k})^{eR_A} \bmod N$ for all $j = 1, \ldots, m$ and $k = 1, \ldots, p$.
Also, for all $i = 1, \ldots, n$, Alice computes $\{a_i\}^{R_A}, i = 1, \ldots, n$ and randomly permutes it using the random permutation ξ_A. Alice then sends $Z_A \parallel M'_{B_j:k}, j = 1, \ldots, m; k = 1, \ldots, p \parallel \xi_A\{a_i\}^{R_A}, i = 1, \ldots, n$ to each individual.

6: Each individual further chooses and stores another random number, $R_j \longleftarrow_r Z_{N/2}$ computes $Z_{B_j:k} = g^{eR_j} \bmod N$ and $M'_{A:i} = (M_{A:i})^{eR_j} \bmod N$ for all $i = 1, \ldots, n$ and $j = 1, \ldots, m$.
For all $j = 1, \ldots, m$ and $k = 1, \ldots, p$, each individual computes $\{b_{jk}\}^{R_j}$ and randomly permutes it using the random permutation ξ_j. Each individual then sends $Z_j \parallel M'_{A:i}, i = 1, \ldots, m \parallel \xi_j\{b_{jk}\}^{R_j}, j = 1, \ldots, m$ and $k = 1, \ldots, p$ to Alice.

7: Alice then encrypts and sends her random number, R_A, to each individual using their public keys. Thus, Alice sends $E_{e_j}(R_A)$ to each individual. Also, each individual encrypts and sends his/her random number, R_j, to Alice using her public key. Thus, each individual sends $E_{e_A}(R_j)$ to Alice.

8: Alice then computes, $\xi_A\{a_1^{R_A R_j}, a_2^{R_A R_j}, \ldots, a_n^{R_A R_j}\}$ and the individuals also compute $\xi_j\{b_{j1}^{R_j R_A}, b_{j2}^{R_j R_A}, \ldots, b_{jp}^{R_j R_A}\}$

9: For all $i = 1, \ldots, n$, Alice computes $D_{A:i} = M'_{A:i} \cdot Z_{B_j:k}^{-R_{A:i}} \bmod N$

10: For all $k = 1, \ldots, p$ each individual computes $D_{B_j:k} = M'_{B_j:k} \cdot Z_{A:i}^{-R_{B_j:k}} \bmod N$

11: Alice computes and outputs $|I_{Aj}| \in A \cap B_j$ if $\exists i, j$ and k s.t. $D_{A:i} = \xi_j\{b_{j1}^{R_j R_A}, b_{j2}^{R_j R_A}, \ldots, b_{jp}^{R_j R_A}\}$. The number of attributes she has in common with each individual is $|I_{Aj}| = \xi_A\{a_1^{R_A R_j}, a_2^{R_A R_j}, \ldots, a_n^{R_A R_j}\} \cap \xi_j\{b_{j1}^{R_j R_A}, b_{j2}^{R_j R_A}, \ldots, b_{jp}^{R_j R_A}\}$.

12: Each individual also computes and outputs $|I_{jA}| \in A \cap B_j$ if $\exists i, j$ and k s.t. $D_{B_j:k} = \xi_A\{a_1^{R_A R_j}, a_2^{R_A R_j}, \ldots, a_n^{R_A R_j}\}$. The number of attributes each individual has in common with Alice is $|I_{Aj}| = \xi_j\{b_{j1}^{R_j R_A}, b_{j2}^{R_j R_A}, \ldots, b_{jp}^{R_j R_A}\} \cap \xi_A\{a_1^{R_A R_j}, a_2^{R_A R_j}, \ldots, a_n^{R_A R_j}\}$.

---

5 Security

In the algorithm, for all the attributes of Alice and the individual(s), the CA computes $S_i = H(ID_A \parallel a_i)^d \bmod N$ and $\sigma_{jk} = H(ID_j \parallel b_{jk})^d \bmod N$ respectively. This computation is to certify the input attributes of the persons in the protocol. By this, the attributes of Alice and the individual(s) in the protocol are bound to them. Hence, they cannot change or modify their attributes so as to gain more information from the others. This facilitates the security of this protocol as the persons in the protocol cannot input attributes they do not possess. Alice and the individual(s) in each step in the algorithm ensure that the other cannot know the actual attributes they possess before the protocol ends. An individual may terminate the protocol before it ends if s/he is able to know the other persons' personal attributes.

In step 5 of the algorithm, Alice computes $\xi_A\{a_i\}^{R_A}, i = 1, \ldots, n$. The computation of $\xi_A\{a_i\}^{R_A}, i = 1, \ldots, n$ makes it computationally impossible for any individual to map $a_i^{R_A}$ to the corresponding attribute in $\xi_A\{a_i\}^{R_A}$ in polynomial time. Hence in step 8, there is no way any individual can know the actual attributes of Alice. Likewise, the computation of $\xi_j\{b_{jk}\}^{R_j}, j = 1, \ldots, m; k = 1, \ldots, p$ makes it computationally impossible for Alice to map $b_{jk}^{R_j}$ to the corresponding attribute in $\xi_j\{b_{jk}\}^{R_j}$ in polynomial time. Hence in step 8, there is no way Alice can know the actual attributes of any of the individual(s).

At the end of the protocol, Alice outputs $|I_{Aj}| \in A \cap B_j$ if $\exists i, j$ and k s.t. $D_{A:i} = \xi_j\{b_{j1}^{R_j R_A}, b_{j2}^{R_j R_A}, \ldots, b_{jp}^{R_j R_A}\}$. Hence, Alice actually outputs:

$D_{A:i} = M'_{A:i} \cdot Z_j^{-R_{A:i}} \bmod N = (M_{A:i})^{eR_j} \cdot Z_j^{-R_{A:i}} \bmod N = (M_{A:i})^{eR_j} \cdot (g^{eR_j})^{-R_{A:i}} \bmod N = (S_i \cdot g^{R_{A:i}})^{eR_j} \cdot (g^{eR_j})^{-R_{A:i}} \bmod N = S_i^{eR_j} \cdot g^{eR_{A:i}R_j} \cdot g^{-eR_j R_{A:i}} \bmod N = S_i^{eR_j} \bmod N = [(H(ID_A \| a_i))^d]^{eR_j} \bmod N = [H(ID_A \| a_i)]^{R_j} \bmod N$. From Alice's output of $D_{A:i} = \xi_j\{b_{j1}^{R_j R_A}, b_{j2}^{R_j R_A}, \ldots, b_{jp}^{R_j R_A}\}$, it can be observed that $H(ID_A \| a_i)^{R_j} \bmod N = \xi_j\{b_{j1}^{R_j R_A}, b_{j2}^{R_j R_A}, \ldots, b_{jp}^{R_j R_A}\}$.

Each individual at the end of the protocol computes and outputs $|I_{jA}| \in A \cap B_j$ if $\exists i, j$ and k s.t. $D_{B:k} = \xi_A\{a_1^{R_A R_j}, a_2^{R_A R_j}, \ldots, a_n^{R_A R_j}\}$. Also, each individual actually outputs; $D_{B:k} = \xi_A\{a_1^{R_A R_j}, a_2^{R_A R_j}, \ldots, a_n^{R_A R_j}\} = M'_{B:k} \cdot Z_A^{-R_j} \bmod N = (M_{B:k})^{eR_A} \cdot (g^{eR_A})^{-R_j} \bmod N = (\sigma_{jk} \cdot g^{R_j})^{eR_A} \cdot (g^{eR_A})^{-R_j} \bmod N = \sigma_{jk}^{eR_A} \cdot g^{eR_A R_j} \cdot g^{-eR_A R_j} \bmod N = \sigma_{jk}^{eR_A} \bmod N = [H(ID_j \| b_{jk})^d]^{eR_A} \bmod N = H(ID_j \| b_{jk})^{R_A} \bmod N$. From the output of the individuals, $D_{B:k} = \xi_A\{a_1^{R_A R_j}, a_2^{R_A R_j}, \ldots, a_n^{R_A R_j}\}$, it can be observed that $H(ID_j \| b_{jk})^{R_A} \bmod N = \xi_A\{a_1^{R_A R_j}, a_2^{R_A R_j}, \ldots, a_n^{R_A R_j}\}$.

Correctness of the Protocol: The protocol is correct since for Alice, as $D_{A:i} = \xi_j\{b_{j1}^{R_j R_A}, b_{j2}^{R_j R_A}, \ldots, b_{jp}^{R_j R_A}\}$, hence $H(ID_A \| a_i)^{R_j} \bmod N = \xi_j\{b_{j1}^{R_j R_A}, b_{j2}^{R_j R_A}, \ldots, b_{jp}^{R_j R_A}\}$. Also, for each individual, since $D_{B:k} = \xi_j\{b_{j1}^{R_j R_A}, b_{j2}^{R_j R_A}, \ldots, b_{jp}^{R_j R_A}\}$, hence $H(ID_j \| b_{jk})^{R_A} \bmod N = \xi_A\{a_1^{R_A R_j}, a_2^{R_A R_j}, \ldots, a_n^{R_A R_j}\}$.

Achievement of Privacy Levels:

Privacy level 1 is achieved in steps 11 and 12 of the algorithm. Alice computes and outputs $H(ID_A \| a_i)^{R_j} \bmod N = \xi_j\{b_{j1}^{R_j R_A}, b_{j2}^{R_j R_A}, \ldots, b_{jp}^{R_j R_A}\}$. Alice is then able to compute the number of attributes she has in common with each individual by computing $\xi_A\{a_1^{R_A R_j}, a_2^{R_A R_j}, \ldots, a_n^{R_A R_j}\} \cap \xi_j\{b_{j1}^{R_j R_A}, b_{j2}^{R_j R_A}, \ldots, b_{jp}^{R_j R_A}\}$. This only allows Alice to know the number of attributes she has in common with each individual. In like manner, each individual also computes and outputs $H(ID_j \|$

$b_{jk})^{R_A} \mod N = \xi_A\{a_1^{R_A R_j}, a_2^{R_A R_j}, \ldots, a_n^{R_A R_j}\}$. Each individual is also able to compute the number of attributes s/he has in common with Alice by computing $\xi_A\{a_1^{R_A R_j}, a_2^{R_A R_j}, \ldots, a_n^{R_A R_j}\} \cap \xi_j\{b_{j1}^{R_j R_A}, b_{j2}^{R_j R_A}, \ldots, b_{jp}^{R_j R_A}\}$. This output also enables each individual to know the number of attributes s/he has in common with Alice.

Privacy level 2 is achieved after they have securely exchanged their random permutations to enable them find the actual attributes they have in common.

Attacks on the Protocol and Countermeasures: In this protocol, attacks by persons outside the protocol is not possible. The structure of the algorithm coupled with the certification of private input attributes further prevents malicious and semi-honest attacks. Alice's private input set and the individuals' private input sets are certified by the CA hence, only certified set of attributes are used in the protocol. This is to ensure that the attributes used by the persons in the protocol, they really possess them. Thus the certification prevents cheating (semi-honest attacks) by the persons in the protocol. Our protocol is collusion resistant. In this protocol, the individuals are unaware of the presence of others. Thus, each individual executes the protocol with the initiator independently without the knowledge of other individuals in the protocol. In lieu of this, the protocol guards against collusion attacks. Also, as the matched-pair will eventually know the type of common attributes s/he has with the other individual, to a large extent user profiling cannot be prevented but minimized. Nonspoofability of the other users' attributes is another characteristic of our protocol. As the attributes of the persons in the protocol are certified, an individual cannot query another's attributes without his/her knowledge.

5.1   Simulation

Simulation for our matchmaking algorithm was conducted in java. We focused only on the execution time without considering the communication time. In this simulation, the execution time is mainly decided by the number of participants and the number of attributes they possess. The prime numbers p and q we chosen to be 1024 bits with RSA modulus of 1024 bits. Also, each attribute was represented by 64 bits. We simulated the algorithm on an i5 PC which has 2.67 GHz processor with 2G RAM. In order to get more accurate execution time, an average of 60 repeated execution times was computed. In the experiment, we considered varying number of users, $j = 1, 5, 10, 15, 20$. The initiator has the same number of attributes whilst the number of attributes each user has equals $k = 10, 15, 20, 25, 30$. Fig. 1 shows the execution times for the different number of user with varying number of attributes. The x-axis shows the number of users and the y-axis shows the execution time. The graph shows the execution times for the varying number of users and users' attributes. It can be observed that the execution times increases as the number of users and attributes increase.

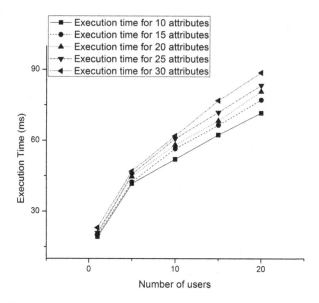

Fig. 1. Comparison of execution time for the number of attributes

6   Conclusion

With the increasing popularity of mobile social networks, it is important to develop secure and privacy-preserving attribute matchmaking protocols that will enable users to effectively find matching-pairs and interact with each other. By this proposed protocol, an initiator can find the best match from among many potential pair-seekers without leaking any private information.

References

1. Ateniese, G., De Cristofaro, E., Tsudik, G. (If) size matters: Size-hiding private set intersection. In: Catalano, D., Fazio, N., Gennaro, R., Nicolosi, A. (eds.) PKC 2011. LNCS, vol. 6571, pp. 156–173. Springer, Heidelberg (2011)
2. Kissner, L., Song, D.: Privacy-preserving set operations. In: Shoup, V. (ed.) CRYPTO 2005. LNCS, vol. 3621, pp. 241–257. Springer, Heidelberg (2005)
3. De Cristofaro, E., Tsudik, G.: Practical private set intersection protocols with linear complexity. In: Sion, R. (ed.) FC 2010. LNCS, vol. 6052, pp. 143–159. Springer, Heidelberg (2010)
4. De Cristofaro, E., Lu, Y., Tsudik, G.: Efficient techniques for privacy-preserving sharing of sensitive information. In: McCune, J.M., Balacheff, B., Perrig, A., Sadeghi, A.-R., Sasse, A., Beres, Y. (eds.) Trust 2011. LNCS, vol. 6740, pp. 239–253. Springer, Heidelberg (2011)
5. Yang, Z., Zhang, B., Dai, J., Champion, A., Xuan, D., Li, D.: Esmalltalker: A distributed mobile system for social networking In physical proximity. In: IEEE, ICDCS, pp. 468–477 (2010)

6. Eagle, N., Pentland, A.: Social Serendipity: Mobilizing Social Software. In: IEEE Pervasive Computing, Special Issue: The Smartphone, pp. 28–34 (2005)
7. Xie, Q., Hengartner, U.: Privacy-Preserving Matchmaking for Mobile Social Networking Secure Against Malicious Users. In: Proc. 9th Int'l. Conf. on Privacy, Security (PST), and Trust 2011, pp. 252–259 (2011)
8. De Cristofaro, E., Kim, J., Tsudik, G.: Linear-complexity private set intersection protocols secure in malicious model. In: Abe, M. (ed.) ASIACRYPT 2010. LNCS, vol. 6477, pp. 213–231. Springer, Heidelberg (2010)
9. Agrawal, R., Evfimievski, A., Srikant, R.: Information Sharing Across Private Databases. In: Proc. of SIGMOD, pp. 86–97 (2003)
10. Camenisch, J., Zaverucha, G.M.: Private intersection of certified sets. In: Dingledine, R., Golle, P. (eds.) FC 2009. LNCS, vol. 5628, pp. 108–127. Springer, Heidelberg (2009)
11. Wang, Y., Zhang, T., Li, H., He, L., Peng, J.: Efficient Privacy Preserving Matchmaking for Mobile Social Networking against Malicious Users. In: IEEE 11th International Conference on Trust, Security and Privacy in Computing and Communications, pp. 609–615 (2012)
12. De Cristofaro, E., Jarecki, S., Kim, J., Tsudik, G.: Privacy-preserving policy-based information transfer. In: Goldberg, I., Atallah, M.J. (eds.) PETS 2009. LNCS, vol. 5672, pp. 164–184. Springer, Heidelberg (2009)
13. De Cristofaro, E., Durussel, A., Aad, I.: Reclaiming Privacy for Smartphone Applications. In: IEEE International Proc. of Pervasive Computing and Communications (PerCom), pp. 84–92 (2011)
14. Freedman, M.J., Nissim, K., Pinkas, B.: Efficient private matching and set intersection. In: Cachin, C., Camenisch, J.L. (eds.) EUROCRYPT 2004. LNCS, vol. 3027, pp. 1–19. Springer, Heidelberg (2004)
15. Kjeldskov, J., Paay, J.: Just-for-Us: A Context-Aware Mobile Information System Facilitating Sociality. In: Proc. 7th International. Conf. on Human Computer Interaction with Mobile Devices and Services, pp. 23–30 (2005)
16. Li, K., Sohn, T., Huang, S., Griswold, W.: PeopleTones: A System for the Detection and Notification of Buddy Proximity on Mobile Phones. In: Proc. 6th Intl. Conf. on Mobile Systems (MobiSys), pp. 160–173 (2008)
17. Hazay, C., Lindell, Y.: Efficient Protocols for Set Intersection and Pattern Matching with Security Against Malicious and Covert Adversaries. Journal of Cryptology 23(3), 422–456 (2010)
18. Jarecki, S., Liu, X.: Efficient oblivious pseudorandom function with applications to adaptive OT and secure computation of set intersection. In: Reingold, O. (ed.) TCC 2009. LNCS, vol. 5444, pp. 577–594. Springer, Heidelberg (2009)
19. Lu, R., Lin, X., Shen, X.: SPOC: A secure and privacy-preserving opportunistic computing framework for mobile-health emergency. IEEE Transactions on Parallel and Distributed Systems 24(3), 614–624 (2013)
20. Dachman-Soled, D., Malkin, T., Raykova, M., Yung, M.: Efficient robust private set intersection. In: Abdalla, M., Pointcheval, D., Fouque, P.-A., Vergnaud, D. (eds.) ACNS 2009. LNCS, vol. 5536, pp. 125–142. Springer, Heidelberg (2009)
21. Liu, M., Lou, W.: FindU: Privacy-preserving personal profile matching in mobile social networks. In: Proc. of Infocom (2011)
22. Lu, R., Lin, X., Liang, X., Shen, X.: Secure handshake with symptoms-matching: the essential to the success of mhealthcare social network. In: Proc. BodyNets (2010)

23. De Cristofaro, E., Tsudik, G.: Practical private set intersection protocols with linear complexity. In: Sion, R. (ed.) FC 2010. LNCS, vol. 6052, pp. 143–159. Springer, Heidelberg (2010)
24. Camenisch, J., Kohlweiss, M., Rial, A., Sheedy, C.: Blind and anonymous identity-based encryption and authorised private searches on public key encrypted data. In: Jarecki, S., Tsudik, G. (eds.) PKC 2009. LNCS, vol. 5443, pp. 196–214. Springer, Heidelberg (2009)
25. Stefanov, E., Shi, E., Song, D.: Policy-enhanced private set intersection: Sharing information while enforcing privacy policies. In: Fischlin, M., Buchmann, J., Manulis, M. (eds.) PKC 2012. LNCS, vol. 7293, pp. 413–430. Springer, Heidelberg (2012)
26. Sarpong, S., Xu, C.: A Secure and Efficient Privacy-preserving Matchmaking for Mobile Social Network. In: International Conference on Computer, Network Security and Communication Engineering (CNSCE), pp. 362–366 (2014)
27. Rabin, M.: How to exchange secrets by oblivious transfer, Tech. Rep. TR-81, Harvard Aiken Computation Laboratory (1981)
28. De Cristofaro, E., Jarecki, S., Kim, J., Tsudik, G.: Privacy-preserving policy-based information transfer. In: Goldberg, I., Atallah, M.J. (eds.) PETS 2009. LNCS, vol. 5672, pp. 164–184. Springer, Heidelberg (2009)
29. Shamir, A.: Identity-based cryptosystems and signature schemes. In: Blakely, G.R., Chaum, D. (eds.) CRYPTO 1984. LNCS, vol. 196, pp. 47–53. Springer, Heidelberg (1985)
30. Hazay, C., Lindell, Y.: Efficient Protocols for Set Intersection and Pattern Matching with Security Against Malicious and Covert Adversaries. Journal of Cryptology 23(3), 422–456 (2010)
31. Lin, H., Chow, S.S.M., Xing, D., Fang, Y., Cao, Z.: Privacy preserving friend search over online social networks. Cryptology EPrint Archive (2011), http://eprint.iacr.org/2011/445.pdf
32. Camenisch, J., Kohlweiss, M., Rial, A., Sheedy, C.: Blind and anonymous identity-based encryption and authorised private searches on public key encrypted data. In: Jarecki, S., Tsudik, G. (eds.) PKC 2009. LNCS, vol. 5443, pp. 196–214. Springer, Heidelberg (2009)
33. Sun, J., Zhu, X., Fang, Y.: A privacy-preserving scheme for online social networks with efficient revocation. In: Proceedings of the IEEE Conference on Computer Communications (INFOCOM 2010), pp. 1–9.
34. Li, M., Cao, N., Yu, S., Lou, W.: FindU: Privacy-preserving personal profile matching in mobile social networks. In: Proc. of IEEE Infocom, pp. 2435–2443 (2011)
35. Pietiläinen, A., Oliver, E., LeBrun, J., Varghese, G., Diot, C.: Mobiclique: middleware formobile social networking. In: Proceedings of the 2nd ACM Workshop on Online Socialnetworks, pp. 49–54. ACM (2009)
36. De Cristofaro, E., Lu, Y., Tsudik, G.: Efficient techniques for privacy-preserving sharing of sensitive information. In: McCune, J.M., Balacheff, B., Perrig, A., Sadeghi, A.-R., Sasse, A., Beres, Y. (eds.) Trust 2011. LNCS, vol. 6740, pp. 239–253. Springer, Heidelberg (2011)
37. Dachman-Soled, D., Malkin, T., Raykova, M., Yung, M.: Efficient robust private set intersection. In: Abdalla, M., Pointcheval, D., Fouque, P.-A., Vergnaud, D. (eds.) ACNS 2009. LNCS, vol. 5536, pp. 125–142. Springer, Heidelberg (2009)
38. Li, M., Yu, S., Cao, N., Lou, W.: Privacy-Preserving Distributed Profile Matching in Proximity-based Mobile Social Networks. IEEE Transactions on Wireless Communications 12(5), 2024–2033 (2013)

A New Filter Approach Based on Generalized Data Field

Long Zhao[1], Shuliang Wang[2], and Yi Lin[1]

[1] State Key Laboratory of Software Engineering, Wuhan University, Wuhan, China
zxcvbnm9515@163.com
[2] School of Software, Beijing Institute of Technology, Beijing, China
slwang2005@gmail.com

Abstract. In this paper, a new feature selection method based on generalized data field（FR-GDF）is proposed. The goal of feature selection is selecting useful features and simultaneously excluding garbage features from a given feature set. It is Important to measure the "distance" between data points in existing feature selection approaches. To measure the "distance", FR-GDF adopts potential value of data field. Information entropy of potential value is used to measure the inter-class distance and intra-class distance. This method eliminates unimportant or noise features of original feature sets and extracts the optional features. Experiments prove that FR-GDF algorithm performs well and is independent of the specific classification algorithm.

Keywords: generalized data field, potential value, feature selection, information entropy.

1 Introduction

In the past 20 years, the dimensionality of the datasets has increased dramatically. Feature selection is choosing useful features from the original feature space. Data classification is a basic task in areas of machine learning, data mining and pattern recognition. The main task of classification is constructing an appropriate classification model based on the known data and tagging new data according by this model. In classification problems, the more number of attributes doesn't mean better decision-making ability.

Duda.R.O pointed out that when the quantity of classifier features is more than a threshold value the error rate of classification model may increases [1].As the increasing of data dimensions, noise features may be introduced to affect the performance of the learning algorithm, thus interfere the learning process. Unfortunately, high-dimensional data leads to dimensionality curse [2]. Therefore, when constructing a classifier, feature selection is very important.

A data sets represented by n features contains 2n feature subsets. When the dimension number n is large, through an exhaustive search to select the optimal feature subset isn't a realistic task. The usual practice is taking a heuristic search strategy in the search process, obtaining optimal or suboptimal feature subsets.

X. Luo, J.X. Yu, and Z. Li (Eds.): ADMA 2014, LNAI 8933, pp. 319–333, 2014.

Feature selection (FS), as it is an important activity in data preprocessing, has been an active research area in the last decade, finding success in many different real world applications[21].We usually divide feature selection methods into three categories: filter approach, wrapper approach and embedded approach (Liu & Yu, 2005) [5]. Filter methods select a subset of features as a preprocessing step which is independent from the induction algorithm. The best feature subset is selected by evaluating some predefined criterion without involving any learning algorithms. However, the wrapper method generally outperforms the filter method in the aspect of the accuracy of the learning machine. Thus, many researchers attempt to speed up the convergence of the wrapper algorithm by using different techniques. Some authors have determined the combination of filter and wrapper methods (Estevez, Tesmer, Perez, & Zurada, 2009; Huang, Cai, &Xu, 2007), which use the knowledge delivered by filter algorithm to guide the classifier [6]. Filter algorithms measure the candidate feature subsets with the evaluation criteria which are independent of the choice of classifiers. Dependency [7], consistency [8], mutual information [9] and distance [10] are some popular alternatives. Among the above evaluation criteria, dependency, consistency and mutual information are all just available in evaluating categorical features. For applying the three criteria to numerical features, a discretization algorithm should be introduced to partition the numerical features into a finite set of intervals and associate each interval with a distinct value [11]. Recently, Wright etc. take recognition problem into solving the classification problem of multiple linear regression model .Sparse Representation-based Classification (SRC) is proposed [22]. Using the ideas of sparse representation, SRC algorithm constructs facial image classification into solving optimization problems of L1 norm minimization. It solved the feature extraction and face cover, camouflage and other face recognition problem.

In this paper, a novel filter method is proposed to select feature for high-dimensional data sets. For high-dimensional data feature selection problem, first calculate the class potential value and inter class potential value of feature vector. The potential value of each original feature will be calculated according to each class label .If the potential value between the different categories S_w is large and the potential value within the category S_b is small, then this feature is important, the opposite feature is noise or a non-important feature.

Consequently, the purpose of clustering is to keep the selected group with the minimal intra-class distance and the maximal inter-class distance .The features with more discriminability will be selected in a higher priority during the whole clustering process. The rest of this paper is organized as follows. Fundamentals are presented in Section 2. The algorithm of feature selection is showed in Section 3.Then, Section 4 is an experimental case to show that the algorithm can effectively reduce dimension and perform the desired task. Finally, conclusion and discussion are discussed in last Section.

2   Generalized Data Field

Inspired by the ideas of the physical field, the interaction between the particles of matter and its description method is introduced to abstract data field space. *D* is the data set; x_i is the feature of D. Ω is data field space. As data set $D = \{x_1, x_2, \cdots, x_n\}$ included in the known space $\Omega \subseteq R^p$.That each object $x \in \Omega$ is a data point in Ω, there is a field of action around it. Any points of this field will be affected by the other points. Thereby a data field is defined on the entire space. Spatial distribution of the data field mainly depends on the interaction range and the radius of the object. Considering the potential function of nuclear field can better highlight the clustering properties of the data distribution, and the Gaussian function has good mathematical properties and universality .This paper use Gaussian potential function to describe the nature of the data field.

The potential of any data field point can be expressed as

$$\varphi(x) = \sum_{i=1}^{n} m_i \times k(\tfrac{x-X_i}{\sigma}) \tag{1}$$

where k(x) is the unit potential function meet $\int k(x)\,dx = 1, \int xk(x)\,dx = 0.\sigma$ describes the interaction process between the control object that is called impact factor. For the quality $m_i \geq 0$ of the object X_i, assumed that satisfy the normalization condition $\sum_{i=1}^{n} m_i = 1$.

Data field is extended to the generalized data field through the impact factor becoming different in multidimensional space. Assumed that the density function and the data set have been known, the potential accuracy of the function primarily depends on factors. Dataset $X^{(A)}$ contains n features $\{X_1^{(A)}, X_2^{(A)}, ...,X_n^{(A)}\}$. Each data point has a particle with certain nature Nat_i. $X_i^{(A)}$is a data point with mass. In Ω every two data points are mutually interacted. $X_i^{(A)}$affects other points, and it is further affected by all the other points. All the fields from different local points $X_i^{(A)}(i =1, 2, ..., n)$ are superposed in global data space Ω. And the superposition enables the field to characterize the interaction among different points. The superposed interaction virtually creates a data field in the whole data space. The potential value of an arbitrary point $x^{(A)}=(x_1,x_2, ..., x_n)^T$in the data field is defined as

$$\varphi(x) = \sum_{i=1}^{n} Nat_i \times k(\tfrac{x-X_i}{\sigma}) \tag{2}$$

where $\varphi(x)$is the potential value of each data point, $K(x)$ is the unit potential function that satisfies $\int k(x)\,dx = 1 \int xk(x)\,dx = 0.\sigma$ is an impact factor. $Nat_i(\sum_{i=1}^{n} Nat_i = 1$, $Nat_i \geq 0$)is the quality of $X_i^{(A)}$.

For the existent multi-dimensional data field, the potential function as Equation (1) shows that the impact factor $\sigma$is stable. That is, the isotropic $\sigma$is supposed to have the same value when measured in different directions. Consequently, the contribution of each observed data point is evenly arranged in all directions. However, the impact factor σ should be anisotropic. The data point has different properties in different dimensions. Different σ values are obtained when measuring $\varphi(x)$in different directions. According to the superposition principle of the potential function, if the quality of the data object is equal, data-intensive areas will have a higher potential values. The potential function can reflect the intensive degree of data distribution. It can be used as an estimate of the overall distribution. Set a unit potential function is K (x), where the proposed nuclear potential function of corresponding field is$K(x) = e^{-\|x\|^k}$. Unit potential function of intended gravitational field can be expressed as$K(x) = \frac{1}{1+\|x\|^k}$.

The superposition potential function equation is$\varphi(x) = \sum_{i=1}^{n}(m_i \cdot K(\frac{x-x_i}{\sigma}))$. Depend-ing on the nature of the probability density function, we can prove that if the value of K(x) in space $\Omega \subseteq R^p$ is limited i.e.$\int_{\Omega} K(x)dx = M < +\infty$. The difference of Potential function and the probability density function is a normalization constant difference.

When the quality of each object is equal, the potential function data field actually gives a physical explanation of kernel density estimation. On the other hand, since the unit potential function K (x) of data field may be not a density function, and the quality of the data object does not require being equal. Furthermore, when the data object has different variations corresponding to different directions or all the data points are almost in a low-dimensional manifold, the estimation in accordance with the above potential function is not appropriate to some extent. Thus, in order to get a better estimation of the potential function, the generalized data field is defined as:

$$\varphi(x) = \sum_{i=1}^{n} Nat_i \times k(H^{-1}(x - X_i)) \tag{3}$$

where H is a positive definite $d \times d$ matrix that is a non-singular constant matrix, σ_j is the j-th impact factor in the j-th dimensional attributes.

3   Feature Selection Method

The importance degree of the feature is measured by the criteria of information measure using the idea of data hierarchical clustering analysis. Two potential values are selected to measure the importance degree of the characteristics for classification. S_w describes the distance or similarity of current samples within the same class. The similarity within class is large when S_w is small. S_b describes the distance or similarity of current samples between the different classes or groups. The higher S_b means that the feature is more dispersed. This feature plays a greater role in the distinction of different classes.

3.1   The Meaning of the Important Feature

The important features will lead to algorithms learning performance degradation. For supervision classification process, the important features are the characters clearly contribute to the separation of each category. The classification algorithms try to make the degree of similarity with the same category larger, rather than make the degree of similarity between different categories smaller. From the perspective of information theory, when the distribution of projected points obeys uniform distribution, every potential value of the data points distribute approximately equal. The value to measure the corresponding feature's importance tends to 0.On the other hand, if the distribution of the points are very asymmetric, the potential value of the points are also very asymmetric. The data field on such asymmetric points will have smaller potential entropy. It measures the corresponding feature's importance. According to the characteristics of selection process, the relevant degree between features in this thesis is measured by information measurements, whose bases are information entropy and its related concepts.

3.2   The Importance Metrics of Characteristics Based on the Generalized Data Field

Let X_i with $i = (1, 2\ldots n)$ be the feature of data set D, F is the dataset of all features, the subset of it $F' \subseteq F$. In the data field, every two data are mutually interacted. One data point affects other, and it is further affected by all the other points. All the fields from different local points are superposed in global data space $\Psi_1, \Psi_2, \cdots, \Psi_n$ and the superposition enables the field to characterize the interaction among different points. The superposed interaction virtually creates a data field in the whole data space the potential value of a feature in the data field is F

$$\text{Im } p(F') = -\sum_{i=1}^{n} \frac{\psi_i}{Z} \cdot \log\left(\frac{\psi_i}{Z}\right) - \sum_{i=1}^{n}\left(1 - \frac{\psi_i}{Z}\right) \cdot log\left(1 - \frac{\psi_i}{Z}\right) \quad (4)$$

$$\psi(x) = \frac{1}{n}\sum_{i=1}^{n}\left\{\prod_{F'} k\left(\frac{x - X_i}{\sigma}\right)\right\} \quad (5)$$

where $Z = \sum_{i=1}^{n} \psi_i$ is a normalization factor, σ is an interaction impact factor. The potential value within class of a feature S_w is large which means that this feature in the class is dispersed, and it is opposite. The potential value within class of a feature S_w and the potential value between different classes of a feature S_b are used to measure a feature's contribution to the current categories. S_w is mainly the potential value of the distance, when the class similarity is bigger. The potential value of S_b is used to measure the original characteristic value between classes. The larger S_b indicates that this feature is more dispersed and play greater role in the distinction different classes. The goal of the feature selection process is to choose the features which have small S_w and large S_b. The importance of the optimized feature subset F*can be obtained with Equation (6).

$$Im(F)_{opt} = Im(F)|_{\sigma=\sigma_{F*}} = \left(\log(n) + \sum_{p=1}^{n} \frac{\varphi_p^{(F)}(x)}{Z} \cdot \log\left(\frac{\varphi_p^{(F)}(x)}{Z}\right)\right)\Bigg|_{\sigma=\sigma_{F*}} \quad (6)$$

Optimal potential is calculated corresponding to minimum entropy

$$\text{Im } p(F') = opt_{\sigma_{F'}}\left(-\sum_{i=1}^{n} \frac{\psi_i}{Z} \cdot \log\left(\frac{\psi_i}{Z}\right) - \sum_{i=1}^{n}\left(1 - \frac{\psi_i}{Z}\right) \cdot \log\left(1 - \frac{\psi_i}{Z}\right)\right) \quad (7)$$

3.3   The Layer-By-Layer Search of Important Feature Subset

Most of the existing characters selection algorithms based on heuristic sequence search strategy are built on a common assumption that feature subsets contain most classification information, and at the same time don't include the information of selected characteristics. It can be seen that the idea of feature selection and cluster analysis is similar. The purpose of cluster is dividing the sample into different groups, make samples from same groups as near as possible, while the samples in different groups as far as possible. Accordingly, the idea of cluster analysis can be naturally applied to the feature selection process, the difference between them is two aspects that the representation method of data points and the distance metrics criterion in clustering analysis.

The Filter model measures the degree of features or subset importance by given evaluating criteria in advance [12]. Wei and Billings uses the squared correlation function describe the dependence between the features, and select the appropriate features with binding sequence search strategy [13].Hall considered a good feature subsets should satisfy such conditions, i.e. it is highly correlated or associated with the classification categories and is very low redundant with the selected features. For this, he proposed a correlation-based feature subset selection algorithm CFS [7].

In this paper the criteria of evaluating the feature subset important are:

$$J(S) = \frac{k \cdot r_{cf}}{(k + k(k-1) \cdot r_{ff})^{\frac{1}{2}}} \tag{8}$$

where k is the number of selected feature subset S, r_{cf} represents the average degree of interrelationship between the single selected feature f and class label C, r_{ff} represents the average degree of interrelationship within the selected characteristics subset S. To solve the problem of measuring the degree of association between the characteristics, if the two features are discrete, then we can use the symmetric uncertainty as measure standard of association degree. Otherwise, the statistical correlation between the characteristics expresses the degree of association between them. In addition, the generation or search of feature subset S in the CFS algorithm can take a variety of search strategies, such as the sequence forward or sequence backward.

The distance between the selected feature subset S and the label subset C, $S_b(C; S)$ can be expressed as the sum of distance between the selected feature s and label class, that is:

$$p(f, s) = \sum_{i=1}^{n} p(C_i, s) \tag{9}$$

$P(f, s)$ is mainly used to indicate the distance between candidate class f with a single selected character s. the calculated method is similar to the distance between classes in pattern recognition, in the process of characteristic cluster analysis, the "distance" between the candidate class f and selected class S can be calculated by summing the "distance" between the candidate class f and each selected characteristics of the select class S, that

$$S(f) = \sum_{s \in S} CU(f, s) \tag{10}$$

The distance within class $S_w(S)$ is calculated according to the following form:

$$S_w(S, f) = S_w(S) + S(f) \tag{11}$$

The size of selected class S also need to be considered in the clustering process aspect the within class distance $S_w(S)$, and between class distance $S_b(S)$. Under normal circumstances, small size of select class S is better, because the less selected features means better robustness in late classification structure. Based on the above analysis, for each candidate class the evaluation function of characteristic is:

$$J(f) = \frac{S_b \cdot (C; S, f)}{|S| + S_w(S, f)} \tag{12}$$

where |S| represents the total number of members in the select class S, which is the number of selected features. The loop is continuing until the number of the characteristics of the candidate class S exceeds a preset threshold value, or the performance of classifier is declining by adding a new feature selection process, and then outputs the obtained feature subset.

Algorithm FR-GDF: Feature Selection based on general data field
Input: A training dataset T= (Ω, F, C);
Output: A selected feature subset F';
1) Initialize relative parameters, e.g. $F' = \Phi$, $S_b=0$, $S_w=0$;
2) For each f in F do
Calculate its mutual information Imp(c, f) with the class labels C
$S_b(C; f) = Imp(C, f)$;
3) $F' = F' \cup \{f\}$; $F' = \{f\}$; $S_w=0; f = argmax \ (Sb(C; f))$;
4) While s<δ do for each f in F do calculate the evaluation criterion $J(f)$ in term Eq.(8);
5) select the feature f with the maximum value of $J \ (f)$;
6) Combine f with F', that is, $F' = F' + \{f\}$; F=F-$\{f\}$;
7) Update the intra-distance S_w and inter-distance S_b of S
$S_w=S_w+S \ (f)$, $S_b=S_b+Imp(C, f)$;
8) return the subset F';
9) End.

4   Experiments

Filter model constructs a score equation based (Score Function) filter type feature selection algorithm. Features scores are mainly based on the Variance Score and Fisher score [23]. Both technologies are the easiest and most widely used feature selection algorithm. Variance score algorithm is an unsupervised feature selection method, the class information is not used Fisher score method takes advantage of the class information of samples is an effective supervised feature selection methods. In recent years, some new scores based feature selection methods have been proposed equation, He et al proposed Laplace scores feature selection method based on the idea of comparing the retention ability of local characteristics [28]. The experiments indicate that Laplace score method can effectively select a reasonable feature subset comparison with variance scores and Fisher scores. Further, Zhang et al based on the idea of semi-supervised learning proposed constraint score feature selection method [25].

This method uses the pairwise constraints for feature selection, by defining constraints scores evaluation criteria. Similar to the Laplace score, constraint score also use the neighbor relationship between the data samples to judge the importance of the feature. The feature selection can be performed efficiently .We used 10 datasets came from the UCI Machine Learning Repository (http://archive.ics.uci.edu/ml). The denotation of the datasets in this paper and their titles in the UCI repository are as follows table1:   [8].

Table 1. Characteristic of Data Sets

Data	Instances	Attributes	Classes
Cancer	198	32	3
Derm	366	33	6
Glass	214	9	6
Heart	270	13	2
Pro	997	20	3
Iris	150	4	3
Sonar	208	60	2
Teach	151	5	3
Wine	178	13	3
Vote	232	16	3

The main characteristics of the datasets are summarized in Table 1. Some basic data preprocessing was done. We filtered instances and attributes with missing values. Since not all classification methods can handle categorical data, each categorical attribute is converted into a vector of attributes. In all experiments, the training inputs are first standardized to zero mean and unit variance, and the test inputs are then standardized using the corresponding training mean and variance.

Most of existing comparisons of different classifiers are based on the same data sets and under the same conditions. Therefore, the previous performance study of FR-GDF simply used a single loop of cross-validation procedure, where classifiers with different hyper-parameters are trained and tested and the performance with the best hyper-parameters is reported. Our purpose in this work is not only to compare FR-GDF methods with traditional classifiers, but also to assess the true accuracy of FS-GDF methods and competing classifiers. In this work, we test the performance of classification algorithms by further separating training data into two parts, which are respectively used for hyper-parameter tuning and validation.

For each of the data sets in Table 1, we employ the ten-fold cross-validation procedure and divide the data into roughly equal-sized parts. Each part is held out in turn as the test set, and the remaining parts are used as the training set. Each training set is further divided into equal-sized parts in a stratified way, where two parts are used to train classifiers with different hyper-parameters and the remaining part is used to validate and determine the best values of hyper-parameters. Once the best hyper-parameters have been decided, the model is applied to the test set and the classification accuracy can be evaluated.

Since the sizes of the last four data sets in Table 1 are very large, it is expensive to simulate using the cross-validation procedure. Therefore, we simply separate them into the training set and test set. The sizes of the training set and test set for each data set are shown in Table 1

The teach and wine data set is as an example, through the above importance measure function, we can calculate their characteristic importance, teach data set consists of 5 feature vectors, each feature importance sorting follows table2:

Table 2. Importance of single feature in teach data

feature	X_5	X_2	X_3	X_1	X_4
IMP	0.3334	0.2090	0.1826	0.1383	0.1367

According to the table given above, we calculate the biggest value of $J(f)$ function, the feature subset is $\{X_2, X_3$ and $X_5\}$, the scatter diagram and potential function value image of this three feature vectors are given as follows:

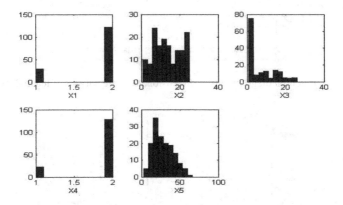

Fig. 1. Frequency histogram on single feature of teach data

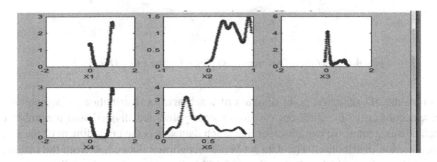

Fig. 2. Frequency histogram on single feature of potential value

Based on the above image we can know, where X_1, X_4 features only have two values, and the distinction degree between different categories is small, compared to it the distinguishing degree of X_2, X_3, X_5 are higher, that is consistent with our obtained results.

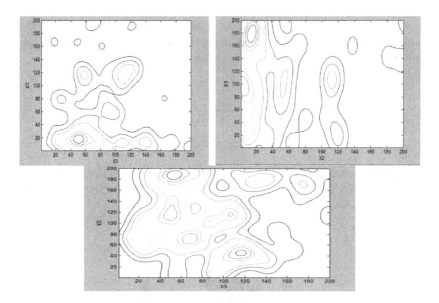

Fig. 3. Potential diagram in two-dimensional feature subsets of teach data

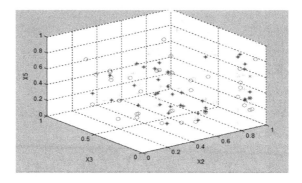

Fig. 4. Scatter plots for three-dimensional feature subset(X_2, X_3, X_5)

From the 3D scattered point diagram of teach and its distribution of equipotential lines, according to the different categories data points are divided into a number of small cluster center, we can observer that teach data sets is the embodiment of the local concentration distribution. The data set Wine recognition data come from the machine learning repository. A chemical analysis of 178 Italian wines from three different cultivars yielded 13 measurements: Alcohol Malic acid Ash Alkalinity of ash, Magnesium, Total phenols, Flavanoids, Nonflavanoid phenols, Proanthocyanins, Color intensity, Hue, OD280/OD315 of diluted wines, Proline, each feature importance are sorting follows:

Table 3. Importance of single feature in Wine data

Feature	X_7	X_{10}	X_1	X_{12}	X_{13}	X_{11}
IMP	0.1593	0.1269	0.0906	0.0898	0.0836	0.0747

Feature	X_6	X_2	X_9	X_5	X_4	X_3	X_8
IMP	0.0784	0.0667	0.0558	0.0492	0.0437	0.0418	0.040

The maximum classification accuracy of this two methods are consistent, the beast number of features is two, but in the other feature number GDF method has higher classification precision than the SRC method.The classification accuracy curve of teach data set is as follows:

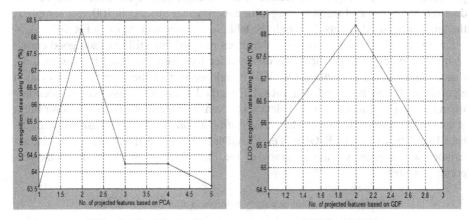

Fig. 5. Classification accuracy of teach used SRC and GDF

The each samples of data source file are complete; its classification accuracy comparison chart is given:

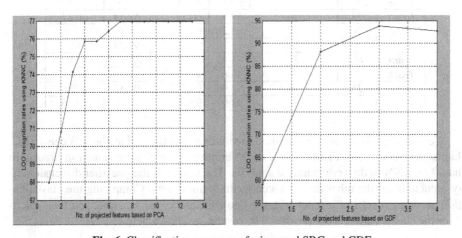

Fig. 6. Classification accuracy of wine used SRC and GDF

Two dimension reduction methods both can effectively improve the classification accuracy, the SRC method starts convergence when the number of selected features is 7, the classification accuracy of it is 77%, the GDF method starts convergence when the number of selected features is 3, the classification accuracy of it is 93.05%, whether from the dimension of choice or the precision of classification it can be seen that the GDF method is obviously better than the SRC [17] method.

In order to validate the independence on the specific classification method of this algorithm, this paper adopts three kinds of typical algorithm to experiment NBC (Naive Bayes Classification) [16], and SVM [18]. Both before and after the dimension reduction classification accuracies were given. Experiment specific process is that, first use data dimensionality reduction algorithm calculates each feature information entropy value, then according to the feature selection algorithm select feature subset S in data set, the number of training sample in experimental process is 100, test sample is the number of full data sets. Through the table we can see that, through FR-GDF algorithm, the number of data dimensions has been reduced, but still maintain good classification accuracy. The performance of classifier has been improved by this method. Results show that this method FR-GDF is independent of the specific classification algorithms. It can be found that FR-GDF method keeps the classification accuracy when reducing the dimensions in 5 data sets. Even if the accuracy becomes lower but it is also in the acceptable range. The classification accuracy has certain ascension in the later five data sets after FR-GDF. The FR-GDF method effectively gets those characteristics which influence the accuracy of the classification and keep down the important features.

Table 4. Classification accuracy before and after feature reduction

Data	Num	Num-FR-GDF	Nbc	Nbc FR-GDF	Svm	Svm-FR-GDF
Cancer	32	16	66.7	64.65	72.73	**72.73**
Derm	33	29	96.71	96.18	98.9	97.27
Glass	9	7	50.47	47.66	75.7	77.57
Heart	13	6	59.46	60.14	62.84	60.81
Pro	20	17	65.26	65.06	69.076	71.3
Iris	4	2	96	96	97.33	96
Sonar	60	23	64.42	67.31	72.115	75.96
Teach	5	3	46.67	48	60	64
Vote	16	7	93.96	95.7	96.55	97.4
wine	13	6	93.97	95.69	47.19	50.56

We take leave-one-out cross method to validate the FR-GDF method. We use SRC, Laplacian Score [24], Fisher score, and Variance Score feature selection algorithm based on KNN to do a comparative analysis. The test set is the one after dimensionality reduction, all the other data is taken as the training sets. Comparing the classification accuracy results of test set with the known category information, and then calculates the classification accuracy rate. The results are shown below:

Table 5. Classification accuracy of tradition method and FR-GDF

Data	Num	Laplacian Score	Src	Fisher Score	Variance Score	FR-GDF
Cancer	32	60.6(18)	**73.3(18)**	68.12(19)	65.7(18)	70.7(16)
Derm	33	93.5(27)	97.27(28)	97.0(30)	95.8(29)	**97.3(29)**
Glass	9	64.5(7)	68.2(8)	73.36(7)	73.72(8)	**73.83(7)**
Heart	13	77.5(10)	80.7(10)	78.00(11)	77.60(11)	**81.41(6)**
Pro	20	87.75	**90.7(11)**	84.7(11)	88.2(10)	90.36(17)
Iris	4	96.40(3)	96.5(3)	96.43(3)	96.47(3)	**98(2)**
Sonar	60	84.2(26)	83.65(25)	82.69(27)	83.20(26)	**89.42(23)**
Teach	5	68(3)	62.67(3)	69.72(4)	68.31(3)	**70.2(3)**
Vote	16	97.07(8)	96.55(9)	95.32(9)	92.1(8)	**97.41(7)**
wine	13	**97.5(8)**	97.9(10)	97.58(10)	97.62(10)	97.82(6)

From Fig7, we can get the information that the FR-GDF algorithm proposed in this paper is superior to the traditional feature selection method in more than one data sets.

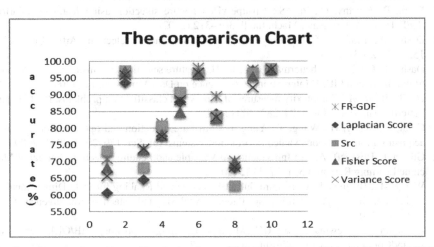

Fig. 7. The comparison Chart of Laplacian Score,SRC, Fisher Score, Variance Score and KNN-GDF classification accuracy

5   Discussion

This method adopts the S_w (potential value within class) and S_b (potential value be-tween different classes) to calculate the information entropy of each feature. Hierar-chical clustering feature selection algorithm is adopted to select these relatively im-portant features. The experiments show that the algorithm FR-GDF effectively reduces the dimensionality of high-dimensional data sets and keeps the classifier performance. The representative features have been selected to structure classifier .Results show that this method is independent of the specific classification algorithm. In most of the classification algorithm it can effectively improve the classification accuracy and reduce the number of dimensions. The overall performance of FR-GDF is better than the other algorithms. The effectiveness of the FR-GDF algorithm has been proved through a series of experiments.

References

1. Duda, R.O., Hart, P.E., Stork, D.G.: Pattern Classification, 2nd edn. John Wiley&Sons, New York (2001)
2. Langley, P.: Selection of relevant features in machine learning. In: Proc of the AAAI Fall Symposium on Relevance, Menlo Park,CA, pp. 140–144 (1994)
3. Jolifie, I.T.: Principal component analysis. Springer, New York (1986)
4. Battiti, R.: Using mutual information for selecting features in supervised neural net learning. IEEE Transactions on Neural Networks 5 (1994)
5. Liu, H., Yu, L.: Toward integrating feature selection algorithms for classification and clustering. IEEE Transactions on Knowledge and Data Engineering 17(3), 1–12 (2005)
6. Estevez, P.A., Tesmer, M., Perez, C., Zurada, J.M.: Normalized Mutual Information Feature Selection. IEEE Transactions on Neural Networks, 1045–9227 (2009)
7. Mitra, P., Murthy, C., Pal, S.: Unsupervised feature selection using feature similarity. IEEE Trans. Pattern Anal. Mach. Intell, 301–312 (2002)
8. Dash, M., Liu, H.: Consistency-based search in feature selection. Artif. Intell. 151, 155–176 (2003)
9. Dash, M., Choi, K., Scheuermann, P., Liu, H.: Feature selection for clustering – a filter solution. In: Second IEEE International Conference on Data Mining, p. 115 (2002)
10. Ho, T., Basu, M.: Complexity measures of supervised classification problems. IEEE Trans. Pattern Anal. Mach. Intell. 24, 289–300 (2002)
11. Guan, Y., Wang, H., Wang, Y., Yang, F.: Attribute reduction and optimal decision rules acquisition for continuous valued information systems. Inf. Sci. 179, 2974–2984 (2009)
12. Guyon, I., Elisseeff, A.: An Introduction to Variable and Feature Selection. Journal of Machine Learning Research 3, 1157–1182 (2003)
13. Wei, H.-L., Billings, S.A.: Feature Subset Selection and Ranking for Data Dimensionality Reduction. IEEE Transactions on Pattern Analysis and Machine Intelligence 29(1), 162–166 (2007)
14. Asuncion, A., Newman, D.J.: UCI Machine Learning Repository [EB/OL]. School of Inf.andCompSci, Univ of California, Irvine (2007), http://www.ics.uci.edu/~mlearn/MLRepository.html
15. Pnevmatikakis, A., Polymenakos, L.: Comparison of Eigenface Based Feature vectors under Different Impairments. In: Proceedings of the17th International Conference on Patten Recognition(ICPR), vol. (l), pp. 296–299 (2004)
16. Hand, D.J., Yu, K.: Idiot's Bayes: Not So Stupid After All Internat. Statist. Rev. 69, 385–398 (2001)
17. Wright, J., Yang, A.Y., Ganesh, A., Sastry, S.S., Ma, Y.: Robust face recognition via sparse representation. IEEE Trans. Pattern Anal. Mach. Intell. 31(2), 210–227 (2009)
18. Vapnik, V.N.: The Nature of Statistical Learning Theory. Springer-Verlag New York, Inc. (1995)
19. Loog, M., Duin, R.P.W., Haeb-Umbach, R.: Multiclass linear dimension reduction by weighted pairwise fisher criteria. IEEE Trans. PAMI 23(7), 762–766 (2001)

20. Yu, K., Ji, L., Zhang, X.G.: Kernel nearestneighbor algorithm. Neural Process.Lett 15, 147–156 (2002)
21. Saari, P., Eerola, T., Lartillot, O.: Generalizability and simplicity as criteria infeature selection: application to mood classification in music. IEEE Transactionson Audio, Speech, and Language Processing 19(6), 1802–1812 (2011)
22. Wright, J., Yang, A.Y., Ganesh, A.: Robust face recognition via sparserepresentation. IEEE Transactions on Pattern Analysis and Machine Intelligence 31(2), 210–227 (2009)
23. Bishop, C.: Neural networks for pattern recognition [M]. Clarendon Press, Oxford (1995)
24. He, X., Cai, D., Niyogi, P.: Laplacian score for feature selection. In: Advances inNeural Information Processing Systems, pp. 507–514 (2006)
25. Zhang, D., Chen, S., Zhou, Z.H.: Constraint Score: A new filter method for featureselection with pairwise constraints. Pattern Recognition 41(5), 1440–1451 (2008)

A Stable Instance Based Filter for Feature Selection in Small Sample Size Data Sets

Afef Ben Brahim[1] and Mohamed Limam[1,2]

[1] LARODEC, ISG, University of Tunis, Tunisia
[2] Dhofar University, Sultanate of Oman

Abstract. In supervised feature selection applications it is common to have high dimensional data, but it is sometimes not easy to collect a large number of examples to represent each pattern or object class. Hence, learning in the small sample case is of practical interest. One reason for this is the difculty in collecting data for each object. We propose a filter approach for feature selection based on instance learning. Its main challenge is that it convert the problem of the small sample size to a tool that allows choosing only a few subsets of features to be combined in order to select the most relevant ones. Each instance proposes a candidate subset of the most relevant features for this instance. Small sample size makes this process feasible. Thus the high dimensionality of data is reduced to few subsets of features which number corresponds to the data sample size and this is when small sample size is of benefit to feature selection process. The combination scheme used for this purpose aims at obtaining a feature selection that yields to good classification performance while beeing stable.

Keywords: Feature selection, stability, Relief, small sample size.

1 Introduction

Data mining is the process of analyzing data from different perspectives and summarizing it into useful information. It allows users to analyze data from many different dimensions or angles, categorize it, and summarize the relationships identified. The major steps in a data mining process are preprocessing, mining, and post-processing. Feature selection is frequently used as a preprocessing step to data mining. It is a process of choosing a subset of original features so that the feature space is optimally reduced according to some evaluation measures. For high-dimensional data with hundreds of thousands of features involved, it is usually intractable [10] to find an optimal feature subset and feature selection could be even considered as a NP-hard problem. But various feature selection algorithms have been proposed with the help of informative feature evaluation metrics and refined searching procedures. They have been also proved both efficient in terms of selection process and effective at improving the performance of learning models built on the selected features. Feature selection is a way of avoiding the curse of dimensionality which occurs when the number of available features significantly outnumbers the number of examples, as is the case

X. Luo, J.X. Yu, and Z. Li (Eds.): ADMA 2014, LNAI 8933, pp. 334–344, 2014.

in Bio Informatics [9]. Feature selection algorithms can be broadly categorized
with respect to the use of a learning method into filter and wrapper models [2]
[10]. The filter model relies on general characteristics of the training data to select some features without involving any mining algorithm. The wrapper model
requires one predetermined mining algorithm in feature selection and uses its
performance to evaluate and determine which features are selected.

Wrapper method could perform better as it finds features better suited to the
predetermined mining algorithm, but would require much more computational
cost than filter methods.

When the number of features becomes very large, data can contain high degree
of irrelevant information which may greatly degrade the performance of mining
algorithms. Therefore, feature selection becomes very necessary for data mining
tasks when facing high dimensional data nowadays and filter model is usually
chosen due to its computational efficiency.

Traditionally, feature selection algorithms are developed with a focus on improving classification accuracy while reducing dimensionality. A relatively
neglected issue is the stability of feature selection which is defined as the insensi-
tivity of a feature selection result to variations in the training data. This issue is
important in many applications with high dimensional data. In microarray data
analysis, a feature selection algorithm may select largely different subsets of features (genes) under variations to the training data [1]. Such instability dampens
the confidence of domain experts in investigating any of the various subsets of
selected features for biomarker identification. It is worthy noting that stability
of feature selection results should be investigated together with classiffication
accuracy, because domain experts are not interested in a strategy that yields
very stable feature sets, but leads to a bad predictive model (e.g., arbitrarily
picking the same set of features under training data variation).

In this work, we aim to develop an efficient filter solution for feature selection in high-dimensional small sample size data which can effectively remove
irrelevant feature while beeing stable.

2   A Filter Based on Instance Learning and Feature Occurrence Frequency

Among feature selection methods, filters rank all variables in terms of relevance,
as measured by a score which depends on the method. They are simple to implement and fast to run. To obtain a signature of size n, one simply takes the top
genes according to the score. A well known algorithm that relies on relevance
evaluation is Relief [3]. Relief algorithm assigns a relevance weight to each feature to denote the relevance of the feature to the target concept. Relief is based
on random selection. For each feature, it samples instances randomly from the
training set and updates the relevance values based on the difference between
the selected instance and the two nearest instances of the same and opposite
class. Then, the feature is scored as the sum of weighted differences in the different class and the same class. The algorithm has been further generalized to

average multiple nearest neighbors, instead of just one, when computing sample margins. It is referred to as Relief-F [13]. Sun et al. showed that Relief-F achieves significant improvement in performance over the original Relief. Sun also systematically proved that Relief is indeed an online algorithm for a convex optimization problem [14]. By maximizing the averaged margin of the nearest patterns in the feature scaled space, Relief can estimate the feature weights in a straightforward and efficient manner. However, Relief based methods suffer from instability, especially in the presence of noisy and high-dimensional outliers.

We propose a filter approach that relies on Relief's feature weighting technique as a way of weighting features according to each instance. Using this weighting process, each instance will propose a feature ranking and the algorithm will then focus on top ranked features for each instance. Lists of ranked features will finally be combined to give a final feature subset. The combination scheme used for this purpose aims at obtaining a feature selection that yields to good classification performance while beeing stable. The superior performance with respect to classification accuracy and excellent robustness to data heavily contaminated by noise make the proposed method promising for using in bioinformatics, where data are severely degraded by background artefacts. The proposed instance based feature selection process is described in the next sections.

2.1   Instance Based Feature Weighting

Let X be a matrix containing m training instances $\mathbf{x}_i = (x_{i1}, \ldots, x_{id}) \in \mathbb{R}^d$, where d is the number of features, and $\mathbf{y}_i = (y_1, \ldots, y_m), i = 1, \ldots, m$ the vector of class labels for the m instances. Let A be the set of features $\mathbf{a}_j = (a_1, \ldots, a_d), j = 1, \ldots, d$, where $d >> m$.

In a preprocessing step of the optimal feature subset selection, the feature space is reduced to m candidate subsets. Each instance of the training data is an expert which proposes a candidate feature subset (CFS) based on an instance feature weighting technique.

Given a distance function, we find two nearest neighbors of each sample x_i for each feature a_j, one from the same class (called nearest hit or NH), and the other from the different class (called nearest miss or NM). The margin of x_{ij} is then computed as

$$W(x_{ij}) = d(x_{ij}, NM(x_{ij})) - d(x_{ij}, NH(x_{ij})) \tag{1}$$

using a distance function. For this paper, we use the Manhattan distance to define a sample's margin and nearest neighbors while other standard definitions may also be used. This weight definition is used in the well-known Relief algorithm (using Euclidean distance) for the feature selection purpose [3]. An intuitive interpretation of this margin is a measure of how much the feature a_j of x_i can be corrupted by noise or how much x_i can "move" in the feature space before being misclassified. These scores are then normalized and we obtain a weighted feature space for each instance x_i.

This weight is then projected on each feature a_j and we get the matrix $\mathbf{W}$ filled with feature weights $w_{j,i}$ as shown in Table 1.

Table 1. Matrix of feature weights

	a_1	a_2	...	a_d
x_1	$w_{1,1}$	$w_{2,1}$	...	$w_{d,1}$
x_2	$w_{1,2}$	$w_{2,2}$	...	$w_{d,2}$
x_3	$w_{1,3}$	$w_{2,3}$	...	$w_{d,3}$
...	...	...	...	...
x_m	$w_{1,m}$	$w_{2,m}$	...	$w_{d,m}$

2.2   Candidate Feature Subsets Construction

Once the algorithm finish the weighting process, features in the space of each instance are ranked based on their weights such as top ranked features are those with highest relevance weight. Note that a feature a_j may have different ranks depending on the instance considered. After this instance based feature ranking step, a candidate subset of cardinality n is chosen from the best ranked features of each instance. This pre-processing step leads to m candidate feature subsets $\{CFS_1, CFS_2, ...CFS_m\}$ each of cardinality n. These candidate subsets must be combined in some manner in order to obtain a final result. The proposed combination technique is detailed in the following section.

2.3   Final Feature Selection by Calculating Feature Occurrence Frequency

In this step, the m candidate subsets component features are gathered together into a single subset S which is the union of all candidates. The resulting subset is projected on the m instances such as for an instance x_i, the feature a_j is assigned 1 if it was selected in CFS_i and 0 otherwise. The final feature selection is obtained by calculating the number of occurrences of each feature over all instances and ranking them based on their occurrence frequency. This ranking technique favors features appearing in the maximum number of candidate feature subsets built based on instances and thus if new instances are tested, there will be strong possibilities that the selected features will also be relevant for classifying them. As the ranking technique is based on aggregating several opinions (candidate subsets), as explained, thus our approach is likely going to improve feature selection stability. The final feature selection step is illustrated in Figure 1.

3   Experiments and Results

In this section we report the experimental setup and results of our proposed filter method and comparison results with four existing methods. The KNN and SVM classifiers are used with all algorithms to evaluate classification performance. Our experimental data consists of seven cancer diagnosis microarray data sets described in section 3.1. Classification performance, stability and final subset cardinality are used as metrics to evaluate our approach.

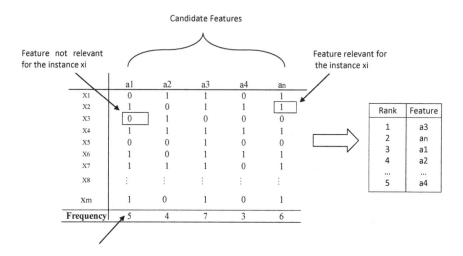

Fig. 1. Feature selection based on calculating feature occurrence frequency

3.1 Data Sets

The experiments are conducted on seven high dimensional low size microarray data sets. The classification in these data sets is binary (two classes) and its task is cancer diagnosis. In diffuse large Bcells (DLBCL) presented in [11], the classification task is the prediction of the tissue types, where genes are used to discriminate DLBCL tissues from Follicular Lymphomas. The task in the Bladder cancer dataset described in [12] is the clinical classification of bladder tumors using microarrays. We consider also another Lymphoma data set which task is to discriminate between two types of Lymphoma based on gene expression measured by microarray technology as in [4]. This dataset contains missing values for numeric attributes that we replace using the KNN imputation method proposed by [5]. Prostate data set described by [8] contains expression level of 12600 genes for 102 samples including prostate tumors and normal samples. Another data set is the Breast cancer data set used by [15]. The Central Nervous System (CNS) by [17] is also considered. It is a large data set concerned with the prediction of central nervous system embryonal tumor outcome based on gene expression. We analyzed also the Lung cancer gene expression, used by [16], which task is to differentiate between malignant pleural mesothelioma (MPM) and lung adenocarcinomas (ADCA). Table 2 summarizes the characteristics of the seven datasets.

Table 2. Datasets characteristics

Dataset	DLBCL	Bladder	Lymphoma	Prostate	Breast	CNS	Lung
No. of samples	77	31	45	102	97	60	181
No. of features	7029	3036	4026	12600	24482	7129	12533

3.2   Evaluation

We use 10-fold stratified cross-validation to evaluate results of feature selection based on the classification performance of KNN and SVM classifiers on the seven data sets. The misclassification error (MCE) of a classifier is defined as the proportion of misclassified instances over all classified instances. This metric is important and always used to evaluate feature selection algorithms for classification tasks. However, it is not sufficient given that there is no best way to evaluate any system and different metrics give different insights into how a feature selection algorithm performs. We evaluate also the stability and the final selected feature subset (SFS) cardinality obtained to compare our proposed method to other existing feature selection methods, namely Relief [3], mRMR [19], t-test and entropy filter methods. Measuring stability requires a similarity measure for feature preferences that will measure to which extent K sets S of s selected features share common features. Those sets can typically be produced by selecting features from different samples of the data. In our work, we use the 10-fold cross- validation for this purpose. In [18], Kuncheva proposed the following stability index:

$$Stab(S_1, .., S_K) = \frac{2}{K(K-1)} \sum_{i=1}^{K-1} \sum_{j=i+1}^{K} (|S_i \cap S_j| - \frac{s^2}{d})/(S - \frac{s^2}{d}), \quad (2)$$

where d is the total number of features, and S_i, S_j are two feature sets built from different partitions of the training samples. This index satisfies $-1 < Stab \leq 1$ and the greater is its value the larger is the number of commonly selected features in various sets. A negative stability index means that feature sets sharing common features are mostly due to chance. We used Kuncheva's stability index in our experiments.

3.3   Performance of Proposed Algorithm

For candidate feature subsets construction, we evaluate subset cardinalities ranging from 1 to 15 features. Our instance based approach is referred to as IB-filter in the experiments. For the proposed method, the number of final selected features depends on the algorithm setting, i.e the candidate subset cardinality and the training data. Given that we use 10-fold cross validation to select features and then record their corresponding misclassification error on each test fold, the number of selected features may vary. For each fold, we record classification performance corresponding to all possible feature subset sizes, i.e for example if 40 features are selected by our approach, classification performance is tested for up to 40 features. Then, to have a general approximation of the optimal number of features to select, we focus only on the 10-fold shared feature subset sizes and calculate the 10-fold cross validation MCE for each possible cardinality. Table 3 and 4 show Min MCE of KNN and SVM classifiers respectively on each data set and the corresponding CFS and SFS cardinalities. Stability is also calculated for the cardinality of the optimal subset over the 10 folds. Only four features give the best classification performance for Prostate cancer data set. Generally, SVM needs a higher number of features than KNN, but SVM classification performance is also higher especially for Prostate cancer data set where KNN gives approximately a MCE of 13% with 4 features, while SVM gives 4, 91% with 85 features. We notice also that high classification performance is generally coupled with high stability which is a good thing in feature selection applications giving a higher confidence on selected features. Stability is high especially for Lung cancer data set. For Breast cancer data set, MCE is about 25% with 85 features and 33% with 95 features respectively for KNN and SVM classifiers. The stability of selected features is also modest. We can deduce that feature selection instability can affect classification performance.

Table 3. Results of IB-filter with KNN classifier on cancer diagnosis data sets

	KNN			
	# CFS	# SFS	Min MCE	Stability
DLBCL	2	27	0.0411	0.7277
Bladder	7	30	0.0333	0.6906
Lymphoma	14	10	0.0000	0.6142
Prostate	14	4	0.1355	0.4443
Breast	8	85	0.2555	0.4231
CNS	8	12	0.2805	0.5138
Lung	2	17	0.0111	0.8567

Table 4. Results of IB-filter with SVM classifier on cancer diagnosis data sets

	SVM			
	# CFS	# SFS	Min MCE	Stability
DLBCL	4	70	0.0125	0.7779
Bladder	11	50	0.0000	0.7347
Lymphoma	4	10	0.0000	0.6958
Prostate	9	85	0.0491	0.7646
Breast	6	95	0.3344	0.5246
CNS	3	36	0.2667	0.6135
Lung	1	32	0.0000	0.9006

3.4 Comparison with Other Algorithms

In this section, we report comparison results of our filter method with four well known filters, Relief, mRMR, t-test and Entropy based feature selection algorithm. The KNN and SVM classifiers are used with all setups to evaluate classification performance. The considered algorithms are applied to several microarray data sets described in Section 3.1. As for our filter method and to not have a local minimum of the cross-validation MCE, we tested the performance of algorithms as a function of the number of features for up to 100 features

Table 5. Compared KNN minimum MCE rates, SFS cardinalities and stability on cancer diagnosis data sets

		IB-filter	Relief	mRMR	t-test	Entropy
DLBCL	Min MCE	**0.0411**	0.0649	0.0779	0.1429	0.1429
	# SFS	27	75	45	55	95
	Stability	0.7277	0.6335	0.5657	0.8791	0.8122
Bladder	Min MCE	**0.0323**	0.0645	0.0645	**0.0323**	**0.0323**
	#SFS	30	60	20	12	10
	Stability	0.6906	0.4650	0.4039	0.6877	0.7057
Lymphoma	Min MCE	**0.0000**	0.0222	**0.0000**	0.0222	0.0667
	# SFS	10	30	12	15	32
	Stability	0.6142	0.4291	0.5951	0.7234	0.7578
Prostate	Min MCE	0.1355	0.2157	**0.0784**	0.1078	0.0980
	# SFS	4	38	25	4	30
	Stability	0.4443	0.6275	0.7096	0.7944	0.6852
Breast	Min MCE	**0.2555**	0.4433	0.2784	0.5258	0.5258
	# SFS	85	6	30	2	2
	Stability	0.4231	0.3628	0.3355	1.0000	1.0000
CNS	Min MCE	**0.2805**	0.3667	0.3000	0.3667	0.3333
	# SFS	12	90	50	25	4
	Stability	0.5138	0.4974	0.3792	0.3970	0.5664
Lung	Min MCE	0.0110	0.0110	**0.0055**	0.0166	**0.0055**
	# SFS	17	20	4	15	17
	Stability	0.8567	0.8409	0.6610	0.8368	0.8900

and recorded the minimum MCE rate and the corresponding SFS cardinality for each algorithm. Stability is also used as metric to evaluate and compare our approach.

Table 5 shows results of the five feature selection algorithms with KNN classifier for the seven microarray data sets. In terms of classification performance, we observe that our algorithm often outperforms other filters. mRMR follows in the second place. The good performance of mRMR is explained by its selection of relevant features while removing redundancy which results on good classification performance. However, stability results of mRMR are modest for most cases. This is also the case for Relief algorithm, which gives modest results for both classification performance and stability. For simple filters like t-test and Entropy based filter, classification performance vary depending on the data set considered. They may give good (Bladder, Lung data sets), modest (Lymphoma, Prostate, CNS) or poor (DLBCL, Breast) classification results comparing to other algorithms. Especially for Breast data set, MCE exceeds 50% both for the two filters. However, both for the two filters also, stability of selected features that resulted on this poor performance is perfect 100%. Most often, high stability is required for good feature selection. However in this case, it is coupled with poor classification performance and as it was argued, stability of feature selection is not enough but it should be considered together with classification

Table 6. Compared SVM minimum MCE rates, SFS cardinalities and stability on cancer diagnosis data sets

		IB-filter	Relief	mRMR	t-test	Entropy
DLBCL	Min MCE	**0.0125**	0.0390	0.0519	0.1299	0.1818
	# SFS	70	70	75	85	85
	Stability	0.7779	0.6627	0.6089	0.8476	0.8164
Bladder	Min MCE	**0.0000**	0.0968	0.0968	0.0968	0.0968
	# SFS	50	38	95	20	10
	Stability	0.7347	0.4445	0.5982	0.7841	0.6923
Lymphoma	Min MCE	**0.0000**	0.0444	0.0222	**0.0000**	**0.0000**
	# SFS	10	21	20	27	75
	Stability	0.6958	0.4490	0.6460	0.7986	0.7509
Prostate	Min MCE	**0.0491**	0.0784	0.0784	0.0980	0.1176
	# SFS	85	65	65	45	65
	Stability	0.7646	0.5838	0.7718	0.8365	0.6722
Breast	Min MCE	0.3344	0.4021	**0.2680**	0.5052	0.5155
	# SFS	95	80	10	2	2
	Stability	0.5246	0.2609	0.3086	1.0000	1.0000
CNS	Min MCE	**0.2667**	0.3500	0.3667	0.3500	0.2833
	# SFS	36	80	65	23	75
	Stability	0.6135	0.5112	0.4442	0.4223	0.7092
Lung	Min MCE	**0.0000**	**0.0000**	0.0055	0.0166	0.0055
	# SFS	32	80	50	80	8
	Stability	0.9006	0.7702	0.8193	0.8831	0.8944

accuracy, because domain experts are not interested in a strategy that yields very stable feature sets, but leads to a bad predictive model. Accordingly, t-test and Entropy methods are not reliable for data sets for which they give good feature selection stability but high MCE.

We compared our approach with the four algorithms based on SVM classifier results also. This gives us the possibility to test the generalization capabilities of all considered algorithms and to know if we can obtain the same conclusions concerning their results, especially classification accuracy, for different classifiers. Table 5 shows SVM results for the five feature selection approaches. According to observed results, our approach is independent of the classification algorithm. When used with SVM classifier, our algorithm gives slightly better results than KNN in most cases. As already noticed in the preceding section, the only exception is for Breast cancer data set, where our feature selection algorithm gives a MCE of about 25% with KNN classifier, i.e outperforming SVM of about 8%. The opposite is noticed for Prostate cancer algorithm, where our approach with SVM outperforms KNN of about 9%. The experimented algorithms have the same behaviour with SVM classifier for most data sets. This is expected from filters which select features independently of the classification algorithm. Thus, they are likely to have a good generalization ability. This is confirmed in our experiments.

In addition to classification performance and stability, a good feature selection method will select the minimum number of features. Thus, it is important to compare selected feature subset cardinality of our approach to those of compared methods. We can say that our approach is also outperforming other methods. A smaller number of relevant features selected by our approach gives better results than larger subsets of relevant features selected by other approaches. For KNN classifier, the only exception where the selected feature subset cardinality is much higher than other selected subsets is for Breast cancer data set with 85 features. However, the corresponding MCE is smaller $(25, 55\%)$. mRMR gives a MCE equal to $27, 84\%$ with only 30 features. Since it is always better to have a minimum number of selected features, with a small classification performance difference, mRMR may be preferred to our approach for Breast cancer data set if this evaluation criterion is considered. For the same data set and with SVM classifier, the same thing is observed concerning selected feature subset cardinality. However, this time the classification performance is in favor of mRMR algorithm with a difference of about 7% with only 10 features against 95 for our approach.

4   Conclusion

In this work we proposed a new filter approach based on instance learning with the main goal of obtaining an efficient and stable feature selection. The main challenge in this approach is that it converts the problem of the small sample size to a mean for choosing only a few subsets of variables to be analyzed. Results show that our proposed method is outperforming existing methods in terms of classification accuracy and stability. We expect investigating research on using small sample size to create hybrid feature selection methods which take as input candidate feature subsets obtained as described in our method and involve them in a wrapper process for the optimal feature subset research.

References

1. Kalousis, A., Prados, J., Hilario, M.: Stability of feature selection algorithms: a study on high-dimensional spaces, Knowl. Inf. Syst. 12(1), 95–116 (2007)
2. Guyon, I., Elisseff, A.: An Introduction to Variable and Feature Selection. Journal of Machine Learning Research 3, 1157–1182 (2003)
3. Kira, K., Rendell, L.: A practical approach to feature selection. In: Sleeman, D., Edwards, P. (eds.) International Conference on Machine Learning, pp. 368–377 (1992)
4. Alizadeh, A.A., Eisen, M.B., Davis, R.E., Ma, C., Lossos, I.S., Rosenwald, A., Boldrick, J.C., Sabet, H., Tran, T., Yu, X., Powell, J.I., Yang, L., Marti, G.E., Moore, T., Hudson Jr., J., Lu, L., Lewis, D.B., Tibshirani, R., Sherlock, G., Chan, W.C., Greiner, T.C., Weisenburger, D.D., Armitage, J.O., Warnke, R., Levy, R., Wilson, W., Grever, M.R., Byrd, J.C., Botstein, D., Brown, P.O., Staudt, L.M.: Distinct types of diffuse large B-cell lymphoma identified by gene expression profiling. Nature 403(6769), 503–511 (2000)

5. Troyanskaya, O.G., Cantor, M., Sherlock, G., Brown, P.O., Hastie, T., Tibshirani, R., Botstein, D., Altman, R.B.: Missing value estimation methods for DNA microarrays. Bioinformatics 17(6), 520–525 (2001)
6. Cover, T., Hart, P.: Nearest neighbor pattern classification. IEEE Transactions on Information Theory 13, 21–27 (1967)
7. Sun, Y., Todorovic, S., Goodison, S.: Local Learning Based Feature Selection for High Dimensional Data Analysis. IEEE Trans. on Pattern Analysis and Machine Intelligence (TPAMI) 32, 1610–1626 (2010)
8. Singh, D., Febbo, P.G., Ross, K., Jackson, D.G., Manola, J., Ladd, C., Tamayo, P., Renshaw, A.A., D'Amico, A.V., Richie, J.P., Lander, E.S., Loda, M., Kantoff, P.W., Golub, T.R., Sellers, W.: Gene Expression Correlates of Clinical Prostate Cancer Behavior. Cancer Cell 1(2), 203–209 (2002)
9. Jain, A., Zongker, D.: Feature Selection: Evaluation, Application, and Small Sample Performance. IEEE Trans. on Pattern Analysis and Machine Intelligence (TPAMI) 19, 153–158 (1997)
10. Kohavi, R., John, G.H.: Wrappers for feature subset selection. Artificial Intelligence 97, 273–324 (1997)
11. Shipp, M.A., Ross, K.N., Tamayo, P., Weng, A.P., Kutok, J.L., Aguiar, R.C., Gaasenbeek, M., Angelo, M., Reich, M., Pinkus, G.S., Ray, T.S., Koval, M.A., Last, K.W., Norton, A., Lister, T.A., Mesirov, J., Neuberg, D.S.: Diffuse large b-cell lymphoma outcome prediction by gene-expression profiling and supervised machine learning. Nature Medicine 9, 68–74 (2000)
12. Dyrskjot, L., Thykjaer, T., Kruhoffer, M., Jensen, J.L., Marcussen, N., Hamilton-Dutoit, S., Wolf, H., Orntoft, T.F.: Identifying distinct classes of bladder carcinoma using microarrays. Nat Genetics 33, 90–96 (2003)
13. Kononenko, I.: Estimating attributes: analysis and extensions of RELIEF. In: Bergadano, F., De Raedt, L. (eds.) ECML 1994. LNCS, vol. 784, pp. 171–182. Springer, Heidelberg (1994)
14. Sun, Y.: Iterative relief for feature weighting: Algorithms, theories, and applications. IEEE Trans Pattern Anal Mach Intell 29(6), 1035–1051 (2007)
15. vant Veer, L.J.: Gene expression profiling predicts clinical outcome of breast cancer. Nature 415, 530–536 (2002)
16. Gordon, G.: Translation of microarray data into clinically relevant cancer diagnostic tests using gene expression ratios in lung cancer and mesothelioma. Cancer Research 62, 4963–4967 (2002)
17. Pomeroy, S.L.: Prediction of central nervous system embryonal tumour outcome based on gene expression. Nature 415, 436–442 (2002)
18. kuncheva, L.: A stability index for feature selection. In: Proceedings of the 25th IASTED International Multi-Conference: Artificial Intelligence and Applications, pp. 390–395 (2007)
19. Peng, H., Long, F., Ding, C.: Feature selection based on mutual information: criteria of max-dependency, max-relevance, and min-redundancy. IEEE Transactions on Pattern Analysis and Machine Intelligence 27, 1226–1238 (2005)

Multi-label Feature Selection
via Information Gain

Ling Li[1], Huawen Liu[1,2,*], Zongjie Ma[1], Yuchang Mo[1], Zhengjie Duan[1],
Jiaqing Zhou[1], and Jianmin Zhao[1]

[1] Department of Computer Science, Zhejiang Normal University, China
hwliu@zjnu.edu.cn
[2] NCMIS, Academy of Mathematics and Systems Science, CAS, China

Abstract. Multi-label classification has gained extensive attention recently. Compared with traditional classification, multi-label classification allows one instance to associate with multiple labels. The curse of dimensionality existing in multi-label data presents a challenge to the performance of multi-label classifiers. Multi-label feature selection is a powerful tool for high-dimension problem. However, the existing feature selection methods are unable to take both computational complexity and label correlation into consideration. To address this problem, a new approach based on information gain for multi-label feather selection (IGMF) is presented in this paper. In the process of IGMF, Information gain between a feature and label set is exploited to measure the importance of the feature and label corrections. After that, the optimal feature subset are obtained by setting the threshold value. A series of experimental results show that IGMF can promote performance of multi-label classifiers.

Keywords: Multi-label classification, High dimension, Feature selection Information gain.

1 Introduction

Unlike single-label classification, each instance in multi-label data may have more than one labels. Multi-label problems are more common in real life. For example, a movie can be associated with both the label action and the label romantic, a video about Windows 8 can be categorized as technology and computer simultaneous. Multi-label classification has been applied to many areas, such as bioinformatics technology [1], text mining [2] and multi-media [3].

Attracted by the application value of multi-label classification, many approaches have been proposed to handle multi-label problems. For instance, BRkNN [4] is algorithm based on the k-Nearest Neighbor (kNN). The first step of BRkNN is to identify the k nearest neighbors, and it is a lazy learning approach. However, BRkNN ignores the label correlations. AdaBoostMH [5] is a extension of the AdaBoost algorithm, and the main purpose is to minimize the

---

* Corresponding author.

X. Luo, J.X. Yu, and Z. Li (Eds.): ADMA 2014, LNAI 8933, pp. 345–355, 2014.
© Springer International Publishing Switzerland 2014

Hamming loss by focusing on the prediction of all correct labels. In order to explore label corrections, IBLR_ML [6] combines the instance-based learning and logistic regression as special cases. However, this method based on kNN involves the choice of the optimal neighborhood and interference of the noise. Besides, MLStacking [7] constructs the multi-label classifier by exploiting the pair-wise correlation between the labels, and then transform the multi-label problem into two categories of single-label problems. MLStacking reduces the complexity of data. However, it has relatively high algorithm complexity.

Similar to traditional classification, multi-label classification also suffers from the so-called curse of dimensionality. Dimensionality reduction, including feature extraction and feature selection, is an effective technology for high-dimension problems. Feature extraction is a method that converts high-dimensional feature space into a low-dimensional space though mapping or transforming. There are many feature extraction methods [8], such as PCA [9], MDDM [11] and LDA [12]. This kind of approaches can help to promote the performance of classifier effectively. However, feature extraction blurs the information of original feature and loses physical significance of data. Therefore, the results of feature extraction are lack of interpretation.

In contrast to feature extraction, based on certain evaluation standards, feature selection tries to select an optimal or most efficient subset of features from the original feature space instead of the original feature space, so as to cut down the dimension of feature space. Traditional feature selection methods handle multi-label data transforming multi-label problems into a set of single-label problems firstly, and then perform feature selection. For example, Spolaor proposed a multi-label feature selection approach based on ReliefF [13]. A strategy of two variable correlation is exploited and ReliefF is used to measure the relevance between features and labels. In [14], a multivariate mutual information-based feature selection for multi-label classification is proposed. In this algorithm, label interactions without resorting to problem transformation have been considered.

However, the mentioned multi-label feature approaches, which involve transfer-based strategies, cause damage to the original label structures. Resulting in the lose of correlation information among partial labels and limiting the capacity of feature selection on promoting the performance of classifiers. Motivated by this, this paper proposes a multi-label feature selection algorithm based on information gain(IGMF), which can perform feature selection at multi-label directly. In the process of IGMF, the optimal features can be obtained by exploring label corrections simultaneously. At the first step, the information gain between a feature and label set is exploited to measure the importance of the feature and label corrections. After that, the optimal feature subset are obtained by eliminating irrelevant or redundant features through comparing with a threshold value. The experimental results show that IGMF can improve the performance of multi-label classifiers.

The rest of the paper is divided into the following sections: Section 2 briefly reviews related work. Our proposed method, IGMF, is presented in Section 3.Section 4 is the experimental part. Finally, the work is concluded.

2   Related Work

Feature selection, which has been widely studied in the field of machine learning, aims at reducing the dimensionality of raw feature space by identifying discriminative features and decreasing the computational complexity. Multi-label feature selection is special technology for multi-label data. Recently, many related algorithms have been proposed. According to the way these algorithms combine the feature selection search with the construction of the classifiers, they can be organized into three categories [15], including embedded methods, wrapper methods, and filter methods.

Embedded and wrapper methods generally exploit a specific classifier to evaluate a specific subset of features. Embedded methods such as MEFS [17], can carry out feature selection and classification learning at the same time, this is because the search for an optimal subset of features is help to build the classifier construction.Namely, the search for an optimal subset of features is built into the classifier construction. Wrapper methods, such as HOML [18], regards the feature selection only as a part of learning algorithms. The interaction between feature subset search and model selection, and feature dependencies are taken into consideration in embedded methods. However, embedded methods and wrapper methods are computationally intensive and rely on classifiers.

In contrast to embedded methods and wrapper methods, filter methods are independent of classifiers and computationally simple and fast. Filter methods firstly assess features according to a certain valuation criteria, and then the optimal feature subset is obtained by search strategies. A Multi-label feature selection algorithm via Label-Specific is proposed in [19]. In this algorithm, feature density on the positive and negative instances set of each label is computed firstly, and then m features of high density from the positive and negative instances set of each label are selected, respectively. The intersection is regarded as the label-specific features of the related label. Multi-label data are classified base on label-specific features. However, this method ignores the label correlations. In order to explore label correlations, a feature selection approach base on F-statistic is presented in [20]. This algorithm firstly calculated values of F of multi labels for each feature. After that, features with higher value than a certain threshold are selected to form the optimal feature subset. This algorithm takes the label correlations into account to some extent, but fails to explore the feature correlations.

3   Multi-label Feature Selection via Information Gain

In this section, the formal concepts of information gain and interaction information are given firstly, and then correlations between labels and features are introduced.Finally, we propose a new feature selection named IGMF for the multi-label data.

3.1   Information Gain and Interaction Information

In information theory, information gain is a measurement that quantifies the degree of association between random variables S and Z. It can be computed as the following formula [21]:

$$IG(S; Z) = H(S) + H(Z) - H(S, Z) \tag{1}$$

where $H(S)$ means information entropy of variable S, and $H(Z)$ indicates information entropy of variable Z. $H(S, Z)$ stands for joint entropy of S and Z.

Interaction information is a generality of amount of information in information theory. Given a set T, and its interaction information $I(T)$ can be computed by [22]:

$$I(T) = -\sum_{X \in T'} (-1)^{|X|} H(X) \tag{2}$$

where T' is a set consisting of all the subsets of T. $|X|$ represents the number of variables in X. $H(X)$ means information entropy of variable X.

If $T = \{t_1, t_2\}$, t_1 and t_2 are random variables, we have:

$$T' = \{\varnothing, t_1, t_2, \{t_1, t_2\}\} \tag{3}$$
$$I(T) = H(t_1) + H(t_2) - H(t_1, t_2) \tag{4}$$

then we can get:

$$I(\{t_1, t_2\}) = IG(t_1; t_2) \tag{5}$$

If $T = \{t, T_1\}$, and t is a random variable, T_1 is a set of variables. $T_1' = \{t_1, t_2, t_3, ..., t_n\}$ is a set consisting of all the subsets of T_1, n is the total number of subsets T_1, we have:

$$T' = T_1' \cup \{\{t\}, \{t, t_1\}, \{t, t_2\}, ..., \{t, t_n\}\} \tag{6}$$

$$
\begin{aligned}
I(T) &= -\sum_{X \in T'} (-1)^{|X|} H(X) \\
&= -\sum_{i=1}^{i=n} (-)^{|t_i|} H(t_i) - \sum_{i=1}^{i=n} (-)^{|t, t_i|} H(\{t, t_i\}) \\
&= \sum_{i=1}^{i=n} (-)^{|t_i|} [H(\{t, t_i\}) - H(t_i)]
\end{aligned} \tag{7}
$$

3.2 Correlations between Labels and Features

Given a feature f and label set Y, information gain $IG(f;Y)$ can represent the degree of corrections between f and Y. It can be decomposed into interaction information of multiple variables as follows:

$$IG(f;Y) = \sum_{i=1}^{i=m} \sum_{y \in Y_i} I(\{f,y\})$$

$$= \sum_{y \in Y_1} I(\{f,y\}) + \sum_{i=2}^{i=m} \sum_{y \in Y_i} I(\{f,y\})$$

$$= \sum_{i=1}^{i=m} I(\{f,y_i\}) + \sum_{i=2}^{i=m} \sum_{y \in Y_i} I(\{f,y\}) \qquad (8)$$

where m is the number of labels, $Y_i = \{y_1, y_2, ..., y_k\}$, where $|y_k| = i, k = 1, 2, 3, ..., m$, according to Eq.(5) and (7), Eq.(8) can be adapted as:

$$IG(f;Y) = \sum_{i=1}^{i=m} IG(f;y_i) +$$

$$\sum_{i=2}^{i=m} \sum_{y \in Y_i} \sum_{i=1}^{i=n} (-1)^{|y_1|} [H(\{f,y_1\}) - H(y_1)] \qquad (9)$$

where n is the total number of subsets belong to y, y_i is the ith subset of y. Therefore, in the case without damaging the associated structures of original labels, $IG(f;Y)$ can not only represent the correlation degree between features and labels, but also represent the effects of feature y on all of the label combinations of label set Y.

3.3 The IGMF Methodology

Recently, numerous multi-label feature selection methods have been proposed. However, most of them failed to explore the existing correlations among labels. In fact, label correlations can help to promote the performance of classifiers. This means that original label set contains the most important information to reflect the label correlation. Existing multi-label feature selection approaches, which can perform feature selection directly on original label set, often rely heavily on classification capacity of classifiers and accompany with high computational complexity. These approaches such as SAML [16], HOML [18],GAFS [18], SFFS [18], MEFS [17] can achieve better classification results, but they are not suitable for large-scale data. Motivated by these problems, this paper proposes a new filter method named IGMF, which is independent of the classifier and fast in calculation.

In filter methods, variable correlations are described by metrics, and widely-used metrics can divided into four groups [24]: distance measures, information measures, dependency measures, and consistency measures. Information gain is a common information metric, which can measure the capacity of nonlinear correlation and quantify the degree of uncertainty of the variables [14]. Therefore, information gain are used as the metric in this paper.

It has been proofed in chapter 3.2 that in the case without damaging the associated structures of original labels, $IG(f;Y)$ can represent the correlation degree between features and labels. In this paper, information gain $IG(f;Y)$ between features and whole label set is used to measure the importance of features. A bigger value stands for greater correlation degree between the feature f and label Y. If set Y and the feature f are independent of each other, the $IG(f;Y)$n gets the minimum value. If the set Y depends on the feature f completely, the $IG(f;Y)$n gets the maximum value. Therefore, the $IG(f;Y)$n can distinguish relationships between features and labels efficiently. In fact, the information gain of each feature and whole label set is not certainly in the same range of measurement. For comparison, the normalized processing was performed on information gain $IG(f;Y)$, followed as:

$$SU(f,Y) = \frac{2 * IG(f;Y)}{H(f) + H(Y)} \tag{10}$$

where $SU(f,Y) \in [0,1]$, a larger value of $SU(f,Y)$ means a greater correlation. $SU(f,Y) = 0$ stands for f and Y are independent from each other. $SU(f,Y) = 1$ means either f or Y can be computed by the given one.

In general, IGMF firstly exploits information gain to measure the degree of correlations between each feature and the whole label set, and then perform normalized processing on their information gain to ensure them in the same range of measurement. After that, the average value of all of the normalized information gain of every feature, and this average value is set as the threshold. Finally, the optimal feature subset are obtained by eliminating irrelevant features through comparing with the threshold value. Therefore, IGMF can adapt to original data directly, and the label correlations are maintained by taking the original label set as a whole. The processing of IGMF algorithm works as Algorithm 1.

It is noted that Algorithm 1 consists of two loops and a mean statement. The first loop calculates the normalized information gain of features and label set, respectively, the computation complexity is $O(n)$. After that, the threshold value is computed with the computation complexity of $O(n)$. The aim of the second loop is to delete the irrelevant features according to the threshold value, the computation complexity is $O(n)$. As a result, the computation complexity of Algorithm 1 is $O(n)$, where n is the number of features. This means that the computation complexity of Algorithm 1 is only associated with the number of features, and irrelevant to the number of labels and independent of any classifiers.

4   Experiments

In this section, in order to evaluate the performance of the proposed algorithm, we compared our method with the filter methods and wrapper methods, re-

---

Alg. 1 The IGMF Algorithm

Input:
  F: $\{f_1, f_2, f_3, ..., f_n\}$
  Y: $\{y_1, y_2, y_3, ..., y_q\}$
Output:
  Best feature set($BestFeature$)
  Initialize IGS=ϕ
  For i=1 **to n do**
    $IG(f_i; Y) = H(f_i) + H(Y) - H(f_i, Y)$
    $SU(f_i, Y) = \frac{2*IG(f;Y)}{H(f)+H(Y)}$
    $IGS_i = SU(f_i, Y)$
    $IGS = IGS \cup IGS_i$
  EndFor
    $\mu = \frac{1}{n}\sum_{i=1}^{n} IGS_i$
    $BestFeature = F$
  for $i = 1$ **to** n **do**
    if $IGS_i < \mu$ then $BestFeature = BestFeature - \{f_i\}$
  endfor

---

spectively. Five commonly used evaluation metrics are employed to evaluate the performance of algorithms, including *Ranking Loss, Hamming Loss One-Error, Coverage* and *Average Precision* [10].

4.1 Data Sets

In our experiments, eight data sets from the real-world applications were used. They are *Cal500, Emotion, Image, Society, Education, Entertainment, Health,* and *Reference* [1]. These data sets are standard public data sets, and widely used in the area of multi-label classification.

Table 1 shows the general information of the benchmark data sets used in experiments, where ♯*Instances* and ♯*Variables* denote the number of instances and the dimensionality of data for each data set respectively.

4.2 IGMF Compared with Filter Methods

On the data sets *Education, Society, Reference,* and *Health*, IGMF has been compared with two approaches, including max [23] and avg [23]. MLNB-Basic[25] and RandSvm [1] are used as the classifier to verify the effectiveness of feature selection methods.

Table 2 shows the experimental results based on MLNB-Basic classifier. On the data sets *Health* and *Education,* our method IGMF obtains the best result on each metric. On *Society* data set, avg gets the best value on *Hamming Loss* and *Ranking Loss,* and IGMF performs best on the other metrics. On *Reference* data set, neither of the three method has significant advantage.

---

[1] http://sourceforge.net/projects/mulan/files/datasets/

Table 1. General information of the experimental data sets

Datasets	♯Instances	♯Variables	♯Labels
Cal500	502	68	174
Education	5000	550	33
Emotion	593	72	6
Entertainment	5000	640	21
Health	5000	612	32
Image	2000	294	5
Reference	5000	793	33
Society	5000	636	27

Table 2. Experimental results based on MLNB-Basic classifier , where ↓ means the smaller, the better, and ↑ means the larger, the better. Bold value shows the winner on each dataset

	Average Precision(↑)	Coverage(↓)	Hamming Loss(↓)	One-Error(↓)	Ranking Loss(↓)
			Education		
IGMF	**0.055302**	**24.098400**	**0.138139**	**0.620000**	**0.092138**
avg	0.053546	25.016800	0.147103	0.653400	0.094093
max	0.052447	25.630200	0.152333	0.634200	0.093636
			Society		
IGMF	**0.101732**	**14.801000**	0.250074	**0.640400**	0.189134
avg	0.098782	15.720600	**0.249615**	0.655800	**0.166067**
max	0.090129	16.561600	0.279785	0.665000	0.188913
			Health		
IGMF	**0.057436**	**23.721800**	0.115588	**0.456600**	**0.065672**
avg	0.054604	25.865000	0.134381	0.497600	0.075810
max	0.055665	25.059800	0.132388	0.510400	0.077735
			Reference		
IGMF	0.051465	**22.260800**	0.154588	**0.451200**	0.092693
avg	0.050895	25.644800	**0.123285**	0.529800	**0.065899**
max	**0.051544**	23.627800	0.165448	0.508200	0.094963

Table 3 shows the experimental results based on RandSvm classifier. On the data sets *Education*, *Society* and *Reference*, our method IGMF obtains the best performance on five metrics. On *Health* data set, max gets the best value on *Coverage* and *Ranking Loss*, at 3.052400 and 0.052316, respectively.

4.3 IGMF Compared with Wrapper Methods

On the data sets *Emotion*, *Image*, *Cal500* and *Entertainment*, IGMF has been compared with three approaches, including SFFS [26], SAFS [27] and GAFS [28]. ML-kNN [29] is used as the classifier to verify the effectiveness of feature selection methods.

Table 4 shows the experimental results. IGMF obtains the best results on three data sets, including *Emotion*, *Image* and *Entertainment*, in terms of five evaluation metrics. On *Cal500* data set, GAFS gets the lowest value, at 0.966690, on the metric of *Hamming Loss*, and it should be noted that the figure for IGMF is 0.9680000, which is slightly lower than GAFS.

Table 3. Experimental results based on RandSvm classifier , where ↓ means the smaller, the better, and ↑ means the larger, the better. Bold value shows the winner on each dataset.

	Average Precision(↑)	Coverage(↓)	Hamming Loss(↓)	One-Error(↓)	Ranking Loss(↓)
		Education			
IGMF	**0.600652**	**3.593400**	**0.045521**	**0.531800**	**0.077973**
avg	0.581765	3.763000	0.048085	0.560000	0.082369
max	0.578220	3.676600	0.048497	0.571200	0.081268
		Society			
IGMF	**0.616479**	**5.124400**	**0.059659**	**0.439000**	**0.121394**
avg	0.609695	5.224800	0.059896	0.445200	0.125184
max	0.613083	5.145200	0.060015	0.446800	0.122190
		Health			
IGMF	**0.749146**	3.136400	**0.039094**	**0.315400**	0.052453
avg	0.723463	3.143800	0.043238	0.356200	0.054234
max	0.730868	**3.052400**	0.041325	0.352400	**0.052316**
		Entertainment			
IGMF	**0.660226**	**2.618000**	**0.033388**	**0.460000**	**0.064784**
avg	0.634178	2.926200	0.034127	0.487000	0.073983
max	0.643944	2.719400	0.034848	0.484800	0.067836

Table 4. Experimental results based on ML-kNN classifier , where ↓ means the smaller, the better, and ↑ means the larger, the better. Bold value shows the winner on each dataset.

	Average Precision(↑)	Coverage(↓)	Hamming Loss(↓)	One-Error(↓)	Ranking Loss(↓)
		Emotion			
IGMF	**0.760334**	**1.937288**	**0.234746**	**0.345763**	**0.197750**
SFFS	0.724609	2.162712	0.256780	0.393220	0.241629
SAFS	0.741229	2.071186	0.239266	0.355932	0.222721
GAFS	0.737034	2.154237	0.243785	0.371186	0.235753
		Image			
IGMF	**0.790722**	0.972500	**0.170700**	0.321500	**0.176375**
SFFS	0.717640	1.226500	0.212500	0.437000	0.241458
SAFS	0.783654	**0.984500**	0.175200	**0.337000**	0.179042
GAFS	0.775878	1.018500	0.180800	0.346000	0.186167
		Cal500			
IGMF	**0.493687**	129.660000	0.968000	**0.118000**	0.182511
SFFS	0.485920	130.926000	0.967874	0.120000	0.186387
SAFS	0.493123	130.358000	0.967069	0.122000	0.183297
GAFS	0.493617	**129.682000**	**0.966690**	0.122000	0.182789
		Reference			
IGMF	**0.584209**	**3.279000**	**0.059314**	**0.554200**	**0.120656**
SFFS	0.500024	3.888000	0.066248	0.673400	0.148980
SAFS	0.579834	3.287800	0.060857	0.563800	0.121585
GAFS	0.576466	3.331000	0.060305	0.565400	0.123518

5   Conclusions

The high-dimension problem existing in multi-label data presents a challenge to the performance of multi-label classifiers. In this paper, we proposes a new feature selection method for multi-label data, named IGMF. IGMF can obtain optimal feature subset by exploiting label corrections fully. The experimental results show that IGMF is an effective approach compared with other feature selection.

Acknowledgements. This work is partially supported by the National NSF of China (61100119, 61272130, 61272468, 61170108 and 61170109), the Open Project Program of the National Laboratory of Pattern Recognition (NLPR) (201204214), Postdoctoral Science Foundation of China (2013M530072), and the NSF of Zhejiang province (LY14F020012, LY13F020016).

References

1. Elisseeff, A., Weston, J.: A kernel method for multi-labelled classification. Advances in Neural Information Processing Systems, pp. 681–687 (2001)
2. Srivastava, A.N., Zane-Ulman, B.: Discovering recurring anomalies in text reports regarding complex space systems. In: Aerospace Conference, pp. 3853–3862. IEEE (2005)
3. Turnbull, D., Barrington, L., Torres, D., et al.: Semantic annotation and retrieval of music and sound effects. IEEE Transactions on Audio, Speech, and Language Processing 16(2), 467–476 (2008)
4. Spyromitros, E., Tsoumakas, G., Vlahavas, I.P.: An empirical study of lazy multi-label classification algorithms. In: Darzentas, J., Vouros, G.A., Vosinakis, S., Arnellos, A. (eds.) SETN 2008. LNCS (LNAI), vol. 5138, pp. 401–406. Springer, Heidelberg (2008)
5. Schapire, R.E., Singer, Y.: Boostexter: a boosting-based system for text categorization. Machine Learning 39, 135–168 (2000)
6. Cheng, W., Hullermeier, E.: Combining instance-based learning and logistic regression for multilabel classification. Machine Learning 76, 211–225 (2009)
7. Tsoumakas, G., Dimou, A., Spyromitros, E., Mezaris, V., Kompatsiaris, I., Vlahavas, I.: Correlation-based pruning of stacked binary relevance models for multi-label learning. In: Proceedings of the Workshop on Learning from Multi-Label Data (MLD 2009), pp. 101–116. Springer Press, Berlin (2009)
8. Liu, H., Motoda, H., Setiono, R., et al.: Feature Selection: An Ever Evolving Frontier in Data Mining. FSDM, 4–13 (2010)
9. Jolliffe, I.: Principal Component Analysis. Springer-Verlag, New York (1986)
10. Tsoumakas, G., Katakis, I., Vlahavas, I.: Mining multi-label data. In: Maimon, O., Rokach, L. (eds.) Data Mining and Knowledge Discovery Handbook, 2nd edn. Spring (2010)
11. Zhang, Y., Zhou, Z.H.: Multilabel dimensionality reduction via dependence maximization. ACM Transactions on Knowledge Discovery from Data (TKDD) 4(3), 14 (2010)
12. Fisher, R.: The use of multiple measurements in taxonomic problems. Annals of Eugenics 7, 179–188 (1936)
13. Spolaor, N., Cherman, E.A., Monard, M.C.: Using ReliefF for Multilabel feature selection. In: Conferencia Latinoamericana de Informatica, pp. 960–975 (2011)
14. Lee, J., Kim, D.W.: Feature selection for multi-label classification using multivariate mutual information. Pattern Recognition Letters 34(3), 349–357 (2013)
15. Saeys, Y., Inza, I., Larranaga, P.: A review of feature selection techniques in bioinformatics. Bioinformatics 23, 2507–2517 (2007)
16. Zhang, Y., You, L., Chen, J.X.: Feature selection for multi-label data by using simulated annealing. Computer Engineering and Design 32(7), 2494–2500 (2011)
17. You, M., Liu, J., Li, G.Z., et al.: Embedded feature selection for multi-label classification of music emotions. International Journal of Computational Intelligence Systems 5(4), 668–678 (2012)

18. Shao, H., Li, G.Z., Liu, G.P., et al.: Symptom selection for multi-label data of inquiry diagnosis in traditional Chinese medicine. Science China Information Sciences 56(5), 1–13 (2013)
19. Qu, H., Zhang, S., Liu, H., et al.: A multi-label classification algorithm based on label-specific features. Wuhan University Journal of Natural Sciences 16(6), 520–524 (2011)
20. Kong, D., Ding, C., Huang, H., et al.: Multi-label relieff and f-statistic feature selections for image annotation. In: 2012 IEEE Conference on Computer Vision and Pattern Recognition (CVPR), pp. 2352–2359. IEEE (2012)
21. Cover, T.M., Thomas, J.A.: Elements of information theory. John Wiley and Sons (2012)
22. Brown, G.: A new perspective for information theoretic feature selection. International Conference on Artificial Intelligence and Statistics, 49–56 (2009)
23. Trohidis, K., Tsoumakas, G., Kalliris, G., Vlahavas, I.: Multi-label classification of music into emotions. In: 9th International Conference on Music Information Retrieval (ISMIR 2008), Philadelphia, pp. 325–330 (2008)
24. Liu, H., Yu, L.: Toward integrating feature selection algorithms for classification and clustering. IEEE Transactions on. Knowledge and Data Engineering 17(4), 491–502 (2005)
25. Zhang, M.L., Pena, J.M., Robles, V.: Feature selection for multi-label naive Bayes classification. Information Sciences 179(19), 3218–3229 (2009)
26. Pudil, P., Novovicov, J., Kittler, J., et al.: Floating search methods in feature selection. Pattern recognition letters 15(11), 1119–1125 (1994)
27. Ronen, M., Jacob, Z.: Using simulated annealing to optimize feature selection problem in marketing applications. European Journal of Operational Research 171(3), 842–858 (2006)
28. Yang, J., Honavar, V.: Feature subset selection using a genetic algorithm, Feature extraction. Construction and Selection, pp. 117–136. Springer, US (1998)
29. Zhang, M.-L., Zhou, Z.-H.: ML-kNN: a lazy learning approach to multi-label learning. Pattern Recognition 40(7), 2038–2048 (2007)

Efficient kNN Algorithm
Based on Graph Sparse Reconstruction

Shichao Zhang*, Ming Zong, Ke Sun, Yue Liu, and Debo Cheng

College of Computer Science & Information Technology,
Guangxi Normal University, Guilin, Guangxi, 541004, China
zhangsc@mailbox.gxnu.edu.cn

Abstract. This paper proposes an efficient k Nearest Neighbors (kNN) method based on a graph sparse reconstruction framework, called Graph Sparse kNN (GS-kNN for short) algorithm. We first design a reconstruction process between training and test samples to obtain the k value of kNN algorithm for each test sample. We then apply the varied kNN algorithm (*i.e.,* GS-kNN) for the learning tasks, such as classification, regression, and missing value imputation. In the reconstruction process, we employ a least square loss function for achieving the minimal reconstruction error, use an ℓ_1-norm to generate different k values of kNN algorithm for different test samples, design an ℓ_{21}-norm to generate the row sparsity for the removal of the impact of noisy training samples, and utilize Locality Preserving Projection (LPP) to preserve the local structures of data. With such an objective function, the GS-kNN obtains the correlation between each test sample and training samples, which then is used to design new classification/regression/missing value imputation rules for real applications. Finally, the proposed GS-kNN method is evaluated with extensive experiments, including classification, regression and missing value imputation, on real datasets, and the experimental results show that the proposed GS-kNN algorithm outperforms the previous kNN algorithms in terms of classification accuracy, correlation coefficient and root mean square error (RMSE).

Keywords: k Nearest Neighbors, sparse, reconstruction process, Locality Preserving Projection.

1 Introduction

K Nearest Neighbor (kNN) algorithm is a classical method in data mining, machine learning, pattern recognition and information retrieval [10][11]. Recently kNN algorithm has been widely applied in classification, regression and missing value imputation [14][21] due to its characteristics, such as simplicity and efficiency.

In kNN algorithm, both the similarity calculation between two samples and the selection of k values are open issues [16]. In metric learning, the methods for computing the distance between samples include Euclidean distance, Minkowsky distance, Mahalanobis distance, and so on. Different distance measures can lead to different results,

---

* Corresponding author.

X. Luo, J.X. Yu, and Z. Li (Eds.): ADMA 2014, LNAI 8933, pp. 356–369, 2014.
© Springer International Publishing Switzerland 2014

while different data distributions require various different distance measures. For simplicity, in this paper, we employ Euclidean distance for calculating the distance (or similarity) between samples. The previously kNN algorithm selected the k value by either a fixed constant for all test samples or conducting the cross-validation method to select the k value for each test sample. This often results in a number of problems in real applications. Here we take an example for illustrating the problem of the selection of the k value in kNN algorithm in Fig. 1.

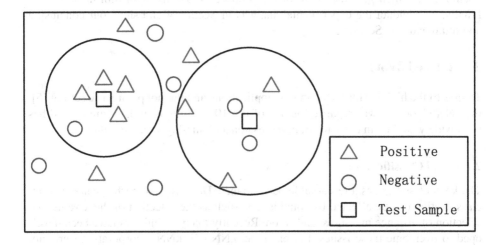

Fig. 1. An illustration on a binary classification for the selection of k value with k = 5

In Fig. 1, by setting k = 5, both the two test samples will be assigned to the positive class according to the kNN rule. Actually, regarding the distribution of data in Fig. 1, it is reasonable for the predicted class label of the left test sample. However, we can find that the right test sample should be assigned to the negative class (*i.e.*, k = 3) rather than the positive class, *i.e.*, k = 5. Besides, we also find the similar scenarios in regression and missing value imputation. The example indicates that using a fixed value of k in kNN algorithm is not always reasonable in real applications. That is, the k value in kNN algorithm should be different for each test sample. Moreover, the different k values should be learnt from the data, *i.e.*, decided by the distribution of data.

To overcome the drawback of kNN algorithm, in this paper, we learn the different k values in kNN algorithm according to the distribution of data, by proposing a sparse learning based kNN framework, called as Graph Sparse kNN (GS-kNN for short). The goal of our method is to make the best use of prior knowledge of data, such as the correlation between two samples, the removal of the noisy samples, and the preservation of the local structures of data. The rationale of the proposed GS-kNN algorithm is that making the best of prior knowledge of data leads to learn the real distribution of data. To do this, we first reconstruct test samples with training samples to obtain the optimal k value for each test sample, and then conduct classification, regression or missing value imputation task using the efficient kNN algorithm with the learnt k values. During the reconstruction process, we use a ℓ_1-norm regularization term to result in the element-wise sparisity (*i.e.,* the sparsity of a matrix is found in the elements) for generating various k values for different test samples. We also employ a ℓ_{21}-norm regularization term to generate the row sparsity, for the removal of the impact of the noisy samples [18][26]. Besides, we utilize a Locality Preserving Projection (LPP) [12] regularization term for preserving the local structures of data during the reconstruction process.

The rest of this paper is organized as follows: we briefly review the applications of kNN algorithm (such as classification, regression, and missing value imputation) in Section 2. We then give the detail of the proposed GS-kNN algorithm in Section 3. Finally, we conduct the experimental analysis in Section 4 and show our conclusion and future work in Section 5.

2   Related Work

Thanks to the high performance in real applications and the nonparametric setting [5], the kNN algorithm was regarded as one of top 10 algorithms in data mining and has been widely applied in classification, regression and missing value imputation.

2.1   kNN Classification

The kNN classifier has a remarkable performance. However, the performance of kNN classification can be affected by some issues, such as the selection of the k value, the selection of distance measures, and so on. Recently many techniques have been developed to overcome these issues. For example, kNN-CF (kNN incorporating Certainty Factor) classification method considered incorporating certainty factor measure into conventional kNN algorithm so that it can be applied on the beginning of the kNN classification, for meeting the need of imbalanced learning [16]. Moreover, the kNN-CF classification method can be easily extended to the dataset with skewed class distribution [16]. Song *et al.* proposed two novel kNN approaches, *i.e.,* Locally Informative-kNN (LI-kNN) and Globally Informative-kNN (GI-kNN), respectively, via designing new measure metrics, for selecting a subset of most informative sample from neighborhoods [8]. Wang proposed a new measure to define the similarity between two data points using the number of neighborhoods, for conducting a new kNN classifier [9].

2.2   kNN Regression

Conventional kNN regression usually encounters for the drawbacks, such as low efficiency, ignoring feature weights in distance calculation, and so on. To address these issues, many approaches have been designed. For example, Hamed *et al.* proposed an interval regression method based on the conventional kNN algorithm, by taking the advantage of the possibility distribution to choose the value of k of kNN algorithm due to the limited sample size [4]. Based on the observation that conventional kNN regression is sensitive to the selection of similarity metric, Yao proposed a general kNN framework to infer the similarity metric as a weighted combination of a set of base similarity measures [13]. Navot *et al.* proposed a new nearest neighbor method to capture complete dependency of the target function [6].

2.3 kNN Missing Value Imputation

Recently, missing value imputation has been shown to be a very important solution to deal with missing data [7][20][15][32], especially kNN based methods. For example, Based on nearest-neighbor imputation, Chen and Shao proposed some jackknife variance estimators, which are asymptotically unbiased and consistent for the sample means [2]. García-Laencina *et al.* proposed to employ mutual information to design a feature-weighted distance metric for conducting kNN [3]. Most recently, Zhang proposed a Shell Neighbors imputation method to select the left and right nearest neighbors of missing data for imputing missing data [17].

However, the previous studies on kNN methods only separately focused on classification, regression and missing value imputation. In this paper, we first revise the drawbacks of conventional kNN method (such as the fixed k value in kNN algorithm, the removal of noisy samples, and the preservation of the local structures of data), and then apply it to simultaneously conduct classification, regression and missing value imputation.

3 Approach

3.1 Notation

In this paper, we denote matrices as boldface uppercase letters, vectors as boldface lowercase letters, and scalars as normal italic letters, respectively. For a matrix $\mathbf{X} = [x_{ij}]$,its i-th row and j-th column are denoted as $\mathbf{x}^i$ and $\mathbf{x}_j$, respectively. Also we denote the Frobenius norm, ℓ_2-norm, ℓ_1-norm and $\ell_{2,1}$-norm of a matrix $\mathbf{X}$ respectively as $||\mathbf{X}||_F = \sqrt{\sum_i ||\mathbf{x}^i||_2^2} = \sqrt{\sum_j ||\mathbf{x}_j||_2^2}$, $||\mathbf{X}||_2 = \sqrt{\sum_i \sum_j |x_{ij}|^2}$, $||\mathbf{X}||_1 = \sum_i \sum_j |x_{ij}|$ and $||\mathbf{X}||_{2,1} = \sum_i ||\mathbf{x}^i||_2 = \sum_i \sqrt{\sum_j \mathbf{x}_{ij}^2}$. We further denote the transpose operator, the trace operator, and the inverse of a matrix $\mathbf{X}$ as $\mathbf{X}^T$, tr(X), and $\mathbf{X}^{-1}$, respectively.

3.2 Reconstruction

Let $\mathbf{X} \in R^{n \times d}$ denotes the set of training samples, where n is the number of training samples and d is the dimensionality of features, and assume $\mathbf{Y} \in R^{d \times m}$ denotes the matrix of test samples, where m is the number of test samples, in this paper, we use training samples $\mathbf{X}$ to reconstruct each test sample $\mathbf{y}_i$ so that the distance between $\mathbf{X}^T \mathbf{w}_i$ and $\mathbf{y}_i$, where $\mathbf{w}_i \in R^n$ denotes the reconstruction weights of training samples, is as small as possible. This leads to the least square loss function [22]:

$$\min_{\mathbf{W}} \sum_{i=1}^m ||\mathbf{X}^T \mathbf{w}_i - \mathbf{y}_i||_2^2 = \min_{\mathbf{W}} ||\mathbf{X}^T \mathbf{W} - \mathbf{Y}||_F^2 \qquad (1)$$

where $||.||_F$ denotes the Frobenius matrix norm and $\mathbf{W} \in R^{n \times m}$ denotes reconstruction weight matrix or the correlations between training samples and test samples.

Obviously, the optimization function in Eq. (1) is convex and smooth, so the optimal weight matrix $\mathbf{W}$ is obtained as $\mathbf{W}^* = (\mathbf{X}\mathbf{X}^T)^{-1}\mathbf{X}\mathbf{Y}$. However, $\mathbf{X}\mathbf{X}^T$ is not always invertible in practical applications. To this end, the conventional objection function is added a smooth regularization term, *e.g.*, an ℓ_2-norm,

$$\min_{\mathbf{W}} ||\mathbf{X}^T\mathbf{W} - \mathbf{Y}||_F^2 + \rho||\mathbf{W}||_2^2 \qquad (2)$$

where $||\mathbf{W}||_2$ is an ℓ_2-norm regularization term and ρ is a tuning parameter. Usually, Eq. (2) is called ridge regression, and its close solution is $\mathbf{W}^* = (\mathbf{X}\mathbf{X}^T + \rho\mathbf{I})^{-1}\mathbf{X}\mathbf{Y}$. Although the ridge regression in Eq. (2) can solve the singular issue in Eq. (1), the solution $\mathbf{W}^*$ does not satisfy our requirement because each element in $\mathbf{W}^*$ may be non-zero value. However, we expect to generate the sparsity for the correlation between training samples and test samples [27]. In this way, each test sample can be represented by partial training samples. To this end, we propose the following objective function for reconstructing test samples by training samples:

$$\min_{\mathbf{W}} ||\mathbf{X}^T\mathbf{W} - \mathbf{Y}||_F^2 + \rho_1||\mathbf{W}||_1 + \rho_2||\mathbf{W}||_{2,1} + \rho_3 Tr(\mathbf{W}^T\mathbf{X}\mathbf{L}\mathbf{X}^T\mathbf{W}) \qquad (3)$$

where $||\mathbf{W}||_1$ and $||\mathbf{W}||_{2,1}$, respectively, are an ℓ_1-norm regularization term and an $\ell_{2,1}$-norm regularization term. $\mathbf{L} \in R^{d \times d}$ is a Laplacian matrix. Different from Eq. (2), we use the ℓ_1-norm to replace the ℓ_2-norm because the ℓ_1-norm has been proved to lead the optimal sparse $\mathbf{W}^*$ [26][30][29]. Moreover, the derived objective function in Eq. (3) (*i.e.*, $\rho_2 = 0$ and $\rho_3 = 0$ in Eq. (3)) is also called the Least Absolute Shrinkage and Selection Operator (LASSO) [22][24][23], which generates element-wise sparsity in the optimal $\mathbf{W}^*$, *i.e.*, irregular sparsity in the elements of the matrix $\mathbf{W}^*$. The larger the value of ρ_1, the more sparse the $\mathbf{W}$.

In Eq. (3), $||\mathbf{W}||_{2,1}$ is an $\ell_{2,1}$-norm regularization term and ρ_2 is a tuning parameter. Moreover, the $\ell_{2,1}$-norm is used to remove the noisy samples which are almost irrelevant to all test samples during the reconstruction process. Specifically, the $\ell_{2,1}$-norm $||\mathbf{W}||_{2,1}$ consists of an ℓ_1-norm embedded by an ℓ_2-norm. Moreover, the ℓ_2-norm regularization term enforces the selection of the same rows across all columns, and the ℓ_1-norm imposes the row sparseness in the linear combination [26]. Therefore, the $\ell_{2,1}$-norm regularization term leads the reconstruction process in Eq. (3) to generate the sparseness through the whole rows of $\mathbf{W}$, row sparsity for short [21]. Moreover, the larger the value of ρ_2 , the more sparse the $\mathbf{W}$.

Besides, this paper considers preserving the local consistency of the structures of the data during the reconstruction process. To this end, a Locality Preserving Projection (LPP) regularization term is added into the objective function Eq. (3). LPP is a nonlinear dimensionality reduction method and can obtain an explicit mapping function [12]. The goal of LPP is to ensure that k nearest neighbors of original data is correspondingly preserved in the new space after conducting dimensionality reduction. Mathematically, assuming that there is a set of data containing n d-dimensional sample set $\mathbf{X} = [\mathbf{x}_1, \mathbf{x}_2, ..., \mathbf{x}_n]$ $(\mathbf{X} \in R^{d \times n})$, LPP is designed to find a projection matrix $\mathbf{W} = [\mathbf{w}_1, \mathbf{w}_2, ..., \mathbf{w}_l]$ $(\mathbf{w}_i \in R^d, \mathbf{W} \in R^{d \times l})$ mapping $\mathbf{X}$ to be $\mathbf{Y} = [\mathbf{y}_1, \mathbf{y}_2, ..., \mathbf{y}_n]$

$(\mathbf{y}_i \in R^l, \mathbf{Y} \in R^{l \times n})$, so that the distance between $\mathbf{W}^T \mathbf{x}_i$ and $\mathbf{y}_i$ is as small as possible. Therefore, the objective function of LPP [12] is defined as:

$$\min_{\mathbf{Y}=\mathbf{W}^T\mathbf{X}} \left(\sum_{i,j=1}^{N} ||\mathbf{y}_i - \mathbf{y}_j||^2 s_{ij} \right) = \underset{\mathbf{w}^T\mathbf{XDX}^T\mathbf{w}=1}{\arg} \min_{\mathbf{w}} \mathbf{w}^T \mathbf{XLX}^T \mathbf{w} \qquad (4)$$

where s_{ij} denotes the correlation between the samples $\mathbf{x}_i$, $\mathbf{D}$ is a diagonal matrix, i.e., $\mathbf{D}_{ii} = \sum_{j=1}^{n} s_{ij}$. Hence, in Eq. (3), $\mathbf{L} \in R^{n \times n}$ is a Laplacian matrix $(\mathbf{L} = \mathbf{D} - \mathbf{S})$. ρ_3 is a regularization term and is used to control the order of magnitude of the LPP part, i.e., controlling the consistency of the magnitude between the least square loss function and the LPP regularization term.

After optimizing Eq. (3) with the proposed optimization method shown in Section 3.3, we obtain the optimal of $\mathbf{W}^*$, i.e., the reconstruction weights or the correlations between training samples and test samples. The element w_{ij} of $\mathbf{W}^*$ denotes the correlation between the i-th training sample and the j-th test sample. The positive weight (i.e., $w_{ij} > 0$) indicates that there is positive correlation between the i-th training sample and the j-th test sample, while the negative weight (i.e., $w_{ij} < 0$) means negative correlation. In particular, the zero weight (i.e., $w_{ij} = 0$) means that there is no correlation between the i-th training sample and the j-th test sample. In this case, the i-th training sample should not be used for predicting the j-th test sample. That is, we use the related training samples (i.e., the training samples with nonzero coefficient) to predict each test sample, rather than using all training samples to predict test samples. In this way, Eq. (3) takes the distribution of data and prior knowledge into account for selecting the k value for each test sample.

To better understand the characteristics of the proposed method, we assume the optimal $\mathbf{W}^* \in R^{5 \times 3}$ as follows.

$$\mathbf{W}^* = \begin{bmatrix} 0 & 0.2 & 0 \\ 0 & 0 & 0 \\ 0 & 0.7 & 0.6 \\ 0.3 & 0.9 & 0 \\ 0.4 & 0 & 0 \end{bmatrix}$$

In this example, we have five training samples and three test samples. According to the proposed method, the values in the first column of $\mathbf{W}^*$ indicates the correlations between the first test sample and five training samples. Due to that there are only two non-zero values in the first column, i.e., w_{14} and w_{15}. This indicates that the test sample is only related to the last two training samples, i.e., the fourth training sample and the fifth training sample. More specifically, in kNN algorithm, we only need to regard the last two training samples as the nearest neighbors of the first test sample, i.e., the corresponding value of k as 2. Meanwhile, according to the values of the second column of $\mathbf{W}^*$, we only need to regard three training samples as the nearest neighbors of the second test sample, i.e., the corresponding value of k is 3. Obviously, for the third test sample, it should be predicted by the third training sample. The corresponding value of k is 1. In this way, the nearest neighbors of each test sample are obtained. Moreover, the

---

Algorithm 1: The Pseudo of Objective Function Equation (3).

Input: $\mathbf{X}$, $\mathbf{Y}$;

Output: $\mathbf{W}^{(t)} \in R^{n \times m}$;

1 Initialize $\mathbf{W}^1 \in R^{n \times m}$, $t = 1$;

2 **while** *not converge* **do**

3   Calculate the diagonal matrices $\mathbf{D}_i^{(t)} (1 \leq i \leq m)$ and $\tilde{\mathbf{D}}^{(t)}$, where the k-th diagonal
   element of $\mathbf{D}_i^{(t)}$ is $\frac{1}{2|w_{ki}^{(t)}|}$ and the k-th diagonal element of $\tilde{\mathbf{D}}^{(t)}$ is $\frac{1}{2||(\mathbf{w}^{(i)})^k||_2}$;

4   For each $i(1 \leq i \leq m)$, $\mathbf{w}_i^{(t+1)} = (\mathbf{X}\mathbf{X}^T + \rho_1 \mathbf{D}_i^{(t)} + \rho_2 \tilde{\mathbf{D}}^{(t)} + \rho_3 \mathbf{L})^{-1} \mathbf{X}\mathbf{y}^i$;

5      $t = t + 1$;

6 **end**

---

value of k in the kNN algorithm is different and is learnt according to the distribution of data.

Furthermore, on one hand, we find the sparsity of $\mathbf{W}^*$ is irregular, *i.e.,* sparsity in elements of the matrix $\mathbf{W}^*$. On the other hand, we also find that the values in the second row of $\mathbf{W}^*$ are all zero, this indicates that the second training sample is unrelated to all test samples. We can regard the second training samples as the noisy training sample. Actually, the ℓ_1-norm in Eq. (3) ensures to product zero values in elements, while the $\ell_{2,1}$-norm in Eq. (3) ensures to remove the impact of the noisy training samples. Furthermore, the LPP term in Eq. (3) ensures to further improve the performance of kNN algorithm.

3.3 Optimization

The objective function Eq. (3) is convex but non-smooth, this paper proposes an efficient solution as follows:

First, after taking the derivative with respect to each row $\mathbf{w}_i(1 \leq i \leq m)$, and then setting it to zero, we can get the following equation:

$$\mathbf{X}\mathbf{X}^T\mathbf{w}_i - \mathbf{X}\mathbf{y}^i + \rho_1 \mathbf{D}_i\mathbf{w}_i + \rho_2 \tilde{\mathbf{D}}\mathbf{w}_i + \rho_3 \mathbf{L}\mathbf{w}_i \tag{5}$$

where $\mathbf{D}_i(1 \leq i \leq m)$ is a diagonal matrix with the k-th diagonal element as $\frac{1}{2|w_{ki}|}$, and $\tilde{\mathbf{D}}$ is a diagonal matrix with the k-th diagonal element as $\frac{1}{2||\mathbf{w}^k||_2}$. Therefore,

$$\mathbf{w}_i = (\mathbf{X}\mathbf{X}^T + \rho_1 \mathbf{D}_i + \rho_2 \tilde{\mathbf{D}} + \rho_3 \mathbf{L})^{-1} \mathbf{X}\mathbf{y}^i \tag{6}$$

We can find that both $\mathbf{D}_i$ and $\tilde{\mathbf{D}}$ are unknown and depend on $\mathbf{W}$. According to [28][21][25], we propose an iterative algorithm in Algorithm 1 to optimize Eq. (3).

Theorem 1. *Algorithm 1 decreases the objective value in Eq. (3) in each of iteration.*

Proof: According to Step 2 in Algorithm 1, we have

$$\mathbf{W}^{(t+1)} = \min_{\mathbf{W}} Tr(\mathbf{X}^T\mathbf{W} - \mathbf{Y})^T + \rho_1 \sum_{i=1}^{m} \mathbf{w}_i^T \mathbf{D}_i^{(t)} \mathbf{w}_i + \rho_2 Tr\mathbf{W}^T \tilde{\mathbf{D}}^{(t)} + \rho_3 \mathbf{L}$$

Therefore, we have

$$Tr(\mathbf{X}^T\mathbf{W}^{(t+1)} - \mathbf{Y})^T(\mathbf{X}^T\mathbf{W}^{(t+1)} - \mathbf{Y}) + \rho_1 \sum_{i=1}^{m} (\mathbf{w}_i^{(t+1)})^T\mathbf{D}_i^{(t)}\mathbf{w}_i^{(t+1)}$$

$$+\rho_2 Tr(\mathbf{W}^{(t+1)})^T\tilde{\mathbf{D}}^t\mathbf{W}^{(t+1)} + \rho_3\mathbf{L}$$

$$\leq Tr(\mathbf{X}^T\mathbf{W}^{(t)} - \mathbf{Y})^T(\mathbf{X}^T\mathbf{W}^{(t)} - \mathbf{Y}) + \rho_1 \sum_{i=1}^{m} (\mathbf{w}_i^{(t)})^T\mathbf{D}_i^{(t)}\mathbf{w}_i^{(t)}$$

$$+\rho_2 Tr(\mathbf{W}^{(t)})^T\tilde{\mathbf{D}}^t\mathbf{W}^{(t)} + \rho_3\mathbf{L}$$

$$\Rightarrow Tr(\mathbf{X}^T\mathbf{W}^{(t+1)} - \mathbf{Y})^T(\mathbf{X}^T\mathbf{W}^{(t+1)} - \mathbf{Y}) + \rho_1 \sum_{i=1}^{d}\sum_{j=1}^{m} \left(\frac{(\mathbf{w}_{ij}^{(t+1)})^2}{2||\mathbf{w}_{ij}^{(t)}||} - ||\mathbf{w}_{ij}^{(t+1)}|| \right.$$

$$+||\mathbf{w}_{ij}^{(t+1)}|| \Bigg) + \rho_2 \sum_{k=1}^{d} \left(\frac{||(\mathbf{w}^{(t+1)})^k||_2^2}{2||(\mathbf{w}^{(t)})^k||_2} - ||(\mathbf{w}^{(t+1)})^k||_2 + ||(\mathbf{w}^{(t+1)})^k||_2 \right) + \rho_3\mathbf{L}$$

$$\leq Tr(\mathbf{X}^T\mathbf{W}^{(t)} - \mathbf{Y})^T(\mathbf{X}^T\mathbf{W}^{(t)} - \mathbf{Y}) + \rho_1 \sum_{i=1}^{d}\sum_{j=1}^{m} \left(||\mathbf{w}_{ij}^{(t)}|| + \frac{(\mathbf{w}_{ij}^{(t)})^2}{2||\mathbf{w}_{ij}^{(t+t)}||} - \right.$$

$$||\mathbf{w}_{ij}^{(t)}|| \Bigg) + \rho_2 \sum_{k=1}^{d} \left(||(\mathbf{w}^{(t)})^k||_2 + \frac{||(\mathbf{w}^{(t)})^k||_2^2}{2||(\mathbf{w}^{(i)})^k||_2} - ||(\mathbf{w}^{(t)})^k||_2 \right) + \rho_3\mathbf{L}$$

$$\Rightarrow Tr(\mathbf{X}^T\mathbf{W}^{(t+1)} - \mathbf{Y})^T(\mathbf{X}^T\mathbf{W}^{(t+1)} - \mathbf{Y}) + \rho_1 \sum_{i=1}^{d}\sum_{j=1}^{m} ||\mathbf{w}_{ij}^{(t+1)}||$$

$$+\rho_2 \sum_{K=1}^{d} ||(\mathbf{w}^{(t+1)})^k||_2 + \rho_3\mathbf{L}$$

$$\leq Tr(\mathbf{X}^T\mathbf{W}^{(t)} - \mathbf{Y})^T(\mathbf{X}^T\mathbf{W}^{(t)} - \mathbf{Y}) + \rho_1 \sum_{i=1}^{d}\sum_{j=1}^{m} ||\mathbf{w}_{ij}^{(t)}|| + \rho_2 \sum_{k=1}^{d} ||(\mathbf{w}^{(t)})^k||_2$$

$$+\rho_3\mathbf{L}$$

According to the literature [31], for any vector $\mathbf{w}$ and $\mathbf{w}_0$, we have $||\mathbf{w}_2|| - \frac{||\mathbf{w}||_2^2}{2||\mathbf{w}_0||_2} \leq ||\mathbf{w}_0||_2 - \frac{||\mathbf{w}_0||_2^2}{2||\mathbf{w}_0||_2}$. So the Algorithm 1 decreases the objective value in each of iteration. $\mathbf{W}^{(t)}$, $\mathbf{D}_i^{(t)}(1 \leq i \leq m)$ and $\tilde{\mathbf{D}}^{(t)}$ will satisfy Eq. (6) at the convergence. As the objective function Eq. (3) is convex, the $\mathbf{W}$ satisfying Eq. (3) is a global optimum solution. Therefore, Algorithm 1 will converge to the global optimum of the objective function Eq. (3).

3.4 Algorithm

In the proposed method, we first use the learnt k values for kNN algorithm, *i.e.*, the GS-kNN algorithm, and then apply it for different tasks, such as classification, regression and missing value imputation. The pseudo of GS-kNN is presented in Algorithm 2.

Algorithm 2: The pseudo of GS-kNN algorithm.

   Input: **X**, **Y**;
   Output:
   switch *task* **do**
      case *1*
         | Class labels;
      endsw
      case *2*
         | Predicted value;
      endsw
      case *3*
         | Imputation value;
      endsw
   endsw
 1 Normalizing **X** and **Y**;
 2 Optimizing Eq. (3) to obtain the optimal solution **W**;
 3 Obtaining the optimal k value for test samples based on **W**;
 4 **switch** *task* **do**
 5   **case** *1*
 6 | Obtaining class labels via majority rule;
 7   **endsw**
 8   **case** *2*
 9 | Obtaining prediction value via Eq. (7);
10   **endsw**
11   **case** *3*
12 | Obtaining imputation value via Eq. (7);
13   **endsw**
14 **endsw**

Firstly, the proposed GS-kNN algorithm employs the majority rule for predicting the class label of the test sample. Secondly, in both the regression task and the missing value imputation task, the proposed GS-kNN algorithm considers, the bigger the correlation between the test sample and its nearest neighbor, the larger the contribution of this nearest neighbor to the test sample. Therefore, this paper proposes a weighted method for the prediction of both the regression task and the task of missing value imputation. Specifically, the weighted predictive value of the j-th test sample is defined as:

$$predictValue\_weight = \sum_{i=1}^{n} \left(\frac{\mathbf{w}_{ij}}{\sum_{i=1}^{n} \mathbf{w}_{ij}} \times \mathbf{y}_{train(i)} \right) \tag{7}$$

where n is the number of training samples, and $\mathbf{y}_{train(i)}$ represents the true value of the i-th training sample.

Table 1. Comparison on Classification Accuracy (mean±STD) of all algorithms

Dataset	kNN	L-kNN	GS-kNN
Climate	0.8944 ± 0.0339	0.9259 ± 0.0360	**0.9407 ± 0.0336**
German	0.6660 ± 0.0615	0.7090 ± 0.0617	**0.7240 ± 0.0587**
Blood	0.7365 ± 0.0605	0.7689 ± 0.0553	**0.7878 ± 0.0537**
Australian	0.7826 ± 0.0538	0.8130 ± 0.0518	**0.8420 ± 0.0531**

4 Experiments

We evaluate the proposed GS-kNN algorithm with the state-of-the-art kNN methods on twelve datasets in three applications, such as classification, regression, and missing value imputation.

4.1 Experimental Setting

In our experiments, we regard the standard kNN algorithm (with k = 5) as the first comparison algorithm. The second comparison algorithm is the L-kNN method, *i.e.,* Eq. (3) with the setting $\rho_2 = 0$, on which we would like to show the importance of removing the noisy samples.

The used datasets mainly came from UCI dataset and LIBSVM website [1]. We conducted experiments on four datasets for classification, regression, and missing value imputation, respectively. Note that there are not missing values in the original datasets, we randomly selected some independent values to be missed according the literatures on missing value imputations [19].

In our experiments, we repeated the experiments on each dataset ten times, and regarded the average performance as the reported results. We used classification accuracy as the evaluation for the classification task. The higher accuracy the algorithm is, the better performance of classification it is. We used correlation coefficient and root mean square error (RMSE) to evaluate the performance of both regression and missing value imputation. Correlation coefficient indicates the correlation between prediction and observation. Generally, the larger the correlation coefficient, the more accurate the prediction.

Experiments on Classification. We listed the classification performance (including the mean of classification accuracy in ten iterations and the corresponding STandard Deviation (STD)) of all algorithms on the four datasets in Table 1.

According to Table 1, we have the following observations:

– The proposed GS-kNN algorithm achieved the best classification accuracy by comparing with the kNN algorithm and the L-kNN algorithm. For example, the GS-kNN algorithm improved 5.37% than the kNN algorithm, and 1.87% than the L-kNN algorithm on average of classification accuracy.
– The GS-kNN algorithm performed better than the L-kNN algorithm because our method used an $\ell_{2,1}$-norm to remove noisy samples. For example, the proposed

Table 2. Comparison on RMSE (mean±STD) of all algorithms

Dataset	kNN	L-kNN	GS-kNN
Mpg	4.0474 ± 0.4154	3.6045 ± 0.2924	**3.4370 ± 0.3063**
Housing	6.1950 ± 1.4083	5.3082 ± 1.1449	**5.1513 ± 1.0959**
Bodyfat	0.0049 ± 0.0010	0.0044 ± 0.0010	**0.0026 ± 0.0008**
ConcreteSlump	5.8989 ± 1.2593	4.9525 ± 1.0956	**4.5109 ± 1.1983**

Table 3. Comparison on Correlation Coefficient (mean±STD) of all algorithms

Dataset	kNN	L-kNN	GS-kNN
Mpg	0.8516 ± 0.0379	0.8733 ± 0.0285	**0.8904 ± 0.0237**
Housing	0.7612 ± 0.0985	0.8386 ± 0.0648	**0.8574 ± 0.0604**
Bodyfat	0.9777 ± 0.0068	0.9865 ± 0.0076	**0.9943 ± 0.0050**
ConcreteSlump	0.7266 ± 0.0571	0.7900 ± 0.0654	**0.8304 ± 0.0463**

method improved by 2.9% and 1.48%, respectively, than the L-kNN, on Australian dataset and Climate dataset. This indicates that both Australian dataset and Climate dataset may contain noisy samples. Moreover, Australian dataset might have more noisy samples.

– Both the GS-kNN and L-kNN are better than the kNN algorithm. This indicates that using different k values in kNN algorithm (such as the GS-kNN algorithm and the L-kNN algorithm) can achieve better classification performance than the method with fixed k value for all test samples, such as the conventional kNN algorithm.

Experiments on Regression. Tables 2 and 3 showed the results of both RMSE (mean±STD) and correlation coefficient (mean±STD)of all algorithms on four datasets.

From Tables 2 and 3, we found that the proposed GS-kNN algorithm had the best prediction performance. For example, the proposed algorithm improved by 1.388 on the ConcreteSlump dataset in terms of RMSE, and 10.38% on the ConcreteSlump dataset in terms of correlation coefficient, than the kNN algorithm. Meanwhile, the proposed algorithm improved by 0.4416 on the ConcreteSlump dataset in terms of RMSE than the L-kNN algorithm, while our GS-kNN improved by 4.04% on the ConcreteSlump dataset in terms of correlation coefficient comparing.

Experiments on Missing Value Imputation. We listed the imputation performance of all algorithms on the four datasets in Tables 4 and 5.

According to Tables 4 and 5, the proposed GS-kNN algorithm achieved similar results to both the classification task and the regression task, than the comparison algorithms on four datasets. That is, our method achieved the best imputation performance, followed by the L-kNN algorithm, and the kNN algorithm ranks the last one. Specifically, the proposed algorithm improved by 3.51% and 10.34%, respectively, than the L-kNN algorithm and the standard kNN algorithm in terms of correlation coefficient, while our method reduced 0.0346 and 0.0625, respectively, than the L-kNN algorithm and the kNN algorithm in terms of RMSE. In particular, the maximal reduce between

Table 4. Comparison on RMSE (mean±STD) of all algorithms

Dataset	kNN	L-kNN	GS-kNN
Abalone	2.9424 ± 0.1710	2.8659 ± 0.1837	**2.7536 ± 0.1907**
Pyrim	0.0692 ± 0.0151	0.0603 ± 0.0149	**0.0523 ± 0.0123**
Triazines	0.1445 ± 0.0324	0.1254 ± 0.0326	**0.1108 ± 0.0250**
Mg	0.1487 ± 0.0163	0.1415 ± 0.0136	**0.1380 ± 0.0135**

Table 5. Comparison on Correlation Coefficient (mean±STD) of all algorithms

Dataset	kNN	L-kNN	GS-kNN
Abalone	0.6943 ± 0.0225	0.7292 ± 0.0277	**0.7429 ± 0.0257**
Pyrim	0.8030 ± 0.0778	0.8865 ± 0.0482	**0.9193 ± 0.0400**
Triazines	0.5593 ± 0.0740	0.6958 ± 0.0827	**0.7704 ± 0.0519**
Mg	0.7569 ± 0.0421	0.7751 ± 0.0356	**0.7944 ± 0.0375**

the proposed method and the kNN algorithm is 0.1888 on Abalone dataset on the evaluation of RMSE, while the maximal improvement for the proposed GS-kNN to the kNN algorithm is 21.11% on Triazines dataset on the evaluation of correlation coefficient.

5   Conclusion and Future Work

In this paper, we have proposed an improved kNN algorithm for three tasks, such as classification, regression and missing value imputation. The proposed method has been designed to revise the conventional kNN algorithm with the following points. Firstly, we learnt different k value for each test sample by making the best use of prior knowledge of data. Secondly, we applied the new kNN algorithm to simultaneously conduct three types of tasks. To this end, we designed a sparse reconstruction framework with three regularization terms to learn the optimal k values for test samples. Specifically, we used the ℓ_1-norm to generate element-wise sparsity for selecting different k values for different test samples. To improve the reconstruction performance, we employed the $\ell_{2,1}$-norm to generate row sparsity for the removal of noisy samples, and further used the LPP regularization term to preserve the local structures of samples. Finally, the results on real datasets experimentally verified the benefit of the proposed method than the state-of-the-art kNN algorithms.

In future, we will focus on designing a nonlinear transformation matrix to learn the correlation between test samples and training samples.

Acknowledgements. This work was supported in part by the National Natural Science Foundation of China under grants 61170131, 61263035 and 61363009, the China 863 Program under grant 2012AA011005, the China 973 Program under grant 2013CB329404, the Guangxi Natural Science Foundation under grant 2012GXNS-FGA060004, the Key Project for Guangxi Universities' Science and Technology Research under grant 2013ZD041.

References

1. Chang, C.-C., Lin, C.-J.: LIBSVM: A library for support vector machines. ACM Transactions on Intelligent Systems and Technology 2, 1–27 (2011)
2. Chen, J., Shao, J.: Jackknife variance estimation for nearest-neighbor imputation. Journal of the American Statistical Association 96(453), 260–269 (2001)
3. García-Laencina, P.J., Sancho-Gómez, J.-L., Figueiras-Vidal, A.R., Verleysen, M.: k nearest neighbours with mutual information for simultaneous classification and missing data imputation. Neurocomputing 72(7), 1483–1493 (2009)
4. Hamed, M.G., Serrurier, M., Durand, N.: Possibilistic KNN regression using tolerance intervals. In: Greco, S., Bouchon-Meunier, B., Coletti, G., Fedrizzi, M., Matarazzo, B., Yager, R.R. (eds.) IPMU 2012, Part III. CCIS, vol. 299, pp. 410–419. Springer, Heidelberg (2012)
5. Mary-Huard, T., Robin, S.: Tailored aggregation for classification. IEEE Transactions on Pattern Analysis and Machine Intelligence 31(11), 2098–2105 (2009)
6. Navot, A., Shpigelman, L., Tishby, N., Vaadia, E.: Nearest neighbor based feature selection for regression and its application to neural activity. In: Advances in Neural Information Processing Systems (NIPS), vol. 19 (2005)
7. Qin, Y., Zhang, S., Zhu, X., Zhang, J., Zhang, C.: Semi-parametric optimization for missing data imputation. Applied Intelligence 27(1), 79–88 (2007)
8. Song, Y., Huang, J., Zhou, D., Zha, H., Giles, C.L.: Iknn: Informative k-nearest neighbor pattern classification. In: Kok, J.N., Koronacki, J., Lopez de Mantaras, R., Matwin, S., Mladenič, D., Skowron, A. (eds.) PKDD 2007. LNCS (LNAI), vol. 4702, pp. 248–264. Springer, Heidelberg (2007)
9. Wang, H.: Nearest neighbors by neighborhood counting. IEEE Transactions on Pattern Analysis and Machine Intelligence 28(6), 942–953 (2006)
10. Wu, X., Zhang, C., Zhang, S.: Efficient mining of both positive and negative association rules. ACM Transactions on Information Systems (TOIS) 22(3), 381–405 (2004)
11. Wu, X., Zhang, C., Zhang, S.: Database classification for multi-database mining. Information Systems 30(1), 71–88 (2005)
12. Wu, X., Zhang, S.: Synthesizing high-frequency rules from different data sources. IEEE Transactions on Knowledge and Data Engineering 15(2), 353–367 (2003)
13. Yao, Z., Ruzzo, W.L.: A regression-based k nearest neighbor algorithm for gene function prediction from heterogeneous data. BMC Bioinformatics 7, S11 (2006)
14. Zhang, S.: Cost-sensitive classification with respect to waiting cost. Knowledge-Based Systems 23(5), 369–378 (2010)
15. Zhang, S.: Estimating semi-parametric missing values with iterative imputation. International Journal of Data Warehousing and Mining 6(3), 1–10 (2010)
16. Zhang, S.: KNN-CF approach: Incorporating certainty factor to knn classification. IEEE Intelligent Informatics Bulletin 11(1), 24–33 (2010)
17. Zhang, S.: Shell-neighbor method and its application in missing data imputation. Applied Intelligence 35(1), 123–133 (2011)
18. Zhang, S.: Decision tree classifiers sensitive to heterogeneous costs. Journal of Systems and Software 85(4), 771–779 (2012)
19. Zhang, S.: Nearest neighbor selection for iteratively kNN imputation. Journal of Systems and Software 85(11), 2541–2552 (2012)
20. Zhang, S., Jin, Z., Zhu, X.: Missing data imputation by utilizing information within incomplete instances. Journal of Systems and Software 84(3), 452–459 (2011)
21. Zhang, S., Qin, Z., Ling, C.X., Sheng, S.: "missing is useful": missing values in cost-sensitive decision trees. IEEE Transactions on Knowledge and Data Engineering 17(12), 1689–1693 (2005)

22. Zhang, S., Zhang, C., Yan, X.: Post-mining: maintenance of association rules by weighting. Information Systems 28(7), 691–707 (2003)
23. Zhao, Y., Zhang, S.: Generalized dimension-reduction framework for recent-biased time series analysis. IEEE Transactions on Knowledge and Data Engineering 18(2), 231–244 (2006)
24. Zhu, X., Huang, Z., Cheng, H., Cui, J., Shen, H.T.: Sparse hashing for fast multimedia search. ACM Transactions on Information Systems (TOIS) 31(2), 9 (2013)
25. Zhu, X., Huang, Z., Cui, J., Shen, H.T.: Video-to-shot tag propagation by graph sparse group lasso. IEEE Transactions on Multimedia 15(3), 633–646 (2013)
26. Zhu, X., Huang, Z., Shen, H.T., Zhao, X.: Linear cross-modal hashing for efficient multimedia search. In: Proceedings of the 21st ACM International Conference on Multimedia, pp. 143–152 (2013)
27. Zhu, X., Huang, Z., Shen, H.T., Cheng, J., Xu, C.: Dimensionality reduction by mixed kernel canonical correlation analysis. Pattern Recognition 45(8), 3003–3016 (2012)
28. Zhu, X., Huang, Z., Yang, Y., Shen, H.T., Xu, C., Luo, J.: Self-taught dimensionality reduction on the high-dimensional small-sized data. Pattern Recognition 46(1), 215–229 (2013)
29. Zhu, X., Suk, H.-I., Shen, D.: Matrix-similarity based loss function and feature selection for alzheimer's disease diagnosis. In: 2014 IEEE Conference on Computer Vision and Pattern Recognition (CVPR), pp. 3089–3096 (2014)
30. Zhu, X., Suk, H.-I., Shen, D.: A novel matrix-similarity based loss function for joint regression and classification in ad diagnosis. NeuroImage (2014)
31. Zhu, X., Zhang, L., Huang, Z.: A sparse embedding and least variance encoding approach to hashing (2014)
32. Zhu, X., Zhang, S., Jin, Z., Zhang, Z., Xu, Z.: Missing value estimation for mixed-attribute data sets. IEEE Transactions on Knowledge and Data Engineering 23(1), 110–121 (2011)

On Dynamic Selection of Subspace for Random Forest

Md Nasim Adnan

Centre for Research in Complex Systems (CRiCS)
School of Computing and Mathematics
Charles Sturt University
Bathurst, NSW 2795, Australia
madnan@csu.edu.au

Abstract. Random Forest is one of the most popular decision forest building algorithms that use decision trees as the base classifiers. The splitting attributes for decision trees of Random Forest are generally determined from a predefined number of randomly selected attribute subset of the original attribute set. In this paper, we propose a new technique that randomly determines the size of the attribute subset between a dynamically determined range based on the relative size of current data segment to the bootstrap samples at each node splitting event. We present elaborate experimental results involving five widely used data sets from the UCI Machine Learning Repository. The experimental results indicate the effectiveness of the proposed technique in the context of Random Forest.

Keywords: decision tree, decision forest, prediction accuracy, random forest.

1 Introduction

From 2005 to 2020, the "Digital Universe" will expand by a factor of 300, from 130 Exabyte to 40,000 Exabyte, or 40 trillion gigabytes (more than 5,200 gigabytes for every man, woman, and child in 2020) [1]. Hence it is almost impossible for domain experts to infer any useful knowledge or pattern manually from such a huge volume of data. Data mining is the method of automatically discovering useful information from large data sets [2]. Classification and clustering are two widely used data mining tasks that are applied for knowledge discovery and pattern understanding.

Classification aims to generate a function (commonly known as the classifier) that maps the set of non-class attributes $\{A_1, A_2, ... A_m\}$ to a predefined class attribute C [2]. There are different types of classifiers including Decision Trees [3], [4], Bayesian Classifiers [5], [6], Artificial Neural Networks [7], [8], [9], and Support Vector Machines [10]. Among these classifiers, decision trees are very popular to the real world users as they can be easily broken down to generate logic rules to infer valuable knowledge [11]. Due to their immense popularity, decision trees with better prediction accuracy can render huge impact on many sensitive application areas such as medical diagnosis.

There are many decision tree induction algorithms such as CART [3], ID3 [4], C4.5 [4], [12], SLIQ [13], Fuzzy SLIQ [14], SPRINT [15] and Explore [16]. Most of

X. Luo, J.X. Yu, and Z. Li (Eds.): ADMA 2014, LNAI 8933, pp. 370–379, 2014.
© Springer International Publishing Switzerland 2014

these algorithms follow the structure of Hunt's algorithm [2]. In Hunt's algorithm a decision tree is induced in a recursive fashion from a training data set. The induction process starts by computing an attribute test measure (such as the Gini Index [3] of the CART algorithm) for all non-class attributes. Then the attribute having the highest test value is selected as a splitting attribute to divide the training data set D into a set of mutually exclusive horizontal data segments [4], [12], [16], [17]. The purpose of this splitting is to create a purer distribution of class values in the succeeding data segments than the distribution in D. The process of selecting the splitting attribute continues recursively in each succeeding data segment D_i until either every data segment gets the purest class distribution or another stopping criterion is satisfied. By "purest class distribution" we mean the presence of a single class value for all records.

It is worthy to mention that ensemble methods work better with unstable classifiers such as a decision tree [2]. Decision forest is an ensemble of decision trees where an individual decision tree acts as the base classifier and the classification is performed by taking a vote based on the predictions made by each decision tree of the decision forest [2].

A decision forest overcomes some of the shortcomings of a decision tree. A decision tree is entirely formed from the training data set. This enables a decision tree to have remarkable classification performance on the examples (records) of the training data set. However, the classification performance on the seen examples does not necessarily get translated into predicting the class values of the unseen (unlabelled) records of the test data set. Decision tree in particular, lacks in generalization performance. Nevertheless, different decision trees have different generalization errors. So, the combination of several decision trees can help mitigating the generalization errors of a single decision tree. As a result, generally a decision forest can deliver better prediction accuracy for unseen records [18].

In order to achieve better ensemble accuracy a decision forest needs both accurate and diverse individual decision trees as the base classifier [19], [20]. An accurate individual decision tree can be obtained by feeding a data set into a decision tree algorithm such as CART [3]. However, if all the individual decision trees generate similar classification results, there is no purpose of constructing a decision forest (as they will commit similar generalization errors). So, it is expected that in a decision forest individual decision trees would be as diverse as possible in terms of classification errors. We already know that decision trees are unstable classifiers. That is – if the training data set is slightly perturbed by adding or removing some records or attributes, the resultant decision tree can be very different from that generated from the unperturbed training data set. In literature we find many forest building algorithms that modify training data sets in different ways. We introduce some of the popular forest building algorithms as follows.

a) ***Bagging*** [21]: Bagging generates new training data set D_i iteratively where the records of D_i are chosen randomly from the original training data set D. D_i contains the same number of records as in D. Thus some records of D can be chosen multiple times and some records may not be chosen at all. This approach of generating a new training data set is known as bootstrap sampling. On an average 63.2% of the original records are present in a boot-strap sample [22]. A decision tree building algorithm is then applied on each bootstrap sample D_i (i = 1, 2, ..., k) in order to generate T number of trees for the forest.

b) **Random Subspace** [20]: The Random Subspace algorithm randomly draws a subset of attributes (subspace) f from the original attribute space m in order to determine the splitting attribute for each node splitting event of a decision tree. The Random Subspace algorithm uses the original training data set in building every decision tree.

c) **Random Forest** [23]: Random Forest is the combination of Bagging and Random Subspace algorithms. In Random Forest, the Random Subspace algorithm is applied on bootstrap samples instead of the original training data set. Random Forest is considered as an improvement over its components Bagging and Random Subspace.

In the next section we present how the size of subspace f can affect the performance of Random Forest and our motivation towards determining more appropriate size of f.

2   Motivation

The Random Forest algorithm randomly draws a subspace f from the entire attribute space m in order to determine the splitting attribute for each node of a decision tree. The individual tree accuracy and the diversity among the decision trees depend on the size of f. If f is sufficiently small then the trees tend to become less correlated thus making them more diverse. However, a sufficiently small f may not guaranty the presence of adequate number of attributes with good classification capacity which may decrease individual tree accuracy. On the other hand, the individual tree accuracy can be increased with relatively larger number of attributes in f by sacrificing diversity.

In literature a lot of variations have been noticed around the size of f [23], [18], [24], [25], [26]. In the Random Forest algorithm, the number of attributes selected for f was chosen to be $|f| = int$ ($\log_2|m|$) + 1 [23]. In both [24] and [25], $|f|$ was selected randomly between 1 to $|m|$. In the extreme cases of [24] and [25], when $|f| = 1$, the method randomly selects an attribute as the splitting attribute and when $|f| = |m|$, the entire attribute space is used as the subspace. In [18], the authors suggested to determine the most appropriate $|f|$ by calculating the forest accuracy of variously sized random subspace. They suggested that the most appropriate $|f|$ may differ for different data sets. However, we argue that this approach is very computation intensive and are not suitable for real world applications. Recently, in [26] the authors implemented $|f| = 2 \times \log_2|m|$ for Random Forest.

For both the original Random Forest and the recent variations of Random Forest, the number of attributes selected for f remains the same for the root node as well as any node down the tree. The root node of a tree contains the entire bootstrap sample

(D_i). Let us assume that we have a data set with 30 attributes. For the original Random Forest the splitting attribute is determined from randomly selected *int* ($\log_2 30$) + 1 = 5 attributes. It is reasonable that this small subset of original attributes may not contain sufficient number of attributes with good classification capacity. Therefore the root node can possibly be split by a weak attribute resulting in improper partitions. The heinous impact of improper partitioning at the beginning may not be mitigated by selecting better attributes down the tree. From the root to down the tree, the partitions (data segments) become smaller and smaller. As a result, the negative impact of selecting a weak attribute gradually decreases from the root to down the tree. On the other hand, for the recent implementation [26] the splitting attribute is determined from randomly selected $2 \times \log_2 30 = 10$ attributes. Obviously, this subset is better than the previous one to contain more attributes with good classification capacity. For this subset, we may have high chance to get better splitting attribute at the root node, yet when splitting a node down the tree the subset may be oversized. Thus the tree may get common strong attributes (as the splitting attribute) down the tree resulting in decreased diversity. Finally, selecting the number of attributes for the subspace randomly between 1 to |*m*| may not direct the characteristics of the trees in any defined path. We may assume that on an average the subspace size would be |*m*|/2. Thus the subspace size becomes larger compared to other proposed subspaces especially for high dimensional data set. Yet, selecting the size of subspace randomly between 1 to |*m*| may induce diversity as the subspace size may differ for many trees. In order to mitigate all these problems, we propose the following technique to determine the size of the subspace for Random Forest.

3   Our Technique

In our technique, the number of attributes for f is randomly selected from the following range:

$$int\,(\log_2|m|) + 1 \text{ and } e^{\frac{|D_i'|}{|D_i|}} \times (int\,(\log_2|m|) + 1)$$

Here, $|D_i'|$ denotes the number of records present in the current data segment whereas $|D_i|$ denotes the number of records present in the entire (initial) bootstrap sample (also equal to the original training data set).

In this way, we propose to determine the size of f by randomly selecting a number from a range which is calculated based on the relative size of the current data segment to the initial bootstrap sample. The upper value of our proposed range is composed of the product of two terms. The first term $e^{\frac{|D_i'|}{|D_i|}}$ establishes the influence of relative size of the current data segment to the initial bootstrap sample on the second term *int* ($\log_2|m|$) + 1 (derived from the original Random Forest algorithm). To explain how the proposed range is calculated and how the number of attributes for f is determined from the proposed range, we present the following example.

Let, |*m*| =30 and thus *int* ($\log_2|m|$) + 1 = 5. Let, $|D_i|$=1000 and at the root node D_i is divided into two subsequent partitions (data segments) D_i^1 and D_i^2. Let $|D_i^1|$=750 and $|D_i^2|$=250. For more comprehensibility, we present the all associated calculations in Table 1.

Table 1. Calculation of the Proposed Range

Current Data Segment	No. of records	$\dfrac{\|D_i'\|}{\|Di\|}$	$e^{\frac{\|D_i'\|}{\|D_i\|}}$	$int\,(\log_2\|m\|) + 1$	$e^{\frac{\|D_i'\|}{\|D_i\|}} \times (int\,(\log_2\|m\|) + 1)$
D_i	1000	1	2.71828	5	13.5914
D_i^1	750	0.75	2.11700	5	10.585
D_i^2	250	0.25	1.10517	5	5.52585

From Table 1, we see that $\|f\|$ can be any number between 5 and 14 for the root node. In the next level of the tree, when D_1^i becomes the current data segment $\|f\|$ can be any number between 5 and 11 and for D_2^i, $\|f\|$ can be any number between 5 and 6. In this way the proposed range is funneled down based on the relative size of the current data segment to the original bootstrap sample. By selecting $\|f\|$ from the proposed range we can ensure strong diversity among the trees as the size of the subspace has strong chance to be different for many trees.

Next, we present the algorithm for our technique of generating the subspace f.

Algorithm 1: Subspace Generation

Input: Initial Bootstrap Sample D_i, Original Attribute Space m, Current Data Segment D_i'.

Output: Subspace f.

Required: $f \leftarrow \emptyset$.

$lower\_range = int\,(\log_2\|m\|) + 1$;

$upper\_range = e^{\frac{\|D_i'\|}{\|D_i\|}} \times (int\,(\log_2\|m\|) + 1)$;

$num\_attr = generate\_random\_number\,(lower\_range, upper\_range)$;

for $i = 1$ to num_attr **do**

    $curr\_attr\_pos = generate\_random\_number\,(1, \|m\|)$;

    $curr\_attr = m[curr\_attr\_pos]$;   /* picks an attribute from m */

    if $curr\_attr \notin f$ **then**

        $f \leftarrow f \cup curr\_attr$;

    end if

end for

$return\ f$;

---

4   Experimental Results

We conduct an elaborated experimentation on five (05) well known data sets that are publicly available from the UCI Machine Learning Repository [27]. The data sets used in the experimentation are listed in Table 2.

Table 2. Description of the Data Sets

Data Set Name	Number of non-class Attributes		Number of Records	Domain Size of the Class Attribute
	Numerical	*Categorical*		
Breast Cancer Wisconsin	33	00	194	2
Dermatology	01	33	366	6
Glass Identification	10	00	214	7
Ionosphere	34	00	351	2
Lung Cancer	56	00	27	2

We maintain the following settings for the entire experimentation:

1. We remove missing values from each applicable data set.
2. We remove the identifier attributes such as *Transaction_ID* from each applicable data set.
3. We apply .632 bootstrap [22] to generate bootstrap samples.
4. Each leaf node of a tree contains at least two records.
5. All trees are un-pruned.
6. We generate exactly 100 trees for every decision forest.
7. We use majority voting to aggregate results for every decision forest.

The experimentation is conducted by a machine with Intel(R) 3.4 GHz processor and 4GB Main Memory (RAM) running under 64-bit Windows 8 Operating System. All the results presented in this paper are obtained using 10-fold-cross-validation (10-CV) for every data set. All the prediction accuracies and timings reported in this paper are in percentage and second respectively. The best results are emphasized through **bold-face.** In the rest of the paper, we call all the variants of the existing Random Forest according to their subspace size.

Ensemble accuracy is the most important performance indicator for any decision forest algorithm. In Table 3 we present the average ensemble accuracies for all the above mentioned variants of Random Forest for all the data sets considered.

Table 3. Average Ensemble Accuracies

Data Set Name	$int(\log_2|m|) + 1$	$2 \times \log_2|m|$	1 to $|m|$	Proposed Technique
Breast Cancer Wisconsin	77.9170	79.4950	79.4030	**81.0740**
Dermatology	86.9630	85.8210	86.7310	**87.0710**
Glass Identification	74.1150	73.6390	72.6860	**74.5910**
Ionosphere	93.7310	91.2150	93.4460	**93.7310**
Lung Cancer	68.8890	73.8890	68.8890	**73.8890**
Average	80.3230	80.8118	80.2310	**82.0712**

From Table 3 we see that our proposed technique can significantly improve the ensemble accuracy compared to other prominent variants. We also compute the prediction accuracy of each individual tree of every forest in order to compute the Average Individual Accuracy (AIA) for all the variants of Random Forest. We report the results in Table 4.

Table 4. Average Individual Accuracies

Data Set Name	$int (\log_2|m|) + 1$	$2 \times \log_2|m|$	1 to $|m|$	Proposed Technique
Breast Cancer Wisconsin	68.3911	**69.0137**	69.0060	68.5746
Dermatology	64.4539	67.7791	**69.0416**	65.7722
Glass Identification	60.5181	**62.0791**	59.3410	60.8784
Ionosphere	88.0171	**89.7960**	88.5511	88.2304
Lung Cancer	61.8891	**62.2835**	61.8891	61.8057
Average	68.6539	**70.1903**	69.5657	69.0523

From Table 4 we see that $2 \times \log_2|m|$ Random Forest has the AIA value among all other variants of Random Forest. However, Table 3 shows that the most important performance indicator ensemble accuracy is the highest for our proposed technique. In line with our previous discussions, we assume that for $2 \times \log_2|m|$ Random Forest diversity of the individual trees may decrease. To verify our assumption, we compute the diversity of a single tree using Kappa [28] as was done in literature [26]. Kappa (K) actually evaluates the level of adjusted agreement between two classifier outputs in the following way.

$$K = \frac{\Pr(a) - \Pr(e)}{1 - \Pr(e)}$$

Here $\Pr(a)$ is the observed agreement between two classifiers and $\Pr(e)$ is the random agreement. As the Kappa value indicates agreement between two classifiers the higher the Kappa value, the lower the diversity. Because the Kappa is defined for only two classifiers, we compute the Kappa for a decision tree with respect to the forest except the used tree. After computing the Kappa value for each decision tree we compute Average Individual Kappa (AIK) for the forest. We next present the Average Individual Kappa (AIK) value as the measure of diversity in Table 5.

Table 5. Average Individual Kappa

Data Set Name	$int (\log_2|m|) + 1$	$2 \times \log_2|m|$	1 to $|m|$	Proposed Technique
Breast Cancer Wisconsin	**0.0704**	0.1127	0.1608	0.1605
Dermatology	**0.5773**	0.6267	0.6410	0.5956
Glass Identification	0.5338	0.5633	**0.5173**	0.5260
Ionosphere	**0.7884**	0.8404	0.8138	0.7980
Lung Cancer	0.1590	0.1620	0.1590	**0.1030**
Average	**0.4258**	0.4610	0.4584	0.4366

We see from Table 5 that *int* ($\log_2 |m|$) + 1 Random Forest achieves the highest diversity in terms of AIK. However, $2 \times \log_2 |m|$ Random Forest has the lowest diversity. This validates our assumption that in turn negatively impacts on ensemble accuracy for $2 \times \log_2 |m|$ Random Forest. These results confirm the fact that diversity is a very important component to achieve better ensemble accuracy. For our proposed technique, both AIA and AIK had a good balance that resulted in better ensemble accuracy. We present the summary of comparative analysis among all the variants of Random Forest in Fig. 1. To make the AIK value more presentable along with ensemble accuracy value and AIA value in the figure we scale up the AIK value by multiplying it with 170.

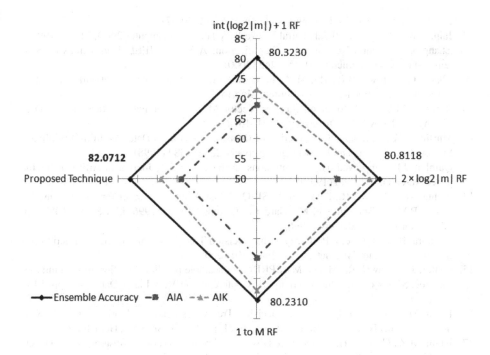

Fig. 1. Summary of Comparative Analysis

5   Conclusion

In this paper, we propose a new technique to find a suitable number of attributes for the subspace to determine the splitting attribute for each node of a decision tree based on the relative size of the current data segment to the initial bootstrap sample. The results presented in this paper show great potential of our technique. In future, we intend to extend our work by including more data sets.

References

1. EMC^2, http://www.emc.com/leadership/programs/digital-universe.htm (last accessed: June 11, 2014)
2. Tan, P., Steinbach, M., Kumar, V.: Introduction to Data Mining. Pearson Education, Boston (2006)
3. Breiman, L., Friedman, J., Olshen, R., Stone, C.: Classification and Regression Trees. Wadsworth International Group, CA (1985)
4. Quinlan, J.R.: C4.5: Programs for Machine Learning. Morgan Kaufmann Publishers, San Mateo (1993)
5. Bishop, C.M.: Pattern Recognition and Machine Learning. Springer-Verlag New York Inc., NY (2008)
6. Mitchell, T.M.: Machine Learning. McGraw-Hill, NY (1997)
7. Jain, A.K., Mao, J.: Artificial Neural Network: A Tutorial. Computer 29(3), 31–44 (1996)
8. Zhang, G.P.: Neural Networks for Classification: A Survey. IEEE Transactions on Systems, Man, and Cybernetics 30, 451–462 (2000)
9. Zhang, G., Patuwo, B.E., Hu, M.Y.: Forecasting with artificial neural networks: The state of the art. International Journal of Forecasting 14, 35–62 (1998)
10. Burges, C.J.C.: A Tutorial on Support Vector Machines for Pattern Recognition. Data Mining and Knowledge Discovery 2, 121–167 (1998)
11. Murthy, S.K.: Automatic Construction of Decision Trees from Data: A Multi-Disciplinary Survey. Data Mining and Knowledge Discovery 2, 345–389 (1998)
12. Quinlan, J.R.: Improved use of continuous attributes in C4.5. Journal of Artificial Intelligence Research 4, 77–90 (1996)
13. Mehta, M., Agrawal, R., Rissanen, J.: SLIQ: A fast scalable classifier for data mining. In: Apers, P.M.G., Bouzeghoub, M., Gardarin, G. (eds.) EDBT 1996. LNCS, vol. 1057, pp. 18–32. Springer, Heidelberg (1996)
14. Chandra, B., Varghese, P.P.: Fuzzy SLIQ Decision Tree Algorithm. IEEE Transactions on Systems, Man, and Cybernetics 38, 1294–1301 (2008)
15. Shafer, J., Agrawal, R., Mehta, M.: SPRINT: A scalable parallel classifier for data mining. In: Proceedings of the 22nd International Conference on Very Large Databases, pp. 544–555 (1996)
16. Islam, M.Z.: EXPLORE: A Novel Decision Tree Classification Algorithm. In: MacKinnon, L.M. (ed.) BNCOD 2010. LNCS, vol. 6121, pp. 55–71. Springer, Heidelberg (2012)
17. Islam, M.Z., Giggins, H.: Knowledge Discovery through SysFor – a Systematically Developed Forest of Multiple Decision Trees. In: Proceedings of the 9th Australian Data Mining Conference, Ballarat, Australia, pp. 195–204 (2011)
18. Bryll, R., Osuna, R.G., Quek, F.: Attribute Bagging: improving accuracy of classifier ensembles by using random feature subsets. Pattern Recognition 36, 1291–1302 (2003)
19. Polikar, R.: Ensemble Based Systems in Decision Making. IEEE Circuits and Systems Magazine 6, 21–45 (2006)
20. Ho, T.K.: The Random Subspace Method for Constructing Decision Forests. IEEE Transaction on Pattern Analysis and Machine Intelligence 20(1), 832–844 (1998)
21. Breiman, L.: Bagging predictors. Machine Learning 24(2), 123–140 (1996)
22. Han, J., Kamber, M.: Data Mining Concepts and Techniques, 2nd edn. Morgan Kaufmann, San Francisco (2006)
23. Breiman, L.: Random Forests. Machine Learning 45(1), 5–32 (2001)
24. Geurts, P., Ernst, D., Wehenkel, L.: Extremely randomized trees. Machine Learning 63, 3–42 (2006)

25. Bernard, S., Heutte, L., Adam, S.: Forest-RK: A New Random Forest Induction Method. In: Huang, D.-S., Wunsch II, D.C., Levine, D.S., Jo, K.-H. (eds.) ICIC 2008. LNCS (LNAI), vol. 5227, pp. 430–437. Springer, Heidelberg (2008)
26. Amasyali, M.F., Ersoy, O.K.: Classifier Ensembles with the Extended Space Forest. IEEE Transaction on Knowledge and Data Engineering 26, 549–562 (2014)
27. UCI Machine Learning Repository,
    http://archive.ics.uci.edu/ml/datasets.html
    (last accessed: Feb 15, 2014)
28. Margineantu, D.D., Dietterich, T.G.: Pruning Adaptive Boosting. In: Proceedings of 14th International Conference on Machine Learning, pp. 211–218 (1997)

Dimensionally Reduction:
An Experimental Study

Xin Song[1], Zhihua Cai[1], and Wei Xu[2]

[1] Dept. of Computer Science,
China University of Geosciences, Wuhan, 430074 China
[2] School of Information,
Renmin University of China, Beijing, 100872, China
songxin066@gmail.com, zhcai@cug.edu.cn, weixu@ruc.edu.cn

Abstract. Dimensionality reduction as the available method to overcome the "curses of dimensionality" has attracted wide attention. However, the pervious studies treat the visualization and the subsequent classification performance separately. In this case, we do not know whether there is a underlying relationship (*i.e.,* direct proportion) between visualization and the followed classification performance.

In this paper we compare several dimensionality reduction techniques on three different types of data sets: 1) Benchmark, 2) Image, and 3) Text data. Specifically, to intuitively evaluate the quality of the dimension reduced data, the visualization analysis is carried out, in which we use a covariance matrix related criteria to quantify the information content of the data. Moreover, we also consider the classification accuracy in different latent spaces (*i.e.,* dimensions) as another performance criterion to further analysis the relationship between the information content and the subsequent classification task. The experimental results show that there is no direct proportion relationship between the information content and the further classification performance.

Keywords: dimensionality reduction, visualization, classification.

1 Introduction

Dimensionality reduction approaches, of which the overall framework is shown in Figure 1, aim to find the low dimensional representation of the data sets in high dimensional space without significant information loss. Dimensionality reduction facilitates the classification, visualization, communication, and storage of high dimensional data and has become the most crucial pre-processing steps in data analysis. Traditionally, dimensionality reduction methods can be divided into two categories, which are linear and non-linear states.

A simple and widely used method is Principle Components Analysis [1] which aims to maximizing variance of the projected low dimensional data. Other linear dimensionality reduction methods, such as Classical Multidimensional Scaling [2], Linear Discriminant Analysis (LDA) [3], are also aim to learn a linear mapping from high dimensional observation to the lower dimension space.

X. Luo, J.X. Yu, and Z. Li (Eds.): ADMA 2014, LNAI 8933, pp. 380–393, 2014.
© Springer International Publishing Switzerland 2014

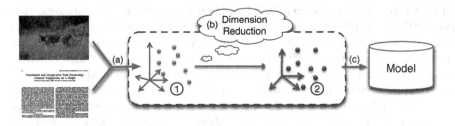

Fig. 1. A conceptual view of dimensionality reduction: The overall framework is to find optimal feature set ② to re-represent the original data ① in the new low-dimensional space by maintaining the most information content.

However, there are always complicated non-linear structures in high dimensional data sets. To drive this challenge, a large number of non-linear dimensionality reduction techniques have been proposed, such as Isometric Mapping (ISOMAP) [4], Maximum Variance Unfolding (MVU) [5], Locally Linear Embedding (LLE) [6], Stochastic Neighbor Embedding (SNE) [7], Gaussian Process Latent Variable Model (GPLVM) [8].

Fodor described several dimensionality reduction methods from those mentioned above in [9]. However, there illustrations are only on text description level, without any experiment. Maaten presented a review and comparative study of techniques for dimensionality reduction [10]. Although an empirical comparison of the classification performance of the techniques for dimensionality reduction is performed, the experimental data mainly forces on artificial data sets with a small quantity of natural data sets. Moveover, it lacks of the analysis of the visualization for each dimensionality reduction approach on different data sets.

According to the experimental results on dimensionality reduction in [11], the author argues that despite the strong performance of those previous techniques (*i.e.,* PCA, LLE, SNE et. al) on artificial data sets, they are often not very successful at visualizing real, high dimensional data. In particular, most of the techniques are not capable of retaining both the local and the global structure of the data in a single map. Motivated by the above discussion, a new technique called t-distributed Stochastic Neighbor Embedding (t-SNE) [11], a variation of SNE, produces significantly better visualizations by reducing the tendency to crowd points together in the center of the map.

The above t-SNE method for dimensionality reduction has achieved good performance on visualization. However, for its classification performance receives very little attention, conversely for the pervious review study in [10]. In this case, there is no comprehensive understanding on that whether there is a underlying relationship (*i.e.,* direct proportion) between visualization and the followed classification performance. In other words, does the good visualization can bring to the superiority on classification results?

In this paper, we choose two linear methods (PCA and LDA) and three nonlinear methods (LLE, SNE, and t-SNE) as the selected techniques to be further analyzed. Then those dimensionality reduction techniques are applied to three different kinds of data sets, including benchmark, image, and text data sets. More specifically, to intuitively evaluate the quality of the dimension reduced data, the visualization analysis is firstly carried out. And then we use a covariance matrix related criteria to quantify the information content of the data. Moreover, we also consider the classification accuracy in different latent spaces (*i.e.*, dimensions) as another performance criteria to further analysis the relationship between the information content (*i.e.*, visualization) and the subsequent classification task. The experimental results show that though t-SNE presents the superiority on visualization compared with other four dimensionality reduction algorithms, it performs not good performance during classification process. In other words, there is no direct proportion relationship between visualization and the further classification performance. At last, we systematically analyze the time complexity of each dimensionality reduction approach.

The rest of the paper is organized as follows. In section 2 to 6 we briefly review the selected dimensionality reduction methods. Section 7 presents an empirical comparisons , followed by the conclusions in Section 8.

2   Principal Components Analysis

Principal Components Analysis (PCA) [1] is a most frequently used linear technique for dimensionality reduction. Its objective is to map high dimensional data to a linear subspace of lower dimensionality, in which the amount of variance in the data is maximal. Therefore, the characteristics of the original data can be preserved as much as possible by using less dimension. It is demonstrable that the loss of the original information content is less than other linear dimensionality reduction method. However, PCA doesn't make an attempt on exploring internal data structure.

Throughout the whole paper, $X = [x_1, ..., x_N]^T$ are defined to represent observed data in the high dimensional space R^D. $Y = [y_1, ..., y_N]^T$ are data in the low dimensional space R^d with $d \ll D$. Hence, we can treat X as an $N \times D$ matrix and Y an $N \times d$ matrix.

Mathematically, PCA attempts to find a linear mapping W that maximizes the cost function trace.

$$\max tr(W^T cov(X)W), s.t. W^T W = I \tag{1}$$

where $cov(X)$ is the covariance matrix of the data X. $W^T W = I$ means that each feature is orthogonal, so that there is no redundant information between each dimension. The optimal W is composed with d eigenvectors, which corresponds to top-d eigenvalues of the data covariance matrix. These eigenvectors form a set of orthogonal basis and best keep the information of the original data.

The low-dimensional data representations Y of the data points X are computed by mapping them onto the linear basis W

$$Y = W'X \tag{2}$$

3   Linear Discriminant Analysis

Linear Discriminant Analysis (LDA)[3], also known as Fisher Linear Discriminant, is a supervised linear dimensionality reduction technique. Unlike PCA trying to preserve information of observed data, LDA finds the low dimensional data that best discriminate among classes. For all samples, the between-class scatter matrix S_b and the within-class scatter matrix S_w are defined by:

$$S_b = \sum_{i=1}^{c} N_i \cdot (\mu_i - \mu) \cdot (\mu_i - \mu)^T \tag{3}$$

$$S_w = \sum_{i=1}^{c} \sum_{x \in X_i} (x - \mu_i) \cdot (x - \mu_i)^T \tag{4}$$

$$S_b w_i = \lambda S_w w_i \tag{5}$$

where c is the number of classes, N_i is the number of data points in class i, μ_i is the mean vector of samples belonging to class i, μ is the global mean and X_i represents the ith class in the dataset.

Eq. (5) is also a generalized eigenvalue problem, which is similar to PCA. The optimal W is composed with d eigenvectors, which corresponds to top-d eigenvalues of the data covariance matrix. The low dimensional data can be calculated according to Eq. (2). In this processing, the only difference with PCA is the calculation of W.

4   Local Linear Embedding

Local Linear Embedding (LLE) [6] is a non-linear technique which constructs a graph representation of the data points and attempts to preserve solely local properties of the data. Because of the property of neighborhood preservation, LLE is widely used in classification and clustering of image data, character recognition, multi-dimensional visualization of data and bioinformatics.

The procedures of LLE can be listed as follow: 1) LLE tries to find the k nearst neighbors of each data point x_i and writes the data point as a linear combination w_{ij}(the so-called reconstruction weights) of those neighbors; 2) Reconstruction errors are measured by the following error function.

$$\psi(W) = \sum_{i=1}^{N} |x_i - \sum_{j=1}^{k} w_{ij} x_j|^2 \quad s.t. \quad \sum_{j=1}^{k} w_{ij} = 1 \tag{6}$$

By minimizing Eq. (6), the weight w_{ij} that combined with the neighbors best approximate each data point x_i can be computed; 3) The last step is mapping each data point x_i into the low dimensional space, such that the cost function is minimized as

$$\phi(Y) = \sum_{i=1}^{N} |y_i - \sum_{j=1}^{k} w_{ij} y_j|^2 \quad s.t. \quad \frac{1}{N} \sum_{i=1}^{N} y_i y_i^T = I, \sum_{i=1}^{N} y_i = 0 \tag{7}$$

5   Stochastic Neighbor Embedding

The main idea behind Stochastic Neighbor Embedding (SNE) [7] is to preserve neighbor similarities of high dimensional space in low dimensional space. In high dimensional space, the asymmetric probability $p_{i|j}$ is

$$p_{i|j} = \frac{exp(-\parallel x_i - x_j \parallel^2 / 2\sigma_i^2)}{\sum_{k \neq i} exp(-\parallel x_i - x_k \parallel^2 / 2\sigma_i^2)} \tag{8}$$

where σ_i is a Gaussian kernel parameter. SNE does a binary search for the value of σ_i which makes the entropy of distribution over neighbors equal to $\log k$. Here k is "perplexity" which is specified by users. In low dimensional space, σ is set to a fixed value $1/\sqrt{2}$. It is described by

$$q_{i|j} = \frac{exp(-\parallel y_i - y_j \parallel^2)}{\sum_{k \neq i} exp(-\parallel y_i - y_k \parallel^2)} \tag{9}$$

SNE tries to minimize the dissimilarities between the probabilities p_{ij} and q_{ij}. The Kullback-Leibler divergence exactly calculates the difference between two probability. Then, the cost function is given by

$$J = \sum_i \sum_j p_{j|i} \log \frac{p_{j|i}}{q_{j|i}} \tag{10}$$

The minimization of the cost function in Eq. (10) is performed using gradient descent approach. The gradient is given by

$$\frac{\partial J}{\partial y_i} = 2 \sum_j (y_i - y_j)(p_{j|i} - q_{j|i} + p_{i|j} - q_{i|j}) \tag{11}$$

To be stuck in local optima, a momentum term is added to the equation. A exponentially decay of the sum of previous gradients is also added for determining the changes in the coordinates of the map points at each iteration of gradient search. Mathematically, the gradient with a momentum term is given by

$$y^{(t)} = y^{(t-1)} + \eta \frac{\partial J}{\partial y_i} + \alpha(t)(y^{(t-1)} - y^{(t-2)}) \tag{12}$$

Where $y^{(t)}$ indicates the solution at iteration t, η indicates the learning rate, and $\alpha(t)$ represents the momentum at iteration t.

6   t-Distributed Stochastic Neighbor Embedding

Although SNE gets good visualizations, a so-called "crowding problem" may occured on some data sets, which will leads to dissimilar points gathering together in the center of the map. Therefore, in [11] van der Maaten and Hinton presented a technique called t-SNE, which is a variation of SNE considering Student's t-distribution with one degree of freedom for data distribution.

The t-SNE extends SNE in two ways: 1)A symmetric joint probability distribution is defined in high dimensional space, which leads to simpler gradient computation in optimization. Equation 9 is used to form this joint probabilities.

$$p_{ij} = \frac{p_{j|i} + p_{i|j}}{2n} \tag{13}$$

where $p_{ii} = 0$ and $\sum_{i,j} p_{ij} = 1$; 2) A Student's t-distribution with one degree of freedom is used in low dimensional space rather than a Gaussian distribution, which can avoid the "crowd problem". The joint probabilities in low dimensional space is defined as

$$q_{ij} = \frac{(1 + \parallel y_i - y_j \parallel^2)^{-1}}{\sum_{k \neq l}(1 + \parallel y_k - y_l \parallel^2)^{-1}} \tag{14}$$

Then, Eq. (14) and Eq. (15) are taken into the Kullback-Leibler divergence to find the optimal low dimensional embedding.

$$C = KL(P \parallel Q) = \sum_{i \neq j} p_{ij} \log \frac{p_{ij}}{q_{ij}} \tag{15}$$

Since a symmetric joint probability distribution is defined in high dimensional space, and a student's t-distribution is used in low dimensional space, the gradient is simpler.

$$\frac{\delta C}{\delta y_i} = 4 \sum_j (p_{ij} - q_{ij})(1 + \parallel y_i - y_j \parallel^2)^{-1}(y_i - y_j) \tag{16}$$

Aside from the basic gradient descent, a number of strategies are employed to improve optimization [11].

7   Experiments

7.1   Data Sets

Benchmark Data: We first compare the dimensionality reduction algorithms on four benchmark data sets, including one "oil flow" data [12,13] and other three UCI data sets ("ionosphere","colic.ORIG", and "wine")[1]. The "oil flow" data set, which has 3 phases of flow (*i.e*, 3 classes) and 1000 samples, does the measurements of oil flow within a pipeline via 12 dimensions. In general, the "ionosphere" data set (with 351 instances and 35 attributes) describes a binary classification task. Meanwhile, the horse data set "colic.ORIG" (with 368 instances and 27 attributes) is used to predict the fate of a horse with colic. And the "wine" data has three types of wine with 178 instances.

---

[1] https://archive.ics.uci.edu/ml/s/

Image Data: The "segment" data set (2310 instances and 19 attributes) [1],which were drawn randomly from a database of 7 outdoor images. The original "mnist" data set[2] has 60,000 handwritten with 28×28 pixels (*i.e.*, 784-dimensional space) for comparison purpose. Besides, the original "USPS" data set consists of 7291 training and 2007 test examples. Moreover,the "India Pine" [14] is a type of remote sensing image data with 200 bands (*i.e.*, attributes), and 21025 pixels (*i.e.*, instances). In this experiment, we randomly selected 1000 instances of these four data sets.

Text Data: Furthermore, we also use four text data sets ("oh0","oh5","oh10", and "oh15") to do the further analysis. Specifically, these data sets are from OHSUMED colection [15] with "oh0" (1003 docs and 3182 words), "oh5" (918 docs and 3012 words), "oh10" (1050 docs and 3238 words), and "oh15" (913 docs and 3100 words). Among them, doc represents the instances with word representing the attribute (*i.e.*, dimension) of data.

7.2   Experimental Setup

For the purpose of evaluating the performance of the selected representative dimensionality reduction algorithms (PCA, LLE, SNE, t-SNE, and LDA), of which the versions are taken form the matlab toolbox for dimensionality reduction[3] with default parameter setting, we first carry out a visualization analysis to intuitively evaluate the quality of the data via different dimensionality reduction approaches. And then, we formulate the quality of the data using the trace of covariance matrix. According to our knowledge, most of the dimensionality reduction approaches try to maintain the information content of the original data. However, the subsequent tasks (*i.e.*, classification) is more important than just considering the quality of the dimension-reduced data. In our experiment, we also consider the classification accuracy in different latent spaces (*i.e.*, dimensions) as another performance criteria, in which a very popular classifier LibSVM [16] is employed to do the subsequent classification. Finally, the time complexity of each selected dimensionality reduction algorithm is also executed.

7.3   Visualization Analysis

In this subsection, Figures 2-4 report the visualization of "oil" benchmark data set, "usps" image data, and "oh0" text data, respectively. For the oil benchmark data as shown in Figure 2, the separation between the classes is almost perfect via t-SNE (Figure 2(d)) and LDA (Figure 2(e)). In other words, these two approaches have very good visualization performance on "oil" data. The LLE approach has the worst visualization performance. Considering the "usps" image data from Figure 3, t-SNE (Figure 3(d)) and LDA (Figure 3(e)) also show the

---

[2] http://yann.lecun.com/exdb/mnist/index.html
[3] http://homepage.tudelft.nl/19j49/Matlab_Toolbox_
for_Dimensionality_Reduction.html

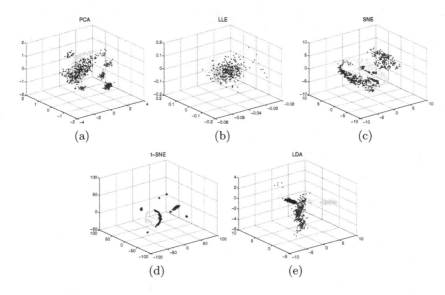

Fig. 2. Visualization of **oil benchmark data set** via different algorithms (a) PCA, (b) LLE, (c) SNE, (d) t-SNE and (e) LDA

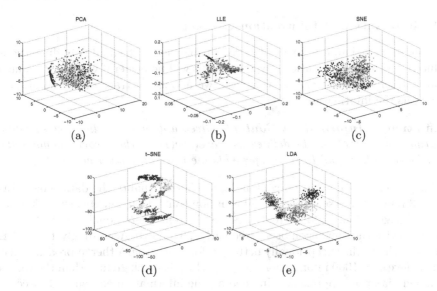

Fig. 3. Visualization of **usps image data set** via different algorithms (a) PCA, (b) LLE, (c) SNE, (d) t-SNE and (e) LDA

better visualization performance than the other three approaches. When it comes to "oh0" text data, which can be considered as a sparse matrix, t-SNE presents the superiority compared with other four dimensionality reduction algorithms.

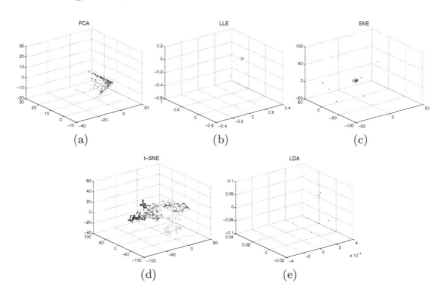

Fig. 4. Visualization of **oh0 image data set** via different algorithms (a) PCA, (b) LLE, (c) SNE, (d) t-SNE and (e) LDA

7.4   Reservation of Information Content

As been discussed above, we can do a rough analysis on the quality of dimension-reduced data via visualization. By contrast, it will be better if we can do a much more subtler measurement. In this case, we adopt the trace of the covariance matrix as the information content.

Definition 1. *Information Content: Given a data set D, the corresponding information content can be defined as $trace(cov(D))$, where $cov(\cdot)$ denotes the covariance matrix, and $trace(\cdot)$ represents the trace of a given matrix.*

The larger the information content is, the better quality the data shows. Tables 1-3 report the reservation for information content on "oil" benchmark data, "usps" image data and "oh0" text data via different dimensionality reduction algorithms with the dimension been fixed with 1, 5, and 10. Accordingly, the t-SNE approach shows the exceptionally better performance than other approaches as it can achieve over 1000 information content, which is much greater than the values obtained via other approaches. In general, the information content will become larger when the dimension of the data increases. For instances, the information content of PCA increase is from 13.987 to 49.513 on "oh0" text data. Naturally, one may argue that does the t-SNE with the best information content (visualization performance) will correspond to the high classification accuracy? According to the following experimental part about the classification performance of the each dimensionality reduction algorithms, the answer is no.

Table 1. Reservation for information content (d=1)

	PCA	LLE	SNE	t-SNE	LDA
oil (Benchmark)	1.004	0.000	51.919	2731.075	16.503
usps (Image)	15.994	0.001	13.028	1635.889	15.193
oh0 (Text)	13.987	0.025	523.213	1406.195	0.000

Table 2. Reservation for information content (d=5)

	PCA	LLE	SNE	t-SNE	LDA
oil (Benchmark)	2.423	0.006	22.827	2363.235	26.542
usps (Image)	35.776	0.005	17.832	3202.030	40.042
oh0 (Text)	37.301	0.125	38.012	2231.455	0.026

Table 3. Reservation for information content (d=10)

	PCA	LLE	SNE	t-SNE	LDA
oil (Benchmark)	2.588	0.013	15.885	2706.304	31.527
usps (Image)	46.135	0.010	15.274	3299.663	52.324
oh0 (Text)	49.513	0.250	42.458	2265.463	0.056

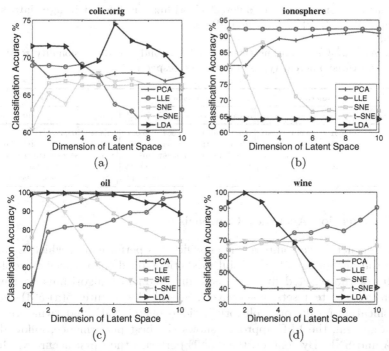

Fig. 5. Classification accuracy of different algorithms on different **benchmark data sets**: (a) colic.orig, (b) ionosphere, (c) oil and (d) wine

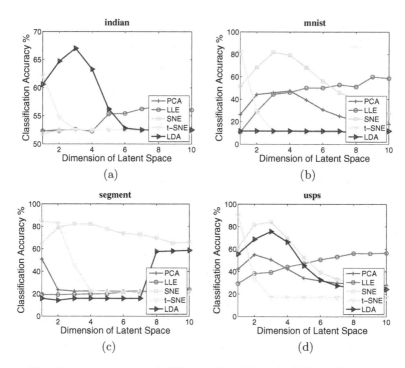

Fig. 6. Classification accuracy of different algorithms on different **image data sets**: (a) indian, (b) mnist, (c) segment and (d) usps

Table 4. Average CPU runtime (measured by second) for dimensionality reduction algorithms on three different types of data sets, respectively

	Benchmark				Image				Text			
	colic.orig	ionosphere	oil	wine	indian	mnist	segment	usps	oh0	oh5	oh10	oh15
PCA	0.001	0.001	0.001	0.001	0.021	0.490	0.001	0.034	1.174	0.965	1.318	0.960
LLE	0.036	0.068	0.274	0.045	0.474	0.748	1.944	0.567	0.404	0.337	0.465	0.339
SNE	17.453	16.683	107.104	6.144	99.678	107.643	106.467	108.153	112.683	87.700	110.782	96.234
t-SNE	5.193	3.560	40.326	1.253	41.549	40.473	40.560	40.155	42.081	35.206	46.359	35.196
LDA	0.002	0.012	0.001	0.001	0.030	3.167	0.002	0.048	321.317	271.010	338.463	293.751

7.5   Classification Accuracy Comparison

In our experiments, we compare the classification performance, which is calculated by the percentage of successful predictions on domain specific problems [17,18,19], of the selected five dimensionality reduction algorithms on benchmark, image, and text sets, respectively. Specifically, Figures 5(a)-5(d) report the classification results on "colic.orig", "ionosphere","oil", and "wine' respectively. In general, the LLE approach shows the best performance, followed by the PCA and SNE. By contact, the t-SNE performs the worst accuracy. Moreover, when the dimension increases, the accuracy of some approaches decreases. This is mainly because that the high dimension may contain the redundant

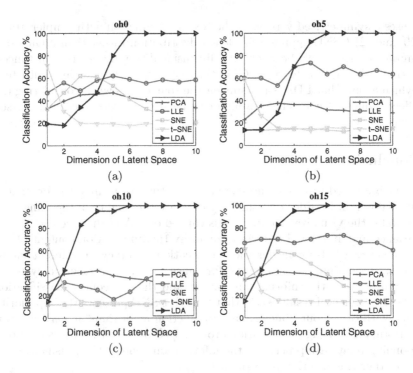

Fig. 7. Classification accuracy of different algorithms on different **text data sets**: (a) oh0, (b) oh5, (c) oh10 and (d) oh15

information, which can lead to the worse classification accuracy. For image data sets, the SNE performs the best in most cases. However, the method with superiority will become LDA when facing the text data. Notice that, the LDA algorithm will achieve the 100% accuracy, which mean that LDA has the perfect classification performance when dealing with the text data (*i.e.*, sparse data).

In summary, t-SNE could not achieve a high classification performance on all the three types of data sets (especially on the real-world image and text data sets), through it has a very good visualization performance.

7.6   Time Complexity

In this subsection, we compare the time complexity for each dimensionality reduction algorithm to systematically analyze its efficiency. Table 4 reports the average CPU runtime (with dimension from 1 to 10) for dimensionality reduction algorithms on three different types of data sets, respectively. For benchmark data sets in low-dimensional space, all the dimensionality reduction approaches have good runtime performance. Among them SNE needs the most CPU time to proceed the dimensionality reduction. The same observation could also be found in the other two types of image and text data sets. Its variation t-SNE shows better runtime performance than SNE, which is mainly attributed to: (1)

t-SNE uses a symmetrized version of the SNE cost function with simpler gradients [20] and (2) t-SNE applies a Student-t distribution rather than a Gaussian to compute the similarity between two dimensionality reduced data instances. Notice that, LDA takes the most CPU time ($\geq 250s$) when dealing with text data, which means that LDA is not suit for the sparse data when considering the time effect thought it can achieve a high classification performance as discussed in Section 7.5.

8   Conclusions

The paper presented a review and comparative study of techniques for dimensionality reduction. In order to intuitively evaluate the quality of the dimension reduced data, the visualization analysis is carried out. And then we use a trace of covariance matrix related criteria to quantify the information content of the data. Moreover, we also consider the classification accuracy in different latent spaces (*i.e.*, dimensions) as another performance criteria to further analysis the relationship between the information content and the subsequent classification task. The experimental results show that though t-SNE presents the superiority compared with other four dimensionality reduction algorithms, it performs not good performance during classification process. In other words, there is no direct corresponding relationship between the information content (*i.e.*, visualization) and the further classification performance.

Acknowledgments. The work was supported by the Key Project of the Natural Science Foundation of Hubei Province, China (Grant No. 2013CFA004), and the National Scholarship for Building High Level Universities, China Scholarship Council (No. 201206410056)and Chinese National "111" Research Project "Attracting International Talents in Data Engineering and Knowledge Engineering Research" of the Renmin University, China, and National Natural Science Foundation of China (Grant No.71001103, 91224008, 91324015), and Humanities and Social Sciences Foundation of the Ministry of Education (No. 14YJA630075).

References

1. Hotelling, H.: Analysis of a complex of statistical variables into principal components. J. Educ. Psych. 24 (1933)
2. Torgerson, W.: Multidimensional scaling: I. theory and method. Psychometrika 17(4), 401–419 (1952)
3. Belhumeur, P.N., Hespanha, J.P., Kriegman, D.J.: Eigenfaces vs. fisherfaces: Recognition using class specific linear projection. IEEE Trans. Pattern Anal. Mach. Intell. 19(7), 711–720 (1997)
4. Tenenbaum, J.B., Silva, V., Langford, J.C.: A Global Geometric Framework for Nonlinear Dimensionality Reduction. Science 290, 2319–2323 (2000)
5. Weinberger, K.Q., Sha, F., Saul, L.K.: Learning a kernel matrix for nonlinear dimensionality reduction. In: Proceedings of the Twenty-First International Conference on Machine Learning, ICML 2004, pp. 839–846 (2004)

6. Roweis, S.T., Saul, L.K.: Nonlinear dimensionality reduction by locally linear embedding. SCIENCE 290, 2323–2326 (2000)
7. Hinton, G.E., Roweis, S.T.: Stochastic neighbor embedding. In: Advances in Neural Information Processing Systems 15 Neural Information Processing Systems, NIPS 2002, Vancouver, British Columbia, Canada, December 9-14, pp. 833–840 (2002)
8. Lawrence, N.: Probabilistic non-linear principal component analysis with gaussian process latent variable models. J. Mach. Learn. Res. 6, 1783–1816 (2005)
9. Fodor, I.K.: A survey of dimension reduction techniques (2002)
10. van der Maaten, L., Postma, E.O., van den Herik, H.J.: Dimensionality reduction: A comparative review (2008)
11. van der Maaten, L., Hinton, G.: Visualizing data using t-sne. J. Mach. Learn. Res. 9, 2579–2605 (2008)
12. Jiang, X., Gao, J., Hong, X., Cai, Z.: Gaussian processes autoencoder for dimensionality reduction. In: Tseng, V.S., Ho, T.B., Zhou, Z.-H., Chen, A.L.P., Kao, H.-Y. (eds.) PAKDD 2014, Part II. LNCS, vol. 8444, pp. 62–73. Springer, Heidelberg (2014)
13. Jiang, X., Gao, J., Wang, T., Zheng, L.: Supervised latent linear gaussian process latent variable model for dimensionality reduction. IEEE Transactions on Systems, Man, and Cybernetics, Part B 42(6), 1620–1632 (2012)
14. Arzuaga-Cruz, E., Jimenez-Rodriguez, L.O., Velez-Reyes, M., Kaeli, D., Rodriguez-Diaz, E., Velazquez-Santana, H.T., Castrodad-Carrau, A., Santos-Campis, L.E., Santiago, C.: A matlab toolbox for hyperspectral image analysis. In: Proceedings 2004 IEEE International Geoscience and Remote Sensing Symposium, IGARSS 2004, vol. 7, pp. 4839–4842 (September 2004)
15. Hersh, W., Buckley, C., Leone, T.J., Hickam, D.: Ohsumed: An interactive retrieval evaluation and new large test collection for research. In: Proceedings of the 17th Annual International ACM SIGIR Conference on Research and Development in Information Retrieval, SIGIR 1994, pp. 192–201. Springer-Verlag New York, Inc., New York (1994)
16. Chang, C.-C., Lin, C.-J.: Libsvm: A library for support vector machines. ACM Trans. Intell. Syst. Technol. 2(3), 27:1–27:27 (2011)
17. Wu, J., Zhu, X., Zhang, C., Yu, P.S.: Bag constrained structure pattern mining for multi-graph classification. IEEE Transactions on Knowledge and Data Engineering 26(10), 2382–2396 (2014)
18. Wu, J., Cai, Z., Zhu, X.: Self-adaptive probability estimation for naive bayes classification. In: IJCNN, pp. 1–8 (2013)
19. Wu, J., Cai, Z.: Attribute weighting via differential evolution algorithm for attribute weighted naive bayes (wnb). Journal of Computational Information Systems 7(5), 1672–1679 (2011)
20. Cook, J., Sutskever, I., Mnih, A., Hinton, G.: Visualizing similarity data with a mixture of maps. In: Society for Artificial Intelligence and Statistics AI and Statistics (2007)

Efficient Deep Learning Algorithm with Accelerating Inference Strategy

Junjie Wang[1,2] and Xiaolong Zhang[1,2]

[1] School of Computer Science and Technology,
Wuhan University of Science and Technology, Wuhan 430065, China
[2] Intelligent Information Processing and Real-time Industrial Systems Hubei Province
Key Laboratory, Wuhan 430065, China
wangjunjie_wust@163.com, xiaolong.zhang@wust.edu.cn

Abstract. In this paper, we present an efficient learning algorithm for Deep Boltzmann Machine (DBM) to get the data-dependent expectation quickly. The algorithm adopts a layer-wise accelerating inference strategy to compute the mean values of all hidden layers, instead of the mean values by repeatedly running the equations of mean-field fixed-point until convergence. By taking advantage of layer-wise inference strategy, we can rapidly get the approximate mean values in a few iterations. This strategy also could learn efficiently a high performance model for high-dimensional high-structured sensory inputs. The proposed algorithm with layer-wise accelerating inference performs well compared to original DBM with given learning tasks.

Keywords: Layer-wise, Accelerating inference strategy, Deep learning.

1 Introduction

Hinton [1, 2] (2006) in DBN (Deep Belief Network) algorithm introduced a multi-hidden-layers network that had an excellent ability to learn the characteristics of input data and could effectively approximate the complex problem or function. He also proposed a fast unsupervised layer-wise training approach to overcome the difficulty of training deep neural network, which improved the variational lower bound of multilayer model based on likelihood's probability and produced the model parameters in accordance with the characteristics of the data. After the pretraining stage, the deep network was composed of model parameters produced by layer-by-layer training, then the entire model was globally fine-tuned. However, this DBN algorithm only adopted bottom-up pass to approximate inference, without considering the top-down influence. The structure of model may lead to fail in disposing the uncertain ambiguous sensory inputs. The wake-sleep algorithm [6] was used to overcome this defect, but it was time-consuming and inefficient.

Salakhutdinov [5, 8] proposed a deep structure of the neural network with all undirected connections, instead of DBN (only the top two layers invoked undirected connection, the rest of the lower layers of the structure invoked top-bottom connections),

X. Luo, J.X. Yu, and Z. Li (Eds.): ADMA 2014, LNAI 8933, pp. 394–405, 2014.
© Springer International Publishing Switzerland 2014

called DBM (Deep Boltzmann Machine). In the pretraining phase, usual bottom-up inference process was added to the feedback of top-down pass, which made DBM better learn uncertainly fuzzy inputs and produce better model of data than the DBN. But the approximate inference process of DBM using the mean-field equations (about 25-50 times to convergence) was much slower than DBN using only a single bottom-up to infer. The training of DBM would be very slow on large scale dataset.

Larochelle [6] presented an approximate inference method called DBM_recnet to overcome the slow convergence problem of DBM inferring mean values, with a separate recognition model. In the DBM_recnet algorithm, recognition model used only a single bottom-up operation to quickly initialize the values of all the hidden layers, then applied the single top-down operation to update recognition model to approximate the mean values, so that the process was relatively much quicker and learned better discriminative model than DBM. However, in the case of multi-layer neural network, DBM_recnet may not produce better recognition model by the feedback of a single top-down operation. Therefore we make use of the features of the model parameters' initialization in layer-wise pretraining process. In addition, we propose the layer-wise accelerating inference strategy that effectively avoids one time's feedback trapped in the local minimum in multiple layers, and get better generative model than that generated by DBM_recnet.

2   Deep Boltzmann Machine

Deep Boltzmann Machine [5] is composed of multi-layers with symmetric random binary neuron units. It contains one input layer V(visible layer) with input neuron units $v \in \{0,1\}$, and a series of hidden-layers with hidden neuron units $h^1 \in \{0,1\}$, $h^2 \in \{0,1\}$, ..., $h^l \in \{0,1\}$. Connection of neuron units only existed in adjacent layers. The DBM network structure is shown as Fig. 1.

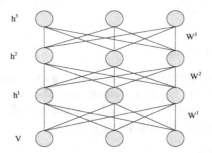

Fig. 1. Structure of Deep Boltzmann Machine (DBM)

W=$\{W^1, W^2, W^3\}$ is a set of model parameters of DBM, where W^1 is weight matrix of neuron units between the visible layer V and the first hidden layer h^1, and W^2, W^3 are respectively the weight matrix of neuron units of hidden layers between h^1 and h^2, and hidden layers between h^2 and h^3. The energy of DBM is defined as:

$$E(v, h, W) = -V^T W^1 h^1 - h^1 W^2 h^2 - h^2 W^3 h^3 \tag{1}$$

where v is the states of visible units, h= $\{h^1, h^2, h^3\}$ is a set of states of three hidden layers' units. The probability that the DBM assigns to a visible vector v is:

$$P(v;W) = \frac{P^*(v;W)}{Z(W)} = \frac{1}{Z(W)} \sum_h \exp(-E(v,h,W)) \tag{2}$$

where $Z(W) = \sum_v \sum_h \exp(-E(v,h,W))$ is the partition function.

In DBM, the conditional distribution of the hidden layers' units and the visible layer's units can be obtained by the logistic function as follows:

$$p(h_j^1 = 1|v, h^2) = \sigma(\sum_i W_{ij}^1 V_i + \sum_k W_{kj}^2 h_k^2) \tag{3}$$

$$p(h_k^2 = 1|h^1, h^3) = \sigma(\sum_j W_{jk}^1 h_j^1 + \sum_m W_{km}^3 h_m^3) \tag{4}$$

$$p(h_m^3 = 1|h^2) = \sigma(\sum_k W_{km}^3 h_k^2) \tag{5}$$

$$p(v_i = 1|h^1) = \sigma(\sum_j W_{ij}^1 h_j^1) \tag{6}$$

where $\sigma = \frac{1}{1 + e^{-(x)}}$ is a logistic function

If we set the values of the model parameters W^2 and W^3 are all zeros, the model recovers the Restrict Boltzmann Machine (RBM) model [3, 7]. If we set the direction of two lower layers are top-down directed connections, the model recovers the Deep Belief Network (DBN) model.

2.1   Approximate Learning

The main goal in the pretraining stage is maximizing the value of P (v;W) that generates the model efficiently to learn the training sets. The max value of P (v;W) is obtained by the derivative of P (v;W). The log-likelihood of P (v;W) with respect to weight matrix W^1 takes the following form:

$$\frac{\partial \log(P(v;W))}{\partial W^1} = E < vh^{1^T} >_{data} - E < vh^{1^T} >_{\text{model}} \tag{7}$$

where $E < vh^{1^T} >_{data}$ denotes the expectation of completed data distribution $< vh^{1^T} >_{data} = P(h/v;W)P(v)$ with $P(v) = (1/N)\sum_n \delta(v - v_n)$ representing the empirical distribution, $E < vh^{1^T} >_{model}$ is the expectation of generative model designed by $< vh^{1^T} >_{model}$. The log-likelihoods with respect to parameters W^2 and W^3 take similar procedure instead of the involved dataset $< h^1 h^{2^T} >$ and $< h^2 h^{3^T} >$ respectively.

The cost time of exact computation of the data-dependent expectation is $O(2^{hidnums})$, and the cost time of exact computation of the model expectation is $O(2^{hidnums+visnums})$, the exact maximum likelihood learning is expensive and complex.

In DBM, a mean-field inference is used to approximately estimate the expectation of entire data $< vh^{1^T} >_{data}$ and Markov Chain Monte Carlo (MCMC) approximate stochastic approach is used to estimate the expectation of model $< vh^{1^T} >_{model}$[4]. And $p(h/v;W)$ denotes the true posterior distribution over all hidden units for the input data v, and any approximate the posterior distribution is denoted as $Q(h/v;W)$. The variational lower bound of the DBM model's log-probability is shown as follows:

$$\log P(v;W) \geq \sum_h Q(h\,|\,v;\mu) \log P(v,h,W) + H(Q)$$

$$=\mathrm{logP}(v;W) - \mathrm{KL}(Q(h\,|\,v;\mu)\,\|\,P(h\,|\,v;W)) \tag{8}$$

where H() is the entropy function. If and only if $Q(h|v;\mu) = P(h|v;W)$ the lower bound of the model can be maximized. The true posterior distribution is approximated by approximating a fully factorized distribution over all the hidden layers: $Q(h\,|\,v;\mu) = \prod_{j=1}^{h1num} \prod_{l=1}^{h2num} \prod_{k=1}^{h3num} q(h_j^1\,|\,v) q(h_l^2\,|\,v) q(h_k^3\,|\,v)$, with $q(h_i^l = 1) = \mu_i^l$ for l=1,2,3 where $\mu = \{\mu^1, \mu^2, \mu^3\}$ is the mean-field parameters called mean values. The lower bound on the log-probability of the data takes the form:

$$\log P(v;W) \geq v^T w^1 \mu^1 + \mu^{1^T} w^2 \mu^2 + \mu^{2^T} w^3 \mu^3 - \log Z(W) + H(Q) \tag{9}$$

From the above inequality, μ_i^l can maximize the lower bound of the model with respect to the current model parameters W, which should satisfy the conditions of mean-field fixed-point equations:

$$\mu_j^1 \leftarrow \sigma(\sum_i W_{ij}^1 V_i + \sum_k W_{jk}^2 \mu_k^2) \tag{10}$$

$$\mu_k^2 \leftarrow \sigma(\sum_j W_{jk}^1 \mu_j^1 + \sum_m W_{km}^3 \mu_m^3) \tag{11}$$

$$\mu_m^3 \leftarrow \sigma(\sum_k W_{km}^3 \mu_k^2) \tag{12}$$

2.2   Greedy Layer-Wise Pretraining of DBM

Hinton introduced a greedy, layer-by-layer unsupervised algorithm that contains a list of RBMs [9, 11]. After the greedy learning, the weights form the single overall model parameters, and then fine-tuning the parameters on the entire model. As shown in Fig. 2, DBN's model contains top two undirected connection layers and the rest of the lower layers which are top-bottom connections formed directed generative model.

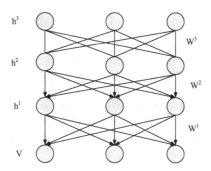

Fig. 2. Structure of Deep Belief Network (DBN)

After learning the first RBM in DBN, the generative model can be written as:

$$P(v;w) = \sum_{h^1} P(h^1;W^1)P(v \mid h^1;W^1) \tag{13}$$

where $P(h^1;W^1) = \sum_{v} P(h^1,v;W^1)$, if the second RBM is initialized correctly, $P(h^1;W^2) = \sum_{h^2} P(h^1,h^2;W^1)$ substitutes for $P(h^1;W^1)$ to make the entire model efficient and accurate, and it is the reason that the lower level layers are directed graph in the structure of DBN as shown in the Fig. 2. In DBM, $P(h^1;W^2)$ is also considered to produce better models than $P(h^1;W^1)$. However, the DBM adopts averaging the two models of h^1, which can be composed with 1/2 bottom-up and 1/2 top-down to infer $P(h^1;W^1,W^2)$ approximately. Since h^2 is also generated from V, in which means that there are two evidences to verify the probability of h^1, so DBM is more robust and better performed than DBN.

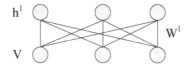

Fig. 3. DBM layer-by-layer training (input layer-first hidden layer)

For example, the DBM of two hidden layers, greedy layer-wise pretraining is to initialize model parameters of DBM. We double the input data V to compensate for the lack of values of top-down pass at the beginning of learning. The model's parameter W^1 is updated by Contrastive Divergence learning [3]. The initialization of values on V and h^1 in Fig. 3 is defined as:

$$p(h^1_j=1|v)=\sigma(\sum_i W^1_{ij}V_i + \sum_i W^1_{ij}V_i) \tag{14}$$

$$p(v_i=1|h^1)=\sigma(\sum_j W^1_{ij}h_j) \tag{15}$$

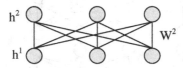

Fig. 4. DBM layer-by-layer training (first hidden layer-the second layer)

After learning the first RBM, we can get the states of the first hidden layer, which can be treated as the raw data of the second RBM. We double the top-down level of the second RBM to compensate h^1 without bottom-up pass. The conditional distributions for this model in Fig. 4 take the form:

$$p(h^2_m=1|h^1)=\sigma(\sum_j W^1_{jm}h_j) \tag{16}$$

$$p(h^1_j=1|h^2)=\sigma(\sum_m W^1_{jm}h^2_m + \sum_m W^1_{jm}h^2_m) \tag{17}$$

After layer-by-layer learning, the above two models form a single model, which is defined as:

$$p(h^1_j=1|v,h^2)=\sigma(\sum_j W^1_{ij}V_i + \sum_k W^2_{kj}h^2) \tag{18}$$

3   Layer-Wise Accelerating Inference Strategy

In DBM pretraining process, the main problem is to maximize the approximate learning to get a model in line with the characteristics of the data, which requires mean-field inference for updating the parameters of DBM. But in the mean-field inference process, the average induce local domain of each neuron by the mean-field fixed-point equations replace the random fluctuations of the neuron units, which requires mean values converging to the steady states, and convergence requires continuous loop iterations. So the computation of updating DBM parameters by invoking the mean-field fixed-point equations is more expensive than DBN using only a single bottom-up to infer especially on large amount of data. It is necessary to accelerate the inference.

In the process of getting the mean values, two problems should be noticed: First, there are existed mapping relationships between the inputs V and the mean values called u, which can be denoted as f (V, W)=u. We should not change the model parameters W to obtain mean values in the existing model. Then, if there is nothing to change, we only make use of mean-field fixed-point equations to update mean values until convergence, but the process is expensive and should be avoided. Second, after the stochastic approximation procedure (SAP) [10], the model parameters are updated, and the mean values should be changed dynamically as the model parameters are updated.

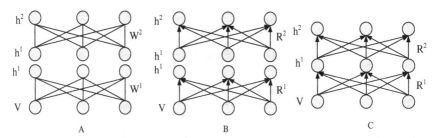

Fig. 5. A) DBM pretraining structure; B) A separate recognition model with layer-wise accelerating inference; C) A separate recognition with globally fine-tuning

Responding to the above problems, DBM_recnet algorithm proposed a separate recognition model as shown in Fig.5C. The recognition model used only a single bottom-up operation to approximate the mean values. However, if neural network is multi-layer, DBM_recnet may not produce better recognition model by the feedback of a single top-down operation. The single feedback may be trapped in the local minimum and the recognition weights are updated inaccurately to make the recognition model invalid. Thus, we establish a simple separate model like DBM_recnet to accelerate the inference of mean values.

Compared to DBM_recnet with a simple globally inference, we propose layer-wise accelerating inference strategy to approximate more accurately and quickly the mean values. There are two reasons to show the advantages of our strategy. First, the structure of layer-wise accelerating inference is similar to the structure of DBM pretraining as shown in the Fig. 5 (A and B). We can make use of the characteristic of weights updated by layer to adjust the each recognition weight. Making the each layer's prediction effectively approximate the each layer's mean values, we can obtain relative accurate initial recognition model. Second, if the layers are too many, use the only simple globally inference to guess the prediction of mean values, then make the prediction close to the target of K-steps' mean-field inference, finally globally fine-tune with using the back-propagation algorithm one time, which may lead to be trapped in the local minimum. To avoid this situation, our strategy with layer by layer fine-tuning the recognition weight can progressively minimize the error between prediction and target as shown in Fig. 5B, then globally fine-tune as shown in Fig. 5C makes the prediction more precisely approximate the target, and obtains faithful recognition model.

After the stage of DBM pretraining as shown in Fig. 5A, we get the parameters of model $\{W^1, W^2\}$. A simple layer-wise accelerating inference strategy is presented as follows.

Firstly, in the stage of the first layer recognition model learning, given a visible vector V, the recognition weight R^1 of the first layer that initialized by W^1, is used to compute the initial guess h^1 of the approximating posterior $p(h^1|v)$ by the equation 19. Then we apply K cycles' mean-field fixed-point equation, initialized by $u^1 = h^1$, to obtain the new mean-field parameter u^1. Finally, make use of the cross-entropy error between u^1 and h^1 to update the first recognition weight R^1. The cross-entropy error is defined in the equation 22. The gradient of cross-entropy error (u^1, h^1) with respect to the first layer recognition R^1, can be computed by back-propagation algorithm. Because there is only one layer, back-propagation of error can be corrected very quickly.

$$p(h^1_j = 1) = \sigma(\sum_i 2R^1_{ij}V_i) \tag{19}$$

$$h^1_j = \sigma(\sum_i 2R^1_{ij}V_i) \tag{20}$$

$$h^2_l = \sigma(\sum_j R^2_{jl}h_j) \tag{21}$$

$$\text{cross-entropy}(u^1, h^1) = [-\sum_j h_j \log(u_j) - \sum_j (1-h_j)\log(1-u_j)] \tag{22}$$

Secondly, after updating the first layer recognition model R^1, we can get the new parameter R^1. In the same way, the second recognition weight R^2 is initialized by $R^2 = W^2$, and the first hidden h^1 is initialized by the equation 20 with updated R^1. We can compute the initial guess h^2 of the approximating posterior $p(h^2|h^1)$ with the equation 21. The remaining operations are the same with process of the first layer recognition model.

Finally, after obtaining each layer's new recognition model parameters $\{R^1, R^2\}$, the overall recognition model can be simply adjusted. Each layer's guess (h^1, h^2) of approximating posterior distribution can be computed by the equation 20 and 21. Then make use of K (K=1 or 5) steps' mean-field fixed-point equations, with initialization of $u^1 = h^1$ and $u^2 = h^2$, to obtain the new entire mean values (u^1, u^2). At last, apply the cross-entropy error to update the whole recognition parameters by minimizing the cross-entropy error with the back-propagation algorithm.

The above process is layer-wise accelerating inference strategy. Although we add some operations, we can approximate effectively mean values and produce better desired model in the fewer iterations of mean-field inference.

The detail of the efficient deep learning algorithm with accelerating inference strategy, which can be called as LAIDBM (Layer-wise Accelerating Inference Deep Boltzmann Machine), is presented as follows:

Layer-wise Accelerating Inference Deep Boltzmann Machine Algorithm

Inputs: training set $<V_1, V_2, ..., V_n>$ as input-data

Steps :
1. Greedy layer-wise DBM pretraining uses Equations(14-17) to get the initial weights $W=\{W^1, W^2\}$
2. The initialization of SAP's process and Recognition model: randomly initialize the M fantasy particles$\{v^F, h^F\}$,where $h^F=\{h^{F1}, h^{F2}\}$;initialize the recognition model parameters R=W,where $R=\{R^1, R^2\}$;

For minibatch-data in input-data

    Data=minibatch-data

3. Layer-wise accelerating inference strategy
    For i=1 to hidden_layer_num

        Using Recognition model R^i and Data to obtain the prediction h^i.

        Set $u^i=h^i$, run the mean-field equation for K-iterations to obtain new approximate mean value u^i.

        Adjust the R^i by taking the gradient steps: $R^i = R^i + \alpha \dfrac{\partial cross\_entory(u^i, h^i)}{\partial R^i}$.

        Update h^i by using the new recognition model R^i.

        Data =h^i.

    End for i;

4. Globally fine-tune the   recognition weight
    Get the new prediction $h=\{h^1, h^2\}$ after the above process, then set u=h, run the mean-field equations to obtain new approximate mean value u where $u=\{u^1, u^2\}$;finally, adjust R by taking the equation as follows:

$$R = R + \alpha \frac{\partial cross\_entory(u, h)}{\partial R}$$

5. SAP process
Using the Gibbs Sampler to get new fantasy particles $\{v^F, h^F\}$

6. Update the DBM's parameter

$$W^1 = W^1 + \alpha(\frac{1}{N}\sum v(u^1)^T - \frac{1}{M}\sum v^F(h^{F1})^T)$$

$$W^2 = W^2 + \alpha(\frac{1}{N}\sum u^1(u^2)^T - \frac{1}{M}\sum h^{F1}(h^{F2})^T)$$

End for minibatch-data

Outputs: The model parameter of LAIDBM $\{W^1, W^2\}$

4   Experimental Results

In the experiments we use two common datasets: the mnist handwritten digits dataset and OCR English letters dataset, and report the classification performance compared to DBM, DBM_recnet and other algorithms. To speed up learning, the datasets are divided into mini-batches, each containing 100 cases, and the parameters of model are updated after each mini-batch.

The mnist dataset contains 60000 training and 10000 testing images of handwritten digits (0 to 9), and each data is 28*28 pixels image. OCR English letters dataset contains 42152 training and 10000 testing binary pixels images of 26 English letters (a to z), with 16*8 pixels.

The experiment adopt the same structure with DBN, DBM, and DBM_recnet algorithm to compare with those algorithm. The structure of DBM on mnist dataset is 784-500-1000-10, which contains the input layer(visible layer) with 784 visible units, two hidden layers with 500 and 1000 hidden units and the last layer(output layer) with 10 units that are numbers of classification. The procedure of the algorithm can be divided into three stages: the stage of RBM's layer-wise training, the stage of DBM integration and the stage of classified fine-tuning. The iterations of training in each stage are 100 times. The structure of OCR dataset's experiment is 128-2000-2000-26, which has the same process with mnist's experiment.

Table 1. Classification error rate on the test set for different inference strategies

	mnist	ocr_letter
MF-0	1.38%	8.68%
MF-1	1.15%	8.54%
recnet_1	1.01%	8.45%
LAIDBM_1	0.96%	8.43%
MF-5	1.01%	8.50%
recnet_5	0.96%	9.39%
LAIDBM _5	0.87%	8.25%
MF-Full	0.95%	8.58%

To compare the different inference approaches, we train several DBMs using different inference procedures. The algorithm in this paper improves on the basis of DBN, DBM and DBM_recnet algorithm. Our first two DBMs using layer-wise accelerating strategy with the numbers of running mean_field steps set to 1 and 5. These models are called as LAIDBM _1 and LAIDBM _5. Our second two DBMs using a separate recognition model with iterations of mean_field set to 1 and 5. We call these models recnet_1 and recnet_5. Our third DBMs are original without the recognition model. We call these models MF_1, MF_5, MF_0, MF_Full with the underline indicated the steps of mean_field. MF_Full means that the mean_field's updates will run until convergence (25 iterations). MF_0 means approximate inference do not apply the mean_field inference.

Table 2. Iterations of training on the train set to reach the classification performance

	mnist	ocr_letter
DBM	40 (0.95%)	71 (8.50%)
recnet	28 (0.95%)	100 (8.45%)
LAIDBM	16 (0.87%)	83 (8.25%)

Table 3. Classification error rate on the test set for different structures

	mnist	ocr_letter
DBN	1.17%	9.68%
SVM	1.40%	9.70%
RBM	14.00%	33.47%
BPNN	4.00%	13.59%
LAIDBM _5	0.87%	8.25%

From Table 1, our algorithm (LAIDBM) with 5 iterations of mean_field on the mnist dataset achieve the best test error rate of 0.87%, and perform better than DBM, DBM_recnet under the same conditions. On the ocr_letter dataset our algorithm also achieved the minimum error rate of 8.25%. The experiment verified that the layer-wise accelerating inference method can effectively improve the DBM's model and achieve better results. From Table 2, our algorithm can reach the classification performance in a few of iterations. On the mnist dataset, LAIDBM reach the minimum error rate by 16 iterations of training. The iterations of training are fewer than the iterations of DBM and DBM_recnet. On the ocr_letter dataset, our algorithms achieve the best performance by training 83 times, which is better than DBM_recnet. From Table 3, compared with SVM, RBM, BPNN (Back Propagation Neural Network with one hidden layer) shallow networks, the algorithm proposed in this paper which enhances the deep structure of DBM is more effective in the classification tasks.

5   Conclusion

The algorithm with layer-wise accelerating inference strategy can efficiently improve the ability of computing mean values, which is better than the original DBM with carrying out the mean-field fixed-point equations until the mean values are convergence. The used accelerating inference strategy speeds up the inference of the mean values, allowing for our algorithm to perform well on large scale dataset. The strategy makes back propagation be out of the local minimum, the recognition weights be updated accurately. The experiment results demonstrate that our algorithm (LAIDBM) can efficiently learn the model and has advantage over the exist algorithm on the mnist and ocr_letter classification tasks.

Acknowledgements. This work was supported in part by National Natural Science Foundation of China (61273225, 61273303, 61373109), the Project (2009CDA) from Hubei Provincial Natural Science Foundation, P.R.China, Program for Outstanding Young Science and Technology Innovation Teams in Higher Education Institutions of Hubei Province (No.T201202), and the Program of Wuhan Subject Chief Scientist (201150530152), as well as National "Twelfth Five-Year" Plan for Science & Technology Support (2012BAC22B01).

References

1. Hinton, G.E., Salakhutdinov, R.R.: Reducing the dimensionality of data with neural networks. Science 313(5786), 504–507 (2006)
2. Hinton, G.E., Osindero, S., Teh, Y.W.: A fast learning algorithm for deep belief nets. Neural Computation 18(7), 1527–1554 (2006)
3. Hinton, G.E.: A practical guide to training restricted boltzmann machines. In: Montavon, G., Orr, G.B., Müller, K.-R. (eds.) Neural Networks: Tricks of the Trade, 2nd edn. LNCS, vol. 7700, pp. 599–619. Springer, Heidelberg (2012)
4. Salakhutdinov, R., Hinton, G.E.: A Better Way to Pretrain Deep Boltzmann Machines. NIPS 3, 2456–2464 (2012)
5. Salakhutdinov, R., Hinton, G.: An efficient learning procedure for deep Boltzmann machines. Neural Computation 24(8), 1967–2006 (2012)
6. Salakhutdinov, R., Larochelle, H.: Efficient learning of deep Boltzmann machines. In: International Conference on Artificial Intelligence and Statistics, vol. 9, pp. 693–700 (2010)
7. Nair V, Hinton G E.: Implicit mixtures of restricted Boltzmann machines. Advances in Neural Information Processing Systems, pp. 1145-1152 (2009)
8. Salakhutdinov, R., Hinton, G.E.: Deep boltzmann machines. In: International Conference on Artificial Intelligence and Statistics, vol. 9, pp. 448–455 (2009)
9. Nair, V., Hinton, G.E.: 3D object recognition with deep belief nets. In: Advances in Neural Information Processing Systems, pp. 1339–1347 (2009)
10. Salakhutdinov, R.: Learning deep Boltzmann machines using adaptive MCMC. In: Proceedings of the 27th International Conference on Machine Learning (ICML 2010), pp. 943–950 (2010)
11. Erhan, D., Bengio, Y., Courville, A., et al.: Why does unsupervised pre-training help deep learning? The Journal of Machine Learning Research 11, 625–660 (2010)

Multi-angle Evaluations of Test Cases Based on Dynamic Analysis

Tao Hu and Tu Peng

School of Software, Beijing Institute of Technology, China
tsyj.hu@gmail.com

Abstract. This paper presents dynamic analysis of test cases. By software mining, we get dynamic call tree to reproduce the dynamic function calling relations of test cases and static call graph to describe the static calling relations. Based on graph analysis, we define some related testing models to evaluate the test cases with the execution of software. Compared with the models of evaluating test cases in static analysis, the models given in this paper can be used on large-scale software systems and the quantization can be completed automatically. Experiments prove that these models of dynamic analysis have an excellent performance in improving testing efficiency and also build a foundation of quantization for the management, selection, evaluation of capability to find software defects of test cases. Even more critical is that they can indicate the directions of improvement and management of the test for testers.

Keywords: software testing, evaluations of test cases, dynamic analysis.

1   Introduction

Software testing is an important method of software quality assurance [1]. Reasonable test cases will be the foundation of an efficient testing. With the increment of software scale, software testing becomes more complicated, and the number of test cases increases further. However, the increased number of test cases doesn't always mean the increment of testing efficiency, because of the redundant cases, which make the testing more expensive. Software testing focuses not only on how many software defects were found, and test manager need to do more about the management of test cases. Here are some problems in front of the test managers. First, how to evaluate and compare these test cases becomes a crucial problem of the management of test cases. Second, when the cost of testing is not sufficient to pay all the test cases sets, how to select the efficient sets for testing is also a crucial problem. Third, the test efficiency of a test case set can also be increased, even though the set has achieved its testing effect. But there are not explicit rules on how to improve the quality of test cases sets.

In a word, the above problems all focus on the accurate evaluations of test cases and sets, namely the value of test cases. There is no clear, unified definition of software testing value, but some studies have presented related concepts, such as the calculation of coverage rate in white-box testing, and test case effectiveness [12].

X. Luo, J.X. Yu, and Z. Li (Eds.): ADMA 2014, LNAI 8933, pp. 406–420, 2014.
© Springer International Publishing Switzerland 2014

However, we still lack deeper analysis on test cases, and most present models of evaluating test cases are based on static analysis, such as the design phase of test cases.

In the design phase of test cases for white-box testing, testers should use static trace to look up the codes and use related testing criteria to evaluate test cases, such as statement coverage rate, prime path coverage rate. In static analysis phase, these testing criteria also express the thought that the higher the coverage rate is, the more valuable a test case is. But static analysis of testing value, such as white-box testing, has some shortages. For example, it is very expensive, it can only be used in a small module, the coverage rate is difficult to calculate, and it has to depend on human intervention. Apart from those, static analysis of testing value focus more on testing criteria, and two test cases sets with the equal coverage rates will be regarded as the same. However, different test cases may have different executions, which mean different test effects.

In order to solve the above problems, this paper presents some dynamic models of evaluating the test cases with the execution of software. In contrast with the above static analysis, these models are called dynamic analysis of test cases. Because of various requirements of testing, the testing criteria are also different, such as white-box testing including many code coverage criteria [10], and this paper also defines a series of testing criteria, which are suitable for dynamic analysis. The application, in section 4 of the paper, illustrates that these models can achieve multi-angle measurements and accurate evaluations of test cases and can further build the foundation of the management of test cases. In addition, compared with static analysis, dynamic analysis has the following advantages, such as quantization, automation, wide application, high efficiency, the ability to indicate the directions of improvement and management of test cases.

The rest of the paper is organized as follows: Section 2 is an overview of our approach. Section 3 defines static call graph and dynamic call graph, and the transformation of dynamic call graph to tree. Section 4 introduces testing criteria models of dynamic analysis as well as related applications. Section 5 introduces related work. Finally, section 6 concludes this paper and proposes extensions.

2   Approach Overview

The approach overview is shown in Fig.1. Our implementation includes five modules: Egypt, AspectC++, Miner, Visualizer and Data_Analyser. Egypt [12] is used to get static function calling relations from codes, just like Source Insight or Codeviz. As for dynamic call tree, there is no present tool to make it, so we choose AspectC++ [11], a kind of aspect oriented language. AspectC++ can weave scrutiny codes into the source codes, which is a kind of program instrumenter. Once the mixed codes and test cases are put into execution, the function calls are recorded in execution trace. However, the raw calling information represents a graph, and a digger is needed to analyze and transform the raw information into another kind, which represents a tree. The output of digger is written in dot language in both static and dynamic methods. Visualizer is implemented to translate these information into a graph or a tree, and

Graphviz is an important part of the Visualizer. At this time, we complete software mining and the further work is graph mining. Based on the information dug from the dynamic call tree and static call graph, we define testing criteria and models. All these information including related graph, tree and testing models, will be put into Data_Analyser. After the calculations, Data_Analyser automatically outputs the results which we use to evaluate test cases.

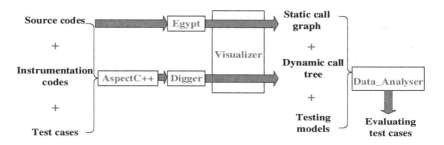

Fig. 1. Approach Overview

Fig.2 shows a static call graph of a program, and Fig.4 show the dynamic call trees of the same program under four test cases. Fig.3 shows a transformation of dynamic call graph to tree.

3 Software Mining

Definition 1. (*Static Call Graph*) A static call graph $G<V, E>$ is a directed graph. Each node represents a function. $E \subset V \times V$, $p \in V, q \in V$, edge $e = (p,q) \in E$ represents that function q is invoked by function p.

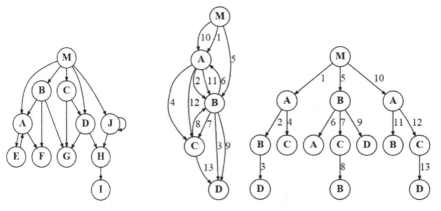

Fig. 2. Static Call Graph   **Fig. 3.** Dynamic Call Graph and Tree

As it is shown in Fig.2, static call graph doesn't contain information about control flow. Each directed edge which can be regarded as a rule of calling. Each node is labeled with a unique function name, and the node labeled "M", whose in-degree are zero, represents function "main" (the same in other figures). From Fig.2, We can know the number of functions in a program, as well as "longest static path" (it will be introduced next).

Definition 2. (*Dynamic Call Graph*) A dynamic call graph $G<V, E, O>$ is a labeled, directed, graph. Each node $v \in V$ represents a function. $E \subset V \times V$, edge $e = (p,q) \in E$ represents that function q is invoked by function p. $O \subset E \times \{0,1,2,3...\}$ and $O(e)=t$, where t is a non-negative calling number and represents the temporal order of the invocation of function q. (Peng has defined dynamic call graph[13].)

In order to separate "longest dynamic path" (it will be introduced next) and improve analysis efficiency, we transformed dynamic call graph to tree.

Definition 3. (*Dynamic Call Tree*) A dynamic call tree $T<V, E, O>$ is a labeled, undirected, connected graph without simple cycles. Each node $v \in V$ represents a function. $E \subset V \times V$ and edge $e = (p,q) \in E$ represents that function q is invoked by function p. $O \subset E \times \{0,1,2,3...\}$ and $O(e)=t$, where t is a non-negative calling number and represents the temporal order of the invocation of function q. (Peng has proposed the transformation of graph to tree [13].)

It is noticeable that different nodes which are labeled with the same string represent the same function.

Fig.3 shows a transformation of dynamic call graph to tree. Fig.4 show the dynamic call trees. From these trees, we can recognize the caller and callee, the order of each call. We can see direct recursion $(J \rightarrow J)$ in Fig.4(a) and simple indirect recursion $(A \rightarrow E \rightarrow A \rightarrow E)$ in Fig. 4(c).

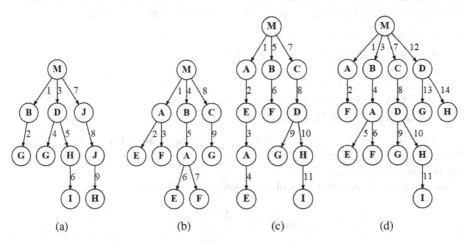

(a)   (b)   (c)   (d)

Fig. 4. Dynamic Call Tree

4   Dynamic Analysis

In this section, in order to achieve multi-angle measurements and evaluations of test cases, we will define some testing criteria suitable for dynamic analysis. For convenience, we will use the labels: TC_4, TC_5, TC_6, and TC_7 to represent the four test cases in Fig.4.

4.1   Testing Criteria about Function Coverage

Definition 4. *(Function Coverage)*: Assume that X is the set of function names of a dynamic call tree and Y is a similar set of a corresponding static call graph.
Let $x = |X|$, $y = |Y|$ then the function coverage rate of the test case is $x/y * 100\%$.

The performance of function coverage in a large-scale system is just like statement coverage of white-box testing in a small module. The former is based on functions, but the latter is based on statements.

Apparently, the higher the function coverage rate is, the higher the likelihood of finding software defects.

Applications:
I. Act as a Kind of Selection Technique to Improve the Efficiency of Regression test. Many effective approaches have been presented to improve the efficiency of regression test, such as regression test selection techniques [2][3]. We can record the functions invoked in the execution of a test case and attach them to the test report. According to these records and software modified log, we easily know whether a function in a test case is modified to decide whether the test case is needed to be re-tested in regression test. In other words, we can use this model to reduce the test cases of regression test without changing the anticipant testing effect.

II. Act as an Evaluation of Test Cases. We can calculate the function coverage rate of single test case, and the rate is an evaluation of the testing execution.

List 1:   The functions sets of Fig.2, 4
    ALL:{ M, A, B, C, D, E, F, G, H, I, J}
    TC_4:{ M, B, D, G, H, I, J}
    TC_5:{ M, A, B, C, E, F, G}
    TC_6:{ M, A, B, C, D, E, F, G, H, I}
    TC_7:{ M, A, B, C, D, E, F, G, H, I}

So the function coverage rates of the four test cases are 63.6%, 63.6%, 90.9%, 90.9%. Apparently, test case 6 or 7 has higher coverage rate, so we can roughly think that they are more valuable.

Jaccard Formula is used to calculate the similarity of two sets:

$$sim(X,Y) = \frac{X \cap Y}{X \cup Y}$$

Then the similarity degree of test case 4 and 5 is the lowest, 27.3%, which means test case 4 and 5 focus on testing different units, but the similarity degree of test case 6 and 7 is 100%. So we can place test case 4 and 5 in the same test case set to achieve a high coverage rate. While using the similarity model, we can set two proper thresholds to help us manage the test cases, and more details of this will be described in the usage of "Degree of Execution Difference" (in Definition 14).

III. Select a Set of Test Cases to Achieve the Highest Coverage Rate

a) Requirements Description

We now have n test cases, and we have got n dynamic call trees. Now we need to choose k test cases to integrate into a test set, and the problem is which cases we should choose to achieve the highest coverage rates. Apparently, this method will be a guide to manage test cases.

b) Abstraction Description

Input: A number k and a collection of sets $S = S_1, S_2, ..., S_n$

Objective: Find a subset $S' \subseteq S$ of sets, such that $|S'| \leq k$ and the number of covered

elements $\left| \bigcup_{S_i \in S'} S_i \right|$ is maximized.

This problem can also be formulated in integer linear program [11]. It belongs to Maximum coverage problem [11] which is NP-hard [5].

c) Algorithm Description

The greedy algorithm for maximum coverage chooses sets according to one rule: at each stage, choose a set which contains the largest number of uncovered elements. It can be shown that this algorithm achieves an approximation ratio of $1-1/e$ [4]. Inapproximability results show that the greedy algorithm is essentially the best-possible polynomial time approximation algorithm for maximum coverage [5].

In this problem, a basic element of set is a function name, a string, so the comparison between two elements will be expensive. Some pretreatment work need to be done, where we use a unique integer Id to represent a function. More details about this algorithm can be found in [4][5].

d) Implementation

From **List 1**, the four test cases TC_4, TC_5, TC_6, TC_7, $n = 4$:

i) $k = 1$: $\{TC_6\}$ and $\{TC_7\}$ can achieve the highest rate 90.9%.

ii) $k = 2$: $\{TC_4, TC_5\}$, $\{TC_4, TC_6\}$, $\{TC_4, TC_7\}$ can achieve the highest rate 100%.

Depending on practical requirements, we can control the size of test cases sets and the coverage rate by setting proper k.

4.2   Testing Criteria about Longest Path Coverage

Definition 5. (*Longest Static Path*): $G(V,E)$ is a static call graph, P is a sequence: $v_0, e_1, v_1, e_2, ... e_k, v_k$, edge $e_i = (v_{i-1}, v_i)$, $v_i \in V, e_i \in E$

If P subjects to the following three rules:

   i) Node v_0 represents function "main"

   ii) If $i \neq j$, then $v_i \neq v_j, e_i \neq e_j$

   iii) $\forall e_{k+1} \in E$, if $e_{k+1} = (v_k, v_{k+1})$, then $v_{k+1} \in \{v_0, v_1, ... v_k\}$

Then we call P a "Longest Static Path" of the static call graph G and the length is $L(P) = |\{e_1, e_2, ... e_k\}| = k$. We can also use its logogram "LSP" for convenience. In addition, a static call graph may have more than one LSP and, from rule 3, we can find that every LSP is unique and is not sub-path of another LSP. For example, in Fig. 4(a), paths $M \rightarrow B \rightarrow G$ and $M \rightarrow D \rightarrow G$ are all LSP. Obviously, when these three rules are satisfied, a LSP will contain a node only once, and not contain loop circuit, which are useful to cut some redundant paths especially when there is recursion in the program. For example, in Fig. 4(a), there are direct recursion ($J \rightarrow J$) about which the path $M \rightarrow J \rightarrow H$ is LSP, and simple indirect recursion ($A \rightarrow E \rightarrow A \rightarrow E$) in Fig. 4(c), about which path $M \rightarrow A \rightarrow E$ is LSP. In a static call graph, the start of a LSP is function "main". After the simplification, we can get the exact number of LSP in a static call graph, and the reason is similar to that we choose prime path testing rather than path testing in white-box testing.

List 2: The whole 12 LSPs in Fig.2:

   1) $M \rightarrow B \rightarrow A \rightarrow F$ 2) $M \rightarrow A \rightarrow E$ 3) $M \rightarrow A \rightarrow F$ 4) $M \rightarrow B \rightarrow F$
   5) $M \rightarrow B \rightarrow A \rightarrow E$ 6) $M \rightarrow B \rightarrow G$ 7) $M \rightarrow C \rightarrow G$ 8) $M \rightarrow D \rightarrow G$
   9) $M \rightarrow C \rightarrow D \rightarrow G$ 10) $M \rightarrow D \rightarrow H \rightarrow I$
   11) $M \rightarrow J \rightarrow H \rightarrow I$ 12) $M \rightarrow C \rightarrow D \rightarrow H \rightarrow I$

Definition 6. (*Longest Dynamic Path*): $T(V,E,O)$ is a dynamic call tree. P is a path of T from root node to a leaf node. After cutting down all the identical subsequences of P, then we get P'. It has a node sequence $(v_0, v_1, ... v_k)$ and every node v_i has a unique name which represents a function and v_0 represents function "main". We call P' a "Longest Dynamic Path" of dynamic call tree T and the length $L(P') = |\{v_0, v_1, ... v_k\}| - 1 = k$ is the number of edges of P'. We can also use its logogram "LDP" for convenience.

   For example, in Fig. 4(a), $P(M \rightarrow J \rightarrow J \rightarrow H)$ is a raw path, and its simplification $P'(M \rightarrow J \rightarrow H)$ is a LDP. In Fig. 4(c), $P'(M \rightarrow A \rightarrow E)$ rather than $P(M \rightarrow A \rightarrow E \rightarrow A \rightarrow E)$ is a LDP.

Definition 7. (*Valid LDP*): Let X is the set of LDPs of a dynamic call tree, Y is the set of LSPs of the corresponding static call graph. Let $x \in X$, $y \in Y$, if $x = y$, then y has an image x in X and y is covered. We call x a valid LDP.

In Fig. 4(a), path (M →D →H →I) is a valid LDP because there is a same LSP in the static call graph, however the path (M →J →H) is not. The valid LDPs of each test case will be seen in *List 3*.

Definition 8. (*Longest Path Coverage Rate*): Assume that there are x valid LDPs in a dynamic call tree (the same valid LDPs are counted only once), and y LSPs in a static call graph. The longest path coverage rate of the test case is $x/y*100\%$.

The performance of longest path coverage in a large-scale system is just like prime path coverage in white-box testing. In the design phase of prime path coverage, we can know how many prime paths should be designed by calculating the circuit complexity, but sometimes the prime paths containing logic errors therefore some of them won't be cover. So does longest path coverage. That is to say, longest path coverage rate may not reach 100% in some programs.

Applications

I. Act as an Evaluation of Test Cases

We can evaluate single test case or test cases sets by calculating longest path coverage rate, just like prime path coverage in white-box testing. For example, in Fig. 4(a), there are 3 valid LDPs: {M →B →G, M →D →H →I, M →D →G}, and there are 12 LSPs(from *List 2*) in the static call graph in Fig.2, so the longest path coverage rate of TC_4 is 25%. The rate of TC_5, TC_6 and TC_7 are 41.7%, 33.3%, 50% respectively. Apparently, test case 7 has the highest coverage rate, so we think that it is more valuable under this testing criterion.

We can also use Jaccard Formula to calculate the similarity of two sets of LDPs. From *List 3*, we get the similarity degree of test case 4, 5 or test case 4, 6 is the lowest, 0, which means test case 4 and 5 focus on testing different paths. The similarity degree of test case 5, 6 is the highest, 37.5%. So we can place test case 4, 5 or 4, 6 in the same test case set to achieve a high coverage rate. While using the similarity model, we can also set two proper thresholds, and more details of this will be described in the usage of "Degree of Execution Difference" (in Definition 14).

II. Select a Set of Test Cases to Achieve the Highest Coverage Rate

a)Requirements Description

We now have n test cases, and we have got n dynamic call trees. Now we need to choose k test cases to integrate into a test set, and the problem is which cases we should choose to achieve the highest rate of longest path coverage. Apparently, we can use this method to manage test cases.

b)Abstraction Description and Algorithm

The problem also belongs to Maximum coverage problem. The approach is also the same with the above solution in function coverage, but pretreatment work is that we should label each path with a unique integer Id.

c) Implementation

List 3: The valid LDPs of TC_4, TC_5, TC_6, TC_7

TC_4:{M→B→G, M→D→G, M→D→H→I}

TC_5:{M → A → E, M → A → F, M → B → A → E, M → B → A → F, M→C→G}

TC_6:{M→A→E, M→B→F, M→C→D→G, M→C→D→H→I}

TC_7:{M→A→F, M→B→A→F, M→B→A→E, M→C→D→G, M→C→D→H→I, M→D→G}

From **List 3**, TC_4, TC_5, TC_6, TC_7, $n = 4$:

ⅰ) $k = 1$:{TC_7} can achieve the highest rate 50%.

ⅱ) $k = 2$:{TC_4, TC_5}, {TC_4, TC_7}, {TC_5, TC_6}, {TC_5, TC_7} {TC_6, TC_7} can achieve the highest rate 66.7%.

ⅲ) $k = 3$:{TC_4, TC_5, TC_6} can achieve the highest rate 100%.

Depending on practical requirements, we can control the size of test cases sets and the coverage rate by setting proper k.

In a word, "function coverage" and "longest path coverage" are two testing criteria suitable for dynamic analysis. These measurement models help us evaluate and manage the test cases. Based on the quantization, we can select and sort test cases.

4.3   Measurement Models about Value of Test Cases

Software practice shows that it is more likely to find defects in the complex parts, so the testers must pay sufficient attention to these sections. The higher the coupling degree of a module is, the more important it will be for the right software execution. The important modules must be given adequate test, which will be useful to improve the test efficiency. Before giving the definition about value of test cases, we will introduce some related models.

Definition 9. (*Function Weight and Calling Times*): In a static call graph, function weight is represented by the corresponding node's degree including indegree and outdegree. In the contrast, calling times of a function is dug from dynamic call tree. Calling times of a function represents how many times a function was invoked in the execution of a test case.

A practical example is that the ranking of Google search results is based on the times, for which a page is referenced. That is to say, the related times represents the importance of a page.

The maximum degree of the graph in Fig. 2 is 5, node A's degree, so function A's weight is 5. We have the reason to give function A sufficient test. In Fig. 4(b), we can see function A is invoked twice, so the calling times of function A is 2.

Definition 10. (*Test Target Vector*): $T_F = (F_1, F_2, ..., F_n)$, F_i is   corresponding to a function in static call graph. The number n is the total number of functions(except "main") in a program. Function "main" is the start of a program, and we don't count it in the vector.

Test target vector indicates the object of testing, and is an important foundation for the management of test process. In Fig. 2, $T_F = (A, B, C, D, E, F, G, H, I, J)$.

Definition 11. (*Test Unit Weight Vector*): $W = (w_1, w_2, ..., w_n)$, w_i is the function weight of F_i, a component of T_F.

Definition 12. (*Test Density*): $D = (d_1, d_2, ..., d_n)$, $d_i = t_i / w_i$, t_i and w_i represent calling times and function weight of F_i separately. D represents the test density vector of a test case, and d_i represents F_i's test density.

Apparently, getting the same calling times, the modules with a heavy function weight will have a lower test destiny.

"*Test Density Average of each test case*": $\quad D_A = \dfrac{1}{n} \sum\limits_{i=1}^{n} d_i$

D_A is used to evaluate the average testing condition of a test target vector. The higher the average is, the larger the comprehensive test scope is.

"*Test Density Average of each function*": $\quad Ka = (k_1, k_2, ..., k_n)$,

$$k_j = \frac{1}{n} \sum_{i=1}^{n} D_i[j]$$

where D_i represents the test density vector of test case i, $D_i[j]$ represents F_j's test density of test case i, n is the number of test case.

Ka can evaluate the average test quality of each unit about n test cases. We can find important information from the components of *Ka*. For example, if the values of components of *Ka* are lower than the test density baseline(which will be introduced next), we believe that the four test cases don't achieve a sufficient test for these parts, and then testers should provide other test solutions, such as building stub modules and driver modules for a sole test.

Definition 13. (*Test Density Baseline*): Test density baseline, a number set by testers, is a guide for testers to evaluate and manage the test cases.

When we use this model, if we find a function module's test destiny is lower than the baseline, we have the reason to believe that the test case can't provide adequate test for this function. Test density baseline model is quantization of the test results, and it enables the testers to stand in a proper position to evaluate the test results quantificationally. It also indicates the testers when and which function should be retested solely.

Definition 14. (*Degree of Execution Difference*): Let x and y represent the test destiny vector of test case 1 and 2 separately. Here we use Squared Euclidean Distance to describe degree of execution difference of two test cases:

$$dif(x, y) = \sum_{i=1}^{n} (x_i - y_i)^2$$

where n is the number of dimensions of T_F.

The basis of many measures of similarity and dissimilarity is Euclidean Distance [8]. The standard Euclidean distance can be squared in order to place progressively greater weight on objects that are farther apart [9]. So this formula focuses more on the greater function weight, which represents our testing center.

While using the degree of execution difference measurement model, testers are asked to set two thresholds: Substitution Distance and Merger Distance. Let SD, MD, and ED represent substitution distance, merger distance, and degree of execution difference separately. SD is smaller than MD in general. If $ED < SD$, we say execution difference between the two test cases is in tolerant distance where we can use one of them to substitute the other to reduce the number of test cases. If $ED > MD$, the execution difference between the two test cases is big, that is to say, the two cases focus on testing different parts of the test target vector, so we have the reason to place them in the same test set to provide a sufficient test for more test units. Of course, proper thresholds will be significant for the management of test cases.

Definition 15. (*Value of Test Cases*): V is the value of a test case:

$$V = (w_1 \quad w_2 \quad \Lambda \quad w_n) \times (t_1 \quad t_2 \quad \Lambda \quad t_n)^T$$

where w_i and t_i are the function weight and calling times of F_i, a component of test target vector T_F.

From the definition, we see that the value of test cases focuses more on the greater function weight. The quantization value provides us a direct evaluation of test cases.

4.4   Experiment of Dynamic Analysis

In part **4.1** and **4.2** of this section, we have evaluated the four test cases from two testing criteria: function coverage and longest path coverage, and also introduced their applications. In this part, we will further evaluate the four test cases using other measurement models defined in part **4.3**.

For convenience, we still use the labels: TC_4, TC_5, TC_6 and TC_7 to represent the four test cases in Fig.4. In the same way, calling times vectors are E_4, E_5, E_6 and E_7. Test density vectors are D_4, D_5, D_6 and D_7. Test density average vectors are D_{A4}, D_{A5}, D_{A6} and D_{A7}.   The values of the four test cases are V_4, V_5, V_6 and V_7. Ka represents the average test destiny of each component about the four test cases. We use $dif(x, y)$ to represents the degree of execution difference, where x and y are two test cases.

From static call graph Fig.2, we can get the following information:

Test target vector: $T_F = (A, B, C, D, E, F, G, H, I, J)$
Test unit weight vector: $W = (5, 4, 3, 4, 2, 2, 3, 3, 1, 4)$

With the help of my Data_Analyser, we can easily get following results of statistics:

Calling times vectors:

$$E4= (0 \quad 1 \quad 0 \quad 1 \quad 0 \quad 0 \quad 2 \quad 2 \quad 1 \quad 2 \quad)$$
$$E5= (2 \quad 1 \quad 1 \quad 0 \quad 2 \quad 2 \quad 1 \quad 0 \quad 0 \quad 0 \quad)$$
$$E6= (2 \quad 1 \quad 1 \quad 1 \quad 2 \quad 1 \quad 1 \quad 1 \quad 0 \quad)$$
$$E7= (2 \quad 1 \quad 1 \quad 2 \quad 1 \quad 2 \quad 2 \quad 2 \quad 1 \quad 0 \quad)$$

Test density vectors:

$$D4= (0 \quad 0.25 \quad 0 \quad 0.25 \quad 0 \quad 0 \quad 0.67 \quad 0.67 \quad 1 \quad 0.5 \quad)$$
$$D5= (0.4 \quad 0.25 \quad 0.33 \quad 0 \quad 1 \quad 1 \quad 0.33 \quad 0 \quad 0 \quad 0 \quad)$$
$$D6= (0.4 \quad 0.25 \quad 0.33 \quad 0.25 \quad 1 \quad 0.5 \quad 0.33 \quad 0.33 \quad 1 \quad 0 \quad)$$
$$D7= (0.4 \quad 0.25 \quad 0.33 \quad 0.5 \quad 0.5 \quad 1 \quad 0.67 \quad 0.67 \quad 1 \quad 0 \quad)$$

Test Density Average of each function:

$$Ka = (D_4 + D_5 + D_6 + D_7)/4 = (0.3 \quad 0.25 \quad 0.25 \quad 0.25 \quad 0.63 \quad 0.63 \quad 0.5 \quad 0.42 \quad 0.75 \quad 0.13)$$

Test Density Average of each test case:

$$D_{A4} = 0.33 \qquad\qquad D_{A5} = 0.33$$

$$D_{A6} = 0.44 \qquad\qquad D_{A7} = 0.53$$

Value of test cases:

$$V_4 = W \times E_4 = 29 \qquad\qquad V_5 = W \times E_5 = 28$$
$$V_6 = W \times E_6 = 34 \qquad\qquad V_7 = W \times E_7 = 44$$

Degree of execution difference:

$$dif\ (D_4, D_5) = 4.13917 \qquad\qquad dif\ (D_4, D_6) = 1.99333$$
$$dif\ (D_4, D_7) = 1.83361 \qquad\qquad dif\ (D_5, D_6) = 1.42361$$
$$dif\ (D_5, D_7) = 2.05556 \qquad\qquad dif\ (D_6, D_7) = 0.78472$$

Here we provide the thought of dynamic analysis:

i. Evaluate Single Test Case
In the vector of test unit weight, W, we can easily find function A has the highest degree of coupling, so the testers need to have a sufficient test practice for it. From the results of the values of test cases, we can clearly get that TC_7 has the best test effect. At the same time, TC_7 also has the largest test destiny average, which seems that it has a good general test for all the functions.

ii. Indicate the Directions of Improvement
Assume test density baseline be 0.3, that is to say, the test of a unit is eligible if its test density is higher than 0.3. On the one hand, from the test destiny vector, we can easily find if a test unit get sufficient test from a test case, depending on which we can eval-uate a test case. For example, from D_4, D_5, D_6 and D_7, every test case has some com-ponents whose test density is lower than 0.3 or even 0, which means these test cases can't achieve perfect test for every component and some sole test are needed. On the

other hand, from these four test cases, we get the average test destiny of each component, as it is shown in vector Ka. If the values of components of Ka are lower than the baseline 0.3, we have the reason to believe that the four test cases don't achieve a sufficient test for these parts, and then testers should provide other test solutions, such as building stub modules and driver modules for a sole test.

iii. Manage Test Cases Sets

From test density vectors: D_4, D_5, D_6 and D_7, we can easily find that every test case focuses on different components of the test target vector. The difference is quantified by the measurement model, degree of execution difference. Let SD and MD represent substitution distance and merger distance separately. If we set $SD=1.0$, $MD=4.0$, then $dif(D_6, D_7) < SD$, so we have the reason to think that TC_6 and TC_7 have similar test efficiency, so we can choose one of them to substitute the other to reduce the number of test cases. It is worth mentioning that this similar relation isn't transitive. Because this sort of similar relation is not exact congruence, the deviation will be enlarged in the transferring process so as to go beyond the threshold.

At the same time, we get $dif(D_4, D_5) > MD$, that is to say, the two test cases focus on different testing parts in the program. So we can place the two test cases in the same test case set to enable the set to implement an all-sided test for the program.

4.5   Conclusions of Dynamic Analysis

Though we defined different testing criteria and used different models to evaluate the test cases, we found that TC_7 had a good performance in many measurement models, such as function coverage, longest path coverage, and value of test case, which illustrates that using the dynamic analysis as well as the testing models to evaluate the test cases are credible.

From the measurement of these quantization models, we find many problems that can't be detected by general testing, which verifies the importance of these models as well as the validity of these definitions.

Dynamic analysis is based on both static and dynamic function calling relations, so we defined static call graph and dynamic call tree to represent these relations and displayed them in the figures. After building the foundation of analysis, we described these models in detail. Overall, depending on whether the values of these models are constant, these models can be divided into two categories: static models and dynamic models. Static models dug from static call graph include test target vector and test unit weight vector, which are constant and the foundation of dynamic models. Based on different application purposes, dynamic models also have two categories: value models for evaluating test cases, guide models for managing test cases. On the one hand, value models, including function coverage, longest path coverage, value of test case, test density, and test density average of each test case, provide an exact quantization evaluation for each test case. On the other hand, Guide models include the following models: the similarity of function coverage or longest path coverage, degree of execution difference, test density baseline, and degree of execution difference, which can be a guide indicating the testers which problems the current test has, how to manage test cases, and the directions of improvement of test cases.

5   Related Work

There has been a considerable amount of studies on software testing, evaluation of test cases, graph mining and analysis, and aspect oriented programming.

Ostrand has talked about education, training, experience of software testing [19]. Bertolino redefined software testing [20]. Many studies have presented related concepts about effective software testing and the evaluations of test cases. Of course, the aim of evaluating test cases is to improve testing efficiency. Here are some present concepts about evaluations of test cases. White-box test design techniques include the following code coverage criteria: statement coverage, prime path coverage, branch testing and so on [8]. We evaluate the test cases by calculating the coverage rates. However this kind of evaluations is based on static analysis of test cases, which has many shortages. In addition the topic about test case effectiveness is also an evaluation of test cases. Chernak defined test case effectiveness as the ratio of defects found by test cases to the total number of defects reported during the function test cycle [12]. Chernak suggested 75 percent to be an acceptable baseline value, and intended to improve test-case effectiveness with this quantization base [12].

Graph brings intuitions for program comprehension. Many studies have presented the methods of getting static call graph and dynamic call graph. Especially for static call graph, some present software, such as Source Insight, CodeViz, all provide related functions. Dynamic call graph is obtained by execution trace mining [14][15]. Peng has proposed the transformation of graph to tree, and he has deeply studied the simplification of dynamic call tree [13].

AspectC++ is an aspect-oriented extension of C and C++ languages [16].The aspect-oriented extension of Java is AspectJ [17]. Many studies have focused on aspect-oriented programming and its applications. Feng has used AspectJ in reverse engineering [18].   Zhao used AspectJ to get dynamic call graph from a program [15].

6   Conclusions and Extensions

Software testing focuses not only on how many software defects were found. For the shortcomings of evaluating test cases based on static analysis, this paper presented dynamic analysis and some relevant testing criteria and measurement models   to achieve multi-angle evaluations of test cases. At last, we further introduced the thought and significance of dynamic analysis by analyzing four test cases. Experiments prove that dynamic analysis can successfully solve the shortages of static analysis, and they have the following advantages, such as quantization, automation, wide application, high efficiency, the ability to indicate the directions of improvement and management of test cases.

Dynamic call trees and static call graphs are dug from program, and the above testing models are derived from these trees and graphs. Software testing should not be separated from software development or coding. Software is a big treasure, so we can dig useful information and make reasonable testing criteria to apply to software test.

Acknowledgment. This work is supported by the Natural Science Foundation of Beijing (4133088), the Foundation for Doctoral Program by the Ministry of Education (3080036621203), the Foundation for Fundamental Research by Beijing Institute of Technology(20120842004), and the Foundation for Overseas Students of China.

References

[1] Pressman, R.S.: Software Engineering, A Practitioner's Approach, 4th edn. McGraw-Hill, New York (1997)

[2] Biswas, S., Mall, R., Satpathy, M., Sukumaran, S.: A model-based regression test selection approach for embedded applications. ACM SIGSOFT Software Engineering Notes 34(4), 1–9 (2009)

[3] Rothermel, G., Harrold, M.J.: Analyzing regression test selection techniques. IEEE Transactions on Software Engineering 22(8), 529–551 (1996)

[4] Feige, U.: A threshold of ln n for approximating set cover. J. ACM 45, 634–652

[5] Hochbaum, D.S.: Approximating covering and packing problems: Set cover, vertex cover, independent set, and related problems. In: Approximation Algorithms for NP-hard Problems, pp. 94–143. PWS Publishing Company, Boston (1997)

[6] http://www.analytictech.com/mb876/handouts/distance_and_correlation.htm (June 2014)

[7] http://en.wikipedia.org/wiki/Euclidean_distance#Squared_Euclidean_distance (July 2014)

[8] http://en.wikipedia.org/wiki/White-box_testing (May 2014)

[9] http://www.aspectc.org/ (June 2014)

[10] http://www.gson.org/egypt/ (July 2014)

[11] http://en.wikipedia.org/wiki/Maximum_coverage_problem (July 2014)

[12] Chernak, Y.: Validating and Improving Test-Case Effectiveness. IEEE Software 18(1) (January-February 2001)

[13] Peng, T.: Program Verification by Reachability Searching over Dynamic Call Tree. In: ADMA 2014 (2014)

[14] Behrmann, G., David, A., Larsen, K.G.: A Tutorial on Uppaal: Toolbox for Verification of Realtime System, Department of Computer Science, Aalborg University, Denmark

[15] Zhao, C., Kong, J., Zhang, K.: Program Behavior Discover and Verfication: A Graph Grammar Appraoch. IEEE Transaction on Software Engineering (2010)

[16] http://en.wikipedia.org/wiki/AspectC%2B%2B (June 2010)

[17] Kiczales, G., Hilsdale, E., Hugunin, J., Kersten, M., Palm, J., Griswold, W.G.: An Overview of AspectJ. In: Brusilovsky, P., Corbett, A.T., de Rosis, F. (eds.) UM 2003. LNCS, vol. 2702, pp. 327–353. Springer, Heidelberg (2003)

[18] Feng, X.: Analysis of AspectJ and its Applications in Reverse Engineering. Master Thesis of Software Engineering, Xian Electrical Science and Technology University

[19] Ostrand, T., Weyuker, E.: Software testing research and software engineering education. ACM, New York (2010)

[20] Bertolino, A.: Software Testing Research and Practice. In: Börger, E., Gargantini, A., Riccobene, E. (eds.) ASM 2003. LNCS, vol. 2589, pp. 1–21. Springer, Heidelberg (2003)

An Interest-Based Clustering Method for Web Information Visualization

Shibli Saleheen and Wei Lai

Faculty of Science, Engineering and Technology
Swinburne University of Technology
Hawthorn, Victoria, Australia
{ssaleheen,wlai}@swin.edu.au

Abstract. Web graph is a tool to visualize web information as network. It unfurls inherent connectivity of the web for end users from a different viewpoint. The enlarged size of the web causes the information overload problem and forces the wide use of compression techniques such as filtering and clustering on graphs during presentation of web information. In addition, the Internet users, their intentions and activities on the web differ. User interest-based web graph, which is modulated by user interests during construction, is used to accommodate differences over end users and/or their needs. However, user interest-based web graph features an unorthodox way to present connectivities among nodes by utilizing edge labels. This complicates further operations such as clustering and focused visualization on web graphs. This paper introduces a novel approach to cluster user interest-based web graphs by adopting the divide and conquer strategy. It is demonstrated that, this approach can effectively cluster the user interest-based web graph.

Keywords: Clustering, Web Networks, Visualization, User Interests.

1 Introduction

Because of its rapid growth, the web is getting more complicated as a source for end users to extract relevant information. Search engines such as Google, Bing and Yahoo provide information as lists to satisfy the needs of users. However, these lists fail to expose structural and/or semantic relationships of the information. To address the problem concerned with revealing the inherent connectivity of web information to end users, researchers have introduced web graph to deliver web information. According to Manning et al.[8], the web can be viewed as a directed graph where static HTML pages are treated as nodes and hyperlinks among them are as directed edges. Various forms of web graph are used to present information sourced from the Internet such as search results, social networking structure and micro-blog relations for ease of navigation and exploration.

The context of this paper describes the web graph as a medium to visualize web information in the form of a graph $G = (V, E)$, where V is a set of nodes and E is a set of edges, similar to the general graph representation. Nevertheless, the web graph defines elements of V and E differently than both traditional

X. Luo, J.X. Yu, and Z. Li (Eds.): ADMA 2014, LNAI 8933, pp. 421–434, 2014.

graphs and the definition by Manning et al.[8]. Structure-based web graph unfolding structural relationships, represents V (the set of nodes) as web pages and E as hyperlink relations, whereas semantic-based web graph which is the outcome of a content-based method, defines V similar to that of a structure-based approach or as a set of terms and E as a set of scores of a similarity function $SIM(v_i, v_j)|v \in V$ on nodes. An edge, $e \in E$, can be undirected for two nodes v_i and v_j when the similarity function holds same score in either direction, $SIM(v_i, v_j) = SIM(v_j, v_i)$.

Introduction of web graphs to visualize web information benefits end users in finding information quickly and effectively and in getting an insight into how chunks of web information are related. Despite these benefits, web graph poses potential challenges regarding feasibility. The problem associated with the web graph visualization is two-fold. First, for its static nature, the same web graph is presented to various end users regardless of having dynamic information need. Second, information overload makes the web graph difficult to be visualized to end users and to be served as a source to navigate and explore interested information. The first problem, i.e., lack of diversity in web graphs can be addressed if user related information is used to influence graph construction process. User interests are incorporated into graph generation processes: selecting nodes and defining edges. In our previous work[11] such a web graph is presented which labels edges according to user interests exposed by nodes. It is called the user interest-based web graph(UIWG hereinafter). To address the second problem, i.e., to reduce the size of the web graph and exhibit clarity in visualization, compression methods like filtering irrelevant nodes and/or edges and clustering similar nodes are widely used. However, because of the unorthodox feature of having edge labels, the demonstrated web graph[11] cannot be effectively filtered and clustered by existing methodologies. As a consequence, an atypical method is required to resolve these and introduced in this paper, namely Divide and Conquer(D&C), to accomplish clustering on UIWG. This paper also demonstrates that the D&C clustering can achieve a better clustered graph from UIWG.

The rest of the paper is organized as: Section 2 presents the motivation of this work; Section 3 discusses the components of UIWG briefly and formulates the problem; the D&C clustering is described in Section 4; experimentation by case study is provided in Section 5; Section 6 reports the related work; and Section 7 concludes the paper stating contributions and future work.

2   Motivation

Clustering plays an important role in reducing the size and is very effective for the web graph visualization. However, most clustering methods consider that all edges of a graph resemble the same context, i.e., similarities among nodes are constituted based on identical relationships of nodes. Structure-based web graphs treat an edge as a hyperlink relation between two web pages. In semantic-based web graphs, content similarity measures establish the sole property in connecting two nodes. Available clustering algorithms exploit structural or contextual

Table 1. User Interests

Interest	Weight
I_1	0.90
I_2	0.95
I_3	0.60
I_4	0.60

Table 2. Interest Term Similarity

Interest	Term	Similarity Score
I_1	a	0.85
I_1	d	0.95
I_2	b	0.75
I_2	c	0.80
I_2	e	0.85

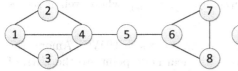

Fig. 1. Traditional Web Graph

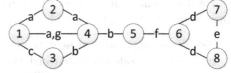

Fig. 2. User Interest-based Web Graph

relationships of edges. The edge representation in UIWG is different because it includes the ground on which nodes are connected in addition to the degree of similarity. This makes existing clustering approaches less effective on UIWG.

Let us consider a web graph that consists of 8 nodes and 10 edges. Traditional representations connect the nodes of the web graph based on hyper-link/contextual relationships whereas UIWG connects them differently. Let us also consider an end user having four interests shown in Table 1 with their corresponding weight values for this user. Table 2 presents pairwise similarity scores among user interests and document(web page) terms. Pairs such as (I_1, b), (I_2, a) and (I_3, f) which do not appear in the table are assumed to have 0 similarity scores. Figures 1 and 2 depict the difference between traditional web graph and UIWG. It is observed that unlike in Figure 1, in Figure 2, an edge shows the information on which it is established. For example, edge-labels 'a' and 'b' represent two distinct terms extracted from the document corpus. Some edges are labelled with multiple terms. For example, if an edge with label 'a,g' connects two documents, then 'a' and 'g' are assumed to be prominent terms in both documents. An edge can have a weight score in addition to linking information.

The purpose of user interest-based visualization is to portray the web graph tailored for an end user. Typically, needs of an individual differ from those of other users. Even needs of a specific end user vary depending on context such as time and location. Therefore, it is useful for end users to have the web graph clustered according to their needs. This can be achieved by grouping and tagging nodes resembling similar topics imaged by the end user's mind map.

Most existing clustering methods will produce two clusters as $C_1 = \{1, 2, 3, 4\}$ and $C_2 = \{6, 7, 8\}$ from the web graph of Figure 1. Because of the difference in edge presentations(divided into seven categories {'a','b',$\cdots$,'g'}), the above generated clusters are less meaningful for UIWG of Figure 2 though the degree of each node is exactly equal to its counterpart in Figure 1. This calls for different techniques to cluster UIWG. For further enhancement, the new clustering should adhere to hierarchical structure which produces the most compact[3] web graph.

3   Problem Definition

A brief review of the components and the representation of UIWG[11] is presented before formulating the problem. Documents downloaded from the Internet have gone through several text-processing methods such as stop-word removal, stemming and term extraction to generate vectors for representing documents. As a consequence, the information representation model is considered as a vector space model. The set of terms is denoted as $K = \{t_1, t_2, ..., t_n\}$ whereas a vector v for a document is denoted as $v = \{w_1 t_1, w_2 t_2, ..., w_n t_n\}$, where weight scores range from 0 to 1, i.e., $0 \leq w \leq 1$.

- **Nodes** are used to represent the document vectors in UIWG. Among the two types of node presented in UIWG, the leaf node points to the URL of the corresponding web page. Clicking on such nodes results in accessing the linked web pages. The rest are abstract representations, known as cluster nodes, which are sets of nodes related to each other based on common characteristics. A cluster node can contain another cluster in the set of members.
- **Edges** are calculated based on similarity measurements such as Cosine similarity and Jaccard index in UIWG. In[11] the cosine similarity measure is used to calculate an edge. Because a term-based vector is used to represent a node, the similarity measure accounts all terms of two vectors for calculation. However, the most prominent term(s) is treated as the base for connecting nodes. That means, if two documents are represented with the vectors v_i and v_j respectively, then the edge connecting them in UIWG can be written as a function $E_{Tij}|T \in \mathcal{P}(K)$ where $\mathcal{P}(K)$ is the power set of K. $(T = \phi) \cong (E_T = null)$ means vectors v_i and v_j have no connecting edge.
- **Graph Representation** reveals node to node relationships of UIWG by an adjacency matrix as follows:

$$G_A = \begin{array}{c} \\ v_1 \\ v_2 \\ \vdots \\ v_r \end{array} \begin{array}{cccc} v_1 & v_2 & \cdots & v_r \\ \left(\begin{array}{cccc} 0 & E_{T12} & \cdots & E_{T1r} \\ & 0 & \cdots & E_{T2r} \\ & & \ddots & \vdots \\ & & & 0 \end{array} \right) \end{array} \qquad (1)$$

where E_T is a function which exhibits two properties: (1) the measure how nodes are similar to each other numerically and (2) a set of terms reflecting how nodes are related. Therefore, E_T can be expressed as wT where w represents the numerical weight score and $T \subset \mathcal{P}(K)$ represents the set of common terms. In other words, the similarity score for a node pair is w and they are connected by the elements of T. Because UIWG is undirected, the same weight score appears when nodes come in reverse order, i.e., $E_{T12} = E_{T21}$. Therefore, it is sufficient to store the values situated one side of the diagonal only, which in this case is the upper left of the adjacency matrix.

The aim of this work is to generate a set of clusters from a given UIWG to reduce the size thus the end user with profile $P(I_1, I_2, ..., I_l)$ gets a better

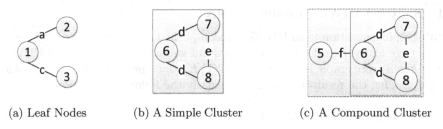

(a) Leaf Nodes   (b) A Simple Cluster   (c) A Compound Cluster

Fig. 3. Types of Node in a Clustered UIWG

visualization to navigate and explore the web and conducts an effective search for interested information. The generated clusters should confront hierarchical structure, i.e., either a cluster is a complete sub-cluster of another cluster or it is mutually exclusive from other clusters. In other words, clusters do not overlap if no super/sub relation is present. This yields the following types of cluster:

Simple Cluster is denoted by C_p which consists only of leaf nodes of UIWG. Therefore, C_p is an element of the power set of V, i.e., $C_p \in \mathcal{P}(V)$ where $|C_p| \geq 2$ which means C_p contains at least 2 leaf nodes.
The set of simple clusters is denoted by C_P. Simple clusters do not share members among themselves and hence are mutually exclusive from each other. For two simple clusters C_p and C_q it is written, $C_p \cap C_q = \phi$.
Compound Cluster is denoted by C_c which consists of at least one simple or compound cluster, plus zero or more leaf nodes that are not members of any cluster, and has a cardinality greater than 2. A compound cluster C_c from the set of compound clusters C_C, is written as $C_c \in \{\mathcal{P}(C_C \cup C_P \cup V) : \forall C_p \not\subset \exists C_C \ and \ \forall v \not\subset \exists(C_P \cup C_C)\}$ where $|v \in C_c| < |C_c| \geq 2$.

Representations of nodes and clusters are presented in Figure 3. Simple and compound clusters are portrayed in Figures 3b and 3c respectively whereas leaf nodes and connecting edges are shown in Figure 3a. Given a UIWG $G = (V, E)$ where $v \in V$ denotes a leaf node and $e = E_T \in \mathcal{P}(K)$ denotes an edge, a cluster set $C = \{C_1, C_2, ..., C_m\}$ is to be generated where $C_i \in (C_C \cup C_P)$.

4 Divide and Conquer(D&C) Clustering

Existing graph clustering methods cannot cluster a UIWG effectively for end users because edges of UIWG are multivariate, i.e., heterogeneous relation types are present to describe different edges and multivalued, i.e., multiple instances of homogeneous information are combined to present one edge.

Reduction of edge diversity is necessary to enhance the effectiveness of clustering processes. The problem defined in Section 3 has been approached in this work by adopting the divide and conquer strategy in three steps. Firstly, UIWG is split into multiple sub-graphs. Secondly, an elementary cluster set for each sub-graph is calculated. Finally, clusters of all elementary sets are combined to form the final cluster set. The following sections describe these steps in detail.

4.1   Splitting UIWG into Sub-graphs

A sub-graph generation from UIWG is accomplished by performing unary operations: edge and vertex deletion. The resultant sub-graph is a minor[1] of UIWG.
If the given web graph is denoted as $G(V, E)$, for the end user with profile
$P(I_1, I_2, ..., I_l)$, the resultant sub-graphs satisfy the following equation:

$$G = G_1 + G_2 + \cdots + G_l \tag{2}$$

where, G_x represents the x-th sub-graph and the adjacency matrix representation
for G_x, G_{Ax} is achieved by applying I_x to G_A, i.e., $G_{Ax} \cong I_x \cdot G_A$ where the
value of x is bounded in $1 \le x \le l$; l is the number of user interests. Therefore,
using Equation 1, $I_x \cdot G_A$ is calculated as follows:

$$I_x \cdot G_A = I_x \cdot \begin{pmatrix} 0 & E_{T12} & \cdots & E_{T1r} \\ & 0 & \cdots & E_{T2r} \\ & & \ddots & \vdots \\ & & & 0 \end{pmatrix} = \begin{pmatrix} 0 & I_x \cdot E_{T12} & \cdots & I_x \cdot E_{T1r} \\ & 0 & \cdots & I_x \cdot E_{T2r} \\ & & \ddots & \vdots \\ & & & 0 \end{pmatrix} \tag{3}$$

The function E_{Tpq} is a combination of weight ω_{pq} and term set $T_{pq} \in \mathcal{P}(K)$,
i.e., $\omega_{pq} T_{pq}$. As I_x is an interest term, the dot product of $I_x \cdot E_{Tpq}$ is written as:

$$I_x \cdot E_{Tpq} = \omega_{pq} \sum_{i=1}^{|T_{pq}|} I_x \cdot t_{pqi} \tag{4}$$

Because I_x and t_{pqi} are terms, $I_x \cdot t_{pqi}$ is defined as a similarity score between
two terms which is obtained by a standard measure. The Lin measure[7] of
WordNet is used to calculate the similarity score λ between these two terms:

$$I_x.t_{pqi} = \lambda_i \cdot I_x \tag{5}$$

Now, $I_x.t_{pqi}$ of Equation 4 is substituted by $\lambda_i \cdot I_x$ using Equation 5 to eliminate
multi-term relationships among nodes. Therefore, Equation 4 becomes:

$$I_x \cdot E_{Tpq} = \omega_{pq} \sum_{i=1}^{|T_{pq}|} I_x \cdot \lambda_i = I_x \cdot \omega_{pq} \sum_{i=1}^{|T_{pq}|} \lambda_i = I_x \cdot \Omega_{pq} \tag{6}$$

as similarity score λ_i is scalar, $\sum \lambda_i$ and hence Ω_{pq} are also scalar. At this point,
the set of terms T is replaced with the term I_x. This is used in Equation 3 to
achieve a scalar matrix representation for sub-graph G_x from UIWG, G, i.e., the
adjacency matrix of G_x. The representation is given below:

$$I_x \cdot G_A = \begin{pmatrix} 0 & I_x \cdot \Omega_{12} & \cdots & I_x \cdot \Omega_{1r} \\ & 0 & \cdots & I_x \cdot \Omega_{2r} \\ & & \ddots & \vdots \\ & & & 0 \end{pmatrix} = I_x \cdot \begin{pmatrix} 0 & \Omega_{12} & \cdots & \Omega_{1r} \\ & 0 & \cdots & \Omega_{2r} \\ & & \ddots & \vdots \\ & & & 0 \end{pmatrix} \cong G_{Ax} \tag{7}$$

---

[1] An undirected graph H is a minor of another graph G if H can be formed from G
by removal and contraction of edges and by deletion of vertices.

Algorithm 1. Splitting UIWG into Sub-graphs

Input : $G(V, E), P(I_1, I_2, \cdots, I_l), \sigma$
Output: $G_1(V, E_1), G_2(V, E_2), \cdots, G_l(V, E_l)$

```
1  begin
2      for q = 1 to l do
3          set G_Aq ← Φ
4          forall the element e of adjacency matrix G_A do
5              set index ← index(e, G_A); s ← 0; T ← termset(e); w ← weight(e); score ← 0
6              if T ≠ Φ then
7                  forall the element t of T do
8                      set s ← s + sim(t, I_q)
9                  if s * w > σ then
10                     set score ← s * w
11             set index(G_Aq) ← score
12         generate G_q(V, E_q) from G_Aq
```

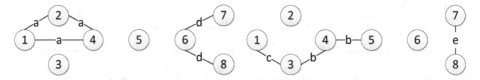

Fig. 4. Graph with Related Edges of I_1 **Fig. 5.** Graph with Related Edges of I_2

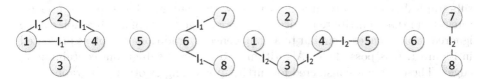

Fig. 6. Sub-graph from Figure 4   **Fig. 7.** Sub-graph from Figure 5

From Equation 7 it is clear that the sub-graph G_x can be obtained from the deduced adjacency matrix, G_{Ax}. According to Equation 2, a sub-graph for each interest of profile P is generated if the interest score(updated on explicit and implicit feedback captured from usage) is greater than a predefined threshold, γ. Algorithm 1 describes the methodology for splitting UIWG into sub-graphs.

Figures 4 and 5 show examples of edges related to user interests. Figure 4 shows edges with labels 'a' and 'd'. It is noticed from Table 2 that 'a' and 'd' are related to I_1. However, In Figure 5, the terms 'b', 'c' and 'e' are present as they are related to I_2 according to Table 2. However, to make the sub-graphs similar to traditional graphs, the above stated situations must be avoided. Algorithm 1 makes sure that no such characteristic arises by replacing them with interest terms. The final score of an element in the deduced adjacency matrix is set to 0 if it falls below a threshold σ, i.e., if the final score does not satisfy the condition given in line 9. The resultant sub-graphs for the interests I_1 and I_2 are presented in Figures 6 and 7 which are obtained from Figures 4 and 5 respectively.

Algorithm 2. Cluster-set and Topic Specific Clustered Sub-graph

---

Input   : $G(V, E), G_x(V, E_x), G_A, G_{xA}, \tau$
Output: $C_1, C_2, \cdots, C_s, G(V, E')$

1   **begin**
2 | set $avg \leftarrow 2|E_x|/|V| + \tau;\ S_x \leftarrow \Phi$
3 | **forall the** v *of* V **do**
4 | | calculate $deg(v)$
5 | | **if** $deg(v) > avg$ **then**
6 | | | set $S_x \leftarrow S_x \cup v$

7 | set $s \leftarrow |S_x|;\ C_1, C_2, \cdots, C_s \leftarrow \Phi$
8 | apply $Kmeans$ algorithm to the seed nodes set S_x and reconstructed adjacency matrix G_{xA} to construct the clusters $C_1, C_2, \cdots, C_s$
9 | calculate adjacency matrix of difference graph$(G - G_x)$ using Equation 8
10 | **forall the** *element* $e_{(i,j)} = 1$ *of* $(G - G_x)_A$ **do**
11 | | set $x, y \leftarrow no$
12 | | **if** $v_i \in C_1 \cup C_2 \cup \cdots \cup C_s$ **then**
13 | | | set $x \leftarrow yes$

14 | | **if** $v_j \in C_1 \cup C_2 \cup \cdots \cup C_s$ **then**
15 | | | set $y \leftarrow yes$

16 | | **if** $x \neq y$ **then**
17 | | | set abstract edge E_a between v_i and the cluster containing v_j or vice versa

---

4.2   Cluster-Set Generation and Topic Specific Clustered Sub-graph

Each sub-graph generated in Section 4.1 is processed for generating a clustered sub-graph. Because no term is associated with a sub-graph except one user interest term that contributed to its generation, edge labels of a sub-graph can be ignored, which makes it similar to a homogeneous/traditional graph as presented in Figure 1. It is possible for Algorithm 1 to generate sub-graphs containing no edge[2]. These are not considered for further processing as other sub-graphs have the same set of nodes. As a result, no potential adverse effect on clustering is present. Hence, the number of sub-graphs is less than or equal to total interests.

Clustering algorithms that operate on graphs with homogeneous edges can be applied to a sub-graph. However, the K-means clustering algorithm is adopted in this work, because the K-means algorithm groups nodes, based on their similarity. Some terms related to clustering and clustered graphs are reviewed below:

Seed Nodes. The members of a set of nodes(S), having degree greater than the minimum average degree $\mu + \tau$, are seed nodes where μ is the average degree of a graph G and τ is a threshold predefined by the end user.
The degree of a node v, $deg(v)$, in a graph is defined by the number of incident edges of v. Because all graphs(UIWG and sub-graphs) in this context are undirected, the degree of a graph is two times its number of edges, i.e., $2|E|$. This calculates the minimum average degree as $2|E|/|V| + \tau$.

Abstract Nodes. The super-nodes, representing a portion of the graph, whose members are highly connected to each other by common characteristics. Grouping nodes to form super-nodes is achieved using node similarity scores.

---

[2] This occurs when no term in the actual UIWG is related to a user interest, and applying this interest to G_A produces an adjacency matrix of all elements as 0.

(a) Clustered Graph for I_1 (b) Clustered Graph for I_2

Fig. 8. Clustered Graphs

Abstract Edges. The edges that do not appear in the actual graph but link the abstract nodes of the clustered graph after a collapse is imposed on an abstract node. An abstract edge generally connects two super-nodes or a super-node and a leaf node whose member sets are mutually exclusive. Generally, an abstract edge is only established when an edge exists between two nodes of the actual graph(UIWG) and at least one node is included in a super-node or both of them are included in different super-nodes.

The adjacency matrix(from Section 4.1) of a sub-graph is reconstructed by excluding isolated nodes. Most clustering algorithms utilize a similarity matrix. However, the number of seed nodes needs to be defined before the K-means is applied. Finding the number of centroids has been described in many ways[4][6]. Rather than adopting the commonly used random seeding, degrees of nodes are used to select seed nodes. From the adjacency matrix, nodes which have more edges than most other nodes, i.e., $deg(v) > avg(deg(G)) + \tau$, are treated as centroids. Algorithm 2 presents the steps of sub-graph clustering. Using the UIWG, G, one of generated sub-graph, G_x, their corresponding adjacency matrices, G_A for G and G_{xA} for G_x, and the threshold τ, Algorithm 2 produces the s number of clusters where s is the cardinality of the seed node set, S_x, of the sub-graph.

The clustered UIWG can be constructed from clustered sub-graphs at this point. However, edges that are not related to an applied user interest(I_x) have been omitted from a sub-graph, G_x. For the sake of visualization, it is important to construct a sub-graph that resembles the actual UIWG as well as topic specific clustering. Therefore, to provide the opportunity of visualizing topic specific clustered UIWG to the end user, reconsideration of omitted edges is necessary to rebuild the topic specific clustered sub-graph. The graph difference[3] technique is invoked to produce the topic specific clustered sub-graph in this context. The goal is to compute the remaining edges which have not been included in the sub-graph clustering process. For this, the (i, j)-th element of the adjacency matrix of $(G - G_x)$ is computed using the following equation:

$$(G - G_x)_{A(i,j)} = \begin{cases} 0 & \text{if both } G_{A(i,j)}, G_{Ax(i,j)} > 0 \text{ or } i = j \\ 1 & \text{if } G_{A(i,j)} - G_{Ax(i,j)} > 0 \end{cases} \tag{8}$$

Figure 8 shows generated clustered graphs for both I_1 and I_2. During visualization, if the end user chooses to view the topic specific graph with interest I_1 or I_2, it is now possible to show Figure 8a or 8b respectively.

---

[3] Graph difference of graphs G and H of same degree is another graph denoted by $(G - H)$ whose adjacency matrix is calculated from the difference of adjacency matrices of G and H.

Algorithm 3. Combining Elementary Clusters to Form Clustered UIWG

Input : $CS\{C_1, C_2, \cdots, C_s\}, FC, P(I_1, I_2, \cdots, I_l)$
Output: $FC\{C_1, C_2, \cdots, C_{s'}\}$
1 **begin**
2   set $CS_{temp} \leftarrow \Phi$; $f(C) \leftarrow 0$
3   **forall the** $C \in CS$ **do**
4   set relevance, $C_r \leftarrow \sum_{t \in TS(C)} \sum_{I \in P} SIM(t, I)$ // TS(C)=all terms of a cluster

5   **forall the** $v \in C_1 \cup C_2 \cup \cdots \cup C_s$ **do**
6   list the clusters in which v is a member
7   find the candidate C_{can} by $max(|C|)$ and then by $max(C_r)$
8   set $f(C_{can}) \leftarrow f(C_{can}) + 1$

9   sort CS by $f(C)$ and then by the w of the applied I to this C
10   **while** $CS \neq \Phi$ **do**
11   set $C_t \leftarrow$ the first element of CS
12   set $FC \leftarrow FC \cup C_t$; $CS \leftarrow CS - C_t$
13   **forall the** $C \in CS$ **do**
14   **if** $(C \cap C_t) \neq \Phi$ **then**
15   **if** $|C \cap C_t| > 1$ **then**   // clusters with single member are not counted
16   set $CS_{temp} \leftarrow CS_{temp} \cup (C \cap C_t)$
17   **if** $|C - (C \cap C_t)| > 1$ **then**
18   set $CS_{temp} \leftarrow CS_{temp} \cup (C - (C \cap C_t))$

19   set $CS \leftarrow CS - C$

20   **if** $CS_{temp} \neq \Phi$ or for at least one $CS_i, CS_j \in CS_{temp}$; $CS_i \cap CS_j \neq \Phi$ **then**
21   repeat Algorithm 3 with inputs CS_{temp} as CS, current FC, and the P

4.3 Combining Elementary Clusters to Form Clustered UIWG

Combining elementary clusters is the last phase of the hierarchical clustered UIWG generation. Because the hierarchy does not support cluster overlapping, a node in the final UIWG cannot be a member of two clusters C_x and C_y unless they pose a super/sub-cluster relationship, i.e., $C_x \cap C_y \neq \phi$ and ($C_x \subset C_y$ or $C_y \subset C_x$). If two clusters overlap and they have at least one non-shared node, one is kept intact and the other is disregarded as described in Algorithm 3 which works iteratively. For all elementary clusters, nodes that fall inside two or more clusters are listed and one potential cluster for each of them is computed. The potential cluster that appears most times is considered to be the final cluster. Remaining clusters are decomposed for further iterations till an overlap exists.

For this example, Table 3 is constituted from Figure 8. Overlapping clusters with their members are listed in Table 3a. Table 3b lists participating nodes of overlapping clusters and their candidate clusters. An overlapped cluster with the maximum number of appearances is selected as the candidate cluster. For example, in Table 3c, clusters C_3 and C_2 have the highest frequency in respective overlapped sets, $\{C_1, C_3\}$ and $\{C_2, C_4\}$. For the first iteration, they are selected as final clusters. In cases where more than one cluster share the highest frequency, the cluster that is generated by the most weighted user interest is selected.

The remaining clusters, in this case, C_1 and C_4, are decomposed again for next iteration. The re-computation of the set of overlapping clusters is performed as described in Algorithm 3(lines 12 to 19). This removes the selected final cluster(e.g., C_3) from the candidate list and takes the intersection of a set and its

Table 3. Combining Overlapped Clusters in Iteration 1

(a) Clusters

Clusters	Members
C_1	1, 2, 4
C_2	6, 7, 8
C_3	1, 3, 4, 5
C_4	7, 8

(b) Candidate Clusters

Node	Clusters	Candidate
1	C_1, C_3	C_3
2	C_1	C_1
3	C_3	C_3
4	C_1, C_3	C_3
5	C_3	C_3
6	C_2	C_2
7	C_2, C_4	C_2
8	C_2, C_4	C_2

(c) Cluster Frequency

Clusters	Frequency
C_1	1
C_2	3
C_3	4
C_4	0

Table 4. Clusters for Iteration 2

Clusters	Members
$(C_3 \cap C_1)$	1, 4
$C_1 - (C_3 \cap C_1)$	2
$(C_2 \cap C_4)$	7, 8
$C_4 - (C_2 \cap C_4)$	6

Table 5. Merged Clusters

Clusters	Members
$(C_3 \cap C_1)$	1, 4
C_3	1, 3, 4, 5
$(C_2 \cap C_4)$	7, 8
C_2	6, 7, 8

 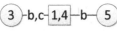

(a) Final Graph   (b) Expansion of C_3 (c) Expansion of C_2

Fig. 9. Final Clustered Graphs

overlapped selected final set(e.g., $C_3 \cap C_1$) and difference of this and intersection sets, i.e., $C_1 - (C_3 \cap C_1)$. The recalculation is performed for the set C_1 with respect to selected final cluster C_3. Clusters recomputed for next iteration are given in Table 4. There remains no overlapped cluster and the process exits. It is possible for the Algorithm 3 to generate clusters containing a single node. Excluding such clusters, Table 5 presents the final set of clusters for the hierarchical structure.

Figure 9 shows final visualizations for the end user. Figure 9a presents the visualization where top level clusters are collapsed. In Figure 9b and 9c, expanded views of clusters C_3 and C_2 are shown respectively. Clusters C_3 and C_2 can be labelled as I_2 and I_1 respectively to make them more understandable to the end user, as these labels reflect the interests of the user profile.

5   Experimentation

To evaluate the effectiveness of proposed approach, we develop a user interface using the JGraph[4] framework. In this case study, we choose the web site of the British Broadcasting Corporation[5]. We collect 39 pages that are used to generate the initial graph. We apply the process described in[11] to generate UIWG

---

[4] http://www.jgraph.com/mxgraph.html
[5] http://www.bbc.co.uk

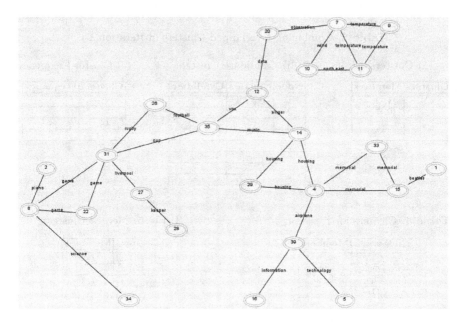

Fig. 10. The Actual User Interest-based Web Graph

Table 6. Allocation of Nodes by Different Clustering Approaches

(a) Content-based Approach

Clusters	Members
north-east-ene-temperature	11, 10
singer-licence-memorial-manchester	14, 15, 33, 1, 34, 35, 12, 29, 16, 20, 26, 28
technology-airplane-information	5, 39
game-liverpool-celtic-season	31, 6, 22, 27

(b) D&C Approach

Clusters	Members
Football	6, 22, 26, 27, 28, 31, 35
Technology(1)	7, 12, 20
Technology(2)	4, 39, 16, 5
Music(1)	6, 22, 31
Music(2)	1, 15

presented in Figure 10. We carefully remove the isolated nodes, which are not connected to other nodes, produced during UIWG generation. It is observed from Figure 10 that nodes with similar topics are scattered over various locations of UIWG. For instance, nodes 39 and 34 have topics related to `technology` but are not close to each other. Should existing content-based clustering be applied, it would generate overlapping clusters and would make the visualization difficult for end users. Clusters and their members obtained by the content-based approach are presented in Table 6a. For D&C clustering, we exploit a user profile with three interests - `technology, football, music`. Table 6b presents the organization of clusters produced by D&C clustering. Corresponding visualizations are presented in Figure 11 where edge labels are intentionally deleted for clarity.

The D&C clustering has benefits from both an end-user and a structural perspective. The clusters of Table 6a suggest that a user with those three interests has difficulty in finding relevant information as cluster tags do not reflect user interests. In contrast, Table 6b shows clusters that are more related to user

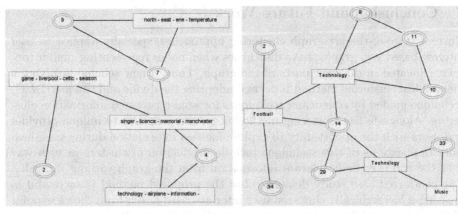

(a) Clustering by Content-based Approach   (b) Clustering by D&C Approach

Fig. 11. Clustered UIWG by Different Approaches

interests. Therefore, it is evident that D&C clustering produces better visualization for end users. From a structural perspective, if we look at the 'singer-licence-memorial-manchester' cluster of Table 6a, we find that node 34 is included whereas node 6 is in a separate cluster though the connectivity of node 34 is only to node 6. On the other hand, this situation is nicely handled in the D&C approach by creating separate clusters with the same tags in different locations of UIWG. The D&C approach also constitutes hierarchical structure as we observe that cluster 'Music(1)' is a complete subset of 'Football'. This implies that the nodes of the 'Music(1)' cluster are related to 'sports' after 'music'.

6   Related Work

Graph clustering is widely used in various domains[13] such as data transformations, information networks and usage and database systems. Clustering techniques broadly fall into two sub-categories: structure-based and content-based. Structure-based clusterings account for hyperlink relationships between chunks of web information. PageRank-based approaches are found to cluster web documents[1], to cluster web graphs by inducing node similarities[5] and by network indices[10]. Content-based clusterings consider the semantic similarity of documents. Some apply heuristics on web graphs[2] to cluster web documents semantically. Major issues in graph clustering include reduction of size, overlapping clusters, time complexity, etc. An attempt to reduce overlapped clusters in multi-level web graph clustering is presented in[9]. An approximation-based approach for faster clustering of huge graphs like web graph is proposed in[12]. A clustering algorithm for information networks consisting heterogeneous objects is presented in[14] using a star-network schema where a ranking-based probabilistic model is applied to each partition achieved from the heterogeneous network.

7   Conclusion and Future Work

Current state-of-the-art graph clustering approaches, specially applied to user interest-based web graphs, have difficulties when nodes representing similar topics are located in different parts of the graph. Thus, a new scheme involving a hierarchical clustering method is devised adopting the divide and conquer(D&C) technique guided by concurrent techniques for semi-structured information clustering. Alongside the clustering methodology, the proposed technique provides end users with the opportunity to explore focused information during visualization. Implications of this technique include production of clusters in such way that the end user finds relevant information from the graph without difficulty. The conducted case study demonstrates: the proposed method is successful in creating a hierarchically clustered user interest-based graph from the view-point of the end user while preserving characteristics of the graph.

The future directions of this work include enhancements of time constraints and picking up user interests accurately. To make clustering faster and more accurate in real-time visualization, we are working in the direction of accomplishing probabilistic interest-based clustering and designing a better user model.

References

1. Avrachenkov, K., Dobrynin, V., Nemirovsky, D., Pham, S.K., Smirnova, E.: Pagerank based clustering of hypertext document collections. In: SIGIR, pp. 873–874. ACM (2008)
2. Bekkerman, R., Zilberstein, S., Allan, J.: Web page clustering using heuristic search in the web graph. In: IJCAI, pp. 2280–2285. MORGKAUF (2007)
3. Brisaboa, N.R., Ladra, S., Navarro, G.: Compact representation of web graphs with extended functionality. Inf. Syst. 39, 152–174 (2014)
4. Hartigan, J.A.: Clustering Algorithms, 99th edn. John Wiley & Sons, Inc., New York (1975)
5. Huang, X., Lai, W.: Clustering graphs for visualization via node similarities. J. Vis. Lang. Comput. 17(3), 225–253 (2006)
6. Jain, A.K.: Data clustering: 50 years beyond k-means. Pattern Recogn. Lett. 31(8), 651–666 (2010)
7. Lin, D.: An information-theoretic definition of similarity. In: ICML, pp. 296–304. ACM (1998)
8. Manning, C.D., Raghavan, P., Schütze, H.: Introduction to information retrieval, vol. 1. Cambridge University Press, Cambridge (2008)
9. Meyerhenke, H., Sanders, P., Schulz, C.: Partitioning complex networks via size-constrained clustering. CoRR abs/1402.3281 (2014)
10. Rattigan, M.J., Maier, M., Jensen, D.: Graph clustering with network structure indices. In: ICML, pp. 783–790. ACM (2007)
11. Saleheen, S., Lai, W.: A new type of web graph for personalized visualization. In: PacificVis, pp. 238–242. IEEE (2014)
12. Savas, B., Dhillon, I.S.: et al.: Clustered low rank approximation of graphs in information science applications. In: SDM, pp. 164–175. SIAM (2011)
13. Schaeffer, S.E.: Survey: Graph clustering. Comput. Sci. Rev. 1(1), 27–64 (2007)
14. Sun, Y., Yu, Y., Han, J.: Ranking-based clustering of heterogeneous information networks with star network schema. In: KDD, pp. 797–806. ACM (2009)

A Serial Sample Selection Framework for Active Learning

Chengchao Li, Pengpeng Zhao*, Jian Wu,
Haihui Xu, and Zhiming Cui

School of Computer Science and Technology, Soochow University,
Suzhou, 215006, P.R. China
{20124227051,ppzhao,jianwu,20124227021,szzmcui}@suda.edu.cn

Abstract. Active Learning is a machine learning and data mining technique that selects the most informative samples for labeling and uses them as training data. It aims to obtain a high performance classifier by labeling as little data as possible from large amount of unlabeled samples, which means sampling strategy is the core issue. Existing approaches either tend to ignore information in unlabeled data and are prone to querying outliers or noise samples, or calculate large amounts of non-informative samples leading to significant computation cost. In order to solve above problems, this paper proposed a serial active learning framework. It first measures uncertainty of unlabeled samples and selects the most uncertain sample set. From which, it further generates the most representative sample set based on the mutual information criterion. Finally, the framework selects the most informative sample from the most representative sample set based on expected error reduction strategy. Experimental results on multiple datasets show that our approach outperforms Random Sampling and the state of the art adaptive active learning method.

Keywords: Data Mining, Active Learning, Sampling Strategy, Uncertainty, Representativeness.

1 Introduction

In many machine learning classification tasks, to learn a high performance classifier, the learners require a sufficient number of labeled training data. However,in many circumstances, the reality is that unlabeled data are in large number and easy to obtain, while labeling them is expensive or time consuming. For example, it is easy to crawl a large number of webpages, however, it typically requires manual effort to annotate these pages. In order to overcome the labeling bottleneck, active learning is a very useful tool with the goal of reducing the overall labeling effort. Active learning models attempt to overcome the labeling bottleneck by querying as little as unlabeled samples with most information from an oracle (e.g., a human expert) and have been used in many areas of machine learning

---

* Corresponding author.

X. Luo, J.X. Yu, and Z. Li (Eds.): ADMA 2014, LNAI 8933, pp. 435–446, 2014.

and data mining, such as outlier detection [1], collaborative filtering [2,3], text classification [4,5], and image classification [6,7].

From a sample-selection perspective, active learning sampling approaches can be mainly divided into two categories. One merely uses uncertainty to measure informativeness of samples and selects the most uncertain sample relative to current classifier for labeling [6,8,9]. Although in some cases such methods have good effects, they only consider the relationship between current sample and the labeled samples, which ignore the distribution information of unlabeled data set. Thus uncertainty strategies may select isolated noise samples. The other one not only takes the uncertainty but also the interrelation between current sample and unlabeled data into account [7,8,10]. These methods consider both the uncertainty and representativeness of samples and choose samples with larger combined value as higher informative ones, which, to some extent, overcome the disadvantages of uncertainty sampling. However, existing methods measure uncertainty and representativeness simultaneously and deal with all unlabeled data. When the number of unlabeled data is large, the computation cost will undoubtedly be great. Due to the rapid development of storage, sensing, networking, and communication technologies, recent years have witnessed a gigantic increase in the amount of daily collected data. As a result, it is computationally prohibitive to directly deal with all unlabeled collected data. Meanwhile, it is difficult to determine the weight factor of uncertainty and representativeness.

To solve problems above, this paper proposes a serial sample selection framework which based on the *Most Uncertain Sample Set* (MUSS) to select the *Most Representative Sample Set* (MRSS). First, it measures uncertainty of unlabeled data and selects the MUSS based on an uncertainty sampling approach. Then, it selects the MRSS from MUSS based on the mutual information approach. Finally, the framework selects the *Most Informative Sample* (MIS) based on the expected error reduction strategy. Our approach has following advantages: First, it not only considers the relations between current sample and labeled set to ensure the high uncertainty of selected samples but also makes full use of interrelations between current sample and unlabeled data to ensure high representativeness of selected samples. Second, it does not need to consider weight factor of uncertainty and representativeness because it selects MRSS from MUSS, which ensures that selected samples are both with higher uncertainty and representativeness. Third, compared with directly dealing with all unlabeled data, our method can avoid selecting samples with low informativeness, thereby effectively improving the algorithm efficiency. Experimental results on three evaluation data sets demonstrate the effectiveness of the proposed approach.

2   Related Work

In order to reduce the workload and get high performance classifier, sampling strategy has been widely studied. Common approaches are as follows: Active learning based on entropy measure [8] used probability classifier to calculate entropy for each unlabeled sample. The greater the entropy is, the higher uncertainty of the sample is. Joshi [6] found that due to the influence of unimportant class probability values, samples with larger entropy value are not necessarily of higher uncertainty and proposed a new sample selection criteria-Best vs Second-Best (BvSB). This criterion has better performance in practical applications. Another widely used framework is Query by Committee (QBC) [11]. The committee is constituted by a set of group classifiers. Each member is then allowed to vote on the labels of query candidate samples. Then they query samples which they most disagree and label them. In essence, QBC method is also based on uncertainty sampling. However, these methods only consider the influence of labeled samples and ignore the distribution of unlabeled data set. According to literatures [12,13], unlabeled samples have a great influence on classification accuracy. If the current sample can represent the remaining unlabeled samples better, we say that the sample is high representative. Taking both labeled and unlabeled sample set into accounts not only measures the uncertainty of current sample relative to the current classifier but also representativeness relative to the remaining unlabeled sample set, which can effectively avoid selecting isolated samples and noise samples. There have been some researches on combination of uncertainty and representativeness. Settles and Craven [10] proposed information density method. It first used traditional uncertain method to measure "basic" information of current sample. Then it used samples' feature vector cosine similarity to measure average similarity between the current sample and all remaining unlabeled samples, which denoted samples' density or representativeness. Density is multiplied with uncertainty and a fixed threshold is set to control the weight of density items. Literature [7] showed the basic idea of an adaptive active learning method which is similar to the literature [10], and it fixed a set of threshold initially, instead of only a threshold value. Selecting the optimal threshold is equivalent to selecting the most informative sample. Experiments showed that adaptive method is superior to information density method. But methods above all take uncertainty and representativeness into account simultaneously. Therefore these methods have the drawback of dealing with large number of non-informative samples, as we discussed above. In this paper, we develop a serial active learning framework, which overcomes their disadvantages.

3   Proposed Approach

3.1   Serial Sample Selection Framework

Considering using the least amount of labeled samples to obtain high performance classifier model, we mainly start from the following aspects. First, in order to minimize label cost, selected samples for labeling should be the most uncertain ones for the current classifier model and the most representative ones for unlabeled sample set. Second, since the number of unlabeled samples is huge, representative samples should be merit-based selection which would improve the operation efficiency. Based on the two considerations above, we propose a serial active learning framework. Within this framework, it first obtains the MUSS using the BvSB approach. From MUSS, it further generates the MRSS, based on the mutual information criterion. Then, the framework selects the MIS based

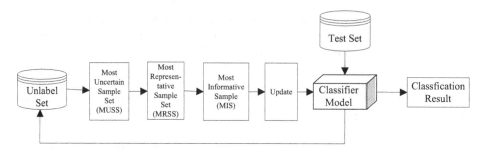

Fig. 1. Our Serial Sample Selection Framework

on expected error reduction strategy and submits the MIS to human experts for labeling. Finally, it updates classifier and labeled and unlabeled sample set. Our algorithm framework is shown in Fig. 1.

3.2   Most Uncertain Sample Set(MUSS) Selection

Literatures [6,14] introduced an uncertainty sampling criteria which considered the difference between the probability values of the two classes having the highest estimated probability value as a measure of uncertainty. From another perspective, we could consider this criteria works as a greedy approximation to entropy [8] for estimating classification uncertainty. The understanding crucially relies on our use of one-versus-one approach for multiclass classification. For details, see literature [6]. Say that our estimated probability distribution for a certain sample x_i is denoted by $\mathbf{P}$, where P_i denotes the membership probability for class y_i. Also suppose that the distribution $\mathbf{P}$ has a maximum value for class y_h. The classification confidence for the classifiers in this set is indicated by the difference in the estimated class probability values, $P_h - P_i$. This difference is an indicator of how uncertain the particular sample x_i is to a certain classifier. Minimizing the probability difference and maximizing the uncertainty, similar with literature [6],we obtain the BvSB measure

$$BvSB = \arg\min_{x_i \in U}(\min_{y_i \in Y_C, y_i \neq y_{Best}}(P(y_{Best}|x_i) - P(y_i|x_i)))$$
$$= \arg\min_{x_i \in U}(P(y_{Best}|x_i) - P(y_{second-Best}|x_i)). \tag{1}$$

where y_{Best} and $y_{Second-Best}$ are the first and second most probable class labels under the current classifier model, respectively, and Y_C represents set of all classes. Samples with small $BvSB$ values are more ambiguous, thus knowing the true label of which would help the classifier model discriminate more effectively between them. Based on this criterion, the influence of uncertainty of all unlabeled samples to the current classifier is measured. Then top MU(a predefined constant-size of MUSS) samples from the unlabeled data set, in terms of the BvSB criterion, are constituted for the MUSS. But uncertainty sampling does not consider the informativeness of current samples in unlabeled data set and

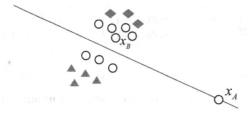

Fig. 2. An illustration of when uncertainty sampling can be a poor strategy for classification

may select noisy samples, which have high uncertainty but cannot provide much help to the classifier. To consider the influence of unlabeled data and avoid the most uncertain samples being noisy points, we also need to consider the representativeness of samples.

3.3 Most Representative Sample Set(MRSS) Selection

In many cases, the most uncertain instances or samples may be isolated or noise samples. As shown in Fig.2, triangles and diamonds represent labeled samples in labeled set and circles represent unlabeled samples in unlabeled set. Since sample x_A is on the decision boundary, it would be queried as the most uncertain. However, querying sample x_B has more information thus could represent the data distribution as a whole better.

In order to solve the above problem of noise samples, representativeness of samples should be considered. Existing methods utilize uncertainty and representativeness at the same time. But our method considers representativeness of samples among the MUSS. According to information theory, mutual information [15] is used to measure relationships between two random variables.

$$I(X, Y) = E(X) - E(X|Y) \tag{2}$$

where $E(X)$ and $E(X|Y)$ respectively represent entropy and conditional entropy of the set variable X. The larger the value $I(X, Y)$ is, the greater the association between X and Y is. This criterion is used to measure representativeness of current sample. We get the definition of representativeness for a candidate sample x_i based mutual information as below.

$$R(x_i) = I(x_i) = E(x_i) - E(x_i|X_{U_i}) \tag{3}$$

X_{U_i} represents the set of unlabeled samples after removing sample x_i from unlabeled sample set. We use Gaussian Process framework [16,17] to compute $E(x_i)$ and $E(x_i|X_{U_i})$ in equation (3). In order to use Gaussian distribution, we associate a random variable with each sample x_i. A symmetric positive definite Kernel function $G(\cdot, \cdot)$ is then used to produce the covariance matrix.

$$\sigma_i^2 = G(x_i, x_i) \tag{4}$$

$$\Sigma_{U_iU_i} = \begin{pmatrix} G(x_1,x_1) & \cdots & G(x_1,x_{i-1}) & G(x_1,x_{i+1}) & \cdots & G(x_1,x_u) \\ G(x_2,x_1) & \cdots & G(x_2,x_{i-1}) & G(x_2,x_{i+1}) & \cdots & G(x_2,x_u) \\ \vdots & \cdots & \vdots & \vdots & \cdots & \vdots \\ G(x_u,x_1) & \cdots & G(x_u,x_{i-1}) & G(x_u,x_{i+1}) & \cdots & G(x_u,x_u) \end{pmatrix} \quad (5)$$

where σ_i^2 and $\Sigma_{U_iU_i}$ represent covariance matrix of unlabeled samples and we assume $U_i = \{1, 2, \cdots, u\}$. One commonly used kernel function is the Gaussian kernel $G(x_i, x_j) = e^{-\lambda(x_i-x_j)^2}$ where is λ a constant. According to the conditional covariance,

$$\Sigma_{i|U_i} = \Sigma_{ii} - \Sigma_{iU_i}\Sigma_{U_iU_i}^{-1}\Sigma_{U_ii} \quad (6)$$

$\Sigma_{ii} = \sigma^2$, Σ_{iU_i}, Σ_{U_ii} and $\Sigma_{U_iU_i}^{-1}$ are calculated by equation (5). Closed-form solutions exist for the entropy of multivariate Gaussian distributions such that

$$E(x_i) = \frac{1}{2}\ln(2\pi e\Sigma_{ii}), E(x_i|X_{U_i}) = \frac{1}{2}\ln(2\pi e\Sigma_{i|U_i}) \quad (7)$$

Using equation (7), representativeness $R(x_i)$ of sample x_i can finally be rewritten into the following form.

$$R(x_i) = E(x_i) - E(x_i|X_{U_i}) = \frac{1}{2}\ln(\frac{\Sigma_{ii}}{\Sigma_{i|U_i}}) \quad (8)$$

The larger the value $R(x_i)$ of sample x_i is, the higher informativeness of unlabeled sample is. Then top MR(a predefined constant-size of MRSS) samples from MUSS, in terms of this estimation criterion, are constituted for the MRSS.

3.4   Most Informative Sample(MIS) Selection

According to analysis in section 3.3, samples in MRSS are both with higher uncertainty and representativeness. Then the MIS is picked based on expected error reduction strategy. The main idea of this strategy is stated as follows. First, we add each candidate sample into labeled sample set and update classifier so that we could get a new classifier. Then, we use the new classifier to classify the remaining unlabeled sample set and select the samples with minimal expected future error rate. Based on the expected error rate of every sample from MRSS, this strategy selects the sample with minimal expected future error rate and submits it to experts for labeling in each iteration. Our expected future error approach is as follows

$$U_{predict} = \sum_{u=1}^{U}(1 - P_{\Theta_{L+<x_i,y_i>}}(y_{Best}|x_u)) \quad (9)$$

where $\Theta_{L+<x_i,y_i>}$ refers to the new classifier model which had sample x_i from label y_i added to labeled set L and y_{Best} represents the class label with the highest posterior probability. $U_{predict}$ represents the prediction loss of the new

---

Algorithm 1. Serial Sample Selection Algorithm

Input: labeled sample set L, unlabeled sample set U
Output: the final classifier model Θ

1 **while** *the stop criterion is not satisfied* **do**
2 | Training on L to learn the classifier model Θ
3 | **for** *each sample x_i in U* **do**
4 | | Compute uncertainty of x_i using Eq.(1)
5 | Select the most uncertain sample set MUSS
6 | **for** *each sample x_i in MUSS* **do**
7 | | Compute representativeness of x_i using Eq.(8)
8 | Select the most representative sample set MRSS
9 | Select sample x^* from MRSS using Eq.(9) and Eq.(10)
10 | Query the true label y^* of sample x^*
11 | Update labeled sample set: $L = L \bigcup < x^*, y^* >$
12 | Update unlabeled sample set: $U = U \backslash < x^*, y^* >$
13 **return** Θ

---

classifier on all unlabeled data. Sample strategy based on equation(9) is defined in equation(10)

$$x^* = \arg \min_{x_i \in MRSS} \sum_{y_i \in Y_C} P_\Theta(y_i|x_i) U_{predict} \tag{10}$$

where Y_C represents set of all classes, and x^* refers to the sample selected for labeling.

Our overall active learning algorithm is given in Algorithm 1. The inputs of the algorithm are: the set of labeled samples, set of unlabeled samples. At the beginning, a SVM classifier model Θ is initialized for the labeled set L. It first obtains the MUSS according to the BvSB approach in section 3.2. From MUSS, it further generates the MRSS based on representativeness criterion in section 3.3. Then the MIS is picked based on expected error reduction strategy in section 3.4. Finally, the MIS is submitted to human experts for labeling and update current classifier, labeled and unlabeled sample set. The process continues until the stop criterion(eg,a target accuracy has been reached) is satisfied.

3.5   Time Complexity Analysis

In this section, we evaluate the time complexity of proposed approach and Adaptive Approach based on the time cost for a query process over the unlabeled data set. We assume that number of unlabeled samples and sample classes are n and k, respectively. Size of MUSS and MRSS are respectively MU and MR. Time cost of two approaches consists of four parts, which are measurement of uncertainty, representativeness, informativeness and selection of MIS based on expected error strategy. Detailed analysis results of four parts of time cost is shown in Table 1.

Table 1. Comparison of time complexity

Process	Adaptive Approach	Proposed Approach	Descriptions
Uncertainty	$1/2 \times n \times (n+1)$	$1/2 \times n \times (n+1)$	Equal
Representativeness	$1/2 \times n \times (n+1)$	$MU \times n$	$MU \ll n$
Informativeness	$10 \times n \times (n+1)$	0	Measured
MIS Selection	$10 \times k \times 1/2 \times n \times (n+1)$	$MR \times k \times 1/2 \times n \times (n+1)$	$MR < 10$

From Table 1 we can know that our approach enhancement on computation rate is mainly reflected in following two aspects. Firstly, time complexity of representative sample selection using Adaptive Approach is $O(n^2)$ while our approach is $O(MU \times n)$, where $MU \ll n$. Secondly, time complexity of informativeness measurement of our approach is 0(combination of the first two steps, namely, representativeness is the value of informativeness) while adaptive approach is $O(n^2)$.

4   Experimental Results

In this section, we present experiments on three different datasets[18,19] to evaluate the effectiveness and efficiency of our proposed approach. Literature [7] shows that Adaptive Active Learning Approach is superior to traditional uncertainty sampling methods [6,8] and ones that based on combination of uncertainty and representation with fixed thresholds [10,11]. So we only compare our approach to Adaptive Approach. The comparison also involves Random Sampling as it is regarded as a baseline. LibSVM [20] is used to train a basic SVM classifier for all approaches in our experiments. For each data set, we report average accuracies across 50 random train/test splits.

4.1   Datasets

Letters Dataset. Letters dataset has 20000 samples and its classes are distributed from A to Z. We randomly selected four subsets containing two 7 classes subsets(A-G,H-N) and two 6 classes subsets(O-T,U-Z) from it. For each category, 200 samples are randomly selected from each class. In Fig. 3, proposed approach outperforms Adaptive Approach, which indicates that the most informative sample indeed exist in MUSS and MRSS with only a fraction of unlabeled samples has been considered. Meanwhile, although Adaptive Approach specifies a set of thresholds and automatically selects the corresponding optimal samples to label, it may still ignore the most informative ones since thresholds themselves have limited range, and selected samples with different thresholds are not necessarily the most informative ones in the whole set.

Pendigits Dataset. The classes are distributed from 0 to 9. Each sample has 16-dimensional features. The number of selected training data set is 340 while that of test samples is 3498, as the whole test set is involved. In Fig.4, proposed

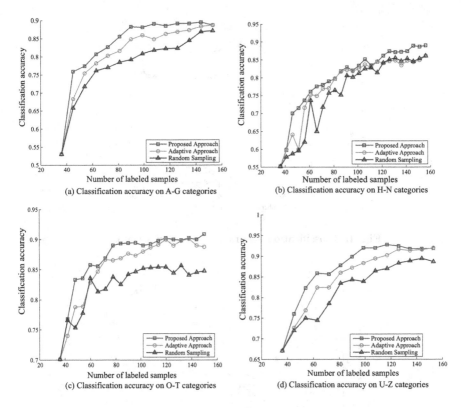

Fig. 3. Classification accuracy on Letters dataset

method and Adaptive method have similar results in early iterations. As the number of labeled samples increases, the classifier achieves higher performance with much fewer labeled samples in proposed approach than in other ones, which demonstrates that the proposed active learning strategy selects more effective samples. Adaptive Approach adjusts the combination of uncertainty and representativeness according to the changing thresholds, which may choose samples of high representativeness and low uncertainty. In this case, our approach is superior to Adaptive Approach for not choosing these samples with little help for improving the classification accuracy. Proposed method is the first to achieve highest classification accuracy and then tends to be stable, which indicates that training samples have been relatively sufficient and is capable of representing the distribution of the whole data set. Newly added training samples in later iterations have little effect on classifier model.

Natural Scene Dataset. Previously selected UCI datasets are simple and here we use complex datasets for validation. We used the 13 Natural Scene Categories dataset, which consists of both natural (coast, forest, mountain, etc.) and man-made scenes (kitchen, tall building, street, etc.), and it is a fairly complete dataset for natural scene dataset. We conducted experiments on the dataset us-

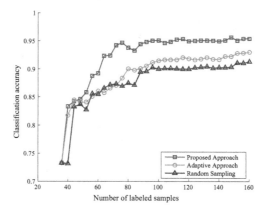

Fig. 4. Classification accuracy on Pendigits Dataset

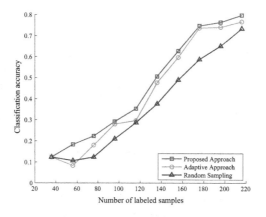

Fig. 5. Classification accuracy on Natural Scene Dataset

ing GIST features and randomly selected ten subsets with 13 classes from it. The average result of ten subsets is shown in Fig. 5. On these subsets, different approaches have advantages in different scenarios and the difference between proposed approach and Adaptive Approach is small. Over these high-dimensional complex datasets, size of MUSS and MRSS have a certain influence over classification accuracy.

4.2   Classification Accuracy

Unlabeled samples can be divided into the following four categories. (1)High uncertainty and high representativeness; (2)High uncertainty and low representativeness; (3)Low uncertainty and high representativeness; (4)Low uncertainty and low representativeness. Both proposed approach and Adaptive Approach will select samples in (1). For (4), all approaches will not select these samples, except for Random Sampling. Random Sampling will not be analyzed for it randomly selects samples for labeling and we just treat it as a baseline. In the following section, we will analyze the differences between the proposed approach and Adaptive Approach on selection of samples for (2) and (3). Our approach selects MRSS from the MUSS, that is, for (2) and (3), proposed approach prefers the latter one, while Adaptive Approach has no preference for (2) and (3). They have the same opportunity to be selected. The most significant difference lies on the selection on (2). For samples in (2), there are two possible cases. One case is that such samples are noise points. Adaptive Approach considers these samples to be high informative ones and select them for labeling. But they have no contribution to the improvement of classification accuracy. In contrast, our approach ignores such samples, so it can reduce label cost and improve classification accuracy. Another case is that those samples are not noise points. In this case, Adaptive Approach selects more effective samples. However, it can be seen from the overall classification results on three datasets that noise samples exist inevitably in reality, which means that our approach is superior to Adaptive approach in most cases.

5   Conclusion and Future Work

In this paper, we present a serial sample selection framework for active learning. It ensures that selected samples are both with higher uncertainty and representativeness and avoids selecting noisy or isolated samples, thereby effectively improving the algorithm efficiency. The method is effective in terms of select one current sample from large unlabeled data set in each iteration. In the future, we will extend this work to the batch mode active learning which allows multiple samples to be selected for manual labeling at one time. Another direction is to further study the definition of a stopping criterion for active learning approaches, because an appropriate stopping criterion also has a great impact on saving the annotation cost and keeping classifier's performance.

Acknowledgments. This work is partially supported by NSFC (No.61003054, No.61170020); Science and Technology Support Program of Suzhou (No.SG2012 57); Science and Technology Support program of Jiangsu province (No.BE20120 75); Open fund of Jiangsu Province Software Engineering R&D Center (SX2012 05); Jiangsu Province Colleges and Universities Natural Science Research Project under grant No.13KJB520021, Suzhou. We would like to thank the anonymous reviewers for their careful reading of the draft and thoughtful comments.

References

1. Abe, N., Zadrozny, B., Langford, J.: Outlier detection by active learning. In: KDD, pp. 504–509 (2006)
2. Koren, Y., Bell, R.: Advances in collaborative filtering. In: Recommender Systems Handbook, pp. 145–186. Springer (2011)

3. Rubens, N., Kaplan, D., Sugiyama, M.: Active learning in recommender systems. In: Recommender Systems Handbook, pp. 735–767 (2011)
4. Nguyen, H.T., Smeulders, A.W.M.: Active learning using pre-clustering. In: International Conference on Machine Learning, pp. 623–630 (2004)
5. Wang, Z., Ye, J.: Querying discriminative and representative samples for batch mode active learning. In: KDD, pp. 158–166 (2013)
6. Joshi, A.J., Porikli, F., Papanikolopoulos, N.P.: Scalable active learning for multiclass image classification. IEEE Trans. Pattern Anal. Mach. Intell. 34(11), 2259–2273 (2012)
7. Li, X., Guo, Y.: Adaptive active learning for image classification. In: CVPR, pp. 859–866 (2013)
8. Settles, B.: Active learning literature survey, vol. 52, pp. 55–66. University of Wisconsin, Madison (2010)
9. Fu, Y., Zhu, X., Li, B.: A survey on instance selection for active learning. Knowl. Inf. Syst. 35(2), 249–283 (2013)
10. Settles, B., Craven, M.: An analysis of active learning strategies for sequence labeling tasks. In: EMNLP, pp. 1070–1079 (2008)
11. Olsson, F.: A literature survey of active machine learning in the context of natural language processing. Sics technical report t2009:06, Swedish Institute of Computer Science (2009)
12. Fujino, A., Ueda, N., Nagata, M.: Adaptive semi-supervised learning on labeled and unlabeled data with different distributions. Knowledge and Information Systems 37(1), 129–154 (2013)
13. Bouguelia, M.R., Belaïd, Y., Belaïd, A.: A stream-based semi-supervised active learning approach for document classification. In: ICDAR, pp. 611–615 (2013)
14. Tuia, D., Ratle, F., Pacifici, F., Kanevski, M.F., Emery, W.J.: Active learning methods for remote sensing image classification. IEEE Transactions on Geoscience and Remote Sensing 47(7), 2218–2232 (2009)
15. Batina, L., Gierlichs, B., Prouff, E., Rivain, M., Standaert, F.X., Veyrat-Charvillon, N.: Mutual information analysis: a comprehensive study. Journal of Cryptology 24(2), 269–291 (2011)
16. Kapoor, A., Grauman, K., Urtasun, R., Darrell, T.: Active learning with gaussian processes for object categorization. In: ICCV 2007, pp. 1–8. IEEE (2007)
17. Rasmussen, C.E., Nickisch, H.: Gaussian processes for machine learning (gpml) toolbox. The Journal of Machine Learning Research 9999, 3011–3015 (2010)
18. Frank, A., Asuncion, A.: Uci machine learning repository (2010), http://archive.ics.uci.edu/ml
19. Li, F.F., Perona, P.: A bayesian hierarchical model for learning natural scene categories. In: CVPR, pp. 524–531 (2005)
20. Chang, C.C., Lin, C.J.: LIBSVM: a library for support vector machines (2001), Software available at http://www.csie.ntu.edu.tw/~cjlin/libsvm.

Map Matching for Taxi GPS Data
with Extreme Learning Machine

Huiqin Li and Gang Wu*

College of Information Science and Engineering, Northeastern University
Shenyang 110004, P.R.China
gloray@163.com, wugang@ise.neu.edu.cn

Abstract. In this work, we explore a new map matching method
through mining historical GPS data collected by taxis. The principle behind is that the map matching can be regarded as a pattern recognition
if there are enough historical GPS points labelled with road network
information. Supervised learning algorithms are feasible for this situation. However, the learning speed of conventional learning techniques is
often not satisfactory, especially facing enumerous classes (road labels).
Considering the matching (classifying) speed and accuracy, we employ
the Extreme Learning Machine (ELM) as a multi-class classifier for its
excellent performance in the learning speed. Furthermore, we propose
MapReduce based GPS trajectories training data preprocessing algorithms and an optimal ELM parameter selection algorithm. Extensive
experimental results show that, compared to the SVM-based approach,
the ELM-based approach achieved faster learning speed and matching
speed, while got close to similar performance on matching accuracy.

Keywords: map matching, pattern recognition, ELM classifier.

1   Introduction

In the vehicle navigation system, the movement track of vehicle on the electronic map reflects the real-time measurement of the GPS devices. However, the
navigation effect is greatly influenced by device error, signal strength, and even
the surrounding geography and weather. Since the GPS data inevitably contain
numerous uncertain information, a vehicle trajectory appeared on the map may
not coincide with the actual track. Therefore, it is quite necessary to detect such
kind of error before any further geographic information mining and analysis.

Map matching is such a process that identify and correct the GPS point to the
corresponding position on the spatial road network through software methods.
Map matching algorithms have been deeply investigated in the past two decades.
But there is much room for improvement. Firstly, not all of the information is
effectively utilized, such as the vehicle direction, the map topology information,
and historical GPS trajectories. Secondly, the adaptability is poor. Even a small

---

* Corresponding author.

X. Luo, J.X. Yu, and Z. Li (Eds.): ADMA 2014, LNAI 8933, pp. 447–460, 2014.

error from the map, e.g. coordinates offset of vector electric map, may cause a strong impact on the precision of the map matching result. Besides, the procedures and rules of the map matching are usually complex and hence not suitable for real-time processing.

Substantially, the map matching is regarded as a pattern recognition process. The patterns are supervised learnt from plentiful historical GPS trajectory data so as to predict new sampling GPS points without tedious matching operations. Considering the large scale of road segments (class label) and GPS trajectory data (training set), we adopt Extreme Learning Machine (ELM) to obtain superior matching accuracy and faster matching speed. ELM is based on the single hidden layer feedforward neural network (SLFNs) [1], which has faster learning speed than traditional neural network algorithms and support vector machine (SVM). With the help of our optimal ELM parameter selection algorithm, the proposed ELM based map matching approach shows significant advantages in the real world experiments.

The contribution of this work are summarized as follows:

- We propose an approach to label a GPS point with corresponding road segment (class) id, which takes various factors into consideration. The factors include the direction of a GPS point, the topological relations between continuous GPS points in the trajectory.
- We propose an ELM based solution to estimate the GPS point's actual position. Map matching is converted into a multi-class classification problem, and hence efficiently solved by ELM algorithm.
- We design a solution to find the optimal ELM parameter. At the same time, we do some adjustment for ELM basic inputs and bring about precision greatly improved.

The paper is structured as follows: Section 2 introduces the related work on map matching. Section 3 defines some concepts appeared in this paper, along with the accuracy criteria and data specification. Section 4 analyzes feature vector extraction for classifier, and introduces learning theories of ELM algorithm. Section 5 presents data processing. Section 6 presents experiments and performance evaluation. The final section gives conclusions.

2   Related Work

The study of map matching is a quite active research field. In this section, we present some of the related work and their main contributions.

According to the information selected from inputed GPS data to use, existing methods can be categorized into four groups: geometric, topological, probabilistic and other advanced techniques [2].

Geometric map matching algorithms are direct and easy, which take advantage of the geometric information of the spatial road network data by considering only the shape of the road without the connectivity between the roads. This kind of methods can often be found in early studies (Bernstein and Kornhauser [3]).

However, the matching quality usually is unstable if urban road networks are complex and density. The context information of GPS trajectory is not taken into account, and it is very sensitive to the offset error of road network.

Topological map matching algorithms utilize the geometry of the roads as well as the connectivity and contiguity of the roads to constrain the candidate position for a sampling GPS point. Greenfeld [4] proposed a weighted topological algorithm. This method makes a topological analysis of a road network and matches using only coordinate information, which neglects any heading or speed information determined from GPS.

Probabilistic algorithms require the definition of an polygon confidence region around a sampling position obtained from a navigation sensor. Choose candidate road segments from confidence region. After determining the matching road, the point is mapped to the closest road according to shortest distance [5].

Advanced map matching algorithms are referred to as those algorithms that use more refined concepts such as the Kalmam filter [6], and the fuzzy logic model [7]. Kalman filter has good filtering performance for signal containing random noise, so it is a preferable tool to process the vehicle trajectory error. But the model of Kalman filter has a higher request for the probability distribution of noise. Yang Li [8] proposed an approach based on joint segment selection. But this method is not suitable for real-time processing, due to high time cost of the joint segment selection procedure. Advanced map matching algorithms require a large amount of data to learn and summarize algorithm parameters.

3   Preliminaries

3.1   Definitions

Definition 1 (Road Segment). *A road segment r is a path between two road nodes. A road segment usually includes other necessary properties, such as r.id denoting the id of r, r.oneway indicating whether the road is one-way or two-way.*

Definition 2 (GPS Point). *A GPS point is a measured point by GPS chip, which consists of taxi ID, timestamp, taxi state, latitude, longitude, taxi speed, taxi driving direction. Driving direction is the angle between the heading direction and north direction where the angle is an integer that ranges within 0~360, clockwise increases, and takes degree as the unit.*

Definition 3 (Taxi Trajectory). *A trajectory is composed of successive GPS points of the same taxi in a continuous period of time. A taxi can have multiple trajectories in one day. We consider three working states for a normal taxi, i.e., empty state (0), carrying passenger state (1), and parking state (3).*

The following two observations help us to design our map matching solution. To avoid influence of sampling errors, only the GPS points with driving directions range up and down 15 degrees around the direction of corresponding located road position are considered.

Observation 1. *On a one-way road, vehicles are allowed to move in only one direction. On this type of roads, the directions of GPS points change within a small scale around certain angle determined by the shape and the location of the road.*

Observation 2. *A two-way road allows vehicles to move on both directions. For a certain road node, assume that the included angle between a two-way road and north direction is α (ranges within $0{\sim}180$), while the direction of another included angle is $\alpha + 180$ degrees. Thus, the directions of GPS points on two-way road change within a small scale around α and $\alpha + 180$ respectively.*

3.2 Problem Description and Accuracy Criteria

Assume that we got taxi A's all trajectories on November 1, 2012. Figure 1 shows one trajectory of taxi A. It is obviously that many GPS points are not on any road segment. It is essential to identify and correct the GPS point to the corresponding position on the road segment of the spatial road network.

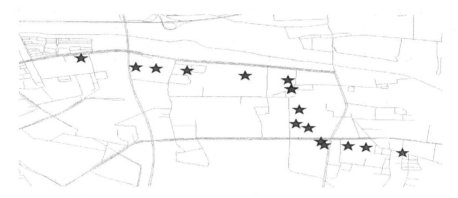

Fig. 1. One trajectory of taxi A

We solve the above problem through our approach which can get the located road segment ID of each GPS point. We adopt *ACCURACY* index to measure the scale of GPS points that are predicted correctly. *CORRECT* means the number of correctly matched GPS points in testing dataset, and *SUM* means the total number of input GPS points in testing dataset. *ACCURACY* index is defined as follows:

$$ACCURACY = CORRECT/SUM \tag{1}$$

3.3 Datasets

Taxi GPS Dataset. The historical GPS dataset used in this work was collected by 12,000 taxis in Beijing in November 2012 (from November 1st, 2012 to

November 30th, 2012), which completely recorded the movement of this city's taxis[1]. The dataset consists of 16,730 time delimited text files, and each file contains over 20,000 discrete GPS point records. The consecutive records are irrelevant in each text file. Table 1 shows a sample of the time delimited text file "20121110035412.txt" in the dataset.

Table 1. Sample data used

Taxi ID	Taxi State	Timestamp	Longitude	Latitude	Direction
486394	0	20121110040854	116.1595001	39.8041000	278
566841	2	20121110040857	116.0809860	39.7710724	114
174973	0	20121110040859	116.6232986	40.3198776	354
194608	1	20121110040859	116.4608459	39.9251099	358
...	...	...	...	...	...

Although the sampling rates for the taxis are quite different, most of the intervals between two consecutive GPS points in one trajectory are less than 10 seconds, which can be considered as high frequency data. In practice, low frequency data is more reasonable because sampling high frequency data is not cost-effective, and results in high redundancy and power consumption. Therefore, it motivates us to create low frequency data from high frequency data by thinning out the plots. We tried five kinds of intervals, i.e. 20s, 30s, 40s, 50s and 60s. Finally, we decided to use 30 seconds to resample the trajectories. It ensures that the distance of continuous two points in one trajectory cannot be too large, and hence can be used to judge the current point's position according to the adjacent point's position. At the same time, points in one trajectory are not too dense, which can reduce the overhead of computation.

Road Network Dataset Another dataset is the road network of Beijing [2], which is a fairly dense and complex network. It is the map matching target in our follow-up experiments. Road network is a system consisting of a various of roads that have different functions, different levels, and different area of the city. A road network offers base for network analysis, for example, finding the best route or creating location-based service, and location recommendation.

4 Map Matching with ELM Algorithm

4.1 Feature Vector Extraction

We analysed the taxi trajectories in different days, and found that these trajectories showed similarity on the map. Figure 2 displays the trajectories of one Taxi in ten days. Different icons indicate the trajectories of different dates. It is obvious that there are many position overlap areas. We can model the road segments of this area through abundant taxi trajectories. In this sense, it is reasonable to use multi-class classification algorithm to learn the models of the roads from different trajectories of different dates.

---

[1] http://www.datatang.com/data/44502
[2] http://www.datatang.com/data/43855

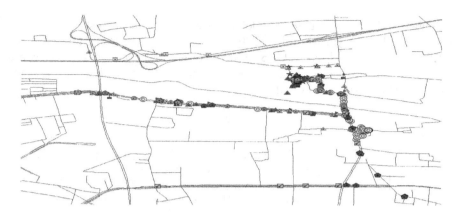

Fig. 2. The trajectories of one Taxi in ten days

Usually, different GPS points that match to the same road segment are gathered together. In the case of accurate electronic map, points distribution density is larger around road centerline. However, it is hard to distinguish the GPS points distributed on two parallel road segments in opposite directions merely according to the location information. Fortunately, parallel road segments in opposite directions are distinguishable with the direction information. Thus, for each GPS point record, the feature vector is defined to be <longitude, latitude, direction>.

4.2   ELM for Map Matching

Extreme learning machine (ELM) is a new learning algorithm based on single hidden layer feedforward neural network (SLFNs). The structure of SLFNs is shown as Figure 3. Compared with general feedforward neural networks, ELM learning algorithm looks much simpler. In ELM, the input weights a_i (linking the ith Input Neurons to the Hidden Neurons) and the hidden nodes biases b_i(the threshold of the ith hidden node) are randomly chosen. The count of hidden neurons L is the only parameter that need to be tuned. β_i is the weight vector connecting the ith hidden node and the output nodes, which is determined analytically through SLFNs.

For N arbitrary diverse samples (x_i, t_i), where $x_i = [x_{i1}, x_{i2}, ..., x_{in}]^T \in R^n$ and $t_i = [t_{i1}, t_{i2}, ..., t_{im}]^T \in R^m$. The output of hidden nodes are defined by function $G(a_i, b_i, \mathbf{x})$ as follow:

$$G(a_i, b_i, \mathbf{x}) = g(a_i \cdot \mathbf{x} + b_i) \tag{2}$$

where g(x) is an activation function. In this work, we adopt sigmoid function $g(x) = 1/(1 + exp(-x))$ as the activation function.

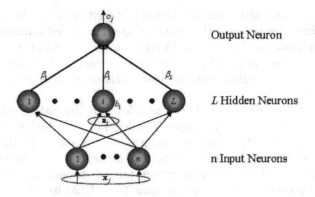

Fig. 3. The SLFNs architecture[9]

The output of SLFNs is:

$$o_j = \sum_{i=1}^{L} \beta_i G(a_i, b_i, \mathbf{x}) = \sum_{i=1}^{L} \beta_i g(a_i \cdot x_j + b_i) \, , j = 1, \cdots, N \qquad (3)$$

There exist β_i, a_i and b_i such that:

$$\sum_{i=1}^{L} \beta_i g(a_i \cdot x_j + b_i) = t_j, j = 1, \cdots, N \qquad (4)$$

In ELM, the input weights a_i and the hidden layer biases b_i are chosen randomly, so SLFNs can be simply regarded as a linear system. The above N equations can be written as follow:

$$\mathbf{H}\beta = \mathbf{T} \qquad (5)$$

where

$$\mathbf{H}(a_1, \cdots, a_N, b_1, \cdots, b_N, x_1, \cdots, x_N) = \begin{pmatrix} g(a_1 \cdot x_1 + b_1) & \cdots & g(a_N \cdot x_1 + b_N) \\ \vdots & \ddots & \vdots \\ g(a_1 \cdot x_N + b_1) & \cdots & g(a_N \cdot x_N + b_N) \end{pmatrix}_{N \times N},$$

$$\beta = \begin{pmatrix} \beta_1^T \\ \vdots \\ \beta_N^T \end{pmatrix}_{N \times m} \quad and \quad \mathbf{T} = \begin{pmatrix} t_1^T \\ \vdots \\ t_N^T \end{pmatrix}_{N \times m}$$

For the linear system $\mathbf{H}\beta = \mathbf{T}$, the smallest norm least-squares solution is:

$$\hat{\beta} = \mathbf{H}'\mathbf{T} \qquad (6)$$

where $\mathbf{H}'$ is the Moore-Penrose generalized inverse of hidden layer output matrix $\mathbf{H}$. The special solution $\hat{\beta} = \mathbf{H}'\mathbf{T}$ tends to reach not only the smallest training error but also the smallest norm of weights (detailed proof can be found in [1]). This salient feature is superior to traditional gradient-based learning algorithms which easily fall into local minimum. In addition, the network has better performance if the weights are smaller, while the gradient-based learning algorithms only try to reach the smallest training errors without considering the magnitude of the weights[10].

Different from traditional gradient-based learning algorithms which need to consider some problems like improper learning rate, overfitting and intensive human intervene, etc., ELM tends to reach the solutions directly. Unlike feedforward neural networks whose all parameters of the networks are tuned iteratively, ELM could generate the hidden node parameters (a_i, b_i) randomly. The learning speed of ELM is extremely fast. At the same time, ELM have better scalability and achieve similar (for regression and binary class cases) or much better (for multi-class cases) performance than conventional feedforward neural network algorithms like the conventional back-propagation (BP) algorithm, and support vector machines(SVM) [11].

5   Data Preprocessing

5.1   Trajectory Acquisition

As we know, the more the number of class labels, the longer the time required for training. Since there are 433,391 road segments in our road network dataset, the training time will be extremely long. Therefore, we introduce a grid partition method to preprocess the data, so that the trainings on the partitions can be paralleled. Suppose that the length of map is L, and the height is H. If we divide the map into $N \times N$ same size partitions, each grid partition will have a length $l = L/N$ and a height $h = H/N$. Assume that the coordinate of upper left corner of the map is $p_0(lat_0, lon_0)$, and the coordinate of any GPS point p in the map is $p(lat, lon)$. Then the ID of the grid partition that p belongs to can be determined by the following formula:

$$ID = floor(N(lat_0 - lat)/h) + floor((lon - lon_0)/l) + 1 \qquad (7)$$

where $floor()$ denotes rounding function.

Figure 4 illustrates one of the grid partitions. The coordinate of upper left corner is $C_0(116°42'0''E, 39°56'0''N)$ and the coordinate of lower right corner is $C_1(116°45'0''E, 39°54'0''N)$.

We adopt MapReduce computing framwork to process historical GPS points. MapReduce is a highly effective and efficient tool for large-scale data processing [12]. Map function takes an input pair and produces a set of intermediate key/value pairs. The MapReduce library groups together all intermediate values associated with the same intermediate key and passes them to the reduce function. Our map function emits taxi ID plus associated GPS point records. The reduce function merges all GPS point records emitted for a particular taxi. In reduce function, firstly, we separate GPS data in accordance with the grid

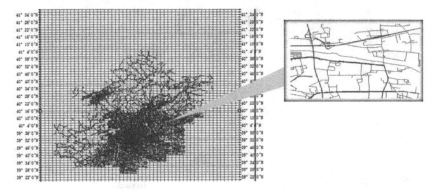

Fig. 4. The grid partition result

partitions of the road network, and extract points records whose taxi state are 0 or 1. Secondly, we sort the discrete GPS points in chronological order. Finally, we get point records which are from the same taxi in the same date and grid.

5.2 Class Label Acquisition

5.2.1 Points from GPS Data

As described in Section 5.1, we have got every taxi's trajectories of November 2012 in the target grid partition. We converted these trajectories into such format that can be displayed on the electronic map intuitively. To facilitate the labelling process, the road network grid and the trajectories within the grid are put together on an electronic map [3]. According to the coordinates, time order, and the taxi driving direction, we can label the located road segment ID for each GPS point. First, we roughly estimate current GPS point's position according to its adjective points' positions. In this way, the spatial locality is taken into account. Then, we determine the located road segment ID in terms of driving direction according to two observations in Section 3.1.

Figure 5 shows the procedure of class label acquisition. In this example, all road segments are one-way roads. The dotted arrow line starting from the star icon edge represents the taxi driving direction.

In Figure 5, there are four time-continuous GPS points, i.e., P1, P2, P3, and P4, from which we can determine the movement trend of the trajectory as shown by gray arrow line. Consequently, we obtain candidate road segments of each GPS points. For P1, there are two candidate road segments, Road1 and Road2. Considering that the driving direction is kept to the right side of the road in China, if P1 drove on Road1, the direction of P1 should be an obtuse angle, while the direction of P1 is an acute angle. Thus, P1 drove on Road2, as shown in S1. For P2, the candidate roads are Road3, Road4 and Road5. We

---

[3] Since there exists certain deviation in the original coordinate of our road network, we corrected the deviation with a tool named GoodyGIS which can be downloaded from http://www.goodygis.com/

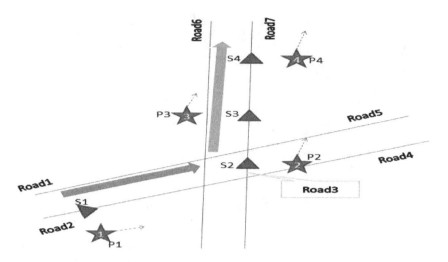

Fig. 5. The procedure of point tags acquisition

assume the angle between $Road_i$ and north direction is $\angle X_i$. Then $\angle X_3 = 2°$, $\angle X_4 = 70°$, and $\angle X_5 = 260°$. Since the driving direction of P2 is 15 degrees, it is more reasonable that P2 drove on Road3 as shown in S2. The rest can be done in the same way. P3 and P4 drove on Road7, as shown in S3 and S4 respectively.

We use the above method to label each GPS point with overlaping satellite map and vector electronic map. It effectively avoids the matching inaccuracy caused by the error, e.g. offset error, originated from the road network.

5.2.2   Points from Road Network

We found that the road network itself can provide a large number of GPS point coordinates data. These points are key points of each road that ensure the shape and the location of one road. Figure 6 shows the key points of one road in the our road network dataset. There are a few key points on straight segment while many key points on winding segment.

As these GPS points are from the road network, the road segments ID can be got easily. However, they lack *direction* feature to be used as training data. To solve the problem, we devise a method of estimation. Given a set of continuous key points, we can calculate the direction for each point from their coordinates. Assume the current point which is on one-way road is $P_0(lon_0, lat_0)$, whose subsequent point is $P_1(lon_1, lat_1)$. Assume the direction of P_0 is α. The arctan value is calculated as follow:

$$\alpha = \arctan((lon_1 - lon_0)/(lat_1 - lat_0)) \tag{8}$$

where arctan() is the arctan function.

Then, we can get final direction of current point through quadrant processing. If the point is on a two-way road, the same coordinate is corresponding to two point records which are in opposite directions, α for one and $\alpha + 180$ for the

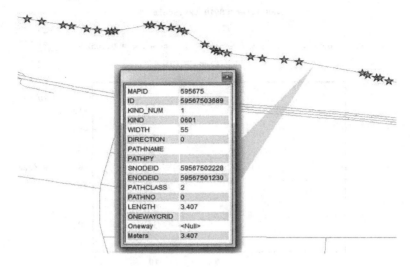

Fig. 6. Key points of one road

other. By this means, we obtain thousands more training data records in one grid partition, which greatly improving the accuracy of matching.

6 Experiments and Performance Evaluation

6.1 Experiments

The training data set used in the experiments contains 8,678 GPS point records, which consists of two parts. One part includes 3,227 labelled GPS point records from 5 days in the targeting grid partition through the method introduced in section 5.2.1. Another part of training data includes 5,451 labelled point records which are generated from the road segments' key points in the grid partition. The testing data includes 1,334 point records from the historical GPS dataset, which have no intersection with the training data.

In order to analyze the performance of ELM, we exploit the Support Vector Machine(SVM) algorithm to do comparative experiments on the same datasets. In this paper, the kernel function used in SVM is radial basis function. SVM algorithm has two important parameters that are the penalty factor c and the kernel function parameter g. We get the optimal parameters that are $c = 32768$ and $g = 0.5$ with the grid search method [13]. It costs 220 minutes to find the best parameters.

We design a solution to get best hidden layer nodes of ELM.

step 1: Choose a smaller number of hidden layer nodes (in our solution is 20), denoted as base.

step 2: Set a larger increment, denoted as the first increment (in our solution is 100). At first increment interval, increase the number of hidden lay nodes from base until the test accuracy decreases. Record the nodes number that reaches highest testing accuracy at this interval as A_1, and the last nodes number as B_1.

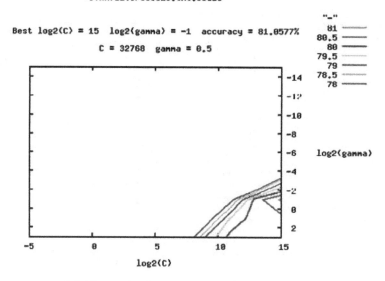

(a) The optimal parameters of SVM.

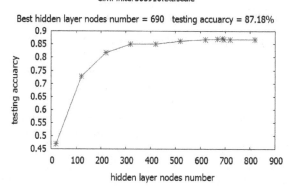

(b) The optimal parameters of ELM.

Fig. 7. The optimal parameters of SVM and ELM

step 3: Set a smaller increment than first increment, denoted as the second increment (in our solution is 50). At second increment interval, increase the number of hidden lay nodes from A_1 until the test accuracy decreases or hidden layer nodes go beyond $B1$. Record the nodes number that reaches highest testing accuracy at this interval as A_2, and the last nodes number as B_2.

step 4: go to step 3 until the increment is 1.

Through above solution, the best hidden layer nodes for this dataset is 690. It takes 37 minutes to find the best hidden layer nodes.

Figure 7 shows the optimal parameters of SVM and ELM.

6.2   Performance Evaluation

After setting the hidden layer nodes of ELM, we directly learnt the ELM model with the training data. But the testing accuracy is not high while training speed is fast. Then, we make normalization which all the input attributes (except expected label) are normalized into the range [-1, 1]. After normalization, the testing accuracy is greatly improved and results are more stable than before. Table 2 shows the non normalized and normalized results of ELM.

Table 2. The normalized and non normalized results of ELM

Type	Hidden nodes	Training time	Testing time	Testing accuracy
non normalization	690	183.9265s	1.0367s	66.12 %
normalization	690	203.1113s	3.2739s	87.18 %
non normalization	100	5.4943s	0.2580s	39.81%
normalization	100	14.18058s	0.6082s	70.31%

Table 3 shows the performance difference of two algorithms. The learning speed (the sum of parameter adjustment time and training time) of ELM is more than 5 times faster than SVM for this case. The testing accuracy obtained by the ELM algorithm is very close to the SVM. More delightfully, the forecasting time of ELM for each GPS point is 0.002454s, while SVM is 0.010714s. Thus, our approach is suitable for real-time applications.

Table 3. The performance of ELM and SVM

Type	Features	Classes	Parameter adjustment	Training time	Testing time	Testing accuracy
ELM	3	583	37min	203.1113s	3.2739s	87.18 %
SVM	3	583	220min	123s	14.293s	91.00%

7   Conclusions

This paper presents a pattern recognition based method to solve the map matching problem. The map matching problem is regarded as a classification problem, and solved using Extreme Learning Machine which can reach a high accuracy at extremely fast learning speed for multi classification learning. In order to reduce the computation time, we make the learning process parallelizable by performing grid partition. In addition, we further optimize the ELM algorithm by proposing an algorithm for quickly finding the optimal number of hidden layer nodes, and improving the matching accuracy through data normalization that makes the results more stable than before.

Acknowledgements. This project is supported by NSFC (Grant No. 61332006, 61370154, 61025007, 61328202); State Key Development Program for Basic Research of China (Grant No. 2011CB302200-G); National High Technology Research and Development Program of China (Grant No.2012AA011004).

References

[1] Huang, G.B., Zhu, Q.Y., Siew, C.K.: Extreme learning machine: a new learning scheme of feedforward neural networks. In: Proceedings of the 2004 IEEE International Joint Conference on Neural Networks, vol. 2, pp. 985–990 (2004)

[2] Quddus, M.A., Ochieng, W.Y., Noland, R.B.: Current map-matching algorithms for transport applications: State-of-the art and future research directions. Transportation Research Part C: Emerging Technologies 15(5), 312–328 (2007)

[3] Bernstein, D., Kornhauser, A.: An introduction to map matching for personal navigation assistants (1998)

[4] Greenfeld, J.S.: Matching GPS observations to locations on a digital map. In: Transportation Research Board 81st Annual Meeting (2002)

[5] Chen, F., Shen, M., Tang, Y.: Local path searching based map matching algorithm for floating car data. Procedia Environmental Sciences 10, 576–582 (2011)

[6] Yang, D., Cai, B., Yuan, Y.: An improved map-matching algorithm used in vehicle navigation system. In: Proceedings of the 2003 IEEE Intelligent Transportation Systems, vol. 2, pp. 1246–1250. IEEE (2003)

[7] Quddus, M.A., Noland, R.B., Ochieng, W.Y.: A high accuracy fuzzy logic based map matching algorithm for road transport. Journal of Intelligent Transportation Systems 10(3), 103–115 (2006)

[8] Li, Y., Huang, Q., Kerber, M., et al.: Large-scale joint map matching of GPS traces. In: Proceedings of the 21st ACM SIGSPATIAL International Conference on Advances in Geographic Information Systems, pp. 214–223. ACM (2013)

[9] Huang, G.B.: Extreme learning machine: learning without iterative tuning. School of Electrical and Electronic Engineering. NTU, Singapore (2010)

[10] Huang, G.B., Zhou, H., Ding, X., et al.: Extreme learning machine for regression and multiclass classification. IEEE Transactions on Systems, Man, and Cybernetics, Part B: Cybernetics 42(2), 513–529 (2012)

[11] Huang, G.B., Zhu, Q.Y., Siew, C.K.: Extreme learning machine: theory and applications. Neurocomputing 70(1), 489–501 (2006)

[12] Dean, J., Ghemawat, S.: MapReduce: simplified data processing on large clusters. Communications of the ACM 51(1), 107–113 (2008)

[13] Chang, C.-C., Lin, C.-J.: LIBSVM: a library for support vector machines. ACM Transactions on Intelligent Systems and Technology 2, 27:1–27:27 (2011), http://www.csie.ntu.edu.tw/

A Segment-Wise Method for Pseudo Periodic Time Series Prediction*

Ning Yin, Shanshan Wang, Shenda Hong, and Hongyan Li**

School of Electronics Engineering and Computer Science, Peking University, Beijing 100871
Key Laboratory of Machine Perception, Ministry of Education, Beijing 100871
{yinning,lihy}@cis.pku.edu.cn

Abstract. In many applications, the data in time series appears highly periodic, but never exactly repeats itself. Such series are called pseudo periodic time series. The prediction of pseudo periodic time series is an important and non-trivial problem. Since the period interval is not fixed and unpredictable, errors will accumulate when traditional periodic methods are employed. Meanwhile, many time series contain a vast number of abnormal variations. These variations can neither be simply filtered out nor predicted by its neighboring points. Given that no specific method is available for pseudo periodic time series as of yet, the paper proposes a segment-wise method for the prediction of pseudo periodic time series with abnormal variations. Time series are segmented by the variation patterns of each period in the method. Only the segment corresponding to the target time series is chosen for prediction, which leads to the reduction of input variables. At the same time, the choice of the value highly correlated to the points-to-be-predicted enhances the prediction precision. Experimental results produced using data sets of China Mobile and bio-medical signals both prove the effectiveness of the segment-wise method in improving the prediction accuracy of the pseudo periodic time series.

Keywords: pseudo periodic time series, entropy, time series segment-wise method, time series prediction.

1 Introduction

In many practical applications, the data in time series appears highly periodic, but never exactly repeats itself. Such time series are called pseudo periodic time series. Typical examples can be found in voiced speech signal, gait signal, electrocardiogram signal [1] (e.g. Figure 1) and economic times-series (e.g. Figure 2).

Predictions on pseudo periodic time series are of great importance in many fields. An accurate prediction of bio-medical signals, for example, can help to take timely precautions before the outbreak of a disease.

---

* This work was supported by Natural Science Foundation of China (No.60973002 and No.61170003), the National High Technology Research and Development Program of China (Grant No. 2012AA011002), and MOE-CMCC Research Fund(MCM20130361).
** Corresponding author.

X. Luo, J.X. Yu, and Z. Li (Eds.): ADMA 2014, LNAI 8933, pp. 461–474, 2014.

Time series often show trends of variation. Some variations have abnormal values and recur within every period. These abnormal variations add to the difficulty of making predictions since they imply fundamental changes in the underlying objects and possess a high domain significance. For example, in Figure 2, daily revenues are outstandingly high on the first and last few days of the month: between two and five times higher than the mean value, which is around ten million. Since the majority of the revenue is generated on these few days, the variations on these days are of greater value than those on other days in business analysis and monitoring.

However, it is not a trivial task to analyze and accurately predict this kind of time series. Any prediction model would be flawed if it ignored periodicity. Furthermore, since a pseudo period never exactly repeats itself and shows variations as time elapses, the periodic prediction method which highly depends on the fixed periodicity and is estimated by a deterministic periodic function would not perform satisfactorily. Once the periodic prediction method is applied to pseudo periodic time series, errors will accumulate and lead to a poor prediction.

Most of the existing prediction methods assume that the time series is stationary and usually view the seasonality as an unobserved component with constant variance and zero sum over seasons. Hence, any abnormal variations have to be smoothed out before modeling with most of the existing methods. However, these abnormal variations can actually be of great value in certain domains. Smoothening will not only increase the deviations in fitting and prediction, but also lower the significance of a model. In pseudo periodic time series, deviating points cannot be compared with neighboring points directly, but only with the corresponding points in the last period, which leads to two main problems. First, it is difficult to make comparison between two periods of different time span. Second, it is even harder to identify the corresponding variation points in two pseudo periods.

The main contribution of the paper is that we propose an entropy-based segment-wise prediction method to solve the prediction problem of pseudo periodic time series

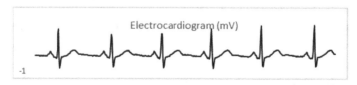

Fig. 1. Data evolution over an electrocardiogram (ECG) stream

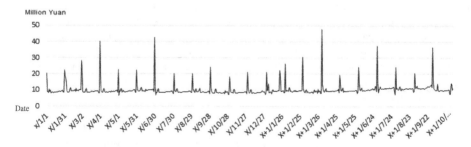

Fig. 2. AD04. ("AD04" refers to a China Mobile KPI. "X" in date refers to a particular year.)

with abnormal variations. Although our focus here is on pseudo periodic time series with abnormal variations, the approach proposed in the paper is highly generalizable and can be applied to many other types of periodic time series or streams. "Segment-wise" denotes the decomposition of a time series into subsequences sharing similar variation patterns, paving roads for later process. Only the segment corresponding to the target time series is chosen for prediction, which leads to a reduction in the number of input variables. At the same time, choosing a value highly correlated to the points-to-be-predicted enhances the prediction precision. The experimental results produced using data sets of China Mobile and bio-medical signal have both proved the effectiveness of the segment-wise method in improving the prediction accuracy of the pseudo periodic time series.

2 Related Work

A vast amount of research has been devoted to the analysis and prediction of time series, some of which performing quite well. Yet, universal as these methods might be, they are not suited for all applications. When dealing with pseudo periodic time series with abnormal variations, these methods tend not to fulfill our expectations. Therefore, more effective methods need to be investigated.

Linear predication methods such as Moving Average (MA), Auto Regressive (AR), and ARIMA [2], are only suitable for stationary time series. To ensure the stability of the above methods, all abnormal variations in the pseudo periodic time series have to be smoothed to avoid a likely recurrence of the abnormal values in the following periods, which leads to an even greater prediction error. Given the importance that this paper assigns to these abnormal variations, linear methods might therefore not be a suitable solution.

Similarly, methods for seasonal time series prediction are proposed quite often but fail to estimate the recurring but persistently changing patterns within a given year [3]. The nature of pseudo periodicity makes it impossible for these methods to estimate seasonal changes precisely.

Nonlinear prediction methods have also been widely used in time series prediction, including artificial neural networks [4, 5], hybridization of traditional forecasting methods and intelligent techniques [6, 7] and Garch models [12]. Although they work unexpectedly well in certain cases of nonlinear time series, the majority of the methods are like a "black box", which means they hardly tell us anything about the domain knowledge implied by time series.

3 Preliminaries

3.1 Notation

Some of the key terms used in the paper include:

Definition 1: *Time Series.* A time series Y is an ordered set of n variables. $y_1, \ldots y_n$ are typically arranged by temporal order separated by a constant time interval.

A pseudo time series is a series that appears repetitive but displays differences between pairs of consecutive periods, either in the key value or the time interval [9]. Daily pseudo periodic time series are the main focus in this paper.

Definition 2: *Subseries.* Given a time series Y, a subseries S of Y is a sampling of contiguous sequences derived from Y. The size of a subseries S is the number of points contained in it and is denoted as $|S|$.

Definition 3: *Subsequence.* A subsequence is a sequence that can be derived from another sequence by deleting some of its elements without changing the order of the remaining elements. Given a time series, $Y = y_1, y_2 \dots y_n$, the subsequence X is a subset of Y whose members maintain their original order. The size of a subsequence R is the number of points contained in it and is written as $|X|$.

3.2 Task Specification

The purpose of time series predictions is to estimate future value patterns as precisely as possible on the basis of previous time series. When dealing with time series, several techniques pay particular attention to the last values of the series. One example is the "one-step ahead" prediction:

$$\hat{y}_t = f(y_{t-1}, y_{t-2}, \dots, y_{t-M}) \tag{1}$$

where M denotes the number of inputs, $y_{t-1}, y_{t-2}, \dots, y_{t-M}$ denotes the observed values.

If values in the series are treated in order as is shown in example (1), we would be confronted with three problems. First, in order to make an accurate prediction, the value of M must be large enough to take several periods into consideration. If the input dimensionality is too large, one can be confronted with the "curse of dimensionality" problem [10]. Moreover, the computational complexity and memory requirements for the learning model will undoubtedly increase. Second, despite its comparably good performance, a complex model with too many inputs is obviously more difficult for people to fully grasp than a simple model with fewer inputs. Finally, since the abnormal variations are always caused by changes in underlying or external factors, they have nothing to do with the adjacent points but rather the last periods of the series. Additional irrelevant inputs will lead to poorer models.

The prediction function of segment-wise method is defined as:

$$\hat{y}_t = f\left(y_{p_1(t)}, y_{p_1(t)-1}, \dots, y_{p_1(t)-M_1}, \dots, y_{p_i(t)}, y_{p_i(t)-1}, \dots, y_{p_i(t)-M_i}, \dots\right), i = 1, \dots n \tag{2}$$

where $p_i(t)$ is a set of functions mapping index t to the corresponding index in the previous i^{th} period which sharing the closest variation pattern to the current one, and M_i is the time span of the coordinate variation in the previous period. Finding $p_i(t)$ and M_i can therefore be defined as the main task. After this, a prediction model can be built based on Eq. (2). In doing so, the segment-wise prediction method can track the evolution of different variation patterns and give an accurate prediction based on previous coordinate variations.

4   Segment-Wise Prediction Method

The prediction method of pseudo periodic time series with abnormal variations intro-duced in the paper can be divided into three main steps. First, the pseudo periodic time series are split into several periods. Then, each period is segmented based on entropy in order to extract different variation patterns. Finally, the prediction model based on Eq. (2) mentioned above can be built based on this segment-wise model. The medical signal ECG is used as an example to illustrate the process of build seg-ment-wise model in the rest of this paper. Daily time series like China Mobile KPIs, as illustrated in Figure 2, have predicable period length (the month of the year), and is much easier to predict. The paper will end on a further discussion of this case at the end of the section.

4.1   Period Splitting

Periods in a pseudo periodic time series share a similar pattern despite the differences in their value and time interval. To identify the implicit pattern, all of the series' peri-ods should be checked. Due to the nature of pseudo periodic time series, merely di-viding the time series into intervals with a fixed length would not yield the desired result. A more effective method should be carefully designed in order to achieve a precise period splitting. Several period detection methods, such as [11, 12] are widely used already. However, both time and space costs are high over pseudo periodical time series, especially in an online environment where large amounts of data are col-lected within a short period of time.

In fact, if we pay a closer attention to the time series with abnormal variations, it is not hard to notice that these variations follow certain basic rules in a specific field, which means that efficient splitting methods could be devised on the basis of careful study. As is shown in Figure 1, the medical signal streams are composed of waves with various time lengths and key values. The waves start and end at valley points below a certain value, which enables us to divide the time series into periods at these valley points. Since the upper bound value may change as the time series evolves, we can automatically update it using the average value of the past valley points [9]: $U_b = \propto (\sum_{i=1}^{N} T_i)/N$, where N is the number of the past valley points and $\propto$ is an adjustment factor depending on the evolution of the data. In the ECG experiment, the best split effect was yielded when $\propto = 1.1$. For other daily time series like the China Mobile KPIs in Figure 2, each month could be used as a period, since both the busi-ness rules and consumption habits recur in monthly intervals.

4.2   Time Series Entropy

Entropy is a widely-used concept when measuring the complexity of time series, be-cause it captures the intensification of the variations in time series. We employed a concept very similar to the information gain used in the traditional decision tree [13].

With the help of this concept, different variation patterns in time series can be distinguished efficiently, paving the way for the following model-building and prediction.

The entropy is calculated after the symbolization and coding of the time series. The basic idea guiding the symbolization is to reduce the large amount of possible continuous values into fewer discrete values. This is called the coarse-grained process.

Definition 4: *The Symbol Representation of a Time Series.* Given a time series Y, a discrete symbolic series $R = r_1, ... r_n$ $(r_i \in A_T)$, which is defined by a specific symbolic approach, can be attained after the symbolization of Y. A_T is the alphabetical set used to represent Y.

This process is capable of capturing the large-scaled features which are sufficient for building up the segment-wise model of the time series. The symbolized time series has clear advantages, such as the prevention of noise and the reduction of data. Meanwhile, it effectively saves computational resources over the too-slight variations, boosting the performance of the algorithm and is adaptable to stream environment.

Conventional ways of symbolizing the time series are histograms, first-order differences, Move Average etc. The number of symbol sets has an impact on the accuracy of entropy. Abnormal and normal variations have to be distinguished in this paper, so the size of the symbol set is set to 2. Given that the time complexity would be increased by too many symbols, and despite these costs, cannot certainly bring about the desired effect, we symbolize the time series in the following way:

For each t_i in T, $r_i = \begin{cases} 0, if\ t_i < \mu + \propto \delta \\ 1, if\ t_i \geq \mu + \propto \delta \end{cases}$

μ and δ are the mean value and standard variance of T. $\propto$ is an adjustment factor depending on the proportion of abnormal variations, which can be easily observed from the past values. In the application of ECG and China Mobile KPIs shown in Figure 1 and Figure 2, the method yields the best effect when $\propto=1.2$.

Time series entropy could be calculated easily after the symbolization of Y.

Definition 5: *Time Series Entropy.* Let R be the symbolic series of Y and A_Y be the alphabetical set. For each symbol a in the alphabetical set A_T, the probability of appearance is defined as $p(a)$. Thus, the entropy of time series Y can be defined as: $I(Y) = \sum_{a \in A_Y} -p(a) \log (p(a))$.

The symbolization of subseries $S_{m,y}$ is the corresponding subseries of R. Notably, the alphabetical set $A_{S_{m,y}}$ only contains symbols that have appeared in the symbolic representation of $S_{m,y}$, and thus $A_{S_{m,y}} \subseteq A_T$. The subseries entropy can be defined as:

$$I(S_{m,y}) = \sum_{a \in A_{S_{m,y}}} -p(a) \log(p(a))$$

4.3   Period Queues

One of the main difficulties in the analysis of pseudo periodic time series is the difference in the time span of each of its periods. To address the problem, some transformation methods have been proposed. For example, when predicting the daily tax revenue, the time span between two months has been made constant by a transformation function, which enables analyzers to model the data for months with varying numbers and spacing of bank days in a parsimonious way [3]. However, this method could hardly be employed to handle the ECG data or other pseudo periodic time series. Different from daily time series, the time spans of each of its periods in ECG data or other pseudo periodic time series are totally unpredictable.

In this paper, after period splitting, periods of pseudo periodic time series are stored in a data structure called period queues. Periods sharing similar time spans are put in the same period queue, with the help of which the mutual variation structure can be identified more easily.

Definition 6: *Period Queue*. Let Y be a time series and $P_i, i = 1, \dots, n$ be the periods generated from the period splitting. A period queue Q_i is defined as a queue containing several periods of Y, and $Q_i = \{P_{i_1}, P_{i_2}, \dots P_{i_c}\}$.

The strategy for building up a set of period queues is specified as follows:

τ is set as the tolerance factor of difference of periodic length. This value is generally taken to be 2% of the periodic length. Next, all the periods acquired are clustered and the periods with a difference of periodic length within τ are located in the same period queue in a chronological order. In this way, several period queues could be gained and we denote the number of the queues as m. The capacity of each period queue is written as c. To update period queue when new data forms a complete period, the period is placed in the corresponding period queue. When the quantity of periods stored in the queue exceeds its ceiling, the head period of the queue is moved out. The extra storage cost is thought to be O(mc), since the periods in the period queue are saved as indices of original time series.

The variation within each period queue is defined by entropy of period queue.

Definition 7: *period queue entropy*. Let Y be a time series and $Q_i = \{P_{i_1}, P_{i_2}, \dots\}$ be a period queues. The entropy of a certain period queue Q_i is the weighting sum of its subseries entropy and defined as $I(Q_i) = \sum_{p \in Q_i} \frac{len(p)}{\sum_{q \in Q_i} len(q)} I(p)$, and $len(p)$ denotes the periodic length of period p.

The division based on period queue is of advantages when we try to find out the influence that periodic rules have on the time series variations, and it reduces the effects caused by certain non-repeatable variations.

4.4   Entropy Based Segmentation

Affected by various factors, different variations could appear in a single period, and the pattern of these variations is of limited values. To reduce the number of inputs and pick out the data most valuable to the prediction, the time series needs certain division. Meanwhile, $p_i(t)$ and M_i in Eq. (2) can be obtained during the process of the division. We denote k as the segmentation number, depending on the field the problem is in. Generally speaking, before reaching a certain threshold, the descriptive capability and predictive accuracy of the model increase as the value of k increases.

So does the time complexity. In the cases of ECG and China Mobile KPI, $k = 3$ turns out to be sufficient.

The segmentation strategy can be described as a set containing k-1 segmentation points, which is introduced to divide each period of the period queue into k segments.

Definition 8: *Period Queue Segmentation Strategy.* Given a period queue Q_i, the employed period queue segmentation strategy is defined as $D_i = \{d_1, \ldots d_{k-1}\}$. A set of integers $1, \ldots, k - 1$ indicates the offset from the start of a period and divides all the periods of Q_i into smaller subseries.

To find out the best segmentation strategy, we use entropy as the main criterion to evaluate its performance. The introduction of entropy brings a number of advantages. First, the strategy avoids parameters, so that the segmentation algorithm is of high universality and can be used in different fields. Second, the strategy is of excellent extensibility which means that the algorithm is always adaptable, even as the value of k increases. Third, the introduction of entropy can always achieve the best results of the segmentation. Furthermore, information gain is employed to evaluate the quality of the segmentation points.

Information gain and the best segmentation strategy are defined as follows:

Definition 9: *Information Gain.* Given a period queue Q_i and a segmentation gy D_i, the entropy before and after applying D_i on Q_i are denoted as $I(B_i)$ and $\hat{I}(B_i)$. The information gain achieved by the segmentation strategy can then be expressed as $Gain(D_i) = I(B_i) - \hat{I}(B_i)$.

Definition 10: *Optimal Period Queue Strategy (OPQS).* The optimal period queue segmentation strategy on Q_i is the one which can reach the goal of $Gain\left(D_i^{opqs}\right) \geq Gain(D_i)$, for any D_i.

The algorithm used to calculate both the period queue entropy and information gain after segmentation is described as:

```
CalculateEntropy(period queue Q, seg_set)
    entropy_b ← 0
    For each period P in Q
        subseries_set ← segment S by seg_set
        entropy_s ← 0
        For each subseries in subseries_set
```
$$entropy\_s = entropy\_s + \frac{|subseries|}{|P|} I(subseries)$$
```
    EndFor
```
$$entropy\_b = entropy\_b + \frac{|P|}{|Q|} entropy\_s$$
```
    EndFor
    Return entropy_b
InformationGain(period queue Q, seg_set, seg_point)
    entropy ← CalculateEntropy(Q, seg_set)
    entropy_new ← CalculateEntropy(Q, seg_set U
    { seg_point })
    Return entropy - entropy_new
```

The ideal segmentation point in the period queue can be found by calculating the traverse of all possible segmentation points in order and comparing the information gain each time. The algorithm is listed below:

```
GenerateCandidates(period queue Q, seg_set)
    candidates ← Ø
    For (i = 1; i < |Q|; i++)
        If i ∉ seg_set;
            candidates ← candidates U {i}
        EndIf
    EndFor
    Return candidates

FindingBestSeg (period queue Q, seg_set)
    candidates ← GenerateCandidates(Q, seg_set)
    bsf_gain ← 0
    bsf_seg_point ← 0
    For each seg_point in candidates
        gain←InformationGain(Q, seg_set, seg_point)
        If gain>bsf_gain
            bsf_gain ← gain
            bsf_seg_point ← seg_point
        EndIf
    EndFor
    Return bsf_seg_point
```

The first *k-1* optimal segmentation points are chosen successively to constitute the best segmentation strategy. The algorithm is presented in the following table.

```
FindingBestSpiltStrategy(period queue Q, k)
    seg_set ← Ø
    For (i = 1; i < k; i++)
        bsf_seg = FindingBestSeg (Q, seg_set)
        If bsf_seg_point ≠ 0;
            seg_set ← seg_set U {bsf_seg }
        EndIf
    EndFor
    Return bsf_seg
```

Suppose that the capacity of each queue is set to c, and the average period of a pseudo periodic time series is t. Then it is quite obvious that the time complexity of the construction of an optimal segmentation on the basis of the original time series can be expressed as $O(m*c*t*k)$, or $O(l*t*k)$, where l stands for the length of the time series. Given that the value of t, k is far less than that of l, hence the time complexity of the whole process can be expressed as $O(l)$. The whole model is refreshed only if the corresponding period queue is refreshed, which in turn occurs whenever a complete round of periodic waves has been observed lately in a stream environment. Consequently, the cost of its refreshing is estimated as $O(c*t*k)$.

4.5   Segment-Wise Model

In order to make the analysis of time series more intuitive, the corresponding segments in each period is extracted and reconstructed into subsequences in chronological order. Since each period queue has k segments as mentioned in the last step, k subsequences can be formed. The process realizes the mapping of $p_i(t)$ in Eq.(2) and results in M_i. Compared with the original time series, the data within each reconstructed subsequence is affected by similar factors, and displays smaller variations between neighboring points. Therefore, a more accurate prediction can be drawn from a general prediction method. The formalized description of this process can be expressed as follows:

Suppose all the period queues are denoted as $Q_1, Q_2, ..., Q_m$, ordered by periodic length in ascending order and their corresponding periodic length are denoted as $\daleth = \tau_1, \tau_2, ..., \tau_m$. To a certain period queue, all the periods in it are recorded in form of a matrix $\mathbb{Q}$ by time indices. The starting time of each period is saved in a row with $t_{i1} > t_{i2} > \cdots > t_{ic}$, so that the finishing time can be calculated as $t_{ij} + \tau_i$.

$$\mathbb{Q} = \begin{bmatrix} t_{11} & \cdots & t_{1c} \\ \vdots & \ddots & \vdots \\ t_{m1} & \cdots & t_{mc} \end{bmatrix}, \qquad \Delta = \begin{bmatrix} d_{11} & \cdots & d_{1k} \\ \vdots & \ddots & \vdots \\ d_{m1} & \cdots & d_{mk} \end{bmatrix}$$

A $m*k$ matrix Δ of which each row represents the segmentation strategy of a period queue is achieved after the segmentation of all period queues. m stands for the total number of the period queue and k indicates the number of segmentations. d_{ij} is the time offset to the starting point, and $d_{ik} = \tau_i$. Hence, the segment-wise model can be described by a two-tuple $(\mathbb{Q}, \Delta)$.

Suppose the current time is t, and the nearest time point in the set $\mathbb{Q}$ is t_{l1}. Since the data in the range of $(t_{l1} + \tau_l + 1, t - 1)$ have not formed a complete period, the total length of the period cannot be decided yet. Two methods are now available to help deciding which period queue to use for the prediction of the coming period. (1) Given the slight variations between the two neighboring periods in a pseudo periodic time series, the period queue of the last period can be used. (2) Given part of a period time series matching techniques [19 20] can be utilized to find the best fitted period. In this manner, $(t_{l1} + \tau_l, t - 1)$ is used as the target series to match with the period series located in tails of all the period queues, and returning the queue with the most matching tail period. Generally, the match is less precise if $(t - 1) - (t_{l1} + \tau_l)$ is less than 1/3 of the average period. So, only under this circumstance, is the first, rather than the second approach employed.

The period queue is denoted as Q_s here, and we use $t - (t_{l1} + \tau_l)$ to decide which segment to employ. Suppose there is a j that satisfies the requirement $d_{s\,j-1} < t - (t_{l1} + \tau_l) \le d_{sj}$, then the jth segment should be used in the prediction. As to $p_i(t)$ and M_i: $p_i(t) = t_{si} + \tau_s, M_i = d_{sj}$, Eq (2) is thus changed into:

$$\hat{y}_t = f\left(y_{t_{s1}+\tau_s}, y_{t_{s1}+\tau_s-1}, ..., y_{t_{s1}}, ..., y_{t_{si}+\tau_s}, y_{t_{si}+\tau_s-1}, ..., y_{t_{si}}, ...\right), i = 1, ... c \qquad (3)$$

f can be fitted by most time series predicting methods. In the next section, further comparisons will be made between the prediction ability of the segment-wise model and that of the original time series, shedding light on the improvements in prediction accuracy achieved by the segment-wise model.

5   Experimental Evaluation

To evaluate the prediction ability of the segment-wise model, several prediction techniques were used as benchmarks. These methods include Autoregressive Integrated Moving Average (ARIMA), Linear Regression (LR), Artificial Neural Network (ANN) [15], Nonlinear Autoregressive (NAR) [16] and Nonlinear Autoregressive with External Input (NARX) [16]. These methods will be applied to the original time series of Eq. (1) and then to the segment-wise models of Eq. (2).

Real datasets from two domains are used. For the economic domain, 28 KPIs provided by China Mobile are used, each of which is a daily time series spanning three years. Two and a half years of the data are used for training and the rest for prediction validation. Since the series yield similar results, for the sake of brevity, we only illustrate the results of three of them: AD01, AD03 and AD04. The second dataset from the domain of medical applications contains ECG signals downloaded from the abdominal and direct fetal ECG Database[1]. There are 6000 points in this time series. The first 5000 values are used for training and the remaining data for testing. Every prediction method is carried out thrice on each dataset, and the average is used as the final prediction result. All the experiments were conducted on a PC with an Intel Core i3-540 CPU, 4*2 GB RAMS and running window 8. The segment-wise method is conducted using C# under Visual Studio 2010. Other prediction methods are conducted using R (R Software Package 3.0.2) or the environment of SPSS (Statistical Product and Service Solutions 20.0) and MATLAB (Matrix Laboratory 2010B). After a brief introduction to some prediction methods commonly used, we report the experimental results.

ARIMA is an important time series prediction method and has been wildly used in the field of economics. For each dataset, we chose the model of a lower MAPE in two ways: first, we build an ARIMA model by using the automatic modeling process in SPSS; second, we build up a model programmed with R. The second strategy follows the general ARIMA modeling procedure. The ARIMA models are employed for the datasets AD01, AD03 and AD04 which correspond to ARIMA(0,0,0), ARIMA(0,1,1) and ARIMA(0,0,1) respectively. Since there are several segments on each segment-wise model and each segment has a different ARIMA model, we will not specify all models here. ARIMA with smoothening employed T4253H function in SPSS. Seasonal methods such as SARIMA and Holt's yield similar result which is why they were excluded from the exposition.

LR is a wildly used method in statistics [17]. In the prediction of China Mobile KPIs, the explanatory variable date is decomposed into three variables: Year, Month and Day. To predict the response variable, we construct the following model:$\hat{Y} = \hat{\beta}_0 + $ el:$\hat{Y} = \hat{\beta}_0 + \hat{\beta}_1 X_{Year} + \hat{\beta}_2 X_{Month} + \hat{\beta}_3 X_{Date}$ Our goal is to select model parameters (intercept

---

[1] http://www.physionet.org/cgi-bin/atm/ATM

and slopes) so as to minimize the difference between observed response values and those predicted by the model. Specifically, model parameters are selected to minimize the sum of squared residuals:

$$\sum_{i=1}^{n}(Y_i - \hat{Y}_i)^2 = \sum_{i=1}^{n}(Y_i - \hat{\beta}_0 + \hat{\beta}_1 X_{Year} + \hat{\beta}_2 X_{Month} + \hat{\beta}_3 X_{Date})^2 = \sum_{i=1}^{n} \varepsilon^2$$

We choose the gradient descent [23] to estimate parameters that provide the best fit for a training set of input-output pairs. $\beta := \beta - \alpha \nabla F(\beta)$ Here, α is the step of the gradient descent. The result is shown in Figure 5.

ANNs provide a general, practical method for deriving real-valued and vector-valued functions from exemplary data. For a certain type of problems, such as interpreting complex real-world sensor data, artificial neural networks are among the most effective learning methods currently known [15]. For the prediction of China Mobile KPIs, each month in a year or each year is regarded as one node in the input layer. We use gradient descent to manipulate our network parameters to best fit a training set of input-output pairs. Finally, we use the ANN model to predict the output of a given date. The ANN is built in MATLAB and has 46 nodes on the input layer, 4 nodes on the only hidden layer and 1 node on the output layer. A similar structure has been used in [15]. Since NAR and NARX yield similar result, they are not included here.

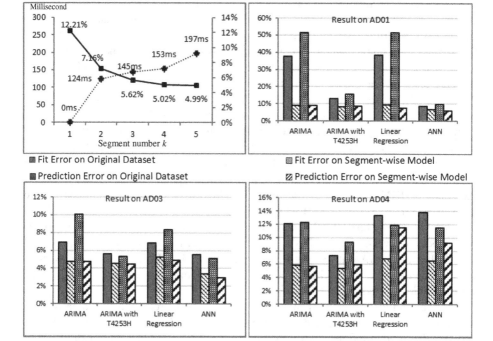

Fig. 3. Top-left: time required to build segment-wise model (dashed) and ARMIA prediction error on this model (solid), for increasing segment number. Top-right and bottom: Fitting and prediction error on original dataset and segment-wise model produced by different methods.

The relation between segment number k and time consuming and prediction accuracy by ARIMA with smoothening has also been explored, as seen in Figure 3 (top left). The result is based on the dataset AD04 but other datasets yield similar result. In Figure 3, we can easily see that $k=3$ strikes a relatively good balance between both computation complexity and prediction accuracy for AD04 and also for the other datasets we tested.

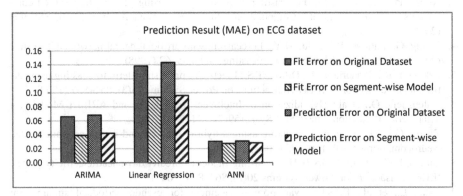

Fig. 4. Prediction Result on ECG dataset, measured by Mean Absolute Error (MAE)

The comparative prediction accuracy of direct modeling (Eq. 1) and segment-wise modeling (Eq. 3) using different methods is illustrated in Figure 3 and Figure 4. Predictions based on segment-wise modeling achieve substantially higher accuracy than those based on original or processed series. Since 1) the process of segmentation ensures data within each segments most likely impacted by homogeneous rules and more smooth and stationary than original series, 2) the segmentation model provides less but more relevant inputs for prediction functions, prediction functions built on segmentation model have lower fit errors and prediction errors.

6   Conclusion

Pseudo periodic time series appear in many practical applications. However, the prediction of pseudo periodic time series is a non-trivial task that has not been studied thoroughly so far. To address these analytic and predictive problems, this paper has introduced a time series segment-wise method. To build the segment-wise model, the entire time series is first split into several periods. Then, each period is divided into segments by the detected variations; these variations are found out through the measurement of entropy. Finally, these segments are used to construct a segment-wise model. Compared with the original time series, the reconstructed subsequences are affected by similar factors, but with slighter variations between neighboring points. Hence, more accurate predictions can be achieved using this general prediction method. Extensive experiments in the fields of finance and medical-care have been conducted and the results have demonstrated both an increase in the accuracy and the efficiency when using the segment-wise model.

References

1. Herzel, H.: Bifurcations and chaos in voice signals. Applied Mechanics Reviews 46(7), 399–413 (1993)
2. Box, G.E.P., Jenkins, G.M.: Time Series Analysis: Forecasting and Control, 3rd edn. Prentice Hall PTR, Upper Saddle River (1994)
3. Koopman, S.J., Ooms, M.: Forecasting daily time series using periodic unobserved components time series models. Computational Statistics & Data Analysis 51(2), 885–903 (2006)
4. Zhang, G., Patuwo, B.E., Hu, M.Y.: Forecasting with artificial neural networks: The state of the art. International Journal of Forecasting 14(1), 35–62 (1998)
5. Szkuta, B.R., Sanabria, L.A., Dillon, T.S.: Electricity price short-term forecasting using artificial neural networks. IEEE Transactions on Power Systems 14(3), 851–857 (1999)
6. Valenzuela, O., et al.: Hybridization of Intelligent Techniques and ARIMA Models for Time Series Prediction. Fuzzy Sets Syst. 159(7), 821–845 (2008)
7. Zhang, G.P.: Time series forecasting using a hybrid ARIMA and neural network model. Neurocomputing 50(0), 159–175 (2003)
8. Garcia, R.C., et al.: A GARCH forecasting model to predict day-ahead electricity prices. IEEE Transactions on Power Systems 20(2), 867–874 (2005)
9. Tang, L., et al.: Effective variation management for pseudo periodical streams. In: SIGMOD (2007)
10. Verleysen, M., François, D.: The curse of dimensionality in data mining and time series prediction. In: Cabestany, J., Prieto, A.G., Sandoval, F. (eds.) IWANN 2005. LNCS, vol. 3512, pp. 758–770. Springer, Heidelberg (2005)
11. Indyk, P., Koudas, N., Muthukrishnan, S.: Identifying Representative Trends in Massive Time Series Data Sets Using Sketches (2000)
12. Vlachos, M., Philip, S.Y., Castelli, V.: On Periodicity Detection and Structural Periodic Similarity (2005)
13. Olshen, L.B.J.F., Stone, C.J.: Classification and regression trees. Wadsworth International Group (1984)
14. Lin, J., et al.: Experiencing SAX: A novel symbolic representation of time series. Data Mining and Knowledge Discovery 15(2), 107–144 (2007)
15. Anderson, J.R., et al.: Machine learning: An artificial intelligence approach, vol. 2. Morgan Kaufmann (1986)
16. Menezes Jr., J.M.P., Barreto, G.A.: Long-term time series prediction with the NARX network: An empirical evaluation. Neurocomputing 71(16), 3335–3343 (2008)
17. Kabacoff, R.: R in Action. Manning Publications Co. (2011)
18. Snyman, J.A.: Practical mathematical optimization: An introduction to basic optimization theory and classical and new gradient-based algorithms, vol. 97. Springer (2005)
19. Chan, K., Fu, A.W.: Efficient time series matching by wavelets (1999)
20. Wu, H., Salzberg, B., Zhang, D.: Online Event-driven Subsequence Matching over Financial Data Streams. In: SIGMOD 2004. ACM, New York (2004)

Constructing Decision Trees
for Unstructured Data

Shucheng Gong and Hongyan Liu*

Department of Management Science and Engineering,
Tsinghua University, Beijing, China
{gongshch.10,liuhy}@sem.tsinghua.edu.cn

Abstract. The volume of unstructured data has been growing sharply
as the era of Big Data arrives. Decision tree is one of the most widely
used classification models designed for structured data. Unstructured
data such as text need to be converted to structured format before being analyzed using decision tree model. In this paper, we discuss how
to construct decision trees for datasets containing unstructured data.
For that purpose, a decision tree construction algorithm called CUST
was proposed, which can directly tackle unstructured data. CUST introduces the use of splitting criteria formed by unstructured attribute
values, and reduces the number of scans on datasets by designing appropriate data structures. Experiments on real-world datasets show that
CUST improves the efficiency of building classifiers for unstructured data
and performs as well as, if not better than existing solutions in classification accuracy.

Keywords: Data mining, Classification, Decision tree.

1 Introduction

The classification techniques in the field of data mining have been widely applied
to various real-world problems such as precise marketing and risk prediction.
With the advent of Big Data era, the volume of unstructured data is exploding.
Therefore, how to perform accurate and efficient classification on unstructured
data has become a problem worth pondering. Popular classification algorithms
include Naïve Bayes, Support Vector Machine (SVM), decision tree and association rule-based classification algorithms, etc. Among these algorithms, decision
tree algorithm is one of the most frequently used classification algorithms, with
the advantages of fast speed, relatively high accuracy, easily understood classification model and no assumption on distribution [15]. However, classical decision
tree classification algorithms can only deal with completely structured data, and
hence when they are used for unstructured data, it is necessary to first convert
the data into structured data. The converted structured data tends to be sparse
with high dimensionality, leading to low algorithm efficiency. Other classical classification algorithms like Naïve Bayes also have this disadvantage. Association

---

* Corresponding author.

X. Luo, J.X. Yu, and Z. Li (Eds.): ADMA 2014, LNAI 8933, pp. 475–487, 2014.

rule-based classification algorithms can directly handle unstructured data, and are able to mine more meaningful rules compared to decision tree classification algorithms, but these algorithms generally run more slowly and would treat structured attributes in the dataset the same way as they treat unstructured attributes, thus wasting some useful information.

For example, Table 1 shows a dataset about advertisement click. Each row records the basic features of a user and the content of the webpages he/she browses. The attribute "Webpage Content" is an unstructured attribute where letters a-f represent the text content of the webpages (each letter could stand for a word or phrase).

Table 1. Initial Training Dataset

Gender	Age	Webpage Content	Click or Not
Male	18	abc	Y
Female	25	ade	Y
Male	38	cf	N
Male	22	e	Y
Female	30	bcf	Y
Female	24	bdef	N
Male	40	de	N
Female	25	cdf	N
Male	33	de	Y
Female	15	acef	N
Female	20	abc	Y
Male	27	ace	N

In order to construct a decision tree model for predicting whether a user would click certain advertisement, we need to convert the dataset into a structured dataset as shown in Table 2. Apparently, the number of attributes increases a lot, which will lower the efficiency of constructing decision trees. Besides, many meaningless values of zero in Table 2 will mislead the building of classifiers.

If using association rule-based classification algorithms, we need to first discretize attribute "Age", then convert attribute "Gender" and "Age" into unstructured attributes like "Webpage Content" before starting to build classifiers. Depending on the minimum support and confidence, maybe too many association rules will be discovered so that there may be several contradictory rules applicable for an instance of unknown class, or maybe too few association rules will be discovered so that they cannot cover all the possible combinations of attribute values and hence we may not be able to find any suitable rules for certain instances of unknown class.

Table 2. Completely Structured Dataset

Gender	Age	a	b	c	d	e	f	Click or Not
Male	18	1	1	1	0	0	0	Y
Female	25	1	0	0	1	1	0	Y
Male	38	0	0	1	0	0	1	N
Male	22	0	0	0	0	1	0	Y
Female	30	0	1	1	0	0	1	Y
Female	24	0	1	0	1	1	1	N
Male	40	0	0	0	1	1	0	N
Female	25	0	0	1	1	0	1	N
Male	33	0	0	0	1	1	0	Y
Female	15	1	0	1	0	1	1	N
Female	20	1	1	1	0	0	0	Y
Male	27	1	0	1	0	1	0	N

As both of the two existing solutions are not quite satisfying, this research tries another way and proposes a decision tree construction algorithm CUST (Constructing Unstructured and Structured data-based decision Trees) which can directly handle unstructured data as well as structured data.

2 Related Work

2.1 Decision Tree Classification Algorithms

The construction of a globally optimal decision tree is generally NP-complete. Therefore, almost all decision tree construction algorithms use a greedy approach to grow the tree in a top-down way [5]. However, various decision tree classification algorithms differ in the form and measurement index of splitting criteria, the data storage structure, etc. According to whether each splitting criterion in a decision tree could involve more than one attribute, we can further break decision tree classification algorithms into two subtypes: univariate decision tree classification algorithms and multivariate decision tree classification algorithms.

Among univariate decision tree classification algorithms, classical algorithms include ID3 [13], C4.5 [12], etc. Later, algorithms good at dealing with large datasets were proposed, such as SPRINT [14], RainForest [5], SPDT [2] and CUDT [11]. SPRINT almost has no requirement for memory size, and it is relatively fast and scalable. RainForest separates the issue of algorithm scalability from the issue of decision tree quality, mainly addressing the data management problem during the construction of decision trees. RainForest is about three times higher in efficiency than that of SPRINT. Both SPRINT and RainForest try to optimize algorithm performance by designing appropriate data structures. SPDT is a streaming parallel decision tree algorithm executed in a distributed environment and it compresses the data to make full use of limited memory.

CUDT is a parallel decision tree algorithm based on CUDA (compute unified device architecture) and it is much faster than SPRINT for large datasets.

Compared to univariate decision tree classification algorithms, multivariate decision tree classification algorithms generally have higher classification accuracy and fewer decision tree nodes, but the splitting criteria might be more complex. Therefore, the constructed decision trees are not necessarily more concise. In addition, although this type of decision tree classification algorithms could mine and utilize the relationships among different attributes to some extent, right now the relationships are mainly limited to linear ones [4]. However, the relationships among unstructured attribute values are not merely simple linear relationships. Hence, other methods to handle unstructured data for decision tree model building are still desirable.

2.2 Association Rule-Based Classification Algorithms

Association rule mining is originally an association analysis technique, which is used to mine the relationships among different items. Association rule mining could date back to market basket analysis [9], of which "beer and diapers" is a famous example. There are already many mature association rule mining algorithms which use different methods to find association rules that satisfy user-defined requirements, the widely used ones including Apriori [1] and FP-Growth [6]. When limiting the consequents of the association rules to be class attribute values, we can mine classification association rules which can be used to build classifiers. In chronological order, typical association rule-based classification algorithms include CBA [8], CMAR [7], CPAR [16], etc. This type of algorithms often needs to examine a large number of candidate association rules, which leads to low efficiency and poor scalability.

Some researchers have tried combining decision tree classification algorithms with association rule-based classification algorithms to mitigate their deficiencies. [10] proposed generalized decision trees (GDT) to better manipulate classification association rules. GDT could be applied to unstructured data because it is indeed an association rule-based classification algorithm. This also means that it shares the aforementioned disadvantages of association rule-based classification algorithms when handling unstructured data.

3   Decision Tree Construction Algorithm

To build decision tree models for data containing unstructured attributes, we proposed a new decision tree construction algorithm, CUST (Constructing Unstructured and Structured data-based decision Trees).

3.1   Data Structures of CUST

CUST learns from the storage structures of RainForest and FP-Growth, mainly AVC-set from the former and FP-Tree from the latter. CUST is able to deal with datasets containing both structured part and unstructured part, to which completely structured datasets and completely unstructured datasets are special cases. In CUST, we design Label Distribution Table (LDT) to store the structured part of the dataset, Unstructured Attribute Tree (UA-Tree) for the unstructured part, and Row Class Table (RCT) for the class information of all the training data instances. CUST also uses pointers to connect these data structures. Here we only discuss the former two major data structures.

Label Distribution Table (LDT). For structured attributes, CUST adapts from the AVC-set storage structure of RainForest and adds a list of pointers for each row, forming a new storage structure called Label Distribution Table (LDT). The LDT of attribute "Gender" in the dataset shown in Table 1 is displayed in Table 3. The pointers in it point to specific rows in the Row Class Table.

Table 3. LDT of attribute "Gender"

Gender	Click	Not Click	Pointer List
Male	3	3	[1][3][4][7][9][12]
Female	3	3	[2][5][6][8][10][11]

UA-Tree. Similar to FP-Tree, UA-Tree is a prefix tree where each node corresponds to an attribute value except for the root node. It is worth noting that we could view any dataset containing unstructured part as a dataset with one unstructured attribute, as long as we could distinguish unstructured attribute values from different sources. Each data instance corresponds to a branch or a section of a branch in the UA-Tree, and the node at the end of this section is connected to certain rows in the Row Class Table via pointers. In addition, along any branch of a UA-Tree, the deeper level a tree node is at, the lower the appearance frequency of its corresponding attribute value. UA-Tree also contains an auxiliary structure called Header Table mainly for indexing. Header Table stores all attribute values in descending order of appearance frequency and each attribute value is connected to tree nodes with the same attribute value successively via pointers. The unstructured part of the dataset shown in Table 1 is the attribute "Webpage Content", and its initial UA-Tree is shown in Fig. 1.

3.2 CUST Algorithm

CUST makes improvements upon the decision tree construction process of Rain-Forest. The major steps of CUST are described in Fig. 2. In the following parts we will introduce how CUST determines splitting criterion and splits the dataset.

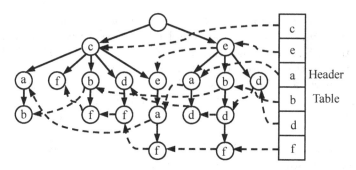

Fig. 1. Example of UA-Tree

```
Input: (Semi-)Unstructured dataset D, Measurement index MI,
       Stopping criterion SC
Output: Classification Decision Tree DT

Schema:
(1)  Scan D, construct initial LDT ldt and UA-Tree T
(2)  DT=createTree(ldt, T, MI, SC)

createTree(LDT ldt, UA-Tree T, Measurement index MI, Stopping
    criterion SC)
(1)  Create a new node pointer root
(2)  if (SC is satisfied)
(3)      return root
(4)  Determine the splitting criterion from ldt and T
     according to MI (producing k child nodes)
(5)  for (i=0; i<k; i++)
(6)      Split ldt and T according to the splitting criterion,
         getting the corresponding LDT ldtᵢ and UA-Tree Tᵢ
(7)      root->child[i]=createTree(ldtᵢ, Tᵢ, MI, SC)
(8)  endfor
(9)  return root
```

Fig. 2. Decision Tree Construction Schema of CUST

Determining Splitting Criterion. Similar to RainForest, CUST adopts gini index [3] as the measurement index of candidate splitting criteria. The lower the gini index, the more effective the splitting criterion. For the structured part, the gini index could be easily computed according to the LDT, and we don't give details here. For unstructured attributes, there are many possible forms of splitting criteria, such as "contain e", "contain a&b", "c appears more than three times", etc. This research looks into the former two forms, namely contain-type splitting criteria involving a single attribute value or two attribute values.

To make it more convenient to collect the class distribution data of all the candidate splitting criteria, we store class distribution information for the purpose of counting in every attribute value node of UA-Tree, i.e., record the numbers of instances that belong to each class. Specifically, when collecting class distribution data, we examine UA-Tree nodes of different attribute values in ascending order of appearance frequency. At the bottom layer of UA-Tree, CUST aggregates the class distribution information contained in the rows of the RCT which are connected to these nodes. Then CUST passes on the class distribution data corresponding to single attribute values and double attribute values to parent nodes, and accumulate these data at the current nodes in the meantime. When we reach the root node of UA-Tree, the class distribution data of all the candidate splitting criteria formed by various single attribute values and double attribute values have been attained, and the root node would have stored the overall class distribution information of the sub-dataset associated with current decision tree node, which is useful for computing gini index and checking whether the stopping criterion is satisfied. After computing the gini indices of all the candidate splitting criteria from both the structured part and the unstructured part, CUST chooses the splitting criterion with the lowest gini index as the splitting criterion of current decision tree node.

Performing Splitting. New child nodes will be created according to the determined splitting criterion. Since the dataset associated with certain decision tree node is actually stored in LDT, UA-Tree and RCT, splitting data instances is in fact splitting these three storage structures. Because there are both pointers to certain rows of RCT in LDT and in the leaf nodes (here leaf nodes refer to nodes at the end of a section of a branch corresponding to a data instance) of UA-Tree, it is pretty simple to split or update LDT, UA-Tree and RCT of child nodes.

In order to avoid overfitting, we set the stopping criterion to be "the data instances associated with current node all belong to the same class or the classification accuracy of current node reaches 80%" in this research.

4   Experiments

To evaluate the performance of the proposed algorithm, we conducted five sets of experiments on a real-world dataset and compared CUST with two existing solutions, RainForest and CPAR. The dataset provided by Wise 2013 Challenge Track 2[1] was extracted from Sina Weibo, involving 1,126,049 users. The dataset contains unstructured attributes such as the weibo content, as well as structured attributes such as gender. We adopted ten-fold cross validation in the experiments.

The first set of experiments was conducted to evaluate the efficiency with respect to different numbers of instances. The datasets used contain eight structured attributes and one unstructured attribute with 263 possible attribute values. This means that the converted completely structured dataset contains 8+263=271 attributes.

Fig. 3 compares the time it took to build classifiers for datasets with different numbers of instances, using CUST, RainForest and CPAR respectively.

---

[1] WISE 2013 Challenge. http://wise2013.njue.edu.cn/wise2013challenge.html

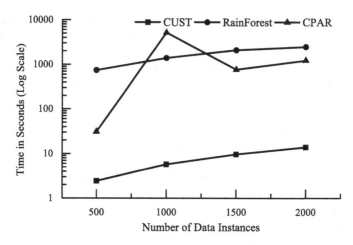

Fig. 3. Efficiency of Three Algorithms for Different Numbers of Data Instances

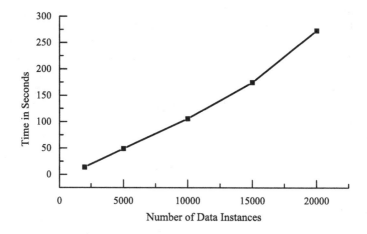

Fig. 4. Efficiency of CUST for Different Numbers of Data Instances

Fig. 3 shows that given the same number of data instances, the efficiency of CUST is apparently higher than that of CPAR and that of RainForest. In addition, CUST seems to have pretty good algorithm scalability in the dimension of the number of data instances.

Fig. 4 compares the time it took to build classifiers using CUST across a wider range of numbers of data instances.

Fig. 4 shows that CUST is able to deal with relatively large datasets, and it does have excellent scalability in this dimension.

The second set of experiments was conducted to examine the classification accuracy for datasets with different numbers of data instances. Fig. 5 compares the results using CUST, RainForest and CPAR respectively.

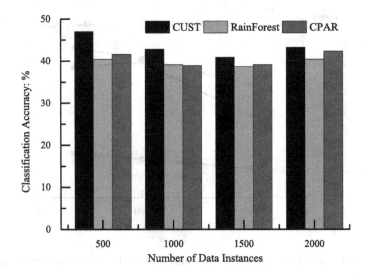

Fig. 5. Accuracy of Three Algorithms for Different Numbers of Data Instances

We can see from Fig. 5 that the ranking of the classification accuracy is CUST>CPAR>RainForest in general, and the latter two sometimes interchange. This shows that CUST at least guarantees classification accuracy as it boosts the efficiency, if it doesn't improve accuracy compared to existing algorithms.

The third set of experiments was conducted to examine the efficiency of dealing with datasets with different numbers of unstructured attribute values. Fig. 6 compares the time of building classifiers with the three algorithms.

Similar to the results of the first set of experiments, given the same number of unstructured attribute values, the efficiency of CUST is significantly higher than that of CPAR and that of RainForest, and CUST is quite scalable in this dimension.

The fourth set of experiments was conducted to examine the classification accuracy for datasets with different numbers of unstructured attribute values. Fig. 7 compares the results of the above three algorithms.

Similar to the results of the second set of experiments, the ranking of the classification accuracy is CUST>CPAR>RainForest, indicating that CUST has a good performance on classification accuracy.

Finally, we have also explored the effect of allowing unstructured splitting criteria formed by double attribute values. This kind of splitting criteria looks like "contain a&b simultaneously". We compared the classification accuracy and the number of decision tree nodes of this version of CUST with that only allowing one unstructured attribute value for splitting criteria. The results on datasets with different numbers of data instances are shown in Fig. 8 and Fig. 9.

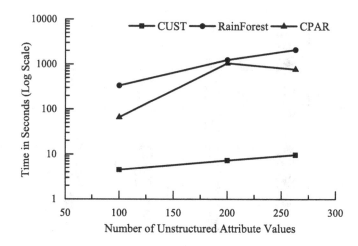

Fig. 6. Efficiency of Three Algorithms for Different Numbers of Unstructured Attribute Values

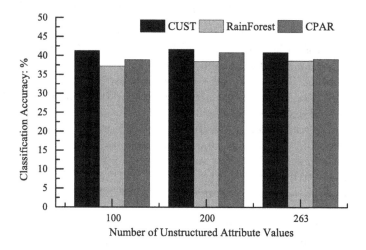

Fig. 7. Accuracy of Three Algorithms of Different Numbers of Unstructured Attribute Values

From Fig. 8 we can see that there is no significant difference of accuracy between the two versions, while Fig. 9 shows that the changes of the number of nodes are not consistent across different datasets.

This indicates that the effect of allowing unstructured splitting criteria formed by double attribute values may depend on the properties of datasets. For example, for sparse datasets like the ones we used in the experiments, the proportion of data instances that simultaneously contain two certain unstructured attribute values is very low, and hence the numbers of data instances associated with the resulting two child nodes are quite imbalanced. Under this condition, many

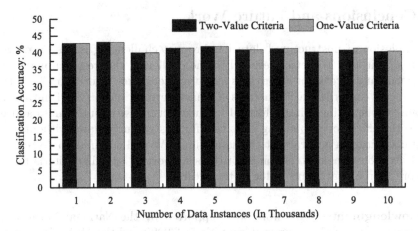

Fig. 8. Accuracy of Two Versions of CUST with/without Unstructured Splitting Criteria Formed by Double Attribute Values

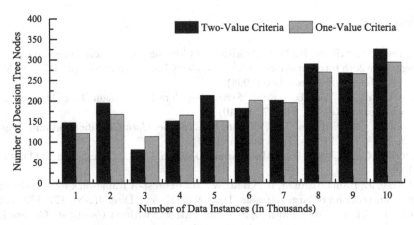

Fig. 9. Number of Decision Tree Nodes of Two Versions of CUST with/without Unstructured Splitting Criteria Formed by Double Attribute Values

times of splitting are still needed from the current node to the leaf nodes. Therefore, the size of the constructed decision tree may be increased rather than be decreased. In a word, the results show that the introduction of unstructured splitting criteria formed by double attribute values doesn't significantly improve the classification accuracy or reduce the size of the constructed decision tree for this particular dataset. It is still possible that more effective and concise decision trees could be produced with this kind of splitting criteria for other datasets. Further research is included in our future work.

5   Conclusions and Future Work

The explosion of unstructured data has brought challenging issues for classification algorithms. To better respond to the challenge, this research makes improvements upon relevant algorithms and proposes a new algorithm CUST that can directly construct decision trees for unstructured data. CUST introduces contain-type splitting criteria formed by unstructured attribute values and enhances the efficiency of building classifiers for unstructured data, without sacrificing classification accuracy at the same time. We will continue to do research about decision tree construction algorithms supporting splitting criteria formed by more than one attribute value in the future.

Acknowledgments. This work was supported by the National Natural Science Foundation of China under Grant No.71272029 and the Major Program of National Social Science Fund of China under Grant No. 13&ZD184.

References

1. Agrawal, R., Srikant, R.: Fast Algorithms for Mining Association Rules. In: Proceedings of the 20th International Conference on Very Large Databases, pp. 487–499. Morgan Kaufmann, San Francisco (1994)
2. Ben-Haim, Y., Yom-Tov, E.: A Streaming Parallel Decision Tree Algorithm. J. Mach. Learn. Res. 11, 849–872 (2010)
3. Breiman, L., Friedman, J.H., Olshen, R.A., Stone, C.J.: Classification and Regression Trees. CRC Press, Boca Raton (1984)
4. Brodley, C.E., Utgoff, P.E.: Multivariate Decision Trees. Mach. Learn. 19(1), 45–77 (1995)
5. Gehrke, J., Ramakrishnan, R., Ganti, V.: RainForest-A framework for fast decision tree construction of large datasets. Data Min. Knowl. Disc. 4(2-3), 127–162 (2000)
6. Han, J., Pei, J., Yin, Y.: Mining Frequent Patterns without Candidate Generation. In: Proceedings of the 2000 ACM SIGMOD International Conference on Management of Data, pp. 1–12. ACM, New York (2000)
7. Li, W., Han, J., Pei, J.: CMAR: Accurate and Efficient Classification Based on Multiple Class-Association Rules. In: Proceedings of the 2001 IEEE International Conference on Data Mining, pp. 369–376. IEEE Computer Society, Washington, DC (2001)
8. Liu, B., Hsu, W., Ma, Y.: Integrating Classification and Association Rule Mining. In: Proceedings of the 4th International Conference on Knowledge Discovery and Data Mining, pp. 80–86. AAAI Press, Menlo Park (1998)
9. Liu, H.: Business Intelligence Techniques and Application. Tsinghua University Press, Beijing (2013) (in Chinese)
10. Liu, H., Yu, J.X., Lu, H.: Unifying Decision Tree Induction and Association Based Classification. In: Proceedings of the 2002 IEEE International Conference on Systems, Man and Cybernetics. IEEE Computer Society, Washington, DC (2002)
11. Lo, W.-T., Chang, Y.-S., Sheu, R.-K., Chiu, C.-C., Yuan, S.-M.: CUDT: A CUDA Based Decision Tree Algorithm. Scientific World Journal 2014, Article ID 745640 (2014)

12. Quinlan, J.R.: C4.5: Programs for Machine Learning. Morgan Kaufmann, San Francisco (1993)
13. Quinlan, J.R.: Induction of decision trees. Mach. Learn. 1(1), 81–106 (1986)
14. Shafer, J., Agrawal, R., Mehta, M.: SPRINT: A Scalable Parallel Classifier for Data Mining. In: Proceedings of the 22nd International Conference on Very Large Databases, pp. 544–555. Morgan Kaufmann, San Francisco (1996)
15. Tan, P.N., Steinbach, M., Kumar, V.: Introduction to Data Mining. Addison-Wesley, Boston (2005)
16. Yin, X., Han, J.: CPAR: Classification Based on Predictive Association Rules. In: Proceedings of the 3rd SIAM International Conference on Data Mining, pp. 331–335. SIAM, San Francisco (2003)

Attack Type Prediction Using Hybrid Classifier

Sobia Shafiq, Wasi Haider Butt, and Usman Qamar

Department of Computer Engineering
College of Electrical & Mechanical Engineering (CEME)
National University of Sciences and Technology (NUST), Pakistan
sobiashafiq786@yahoo.com,
{wasi,usmanq}@ceme.nust.edu.pk

Abstract. Due to the rapid increase in terrorist activities throughout the world, there is serious intention required to deal with such activities. There must be a mechanism that can predict what kind of "attack types" can happen in future and important measures can be taken out accordingly. In this paper, a hybrid classifier is proposed which consists of some existing classifiers including K Nearest Neighbor, Naïve Bayes, Decision Tree, Averaged One Dependence Estimators and BIFReader. The proposed technique is implemented in Rapid Miner 5.3 and it achieves the satisfied level of accuracy. Results reveal the improvement in accuracy for the proposed technique as compare to the individual classifiers used.

Keywords: Classification, Prediction, K-NN, Naive Bayes, Decision Tree, AODE, BIFReader.

1   Introduction

Terrorism has been around the world as long as one can remember but from past one decade there is a huge increase in terrorist activities. Terrorists perform activities like hijacking, murder, kidnapping and bombing in order to achieve an agenda. These activities not only subjected to one region or country but throughout the world. As these incidents increased in past few years the phenomena of terrorism became an important issue to the government authorities [1]. There is a serious need that security agencies or government authorities can detect terrorist activities before they happen and take precautionary measures accordingly. This research is based on prediction of "attack types" using data mining techniques.

The rest of this paper is organized as follows. Section 2 discusses the related work. Section 3 illustrates proposed framework. Section 4 provides the experimentation and implementation of proposed technique. Section 5 analyses experimental results. Section 6 concludes this paper and points out some future work of proposed framework.

2   Related Work

Data Mining is a process in which data analysis is done. In data mining different discovery algorithms are applied on the data which can produce particular patterns or

X. Luo, J.X. Yu, and Z. Li (Eds.): ADMA 2014, LNAI 8933, pp. 488–498, 2014.
© Springer International Publishing Switzerland 2014

models over the data [2]. In data mining different type of techniques and algorithms are used like Classification, Clustering, Artificial Intelligence, Association Rules, decision Trees etc for knowledge discovery from large set of databases. From all of above mentioned techniques, classification is one of the most popular and studied technique which can be used for predicting something. Value of an attribute can be predicted based on the values of other attributes. The attribute which is to be predicted is known as "class" [3]. Classification and prediction are two forms of data analysis which describe about classes or tell about future trends by building models.

This section briefly describes the different techniques of classification in data mining such as K Nearest Neighbor classification (K-NN), Naive Bayes Classification, Decision Tree, Averaged One-Dependence Estimators (AODE) and BIFReader Classification. This section also gives a view of ensemble or hybrid classifiers.

K Nearest Neighbor or K-NN is a classification algorithm that was introduced in early 1950's. N. Suguna, and Dr. K. Thanushkodi [4] state that K-NN is an important pattern recognition algorithm. In K-NN classification rules are produced using only training dataset instead of any additional data. Xingjiang Xiao and Huafeng Ding [5] present that in K-NN classifier, classification rules are generated based on training datasets. When test data is given to KNN classifier it predicts its category based on training dataset which are nearest neighbor to the test data. This algorithm first calculates the distance of the new query and already known samples to find out the K nearest neighbors. Once K nearest neighbors is gathered, the majority of these K nearest neighbors are taken to be the prediction of the query instance. Pratiksha Y. Pawar and S. H. Gawande [6] in their research presents that K-NN classification is outstanding because of its simplicity. This technique is widely used in text classification as well as in classification tasks with multi categorized document because of its simplicity and easy implementation. One drawback of this algorithm is that it takes more time if the training dataset is large. LiuYu and Chen Gui-Sheng [7] presents that K-NN is the most widely used lazy learning method. It is one of most powerful classification algorithm which can deal with complex problems easily. K-NN provides well classified samples when the training dataset is large.

Pat Langley and Stephanie Sage [8] in their research states that naive bayes is an important techniques used for probabilistic induction. This classification technique represents each class with a single probabilistic summary. Jaideep Vaidya et al [9] presents that Naive Bayes is a very useful Bayesian learning technique and it is more concerned with high dimensional tasks. Naive Bayes classifier takes an arbitrary number of continuous or categorical variables and classifies an instance to belong one of several classes. It is based on bayesan theorem with strong independence assumption. A Naive Bayes classifier considers that the absence or presence of a particular feature or an attribute is independent of absence or presence of any other feature. The advantage of this classifier is that it requires small amount of training data to calculate the means and variances of the variables required for classification.

Amany Abdelhalim and IssaTraore [10] presents that decision trees are models that direct the decision making process. Decision trees can be created through the dataset (data-based decision trees) as well as from rules proposed. A decision tree is an efficient technique for guiding a decision process as long as no changes occur in the dataset used to create the decision tree. Decision tree is a flowchart like tree structure which works both on numerical as well as categorical data. Decision Trees are based on recursive portioning. Decision tree is a directed tree like structure having nodes. Its node is of three types. One is Root node that has no incoming edges, second is internal or test node that have outgoing edges and third one is leaves. Each internal node represents a test on an attribute, branch or edges represent result of those test whereas each leave denotes a label class. Decision trees are popular in field of classification because they do not require any domain knowledge thus it is widely used for classification purposes. [11]

Liangxiao Jiang and Harry Zhang [12] in their research presents that Naive Byse is a classification model based on probabilities and on attributes independence assumptions. In real world data this assumption of attribute independence is violated. Researchers have done research regarding this issue and tried to get better the Naive Bayes's accuracy. This is done by fading its attribute independence assumption. Averaged One-Dependence Estimators (AODE) is proposed by a researcher in this regard which helps in weakening the attribute independence assumption of naive bayes. AODE has done this by averaging all the models, from a restricted class of one dependence classifier. AODE is a probabilistic classification method. It deals with the attribute independence assumption of naive bayes by averaging all of the dependence estimators. [13].

BIF reader constructs a description of a Bayes Net classifier which is stored in XML BIF 0.3 format. It accepts input in form of an example dataset, which is training dataset and give model in output.

Nicholas Stepenosky et al state in their research that collection or combination of classifiers is now more popular than individual classifiers because of their better performance and superiority over individual classifier system. Ensemble techniques of classifiers include bagging, boosting, voting techniques. These techniques are quite effective on many applications. The main idea behind combining classifiers is that individual classifiers are diverse and chances of errors are higher while combining classifiers can reduce errors and results in better performance through averaging. [14]

Lior Rokach in his research state that the idea of ensemble classifiers is used to build a model by combining different classifiers for better performance. Different type of techniques can be used for combining classifiers one of which is majority voting. In majority voting technique classification of an unlabeled instance is carried out according to the class that obtains the highest number of votes. [15]

3   Proposed Technique

The proposed framework consists of a hybrid classifier which is shown in figure 1 and has following steps:

1. Data Gathering
2. Data Pre-processing
3. Data Mining
4. Data Deployment and testing

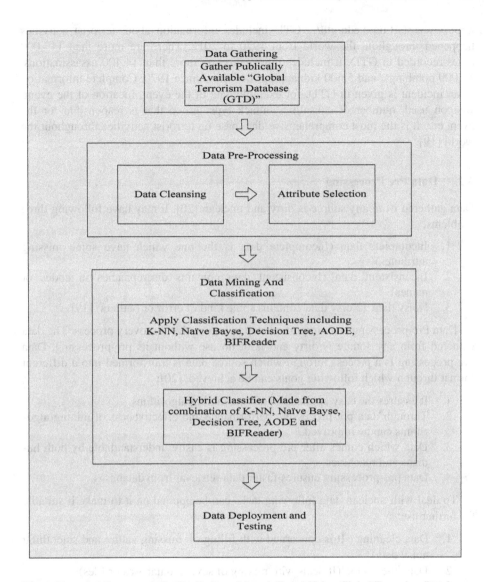

Fig. 1. Proposed Framework for Prediction of Attack Types Using Global Terrorism Database

3.1   Data Gathering

First phase of data mining process is data gathering. This step is concerned with how
and from where data is collected [20]. For proposed framework Global Terrorism
Database (GTD) is gathered from an open source database. The data base has been
obtained from the National Consortium for the Study of Terrorism and Responses to
Terrorism (START) initiative at University of Maryland, from their online interface at

http://www.start.umd.edu/gtd/. GTD includes information about terrorist activities happened throughout the world from 1970 to 2012. There are more than 113,000 cases recorded in GTD. It includes information on more than 14,400 assassinations, 52,000 bombings, and 5,600 kidnappings activities since 1970. Complete information of an incident is given in GTD, for example date of the event, location of the event, weapon used, number of causalities, attack type, group that is responsible for the event etc. It is the most comprehensive data base on terrorist activities throughout the world [19].

3.2   Data Pre Processing

Data gathered from any source is dirty and unclean [20]. It may have following three problems.

1. Incomplete data (Incomplete data is the one which have some missing attributes)
2. Inconsistent data (Inconsistent data contains discrepancies in codes or names)
3. Noisy data (Noisy data contains some kind of error or outliers) [19]

Data Pre-processing is an important step in knowledge discovery process. The data gathered from any source is dirty and is of no use without its pre-processing. Data Pre-processing is a process through which source data is transformed into a different format through which following goals can be achieved: [20]:

1. It ensures the easy application of data mining algorithms.
2. Through data pre-processing performance and effectiveness of mining algorithms can be improved.
3. Data which comes after pre-processing is easily understandable by both humans and machines.
4. Data pre-processing ensures faster data retrieval from databases.

To deal with unclean data following tasks can be applied on it to make it suitable for further use:

1. Data cleaning (It is concerned with filling of missing values and smoothing noisy data)
2. Data integration (It deals with mixing of several databases or files)
3. Data transformation (it is concerned with normalization)
4. Data reduction (it deals with reducing the amount of data but without disturbing the analytical results)
5. Data discritization (it is part of data reduction but with specific importance, particularly for numerical data)

3.2.1   Data Pre Processing of Global Terrorism Database (GTD)

By going through the above mentioned data pre processing steps, the dataset (GTD) is transformed into a form that is suitable for data mining algorithm. Missing values are removed from GTD and required attribute are selected (Data reduction). There are total 134 attributes and 113114 records in GTD with a lot of missing values. For the proposed framework data in GTD is reduced by selecting 8 attributes and 45221 records which have large impact on prediction of attack type. Data obtained after reduction is both in numerical as well as textual format. Attributes selected for the prediction of attack types are listed below:

1. Country (This attribute represents the country or location where the incident has happened)
2. Region (It is a categorical variable and it represents the region in which the incident occurred)
3. Attack type (It is a categorical variable and shows which kind of attack types is happen e.g. assassination, bombing, kidnapping etc. there are total of 9 kind of attack types recorded in GTD)
4. Target type (This attribute represents the target category)
5. Group name (This attribute represents the group that is responsible for attack)
6. Weapon type (This attribute shows the type of weapon used in attack)
7. Property (This attribute shows that any damage to any property happen or not during a terrorist incident)
8. Ransom (This field shows that, is some ransom demanded or not for an incident?)

Through the proposed framework, attack types will be predicted based on other selected attributes which have most impact on attack types.

3.3   Data Mining and Classification

Classification is an important and predictive data mining method. Classification is used to make prediction using known data. K-NN, Naïve Bayes, Decision Tree, AODE and BIFReader are applied on GTD and a hybrid classifier is proposed by combining above mentioned individual classifiers. Implementation of these classification algorithms is discussed in section 4.

3.4   Data Deployment and Testing

Models are created from dataset using classification algorithms and these models can be used for prediction purpose. In data mining act of applying a model on a dataset is known as deployment. For testing of proposed framework, dataset (GTD) was split into two datasets, one for training purpose and other for testing purpose. Models are created from training dataset by applying classification techniques on that dataset and by using those models testing is performed.

4   Experimentation and Implementation

This section elaborates the implementation of proposed framework. Rapid Miner 5.3 is used for the implementation of proposed framework. Rapid miner is an environment used for experimentation of machine learning and data mining techniques. For prediction of attack types GTD is split into two parts. One is training dataset and other is testing dataset. KNN, Naive Byse, Decision Tree, AODE and BIF Reader are applied on training dataset to build models. The implementation of the proposed technique includes classification using individual classifiers as well as a hybrid classifier.

4.1   K-NN Classifier

A model is created using K-NN classifier in rapid miner. For this purpose dataset (GTD) is retrieved using retrieve data operator. After retrieving dataset, split operator is applied on the dataset for splitting data into two parts i.e. for training and testing purpose. K-NN classifier is applied on training dataset and model is created. This model is then applied on the testing dataset using apply model operator after this accuracy and classification error of the model is noticed using performance operator. Accuracy of K-NN Classifier is 80.50% whereas classification error is 19.50%.

4.2   Naïve Bayes Classifier

A model is created using Naïve Bayes classifier. For this purpose dataset (GTD) is retrieved After retrieving dataset, data is split for testing and training purpose and a model is created using training dataset. This model is then used for testing dataset. Accuracy and classification error for naïve bayes classifier are calculated. Accuracy of naive bayes classifier is 81.54% whereas classification error is 18.46%.

4.3   Decision Tree Classifier

A model is created using decision tree classifier. For creation of model, dataset (GTD) is retrieved in rapid miner and then split into two parts i.e. for training and testing purpose. Decision tree classifier is applied on training dataset and a model is created. That model is then used for testing dataset. Accuracy calculated for decision tree classifier is 82.90% whereas classification error is 17.10%.

4.4   AODE Classifier

A model is created using AODE classifier. For this purpose dataset (GTD) is retrieved. After retrieving dataset, data is split into two parts i.e. for training and testing purposes. AODE classifier is applied on training dataset and model is created. This model is then applied on the testing dataset and after this accuracy of the model is noticed which is 84.07 % whereas classification error is of AODE classifier is 15.93%.

4.5   BIFReader Classifier

A model is created using BIFReader classifier. For this purpose dataset (GTD) is retrieved and after retrieving data is split into two parts i.e. for training and testing purpose. BIFReader classifier is then applied on training dataset and model is created. This model is then applied on the testing dataset and accuracy and classification error of the model is noticed. Accuracy of BIFReader classifier is 83.30 % whereas classification error is 16.70%.

4.6   Hybrid Classifier

The proposed hybrid Classifier is built by combining K-NN, Naïve Bayes, Decision Tree, AODE and BIFReader classifiers for better performance and results. Hybrid classifier is made using Vote Operator in Rapid Miner.

The Vote operator has sub processes which must have at least two learners, called base learners. This operator builds a classification model or regression model depending upon the Example Set and learners. This operator uses a majority vote for predictions of the base learners provided in its sub process. While doing classification, all the operators in the sub process of the Vote operator accept the given Training Data Set and generate a model for classification. For prediction of an unknown example set, the Vote operator applies all the classification models included in its sub process and assigns the predicted class with maximum votes to the unknown example.

A hybrid classifier is proposed for better performance. For this purpose dataset (GTD) is retrieved. After retrieving dataset, data is split into two parts i.e. one for training purpose and one for testing purpose. Vote Operator is applied on training dataset and model is created. As the sub processes of vote operator K-NN, Naive Bayes, Decision Tree, AODE, BIFReader classifiers are used. These classifiers are working as sub processes of vote operator. Each of this classifier will get the dataset and generates a classification model, and then vote operator applies all the classification models from its sub processes, and assigns the predicted class with maximum votes to the unknown example. This model is then applied on the testing dataset and its accuracy and classification error is calculated. Accuracy of Hybrid classifier is 85.10 % whereas classification error is 14.90%.

5   Analysis of Results

This section explains the results of proposed hybrid classifier and individual classifiers in graphical form. Accuracy of K-NN, Naive Bayes, Decion Tree, AODE and BIFReader classifier as well as proposed hybrid classifier is shown in figure 2 where classifiers are taken along x axis, and accuracy is plotted along y axis. Graphical representation of results shows that K-NN has least accuracy whereas proposed hybrid classifier has maximum accuracy.

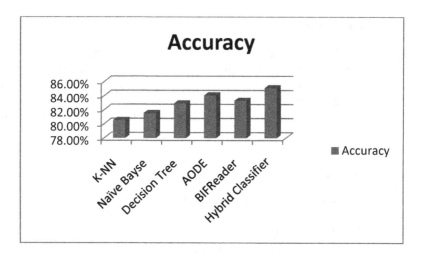

Fig. 2. Graphical Representation of Accuracy of Individual and Hybrid Classifier

Classification error of K-NN, Naive Bayes, Decion Tree, AODE and BIFReader classifier as well as proposed hybrid classifier is shown in figure 3. In figure classifiers are taken along x axis, where as classification error is plotted along y axis. Graphical representation of results shows that proposed hybrid classifier has least classification error whereas K-NN classifier has maximum classification error.

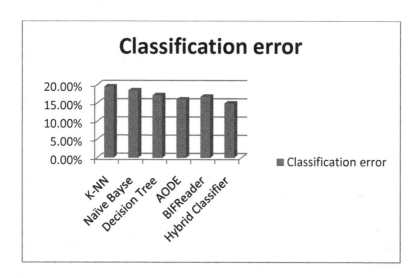

Fig. 3. Graphical Representation of Classification Error of Individual and Hybrid Classifier

6   Conclusion and Future Work

In this research, a hybrid classifier is proposed for predicting terrorist activities using data mining techniques. That hybrid classifier uses five types of classification techniques including K-NN, Naive Bayes, Decision Tree, AODE and BIFReader. Five different classifiers are used for comparison purpose. Accuracy and classification error of the individual classifier as well as hybrid classifier shows that hybrid classifier gives best result for predicting "Attack Types" in future. In future this research work can be extended for different classification algorithms and different techniques for ensemble classifiers.

References

[1] Jenkins, B.M.: The study of terrorism: Definitional Problems. The Rand Corporation, Santa Monica (1980)
[2] Ozer, P.: Data Mining Algorithms for Classification. Radboud University Nijmegen (January 2008)
[3] Bhardwaj, B.K., Pal, S.: Data Mining: A prediction for perform-ance improvement using classification. International Journal of Computer Science and Information Security (IJCSIS) 9(4) (April 2011)
[4] Suguna, N., Thanushkodi, K.: An Improved k-Nearest Neighbor Classifica-tion Using Genetic Algorithm. IJCSI International Journal of Computer Science Issues 4(2) (July 2010)
[5] Xiao, X., Ding, H.: Enhancement of K-nearest Neighbor Algo-rithm Based on Weighted Entropy of Attribute Value. In: 2012 5th International Conference on BioMedical Engineering and Informatics (BMEI 2012), pp. 1261–1264 (2012)
[6] Pawar, P.Y., Gawande, S.H.: A Comparative Study on Different Types of Approaches to Text Categorization. International Journal of Machine Learning and Computing 2(4), 423–426 (2012)
[7] Yu, L., Chen, G.-S.: KNN Algorithm Improving Based on Cloud Model. In: 2010 2nd International Conference on Advanced Computer Control (ICACC), March 27-29, vol. 2, pp. 63–66 (2010)
[8] Langley, P., Sage, S.: Induction of Selective Bayesian Classifiers. In: Proceedings of the 10th Conference on Uncertainty in Artificial Intelligence, Seattle, WA, pp. 399–406. Morgan Kaufmann, San Mateo (1994)
[9] Vaidya, J., Basu, A., Shafiq, B., Hong, Y.: Differentially Private Naive Bayes Classification. In: 2013 IEEE/WIC/ACM International Conferences on Web Intelligence (WI) and Intelligent Agent Technology (IAT), pp. 571–576 (2013)
[10] Abdelhalim, A., Traore, I.: A New Method for Learning Decision Trees from Rules. In: 2009 International Conference on Machine Learning and Applications, pp. 693–698 (2009)
[11] Han, J., Kamber, M.: Data Mining: Concepts and Techniques, 2nd edn.
[12] Jiang, L., Zhang, H.: Weightily Averaged One-Dependence Estimators. In: Proceedings of the 9th Pacific Rim International Conference on Artificial Intelligence, Guilin, China, August 07-11 (2006)
[13] Wu, J., Cai, Z.: Learning Averaged One-dependence Estimators by Attribute Weighting. Journal of Information & Computational Science 8(7), 1063–1073 (2011)

[14] Stepenosky, N., Green, D., Kounios, J., Clark, C.M., Polikar, R.: Majority vote and decision template based ensemble classifiers trained on event related potentials for early diagnosis of Alzheimers's disease. In: Proceedings of the IEEE Int. Conf. on Acoustics, Speech and Signal Processing, pp. 901–904 (2006)

[15] Rokach, L.: Ensemble-based classifiers. Artif. Intell. Rev. 33, 1–39 (2010), doi:10.1007/s10462-009-9124-7

[16] http://www.cecs.louisville.edu/datamining/PDF/0471228524.pdf

[17] http://www.start.umd.edu

[18] http://www.mimuw.edu.pl/~son/datamining/DM/4-preprocess.pdf

[19] http://www.cs.ccsu.edu/~markov/ccsu_courses/DataMining-3.html

[20] Jain, A., Nandakumar, K., Ross, A.: Score normalization in multimodal biometric systems. Pattern Recognition 38(12), 2270–2285 (2005)

*k*NN Algorithm with Data-Driven *k* Value

Debo Cheng, Shichao Zhang*, Zhenyun Deng, Yonghua Zhu**, and Ming Zong

College of Computer Science & Information Technology,
Guangxi Normal University, Guilin, Guangxi, 541004, China
zhangsc@mailbox.gxnu.edu.cn

Abstract. This paper proposes a new *k* Nearest Neighbor (*k*NN) algorithm based on sparse learning, so as to overcome the drawbacks of the previous *k*NN algorithm, such as the fixed *k* value for each test sample and the neglect of the correlation of samples. Specifically, the paper reconstructs test samples by training samples to learn the optimal *k* value for each test sample, and then uses *k*NN algorithm with the learnt *k* value to conduct all kinds of tasks, such as classification, regression, and missing value imputation. The rationale of the proposed method is that different test samples should be assigned different *k* values in *k*NN algorithm, and learning the optimal *k* value for each test sample should be taken the correlation of data into account. To this end, in the reconstruction process, the proposed method is designed to achieve the minimal reconstruction error via a least square loss function, and employ an ℓ_1-norm regularization term to create the element-wise sparsity in the reconstruction coefficient, *i.e.*, sparsity appearing in the element of the coefficient matrix. For achieving effectiveness, the Locality Preserving Projection (LPP) is employed to keep the local structures of data. Finally, the experimental results on real datasets, and the experimental results show that the proposed *k*NN algorithm is better than the state-of-the-art algorithms in terms of different learning tasks, such as classification, regression, and missing value imputation.

Keywords: Sparse learning, k nearest neighbors, locality preserving projection, classification.

1 Introduction

The *k* Nearest Neighbors algorithm (*k*NN for short) is an instance-based, or a lazy learning method. It has been regarded as one of the simplest of all machine learning algorithms [4][13]. The rational of *k*NN is that similar samples belonging to the same class have high probability, while the key idea of *k*NN algorithm is to first select *k* nearest neighbors for each test sample, followed by using the learnt *k* nearest neighbors to predict this test sample. Therefore, *k*NN algorithm was often thought as an algorithm, in which no explicit training step is required.

---

* Corresponding author.
** Yonghua Zhu participated this work when he was a visiting undergraduate student of the Guangxi Key Lab of Multi-source Information Mining & Security (MIMS) of Guangxi Normal University, supervised by Professor Shichao Zhang. He is currently with School of Computer, Electronics and Information, Guangxi University, Nanning, Guangxi, 530004, China.

X. Luo, J.X. Yu, and Z. Li (Eds.): ADMA 2014, LNAI 8933, pp. 499–512, 2014.

A drawback of the kNN algorithm is that it is sensitive to select the value of k. Although studies have been focusing on this topic for a long time, the selection of k value in kNN algorithm is still very difficult and challengeable [8][11]. For example, Lall and Sharama mentioned that the suitable setting of k value should satisfy $k = \sqrt{n}$ for the datasets with a sample size larger than 100 [9]. However, such a setting has been proved to not suitable for all cases of datasets [10].

In this paper, we focus on the setting of k value in kNN algorithm based on the following observations, in which there is a binary classification task (where two classes are marked as the symbol '+' and the symbol '-', respectively) and the test sample with no label is marked with the symbol '?' in Fig. 1.

In Fig. 1, by setting $k = 5$ in the kNN algorithm, both the two test samples will be assigned '+' class. Actually, according to the distribution of data in Fig. 1, it is reasonable for the class label of the first test sample. However, we can find that the second test sample should be assigned as '-' class (*i.e.*, $k = 3$) rather than '+' class (*i.e.*, $k = 5$). Moreover, we can also find the similar scenario in kNN regression or missing value imputation with kNN algorithm. According to our observations, the k value in kNN algorithm should be data-driven, *i.e.*, decided by the distribution of data. Moreover, the k value for each test sample might be different.

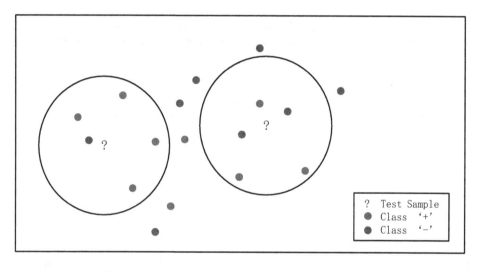

Fig. 1. An example of kNN classification task with $k = 5$

To overcome the above drawback of kNN algorithm, this paper proposes a Sparse learning based kNN method (S-kNN for short) to learn the optimal k values for different test samples when using kNN algorithm. This is carried out by the consideration of both the correlation and the local structure of training samples [37]. Specifically, we first reconstruct test samples with training samples to obtain an optimal k value for each test sample, and then use kNN algorithm with the learnt k to conduct classification or regression or missing value imputation[18][16]. The key step of the proposed method is the reconstruction process, in which, we employ a least square loss function to achieve the minimal reconstruction error, and use an ℓ_1-norm regularization term to result in the element-wise sparsity for generating various k values of different test samples. We also employ Locality Preserving Projection (LPP) [7] for preserving the local structures of data during the reconstruction process, aiming to further improve the reconstruction performance.

Comparing with the previous kNN algorithms, the proposed S-kNN algorithm has the following advantages:

- Different from the previous kNN algorithms that use either a fixed k value for all test samples or the cross-validation method to select the k value without considering the correlation of samples, the proposed S-kNN algorithm learns an optimal k value for each test sample according to the distribution of data. Moreover, during the learning process, the proposed S-kNN algorithm considers the correlation among samples, for further improving the performance. Therefore, the proposed reconstruction process for learning the optimal k values is data-driven.
- This paper proposes a new sparse learning framework for learning k values. Different from conventional Least Absolute Shrinkage and Selection Operator (LASSO) [14][31], our sparse learning framework also takes the local structures of samples into account. Moreover, this paper proposes a novel optimization method to solve the designed objective function.
- Different from the conventional kNN algorithms using for a single task, this paper uses the designed S-kNN method for three tasks, such as classification, regression, and missing value imputation. Moreover, the experimental results on real datasets show the benefit of the proposed method by comparing to the state-of-the-art kNN algorithms.

The remainder of the paper is arranged as follows. We review the applications of kNN algorithm in Section 2 and then give the detail of the proposed method in Section 3. Furthermore, we analyze the experimental results in Section 4 and give our conclusion and future work in Section 5.

2   Related Work

The study of kNN method has been becoming a hot research topic in data mining and machine learning since the algorithm was proposed in 1967 [2][6]. In this section, we briefly review the applications of kNN algorithm in data mining tasks, such as classification, regression, and missing value imputation.

2.1   Classification

The kNN classification was designed to select the k closest training samples to the test samples (*i.e.*, k nearest neighbors) from all the known samples, followed by a simple classifier, *e.g.*, majority classification rule [20][24][17]. Liu proposed an anomaly removal algorithm under the framework of kNN algorithm, for conducting kNN classification [10]. Weinberger *et al.* first demonstrated the reasonability to learn a Mahanalobis distance metric for kNN classification, and then optimized the measure metric with the goal that kNN always belong to the same class while examples form different classes are separated by a large margin [15]. Goldberger *et al.* proposed a neighborhood component analysis (NCA) method, *i.e.*, a new distance metric learning algorithm, to improved kNN classification [6]. Zhang incorporated Certainty Factor (CF) measure to deal with the unsuitability of skewed class distribution in the kNN framework [22].

2.2 Regression

The kNN algorithm for regression has been currently used and studied for many years in pattern recognition and data mining. In regression analysis, Burba *et al.* proposed some asymptotic properties of the kNN based kernel estimator for improving the performance of kNN regression [2]. Moreover, the goal of their paper is to study the non-parametric kNN algorithm. Ferraty and Vieu used the nonparametric characteristics of the kNN algorithm for conducting estimation, classification and discrimination on the high-dimensional data [5].

2.3 Missing Value Imputation

Missing data are often found in the study of data mining and machine learning. The learning task with missing data easily leads to low performance [12][25][38]. Although there are many solutions to deal with missing data, such as missing instance deletion and missing value imputation, and so on, the kNN based algorithms for missing value imputation are the key solutions in real applications [21][23][26][28]. For example, Zhang *et al.* proposed a Grey-Based kNN Iteration Imputation (GBKII) method, replacing Euclidean distance in conventional kNN algorithm with the grey-based distance measure, to improve the performance of kNN algorithm [19]. Recently, Zhang designed to select the left and right nearest neighbors of missing data for improving the performance of the previous kNN algorithm [27].

3   Method

In this section, we first introduce some basic concepts used in this paper. And then the S-kNN method is described. Finally, we give our optimization method to solve the proposed objective function.

3.1 Notation

Throughout the paper, we denote matrices as boldface uppercase letters, vectors as boldface lowercase letters, and scalars as normal italic letters, respectively. For a matrix $\mathbf{X} = [x_{ij}]$, its i-th row and j-th column are denoted as $\mathbf{x}^i$ and $\mathbf{x}_j$, respectively. Also, we denote the Frobenius norm, the ℓ_2-norm, and the ℓ_1-norm, respectively, as $\|\mathbf{X}\|_F = \sqrt{\sum_i \|\mathbf{x}^i\|_2^2} = \sqrt{\sum_j \|\mathbf{x}_j\|_2^2}$, $\|\mathbf{x}_j\|_2 = \sqrt{\sum_i^n x_{i,j}^2}$, and $\|\mathbf{X}\|_1 = \sum_i \sum_j |x_{i,j}|$, respectively. We further denote the transpose operator, the trace operator, and the inverse of a matrix $\mathbf{X}$ as $\mathbf{X}^T$, $tr(\mathbf{X})$, and $\mathbf{X}^{-1}$, respectively.

3.2 Reconstruction and Locality Preserving Projection (LPP)

Given $\mathbf{X} = \{\mathbf{x}_i\}_{i=1}^n \in \mathbb{R}^{n \times d}$ and $\mathbf{Y} = \{\mathbf{y}_i\}_{i=1}^m \in \mathbb{R}^{m \times d}$, where n, m, and d, represents the number of training samples, test samples, and feature dimension, respectively. To reconstruct test samples with training samples for obtaining the reconstruction coefficient matrix $\mathbf{W} \in \mathbb{R}^{n \times m}$, we define the following objective function [32][33][34]:

$$\arg\min_{\mathbf{W}} \sum_i \left\| \mathbf{w}_i^T \mathbf{x}_i - \mathbf{y}_i \right\| \tag{1}$$

where $w_{i,j}$ is used to test the correlation between $\mathbf{y}_i$ and training sample $\mathbf{x}_j$. The larger the value of $w_{i,j}$, the larger correlation between the i-th test sample and the j-th training sample. In particular, the case of $w_{i,j} = 0$ indicates that there is no correlation between $\mathbf{y}_i$ and $\mathbf{x}_j$.

LPP is a nonlinear subspace learning method [7]. The rationale of LPP is to obtain an optimal linear transformation $\mathbf{W}$ so that the local structures of the original data are preserved in the new space, *i.e.*, $\mathbf{W}$ converts the high-dimensional data $\mathbf{X}$ into the low-dimensional data $\mathbf{Y}$ with the following definition:

$$\mathbf{y}_j = \mathbf{W}^T \mathbf{x}_i, i = 1, 2, \cdots, n \tag{2}$$

To this end, the objective function of LPP can be defined as follows:

$$\min_{\mathbf{W}} \sum_{i,j} \left(\mathbf{W}^T \mathbf{x}_i - \mathbf{W}^T \mathbf{x}_j \right)^2 s_{ij} \tag{3}$$

where $\mathbf{S}$ is the weight matrix and each element of $\mathbf{S}$ is defined by a heat kernel $s_{i,j} = exp(\frac{-\|\mathbf{x}_i - \mathbf{x}_j\|^2}{\sigma})^1$.

By plugging Eq. (2) into Eq. (3), and some algebraic transformation operations, we obtain:

$$\frac{1}{2} \sum_{ij} \left(\mathbf{W}^T \mathbf{x}_i - \mathbf{W}^T \mathbf{x}_j \right)^2 s_{ij}$$

$$= \sum_i \left(\mathbf{W}^T \mathbf{x}_i d_{ij} \mathbf{x}_i^T \mathbf{W} \right) - \sum_{ij} \left(\mathbf{W}^T \mathbf{x}_i s_{ij} \mathbf{x}_i^T \mathbf{W} \right)$$

$$= tr(\mathbf{W}^T \mathbf{X} \mathbf{D} \mathbf{X}^T \mathbf{W}) - tr(\mathbf{W}^T \mathbf{X} \mathbf{S} \mathbf{X}^T \mathbf{W})$$

$$= tr(\mathbf{W}^T \mathbf{X} \mathbf{L} \mathbf{X}^T \mathbf{W}) \tag{4}$$

where $\mathbf{D}$ is a diagonal matrix and the i-th diagonal element of $\mathbf{D}$ is defined as $d_{i,i} = \Sigma_j s_{i,j}$. Hence, $\mathbf{L} = \mathbf{D} - \mathbf{S}$ is a Laplacian matrix.

3.3 Approach

We reconstruct test samples $\mathbf{Y}$ with training samples $\mathbf{X}$ to obtain the linear transformation matrix $\mathbf{W}$. We expect to map $\mathbf{X}$ into the space of $\mathbf{Y}$ via $\mathbf{W}$ and make the distance

---

^{1} σ is a tuning parameter. For simplicity, we set $\sigma = 1$ in our experiments.

between $\mathbf{Y}$ and $\mathbf{W}^T\mathbf{X}$ as small as possible. To this end, we employ the least square loss function [35,36]:

$$\left\|\mathbf{W}^T\mathbf{X} - \mathbf{Y}\right\|_F^2 = \left\|\hat{\mathbf{Y}} - \mathbf{Y}\right\|_F^2$$

$$= \sum_{i=1}^{n}\sum_{j=1}^{m}(y_{ij} - \hat{y}_{ij})^2 \tag{5}$$

where $\hat{\mathbf{Y}}$ is the new representation of $\mathbf{X}$ in the space of $\mathbf{Y}$, i.e., $\hat{\mathbf{Y}} = \mathbf{W}^T\mathbf{X}$. $\|\mathbf{W}^T\mathbf{X} - \mathbf{Y}\|_F^2$ represents the reconstruction error. Due to that Eq. (5) is convex, we can easily obtain its global solution $\mathbf{W} = (\mathbf{X}\mathbf{X}^T)^{-1}\mathbf{X}\mathbf{Y}$. However, $\mathbf{X}\mathbf{X}^T$ is not always invertible, so an ℓ_2-norm is often added to remove the issue of invertible. This leads to the ridge regression:

$$\arg\min_{\mathbf{W}} \left\|\mathbf{W}^T\mathbf{X} - \mathbf{Y}\right\|_F^2 + \delta\|\mathbf{W}\|_2^2 \tag{6}$$

where δ is a tuning parameter. The optimal solution of Eq. (6) can be described as a closed solution $\mathbf{W} = (\mathbf{X}\mathbf{X}^T + \delta\mathbf{I})^{-1}\mathbf{X}\mathbf{Y}$, where $\mathbf{I} \in \mathbb{R}^{n \times n}$ is an identity matrix.

The regularization term ℓ_1-norm has been proved to generate sparsity, i.e., zero in the elements of a matrix [30], while studies have shown that the ℓ_2-norm did not surely generate sparse result. In this paper, the element $w_{i,j}$ indicates the correlation between the i-th test sample and the j-th training sample. We expect that each test sample is only represented by part of training samples, i.e., many zero elements on each column in $\mathbf{W}$. Therefore, it makes sense for us to use an ℓ_1-norm to replace the ℓ_2-norm. Meanwhile, we also employ the LPP to preserve the local structures of data after the reconstruction process. To this end, the objective function of the proposed S-kNN method is defined as follows:

$$\arg\min_{\mathbf{W}} \frac{1}{2}\left\|\mathbf{Y} - \mathbf{W}^T\mathbf{X}\right\|_F^2 + \rho_1 tr(\mathbf{W}^T\mathbf{X}\mathbf{L}\mathbf{X}^T\mathbf{W}) + \rho_2\|\mathbf{W}\|_1 \tag{7}$$

where both ρ_1 and ρ_2 are tuning parameters. ρ_1 is designed to balance the magnitude between $tr(\mathbf{W}^T\mathbf{X}\mathbf{L}\mathbf{X}^T\mathbf{W})$ and $\left\|\mathbf{Y} - \mathbf{W}^T\mathbf{X}\right\|_F^2$. Moreover, the larger the value of ρ_1, the larger the contribution of LPP in Eq. (7). In particular, Eq. (7) shrinks to LASSO while setting $\rho_1 = 0$.

Different from LASSO, the proposed S-kNN considers preserving the local structures of data via LPP. Moreover, our S-kNN is used to learn the k value of kNN algorithm. Different from that the conventional kNN algorithms assign the fixed k value for all test samples or learn the k value for each test sample without considering the correlation among test samples, the proposed S-kNN algorithm learns the optimal k value for each test sample via a reconstruction process. During the reconstruction process, the proposed method considers the correlation of test samples and training samples. More specifically, the proposed method considers the correlation of test samples through generating the k values for all test samples once, and considers the correlation of training samples through adding the LPP regularization term in the reconstruction process. Moreover, the proposed S-kNN method is a data-driven method for selecting the optimal k values.

3.4   Optimization

Eq. (7) is a convex but non-smooth function. In this work, we solve it by designing a new accelerated proximal gradient method [37]. We first conduct the proximal gradient method on Eq. (7) by letting:

$$f(\mathbf{W}) = \frac{1}{2} \left\| \mathbf{Y} - \mathbf{W}^T \mathbf{X} \right\|_F^2 + \rho_1 tr(\mathbf{W}^T \mathbf{X} \mathbf{L} \mathbf{X}^T \mathbf{W}) \tag{8}$$

$$\vartheta(\mathbf{W}) = f(\mathbf{W}) + \rho_2 \|\mathbf{W}\|_1 \tag{9}$$

Note that $f(\mathbf{W})$ is convex and differentiable. To optimize $\mathbf{W}$ with the proximal gradient method, we iteratively update it by means of the following optimization rule:

$$\mathbf{W}(t+1) = \arg\min_{\mathbf{W}} \mathbf{G}_{\eta(t)}(\mathbf{W}, \mathbf{W}(t)) \tag{10}$$

where $\mathbf{G}_{\eta(t)}(\mathbf{W}, \mathbf{W}(t)) = f(\mathbf{W}(t)) + \langle \nabla f(\mathbf{W}(t)), \mathbf{W} - \mathbf{W}(t) \rangle + \frac{\eta(t)}{2} \|\mathbf{W} - \mathbf{W}(t)\|_F^2 + \rho_2 \|\mathbf{W}\|_1$, $\nabla f(\mathbf{W}(t)) = (\mathbf{X}\mathbf{X}^T + \rho_1 \mathbf{X}\mathbf{L}\mathbf{X}^T)\mathbf{W}(t) - \mathbf{X}\hat{\mathbf{Y}}^T$, $\langle \cdot, \cdot \rangle$ is an inner product operator, $\eta(t)$ determines the stepsize of the t- iteration, and $\mathbf{W}(t)$ is the value of $\mathbf{W}$ obtained at the t-iteration.

By ignoring the terms independent of $\mathbf{W}$ in Eq. (10), we can rewrite it as follows:

$$\mathbf{W}(t+1) = \pi_{\eta(t)}(\mathbf{W}(t)) = \arg\min \frac{1}{2} \|\mathbf{W} - \mathbf{U}(t)\|_2^2 + \frac{\rho_2}{\eta(t)} \|\mathbf{W}\|_1 \tag{11}$$

where $\mathbf{U}(t) = \mathbf{W}(t) - \frac{1}{\eta(t)} \nabla f(\mathbf{W}(t))$ and $\pi_{\eta(t)}(\mathbf{W}(t))$ is the Euclidean projection of $\mathbf{W}(t)$ onto the convex set $\eta(t)$. Due to the separability of $\mathbf{W}(t+1)$ on each row, i.e., $\mathbf{w}^i(t+1)$, we update the weights for each row individually:

$$\mathbf{w}^i(t+1) = \arg\min_{\mathbf{w}^i} \frac{1}{2} \left\| \mathbf{w}^i - \mathbf{u}^i(t) \right\|_2^2 + \frac{\rho_2}{\eta(t)} \left\| \mathbf{w}^i \right\|_2 \tag{12}$$

where $\mathbf{u}^i(t) = \mathbf{w}^i(t) - \frac{1}{\eta(t)} \nabla f(\mathbf{w}^i(t))$. In Eq. (12), $\mathbf{w}^i(t+1)$ takes a closed form solution [36] as follows:

$$\mathbf{w}^{i*} = max\{|\mathbf{w}^i| - \rho_2, 0\} \cdot sgn(\mathbf{w}^i) \tag{13}$$

Meanwhile, in order to accelerate the proximal gradient method in Eq. (8), we further introduce an auxiliary variable $\mathbf{V}(t+1)$ as follows:

$$\mathbf{V}(t+1) = \mathbf{W}(t) + \frac{\alpha(t) - 1}{\alpha(t+1)}(\mathbf{W}(t+1) - \mathbf{W}(t)) \tag{14}$$

where the coefficient $\alpha(t+1)$ is usually set as $\alpha(t+1) = \frac{1 + \sqrt{1 + 4\alpha(t)^2}}{2}$.

Finally, we list the pseudo of our proposed optimization method in Algorithm 1 and its convergence in Theorem 1.

Algorithm 1. Pseudo code of solving Eq. (7).

    Input: $\eta(0) = 0.01, \alpha(1) = 1, \gamma = 0.002, \rho_1, \rho_2$;
    Output: $\mathbf{W}$;
1 Initialize $t = 1$;
2 Initialize $\mathbf{W}(1)$ as a random diagonal matrix;
3 **repeat**
4   **while** $L(\mathbf{W}(t)) > G_{\eta(t-1)}(\pi_{\eta(t-1)}(\mathbf{W}(t)), \mathbf{W}(t))$ **do**
5 | Set $\eta(t-1) = \gamma\eta(t-1)$;
6   **end**
7   Set $\eta(t) = \eta(t-1)$;
8   Compute $\mathbf{W}(t+1) = \arg\min_{\mathbf{W}} G_{\eta(t)}(\mathbf{W}, \mathbf{V}(t))$;
9   Compute $\alpha(t+1) = \frac{1+\sqrt{1+4\alpha(t)^2}}{2}$;
10   Compute Eq. (14);
11 **until** *Eq. (7) converges*;

Theorem 1. *[36] Assume* $\{\mathbf{W}(t)\}$ *be the sequence obtained by Algorithm 1, then for* $\forall\, t \geq 1$, *the following holds*

$$\vartheta(\mathbf{W}(t)) - \vartheta(\mathbf{W}^*) \leq \frac{2\gamma L \|\mathbf{W}(1) - \mathbf{W}^*\|_{\mathrm{F}}^2}{(t+1)^2} \tag{15}$$

where $\gamma > 0$ *is a predefined constant, L is the Lipschitz constant of the gradient of* $f(\mathbf{W})$ *in Eq. (8), and* $\mathbf{W}^* = \arg\min_{\mathbf{W}} \vartheta(\mathbf{W})$.

Theorem 1 shows that the convergence rate of the proposed accelerated proximal gradient method in Algorithm 1 is $\mathcal{O}(\frac{1}{t^2})$, where t denotes an iteration number.

3.5 Algorithm

In this paper, we first optimize Eq. (7) to obtain the correlation coefficient matrix $\mathbf{W}$, aiming to obtain an optimal k value for each test sample. We then use the selected k to conduct kNN algorithm for different tasks, such as classification, regression, and missing value imputation. The pseudo of S-kNN is presented in Algorithm 2.

In both the regression task and missing value imputation, the bigger the correlation between a test sample and its nearest neighbors[29], the larger the contribution of the nearest neighbors to the test sample. Therefore, this paper proposes a weighted method for the prediction of both the regression task and missing value imputation task. Specifically, the weighted predictive value of the j-th test sample is defined as:

$$\text{predictValue\_weight} = \sum_{i=1}^{n} \left(\frac{w_{i,j}}{\sum_{i=1}^{n} w_{i,j}} \times \mathbf{y}_{train(i)} \right) \tag{16}$$

where n is the number of training samples, and $\mathbf{y}_{train(i)}$ represents the true value of the i-th training sample.

Algorithm 2. The pseudo of S-*k*NN algorithm.

Input: **X**, **Y**;
Output:
switch *task* **do**
    case *1*
      | Class labels;
    endsw
    case *2*
      | Predicted value;
    endsw
    case *3*
      | Imputation value;
    endsw
endsw

1 Normalizing **X** and **Y** (When **Y** is class labels without normalization);
2 Optimizing Eq. (7) to obtain the optimal solution **W**;
3 Obtaining the optimal *k* value for test samples based on **W**;
4 **switch** *task* **do**
5   **case** *1*
6 | Obtaining class labels via majority rule;
7   **endsw**
8   **case** *2*
9 | Obtaining prediction value via Eq. (16);
10   **endsw**
11   **case** *3*
12 | Obtaining imputation value via Eq. (16);
13   **endsw**
14 **endsw**

On the other hand, in the classification task, the proposed S-*k*NN algorithm uses the majority rule of *k* nearest neighbors of each test sample to predict the class label of each test sample.

4 Experimental Analysis

We evaluated the proposed S-*k*NN method in the tasks, such as classification, regression and missing value imputation, by comparing with the state-of-the-art *k*NN algorithms. Note that the classification tasks included binary classification and multi-class classification.

4.1 Experimental Setting

In our experiments, we regarded the standard *k*NN algorithm (with $k = 5$) as the first comparison algorithm. The second comparison algorithm is Eq. (7) with the setting $\rho_1 = 0$, *i.e.*, via LASSO learning different *k* values for test samples. We called this algorithm LASSO based kNN algorithm, L-kNN for short, with which we would like to show the importance of preserving the local structures of data.

The used datasets came from UCI [1], LIBSVM [3] and the website[2]. We conducted experiments on four datasets for classification, regression, and missing value imputation, respectively. We coded all algorithms with MATLAB 7.1 in windows 7 system. We conducted experiments by 10-fold cross-validation method. In this way, we repeated the whole process 10 times to avoid the possible bias in our experiments.

We used classification accuracy as the evaluation for the classification task. The higher accuracy the algorithm is, the better the performance of the classification is.

We employed Root Mean Square Error (RMSE) [31] and correlation coefficient to evaluate the performance of both regression analysis and missing value imputation. Note that there are not missing values in the original datasets, we randomly selected some independent values to be missed according to the literatures on missing value imputation [27].

RMSE is defined as the square root of predicted value and the ground-truth. Its formula is as follows:

$$\text{RMSE} = \sqrt{\frac{1}{n} \sum_{i=1}^{n} (\mathbf{y}_i - \hat{\mathbf{y}}_i)^2} \tag{17}$$

where $\mathbf{y}_i$ indicates the ground-truth, $\hat{\mathbf{y}}_i$ indicates the predicted value. Obviously, the smaller the RMSE is, the better the algorithm is.

Correlation coefficient indicates the correlation between prediction and observation. The range of correlation coefficient is between +1 and -1 inclusive, where 1 is total positive correlation, 0 is no correlation, and -1 is total negative correlation. Generally, the larger the correlation coefficient is, the more accurate the prediction is.

4.2   Results

In this section, we compared the performance of the proposed S-kNN algorithm to two comparison algorithms on real datasets, in terms of three tasks, such as classification, regression, and missing value imputation.

Classification Results. We summarized the classification accuracy results of all algorithms in Table 1. As shown in Table 1, we had the following observations:

- The proposed S-kNN algorithm outperformed the comparison algorithms, such as L-kNN and kNN. More specifically, the proposed S-kNN algorithm averagely improved by 5.97% and 16.32%, respectively, than the L-kNN algorithm and the kNN algorithm, in terms of classification accuracy.
- The S-kNN algorithm performed better than the L-kNN algorithm because the proposed S-kNN method used the LPP regularization term to preserve the local structures of data. In particular, the proposed method improved by 11.48% than the L-kNN, on Heart dataset. Both the S-kNN and the L-kNN outperformed the kNN algorithm. This indicates that using different k values in kNN algorithm (such as

---

[2] http://www.cc.gatech.edu/~lsong/code.html

Table 1. Comparison of classification accuracy

Dataset	kNN	L-kNN	S-kNN
Cleveland	0.7048 ± 0.0029	0.7619 ± 0.0030	**0.8048 ± 0.0038**
Ionnosophere	0.6743 ± 0.0031	0.8286 ± 0.0025	**0.8571 ± 0.0025**
Heart	0.5381 ± 0.0031	0.6741 ± 0.0033	**0.7889 ± 0.0043**
Seeds	0.8048 ± 0.0033	0.8714 ± 0.0015	**0.9238 ± 0.0011**

Table 2. Comparison of RMSE

Dataset	kNN	L-kNN	S-kNN
Mpg	3.8080 ± 0.4417	3.5909 ± 0.3662	**3.4693 ± 0.3158**
Triazines	0.1477 ± 0.0017	0.1357 ± 0.0016	**0.1242 ± 0.0013**
Concreteslup	0.0176 ± 0.000036	0.0151 ± 0.000033	**0.0139 ± 0.0000339**
Bodyfat	0.000021± 3.7921e-09	0.000014 ± 3.3355e-09	**0.000013 ± 3.8892e-09**

the S-*k*NN algorithm and the L-*k*NN algorithm) can achieve better classification performance than the method with fixed k value for all test samples, such as the conventional *k*NN algorithm.

Regression Results. We summarized the results of RMSE and correlation coefficient in Tables 2 and 3, respectively.

From Tables 2 and 3, we found that our proposed S-*k*NN achieved the best performance in terms of both RMSE and correlation coefficient, followed by L-*k*NN and *k*NN. In the evaluation of RMSE, the proposed S-*k*NN averagely reduced by 0.0915 and 0.0336 than the L-*k*NN and the *k*NN, on four datasets. In particular, the proposed S-*k*NN algorithm made the most improvement on Mpg dataset, *i.e.*, reduced by 0.1216 and 0.3387, than the L-*k*NN and the *k*NN. In terms of correlation coefficient, the proposed S-*k*NN averagely increased by 3.66% than L-*k*NN, and by10.42% than the *k*NN, on four datasets. Moreover, the proposed method achieved the maximal increase on Triazines dataset, *i.e.*, 8.8% compared to the L-*k*NN and 22.85% compared to the *k*NN.

Based on the results on regression analysis in our experiments, we can know: On one hand, it is reasonable for the proposed method to take the local structures of data into account in regression analysis. On the other hand, using different *k* values in *k*NN algorithm is practical. Moreover, the *k* values should be learnt from the data.

Table 3. Comparison of correlation coefficient

Dataset	kNN	L-kNN	S-kNN
Mpg	0.8865 ± 0.0019	0.8978 ± 0.0014	**0.9076 ± 0.0011**
Triazines	0.4256 ± 0.0148	0.5661 ± 0.0123	**0.6541 ± 0.0106**
Concreteslup	0.6606 ± 0.0312	0.7719 ± 0.0249	**0.8194 ± 0.0195**
Bodyfat	0.9846 ± 0.000081	0.9918 ± 0.000044	**0.9930 ± 0.000043**

Table 4. Comparison of RMSE

Dataset	kNN	L-kNN	S-kNN
Abalone	2.3894 ± 0.1019	2.1261± 0.1360	**2.0850 ± 0.1207**
Housing	5.2228 ± 1.2315	3.8666 ± 0.2108	**3.7948 ± 0.2175**
Pyrim	0.0673 ± 0.0002	0.0492 ± 0.0003	**0.0484 ± 0.0003**
YachtHydrod	10.5379 ± 3.0492	10.1437 ± 2.6463	**9.4637 ± 3.2310**

Table 5. Comparison of correlation coefficient

Dataset	kNN	L-kNN	S-kNN
Abalone	0.6499± 0.0008	0.7282 ± 0.0024	**0.7376 ± 0.0021**
Housing	0.8285 ± 0.0004	0.9142 ± 0.0004	**0.9197 ± 0.0003**
Pyrim	0.8283 ± 0.0079	0.9140 ± 0.0025	**0.9201 ± 0.0022**
YachtHydrod	0.7200 ± 0.0044	0.7948 ± 0.0039	**0.8162 ± 0.0034**

Imputation Results. We summarized the RMSE and the correlation coefficient, respectively, of the results of missing value imputation on all algorithms in Tables 4 and 5.

From Tables 4 and 5, we found that our proposed S-kNN achieved the best performance in terms of RMSE and correlation coefficient, followed by L-kNN and kNN. Regarding the case that the proposed S-kNN outperformed L-kNN, we knew that the preservation of local structures of data is necessary on missing value imputation with kNN algorithm. By comparing the S-kNN and the L-kNN with the kNN, we found that learning the optimal k value in kNN algorithm is feasible and useful.

In a word, according to the results on three learning tasks, we can make the following conclusion: it might be reasonable to use varied k values in kNN algorithm in real applications, while the optimal k values should be learnt from the data, *i.e.,* the data-driven k value in kNN algorithm.

5 Conclusion

In this work, we have proposed a new kNN algorithm by replacing the fixed k value for all test samples by learning the optimal k value for each test samples according to the distribution of data. To achieve this, we reconstructed test samples by training samples to obtain sparse correlation between each test sample and training samples. Moreover, the sparse correlation decided the k value for each test sample. Furthermore, we used the new kNN algorithm to conduct three learning tasks, such as classification, regression, and missing value imputation. The experimental results on real datasets verified the benefit of the proposed method, by comparing with the state-of-the-art kNN algorithms.

Acknowledgements. This work was supported in part by the National Natural Science Foundation of China under grants 61170131, 61263035 and 61363009, the China 863 Program under grant 2012AA011005, the China 973 Program under grant

2013CB329404, the Guangxi Natural Science Foundation under grant 2012GXNS-FGA060004, the Key Project for Guangxi Universities' Science and Technology Research under grant 2013ZD041.

References

1. Bache, K., Lichman, M.: UCI machine learning repository (2013)
2. Burba, F., Ferraty, F., Vieu, P.: k-nearest neighbour method in functional nonparametric regression. Journal of Nonparametric Statistics 21(4), 453–469 (2009)
3. Chang, C.-C., Lin, C.-J.: LIBSVM: A library for support vector machines. ACM Transactions on Intelligent Systems and Technology 2, 27:1–27:27 (2011), http://www.csie.ntu.edu.tw/~cjlin/libsvm
4. Cover, T., Hart, P.: Nearest neighbor pattern classification. IEEE Transactions on Information Theory 13(1), 21–27 (1967)
5. Ferraty, F., Vieu, P.: Nonparametric functional data analysis: theory and practice (2006)
6. Goldberger, J., Roweis, S.T., Hinton, G.E., Salakhutdinov, R.: Neighbourhood components analysis. In: NIPS (2004)
7. He, X., Niyogi, P.: Locality preserving projections. In: NIPS (2003)
8. Kang, P., Cho, S.: Locally linear reconstruction for instance-based learning. Pattern Recognition 41(11), 3507–3518 (2008)
9. Lall, U., Sharma, A.: A nearest neighbor bootstrap for resampling hydrologic time series. Water Resources Research 32(3), 679–693 (1996)
10. Liu, H., Zhang, S., Zhao, J., Zhao, X., Mo, Y.: A new classification algorithm using mutual nearest neighbors. In: GCC, pp. 52–57 (2010)
11. Meesad, P., Hengpraprohm, K.: Combination of knn-based feature selection and knnbased missing-value imputation of microarray data. In: ICICIC, pp. 341–341 (2008)
12. Qin, Y., Zhang, S., Zhu, X., Zhang, J., Zhang, C.: Semi-parametric optimization for missing data imputation. Applied Intelligence 27(1), 79–88 (2007)
13. Qin, Z., Wang, A.T., Zhang, C., Zhang, S.: Cost-sensitive classification with k-nearest neighbors. In: Wang, M. (ed.) KSEM 2013. LNCS, vol. 8041, pp. 112–131. Springer, Heidelberg (2013)
14. Tibshirani, R.: Regression shrinkage and selection via the lasso. Journal of the Royal Statistical Society. Series B (Methodological), 267–288 (1996)
15. Weinberger, K.Q., Blitzer, J., Saul, L.K.: Distance metric learning for large margin nearest neighbor classification. In: NIPS, pp. 1473–1480 (2005)
16. Wu, X., Zhang, C., Zhang, S.: Efficient mining of both positive and negative association rules. ACM Transactions on Information Systems (TOIS) 22(3), 381–405 (2004)
17. Wu, X., Zhang, C., Zhang, S.: Database classification for multi-database mining. Information Systems 30(1), 71–88 (2005)
18. Wu, X., Zhang, S.: Synthesizing high-frequency rules from different data sources. IEEE Transactions on Knowledge and Data Engineering 15(2), 353–367 (2003)
19. Zhang, C., Zhu, X., Zhang, J., Qin, Y., Zhang, S.: GBKII: An imputation method for missing values. In: Zhou, Z.-H., Li, H., Yang, Q. (eds.) PAKDD 2007. LNCS (LNAI), vol. 4426, pp. 1080–1087. Springer, Heidelberg (2007)
20. Zhang, S.: Cost-sensitive classification with respect to waiting cost. Knowledge-Based Systems 23(5), 369–378 (2010)
21. Zhang, S.: Estimating semi-parametric missing values with iterative imputation. International Journal of Data Warehousing and Mining 6(3), 1–10 (2010)

22. Zhang, S.: KNN-CF approach: Incorporating certainty factor to knn classification. IEEE Intelligent Informatics Bulletin 11(1), 24–33 (2010)
23. Zhang, S.: Shell-neighbor method and its application in missing data imputation. Applied Intelligence 35(1), 123–133 (2011)
24. Zhang, S.: Decision tree classifiers sensitive to heterogeneous costs. Journal of Systems and Software 85(4), 771–779 (2012)
25. Zhang, S.: Nearest neighbor selection for iteratively knn imputation. Journal of Systems and Software 85(11), 2541–2552 (2012)
26. Zhang, S., Jin, Z., Zhu, X.: Missing data imputation by utilizing information within incomplete instances. Journal of Systems and Software 84(3), 452–459 (2011)
27. Zhang, S., Jin, Z., Zhu, X., Zhang, J.: Missing data analysis: A kernel-based multi-imputation approach. In: Gavrilova, M.L., Tan, C.J.K. (eds.) Transactions on Computational Science III. LNCS, vol. 5300, pp. 122–142. Springer, Heidelberg (2009)
28. Zhang, S., Qin, Z., Ling, C.X., Sheng, S.: "Missing is useful": missing values in cost-sensitive decision trees. IEEE Transactions on Knowledge and Data Engineering 17(12), 1689–1693 (2005)
29. Zhao, Y., Zhang, S.: Generalized dimension-reduction framework for recent-biased time series analysis. IEEE Transactions on Knowledge and Data Engineering 18(2), 231–244 (2006)
30. Zhu, X., Huang, Z., Cheng, H., Cui, J., Shen, H.T.: Sparse hashing for fast multimedia search. ACM Transactions on Information Systems 31(2), 9 (2013)
31. Zhu, X., Huang, Z., Cui, J., Shen, H.T.: Video-to-shot tag propagation by graph sparse group lasso. IEEE Transactions on Multimedia 15(3), 633–646 (2013)
32. Zhu, X., Huang, Z., Shen, H.T., Zhao, X.: Linear cross-modal hashing for efficient multimedia search. In: ACM Multimedia, pp. 143–152 (2013)
33. Zhu, X., Huang, Z., Tao Shen, H., Cheng, J., Xu, C.: Dimensionality reduction by mixed kernel canonical correlation analysis. Pattern Recognition 45(8), 3003–3016 (2012)
34. Zhu, X., Huang, Z., Yang, Y., Tao Shen, H., Xu, C., Luo, J.: Self-taught dimensionality reduction on the high-dimensional small-sized data. Pattern Recognition 46(1), 215–229 (2013)
35. Zhu, X., Suk, H.-I., Shen, D.: Matrix-similarity based loss function and feature selection for alzheimer's disease diagnosis. In: CVPR, pp. 3089–3096 (2014)
36. Zhu, X., Suk, H.-I., Shen, D.: A novel matrix-similarity based loss function for joint regression and classification in ad diagnosis. NeuroImage (2014)
37. Zhu, X., Zhang, L., Huang, Z.: A sparse embedding and least variance encoding approach to hashing. IEEE Transactions on Image Processing 23(9), 3737–3750 (2014)
38. Zhu, X., Zhang, S., Jin, Z., Zhang, Z., Xu, Z.: Missing value estimation for mixed-attribute data sets. IEEE Transactions on Knowledge and Data Engineering 23(1), 110–121 (2011)

Topology Potential-Based Parameter Selecting
for Support Vector Machine

Yi Lin[1], Shuliang Wang[2], Long Zhao[1], and DaKui Wang[3]

[1] School of Computer, Wuhan University
[2] School of Software, Beijing Institute of Technology
[3] International School of Software, Wuhan University
slwang2005@gmail.com

Abstract. We present an algorithm for selecting support vector machine's meta-parameter value which is based on ideas from topology potential of data field. By the optimal spatial distribution of topological potential corresponding to minimum entropy potential, it searches so smart that the optimal parameters can be found effectively and efficiently. The experimental results show that it can get almost the same effectiveness with the exhaustive grid search under an order of magnitude lower computational cost. It also can be used to automatically identify kernels and other parameter selection problem.

Keywords : Data field, Topology potential, Potential entropy, Support vector machine, Parameter selection.

1   Introduction

Support Vector Machines (SVM) are supervised learning models proposed by Corinna Cortes and Vapnik in 1995(Cortes and Vapnik 1995). As the generalization capability independent of the input feature dimensions, which overcomes the "curse of dimensionality", SVMs has a very profound impact in the field of data mining and machine learning(Wu, Kumar et al. 2008). But there are three potential drawbacks of SVMs: Uncalibrated class membership probabilities, multi-class SVM and parameters of a solved model (Wikipedia 2014). For the last problem, we presents a parameter selecting algorithm based on topology potential of data field. By the most rational distribution of topology potential in the sense of uncertainty, it effectively and effectively guide the parameter space scaling towards the optimal or near-optimal selecting.

At present, there is no uniform method for SVM's parameter recognition. The most common and reliable approach to parameter selection is to do exhaustive grid search over the whole search space now. Unfortunately, it may result in a large number of evaluations and unacceptably long run times. Although heuristic algorithms such as genetic algorithms and particle swarm etc. could get the feasible solution without traversal, the result is unstable and easy to fall into local extreme points. The vast majority methods use N-fold cross-validation(Kohavi 1995) as the assessment of model performance.

Our method is to start with a very coarse gird covering the whole parameter space and iteratively refine both the grid resolution and search boundaries, keeping the

X. Luo, J.X. Yu, and Z. Li (Eds.): ADMA 2014, LNAI 8933, pp. 513–522, 2014.
© Springer International Publishing Switzerland 2014

scaling to warding regions with high performance and high topology potential values. The results shows that there is 97.02% similarity between topology potential-based approach and grid method, whereas the presented algorithm required only seventh in computation cost. It has a strong practical value in many field, such as kernel selecting and other parameter selecting case.

2   Fundamental

2.1   Topology Potential of Data Field

With the development of field theory, "field" is not only a physical terms, but rather as an abstract mathematical concept, which can be used to describe the distribution of physical or mathematical function in space. Inspired by field theory, data field introduces interaction between material particles and field description method into abstract number field space. Let $D = \{x_1, x_2, ... x_n\}$ be a data set with n objects in space X, where $x_i = (x_{i1}, x_{i2}, ..., x_{ip})'$, $(i = 1, 2, ..., n)$, then each object is equivalent to a particle with a certain quality in a p-dimensional space. There is a data field in the entire space, and any object in the field will be affected by the associative action of other objects(Wang, Gan et al. 2011).

In data field, the field generated by static data can be regarded as a stable active field. It can use a vector intensity function or scalar potential function to describe the spatial distribution of data fields. Because the scalar computation is more simple and intuitive than vector computation, so the scalar potential function is chosen to describe the properties of data field.

In principle, any function who meets the standard morphology of data field potential function can be used to define the data field(Wang, Gan et al. 2011). They can be divided into two categories: long-range field potential functions and short-range field potential functions. The potential value of short-range field decays slower with distance than long-range field function, which is more suitable to describe the interaction between data objects. The most commonly used nuclear field potential function is defined as follow: $\varphi_x(y) = m \times e^{-\left(\frac{\|x-y\|}{\sigma}\right)^k}$, where, $m \geq 0$ represents the field source strength, which can be regarded as the quality of data object; $\sigma \in (0, \infty)$ used to control the interaction range between objects, called impact factor; $k \in N$ is the distance index. So the superposition potential value of data field at any point $x \in \Omega$ in space can be expressed as $\varphi(x) = \varphi_D(x) = \sum_{i=1}^{n} \varphi_i(x) = \sum_{i=1}^{n} \left(m_i \times e^{-\left(\frac{\|x-x_i\|}{\sigma}\right)^2} \right)$, where $\|x - x_i\|$ represents the distance between the object x_i and location x.

2.2   Parameters of Support Vector Machine

The selecting of kernel parameters and penalty factor C has a significant impact on the results of SVMs. How to choose them becomes difficulties. The first problem is how to evaluate the parameter's performance. The N-fold cross-validation(CV)(Kohavi 1995) is the most commonly used method. It converts the original data into some smaller subsets, part as training sets and the other part as test sets. Firstly, the training sets was used to train classifiers, then obtained generated model was tested with test sets, finally, the test accuracy is used to as evaluation index of classification performance.

Unfortunately, there is no uniform solution on the selecting of SVM's optimal parameters internationally. Grid search is still the most common and reliable approach, which can find the highest classification accuracy in sense of CV within whole parameter space. But even grid searches with middling resolution could lead to lots of evaluations and unbearable long run times. In large-scale search, it is more inclined to use heuristic optimization algorithm, such as genetic algorithms(Wu, Tzeng et al. 2009), particle swarm algorithms(Guo, Yang et al. 2008) etc., which can get the feasible solution without traversal. But the results of such heuristic algorithm is unstable and easy to fall into local extreme point.

In many cases, there are more than one parameter setting could achieve the highest verification classification accuracy. To avoid over fitting, it often choose the setting with the smallest C as the optimal choice. If there is a test set, these setting may be tested separately, the parameter setting with the highest test accuracy will be the optimal value.

2.3   Parameter Selecting by Minimizing Topology Potential

The spatial distribution of data field is mainly depends on the impact radius $R = \sigma \times \sqrt[k]{\dfrac{9}{2}} \ (k \in N)$, which describe the interaction between objects, which has little relationship with the specific morphology of potential function or distance index k. For given potential function from, the setting of σ will have a huge impact on the spatial topology of potential field(Wang, Gan et al. 2011). Too small or too large σ are not well reflect the intrinsic distribution relationship of data.

From the view of information theory, potential entropy can be used to measure the uncertainty of original data distribution. When each object's potential value is the same, this source data distribution means the maximum uncertainty and potential entropy. Conversely, if the potential value is very asymmetric, it means the minimal uncertainty and potential entropy. Let $\Psi_1, \Psi_2, ... \Psi_n$ be the potential value of data object $x_1, x_2, ..., x_n$, then the potential entropy is defined as $H = -\sum_{i=1}^{n} \dfrac{\Psi_i}{Z} \log\left(\dfrac{\Psi_i}{Z}\right)$, where $Z = \sum_{i=1}^{n} \Psi_i$ is the normalization factor. The nature of potential entropy minimizing is a single-value nonlinear optimization problem. There are many existing standard algorithms to solve it, such as simulated annealing etc. The obtained impact factor value is the best setting, which give the most reasonable potential field distribution.

According to the superposition principle of data field, if the intensity of data distribution is the same, the region nearby high quality data object will have higher potential values, the quality of data object nearby maximum potential value is also the largest. It means that the potential can reflect the quality distribution of data objects and be used as an estimation of population distribution. Figure 1 shows the comparison of SVM's performance and topology potential simulation on 'zoo'. Left is the 3D graph of SVM model performance, the right is the surface of optimal potential distribution. It is not difficult to find that the topology potential of data field establishes a reasonable smooth population estimation of SVM's performance. This method not only can effectively and smoothly capture various trend of SVM's performance, but also only requires $O(n^2)$ time complexity.

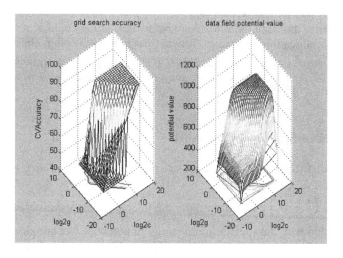

Fig. 1. 3D simulating of data field

Frist, our algorithm evaluates the cross-validation model performance of some parameter in a very coarse sampling covering the whole search space; and then it calculate the optimal topology potential value of data field at grid intersections; finally, it scales the parameter range at center of points with maximum potential value and samples with maximum CV accuracy. This process can be iterated several times until meets termination condition. In last sampling, the sample point with the highest CV accuracy is the optimal meta-parameter setting. The algorithm flow chart is shown in figure 2.

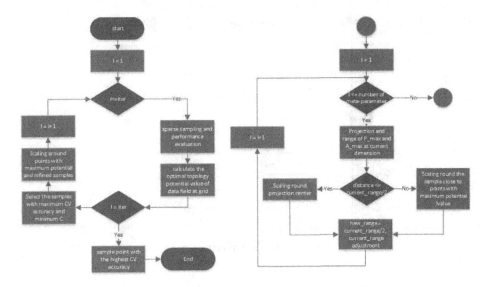

Fig. 2. Algorithm flowchart

This algorithm uses a combination sampling pattern of a standard N-parameter three-level $\{-1, 0, +1\}$ with and N-parameter two-level $\{-1/2, 1/2\}$ form Design of experiments(Montgomery 2008). Figure 3 shows the 13 samples including boundary extremes and key points in two-parameter case. This sampling pattern covers the whole parameter space in a simple and effective way, and the samples includes boundary extremes and key points. It keeps the number of samples at each iteration roughly constant, and easily extended to different dimensional case.

Each sample point is equivalent to a particle with CV accuracy quality in parameter space. There is a data field in whole space, and any object in field was influenced by other objects' associative action. In optimal data field, the parameter space can be scaled at the center of samples with the highest CV accuracy and points with maximum topology potential. First, these two types of extreme points exclude best samples without minimum C are projected on every parameter dimension; then, on the projection direction, if these two types of extreme points is close to each other, parameter range is folded at the center of these points; if they are far apart, parameter range is halved around samples closest to points with maximum potential value; finally, if there are multiple samples meet conditions, it takes the median of these points as shrinkage center. When new range extends outside the initial range, the center point is the point closest to the best point where the new range is contained in the initial range. The algorithm itself is independent of the number of parameters in search space, so it scales naturally to different number of parameters required for different kernel. All evaluated parameter values and their CV accuracy will be saved, and it will directly use the stored value without re-calculation. In our algorithm, the grid resolution is doubled, but it might be increased by other factor.

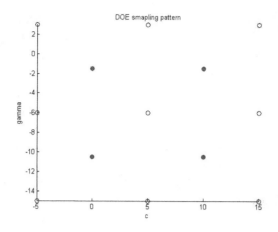

Fig. 3. DOE sampling pattern

3   Experiments and Analysis

3.1   Dataset and Experiment Environments

The 23 datasets used in the experiments is from UCI Machine Learning Repository(K and Lichman.M 2013). They are the most popular datasets in the data mining and machine learning field, which are: balance, breast, dna, german, glass, heart, iono-sphere, iris, landsat, letter, liver, msplice, musk, pima, segment, sonar, spambase, ve-hicle, vote, waveform3, wine, wpbc, and zoo. In experiments, each dataset was divided into training dataset and test dataset with the ration of 3:1, which are normalized to [0, 1]. Table 1 shows the important factor of experiment environments.

Table 1. Experiment Enviorment

Computers	HP xw6600 Workstatoion
Operation System	Windows 7 Ultimate
Software Platform	Matlab R2012a
Toolbox	LIBSVM-3.18

3.2   Results of Topology Potential Algorithm

To many various problems, the Gaussian kernel could get impressive results. For all experiments in this paper, it utilized Gaussian kernel for modeling with

$\log_2 C \in \{-5,15\}$ and $\log_2 \gamma \in \{-15,3\}$. The potential value calculating was done with twenty samples per parameter with uniform resolution in $\log_2$ space.

Figure 4 (a) ~ (e) shows the progress of 5 iterations on "liver" dataset by topology potential method. In each subfigure, the blue hollow circle means the evaluated samples, in which the magenta "* " marked the new evaluation samples in current iteration, magenta mogen david and "*" labeled points with maxima potential value and the samples with great CV accuracy individually, 3D contour shows the trends of potential value, blue lines marks new range. The final selected parameter is the point with blue circle and magenta "*" inside in the last figure. It is clear that overall selecting process is effective along the direction of high performance. The algorithm not only divided a reasonable new area in current range, but also make it out of current range sometimes, which effectively avoided missing the optimal in the process of parameter selecting.

Though the topology potential-based algorithm could efficiently shrink the parameter range, it becomes lack of flexibility in "flat" regions. Figure 4 (f) ~ (j) shows the results on "wbpc" dataset. If the "flat" area be identified earlier, it may identify the optimal faster.

3.3   Results of Grid Search

To further analysis, we compared the quality of final result and cost of topology potential algorithm with the standard grid search method using LIBSVM(Chang and Lin 2011) and other setting. Table 2 lists the final parameter settings, the number of model evaluation and predict performance by two different methods. The grid search sampled 20 points per dimension, for 400 points in two-dimensional space. The "acc" item records the real optimal test accuracy of samples with the highest CV accuracy.

Table 2. Results

dataset	Topological potential				Grid			
	$\log_2 C$	$\log_2 \gamma$	#fun	acc	$\log_2 C$	$\log_2 \gamma$	#fun	acc
balance	12.2862	-7.051	65	98.7261%	12.8947	-7.4211	400	99.3631%
breast	4.8849	-6.1036	65	70%	1.3158	-4.5789	400	72.8571%
dna	3.7993	-4.9194	65	97.2%	1.3158	-4.5789	400	97.2%
german	14.0625	-14.7188	45	75.6%	15	-15	400	75.2%
glass	5.5592	-1.4704	65	1.85185%	1.3158	0.15789	400	0%
heart	1.8092	-4.4309	65	84.2105%	9.7368	-5.5263	400	75%
iono-sphere	-0.29605	-1.4704	65	71.5909%	0.26316	-1.7368	400	71.5909%
iris	3.5855	-0.21217	65	81.5789%	3.4211	0.15789	400	78.9474%
landsat	2.2862	-11.2549	55	70.2%	2.3684	-11.2105	400	70.25%
letter	4.2105	-5.7928	60	93.12%	3.4211	-5.5263	400	92.96%

Table 2. (*continued*)

liver	10.5263	-15	63	74.7126%	9.7368	-15	400	75.8621%
msplice	-1.5789	-6.5921	65	95.3401%	0.26316	-4.5789	400	95.7179%
musk	2.7961	-15	65	100%	1.3158	-15	400	100%
pima	9.6053	-7.0066	60	77.0833%	12.8947	-10.2632	400	77.6042%
segeme nt	8.5526	-12.9572	60	88.5813%	6.5789	-12.1579	400	88.4083%
sonar	5.1316	-2.8026	65	34.6154%	5.5263	-2.6842	400	34.6154%
spamba se	12.0888	-14.7188	55	80.5387%	12.8947	-15	400	80.278%
vehicle	12.5987	-5.5559	60	85.8491%	12.8947	-5.5263	400	86.3208%
vote	3.8158	-3.5132	65	89.9083%	3.4211	-2.6842	400	92.6606%
wave- form3	1.7105	-11.4474	60	87.2%	-1.8421	-5.5263	400	86.96%
wine	11.2829	-15	55	26.6667%	11.8421	-15	400	26.6667%
zoo	0.98684	-3.6908	65	84.6154%	0.2632	-2.6842	400	84.6154%

3.4   Observations

Obviously, topology potential-based method is on par with grid search on performance. The t-test proved that the final accuracy results of these two approaches had
comparable accuracy with 97.02% probability. But the topology potential-based
method used 5 iterations and no more than 65 samples to find the parameter values,
which requires only nearly one-seventh cost of gird search. It is interesting to find that
optimal parameter values by presented approach may differ widely from that found in
grid search, such as the 'pima' dataset.

4   Summary

We have presented an approach based on topology potential that can reliably find
excellent parameter values for SVM with relatively little cost. Using the topology
potential-based algorithm for a good start point and other gradient-descent method for
local search, it should be able to get a better parameter values.

The topology potential establishes a reasonable rough model of SVM's performance
distribution by several limited discrete points' performance. It uses a continuous
single-values potential function about location to reflect parameter's assessment performance, and minimum potential entropy for the most reasonable potential field distribution. Additionally, the contour of topological potential can vividly show the data
field's distribution in low dimension space. This approach not only be a good choice
for SVM's parameter selecting, but also be extended to other parameter selection
problems.

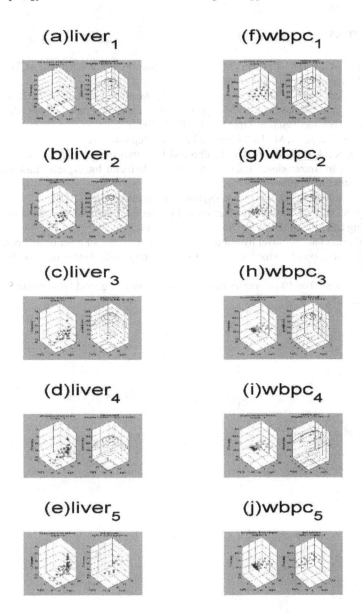

(a)liver$_1$

(f)wbpc$_1$

(b)liver$_2$

(g)wbpc$_2$

(c)liver$_3$

(h)wbpc$_3$

(d)liver$_4$

(i)wbpc$_4$

(e)liver$_5$

(j)wbpc$_5$

Fig. 4. Selecting process

Acknowledgements. The author thank Carl Staelin for many useful Q & A. and Chih_Jen Lin for LIBSVM toolbox(Chang and Lin 2011).This work was supported by National Natural Science Fund of China (61472039, 61173061, and 71201120).

References

1. Chang, C.-C., Lin, C.-J.: LIBSVM: a library for support vector machines. ACM Transactions on Intelligent Systems and Technology (TIST) 2(3), 27 (2011)
2. Cortes, C., Vapnik, V.: Support-vector networks. Machine Learning 20(3), 273–297 (1995)
3. Guo, X., et al.: A novel LS-SVMs hyper-parameter selection based on particle swarm optimization. Neurocomputing 71(16), 3211–3215 (2008)
4. Bache, K., Lichman, M.: UCI Machine Learning Repository (2013)
5. Kohavi, R.: A study of cross-validation and bootstrap for accuracy estimation and model selection. In: International Joint Conference on Artificial Intelligence. Lawrence Erlbaum Associates Ltd. (1995)
6. Montgomery, D.C.: Design and analysis of experiments. Wiley (2008)
7. Wang, S., et al.: Data field for hierarchical clustering. International Journal of Data Warehousing and Mining (IJDWM) 7(4), 43–63 (2011)
8. Wu, C.-H., et al.: A Novel hybrid genetic algorithm for kernel function and parameter optimization in support vector regression. Expert Systems with Applications 36(3), 4725–4735 (2009)
9. Wu, X., et al.: Top 10 algorithms in data mining. Knowledge and Information System 14(1), 1–37 (2008)

Active Multi-label Learning
with Optimal Label Subset Selection

Yang Jiao, Pengpeng Zhao*, Jian Wu, Xuefeng Xian,
Haihui Xu, and Zhiming Cui

School of Computer Science and Technology, Soochow University,
Suzhou, 215006, P.R. China
{20124227038,ppzhao,jianwu,xfxian,20124227021,szzmcui}@suda.edu.cn

Abstract. Multi-label classification, where each instance is assigned with multiple labels, has been an attractive research topic in data mining. The annotations of multi-label instances are typically more difficult and time consuming, since they are simultaneously associated with multiple labels. Therefore, active learning, which reduces the labeling cost by actively querying the labels of the most valuable data, becomes particularly important for multi-label learning. Study reveals that methods querying instance-label pairs are more effective than those query instances, since for each sample, only some effective labels need to be annotated while others can be inferred by exploring the label correlations. However, with the high dimensionality of label space, the instance-label pair selective algorithm will be affected since the computational cost of training a multi-label model may be strongly affected by the number of labels. In this paper we propose an approach that combines instance sampling with optimal label subset selection, which can effectively improve the classification model performance and substantially reduce the annotation cost. Experimental results demonstrate the superiority of the proposed approach to state-of-the-art methods on three benchmark datasets.

Keywords: Multi-label learning, Active learning, Sampling, optimal label subset, Data mining.

1 Introduction

In many machine learning applications, we have plenty of unlabeled data but few labeled data. While labeling is usually expensive since it requires the participation of human experts, training an accurate model with as few labeled data as possible becomes a challenge of great significance. Active learning, which aims on conducting selective instance labeling and reducing the labeling effort of training good prediction models, is a leading approach to this goal [1].

The traditional supervised classification problem is concerned with learning from examples associated with only one label. However, many real-world applications often involve the scenario where each instance can be assigned with

---

* Corresponding author.

X. Luo, J.X. Yu, and Z. Li (Eds.): ADMA 2014, LNAI 8933, pp. 523–534, 2014.

multiple labels. For example, in text categorization, a news document could cover several topics such as sports, London Olympics, ticket sales and torch relay. In social network, a micro-blog can be affixed to the plurality of tags, such as news, economy, entertainment, etc. Similarly, in image annotation, one image can be tagged with a set of multiple words, such as urban, building, and road, which indicate the contents of the image [2–4]. Multi-label learning is a framework which deals with such objects [5]. To label the instance, the annotator must consider every possible label for each instance, even if the positive labels are sparse. In general, the biggest challenge in multi label learning is large output space. With the increase of label space, the quantities of category label set will be exponential growth in output space. Obviously, the labeling process is significantly more expensive and time consuming than single-label problems. Thus, active learning under the multi-label framework has attracted more and more attention.

Despite the value and significance of the problem, studies on multi-label active learning remain in a preliminary state. Most of the active learning researches have focused on the single-label classification problem [6, 7]. The sample selection strategies strictly follow the assumption that each instance has only one label, which are not directly well applicable in multi-label cases, since instance selection decisions in multi-label cases should be based on all labels. A direct way to tackle active learning under multi-label setting is to decompose it into a set of binary classification problems, i.e., each category/label is independently handled by a binary active learning algorithm [8, 9]. Existing multi-label active learning works, such as [10–12], measure the informativeness of an unlabeled instance by treating all labels in an independent way. However, these approaches do not take into account the inherent relationship among multiple labels. To deal with the problem, literature [13] develops a two-dimension active learning algorithm that selects sample-label pairs, which considers both instance and label dimension. Specifically, in each iteration, human annotators are only required to annotate/confirm a selected part of labels of selected samples while the remaining unlabeled part can be inferred according to the label correlations. Although the approach achieved good performance, with the increasing size of sample and label dimension, the learning space will become especial large. Thus the training process is complicated and time consuming.

In this paper, we propose a multi-label active learning classification method, which is called MUSLAP (Max-margin Uncertainty Sampling with Label-set Push). Different from approaches above, this method combines a max-margin-based uncertainty sampling strategy with transductive optimal label set selection that effectively assigns instance multiple labels, which most possibly require annotations by oracles. It considers both the instance and label dimension. An intuitive explanation of this strategy is that both instance and label redundancies exist for multi-label classification. The contributions of different labels to minimizing the classification error are different due to the inherent label correlations. Therefore, annotating an optimal label subset with selected instance provides sufficient information for training the classifiers. Though, different from the approach [13], in each iteration our method queries one instance with suitable label set which is more efficient. Using the idea of batch-mode, the learning rate is greatly improved and the strategy can significantly reduce the required human labor under the same number of queries.

2   Related Work

In single-label classification, active learning is an effective method of reducing labeling effort and cost while training a high-quality prediction model. Given a large pool of unlabeled instances, the active learner iteratively selects the most informative instances for an oracle to label. Numerous widely used sampling strategies are available in the literature. Among them, one of the most commonly used is uncertainty sampling [14], in which an active learner iteratively labels the unlabeled data on which the current hypothesis is most uncertain. Another popular strategy is expected-error reduction [15], which aims to label data to minimize the expected error on the unlabeled data. It typically requires expensive computational effort on estimating the expected error because each of the unlabeled data has to be evaluated. Besides, many other selection criteria are commonly used. For example, diversity measures how different an instance is from the labeled data [16], density measures the representativeness of an instance to the whole data set [17], and so on. There are also some other approaches trying to consider different criteria simultaneously [18, 19].

However, active learning for multi-label classification is still in its infancy. Most multi-label active learning methods decompose multi-label classification into a set of binary classification problems and perform instance selection by each binary classifier independently without considering the label correlation of sample. The literature [10] uses a simple extension of the uncertainty sampling strategy. It decomposes the multi-label classification problem into several binary ones using the one-versus-all scheme and selects the instance that minimizes the smallest support vector machine (SVM) margin among all binary classifiers. In literature [8], an SVM active learning method is proposed for multi-label image classification. The method selects unlabeled data that have the maximum mean loss value over the predicted classes (MML). The multi-label classification problem is also considered as several binary classification tasks. A threshold of loss value is estimated for each binary classifier and is then used to decide the predicted classes for unlabeled data. Literature [11] presents a strategy called maximum loss reduction with maximal confidence (MMC), which uses a multi-class logistic regression to predict the number of labels for an unlabeled instance and then computes the MMC measure by summing up the losses from SVM classifiers on all labels. Different from the studies on the aforementioned methods, literature [12] exploits a multi-label boosting classification method and tests numerous strategies that conduct instance selections by combining measures from each class in an unequally weighted manner. In addition, several other multi-label active learners consider selecting both instance and labels for annotations. For example, literature [13] develops a two-dimensional active learning algorithm that selects sample-label pairs to minimize the Bayesian classification error bound. Literature [20] develops a multi-label multiple-instance active learning approach that selects both an image example and a level of annotation to request. Besides, similar to active learning, the semi supervised/transductive multi label learning method is also based on mining information of unlabeled data,e.g. literature [21] estimates the label sets of the unlabeled instances effectively by utilizing the information from both labeled and unlabeled data.

3 Max-Margin Multi-label Active Learning with Label Set Push

3.1 Multi-label SVM Classification

Our method uses the multi-label SVM for active learning because of its usefulness and effectiveness. Multi-label SVM uses the one-versus-all method to combine predictions of multiple binary SVM classifiers. We let $D = \{x_1, \cdots, x_n\}$ denote the entire dataset, which consists of n instances ($x_i \in \mathbb{R}^d$), where x_i is the input feature vector for the i-th instance and $\mathrm{y} = \{y_1, \cdots, y_n\}$ is the label set of the entire dataset, where $y_i \subseteq LC$ denotes the set of multiple labels assigned to x_i. Here, $LC = \{l_1, \cdots, l_k\}$ is the set of all possible label concepts. The j-th component of y_i corresponds to the output of j-th binary SVM. Given D, each class is separated from all other classes by a binary SVM classifier. Thereafter, k binary SVM classifiers are obtained in total. i-th($i = 1, \cdots, k$) classifier can be written as $\langle w_i^*, x \rangle + b_i^* = 0$. w_i^* and b_i^* can be computed as follows:

$$min \; \Phi(w_i) = \frac{1}{2}\|w_i\|^2 + C\sum_{j=1}^{m}\xi_{ji}$$
$$subject \; to \; y_{ji}(\langle w_i^*, x \rangle + b_i) \geq 1 - \xi_{ji}, \xi_{ji} \geq 0, \forall i \qquad (1)$$

where $\{\xi_{ji}\}$ are the slack variables and C is the parameter controlling the trade-off between function complexity and training error.

3.2 Problem Formulation

In this study, we consider pool-based active learning, which appears to be the most popular scenario for applied research in active learning. The dataset includes both labeled and unlabeled instances. We assume that, without loss of generality, the first n_l instances within D are labeled by $\{y_1, \cdots, y_{n_l}\}$. For convenience, we also denote $L = \{1, \cdots, n_l\}$ as the index set for the labeled instances and $U = \{n_l + 1, \cdots, n\}$ for the unlabeled instances ($n = n_l + n_u$). We aim to design multi-label active learning strategies for learning a satisfactory multi-label SVM classification model with fewer labeled instances and finding an optimal label set y_i for each unlabeled selected instance x_i, thereby leading to lower labeling cost. We will first use a multi-label uncertainty sampling strategy from the perspective of label prediction and then employ the transductive multi-label classification method to select the most likely label set for the oracle to annotate.

3.3 Max-Margin Uncertainty Sampling

Uncertainty sampling is one of the simplest and most effective active learning strategies used in single-label classification. The central idea behind this strategy is that the active learner should query the instance that the current classifier is most uncertain about. Notably, the training processes of multi-label and multi-class classifications via the one-versus-all scheme are exactly the same; however, the prediction values produced by the multiple binary classifiers over the same instance are relevant to each other. Literature [22] argues that the prediction values of binary SVM classifiers trained using the one-versus-all scheme for multi-label classification are directly comparable as well.

We use the max-margin uncertainty sampling method first proposed in literature [23] and proven to be effective. The method uses a global separation margin between the groups of positive and negative label prediction values to model the prediction uncertainty of an instance under the current multi-label SVM classifiers. Given the multi-label SVM classifier $F = [f_1, \cdots, f_k]$, the predicted label vector $\hat{y}_i$ of an unlabeled instance x_i can be determined by the sign of the prediction values, such as $\hat{y}_{ik} = sign(f_k(x_i))$. The separation margin over instance x_i can then be defined as follows:

$$
\begin{aligned}
sep\_margin(x_i) &= \min_{t \in \hat{y}_i^+} f_t(x_i) - \max_{s \in \hat{y}_i^-} f_s(x_i) \\
&= \min_{t \in \hat{y}_i^+} |f_t(x_i)| + \min_{s \in \hat{y}_i^-} |f_s(x_i)|
\end{aligned}
\tag{2}
$$

where $\hat{y}_i^+$ and $\hat{y}_i^-$ denote the sets of predicted positive and negative labels, respectively.

The instance that has the smallest separation margin should be the most uncertain instance under the current classification model. A global multi-label uncertainty measure as the inverse separation margin can then be defined as follows:

$$
u(x) = \frac{1}{sep\_margin(x)}
\tag{3}
$$

3.4 Transductive Label Set Selection

After the max-margin uncertainty sampling, we obtain the instances with the most uncertainty, which will be annotated. For each selected instance, only a few effective labels have to be annotated and the others can be inferred by exploring the label correlations. Thus, we employ a simple and efficient method [21] to determine an optimal label set y_i for unlabeled instance x_i, which improves the performance of multi-label classification by exploiting both labeled and unlabeled data.

First, the label concept composition for a multi-label instance is defined as follows. Suppose that we have a multi-label instance x_i and its label set y_i contains a set of multiple label concepts. Formally, we denote the concept composition for instance x_i as $\alpha_i = (\alpha_{i1}, \alpha_{i2}, \cdots, \alpha_{ik})^T$, where α_{ij} represents the fraction of label concept l_j in instance x_i. Here, we assume that $\alpha_{ij} \geqslant 0$ and $\alpha_i^{\mathrm{T}} 1 = 1 (\forall i)$. In a labeled training instance, all label concepts in its label set have equal weights or importance for concept composition, i.e., the ground-truth concept composition $\bar{\alpha}_i = (\bar{\alpha}_{i1}, \cdots, \bar{\alpha}_{ik})^{\mathrm{T}}$ for a labeled instance x_i, which is defined as follows:

$$\bar{\alpha}_{ij} = \begin{cases} \frac{1}{|y_i|}, & if(\, l_j \in y_i), \\ 0, & otherwise. \end{cases} \quad (i \in L) \tag{4}$$

We assume that the optimal estimation of concept compositions should have the property of smoothness, i.e., similar instances should have similar concept compositions within their label sets. If an unlabeled instance x_i is similar to a labeled instance x_j, the α_i should be similar to $\alpha_j = \bar{\alpha}_j$. Moreover, if two unlabeled instances are similar to each other, their concept compositions should also be similar. Thus it is deemed that we need estimate the concept compositions for all the unlabeled instances jointly/simultaneously in order to find optimal solutions on all the unlabeled data. Our method builds a weighted neighborhood graph $G = (V, E)$ on both labeled and unlabeled instances to characterize the relationship between similar instances. After the k nearest neighbor (kNN) search, we define a sparse matrix W indicating the similarities among neighboring instances, as follows:

$$W_{iz} = \begin{cases} \frac{1}{Z_i} \exp\left(-\frac{\|x_i - x_z\|^2}{2\sigma^2}\right), & if \; z \in \mathrm{N}_i, \\ 0, & otherwise. \end{cases} \tag{5}$$

where N_i is the index set of the kNN of i-th instance. Typically, $\| \cdot \|$ refers to the Euclidean distance. Parameter σ is empirically estimated as the average distance between instances. $Z_i = \sum_{z \in N_i} \exp\left(-\frac{\|x_i - x_z\|^2}{2\sigma^2}\right)$.

Based on the smoothness assumption above, the following general optimization framework is used to estimate the optimal alpha values for unlabeled instances:

$$\min_{\alpha_{n_l+1}, \cdots, \alpha_n} \sum_{i \in U} \sum_{j=1}^{k} \left(\alpha_{ij} - \sum_{z \in N_i} W_{iz} \alpha_{zj}\right)^2 \tag{6}$$

$$s.t. \quad \begin{array}{l} \alpha_{ij} \geq 0, \quad \sum_{j=1}^{k} \alpha_{ij} = 1 \\ \alpha_{ij} = \bar{\alpha}_{ij} (\forall i \in L) \end{array}$$

here the $\bar{\alpha}_{ij}$ is defined as Eq.4.

After solving the optimization problem in Eq.6, we have already derived the optimal alpha values for any unlabeled instance x_i. A sorted list of all potential labels for x_i can be found by ranking all candidate labels using their alpha values in a descending order. Thereafter, we predict the number of labels that should be predicted into the label set of x_i using both labeled and unlabeled data. We let θ_i denote the number of labels in the label set for instance x_i. The θ_i values on the labeled instances are fixed according to the ground truth of their label sets, s.t. $\theta_i = |y_i| \ (\forall i \in L)$. For unlabeled data, the number of labels should be a non-negative integer, here we can relax the $\theta_i \in \mathbf{R}$ and $\theta_i \geq 0 (i \in U)$. Then, the optimal θ_i values can be solved using the following optimization problem.

$$\min_{\theta_1, \cdots, \theta_n} \sum_{i \in U} \left(\theta_i - \sum_{z \in N_i} W_{iz} \theta_z \right)^2 \tag{7}$$

$$s.t. \ \theta_i = |y_i| \ (\forall i \in L)$$

The overall multi-label active learning with label set selection procedure is described in Algorithm 1. First, a subset of the data set D is randomly sampled to initialize the labeled data D_L. Note in the experiments, to be fair, for all algorithms, D_L is initialized with the same set of fully labeled instances rather than instance-label pairs. After the initialization, a multi-label SVM classifiers F^0 is trained on the labeled data D_L. It first obtains the informative sample according to the max-margin uncertainty sampling method in section 3.3. Then the optimal label subset is picked based on the transductive label set selection in section 3.4. Finally, the instance with an optimal label subset is submitted to human experts for labeling and update the model F. This active querying and model updating process is repeated until the number of queries or the required accuracy is reached.

Algorithm 1. Max-Margin Multi-label Active Learning with Label Set Push

---

Input:
$D{:}\{x_1, \cdots, x_n\}$ encoding features of the entire data set;
$Y_L{:}\{y_1, \cdots, y_l\}$ encoding labels of the training set;
Labeled data D_L, unlabeled data D_U;
Initialize:
Train multi-label SVM classifiers F^0 on labeled data D_L;

1: **repeat**
2:    Get predictions for instances in D_U with F;
3:    **for** each x_i in D_U **do**
4:        Use Eq.3 to measure the uncertainty of sample x_i;
5:    **end for**
6:    Select the sample set with the most uncertainty x^*;
7:    Estimate the composition of label concepts within label set of sample based on the information utilized from both labeled and unlabeled data;
8:    Use transductive label prediction method to determine the optimal label set y^* according to the composition of label concepts;
9:    Remove x^* from D_U, query the label set y^*;
10:    Update the model F with (x^*, y^*);
11: **until** the number of queries or the required accuracy is reached

---

4   Experimental Results and Discussion

4.1   Settings

We demonstrate our proposed multi-label active learning approach by conducting experiments on three real-world multi-label classification tasks. We compare MUSLAP with four multi-label active learning approaches: MMC, selects instances that lead to the maximum loss reduction with the largest confidence [11]. MML, selects instances with the mean max loss to query its label [8]. Adaptive, considers both the max-margin prediction uncertainty and the label cardinality inconsistency when selecting query instances [23]. 2DAL, selects instance-label pairs that lead to the maximum reduction of expected error [13].

Experiments are performed on three well-known datasets. The first is yeast [24] for gene functional analysis, which has 2,417 instances of genes and 14 possible class labels; each gene is represented by a 103-dimensional vector and the average number of class labels is 4.24 ± 1.57 for each instance. The second is a text categorization dataset RCV1-V2 [25]. We use a benchmark subset, rcv1v2 (topics; subset), which contains 6,000 documents. We removed words that occurred less than 200 times and topics with less than 50 positive examples, thereby obtaining 497 words and 54 topics. The last is the natural scene classification dataset scene [26], which is relatively small and consists of 2,407 natural scene images belonging to different classes.

For each experiment, we first randomly partitioned the dataset into three parts: labeled set, unlabeled pool, and test set. Thereafter, we ran each comparison approach independently on the same initial setting. The partition settings we used for the three datasets are the following: Yeast has 35 labeled, 1,465 unlabeled, and 917 tests; Rcv1 has 105 labeled, 2,895 unlabeled, and 3,000 tests; and Scene has 45 labeled, 1,166 unlabeled, and 1,196 tests. At each iteration of active learning, one instance or one instance-label pair is selected by the active learning methods based on their own strategy, and then added into the labeled data. After instance-label pairs queried, we train a classification model on the labeled data and evaluate its performance on the holdout test data in terms of two measures: micro-F1 and accuracy. We repeated each experiment 10 times and reported the average results.

To be fair, we use one-versus-all linear SVM (implemented with LIBLINEAR) as the classification model for evaluating all the compared approaches. For MMC, the regression model is also implemented with LIBLINEAR. The other parameters are selected via 5-folds cross validation on the initial labeled data. For the other approaches, parameters are determined in the same way if no values suggested in their literatures.

4.2   Result and Discussion

The experimental results on the three datasets are presented in Figure 1 to 3. Note that two approaches: 2DAL and MUSLAP, which query instance-label pairs, are plotted in solid line, while the other three methods which query the whole label set of instances are plotted in dashed line.

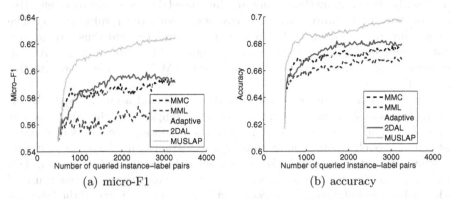

(a) micro-F1   (b) accuracy

Fig. 1. Average results over 10 runs in terms of (a) micro-F1, (b) accuracy on the yeast dataset

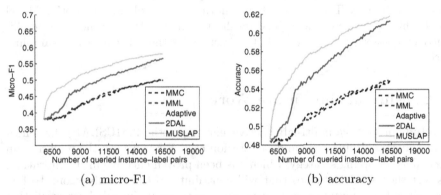

(a) micro-F1   (b) accuracy

Fig. 2. Average results over 10 runs in terms of (a) micro-F1, (b) accuracy on the rcv1 dataset

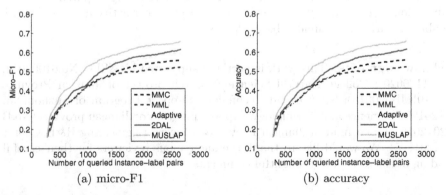

(a) micro-F1   (b) accuracy

Fig. 3. Average results over 10 runs in terms of (a) micro-F1, (b) accuracy on the scene dataset

Generally speaking, methods querying instance-label pairs are more effective than those query instances, which is consistent with the results in [13]. The reason is probably that multiple labels may be correlated and thus redundancy of information may exist among the multiple labels of the same instance. Among the methods query instances only, Adaptive and MMC tend to be more effective than MML.

Figure 1 to 3 show the performance on micro-F1 and accuracy with the increasing number of instance-label pair queries. Compared to the baselines, our proposed method achieves the best performance in most cases. It is reasonable that our approach is superior to those query instances methods, since it also takes into account the optimal label subset which is useful for classification. Different from the former method querying instance-label pairs, we make use of both labeled and unlabeled instances to estimate the cardinality of the label set for each unlabeled instance. After the label set cardinality is estimated, we sort all the labels based on instances concept composition (i.e. the estimated alpha values), and predict the label set with the top ranked labels with the estimated label set cardinality. Then we select the informative sample with its optimal label subset which is more efficient. The results reveal that the instance can be efficiently combined with the suitable label set under the active learning sampling process.

5   Conclusion and Future Work

This study proposes a multi-label active learning method (MUSLAP) that effectively reduces training time and annotation cost. First, we use the max-margin uncertainty sampling strategy, which has been proven effective in the literature, to select the useful instances that will contribute to the classification. To further reduce the annotation cost, we employ a transductive label set prediction method to select an optimal label subset for human experts. Empirical studies on multi-label classification datasets have demonstrated that our method can improve classification performance while significantly reducing required time and annotation cost. In the future, we will try to study other active query strategies combining with the optimal subset selection.

Acknowledgments. This work is partially supported by NSFC(No.61003054, No.61170020); College Natural Science Research project of Jiangsu Province (No.10KJB520018); Science and Technology Support Program of Suzhou(No. SG201257); Science and Technology Support program of Jiangsu province(No.B E2012075); Open fund of Jiangsu Province Software Engineering R&D Center (SX201205). We would like to thank the anonymous reviewers for their careful reading of the draft and thoughtful comments.

References

1. Settles, B.: Active learning literature survey, vol. 52, pp. 55–66. University of Wisconsin, Madison (2010)
2. Carneiro, G., Chan, A.B., Moreno, P.J., Vasconcelos, N.: Supervised learning of semantic classes for image annotation and retrieval. IEEE Transactions on Pattern Analysis and Machine Intelligence 29(3), 394–410 (2007)
3. Qi, G.J., Hua, X.S., Rui, Y., Tang, J., Mei, T., Zhang, H.J.: Correlative multi-label video annotation. In: Proceedings of the 15th International Conference on Multimedia, pp. 17–26. ACM (2007)
4. Yang, Y., Wu, F., Nie, F., Shen, H.T., Zhuang, Y., Hauptmann, A.G.: Web and personal image annotation by mining label correlation with relaxed visual graph embedding. IEEE Transactions on Image Processing 21(3), 1339–1351 (2012)
5. Zhang, M., Zhou, Z.: A review on multi-label learning algorithms (2013)
6. Luo, T., Kramer, K., Samson, S., Remsen, A., Goldgof, D., Hall, L., Hopkins, T.: Active learning to recognize multiple types of plankton. In: Proceedings of the 17th International Conference on Pattern Recognition, ICPR 2004, vol. 3, pp. 478–481. IEEE (2004)
7. Yan, R., Yang, L., Hauptmann, A.: Automatically labeling video data using multi-class active learning. In: Proceedings of the Ninth IEEE International Conference on Computer Vision, pp. 516–523. IEEE (2003)
8. Li, X., Wang, L., Sung, E.: Multilabel svm active learning for image classification. In: 2004 International Conference on Image Processing, ICIP 2004, vol. 4, pp. 2207–2210. IEEE (2004)
9. Brinker, K.: On active learning in multi-label classification. In: From Data and Information Analysis to Knowledge Engineering, pp. 206–213. Springer (2006)
10. Kruse, M.S.R., Nçrnberger, C.B.A., Gaul, W.: From data and information analysis to knowledge engineering
11. Yang, B., Sun, J.T., Wang, T., Chen, Z.: Effective multi-label active learning for text classification. In: Proceedings of the 15th ACM SIGKDD International Conference on Knowledge Discovery and Data Mining, pp. 917–926. ACM (2009)
12. Esuli, A., Sebastiani, F.: Training data cleaning for text classification. In: Azzopardi, L., Kazai, G., Robertson, S., Rüger, S., Shokouhi, M., Song, D., Yilmaz, E. (eds.) ICTIR 2009. LNCS, vol. 5766, pp. 29–41. Springer, Heidelberg (2009)
13. Qi, G.J., Hua, X.S., Rui, Y., Tang, J., Zhang, H.J.: Two-dimensional multilabel active learning with an efficient online adaptation model for image classification. IEEE Transactions on Pattern Analysis and Machine Intelligence 31(10), 1880–1897 (2009)
14. Lewis, D.D., Gale, W.A.: A sequential algorithm for training text classifiers. In: Proceedings of the 17th Annual International ACM SIGIR Conference on Research and Development in Information Retrieval, pp. 3–12. Springer-Verlag New York, Inc. (1994)
15. Roy, N., McCallum, A.: Toward optimal active learning through monte carlo estimation of error reduction. In: ICML, Williamstown (2001)
16. Brinker, K.: Incorporating diversity in active learning with support vector machines. In: ICML, vol. 3, pp. 59–66 (2003)
17. Nguyen, H.T., Smeulders, A.: Active learning using pre-clustering. In: Proceedings of the Twenty-First International Conference on Machine Learning, vol. 79. ACM (2004)

18. Donmez, P., Carbonell, J.G., Bennett, P.N.: Dual strategy active learning. In: Kok, J.N., Koronacki, J., Lopez de Mantaras, R., Matwin, S., Mladenič, D., Skowron, A. (eds.) ECML 2007. LNCS (LNAI), vol. 4701, pp. 116–127. Springer, Heidelberg (2007)

19. Huang, S.J., Jin, R., Zhou, Z.H.: Active learning by querying informative and representative examples. In: Advances in Neural Information Processing Systems, pp. 892–900 (2010)

20. Vijayanarasimhan, S., Grauman, K.: What's it going to cost you?: Predicting effort vs. informativeness for multi-label image annotations. In: IEEE Conference on Computer Vision and Pattern Recognition, CVPR 2009, pp. 2262–2269. IEEE (2009)

21. Kong, X., Ng, M.K., Zhou, Z.H.: Transductive multilabel learning via label set propagation. IEEE Transactions on Knowledge and Data Engineering 25(3), 704–719 (2013)

22. Rifkin, R., Klautau, A.: In defense of one-vs-all classification. The Journal of Machine Learning Research 5, 101–141 (2004)

23. Li, X., Guo, Y.: Active learning with multi-label svm classification. In: Proceedings of the Twenty-Third international Joint Conference on Artificial Intelligence, pp. 1479–1485. AAAI Press (2013)

24. Elisseeff, A., Weston, J.: A kernel method for multi-labelled classification. In: Advances in Neural Information Processing Systems, pp. 681–687 (2001)

25. Lewis, D.D., Yang, Y., Rose, T.G., Li, F.: Rcv1: A new benchmark collection for text categorization research. The Journal of Machine Learning Research 5, 361–397 (2004)

26. Boutell, M.R., Luo, J., Shen, X., Brown, C.M.: Learning multi-label scene classification. Pattern Recognition 37(9), 1757–1771 (2004)

Merging Decision Trees: A Case Study in Predicting Student Performance

Pedro Strecht, João Mendes-Moreira, and Carlos Soares

INESC TEC/Faculdade de Engenharia, Universidade do Porto
Rua Dr. Roberto Frias, 4200-465 Porto, Portugal
{pstrecht,jmoreira,csoares}@fe.up.pt

Abstract. Predicting the failure of students in university courses can provide useful information for course and programme managers as well as to explain the drop out phenomenon. While it is important to have models at course level, their number makes it hard to extract knowledge that can be useful at the university level. Therefore, to support decision making at this level, it is important to generalize the knowledge contained in those models. We propose an approach to group and merge interpretable models in order to replace them with more general ones without compromising the quality of predictive performance. We evaluate our approach using data from the U. Porto. The results obtained are promising, although they suggest alternative approaches to the problem.

Keywords: prediction of failure, decision tree merging, C5.0.

1 Introduction

Interpretable models for predicting the failure of students in university courses are important to support both course and programme managers. By identifying the students in danger of failure beforehand, suitable strategies can be devised to prevent it. Moreover, those models can give clues about the reasons that lead to student attrition, a topic widely studied in educational data mining [1]. Given the vast amount of data available in university information systems, these models are usually created for each course and academic year separately. This means that a very large number of models is generated, which raises problems on how to generalize knowledge in order to have a global view of the phenomena across the university and not only in the context of a single course.

Additionally, it may be expected that there are different groups of models with very different characteristics. For instance, the performance in courses of different scientific areas is likely to be affected by different factors. We propose an approach for the problem of generalizing the knowledge contained in a large number of models that consists of two phases. In the first one, the models are split into groups. The splitting is done based on domain-specific knowledge (e.g. scientific areas) or is data-driven (e.g. by clustering them). In the second phase, the models in a group are aggregated into a single model that generalizes the knowledge contained in the original models, hopefully with small impact on its

X. Luo, J.X. Yu, and Z. Li (Eds.): ADMA 2014, LNAI 8933, pp. 535–548, 2014.

predictive performance. The aggregation method consists mainly of intersecting the decision rules of pairs of models of a group recursively, i.e., by adding models along the merging process to previously merged ones.

In this paper we compare different methods of grouping models and of defining weights of decision rules as part of a strategy to keep the merged models simple and generic. Throughout the merging process we work only with decision rules and delay the merged decision tree creation to the final step. We define an evaluation procedure to compare performance of merged models relatively to the original ones. The case study used for empirical evaluation uses data from the academic management information system of the University of Porto, Portugal. Due to limitations of space, this paper focuses on the process of merging trees. Therefore, some decisions which were based on domain-specific knowledge and preliminary experiments (e.g. variable selection, parameter setting) as well as some aspects of the results have not been discussed in depth.

The main contributions of this paper are: 1) propose the methodology to generalize the knowledge from a large number of models; 2) identify which are the components of the merging process; 3) define different alternatives for these components; and 4) combine them in a series of experiments to assess the impact in the global predictive performance of the merged models.

The remainder of this paper is structured as follows. Section 2 presents approaches to combine decision trees. Section 3 describes the system architecture and methodology. Section 4 presents results and discussion. Section 5 presents the conclusions and future work.

2   Combining Decision Trees

Combining decision tree models is a topic that has been studied with different approaches. Gorbunov and Lyubetsky [2] address the issue from a mathematical point of view, by formulating the problem of constructing a decision tree that is closest on average to a set of trees. Kargupta and Park [3] present an approach in which decision trees are converted to the frequency domain using the Fourier transform. The merging process consists on summing the spectra of each model and then transform the results back into to the decision tree domain.

Concerning data mining approaches, Provost and Hennessy [4,5] present an algorithm that evaluates each model with data from the other models to merge. The merged model is constructed from satisfactory rules, i.e., rules that are generic enough to be evaluated in the other models. A more common approach is the combination of rules derived from decision trees. The idea is to convert decision trees from two models into decision rules by combining the rules into new rules, reducing their number and finally growing a decision tree of the merged model. Parts of this process are presented in the doctoral thesis of Williams [6] and other researchers have contributed by proposing different ways of carrying out intermediate tasks, such as Andrzejak et al. [7], Hall et al. [8,9] or Bursteinas and Long [10]. While each approach is different, we identified a set of phases they share in common, described next.

In the first phase, a decision tree is transformed to a set of rules. Each path from the root to the leaves creates a rule with a set of possible values for variables and a class. These have been called "rules set" [8], "hypercubes" [10] or "sets of iso-parallel boxes" [7]. These designations arise from the fact that a variable can be considered as a dimension axis in a multi-dimensional space. The set of values (nominal or numerical) is the domain for each dimension and each rule defines a region. The representation of the regions is required for the next phase. It is worth noting that all regions are disjoint from each other and together cover the entire space.

In the second phase, the regions of both models are combined using a specific method. Andrzejak *et al.* [7] call this "unification" and propose a line sweep algorithm to avoid comparing every region of each model. It is commonly based on sorting the limits of each region and then analysing where merging can be done. However this method only applies to numerical variables. Hall *et al.* [8] compare all regions with each other. Bursteinas and Long [10] have a similar method but separate disjoint from overlapping regions. One potential problem is that combining regions can lead to a class conflict if overlapping regions have different classes. Andrzejak *et al.* [7] propose three strategies to assign a class to the merged region. The first assigns the class with the greatest confidence, the second, the one with the greater probability and a third strategy, which is the more complex, involves more passes over the data. Hall *et al.* [8] explore the issue in greater detail and propose further strategies, e.g., comparing distances to the boundaries of the variables. However, this approach seems suitable only for numerical variables. Bursteinas and Long [10] use a different strategy by retraining the model with examples for the conflicting class region. If no conflict arises, that class is assigned, otherwise the region is removed from the merged model.

The third phase attempts to reduce the number of regions and it is commonly referred to as "pruning". This is carried out to avoid having models with very complex rules. The most direct approach is to identify adjacent regions, i.e., regions sharing the same class and values of all variables except for one. If that variable is nominal, the values of both regions are included, otherwise the join is only possible if the limits overlap. To further reduce the rules set, Andrzejak *et al.* [7] developed a ranking system retaining only the regions with the highest relative volume and number of training examples. Hall *et al.* [8] only carry out this phase to eliminate redundant rules created during the removal of class conflicts. Bursteinas and Long [10] mention the phase but do not provide details on how it is performed.

The fourth phase consists in growing a decision tree from the decision regions representation. Andrzejak *et al.* [7] attempt to mimic the C5.0 algorithm using the values in the regions as examples. One problem with this method is that it is necessary to divide one region in two to perform the splitting, which increases their number, thus making the model more complex. Hall *et al.* [8] do not perform this phase and the merged model is represented as the set of regions. Bursteinas and Long [10] claim to grow a tree but do not describe the method.

3   Methodology

To carry out the experiments, a system with five processes was developed with the architecture presented in Fig. 1.

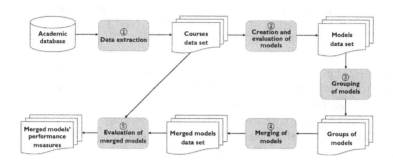

Fig. 1. System architecture

The first process creates the data sets (one for each course in the university) from the academic database. These contain enrollment data (Section 3.1). The second process creates decision tree models for each course, analyses them in order to determine the most important variables and evaluates them to assess their quality (Section 3.2). The third process groups the models according to different criteria (Section 3.3). The models in each group are then merged by the fourth process (Section 3.4). Finally the fifth process evaluates the merged models from a performance improvement point of view (Section 3.5).

3.1   Data Extraction

This process extracts data sets from the academic database of the university information system. The academic database stores a large amount of data on students, program syllabuses, courses, academic acts and assorted data related to a variety of sub-processes of the pedagogical process. The analysis done focuses on the academic year 2012/2013 with the extraction of 5779 course data sets (from 391 programmes), with the variables presented in Table 1.

3.2   Creation and Evaluation of Models

This process has two sub-processes: (1) the models for each course data set are trained and analysed in order to find out the most important variables for prediction; (2) the prediction quality of each model is evaluated.

Model Training and Analysis. The models are decision tree classifiers generated by C5.0 algorithm [11]. Decision trees have the characteristic of not requiring previous domain knowledge or heavy parameter tuning making them

Table 1. Variables for models

Variable	Remarks	Type
Age	age of student at the date of enrollment	Numerical
Sex	male, female	Nominal
Marital status	single, married, divorced, widower, ...	Nominal
Nationality	first nationality of student	Nominal
Displaced	whether the student lived outside the Porto district	Boolean
Scholarship	whether the student has a scholarship	Boolean
Special needs	whether the student has disabilities	Boolean
Type of admission	type of application contest	Nominal
Type of student	regular, mobility, extraordinary	Nominal
Status of student	ordinary, employed, athlete, ...	Nominal
Years of enrollment	# of academic years the student has enrolled in previously	Numerical
Delayed courses	# of courses the student should have completed	Numerical
Type of dedication	full-time, part-time	Nominal
Debt situation	whether there are fees due	Boolean
Approval	whether the student has passed or failed the course	Boolean

appropriate for both prediction, exploratory data analysis and are human interpretable. In this study, students are classified as having passed or failed a course. The C5.0 algorithm measures the importance of variable I_v by determining the percentage of examples tested in a node by that variable in relation to all examples (eq. 1).

$$I_v = \frac{\#examples\ tested\ by\ variable\ v}{\#examples} \tag{1}$$

Model Evaluation. Experimental setup uses k-fold cross-validation [12] with stratified sampling [13]. Failure is the positive class in this problem, i.e. it is the most important class, and thus, we use a suitable evaluation measure $F1$ [14].

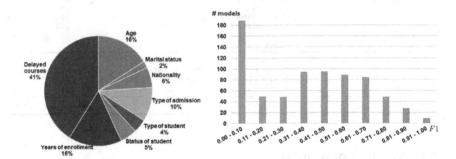

Fig. 2. Distribution of I_v in models   **Fig. 3.** Distribution of $F1$ in models

Training, analysis and evaluation of models is replicated for each course in the data set, however, models were created only for courses with a minimum of 100 students enrolled. This resulted in creating 730 models (12% of the 5779 courses). The variables used in the models are age, marital status, nationality, type of admission, type of student, status of student, years of enrollment, and delayed courses. Fig. 2 shows the average importance I_v of each variable across all models. Delayed courses (41%) is the variable most often used, followed by age (16%) and years of enrollment (16%). The quality of the models varies significantly with only a quarter having $F1$ above 0.60, as presented in Fig. 3.

3.3   Grouping of Models

This process aims to group models according to specific criteria. We grouped models on the basis of scientific area, number of variables, variable importance, and a baseline group containing all models. For creating groups according to variable importance we used the k-means clustering algorithm [15], which created four groups (clusters) using only three of the most important variables identified in Section 3.2, namely age (I_a), years of enrollment (I_{yo}), and delayed courses (I_{dc}). Table 2 presents the four group sets and respective groups, specifying the number of models in each group.

Table 2. Groups sets of models

Group set criteria	Group	# models	%
Scientific areas	1: Architecture & Arts	47	6.44
	2: Computer Science	44	6.03
	3: Engineering	92	12.60
	4: Humanities	38	5.21
	5: Legal Sciences	48	6.58
	6: Mathematics	63	8.63
	7: Medicine	143	19.59
	8: Physical Sciences	98	13.42
	9: Social Sciences	117	16.03
	10: Sport Sciences	40	5.47
Number of variables	1: 0 variables	177	24.25
	2: 1 variable	219	30.00
	3: 2 variables	161	22.05
	4: 3 variables	92	12.60
	5: 4 variables	42	5.75
	6: 5 variables	35	4.79
	7: 6 variables	4	0.56
Importance of variables	1: I_a=17.40, I_{yo}=95.15, I_{dc}=37.56	116	15.89
	2: I_a=6.87, I_{yo}=7.75, I_{dc}=98.98	304	41.64
	3: I_a=97.91, I_{yo}= 6.83, I_{dc}=23.28	102	13.97
	4: I_a=0.06, I_{yo}= 0.00, I_{dc}= 0.00	208	28.50
None	1: Baseline	730	100.00

3.4   Merging of Models

The methodology to merge all models in a group set is done according to the experimental set-up presented in Fig. 4. For the process of merging models, each model must be represented as a set of decision regions. This can take the form of a *decision table*, in which each row is a decision region. Therefore, the first and second models are converted to decision tables and merged, yielding the model a_1, also in decision table form. Then the third model is also converted to a decision table and is merged with model a_1 yielding model a_2.

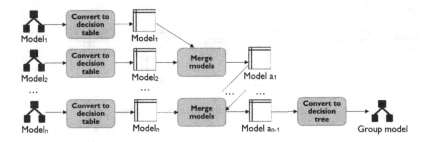

Fig. 4. Experimental set-up to merge models in a group set

The last merged model a_{n-1} is converted to the decision tree form and is evaluated against data of all courses in the group. Each one of these sub-processes and its tasks are detailed in the following sub-sections.

Conversion to Decision Table. A decision table is an alternative way of representing a decision tree. Each path from the top of the tree to the leaves defines a decision region, represented as a row in the decision table. Columns specify the class, weight and set of values of each variable. An example of conversion is presented in Fig. 5. A special case arises when the model is empty, which implies the existence of a single decision region covering the whole space.

```
AGE <= 20: y (95/9)
AGE > 20:
:...MARITAL_STATUS = divorced: n (1)
   MARITAL_STATUS = married: y (3)
   MARITAL_STATUS = single:
   :...DELAYED_COURSES > 0: n (86/28)
      DELAYED_COURSES <= 0:
      :...AGE > 23: y (5/2)
         AGE <= 23:
         :...AGE <= 21: y (16/5)
            AGE > 21: n (11/3)
```

Region #	Class	Weight	Variables		
			AGE	MARITAL_STATUS	DELAYED_COURSES
1	y	44	[0,20]		
2	n	1	[21,+∞)	divorced	
3	y	1	[21,+∞)	married	
4	n	40	[21,+∞)	single	[1,+∞)
5	y	2	[24,+∞)	single	0
6	y	7	21	single	0
7	n	5	[22,23]	single	0

Fig. 5. Example of conversion from decision tree to decision table

In the decision trees generated by the C5.0 algorithm, the leaf nodes show the number of examples used to create the split [11]. This number is used to measure the importance of the region associated with each leaf node. In a model with r decision regions, we define the *weight* of a decision region w_i as the proportion of examples in region i relative to the whole data set. It is a relative measure of the importance of a decision region in the context of a model.

Merge Models. The process of merging two models encompasses three sequential sub-processes: intersection, filtering and reduction as presented in Fig. 6.

Intersection is a sub-process to calculate the cross-product regions of two decision tables, resulting in a decision table for the merged model. The merged

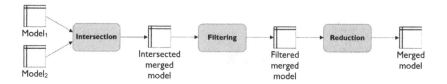

Fig. 6. Process of merging two models

model has the variables of both models. For each pair of regions, the set of values of each variable is intersected. The resulting *merged region* contains all the intersection sets relative to each variable. The intersected merged model is the set of all merged regions.

The intersection of values of each variable from both regions may have the following outcomes:

– If there are common values, then these are assigned to the variable in the merged region.
– If there are no common values for the variable being intersected in both regions, then they are considered *disjoint regions*, regardless of other variables in which the intersection set may not be empty.
– If the variable does not have a value or is not present in one of the models then the set of values in the other model is copied to the merged region (the absence of a variable in a model or its presence with no value for a specific region is considered as a neutral element, i.e., it can take any value without affecting the decision).

Each region of the intersected merged model must also have a weight so that the decision table has the same format as the original models. As weight measures the importance of a region, it is logical to assume that the most immediate choice is to consider the maximum weight of the original regions. However, for evaluation purposes, we think that it is interesting assess the impact on the model's performance by assigning the minimum weight. We also consider a third possibility which is to use the average value of the largest and smallest. To easily compare each possibility, we define a *weight attribution* parameter which can be set to be the *minimum*, *maximum* or *average* value of the original pair of regions, according to each case.

The class to assign to the merged region is straightforward if both regions have the same class, otherwise the class conflict problem arises. The strategies for its resolution are controlled by a *conflict class resolution* parameter. This can be set to assign the same class as the region with *minimum weight* or with *maximum weight*. Fig. 7 presents an example of intersection of decision tables, which illustrates the weight and class assignment to each merged regions. In this example the weight attribution parameter is set to *average* and conflict class resolution parameter to *maximum weight*.

Filtering is the sub-process to remove disjoint regions from the intersected merged model yielding the filtered merged model. Disjoint regions have to be

Model 1				Model 2					Merged model					
Region #	Class	Weight	Variables DELAYED_COURSES	Region #	Class	Weight	Variables AGE	DELAYED_COURSES	Region #	Class	Weight	Variables DELAYED_COURSES	AGE	Region status
1	y	82	[0,1]	1	y	75	[0,20]		1	y	79	[0,1]	[0,20]	OK
1	y	82	[0,1]	2	n	10	[21,+∞)	[1,+∞)	2	y	46	1	[21,+∞)	OK
1	y	82	[0,1]	3	y	13	[21,25]	0	3	y	48	0	[21,25]	OK
1	y	82	[0,1]	4	y	3	[26,+∞)	0	4	y	43	0	[26,+∞)	OK
2	y	15	[2,5]	1	y	75	[0,20]		5	y	45	[2,5]	[0,20]	OK
2	y	15	[2,5]	2	n	10	[21,+∞)	[1,+∞)	6	y	13	[2,5]	[21,+∞)	OK
2	y	15	[2,5]	3	y	13	[21,25]	0	7	y	14	∅	[21,25]	Disjoint
2	y	15	[2,5]	4	n	3	[26,+∞)	0	8	y	9	∅	[26,+∞)	Disjoint
3	n	2	[6,+∞)	1	y	75	[0,20]		9	y	39	[6,+∞)	[0,20]	OK
3	n	2	[6,+∞)	2	n	10	[21,+∞)	[1,+∞)	10	n	6	[6,+∞)	[21,+∞)	OK
3	n	2	[6,+∞)	3	y	13	[21,25]	0	11	y	8	∅	[21,25]	Disjoint
3	n	2	[6,+∞)	4	n	3	[26,+∞)	0	12	n	3	∅	[26,+∞)	Disjoint

Fig. 7. Intersection of decision tables

Region #	Class	Weight	Variables DELAYED_COURSES	AGE
1	y	25	[0,1]	[0,20]
2	y	14	1	[21,+∞)
3	y	15	0	[21,25]
4	y	13	0	[26,+∞)
5	y	14	[2,5]	[0,20]
6	y	4	[2,5]	[21,+∞)
7	y	12	[6,+∞)	[0,20]
8	n	2	[6,+∞)	[21,+∞)

Region #	Class	Weight	Variables AGE	DELAYED_COURSES	Obs.
1	y	51	[0,20]		1;5;7
2	y	18	[21,+∞)	[1,5]	2;6
3	y	28	[21,+∞)	0	3;4
4	n	2	[21,+∞)	[6,+∞)	8

Fig. 8. Filtering a merged decision table   **Fig. 9.** Reducing a merged decision table

removed because they relate to pairs of regions of the original models that have no values in common on at least one variable present in both. Removing regions implies recalculating the weight of each region that remains to obtain a total of 100%. As weights are rounded to integers, the weight of a region less than 0.5 is rounded to zero. Therefore, a possible consequence of the filtering sub-process is that regions with zero weight may arise. Fig. 8 shows the result of filtering out the disjoint region of the merged model of Fig. 7.

The filtered merged model can be empty if all regions of the intersected merged model are disjoint. In such case, the two original models are considered *not mergeable* and the merging process is halted. It is then restarted using the last successfully merged model as model 1 and the next model from the group set as model 2.

Reduction is a sub-process to limit the number of regions in the filtered merged model, to obtain a simpler model. The regions are examined to find out which can be merged. This is possible when a set of regions have the same class and all variables have equal values except for one. In the case of nominal variables, reduction consists on the union of values of that variable from all regions. In the case of numerical variables, currently reduction is only performed if the intervals are contiguous (this procedure will be improved to allow reduction even with non-contiguous intervals).

The weight of the resulting region is the sum of the weights of the regions that are joined. After all regions have been subjected to reduction, they are again examined and those that have zero weight are removed. Another consequence of the reduction is that there may exist variables with the same value in all decision regions. The columns for these variables are removed from the table. Fig. 9 shows the result of reducing the decision table of Fig. 8. The reduction sub-process results in the last successfully merged model of the group so far.

Conversion to Decision Tree. The last merged model of the group is converted to the decision tree representation, yielding the *group model*. For this purpose, examples are generated randomly, bounded by the limits of each variable from the decision table and submitted to the C5.0 algorithm to train a model. Each decision region provides examples which corresponds to a combination of the set of values of each variable with the set of values of other variables and the assigned class to the region. The set of values of numerical variables are bounded by two limits. If the upper limit is missing ($+\infty$), then the maximum observed value from all courses data sets is used. When the lower limit is zero, the lowest observed value across all courses data sets is used (e.g., age). These values are collected as part of the final task of the data extraction process (Section 3.1).

The generation of examples can be controlled with the *examples for numerical variables* parameter. Setting to *limits* only generates two examples (one for each limit) while *samples* generates examples between the limits with a step of 5 (our initial approach used all values but was infeasible due to memory limitations). The weight of the region can also influence the number of examples generated, being controlled by a *weight examples* parameter. If active, the number of generated examples by each region is multiplied by the weight of that region. Generating some examples more frequently than others is a way to preserve the importance of a region over others with less weight in the resulting decision tree.

3.5 Merged Models Evaluation

This process evaluates a group model, using the experimental set-up presented in Fig. 10. After this process, each model has two measures of performance, one ($F1$) from the model obtained from its own data and another ($F1_g$) from the group model. Hence, $\Delta F1$ (eq. 2) allows us to measure changes in predictive power:

$$\Delta F1 = F1_g - F1 \tag{2}$$

If $\Delta F1$ is greater than zero, then there is an improvement in predictive performance by using the group model. If equal to zero, there is no improvement in predictive performance. If lower than zero, then there is loss of predictive performance relative to the original model.

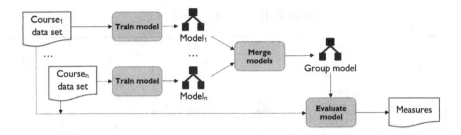

Fig. 10. Experimental set-up to evaluate a group model

The performance of a group model (eq. 3) is the average of $\Delta F1$ of all models of the group. We define the *merging score* (M) of a group (eq. 4) as the number of models that was possible to merge (m) divided by the number of pairs of models in the group $(n-1)$.

$$\overline{\Delta F1} = \frac{\sum\limits_{i=1}^{n} \Delta F1_i}{n} \quad (3) \qquad\qquad M = \frac{m}{n-1} \quad (4)$$

For group set evaluation, we normalized $\overline{\Delta F1}$ of all groups by the number of models (n_k) in each group (eq. 5). Likewise, the merging score of a group set (eq. 6) is the average merging score of all groups normalized by the number of models (g is the number of groups in a group set).

$$\overline{\Delta F1}_{gs} = \frac{\sum\limits_{k=1}^{g} \overline{\Delta F1}_k \times n_k}{\sum\limits_{k=1}^{g} n_k} \quad (5) \qquad\qquad \overline{M}_{gs} = \frac{\sum\limits_{k=1}^{g} M_k \times n_k}{\sum\limits_{k=1}^{g} n_k} \quad (6)$$

4 Results

A set of 24 experiments were run to merge models with the results presented in Table 3. Each experiment is a combination of values of the parameters *weight attribution* (wa), *conflict class resolution* (ccr), *examples for numerical variables* (efnv) and *weight examples* (we). This allows to compare the impact of each parameter on the average merging score $(\overline{\overline{M}})$, improvement in prediction $(\overline{\overline{\Delta F1}})$ and across each group set: scientific areas $(\overline{\Delta F1}_{SA})$, number of variables $(\overline{\Delta F1}_{\#v})$, importance of variables $(\overline{\Delta F1}_{I_v})$, and baseline $(\overline{\Delta F1}_B)$.

Merging Score. The average merging score is 76%, which hardly changes throughout experiments. This implies that none of the parameters has a significant role in the ability to merge models. Fig. 11 shows the average merging score of all experiments across groups in all group sets. The idea behind creating groups of models is to try to bring together models that are similar the most, i.e.,

Table 3. Results of experiments

#	wa	ccr	efnv	we	$\overline{M}$	$\overline{\Delta F1}_{SA}$	$\overline{\Delta F1}_{\#v}$	$\overline{\Delta F1}_{I_v}$	$\overline{\Delta F1}_{B}$	$\overline{\Delta F1}$
1	min	min	limits	no	0.76	0.02	-0.04	0.03	-0.27	-0.06
2	min	min	limits	yes	0.76	0.03	-0.03	0.02	0.04	0.01
3	min	min	samples	no	0.76	0.01	-0.05	0.00	0.06	0.00
4	min	min	samples	yes	0.76	0.01	-0.03	-0.06	0.05	-0.01
5	min	max	limits	no	0.76	-0.02	-0.05	-0.18	0.00	-0.06
6	min	max	limits	yes	0.76	-0.08	-0.05	-0.22	0.00	-0.09
7	min	max	samples	no	0.76	-0.01	-0.04	-0.18	0.00	-0.06
8	min	max	samples	yes	0.76	-0.04	-0.04	-0.20	0.00	-0.05
9	max	min	limits	no	0.75	0.00	-0.06	-0.07	-0.03	-0.05
10	max	min	limits	yes	0.75	-0.02	-0.07	-0.07	0.06	-0.05
11	max	min	samples	no	0.75	-0.03	-0.04	-0.09	-0.03	-0.05
12	max	min	samples	yes	0.75	-0.02	-0.06	-0.13	-0.03	-0.06
13	max	max	limits	no	0.76	0.02	-0.07	-0.08	0.00	-0.03
14	max	max	limits	yes	0.76	0.02	-0.07	-0.07	0.00	-0.03
15	max	max	samples	no	0.76	0.02	-0.08	-0.08	0.00	-0.03
16	max	max	samples	yes	0.76	0.02	-0.07	-0.07	0.00	-0.03
17	avg	min	limits	no	0.76	0.01	-0.03	0.00	0.00	-0.01
18	avg	min	limits	yes	0.76	0.01	-0.04	0.00	0.00	-0.01
19	avg	min	samples	no	0.76	0.03	-0.06	0.00	0.00	-0.01
20	avg	min	samples	yes	0.76	0.02	-0.05	0.00	0.00	-0.01
21	avg	max	limits	no	0.76	0.01	-0.08	-0.06	0.00	-0.03
22	avg	max	limits	yes	0.76	0.02	-0.08	-0.06	0.00	-0.03
23	avg	max	samples	no	0.76	0.03	-0.11	-0.06	0.00	-0.03
24	avg	max	samples	yes	0.76	0.03	-0.08	-0.05	0.00	-0.03
				Avg.	0.76	0.01	-0.06	-0.07	-0.05	

with less likelihood of their merging resulting in disjoint regions. We observe that different group sets affect the merging score. This is particularly noticeable in grouping by the scientific areas in which group #4 (Humanities) has the highest merging score (92%) while group #7 (Medicine) has the lowest (52%). Grouping by number of variables shows that it is not possible to merge models with no variables (group #1), however, from 1 variable onward, it is always possible to merge all models into a single group model. Grouping by variable importance always allow full merging, except in the last group (probably because the models are less similar). The baseline group set has the average value of the experiments (76%). Results show that, from a merging ability perspective, merging by scientific areas is not necessarily the best way to group models while number of variables and importance of variables seem to be more suitable approaches.

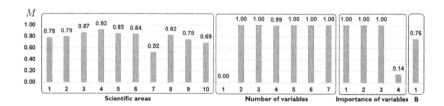

Fig. 11. Merging score distribution by groups of models

Average Improvement in Prediction. Experiments with best results in improvement in prediction are #2 and #3, although average improvement is not very significant (0.01). In all other experiments there is loss of average predictive power, with experience #6 with particularly poor results (-0.09). There is no apparent correlation between any of the four parameters and improvement in predictive performance.

Fig. 12 presents the distribution for experiment #2. Grouping by scientific areas has an average improvement of 0.03, with group #7 (Medicine) showing an improvement of 0.14 while group #3 (Engineering) has a loss of 0.21. Grouping by number of variables, has an average loss of 0.03. Grouping by variable importance has an average loss of 0.02 and it is the group set with highest variance. The baseline group set has an average improvement of 0.04 (the highest of all). Table 3 also shows the average $\overline{\Delta F1}$ across experiments for each group set. Grouping by scientific areas is the only one that has a positive global improvement in prediction with 0.01. The baseline group set has less variation and only had loss in experiment #1.

These results show that, contrary to what happened with the merging score, grouping models by scientific areas yields interesting results in terms of improvement in predictive power. This may be indicative that models obtained from courses with more similar content are more susceptible to generalize knowledge than the ones in which similarity arises from features of the models themselves (such as the number of variables or its importance).

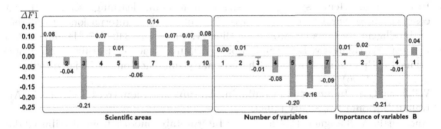

Fig. 12. Average improvement in prediction by groups of models for experiment #2

5   Conclusions and Future Work

This methodology presents an approach to merge decision trees. Its goal is to address the main problems commonly encountered while solving this problem, namely the preservation of the importance of some decision rules throughout all merging process and how to deal with the problem of class conflict of overlapping rules. The generation of the merged decision tree also presents challenges.

To study the impact of these problems, we define four parameters affecting merging and carried out experiments combining them. The case study are decision trees to predict failure of students in courses at the University of Porto. Results show that, on average, it is possible to merge 76% of models in a group. Grouping by scientific areas of courses is the best way to combine courses as the resulting model remains generic without loosing predictive quality. Tuning the four parameters improved predictions in a few cases.

Directions for future work point to improving the process with more elaborate strategies for class conflict resolution. Another important issue is the merge order of models in each group, which has yet to be studied in detail.

Acknowledgments. This work is funded by projects "NORTE-07-0124-FEDER-000059" and "NORTE-07-0124-FEDER-000057", financed by the North Portugal Regional Operational Programme (ON.2 – O Novo Norte), under the National Strategic Reference Framework (NSRF), through the European Regional Development Fund (ERDF), and by national funds, through the Portuguese funding agency, Fundação para a Ciência e a Tecnologia (FCT).

References

1. Dekker, G., Pechenizkiy, M., Vleeshouwers, J.: Predicting students drop out: a case study. In: 2nd International Educational Data Mining Conference (EDM 2009), pp. 41–50 (2009)
2. Gorbunov, K.Y., Lyubetsky, V.A.: The tree nearest on average to a given set of trees. Problems of Information Transmission 47, 274–288 (2011)
3. Kargupta, H., Park, B.: A fourier spectrum-based approach to represent decision trees for mining data streams in mobile environments. IEEE Transactions on Knowledge and Data Engineering 16, 216–229 (2004)
4. Provost, F.J., Hennessy, D.N.: Distributed machine learning: scaling up with coarse-grained parallelism. In: Proceedings of the 2nd International Conference on Intelligent Systems for Molecular Biology, vol. 2, pp. 340–347 (January 1994)
5. Provost, F., Hennessy, D.: Scaling up: Distributed machine learning with cooperation. In: Proceedings of the 13th National Conference on Artificial Intelligence, pp. 74–79 (1996)
6. Williams, G.J.: Inducing and Combining Multiple Decision Trees. PhD thesis, Australian National University (1990)
7. Andrzejak, A., Langner, F., Zabala, S.: Interpretable models from distributed data via merging of decision trees. In: 2013 IEEE Symposium on Computational Intelligence and Data Mining, CIDM (April 2013)
8. Hall, L., Chawla, N., Bowyer, K.: Combining decision trees learned in parallel. Working Notes of the KDD 1997 Workshop on Distributed Data Mining, pp. 10–15 (1998)
9. Hall, L., Chawla, N., Bowyer, K.: Decision tree learning on very large data sets. In: IEEE International Conference on Systems, Man, and Cybernetics, vol. 3, pp. 2579–2584 (1998)
10. Bursteinas, B., Long, J.: Merging distributed classifiers. In: 5th World Multiconference on Systemics, Cybernetics and Informatics (2001)
11. Kuhn, M., Weston, S., Coulter, N., Quinlan, R.: C50: C5.0 Decision Trees and Rule-Based Models. R package version 0.1.0-16 (2014)
12. Stone, M.: Cross-validatory choice and assessment of statistical predictions. Journal of the Royal Statistical Society: Series B 36(2), 111–147 (1974)
13. Kohavi, R.: A study of cross-validation and bootstrap for accuracy estimation and model selection. In: Proceedings of the 14th International Conference on AI (IJCAI), pp. 1137–1145. Morgan Kaufmann, San Mateo (1995)
14. Chinchor, N.: MUC-4 Evaluation Metrics. In: Proceedings of the 4th Message Understanding Conference (MUC4 1992), pp. 22–29. Association for Computational Linguistics (1992)
15. Han, J., Kamber, M., Pei, J.: Data Mining: Concepts and Techniques. Morgan Kaufmann, San Francisco (2011)

Faster MaxScore Query Processing
with Essential List Skipping

Kun Jiang and Yuexiang Yang

College of Computer
National University of Defense Technology
Changsha, Hunan Province, China
deeper@163.com

Abstract. Large search engines process thousands of queries per second over billions of documents, making a huge performance gap between disjunctive and conjunctive text queries in query processing. An important class of optimization techniques called top-k processing is therefore used to narrow this gap. In this paper, we present an improvement to the MaxScore optimization, which is the most efficient known document-at-a-time (DAAT) top-k processing method. Essentially, our approach can speed up MaxScore method by enabling skipping not just in non-essential lists as the original method does but also in essential lists, thus the name Essential List Skipping MaxScore (ELS-MaxScore), and providing more promising candidates for scoring. Experiments with TREC GOV2 collection show that our ELS-MaxScore processes significantly less elements, thus reduces the average query latency by almost 18% over the MaxScore baseline and 84% over the disjunctive DAAT baseline, while still returns the same results as the disjunctive evaluation.

Keywords: top-k processing, document-at-a-time, MaxScore, essential list skipping.

1 Introduction

Due to the rapid growth of the Internet, more and more people are relying on search engines to locate useful information. Large-scale search engines process thousands of queries per second over billions of documents. One major bottleneck in query processing is that the length of the inverted list can easily grow to hundreds of MBs for common terms [1]. Given that search engines need to answer queries within fractions of a second, naively traversing the basic index structure, which could take hundreds of milliseconds, is not acceptable. This problem has long been recognized by researchers, and has motivated a lot of work on optimization techniques [2–4]. An important class of optimization techniques called top-k processing [5, 6] can be used to reduce the query processing latency.

Top-k processing methods do not rank every document in the collection for each query. They manage to rank only the documents that will have a chance to

X. Luo, J.X. Yu, and Z. Li (Eds.): ADMA 2014, LNAI 8933, pp. 549–559, 2014.

enter in the final top-k results. As one of the most famous top-k processing algorithm, MaxScore method has been known for a long time with lots of enhancements, including some auxiliary index structures [7–11]. Essentially, the method distinguishes terms from essential lists and non-essential lists, and drives query processing by skipping in non-essential lists. Different from previous work, we proposed an updated MaxScore by enabling skipping in essential lists, which requires no modification to the underlying index structure. To the best of our knowledge, we are not aware of any previous work that enables skipping in essential lists of MaxScore for better query processing. Experimental results show that our proposed method processes less elements, leading to considerable performance gains over the MaxScore and disjunctive baselines.

The rest of this paper is organized as follows. We provide brief descriptions on inverted index structure, index traversal strategies and MaxScore method in Section 2 and related work in Section 3. In Section 4, we present our ELS-MaxScore method in detail. Experimental comparison of our ELS-MaxScore against a couple of the state-of-the-art methods is demonstrated in Section 5. Finally, Section 6 concludes this paper and presents a prospective of future work.

2   Basic Concepts

2.1   Inverted Index

The inverted index plays a key role in the efficient processing of boolean and ranked queries [12, 13]. It can be seen as an array of lists or postings, where each entry of the array corresponds to a different term or word in the collection, and the lists contain one element per distinct document where the term appears. For each document, the index stores the document identifier (*docid*), the weight of the term in the document(called *frequency*), the exact position of the occurrences and other relevant information (e.g. in the title, in anchor text, or in URLs). The set of terms is called the lexicon of the collection, which is comparatively small in most cases, but the inverted lists of common query terms may consist of many millions of postings. Fig. 1 is an example of inverted index structure, showing the physical storage of lexicon and posting lists. We assume postings have docids and frequencies but do not consider other data such as positions or contexts, thus each posting is of the form (d_i, f_i).

To allow faster access and limit the amount of memory needed, search engines use various compression techniques that significantly reduce the size of the inverted lists. Compression is crucial for search engine performance and various compression techniques have been proposed by researchers, see [4] for some recent work. Because the inverted lists can be very long, we want to skip parts of the lists during query processing. Skiplists divide the inverted lists into blocks of entries and provide pointers for faster access to such blocks, so that a scan in the skiplist determines in which block a document entry may occur, if it does, in the inverted list [14].

Index traversal strategies in query processing that match and score documents in inverted indexes for a query fall into two main categories, scoring by each

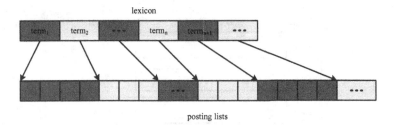

Fig. 1. An example of inverted index structure, showing the physical storage of lexicon and posting lists. Each term of the lexicon points to a posting list consisting of a number of postings describing all places where the term occurs in the collection.

term or scoring by documents, known as Term-At-A-Time (TAAT) and DAAT respectively [7]. For large indexes, TAAT is less efficient than DAAT due to more memory consuming that used to store a large number of accumulate scores [15, 16]. In the following sections, all our algorithms are based on DAAT strategy.

2.2 MaxScore Processing

Top-k processing methods are used to accelerate the basic DAAT index traversal strategy, in which documents that cannot exceed the threshold of the top-k result heap will be ignored. In this section, we detail the method closer to our research, focusing on the most famous top-k processing method, i.e. MaxScore [7, 8]. MaxScore maintains a sorted list $I_{t_i}(t_i \in Q)$ containing the current top-k documents scored so far. The last top-k document score τ is a threshold that documents must overcome to be considered in the top-k documents. Every candidate document d_{cand} with lowest current docid in all posting lists, i.e., $min(I_{t_i}.doc)$, will be considered, and discarded only if the partial score of the term it appears plus the cumulative maximum score of the left terms set Q' is less than the threshold, i.e., $score + \Sigma_{t_i \in Q'} \widehat{s}(I_{t_i}) < \tau$. This version of MaxScore [7] selects candidate with all the docids in the posting lists of query terms. The optimization point lies in the partial scoring of candidate document.

In the description [8], the query terms are sorted from top to bottom by their maxscores $\widehat{s}(I_{t_i})(t_i \in Q)$ and distinguished between essential lists Q_{req} and non-essential lists Q_{nonreq}, as shown in Fig.2. The Q_{nonreq} is defined as a set of lists in which that their maxscores sum up to less than the threshold, by adding lists starting from the list with the smallest maxscore. The other lists are called essential. No document can make it into the top-k results just using postings in Q_{nonreq}, and at least one of Q_{req} terms has to occur in any top-k document. We can safely perform a top-k query by only considering documents in the postings of Q_{req}, thus avoiding the documents that just appear in the postings of Q_{nonreq}. When the next larger docid $min(I_{t_i}.doc)(t_i \in Q_{req})$ in Q_{req} is selected as candidate d_{cand}. The posting list $I_{t_i}(t_i \in Q_{nonreq})$ in Q_{nonreq} should first skip to d_{cand}, then performing lookups into Q_{nonreq} to get the precise document scores. Instead of fully scoring each candidate in all lists, we perform the same

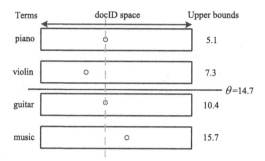

Fig. 2. Selecting non-essential lists in the MaxScore. Here, the lists for piano and violin are non-essential as their maxscores sum up to $5.1 + 7.3 < 14.7$, while guitar and music are essential.

partial scoring technique as recited in the original description [7]. Whenever a new result is inserted into the top-k result heap, we should check if the increase in the threshold means that another list can be added to the non-essential lists. Finally, the algorithm terminates when the essential lists set becomes empty.

3 Related Work

The problem of efficiently computing the ranking of results for a given user query has been largely addressed by top-k processing in [5–11, 15–18] . We can distinguish top-k processing algorithms between safe techniques [5–11, 15, 16] that return exactly the same top-k results as the baseline and unsafe techniques [17, 18] that just return results of equivalent quality. We focus on safe top-k processing algorithms of disjunctive queries, where the most relevant previous techniques are the MaxScore series of methods. In the version firstly proposed by Turtle and Flood [7], every candidate document with lowest current docid in all posting lists is considered, and will be discarded if its partial score in the term plus the cumulative maximum score of the left terms is less than the threshold. A later description of MaxScore by Strohman, Turtle and Croft [8] differs from the original one, in which terms are separated into essential lists and non-essential lists and skipping is occurred only in non-essential lists according to the candidate document in essential lists. Descriptions and evaluations of MaxScore method provided by Lacour et al. [16]and Fontoura [15] are both closer to the original MaxScore. More recently, Jonassen et al. [9] presents a complete combination of the above explanations with efficient skipping by introducing auxiliary an index called self-skipping index.

It is supported by [8] that a basic MaxScore algorithm improves performance of an original DAAT algorithm by 40% and it just scores about 50% of the documents. However, as we just need top-k results from the scoring candidates, the number of mis-scored documents rises with the increase of the index size.

The worst case complexity of the MaxScore is the same as without the optimization. If we can safely make skipping in essential lists, the algorithm can be more efficient by scoring fewer numbers of documents and skipping more postings. As described above, if the score of the candidate is lower than the threshold, it becomes ineffectiveness of the scoring process and useless of skipping to the candidate document in the non-essential lists. Further checking the effectiveness of the candidate can avoid overhead of extra scoring and skipping, thus improve query performance of MaxScore. Suel et al. [10, 11] and Chakrabarti [6] have presented the auxiliary block-max index to explicitly checking the effectiveness of candidate respectively.

As another technique relevant to us, skiplist is the cornerstone of fast skipping in most top-k processing methods. Moffat and Zobel [14] published one of the first papers applying inverted index skipping and presented a method to choose optimal skip-lengths for single- and multiple-level skipping with respect to disk-access and decompression time. Strohman and Croft [18], Chierichetti et al. [19], and Boldi and Vigna [20] presented counterpart methods to estimate optimal skipping distances. In our approach, we adopt the hierarchical index structure and also the skipping method described by Jonassen et al. [9], in which we compress groups of 128 index postings or 128 skipping pointers in chunks, while the pointers corresponding to different skipping-levels are stored in different chunks.

4   ELS-MaxScore

In this section, we present the basic idea of essential lists skipping and the implementation details of the ELS-MaxScore.

4.1   Essential Lists Skipping

The cumulative maxscore of all terms in lists should be larger than the threshold, i.e., $\sum_i^0 \hat{s}(I_{t_i}) > \tau$. When $\sum_{t_i \in q_{essential}} \hat{s}(I_{t_i}) < \tau$, no document can make it into the top-k results just using postings in the essential lists, and at least one of the non-essential terms has to occur in any top-k document. In this case, if the candidate docid in non-essential lists $d_{essential} > d_{nonessential}$, the score of the candidate document in essential lists cannot exceed the threshold. The candidate document in essential lists will be discarded and select the next candidate $d_{essential}.next()$. Through this further checking, we can avoid unnecessary scoring of the candidate document in essential lists.

However, we find that the next candidate document in essential lists will also be mis-scored if $d_{nonessential} > d_{essential}.next()$. For further improvement, the candidate document in essential lists $d_{essential}$ should skip to the candidate document in non-essential lists $d_{nonessential}$ to avoid mis-scoring and skipping of the documents between the candidate document in essential list and candidate document in non-essential lists. In fact, according to the definition of essential lists and the assumption we used here, no document can make it into the top-k results just using postings in essential lists or just using postings in non-essential lists. Thus, candidate documents in essential lists and non-essential lists can skip to each other iteratively until $d_{nonessential} = d_{essential}.next()$, i.e., the top-k result heap and judgment conditions are unchanged.

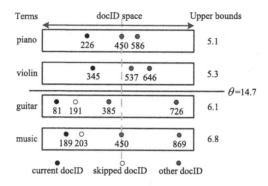

Fig. 3. An example showing how the essential lists skipping works. Assume $d_{essential}$ is the candidate document in essential lists and $d_{nonessential}$ is the candidate document in non-essential lists. In this case, we enable better essential list skipping to $d_{nonessential}$, instead of $d_{essential}.next()$ in essential list.

The idea of essential list skipping is shown in Fig.3. Assume docid 81 is the candidate document in essential lists and docid 226 is the candidate document in non-essential lists. Here, the cumulative maxscore in essential lists sum up to $6.1 + 6.8 < 14.7$, and it is obvious that docid 81 cannot make itself into the result heap. ELS-MaxScore makes the list for guitar skip over docid 191 to docid 226, and reached docid 385. By doing this, skipping is greatly improved compared with the case that using documents one after another as the candidate in essential lists (In the baseline method, docid 189 will be selected as the next candidate). The skipping continues until term piano and music both reached docid 450 in their own lists. Thus, the document with docid 450 is selected as a more promising candidate and sends to final scoring. By doing this, skipping is further improved since docid 385 will be also skipped which is selected as candidate document in the above case. The proof should be obvious.

4.2   The Algorithm

In ELS-MaxScore, we first make some necessary preparation for the query processing, including ordering the iterators by descending maximum score and calculating their cumulative maximum scores from last to first. We maintain a heap to store k result candidates. Here, we do not simply select candidate document with lowest docid within essential lists. Instead, we ignore the candidates using

Algorithm 1. $ELS(\{I_{t_1}, \cdots, I_{t_q}\}, \hat{a}, q)$

Input: $\{I_{t_1}, \cdots, I_{t_q}\}, \hat{a}, q$
Output: candidate document d_{cand}
1: $d_{cand} \leftarrow min_{t_i \leq q_{essential}}(I_{t_i}.doc)$; $d_{essential}, d_{nonessential} \leftarrow Integer.MaxValue$
2: **while** $q_{essential} < q$ **and** $\hat{a}(I_{t_0}) - \hat{a}(I_{t_{q_{essential}}}) < \tau$ **do**
3: $\quad d_{essential} \leftarrow min_{t_i \leq q_{essential}}(I_{t_i}.doc)$
4: $\quad$ **Select** $d_{nonessential} \leftarrow min_{t_i \leq q_{nonessential}}(I_{t_i}.doc)$ **satisfies** $d_{nonessential}.score + \hat{a}(I_{t_0}) - \hat{a}(I_{t_{q_{essential}}}) > \tau$
5: $\quad$ **if** $d_{essential} < d_{nonessential}$ **and** $d_{essential}.skipTo(d_{nonessential}) = false$ **then**
6: $\quad\quad$ remove I_{t_i}, update $\hat{a}, q, q_{essential}$, **continue**
7: $\quad$ **end if**
8: $\quad$ **if** $d_{essential} > d_{nonessential}$ **and** $d_{nonessential}.skipTo(d_{essential}) = false$ **then**
9: $\quad\quad$ remove I_{t_i}, update $\hat{a}, q, q_{essential}$, **continue**
10: $\quad$ **end if**
11: $\quad$ **if** $d_{cand} > d_{essential}$ **then break**
12: $\quad$ **end if**
13: **end while**
14: **return** d_{cand}

our Essential List Skipping (ELS) procedure that listed in Algorithm 1, which can generate more promising candidates.

In Algorithm 1, when non-essential lists exist and the cumulative maximum score $\hat{a}(\hat{a}(I_{t_i}) = \sum_i^0 \hat{s}(I_{t_i}))$ of the terms in essential lists is less than τ(line 2), we iteratively select candidate document of non-essential lists until its score in non-essential lists plus the cumulative maximum score of the terms in essential lists is bigger than τ(line 4). This idea is similar to the description of choosing candidate documents in essential lists in the MaxScore method by Turtle and Flood. For each candidate document in non-essential lists and candidate document in essential lists, they skip to each other iteratively until the two are equal, due to the unchanged top-k result heap and other judgment conditions (line 5-12).

5 Experiments

In this section, we provide our experimental results.

5.1 Experimental Setup

We use the TREC GOV2 collection containing about 25.2 million documents and 32.8 million terms with an uncompressed size of 426GB. We build inverted index structures with 128 docids per block, using PForDelta as the compression algorithm [21], removing English stopword, and applying Porters English stemmer. We adopt the hierarchical index structure and also the skipping method described by Jonassen et al. [9], in which we compress groups of 128 index postings or 128 skipping pointers in chunks, while the pointers corresponding to different skipping-levels are stored in different chunks. The final compressed index size is 7.57GB. We use 10000 queries randomly selected from the TREC2005 Efficiency track queries using distinct numbers of terms with $|q| \geq 2$.

Our experiments were performed on an Intel(r) Xeon(r) E5620 processor running at 2.40 GHz with 8GB of RAM and 12,288KB of cache. All solutions were implemented in JAVA with BM25 [13] as the ranking function. All the codes are openly available by contacting the authors. In every experiment where we report running time, the index was preloaded into memory and the results are averaged over 5 independent runs. We measure the performance by the following criteria: average time per query, decoded blocks, docid evaluated, calls to the scoring function and candidates inserted into the result heap. These criteria are also used in previous work [9, 10].

5.2 Results

One of the main challenges associated with processing of disjunctive queries is that it tends to be much slower than conjunctive queries, as document matching any of the query terms might be returned as a result. In contrast, conjunctive queries requires to return only those documents that match all of the terms, in which the shortest posting list can be used to efficiently skip the longer posting lists and then reduce the amount of data to be processed. Top-k processing methods, including our ELS-MaxScore, are focusing on narrowing the performance gap between the two types of queries. Thus, we compare our algorithm with three baseline algorithms, i.e., disjunctive queries, conjunctive queries and the state-of-the-art MaxScore by Jonassen et al. All of the three methods are using DAAT index traversal strategy. Table 1 shows the query processing time using different algorithms with different number of returned results.

Table 1. Average query processing times in ms of different algorithms with different number of results k, on the TREC 2005 query logs

Algorithms	Avg.	k=10	k=50	k=100	k=500	k=1000
Disjunctive	285.3	276.4	283.8	288.4	288.9	288.9
Conjunctive	24.7	24.7	24.7	24.7	24.8	24.8
MaxScore	54.6	34.6	43.9	49.2	67.5	77.9
ELS-MaxScore	44.7	28.3	33.7	39.3	56.1	66.0
$\Delta(D-E)\%$	84.3%	89.8%	88.1%	86.4%	80.6%	77.2%
$\Delta(M-E)\%$	18.2%	18.2%	23.2%	20.1%	16.9%	15.3%

From Table 1, we can see the existing huge performance gap between disjunctive and conjunctive queries. MaxScore achieves great benefits but still leaves a large room for improvement. This is mainly due to the mis-scoring problem in MaxScore method described in Section 2. ELS-MaxScore provides an 18.2%

further reduction over the MaxScore baseline and 84.3% over the disjunctive baseline in the average query latency by enabling skipping in essential lists. The best case with ELS-MaxScore is 8.8 times faster than the disjunctive evaluation, making the performance between conjunctive and disjunctive queries much closer to each other. Furthermore, we find that the performance for the disjunctive queries is quite stable as it decodes and evaluates all the postings, and it is also for conjunctive queries as the threshold is only used to keep the number of results when each candidate is completely scored. For MaxScore and ELS-MaxScore, the query processing time increases as we increase k. This is mainly because different k results in different thresholds and essential lists when selecting candidates. However, our algorithm always keeps a good improvement over the MaxScore baseline. With k less than 100, the performance of our method is close to the conjunctive evaluation with less than 26.7% difference.

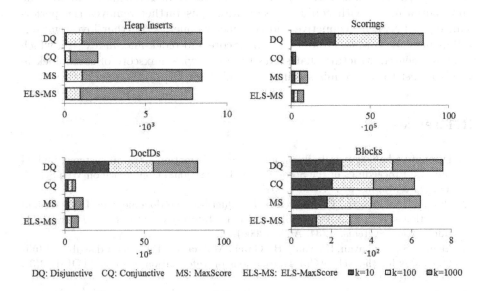

Fig. 4. Average numbers of processed elements of different DAAT methods on TREC GOV2 collection

Fig. 4 shows other criteria for our ELS-MaxScore and the baselines on the TREC 2005 query log with results number k=10, 100, 1000. We can see the huge gap of processed elements between disjunctive and conjunctive queries, which indicates the importance of our optimization work. The number of candidates inserted into the result heap and posting scorings of our ELS-MaxScore is significantly reduced compared with disjunctive and MaxScore baseline. This means that ELS-MaxScore should perform even better when we choose a more complex scoring function than BM25, such as functions with proximity support. We observe also a decrease in the number of evaluated docids and decompressed blocks due to more inverted index skipping. In addition, we find that MaxScore provides a little reduction of the decompressed blocks even compared with conjunctive queries with k less equals 10, and our ELS-MaxScore does it further with all different k. This is mainly because that some long step skipping can even jump over blocks.

6   Conclusion and Future Work

In this paper we have proposed and evaluated an efficient top-k processing method called ELS-MaxScore by skipping in both essential and non-essential lists. Experimental results show that our technique provides additional 18% performance gains on average over the MaxScore baseline and 84% over the disjunctive DAAT baseline, without sacrificing result quality. The best case of our proposed method with the number of returned results less than 100 achieves almost the same performance with conjunctive queries, thus further removes the performance gap between disjunctive and conjunctive queries. Further investigations will include the combination of ELS-MaxScore and block-max index [10], which is an augmented structure that stores the piece-wise maxscore of each block in a posting list to reduce mis-scorings of candidates [11].

References

1. Dean, J.: Challenges in building large-scale information retrieval systems: invited talk. In: Proceedings of the Second ACM International Conference on Web Search and Data Mining, p. 1. ACM (2009)
2. Puppin, D., Silvestri, F., Laforenza, D.: Query-driven document partitioning and collection selection. In: Proceedings of the 1st International Conference on Scalable Information Systems, p. 34. ACM (2006)
3. Melink, S., Raghavan, S., Yang, B., Garcia-Molina, H.: Building a distributed full-text index for the web. ACM Transactions on Information Systems (TOIS) 19(3), 217–241 (2001)
4. Yan, H., Ding, S., Suel, T.: Compressing term positions in web indexes. In: Proceedings of the 32nd International ACM SIGIR Conference on Research and Development in Information Retrieval, pp. 147–154. ACM (2009)
5. Broder, A.Z., Carmel, D., Herscovici, M., Soffer, A., Zien, J.: Efficient query evaluation using a two-level retrieval process. In: Proceedings of the Twelfth International Conference on Information and Knowledge Management, pp. 426–434. ACM (2003)
6. Chakrabarti, K., Chaudhuri, S., Ganti, V.: Interval-based pruning for top-k processing over compressed lists. In: 2011 IEEE 27th International Conference on Data Engineering (ICDE), pp. 709–720. IEEE (2011)
7. Turtle, H., Flood, J.: Query evaluation: strategies and optimizations. Information Processing & Management 31(6), 831–850 (1995)
8. Strohman, T., Turtle, H., Croft, W.B.: Optimization strategies for complex queries. In: Proceedings of the 28th Annual International ACM SIGIR Conference on Research and Development in Information Retrieval, pp. 219–225. ACM (2005)

9. Jonassen, S., Bratsberg, S.E.: Efficient compressed inverted index skipping for disjunctive text-queries. In: Clough, P., Foley, C., Gurrin, C., Jones, G.J.F., Kraaij, W., Lee, H., Mudoch, V. (eds.) ECIR 2011. LNCS, vol. 6611, pp. 530–542. Springer, Heidelberg (2011)

10. Ding, S., Suel, T.: Faster top-k document retrieval using block-max indexes. In: Proceedings of the 34th International ACM SIGIR Conference on Research and Development in Information Retrieval, pp. 993–1002. ACM (2011)

11. Dimopoulos, C., Nepomnyachiy, S., Suel, T.: Optimizing top-k document retrieval strategies for block-max indexes. In: Proceedings of the Sixth ACM International Conference on Web Search and Data Mining, pp. 113–122. ACM (2013)

12. Zobel, J., Moffat, A.: Inverted files for text search engines. ACM Computing Surveys (CSUR) 38(2), 6 (2006)

13. Büttcher, S., Clarke, C., Cormack, G.V.: Information retrieval: Implementing and evaluating search engines. The MIT Press (2010)

14. Moffat, A., Zobel, J.: Self-indexing inverted files for fast text retrieval. ACM Transactions on Information Systems (TOIS) 14(4), 349–379 (1996)

15. Fontoura, M., Josifovski, V., Liu, J., Venkatesan, S., Zhu, X., Zien, J.: Evaluation strategies for top-k queries over memory-resident inverted indexes. Proceedings of the VLDB Endowment 4(12), 1213–1224 (2011)

16. Lacour, P., Macdonald, C., Ounis, I.: Efficiency comparison of document matching techniques. In: Proc. ECIR (2008)

17. Lester, N., Moffat, A., Webber, W., Zobel, J.: Space-limited ranked query evaluation using adaptive pruning. In: Ngu, A.H.H., Kitsuregawa, M., Neuhold, E.J., Chung, J.-Y., Sheng, Q.Z. (eds.) WISE 2005. LNCS, vol. 3806, pp. 470–477. Springer, Heidelberg (2005)

18. Strohman, T., Croft, W.B.: Efficient document retrieval in main memory. In: Proceedings of the 30th Annual International ACM SIGIR Conference on Research and Development in Information Retrieval, pp. 175–182. ACM (2007)

19. Chierichetti, F., Lattanzi, S., Mari, F., Panconesi, A.: On placing skips optimally in expectation. In: Proceedings of the 2008 International Conference on Web Search and Data Mining, pp. 15–24. ACM (2008)

20. Boldi, P., Vigna, S.: Compressed perfect embedded skip lists for quick inverted-index lookups. In: Consens, M.P., Navarro, G. (eds.) SPIRE 2005. LNCS, vol. 3772, pp. 25–28. Springer, Heidelberg (2005)

21. Zukowski, M., Heman, S., Nes, N., Boncz, P.: Super-scalar ram-cpu cache compression. In: Proceedings of the 22nd International Conference on Data Engineering, ICDE 2006, p. 59. IEEE (2006)

Selecting Representative Objects from Large Database by Using K-Skyband and Top-k Dominating Queries in MapReduce Environment

Md. Anisuzzaman Siddique, Hao Tian, and Yasuhiko Morimoto

Graduate School of Engineering,
Hiroshima University,
1-7-1 Kagamiyama, Higashi-Hiroshima, 739-8521, Japan
{siddique,M124671}@hiroshima-u.ac.jp,
{morimoto}@mis.hiroshima-u.ac.jp

Abstract. We consider a problem to select representative distinctive objects in a numerical database, which is an important problem in an early stage of knowledge discovery process. Skyline query and its variants are functions to find such representative objects. Skyline query selects representative objects that are not dominated by any other object in the dataset. Though skyline query is useful function, it cannot control the size of selected objects. In order to solve the problem, "top-k dominating query" and "K-skyband queries" have been introduced. However, conventional algorithms for computing those functions are not well suited for parallel distributed environment. In this paper, we consider a method for computing both queries in a parallel distributed framework called MapReduce, which is a popular framework to handle "big data".

Keywords: Representative Objects, Skyline Query, Top-k Dominating Query, K-Skyband Query, MapReduce.

1 Introduction

In an early stage of knowledge discovery process, to select representative distinctive objects in a database is important to understand the data. Assume that a new hotel database is given. First of all, we often look at representative objects such as the cheapest one, the most popular one, the most convenient one, and so on to understand the data. These objects are examples of representative distinctive objects in the hotel database.

Skyline query [3] and its variants are functions to find such representative objects from a numerical database. Skyline objects in a dataset are objects that are not dominated by any other object in the dataset. Given an m-dimensional dataset DS, an object O_i is said to be in skyline of DS if there is no other object O_j $(i \neq j)$ in DS such that O_j is better than O_i. If there exists such O_j, then we say that O_i is dominated by O_j or O_j dominates O_i.

Figure 1 shows an example of skyline. The table in the figure is a list of hotels, each of which contains two numerical attributes: *distance* and *price*. If we assume

X. Luo, J.X. Yu, and Z. Li (Eds.): ADMA 2014, LNAI 8933, pp. 560–572, 2014.

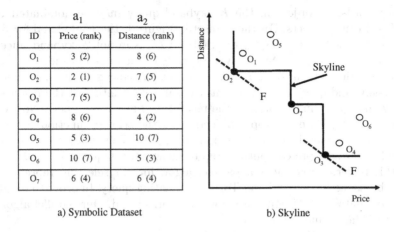

a) Symbolic Dataset   b) Skyline

Fig. 1. Skyline example

that smaller value is better, then the skyline query retrieves objects $\{O_2, O_3, O_7\}$ (see Figure 1 (b)). Where objects O_1 and O_5 are dominated by the object O_2. Similarly, objects O_4 and O_6 are dominated by the object O_3.

Another popular function to find representative objects is "top-k query". The top-k query selects the k objects based on a user specified scoring function. For example, if a user specifies $k = 2$ and "$F = price + distance$" as a scoring function, the top-2 query selects O_2 and O_3.

As illustrated in the above, skyline queries do not require a scoring function and simply find a common subset of non-dominated objects for all linear scoring functions. This intuitive nature of the query formulation has been a key strength of skyline queries, compared to top-k queries that requires users to formulate a function. On the other hand, we cannot control the number of retrieved skyline objects. Skyline queries may retrieve too many objects especially in high dimensional datasets.

Authors of [20] combine the strength of both skyline query and top-k query and propose a new variant of skyline query called "top-k dominating query" to select various k objects. They defined an intuitive score function for modeling the importance of an object $O \in DS$:

$$\mu(O) = |\{O' \in DS \,|O \prec O'\} \,|.$$

The score function $\mu(O)$ return the number of objects dominated by object O. For example, $\mu(O_2) = 2$ since O_2 dominates O_1 and O_5. Similarly, $\mu(O_3) = 2$. Hence, top-2 dominating query of Figure 1 retrieves objects O_2 and O_3. Thus, a top-k dominating query is not only to select representative objects but also to provide importance of each representative object. From a practical perspective, top-k dominating query combines the advantages of top-k and skyline queries without sharing their disadvantages.

K-Skyband query [13] is another variants of skyline query. K-skyband query returns a set of objects, each object of which is not dominated by K objects.

In other words, an object in the K-skyband query may be dominated by at most $K - 1$ other objects. For the dataset in Figure 1, the K-skyband for $K = 2$ retrieves objects $\{O_1, O_2, O_3, O_4, O_7\}$. Object O_5 is in not 2-skyband since it is dominated by O_2 and O_1. Similarly, object O_6 is not in 2-skyband.

Top-k queries return different set of objects if scoring function is different, while K-skyband query always returns the same set. Note that the result set of a K-skyband query must contains the result set of any top-k query. Notice that O_2 and O_3, which are the top-2 objects with the scoring function $F = price + distance$, are contained in the 2-skyband result. Therefore, one can use the skyband query as a pre-computing step to answer top-k query.

In this paper, we consider a parallel algorithm for selecting representative objects by using K-skyband and Top-k dominating query. In order to handle so called "big data", MapReduce framework is introduced. Our parallel algorithm utilize the MapReduce framework.

Contributions of this paper include following aspects:

☐ We have developed a scalable parallel algorithm for K-skyband and Top-k dominating query so that we can handle "big data".

☐ The proposed algorithm compute both K-skyband and Top-k dominating query simultaneously.

☐ We have empirically proved its efficiency through extensive experiments.

The rest of this paper is organized as follows: Section 2 presents the notions and properties of top-k dominating and skyband query computation. We provide detailed examples and analysis of our algorithm in Section 3. Section 4 reviews related work. We experimentally evaluate the algorithm in Section 5 under a variety of settings. Finally, Section 6 concludes the paper.

2   Preliminaries

In this section, we present some definitions and basic properties of proposed algorithm.

Table 1. Database DS

DS_1			DS_2			DS_3		
ID	a_1	a_2	ID	a_1	a_2	ID	a_1	a_2
$O_{1,1}$	3	8	$O_{2,1}$	7	3	$O_{3,1}$	5	10
$O_{1,2}$	2	7	$O_{2,2}$	8	4	$O_{3,2}$	10	5
						$O_{3,3}$	6	6

Given a database DS that is defined by a set of m-attributes $\{a_1, a_2, \cdots, a_m\}$. The database is distributed into n datasets $\{DS_1, DS_2, \cdots, DS_n\}$ on different locations. Without loss of generality, assume that each attribute has non-negative numerical values. We also assume smaller value is preferable in each attribute. We use $O_{i,j}.a_p$ to denote the p-th attribute's value of object $O_{i,j}$ where i represent datasets ID and j represent object ID in the corresponding dataset DS_i. Assume the dataset DS shown in Figure 1 is distributed into three subsets, DS_1, DS_2, and DS_3, each of which has two attributes, a_1 and a_2, as shown in Table 1. In this paper, we assign object id of those distributed databases like $O_{DSID,LOCALOID}$ where $DSID$ denote ID of dataset and $LOCALOID$ is local ID of an object in the dataset. For example, $O_{3,3}$ is an object of DS_3 and its local ID in DS_3 is "3".

Definition 1 (Domination). For objects $O_{i,j}$ and $O'_{i,j}$, an object $O_{i,j}$ is said to *dominate* another object $O'_{i,j}$ with respect to DS, denoted by $O_{i,j} \prec O'_{i,j}$, if $O_{i,j}.a_s \leq O'_{i,j}.a_s$ for all attributes $(s = 1, \cdots, m)$ and $O_{i,j}.a_x < O'_{i,j}.a_x$ for at least one attribute $(1 \leq x \leq m)$. We call such $O_{i,j}$ as *dominant* object and such $O'_{i,j}$ as *dominated* object between $O_{i,j}$ and $O'_{i,j}$. If $O_{i,j}$ dominate $O'_{i,j}$, then $O_{i,j}$ is more preferable than $O'_{i,j}$.

In Table 1 object $O_{1,2}$ dominates object $O_{1,1}$ ($O_{1,2} \prec O_{1,1}$). This is because object $O_{1,2}$ has smaller value in both attributes than objects $O_{1,1}$.

Definition 2 (Skyline). An object $O \in DS$ is in skyline of DS (i.e., a skyline object in DS) if O is not dominated by any other object in DS. The skyline of DS, denoted by $Sky(DS)$, is the set of skyline objects in DS. For dataset DS, objects $\{O_{1,2}, O_{2,1}, O_{3,3}\}$ can dominate all other objects and they are not dominated by each other. Thus skyline query for dataset DS will retrieve $Sky(DS)$ = $\{O_{1,2}, O_{2,1},$ and $O_{3,3}\}$.

Top-k queries can be defined by a scoring function F, which enables the ranking (ordering) of the data objects. The most important and commonly used scoring function is the ranked linear sum function. Each attribute a_i has an associated rank r_i indicating a_i's relative importance for the query. The aggregated score $F_{\mathbf{w}}(O_i)$ for object O_i is defined as a weighted sum of the individual ranks: $F_{\mathbf{w}}(O_i) = \sum_{i=1}^{m} w_i * r_i$, where $\mathbf{w} = (w_1, ..., w_m)$ is a user given weighting vector. The result of a top-k query is a list of the k objects that have the top-k ranking values of $F_{\mathbf{w}}$.

Definition 3 (Top-k query). Given a positive integer k and a weighting vector $\mathbf{w}$, the result set $TOP_k(\mathbf{w})$ of the top-k query is a set of objects such that $TOP_k(\mathbf{w}) \in DB$, $|TOP_k(\mathbf{w})| = k$ and $\forall O_i, O_j$: $O_i \in TOP_k(\mathbf{w})$, $O_j \in DB - TOP_k(\mathbf{w})$ it holds that $F_{\mathbf{w}}(O_i) \leq F_{\mathbf{w}}(O_j)$.

If we set weight vector $[0.5, 0.5]$ and $k = 2$ then Table 1 retrieves objects $O_{1,1}$ and $O_{2,1}$ as top-2 result. We can easily specify the number of retrieved objects by using top-k query. However, to specify a weighting vector is not an easy procedure for an user.

Definition 4 (μ score). The μ score is the number of objects dominated by an object. μ score of an object O is denoted as $\mu(O)$. In Table 1 object $O_{1,2}$ dominates objects $O_{1,1}$ and $O_{3,1}$ so the μ score of $O_{1,2}$ is 2, i. e. $\mu(O_{1,2}) = 2$.

Definition 5 (SB score). The SB score of an object is the number of dominant objects. SB score of an object O is denoted as $SB(O)$. In Table 1 object $O_{3,2}$ is dominated by objects $O_{2,1}$ and $O_{2,2}$ so the SB score of $O_{3,2}$ is 2, i. e. $SB(O_{3,2})$ = 2.

Definition 6 (Top-k dominating query). Given a positive integer k and for a database DS, the top-k dominating query returns k objects that have top-k μ scores from DS.

For dataset shown in Table 1, the top-2 dominating query retrieves $O_{1,2}$ and $O_{2,1}$.

Definition 7 (K-skyband query). Given a positive integer K, K-skyband is a set of objects that are dominated by at most $K - 1$ other objects.

For dataset DS shown in Table 1, the skyband query for $K = 2$ includes objects $\{O_{1,1}, O_{1,2}, O_{2,1}, O_{2,2}, O_{3,3}\}$. Conceptually, K represent the thickness of the skyline. 1-Skyband query corresponds to a conventional skyline. Moreover, top-k result also belongs to the result of K-skyband. One can use K-skyband result as a pre-processing step of skyline as well as top-k query computation.

Definition 8 (Domination Check Set). For an object O assume $r_1(O)$ and $r_2(O)$ be the rank value of attribute a_1 and a_2, respectively. For example, in Figure 1, $r_1(O_4) = 6$ and $r_2(O_4) = 2$. We call the largest $r_s(O)$ ($s = 1, ..., m$) as "the worst rank of O" and a_s as "the worst rank attribute of O". In the example the worst rank of O_4 is 6 and the worst rank attribute is a_1.

Domination Check DC set for an object O is a set of objects that have equal or greater rank than the worst rank of O in the worst rank attribute. For example, O_6 has greater rank than the worst rank, i.e., 6, of O_4 in a_1 (price). Therefore, the DC set of object O_4 is $\{O_6\}$. Similarly, the worst rank of O_3 is 5. So, the DC set of object O_3 is $\{O_4, O_6\}$.

3   Skyband and Top-k Dominating Query Processing

Our parallel distributed algorithm of skyband and top-k dominating queries has following four phases.

Data Map and Ranking: We partition each distributed dataset vertically, i.e., attribute wise and dispatch each partition to the MAP workers. Each MAP worker generates $(Key, Value)$ pairs, where Key is the object ID and $Value$ is the rank of corresponding object in the attribute domain. The output of this phase is $(ID, Rank)$ pairs for each object.

Reduce and Worst Rank Computation: Coordinator collects $(ID, Rank)$ pairs to reduce data access between MAP workers and REDUCE workers. After shuffling, coordinator finds the worst attribute rank for each object. The second map-reduce function is invoked in this stage for skyband and top-k dominating queries computation.

Rank Map and DC Set Computation: Coordinator sends attribute rank to the corresponding map worker and compute domination check DC sets for each object.

Skyband and Top-k Dominating Computation: REDUCE workers perform domination check between DC sets and corresponding object. Next, it sends the result to the coordinator. The coordinator maintains SB score and μ score to compute skyband and top-k dominating query, respectively.

3.1 Data Map and Ranking

We followed a vertical partitioning strategy such that m-dimensional dataset splitted into m partitions. If the number dataset is n then the total number of partitions is equal to $n \times m$, say $\{s_{1,1}, \cdots, s_{1,m}, \cdots, s_{n,1}, \cdots, s_{n,m}\}$. For simplicity, we denote $\{s_{1,1}, s_{2,1}, \cdots, s_{n,1}\}$, $\{s_{1,2}, s_{2,2}, \cdots, s_{n,2}\}$, and $\{s_{1,m}, s_{2,m}, \cdots, s_{n,m}\}$ as $\{S_1\}$, $\{S_2\}$, $\cdots$, and $\{S_m\}$, respectively. In our running example, DS_1 has two attributes a_1 and a_2. We split DS_1 into two partitions called $s_{1,1}$ and $s_{1,2}$. Here, we have two partitions, which we need at least two MAP workers.

Each MAP worker independently operates a non-overlapping partition of the input file and calls the user-defined "map function" to emit a list of $(Key\ Value)$ pairs from its local storage in parallel. In our algorithm, each MAP worker produces $(ID, Rank)$. To calculate the rank value for each key-value pair of $S_l(l = 1, \cdots, m)$, corresponding MAP worker sorts its attribute in ascending order, then replaces the values by their corresponding rank value. Figure 2 shows

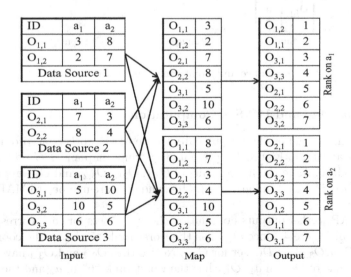

Fig. 2. Data map and ranking process

the "data map and ranking" procedure. It shows that objects $O_{1,2}$ and $O_{2,1}$ have rank "1" for attribute a_1 and a_2, respectively. By replacing each actual value to rank value, we can prevent distribution of some sensitive values.

3.2   Reduce and Worst Rank Computation

Each MAP worker dispatches the $(ID, Rank)$ pairs to the REDUCE workers. After shuffling, each REDUCE worker sends its output to the coordinator. After that, coordinator retrieves the worst rank and the worst rank attribute each object. Figure 3 shows the "reduce and worst rank computation" procedure. In our running example, $O_{1,1}$ has rank value 2 and 6 for attribute a_1 and a_2 respectively. In the object, a_2's rank is the worst (i.e., the worst rank of $O_{1,1}$ is 6 and the worst rank attribute is a_2). Therefore, coordinator generates $(O_{1,1}, < a_2, 6 >)$ for object $O_{1,1}$ as a key-value pair for $O_{1,1}$.

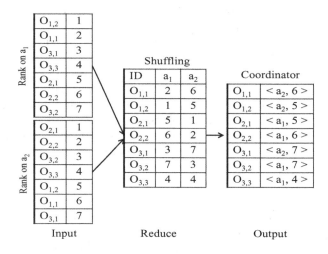

Fig. 3. Reduce and worst rank computation process

3.3   Rank Map and DC Set Computation

Next, coordinator maps the output pairs to the MAP workers according to the worst rank attribute. Figure 4 represents the "rank map and DC set computation" procedure. As in the figure, pairs of $O_{2,1}$, $O_{2,2}$, $O_{3,2}$, and $O_{3,3}$ are mapped to MAP worker for a_1. Similarly, $O_{1,1}$, $O_{1,2}$, and $O_{3,1}$ are mapped to MAP worker for a_2.

Each MAP worker outputs domination check DC sets for each corresponding object. As in Figure 4, since $O_{2,1}$ has the worst rank "5" in a_1, the coordinator outputs $\{O_{2,2}, O_{3,2}\}$ as DC set for $O_{2,1}$. Notice that $O_{2,2}$ and $O_{3,2}$ have greater rank than that of $O_{2,1}$ in a_1. $O_{1,1}$ has the worst rank "6" in a_2 and the DC set member of $O_{1,1}$ is $\{O_{3,1}\}$.

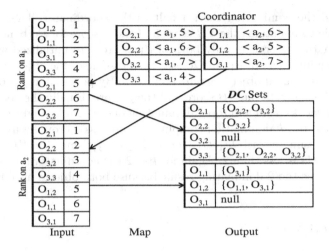

Fig. 4. Rank map and DC set computation

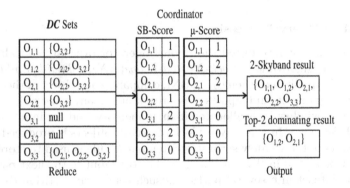

Fig. 5. Skyband and top-k dominating computation

3.4 Skyband and Top-k Dominating Computation

In second reduce phase REDUCE workers perform domination check between an object and its corresponding DC set. This is because here we can apply the following theorem.

Theorem 1. *For two objects* $\{O, O' \in DS\}$, *if* O' *is not in the DC set of object* O *then* O *can not dominate object* O', *i.e.,* $O \nprec O'$.

Proof

Let a_s be the worst rank attribute of O. If O dominates O', O' must be in DC set of O since $O.a_s \leq O'.a_s$. If O' is not in DC set of O, it means $O.a_s > O'.a_s$. Therefore, O can not dominate O'. $\square$

Theorem 1 confirmed that it is sufficient to perform domination check between an object O and the corresponding DC set for object O. Afterward REDUCE workers send the domination check result to the coordinator. The coordinator is responsible to maintains the SB score and the μ score for skyband as well as top-k dominating query. Initially both of the scores are initialized with "0". If an object dominates an object in DC set, then μ score of the object and the SB score of the dominated object incremented by 1. By performing the domination check, coordinator gets the SB score and the μ score. Finally the coordinator is ready to answer skyband and top-k dominating query results. Figure 5 shows the " skyband and top-k domination computation process". From figure for $K = 2$, the skyband result set is $\{O_{1,1}, O_{1,2}, O_{2,1}, O_{2,2}, O_{3,3}\}$, since the SB score value for those objects is less than 2. Again for $k = 2$ the coordinator outputs objects $O_{1,2}$ and $O_{2,1}$ as top-2 dominating result because both objects have the highest μ score.

4   Related Work

Our work is motivated by previous studies of skyline query processing as well as its variants. Those are reviewed in the following sections.

4.1   Skyline Query Processing

Borzsonyi et al. first introduced the skyline operator over large databases and proposed three algorithms: *Block-Nested-Loops(BNL)*, *Divide-and-Conquer* (*D&C*), and B-tree-based schemes [3]. BNL compares each object of the database with every other object, and reports it as a result only if any other object does not dominate it. A window W is allocated in main memory, and the input relation is sequentially scanned. In this way, a block of skyline objects is produced in every iteration. In case the window saturates, a temporary file is used to store objects that cannot be placed in W. This file is used as the input to the next pass. *D&C* divides the dataset into several partitions such that each partition can fit into memory. Skyline objects for each individual partition are then computed by a main-memory skyline algorithm. The final skyline is obtained by merging the skyline objects for each partition. Chomicki et al. improved BNL by presorting, they proposed *Sort-Filter-Skyline(SFS)* as a variant of BNL [6]. Among index-based methods, Tan et al. proposed two progressive skyline computing methods Bitmap and Index [15]. In the Bitmap approach, every dimension value of an object is represented by a few bits. By applying bit-wise *AND* operation on these vectors, a given object can be checked if it is in the skyline without referring to other objects. The index method organizes a set of m-dimensional objects into m lists such that an object O is assigned to list i if and only if its value at attribute i is the best among all attributes of O. Each list is indexed by a B-tree, and the skyline is computed by scanning the B-tree until an object that dominates the remaining entries in the B-trees is found. The current most efficient method is *Branch-and-Bound Skyline(BBS)*, proposed by Papadias et al., which is a progressive algorithm based on the *best-first nearest neighbor* (*BF-NN*) algorithm [13]. Instead of searching for nearest neighbor repeatedly, it directly prunes using the R*-tree structure.

4.2   Skyline Variants

Recently, research community focus has been shifted to the study of queries based on variants of the dominance relationship. Chan et al. identify the problem of computing top-k frequent skyline objects, where the frequency of an object is defined by the number of dimensional subspaces [5]. Li et al. propose a data cube structure for speeding up the evaluation of queries that analyze the dominance relationship of objects in the dataset [10]. However, incremental maintenance of the data cube over updates has not been addressed in [10]. Clearly, it is prohibitively expensive to recompute the data cube from scratch for dynamic datasets with frequent updates. Chan et al. proposed k-dominant skyline, which is based on the k-dominance relationship and developed efficient ways to compute it in high-dimensional space [4]. An object O_i is said to k-dominate another object O_j if O_i dominates O_j in at least one k-dimensional subspace. The k-dominant skyline contains the objects that are not k-dominated by any other object. When k decreases, the size of the k-dominant skyline also decreases. Skyband computation is another kind of query. K-skyband query returns objects that are dominated by at most $K - 1$ other objects [13]. It has been observed that for any increasingly monotone aggregate function, the top-k objects belong to the K-skyband, where $k \leq K$ [8].

More aspects of skyline computation have been explored. Lin et al. proposed n-of-N skyline query to support online query on data streams, i.e., to find the skyline of the set composed of the most recent n elements. In the cases where the datasets are very large and stored distributedly, it is impossible to handle them in a centralized fashion [11]. Balke et al. first mined skyline in a distributed environment by partitioning the data vertically [1]. Vlachou et al. introduce the concept of extended skyline set, which contains all data elements that are necessary to answer a skyline query in any arbitrary subspace [19]. Tao et al. discuss skyline queries in arbitrary subspaces [17]. More skyline variants such as dynamic skyline [12] and reverse skyline [7] operators also have recently attracted considerable attention.

Computing the skyline or its variants are challenging today since there is an increasing trend of applications expected to deal with "big data". Observe that [4,5,10] cannot be directly applied to evaluate top-k dominating queries. Moreover skyband query needs separate algorithm. To compute both type queries in a common method and efficient way this paper proposed an algorithm, which can handle "big data". For such data intensive applications, the parallel distribution frame work or MapReduce [9,2,18] framework has recently attracted a lot of attention. Parallel distributed framework allows easy development of scalable parallel applications to process "big data" on large clusters of commodity machines. An ideal parallel distributed system should achieve a high degree of load balancing among the participating machines, and minimize the space, CPU and I/O time, and network transfer at each machine. There exist some recent works on skyline computation using MapReduce [16,14]. However, to the best of our knowledge there is no such MapReduce algorithm for the k-dominating query and the K-skyband query so far.

5   Performance Evaluation

We set up a cluster of 4 commodity PCs in a high speed Gigabit networks, each of which has an Intel Core i7 3.4GHz CPU, 4GB memory and Windows 8.0 OS. The machines are connected with a Gbps LAN connection. We compile the source codes under JDK 1.6. We conduct a series of experiments with different dimensionalities and data cardinalities to evaluate the effectiveness and efficiency of our proposed method. Since none of the existing methods can handle K-skyband and Top-k dominating queries at a time, as a result we failed to compare proposed method with other algorithms. Each experiment is repeated five times and the average result is considered for performance evaluation. Three data distributions are considered as follows:

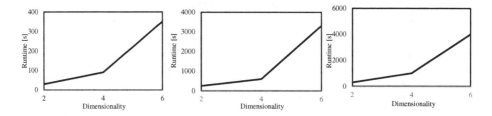

Fig. 6. Performance for different data dimension

Correlated: A correlated dataset represents an environment in which, objects are good in one dimension are also good in the other dimensions. In a correlated dataset, fairly few objects dominate many other objects.

Anti-Correlated: An anti-correlated dataset represents an environment in which, if an object has small coordinates on some dimensions, it tends to have large coordinates on other dimensions or at least another dimension.

Independent: For this type of dataset, all attribute values are generated independently using uniform distribution. Under this distribution, the total number of non-dominating objects is between that of the correlated and the anti-correlated datasets.

Effect of Dimensionality. We study the effect of dimensionality on our MapReduce technique. We fix the data cardinality to 100k and vary dataset dimensionality n ranges from 2 to 6. The run-time results for this experiment are shown in Figure 6(a), (b), and (c). The result shows that as the dimension increases the performance of propose method becomes slower. This is because for high dimension the number of non dominant objects increases and the performance become slower. The result on correlated data dataset is 10 times faster than independent data dataset. Where as it is 12 times faster than anti-coorelated data dataset.

Effect of Cardinality

For this experiment, we fix the data domensionality to 6 and vary dataset cardinality ranges from 75k to 300k. Figure 7(a), (b), and (c) shows the performance on correlated, independent, and anti-correlated datasets. Propose technique is highly affected by data cardinality. If the data cardinality increases then the performances decreases.

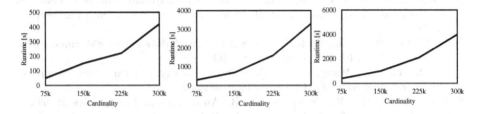

Fig. 7. Performance for different cardinality

6   Conclusion

This paper addresses a problem of selecting representative objects. We consider parallel distributed computation algorithm for K-skyband queries and top-k dominating queries to solve the problem. Extensive experiments demonstrate the efficiency of our algorithm for synthetic datasets

It is worthy of being mentioned that this work can be expanded in a number of directions. First, from the perspective of parallel computing, how to compute both queries from streaming dataset. Secondly, to design an efficient index based (R-tree/B-tree) parallel distributed method is promising research topics.

Acknowledgments. This work is supported by KAKENHI (23500180, 25.03040) Japan.

References

1. Balke, W.-T., Güntzer, U., Zheng, J.X.: Efficient distributed skylining for web information systems. In: Bertino, E., Christodoulakis, S., Plexousakis, D., Christophides, V., Koubarakis, M., Böhm, K. (eds.) EDBT 2004. LNCS, vol. 2992, pp. 256–273. Springer, Heidelberg (2004)
2. Blanas, S., Patel, J.M., Ercegovac, V., Rao, J., Shekita, E.J., Tian, Y.: A comparison of join algorithms for log processing in mapreduce. In: Proceedings of SIGMOD
3. Borzsonyi, S., Kossmann, D., Stocker, K.: The skyline operator. In: Proceedings of ICDE
4. Chan, C.Y., Jagadish, H.V., Tan, K.-L., Tung, A.K.H., Zhang, Z.: Finding k-dominant skyline in high dimensional space. In: Proceedings of ACM SIGMOD
5. Chan, C.-Y., Jagadish, H.V., Tan, K.-L., Tung, A.K.H., Zhang, Z.: On high dimensional skylines. In: Ioannidis, Y., Scholl, M.H., Schmidt, J.W., Matthes, F., Hatzopoulos, M., Böhm, K., Kemper, A., Grust, T., Böhm, C. (eds.) EDBT 2006. LNCS, vol. 3896, pp. 478–495. Springer, Heidelberg (2006)

6. Chomicki, J., Godfrey, P., Gryz, J., Liang, D.: Skyline with presorting. In: Proceedings of ICDE
7. Dellis, E., Seeger, B.: Efficient computation of reverse skyline queries. In: Proceedings of VLDB
8. Gong, Z., Sun, G.-Z., Yuan, J., Zhong, Y.: Efficient top-k query algorithms using K-skyband partition. In: Mueller, P., Cao, J.-N., Wang, C.-L. (eds.) INFOSCALE 2009. LNICST, vol. 18, pp. 288–305. Springer, Heidelberg (2009)
9. Jiang, D., Tung, A.K.H., Chen, G.: Map-join-reduce: Toward scalable and efficient data analysis on large clusters. IEEE Transactions Knowledge Data Engineering, TKDE (2011)
10. Li, C., Ooi, B.C., Tung, A.K.H., Wang, S.: Dada: A data cube for dominant relationship analysis. In: Proceedings of SIGMOD
11. Lin, X., Yuan, Y., Wang, W., Lu, H.: Stabbing the sky: Efficient skyline computation over sliding windows. In: Proceedings of ICDE
12. Papadias, D., Tao, Y., Fu, G., Seeger, B.: An optimal and progressive algorithm for skyline queries. In: Proceedings of SIGMOD
13. Papadias, D., Tao, Y., Fu, G., Seeger, B.: Progressive skyline computation in database systems. ACM Transactions on Database Systems (2005)
14. Park, Y., Min, J., Shim, K.: Parallel computation of skyline and reverse skyline queries using mapreduce. In: Proceedings of VLDB
15. Tan, K.-L., Eng, P.-K., Ooi, B.C.: Efficient progressive skyline computation. In: Proceedings of VLDB
16. Tao, Y., Lin, W., Xiao, X.: Minimal mapreduce algorithm. In: Proceedings of SIGMOD
17. Tao, Y., Xiao, X., Pei, J.: Subsky: Efficient computation of skylines in subspaces. In: Proceedings of ICDE
18. Vernica, R., Carey, M.J., Li, C.: Efficient parallel set-similarity joins using mapreduce. In: Proceedings of SIGMOD
19. Vlachou, A., Doulkeridis, C., Kotidis, Y., Vazirgiannis, M.: Skypeer: Efficient subspace skyline computation over distributed data. In: Proceedings of ICDE
20. Yiu, M.L., Mamoulis, N.: Efficient processing of top-k dominating queries on multi-dimensional data. In: Proceedings of VLDB

Towards Positive Unlabeled Learning for Parallel Data Mining: A Random Forest Framework

Chen Li[1,2] and Xue-Liang Hua[3,*]

[1] Department of Biochemistry and Molecular Biology, Monash University, VIC 3800, Australia
[2] College of Information Engineering, Northwest A&F University, Yangling 712100, China
Chen.Li@monash.edu
[3] Faculty of Information Technology, Monash University, VIC 3800, Australia
Xueliang.Hwa@gmail.com

Abstract. Parallel computing techniques can greatly facilitate traditional data mining algorithms to efficiently tackle learning tasks that are characterized by high computational complexity and huge amounts of data, to meet the requirement of real-world applications. However, most of these techniques require fully labeled training sets, which is a challenging requirement to meet. In order to address this problem, we investigate widely used Positive and Unlabeled (PU) learning algorithms including PU information gain and a newly developed PU Gini index combining with popular parallel computing framework - Random Forest (RF), thereby enabling parallel data mining to learn from only positive and unlabeled samples. The proposed framework, termed PURF (Positive Unlabeled Random Forest), is able to learn from positive and unlabeled instances and achieve comparable classifcation performance with RF trained by fully labeled data through parallel computing according to experiments on both synthetic and real-world UCI datasets. PURF is a promising framework that facilitates PU learning in parallel data mining and is anticipated to be useful framework in many real-world parallel computing applications with huge amounts of unlabeled data.

Keywords: PU information gain, PU Gini index, random forest, parallel data mining.

1 Introduction

Processing of huge amounts of data and high-complexity computations are the two emerging major challenges faced by many real-world applications of data mining in recent years. Accordingly, parallel computing techniques are developed and widely used to address data mining tasks that require time-consuming computation and processing of large databases by dispatching the learning task to several CPUs / jobs to allow separate running. A large number of experiments in different research areas and applications have benefited from the success and effectiveness of these developed parallel algorithms [1][2][3]. Among these, random forest (RF) [4] is one such attractive parallel algorithm that can be readily extended for parallel computing purposes by generating many trees in different CPUs / jobs.

---

* Corresponding author.

X. Luo, J.X. Yu, and Z. Li (Eds.): ADMA 2014, LNAI 8933, pp. 573–587, 2014.

Experiments on many real-world datasets have proved that RF can achieve satisfactory performance and flexible extendibility for parallel computing. However, to date, the majority of parallel algorithms are trained by fully labeled instances, while it is well known that labeling data is time-consuming and requires considerable effort. Prediction of post-translational modification sites (PTMs) provides is a good example that illustrates the limitation and incompatibility of traditional classifications that rely on fully labeled samples: First, millions of protein sequences from different species are available; Second, identifying (labeling) PTMs in each of these proteins requires extensive experimental efforts; and finally, researchers sometimes are more interested in certain types of PTMs (e.g. phosphorylation sites) that might be better regarded as positive samples for data mining purposes.

With the goal of tackling the tasks described above, in this study, we propose a novel computational framework, termed PURF (Positive Unlabeled Random Forest), for parallel computing to learn from positive and unlabeled samples effectively. This framework combines PU learning techniques including widely used PU information gain (PURF-IG) [5] and newly developed PU Gini index (PURF-GI) with an extendable parallel computing algorithm (i.e. RF). Empirical experiments performed on both synthetic and real-world UCI datasets indicate that both these two PU learning techniques perform comparably with RF model trained by fully labeled data, suggesting that in the case of parallel computing, PURF has a strong capability to learn data from large amounts of samples with unknown class and lack of labeled negative samples. To the best of our knowledge, this work represents the first study of positive and unlabeled learning with both PU information gain and PU Gini index for the parallel data mining scenario.

The rest of this paper is organized as follows: Section 2 will review some representative works in terms of PU Learning and parallel computing. A new PU Gini index will be proposed in Section 3. The corresponding algorithms for PURF and its parallel implementation will be described in Section 4. We conduct experiments to compare the performance of supervised RF, PURF-GI and PURF-IG in Section 5, which is followed by conclusions and future work drawn in Section 6.

2   Related Works

We briefly summarize three major directions in this area: **(1) Single classifier for PU learning.** Denis *et al.* developed a new information gain-based method with only positive and unlabeled data and applied it to decision tree [5]. Similar technique was also developed with Naive Bayesian learner in [6]. **(2) Strategies for PU learning (such as two-step strategy and unlabeled sample weighting).** In this direction, traditional classifiers or algorithms will not be changed or modified. Elkan *et al.* developed a method to weigh unlabeled data with the assumption that training data are an incomplete set of positive and unlabeled data [7]. Margin maximization in SVM was also used to incrementally label negative data with the aid of MC (Mapping Convergence) proposed by Yu. [8]. For text classification, a two-step strategy was widely used, with which reliable negative documents were extracted from unlabeled documents and used to train a classifier together with labeled positive documents [9][10][11]. Then previously labeled negative documents could be refined and more reliable negative documents would be retrieved. This process would not stop until a satisfactory threshold was reached.

On the other hand, the research works of parallel data mining techniques can be categorized into following three groups: **(1) Modification of traditional data mining techniques.** For apriori association rules mining, there exist three types of methods, including Count Distribution, Data Distribution and Candidate Distribution [2][12][13]. Zaki *et al.* proposed a parallel algorithm for vertical associate rules mining based on Eclat, in order to aggregate and take advantage of limited computing resources [14]. FP-tree [15] is another classic model to mine association rules. Two algorithms have been developed to grow multiple FP-trees in a parallel manner to different parts of transactions [16][17]. Moreover, Parthasarathy *et al.* proposed a new parallel association rules mining technique in shared-memory system [1]. To tackle imbalanced data mining task, Cheung *et al.* proposed a novel parallel algorithm using a distributed nothing-parallel system to explore association rules [18]. **(2) Parallel computing in computer applications.** For graphics processing, Li *et al.* developed three methods based on Compute Unified Device Architecture (CUDA) to accelerate three data mining algorithms (CU-Apriori, CU-KNN and CU-K-means) in CUDA [2]. For semantic retrieval, Chen *et al.* proposed a parallel computing approach to build engineering concept space, thereby addressing the scalability issue related to large-scale information retrieval [3]. **(3) Parallel computing in bioinformatics.** Parallel computing has been widely used in bioinformatics and computational biology, such as molecular dynamic simulation [19][20] and tool development for biological data analysis [21].

It should be noted that for most of the parallel algorithms, the models are trained by fully labeled data that is difficult to collect in real-world applications.

3 Positive Unlabeled Gini Index

As mentioned in our previous study [22], the PU learning task can be described as follows. Let us write a dataset S ($|S| = n$) that has only positive and unlabeled samples, $S = \{s_1, s_2, \ldots, s_n\}$, where $s_i = <A_i, C_i, y_i>$ is a sample in $S(1 \leq i \leq n)$. Here, $A_i = \{a_{i1}, a_{i2}, \ldots, a_{im}\}$ is an attribute vector with m attributes; C_i denotes the class label of s_i (positive class: $C_i = +1$; negative class: $C_i = -1$); y_i represents whether C_i is known to the model (class known: $y_i = +1$; class unknown: $y_i = -1$). Since for positive unlabeled learning, only positive samples are labeled, if $y_i = +1$, then $C_i = +1$; while if $y_i = -1$, the true class label of s_i is unknown. The PU learning aims at distinguishing positive class ($C_i = +1$) from non-positive class ($C_i \neq +1$) in a given dataset. In the case of binary classification, non-positive class is the negative class. While in the case of multi-class classification, non-positive class can be the set of all classes except the positive class.

Random forest [4] is an ensemble of decision trees built on random bootstrapping of training datasets. Final classification decision of a test sample is determined by the average value of the possibility of all classes. Since RF uses Gini index [23] to decide the split criteria, inspired by estimation method of proportion of positive and negative samples proposed by Denis *et al.* [5], we propose the PU Gini index in this study that explores only positive and unlabeled samples.

Given a learning task with only positive and unlabeled samples, let $PosLevel$ denote the proportion of positive samples in the learning task and S_{node} as the dataset filtered in the current node node in tree T. Note that $PosLevel$ is the proportion of positive samples in this learning scenario. Therefore, $PosLevel$ is unknown and can vary. The details of $PosLevel$ will be discussed in the following section. According to the definition of Gini index, we have

$$PUGini(S_{node}) = 1 - \sum_{i=1}^{|C|} p_i^2. \tag{1}$$

Under the PU learning scenario, there are only two classes: positive class and non-positive class. Let p_1 and p_2 denote the estimates of proportion of positive and non-positive samples in S_{node}, respectively. According to the PU information gain in [5], we have

$$p_1 = \min\{\frac{|POS_{node}|}{|POS|} \times PosLevel \times \frac{|UNL|}{|UNL_{node}|}, 1\}$$

$$p_2 = 1 - p_1, \tag{2}$$

where $|POS|$ and $|UNL|$ are the numbers of positive and unlabeled samples in the training dataset, respectively. $|POS_{node}|$ and $|UNL_{node}|$ are the numbers of positive and unlabeled samples filtered into the current node, respectively. Then for each attribute $a_j (1 \leq j \leq m)$ in A, according to the split point, S_{node} can be divided into two subsets, $S_{node}(a_j \leq splitPoint)$ and $S_{node}(a_j > splitPoint)$. Therefore we have

$$PUGini_{a_j} = \frac{|S_{node}(a_j \leq splitPoint)|}{|S_{node}|} \times$$
$$PUGini(S_{node}(a_j \leq splitPoint)) +$$
$$\frac{|S_{node}(a_j > splitPoint)|}{|S_{node}|} \times$$
$$PUGini(S_{node}(a_j > splitPoint)). \tag{3}$$

Finally, for the current node with the attribute $a_j (a_j \in A)$, the PU Gini can be described as follows

$$\Delta PUGini_{a_j}(S_{node}) = PUGini(S_{node}) - PUGini_{a_j}. \tag{4}$$

Among m attributes, the one with the maximal $\Delta PUGini_{a_j}(S_{node})$ should be chosen as the splitting attribute.

4   Positive Unlabeled Random Forest

4.1   Framework

In this section, we describe the developed framework based on RF to learn from positive and unlabeled samples. In particular, PURF uses bootstrapping (with replacement) to select the training samples for each tree in forest. Therefore after bootstrapping, approximately 2/3 of samples are collected as the training data. In each PU tree T_i in forest, there are nine sub-trees built in accordance with different values of $PosLevel$ (ranging from 0.1 to 0.9 with a step size of 0.1). Then the rest 1/3 of the data will be used to determine the final output of each tree T_i in forest from the nine subtrees. Algorithm 1 shows the framework of the proposed PURF. Step 1 is used to

---

Algorithm 1. PURF Framework

Input:

        training dataset including positive and unlabeled samples, S;

        the feature space of S, A;

        the number of trees in forest, n_{tree};

Output:

        positive and unlabeled random forest F.

1: $F = \phi$;

2: **for** each $i \in [1, n_{tree}]$ **do**

3:     initialize a tree T_i with only one node (the root);

4:     $S_{train} = Bootstrapping(S)$ [4];

5:     $S_{validate} = S - S_{train}$;

6:     **for** each $j \in [1, 9]$ **do**

7:        $PosLevel = \frac{j}{10}$;

8:        $T_{ij} = BuildTree(T_{ij}, PosLevel, S_{train}, A)$;

9:        $T_i = T_i \bigcup T_{ij}$

10:       $T_\theta = GetBestTree(T_i, S_{validate})$

11:    **end for**

12:    $F = F \bigcup T_\theta$;

13: **end for**

14: **return** F;

---

initiate the forest F, $F = \{T_1, T_2, \ldots, T_{n_{tree}}\}$. From steps 2 to 13, a forest F with n_{tree} trees will be generated and returned. As mentioned above, each tree $T_i (1 \leq i \leq n_{tree})$ in F has nine sub-trees with different $PosLevel$ values. The function $BuildTree(T_{ij}, PosLevel, S_{train}, A)(1 \leq j \leq 9)$ is used to generate sub-trees for T_i. Then one of the nine trees will be chosen as the best tree according to the estimate value by $GetBestTree(T_i, S_{validate})$, which will be discussed in the next section. Details of the $BuildTree(T_{ij}, PosLevel, S_{train}, A)$ are shown in Algorithm 2. The key part of Algorithm 2 is to calculate the PU Gini index according to formula (4) or PU information gain [5]. Then the attribute in A with the maximal PU Gini index will be chosen as the splitting attribute. The growth of the tree halts when one of the two criteria is met: (1) the samples filtered into this node are 'pure' (i.e., positive or unlabeled) or (2) the number of samples in the current node is smaller than a pre-set threshold. Then the class of the leaf is determined by p_1 and p_2 from formula (2).

4.2 Tree Selection

As discussed above, each PU decision tree $T = \{T_1, T_2, \ldots, T_9\}$ generated in the forest F has nine sub-trees based on nine values of $PosLevel$ and one sub-tree should

---

Algorithm 2. $BuildTree(T, PosLevel, S_{train}, A)$

// Refer to Algorithm 1 for the details of input parameters;

1: Generate new feature subspace $A' \in A$ by random sampling with replacement;
2: **for** each $A'_m \in A$ **do**
3:    Calculate PU Gini index (formula (4)) or PU information gain according to [5];
4: **end for**
5: Choose attribute A_{split} with the maximal PU Gini index / PU information gain value as splitting attribute;
6: Compute the splitting value $splittingPoint$ for A_{split};
7: Generate child nodes splitting from $splittingPoint$;
8: **for** each child node **do**
9:    $S_{train} = getDataInCurrentNode()$;
10:    $BuildTree(T, PosLevel, S_{train}, A)$;
11: **end for**

---

be chosen as a representative decision tree of T in the forest F. Here, based on [5] and our previous study [22], we use the following four statistics to calculate $e(T)$ to choose the best sub-tree:

1. Number of positive samples in $S_{validate}$, n_{POS};
2. Number of unlabeled samples in $S_{validate}$, n_{UNL};
3. Number of positive samples to be predicted as non-positive, $n_{POS \longrightarrow NP}$;
4. Number of unlabeled samples to be predicted as positive, $n_{UNL \longrightarrow POS}$.

Then for each T_k in T,

$$e(T_k) = \frac{n_{POS \longrightarrow NP}}{n_{POS}} + \frac{n_{UNL \longrightarrow POS}}{n_{UNL}}. \tag{5}$$

Finally, the output of T, T_θ is

$$\theta = \operatorname*{argmin}_{k}(e(T_k)). \tag{6}$$

5   Empirical Study

In order to evaluate the performance of our proposed algorithm, we conducted benchmarking experiments using both synthetic and real-world datasets to assess the ability of PURF to learn from unlabeled samples and compared with RF trained using fully labeled data.

We have implemented PURF with Python[1] based on the scikit-learn package[2] and both two PU techniques including PU Gini index (PURF-GI) and PU information gain (PURF-IG) were tested as split functions. RF models used to compared with PURF-GI and PURF-IG were implemented with Gini index and information gain, respectively. The experiments were conducted on two Apple iMac machines with Intel Core i5 Quad

---

[1] http://www.python.org/
[2] http://scikit-learn.org/stable/

Core CPU, with 12G and 24 G RAM, respectively and one Apple MacBook Pro with Intel Core Duo CPU, and 4G RAM. The machine with Intel Core Duo CPU was only used to conduct the experiments of running time comparison based on different number of cores.

We measured the classification performance of PURF-GI, PURF-IG and RF using Accuracy, F1 and AUC, which are commonly used for measuring the performance of PU classifiers [7][22]. Meanwhile, the time and space complexity were also evaluated by calculating the running time and counting the number of leaves and inner nodes of both algorithms.

5.1 Datasets and Performance Evaluation

Both synthetic and real-world datasets were used to evaluate the performance of our proposed PURF algorithm. We used a waveform generator to construct synthetic data, while the three real-world datasets (Spambase[3], Breast Cancer[4] and Ionosphere[5]) described in Table 1 were downloaded from the UCI website.

Table 1. real-world datasets

Dataset	#Ins.	#Attr.	Positive class	#Positive ins.
Spambase	4601	57	spam	1813
Breast cancer	699	9	malignant	241
Ionosphere	351	34	g	225
			b	126

For transferring from fully labeled data to positive and unlabeled data, we defined the probability $transfer(s)$ that one positive instance s becomes unlabeled with the given percentage of unlabeled data $ratio(UNL)$ and the assumption that $ratio(UNL)$ is larger than the proportion of negative samples in the training dataset:

$$transfer(s) = \frac{|S_{train}| \times (1 - ratio(UNL))}{|S_{train_{pos}}|}, \qquad (7)$$

where $|S_{train}|$ is the number of training samples, while $|S_{train_{pos}}|$ represents the number of positive samples in $|S_{train}|$. If $r > transfer(s)$, s will be transferred to unlabeled sample.

10-fold cross-validation was applied to all experiments conducted in this study. In each fold, we generated PU datasets for $ratio(UNL) = 70\%$, 80% and 90% using the fully labeled dataset with formula (7). Therefore, the performance of PURF-GI/PURF-IG and RF reported in this paper were averaged over the forest generated on these fully labeled and PU datasets of the ten folds. In addition, all the experiments except the

---

[3] http://archive.ics.uci.edu/ml/datasets/Spambase
[4] http://archive.ics.uci.edu/ml/datasets/
Breast+Cancer+Wisconsin+(Original)
[5] http://archive.ics.uci.edu/ml/datasets/Ionosphere

parallel analysis were performed in a parallel manner with quad cores and 10 jobs. The number of jobs is a pre-defined parameter in the Python implementation[6] of RF.

5.2 Portrait of PURF-GI

In this section, we would like to apply both synthetic and real-world UCI datasets to summarise the character of PURF with newly developed PU Gini index in terms of different sizes of training data, tree selection method and time/space complexity.

Number of Samples n_{sample}. The purpose of this group of experiments is to examine the relationship between the number of training samples and the corresponding classification performance of PURF-GI and RF. To do this, we used Waveform Generator to build synthetic datasets. By default, datasets constructed by the waveform generator have three classes: 0, 1 and 2. In this section, every class was considered as positive class once, and the other two were combined as negative class. We constructed three different datasets 20 times whose $n_{sample} = 1,000, 5,000$ and $10,000$, respectively. When conducting experiments, we fixed $n_{tree} = 20$ and $ratio(UNL) = 70\%, 80\%$ and 90%, respectively. The results are shown in Fig. 1. With the increase of n_{sample}, the performance (in terms of Accuracy, AUC and F1) of both PURF-GI and RF generally increased. In general, despite the different $ratio(UNL)$, the performances of PURF-GI and RF were very close to each other, especially for AUC value. For example, when the positive class was 0 and $n_{sample} = 10,000$, the differences of Accuracy, AUC and F1 between PURF-GI and RF were 4.9%, 0.48% and 2.1%, respectively. Note that despite the $ratio(UNL) = 90\%$ (approximately $10,000 \times 0.9 \times 0.9 = 8,100$ samples were unlabeled in 10-fold cross-validation), our proposed PURF-GI still achieved a very close performance to that of RF trained using fully labeled data.

Experiments on Tree Selection. To check whether formulas (5) and (6) are capable of selecting best sub-tree among nine, we conducted a group of experiments based on three real-world UCI datasets with $ratio(UNL) = 80\%$ and $n_{tree} = 1$. For each sub-tree ($PosLevel$ from 0.1 to 0.9), we built and tested PURF-GI on 10 folds. The average Accuracy, F1 and AUC at certain $PosLevel$ are reported in Fig. 2. For each fold at a certain $PosLevel$, we recorded the real selected $PosLevel$ by formulas (5) and (6) and calculated the average values. Then we obtained nine average values of selected $PosLevel$, of which the ranges are shown in the dark areas of Fig. 2. It is clear that the dark areas are all around the $PosLevels$ with best performance, which shows the effectiveness of formulas (5) and (6) for selecting the best sub-tree.

Time and Space Complexity Analysis. In this section, we analysed the time and space complexity of PURF-GI and RF. All the experiments in this section were conducted with the Quad cores and 10 jobs. Fig. 3 shows the time consumed by PURF-GI and RF to generate models on the three real-world datasets with different n_{tree} and $ratio(UNL)$, respectively. It is obvious that, both PURF-GI and RF have linear time complexity. The reason that PURF-GI consumed more time than RF is that in each tree in the forest of PURF-GI, there are a total of nine sub-trees to generate. We also listed the average numbers of the inner nodes and leaves of single tree in both PURF-GI and

---

[6] https://pythonhosted.org/joblib/generated/joblib.Parallel.html

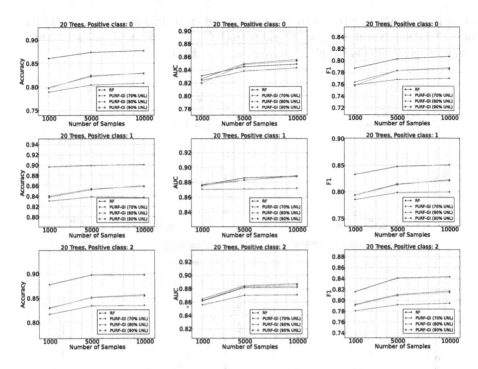

Fig. 1. Experimental results of n_{sample} with different positive classes for PURF-GI

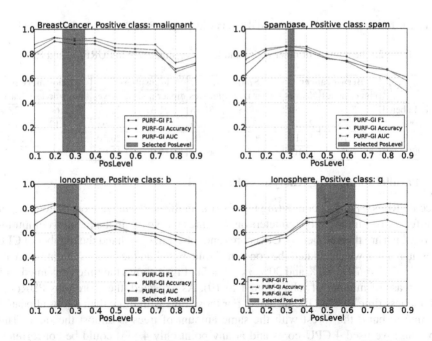

Fig. 2. Experimental results of tree selection on three UCI datasets for PURF-GI

RF in Table 2. All the results were generated based on 10-fold cross-validation of the three UCI datasets with $n_{tree} = 20$ and $ratio(UNL) = 70\%$, 80% and 90%. Since Gini index generates binary splits, $\#inner\ nodes = \#leaves - 1$.

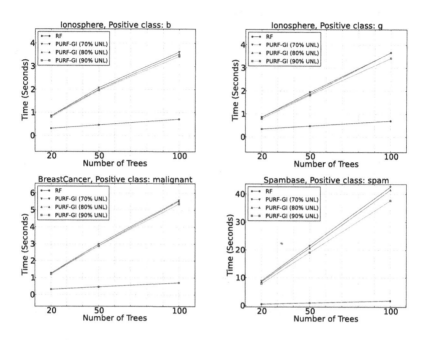

Fig. 3. Time consumed in real-world datasets with 10 jobs and Quad cores for PURF-GI

Table 2. The number of inner nodes and leaves of single tree in PURF-GI and RF

$ratio(UNL)$	Breast cancer		Spambase		Ionosphere_b		Ionosphere_g	
	#inner nodes	#leaves	#inner nodes	#leaves	#inner nodes	#leaves	#inner nodes	#leaves
Fully Labeled	25.50	26.50	260.48	261.48	21.37	22.37	21.64	22.64
70%	20.48	21.48	515.99	516.99	33.65	34.65	13.12	14.12
80%	11.10	12.10	126.80	127.80	19.30	20.30	9.38	10.38
90%	5.16	6.16	65.46	66.46	10.02	11.02	6.42	7.42

5.3   Parallel Computing Analysis of PURF-GI

Since PURF is a parallel computing framework, in this section, we went on to examine the effect of two important parameters on the time complexity in the parallel computing setup, which are the numbers of CPU cores and jobs. First, we fixed the number of CPU cores at 4. Then we calculated the consumed time based on the three UCI datasets with $ratio(UNL) = 70\%$, 80% and 90%, $n_{tree} = 50$. Fig. 4 shows the time consumed with the increase of number of jobs (from 1 to 10). As expected, the more jobs PURF-GI had, the less time PURF-GI consumed. For those runs with more than 4 jobs allocated, the time to build the forest with the same amount of trees is almost the same. This is because we used 4 CPU cores and at any point only 4 jobs could be concurrently processed.

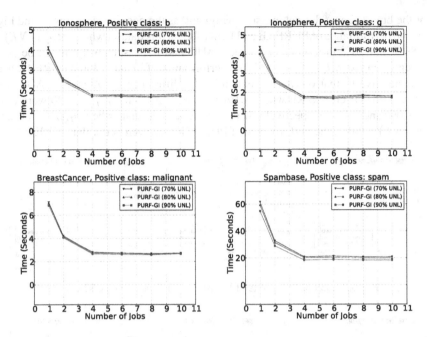

Fig. 4. Time consumed in real-world datasets with different numbers of jobs for PURF-GI

Table 3. Time (second) consumed with different numbers of CPU cores for PURF-GI

#CPU cores	$ratio(UNL)$	Spambase	Breast cancer	Ionosphere_b	Ionosphere_g
Core Duo	70%	59.447	7.557	4.647	4.765
	80%	58.169	7.493	4.535	4.737
	90%	53.652	7.349	4.365	4.567
Quad	70%	21.493	2.830	1.840	1.868
	80%	20.591	2.853	1.794	1.866
	90%	19.085	2.750	1.740	1.773

Then, we fixed the number of jobs at 10 and changed the number of CPU cores from 2 to 4. Then we run PURF-GI on the UCI datasets with $n_{tree} = 50$. The results are listed in Table 3. The first column is the number of CPU cores. The $ratio(UNL)$ is shown in the second column. Time consumed is shown for each dataset (for the ionosphere data, there were two positive classes: b and g.) in columns 3 to 6. These results suggest that PURF can be appropriately applied in the parallel computing scenario.

5.4 PURF-GI/PURF-IG *versus* RF on Real-World Datasets

To illustrate the scalability of PURF to real-world applications, in this section, we performed several experiments using UCI datasets to compare the performance of PURF-GI/PURF-IG to RF. We actually built up separate 10-fold cross-validation sets for the performance comparison of PURF-GI/PURG-IG with RF for each dataset, respectively. The $ratio(UNL)$ ranged from 70% to 90% and $n_{tree} = 20, 50$ and 100.

For the breast cancer dataset, we set 'malignant' as the positive class. Fig. 5 and Fig. 6 show the performance of PURF-GI and PURF-IG on this dataset when $ratio(UNL) = 70\%$, 80% and 90% in terms of Accuracy, AUC and F1, respectively. It can be seen that, even when $ratio(UNL) = 90\%$, the performance of PURF-GI and PURF-IG was still comparative to that of RF. For example, the differences of Accuracy, AUC and F1 between PURF-GI and RF when $ratio(UNL) = 90\%$ and $n_{tree} = 20$ were only 1.2%, 1.0% and 1.8%, respectively. It is also noticeable that, when $n_{tree} = 50$ and $ratio(UNL) = 80\%$, the performance of PURF-GI was even better that of RF.

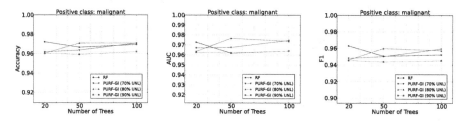

Fig. 5. Experimental results of Breast Cancer dataset for PURF-GI

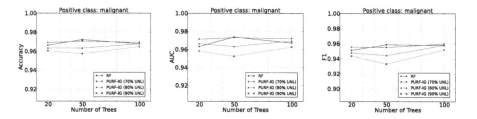

Fig. 6. Experimental results of Breast Cancer dataset for PURF-IG

For the Spambase dataset, 'spam' was set to be the positive class. Fig. 7 and Fig. 8 display the performance of PURF-GI/PURF-IG vs. RF when $ratio(UNL) = 70\%$, 80% and 90%, respectively. Similarly, both PURF-GI and PURF-IG achieved satisfactory performance when compared with RF based on this dataset. For instance, the difference of accuracy between PURF-GI and RF when $n_{tree} = 50$ and $ratio(UNL) = 80\%$ was only 3.1%.

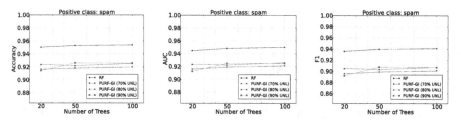

Fig. 7. Experimental results of Spambase dataset for PURF-GI

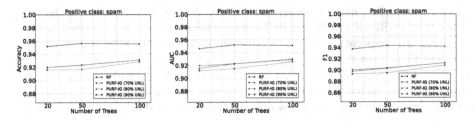

Fig. 8. Experimental results of Spambase dataset for PURF-IG

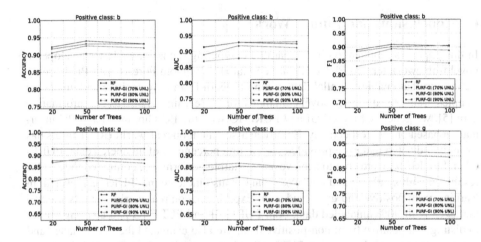

Fig. 9. Experimental results of Ionosphere dataset for PURF-GI

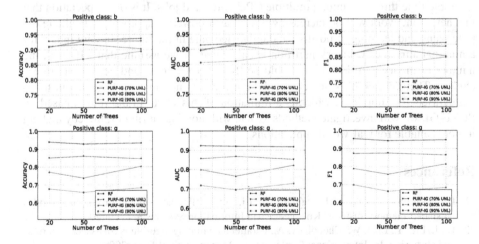

Fig. 10. Experimental results of Ionosphere dataset for PURF-IG

For the ionosphere dataset, we set each class as the positive class in turn. The performance of PURF-GI *vs.* RF and PURF-IG *vs.* RF on this dataset is shown in Fig. 9 and Fig. 10, respectively. When 'b' was set as the positive class, the Accuracy, AUC and F1 of both PURF-GI and PURF-IG were very close to those of RF with $ratio(UNL) = 70\%$ and 80%. When $ratio(UNL) = 90\%$, the performance dropped down. A possible reason for this is that the number of positive samples was very limited in this dataset - there were only a few positive samples that led to poor performance of both PURF-GI and PURF-IG. On the other hand, when 'g' was set to be the positive class, the performance was generally poor.

6   Conclusions and Future Work

In this paper, for application in the parallel computing scenario, we have developed a novel framework termed PURF (Positive unlabeled Random Forest) with PU learning techniques that enables parallel data mining to learn from only positive and unlabeled samples. Using the designed novel Gini index for PU learning as well as PU information gain [5], we generated PURF-GI and PURF-IG with the PU Gini index and PU information based on RF, respectively. Empirical assessment on real-world UCI datasets has strongly indicated that even provided with a high percentage of unlabeled data, PURF-GI and PURF-IG were able to achieve a accpetatble performance compared to RF. Our results on both synthetic and real-world dataset also indicate that PURF is powerful in learning from positive and unlabeled data. Experiments on real-world datasets showed that even with 90% unlabeled data, both PURF-GI and PURF-IG had a strong ability to distinguish positive from non-positive data. We also evaluated the performance and consumed time for calculation of PURF-GI in both parallel and non-parallel situations. As expected, time analysis demonstrated that parallel PURF could indeed save considerable time through running multiple CPU cores and jobs. It is our expectation that this new framework will be increasingly used as a powerful approach to facilitate the processing and learning of positive and unlabeled data in the future. In Gini index, for a nominal attribute A, binary partition is calculated. If we assume that $|A| = n$, the number of binary partition of A should be $2^n - 2$ (power set and empty set are excluded). Therefore in this research, we did not consider nominal attributes in our training datasets for simplicity, which is a limitation of this work. In the future, based on PURF-GI, we will investigate methods to deal with nominal attributes, thereby making it more suitable for real-world applications.

References

1. Parthasarathy, S., Zaki, M.J., Ogihara, M., Li, W.: Parallel data mining for association rules on shared-memory systems. Knowledge and Information Systems 3, 1–29 (2001)
2. Li, J., Liu, Y., Liao, W., Choudhary, A.: Parallel data mining algorithms for association rules and clustering. In: International Conference on Management of Data (2008)
3. Chen, H., Schatz, B., Ng, T., Martinez, J., Kirchhoff, A., Lin, C.: A parallel computing approach to creating engineering concept spaces for semantic retrieval: The Illinois digital library initiative project. IEEE Transactions on Pattern Analysis and Machine Intelligence 18, 771–782 (1996)

4. Breiman, L.: Random forests. Machine Learning 45, 5–32 (2001)
5. Letouzey, F., Denis, F., Gilleron, R.: Learning from positive and unlabeled examples. In: Arimura, H., Sharma, A.K., Jain, S. (eds.) ALT 2000. LNCS (LNAI), vol. 1968, pp. 71–83. Springer, Heidelberg (2000)
6. Calvo, B., Larranaga, P., Lozano, J.A.: Learning Bayesian classifiers from positive and unlabeled examples. Pattern Recognition Letters 28, 2375–2384 (2007)
7. Elkan, C., Noto, K.: Learning classifiers from only positive and unlabeled data. In: Proceedings of the Fourteenth ACM SIGKDD International Conference on Knowledge Discovery and Data Mining (SIGKDD 2008), pp. 213–220 (2008)
8. Yu, H.: Single-Class Classification with Mapping Convergence. Machine Learning 61, 49–69 (2005)
9. Liu, B., Dai, Y., Li, X., Lee, W.S., Yu, P.S.: Building text classifiers using positive and unlabeled examples. In: Proceedings of the Third IEEE International Conference on Data Mining (ICDM 2003), pp. 179–186 (2003)
10. Fung, G.P.C., Yu, J.X., Lu, H., Yu, P.S.: Text classification without negative examples revisit. IEEE Transactions on Knowledge and Data Engineering 18, 6–20 (2006)
11. Yu, H., Han, J., Chang, K.C.C.: PEBL: web page classification without negative examples. IEEE Transactions on Knowledge and Data Engineering 16, 70–81 (2004)
12. Agrawal, R., Shafer, J.C.: Parallel mining of association rules. IEEE Transactions on Knowledge and Data Engineering 8, 962–969 (1996)
13. Han, E., Karypis, G., Kumar, V.: Scalable parallel data mining for association rules, vol. 26. ACM (1997)
14. Zaki, M.J., Parthasarathy, S., Li, W.: A localized algorithm for parallel association mining. In: Proceedings of the Ninth Annual ACM Symposium on Parallel Algorithms and Architectures, pp. 321–330 (1997)
15. Han, J., Pei, J., Yin, Y.: Mining frequent patterns without candidate generation. ACM SIGMOD Record 29, 1–12 (2000)
16. Zaïane, O.R., El-Hajj, M., Lu, P.: Fast parallel association rule mining without candidacy generation. In: Proceedings IEEE International Conference on Data Mining (ICDM 2001), pp. 665–668 (2001)
17. Pramudiono, I., Kitsuregawa, M.: Tree structure based parallel frequent pattern mining on PC cluster. In: Mařík, V., Štěpánková, O., Retschitzegger, W. (eds.) DEXA 2003. LNCS, vol. 2736, pp. 537–547. Springer, Heidelberg (2003)
18. Cheung, D.W., Lee, S.D., Xiao, Y.: Effect of data skewness and workload balance in parallel data mining. IEEE Transactions on Knowledge and Data Engineering 14, 498–514 (2002)
19. Kalé, L., Skeel, R., Bhandarkar, M., Brunner, R., Gursoy, A., Krawetz, N., Phillips, J., Shinozaki, A., Varadarajan, K., Schulten, K.: NAMD2: greater scalability for parallel molecular dynamics. Journal of Computational Physics 151, 283–312 (1999)
20. Sanbonmatsu, K.Y., Tung, C.S.: High performance computing in biology: multimillion atom simulations of nanoscale systems. Journal of Structural Biology 157, 470–480 (2007)
21. D'Agostino, N., Aversano, M., Chiusano, M.L.: ParPEST: a pipeline for EST data analysis based on parallel computing. BMC Bioinformatics 6, S9 (2005)
22. Li, C., Zhang, Y., Li, X.: OcVFDT: one-class very fast decision tree for one-class classification of data streams. In: Proceedings of the Third International Workshop on Knowledge Discovery from Sensor Data (SensorKDD 2009), pp. 79–86 (2009)
23. Steinberg, D., Colla, P.: CART: tree-structured non-parametric data analysis. Salford Systems, San Diego (1995)

Program Verification by Reachability Searching over Dynamic Call Tree

Tu Peng and Kai Wang

School of Software, Beijing Institute of Technology, China
pengtu@bit.edu.cn, kevinkitty7@163.com

Abstract. Dynamic call graph represents runtime calls between entities in a program. Existed studies have used call graph to facilitate program comprehension and verification. However, the dynamic call graph produced by a program execution is complicated, especially when multithreads, loops and recursions are involved. In this paper, we retrieve dynamic call graph from program execution and transform it to call tree, and provide an approach of tree simplification by reducing loops and recursions. We formally define reachability properties over a call tree and reachability based tree isomorphism. We prove the soundness of tree simplification and the applicability to transform safety concerns verification to reachability properties searching. We implement the Dynamic Program Analyzer, and show how the behaviors of multithread programs can be retrieved, comprehended and verified.

Keywords: dynamic call tree, reachability, safety.

1 Introduction

With the wide deployment of software in safety critical industries, improving software reliability has become an important yet challenging task due to increasing software size and complexity, incomplete and incorrect documentation. Studies [15][4] have shown that design flaws and program errors are commonly the main source of security holes that are explored by attackers. Many software testing and verification techniques are thus developed to detect design flaws and program errors.

Many previous studies have focused on the verification of software designs and implementations. Studies [13][4] have shown that pattern based software designs can be formalized and subject to automated verification. UML diagrams represent detailed software designs. Many studies have focused on verification of UML diagrams [7][9][11]. Those studies have aimed at proving correctness of a program by model checking. Model checker CWBNC [2] is used to verifying the CCS [12] specifications. Process Meta language, PROMELA [6] is used to specify UML sequence diagram, and SPIN is used to verifying PROMELA specifications [11].

The above approaches, however, are built based on formal methods which require a deep learning curve and mathematical knowledge for a software developer to comprehend. To alleviate the cognitive loads, call graphs are used as the model of software for analysis and comprehension [13][15]. Dynamic call graph represents calls

X. Luo, J.X. Yu, and Z. Li (Eds.): ADMA 2014, LNAI 8933, pp. 588–601, 2014.

between entities in a program execution, which truthfully represent program execution, while remains a suitable abstracted model subjecting to analysis and comprehend. Dynamic call graph is obtained by execution trace mining [11][5]. Execution trace can be produced by programming instrumentation with AspectJ [8], which is a Java implementation of aspect oriented programming. A study has used [15]graph grammar reduction to verify the runtime call graph against certain expected properties expressed as graph production rules. A method to localize software defect through dynamic call graph mining and matching is proposed [14]. Patterns representing the defect free execution of the target features are identified. Then a dynamic call tree of a failure execution is contrasted with relevant patterns to identify mismatches, which point to calls in source codes that indicate the cause of the failure.

As an enhancement to previous dynamic call graph mining techniques [14][15], we identify thread information of each entity call, hence the dynamic call graph of a multithread program can be truthfully recorded. In addition, this paper alleviates the loads of call graph mining by reducing calls occurred repeatedly inside loops. More precisely, we use an abstracted call graph instead of original dynamic call graph. The abstracted call graph contains fewer nodes while remains indistinguishable to original dynamic call graph in terms of graph searching.

While existed studies aim at call graph analysis, we transform call graph to tree and design efficient reachability searching over the call tree. Successful approaches have been made on program verification by graph grammar reduction[15] and fault localization by graph mining[14]. Graph grammar reduction can verify whether the program behaviors satisfy given patterns represented as graph grammar, but cannot verify reachability properties. In contrast, our study focuses on verifying whether given reachability properties are satisfied. We transform the retrieved call graph to call tree, which has a simplified structure yet maintains complete information of call branches. We formally define a variety of reachability properties over the call tree and develop lemmas and algorithms for efficient searching. Finally, we implement DPA, the integrated environment for Dynamic Program Analyzing.

The rest of the paper is organized as follows: Section 2 is an overview of our approach. Section 3 introduces the transformation of dynamic call graph to tree and tree simplification approaches. Section 4 presents formal definitions and lemmas for reachability properties, tree isomorphism and search algorithms. Section 4 also provides experiments to demonstrate the capability of our approach to find and correct loopholes in programs. Section 5 introduces related work. Finally, section 6 concludes this paper.

2   Approach Overivew

The approach overview is depicted in Fig. 2 left. Our implementation includes four modules: AspectJ, Miner, Visualize and Verifier. In the beginning, we use AspectJ to weave instrumentation codes to the source codes. Then the source codes with test inputs are put into execution and the function calls are recorded in execution trace. The execution trace is processed by Miner to reduce irrelevant information and produce the dynamic call tree presenting the program behaviors. A Visualize is implemented to visualize the dynamic call graph, which helps the user to obtain intuitions of program behaviors. A verifier is implemented to automatically check whether the call tree satisfies certain safety and liveness properties. Verification results are showed by the Visualize.

3   Graph Mining

In order to transform the execution trace of a target program to a dynamic call graph, AspectJ is used to weave code segment to the target program. During every round of program execution, every function invocation will be recorded in execution trace. Each function invocation trace consists of the function caller name, the called function name, current thread identity and temporal order of the invocation (calling number). The execution trace is transformed to a dynamic call graph.

Definition 1. A dynamic call graph $G<V, E>$ is a labeled directed graph.   Each vertex $i \in V$ represents a function. Each edge $(i, j, t) \in E$ where $i, j \in V$ and non-negative integer t represents that function j is invoked by function i at time point t.

Time point t is not the exact time when the function is called; t denotes the order of function invocation in an execution of the program. In a single processor computer, each function has a unique time point in the execution. An example of dynamic call graph is shown in Fig. 1 left graph. Two edges $(M, A, 1)$ and $(M, A, 7)$ represent function M calls A at time 1 and 7 respectively.

Definition 2. (Dynamic Call Tree) Dynamic call graph $G_1<V, E_1>$ can be transformed into call tree $G<N, E>$ by the following rules:

1. $(M, 0) \in N$, where M denotes the main function
2. For $\forall j \in V$ and $(i_1, j, t_1), \cdots, (i_n, j, t_n) \in E$, $(j, t_i) \in N$
3. $E_1 = E$
    $(f, g, t) \in E$ is denoted as $f \mapsto g$.

A call tree has the same edges as the corresponding call graph. If a function is called n times, it is represented as one vertex in the call graph, while as n nodes in the call tree. A call tree can be searched, mined or matched more efficiently than graph.

Definition 3.   Let $G<N, E>$ be a call tree. Let f and g be two functions with $(f, t), (g, t') \in N$. g is reachable from f, denoted as $f \rightarrow g$, with the time range of $[t, t']$ and call depth of n-1, if there exists a sequence of functions $< f_1, f_2, \cdots, f_n >$ subject to

$$(f_i, t_i) \in N, \quad f_1 = f, \quad f_n = g, \quad f_i \mapsto f_{i+1}.$$

The middle graph in Fig. 1 shows a dynamic call tree, and the left graph shows its corresponding dynamic call graph. A call tree is obtained by marking each of the repeated invocations of one function with an individual node, labelled with the function name. For example, the call graph shows that function A is called three times at time 1, 5 and 7 respectively. The corresponding call tree has three nodes labelled as A, which is called at time 1, 5 and 7. The right graph shows a multi-thread program, where red color branch illustrates a second thread.

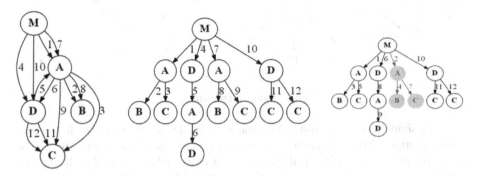

Fig. 1. Dynamic Call Graph

A dynamic call tree including loop and recursions is show in the right graph in Fig. 2. Loop and recursions create redundant nodes and edges in the call tree. To find useful information efficiently, we need to simplify the call tree. The simplification can be done in different level, with respect to different requirements of program verification.

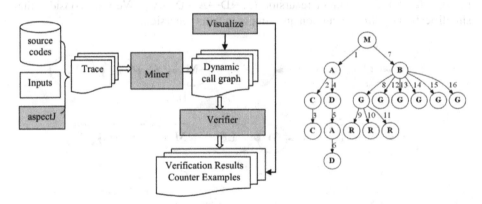

Fig. 2. Over view and Dynamic Call Tree-Loop and Recursion

To simplify loop, we exploit the idea of tree isomorphism: two trees with different number of nodes and edges, while representing the same runtime structure (branch information) and temporal order of invocations. In program verification, one wants to know the runtime structure (branch information) and temporal order of invocations, which must be reserved by the simplified tree. To obtain an isomorphic simplified call tree, we define virtual leaf nodes, which are used to maintain the branch information of the original tree. Fig. 3 *a.* shows examples of virtual leaf nodes which are marked as LF. We omit the time stamp information for simplicity.

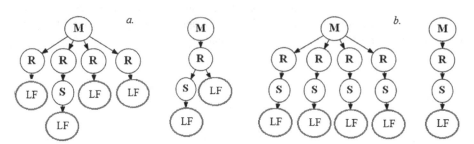

Fig. 3. Loop Simplification

In tree simplification, we use virtual leaf node to identify the call branch of a program execution. To better clarify our idea, an example is presented in Fig. 3. The left graph shows the original call tree, where a loop contains many M→R and one R→S. The right graph shows a simplified call tree, where repeated invocations of R are merged into one node. Meanwhile, the simplified call tree maintains S→LF and R→LF, which represent S is leaf node and some Rs are leaf nodes. Fig. 3 *b.* shows another example of loop simplification, where no R is a leaf node. Without virtual nodes, we will not be able to discriminate the left graphs from *a.* and *b.*, because their simplified trees would have been identical.

Recursion is the self invocations of a function. Recursions include direct recursion (C→C→C→C), and indirect recursion (A→D→A→D→A). We only consider the simplification of direct recursion and simple indirect recursion.

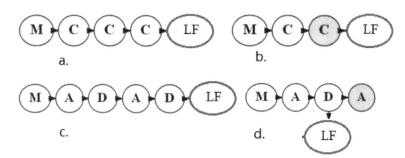

Fig. 4. Recursion Simplification

Fig. 4 illustrates recursion simplification. Direct recursion is simplified in the following way: one node represents the function called (blank node C in graph. b.) and one special node represents the recursion (dark node C in graph. b.). Indirect recursion is simplified in the following way: repeated calling sequence (A→D→A→D) is represented by one calling sequence (A→D) and one special node represents the recursion (dark node A in graph d.).

4   Verification of Program Behaviors

In this section, we formally define reachability properties of dynamic call tree. Reachable functions or call sequences may exist more than once in a call tree, e.g., two threads call the same function. It is important to distinguish different reachability properties in definition. Then we prove the tree simplification preserve reachability properties.

4.1   Reachability Isomorphism

To describe reachability over call tree, we extend propositional logics by introducing quantifiers.

Definition 4. The syntax of reachability formula is

$$R := A\,P\mid E\,P\mid P$$
$$P := G\,T\mid F\,T$$
$$T := f \rightarrow g\mid g\mid \neg g$$

where P is a predicate defining the reachability from function f to g in dynamic call graph. $G\,f \rightarrow g$ is true if and only if g is called on every path of f. $F\,f \rightarrow g$ is true if and only if g is called on at least one path of f. $E\,P$ is true if and only if reachability P is satisfied on at least one sub tree starting from f. $A\,P$ is true if and only if P is satisfied on all sub trees starting from f.

To be more specific, we formulate four types of reachability properties in Fig. 5:

- $A\,G\,f \rightarrow g$: for all calls of f, g is called globally (on all paths of f)   d.
- $A\,F\,f \rightarrow g$: for all calls of f,g is called finally (on at least one path of f)   c.
- $E\,G\,f \rightarrow g$: exist a call of f, g is called globally (on all paths of f)   b.
- $E\,F\,f \rightarrow g$: exist a call of f,g is called finally (on at least one path of f)   a.

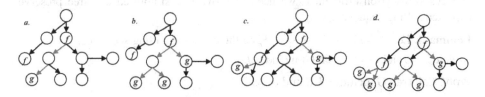

Fig. 5. Reachability Properties

Definition 5. (*Tree isomorphism*)   Let G and G_1 be two call trees. Let $Rp(G)$ and $Rp(G_1)$ be the set of reachability properties satisfied by G and G_1. Let $Rn(G)$ and $Rn(G_1)$ be the set of reachability properties not satisfied by G and G_1. Let $R(G)$ be the set of all reachability properties which can be defined among the nodes of G. Then G and G_1 are reachability *isomorphic tree*, denoted by $G \approx G_1$, only if $Rp(G)=Rp(G_1)$, $Rn(G)=Rn(G_1)$, and $Rp(G) \cup Rn(G) = R(G)$.

Lemma 1. Loop simplification tree illustrated in Fig. 3 *a.* preserves reachability properties over a call.

Proof. shows a general situation of loop, where function R is repeatedly called by M, and S is called sometimes by R. Let the left graph and right graph be denoted by G_L and G_R respectively. It is observable that:

$Rp(G_L)= Rp(G_R)=\{ A\ G\ M \rightarrow R, A\ F\ M \rightarrow R, E\ G\ M \rightarrow R, E\ F\ M \rightarrow R, E\ F\ M \rightarrow S$, $E\ F\ R \rightarrow S\}$
$Rn(G_L)= Rn(G_R)=\{ A\ F\ M \rightarrow S, E\ G\ M \rightarrow S, A\ G\ M \rightarrow S, A\ F\ R \rightarrow S, E\ G\ R \rightarrow S, A\ G\ R \rightarrow S\}$

Hence we have $G_L \approx G_R$, i.e., loop simplification preserves reachability properties.

Lemma 2. Recursion simplification illustrated in Fig. 4   preserves reachability properties over a call tree.

Proof. Let the call tree *a.* and *b.*   be denoted by G_A and G_B. It is observable that:

$Rp(G_A)= Rp(G_B)=\{ A\ G\ M \rightarrow R, A\ F\ M \rightarrow R, E\ G\ M \rightarrow R, E\ F\ M \rightarrow R\}$
$Rn(G_A)= Rn(G_B)=\{\}$

Let the call tree *c.* and *d.* be denoted by G_C and G_D. It is observable that:

$Rp(G_C)= Rp(G_D)=\{A\ G\ M \rightarrow A, A\ F\ M \rightarrow A, E\ G\ M \rightarrow A, E\ F\ M \rightarrow A, A\ G\ M \rightarrow D, A\ F\ M \rightarrow D, E\ G\ M \rightarrow D, E\ F\ M \rightarrow D, A\ G\ A \rightarrow D, A\ F\ A \rightarrow D, E\ G\ A \rightarrow D, E\ F\ A \rightarrow D\}$
$Rn(G_C)= Rn(G_D)=\{\}$

Hence we have $G_A \approx G_B$ and $G_C \approx G_D$, i.e., recursion simplification preserves reachability properties.

4.2   Verification of Expected Call Sequence and Safety Properties

In software verification, the commonly questioned safety concern is whether expected calling sequences exist in a program or not. In order to perform expected calling sequences verification efficiently, we need to prove that simplified call tree preserve expected calling sequence.

Lemma 3. Let $L = F_1 \rightarrow F_2 \rightarrow \cdots \rightarrow F_n$ be the expected calling sequence. If call tree G includes L and $G \approx G_I$ then G_1 includes L.

Proof. Call tree G includes L if and only if

$$Rp(G)= \{EF\ F_i \rightarrow F_j \,|1 \le i < n, 1 < j \le n, i < j\}$$

Since $G \approx G_I$, we have $Rp(G)=Rp(G_I)$, hence G_1 includes L.

Another important safety concern is categorized into safety and liveness properties. Safety properties are in the form: "something bad will never happen"[1]. For instance, in a model of a nuclear power plant, a safety property might be, that the operating temperature is always (invariantly) under a certain threshold, or that a meltdown never occurs. A variation of this property is that "something will possibly never happen". Liveness properties are of the form: something good will eventually happen, e.g. when pressing the on button of the remote control of the television, then eventually the television should turn on. Or in a model of a communication protocol, any message that has been sent should eventually be received. We formulate safety and liveness properties as reachability properties. Definition 4 actually defines four important liveness properties in terms of reachability properties. We summarize some important safety properties in the following lemma.

Lemma 4. *Safety* properties can be expressed as disjunction of negation of *liveness* properties: 1. $A\,Gf \rightarrow \neg g \equiv \neg (E\,Ff \rightarrow g) \lor \neg E\,Ff$ 2. $E\,Gf \rightarrow \neg g \equiv \neg (A\,Ff \rightarrow g) \lor \neg A\,Ff$.

Proof. By swapping negation with quantifiers, it is straightforward to obtain 1 and 2.

Lemma 3 proves the applicability to transform the verification of expected calling sequence against a program to the searching of reachability properties over a simplified call tree. Lemma 4 allows us to transform the verification of safety properties to the verification of liveness properties. Those lemmas prove the soundness of tree simplification and the applicability to transform safety concerns verification to reachability properties searching.

4.3   Reachability Searching Algorithms

We have designed algorithms to search reachability properties over dynamic call tree. In this paper, we present the search algorithm for $AG\,x \rightarrow y$, the most complicated reachability.

Definition 6.(*Algorithm AG*)
Input:
    $G<N, E>$, $(x,t_1),(y,t_2) \in N$
Output:
    whether it is true that for any branch of x, y is called globally.
Varibles:
    Id : the unique identifier of a node.
    Name: the function represented by the node and it may not be unique.
    CntLeaf : the number of children of a given node.
    IndexTable: the index table for nodes of the dynamic call graph, through which one have quick access for a node by referring its name. It is obtained by scanning the dynamic call graph.
Step 1: Seaching for all nodes named x and its *Cntleaf* within *IndexTable*, and recorded in a hash table *xMap= map<Id, CntLeaf>*
    for each node $(n, id\_n, CntLeaf)$ in *IndexTable*
        if (n equals to x)
            xMap=xMap+map<id_n, CntLeaf>
        end if
    end for

Step 2: Seaching for all nodes named y in *IndexTable*, for each node named y:
    Do a backward search from its parent node. If a node named x is met, then subtract *CntLeaf* of y from that of x, i.e., *CntLeaf[x] := CntLeaf[x] - 1*, update *xMap* correspondingly.

for each node (*n*, *id_n*, *CntLeaf*) in *IndexTable*
if   (*n equals to y*)
   repeat
     get the parent node *p* of *id_n*   until *p equals to x*
    end repeat
    if (*p* is not found)   **return** *false*;
   else
     set *xMap* with new value *map<id_p, CntLeaf*-1>
    end if
end if
end for

Step 3: Traverse *xMap*, if there exists a node whose *CntLeaf* value is greater than 0, return false, otherwise return true.
for each map *map<id, CntLeaf>* in *xMap*
   if (*CntLeaf*>0) **return** *false* **end if**
end for
return *true*

Lemma 5. Suppose *G<N, E>* is a call tree with $(x, t_1), (y, t_2) \in N$ and $|N| = n$. The time complexity of *AG x* → *y* is less than $2n + n \times \log_2 n$.

Proof. According to the *Algorithm AG*: in any case, the time complexity of step i) is *n*. Time complexity of step ii) is $n \times d$, where *d* in worst case is (the average depth of the tree) estimated as $\log_2 n$. In worst case, the time complexity of step iii) is *n*. Hence in worst case, the time complexity of *Algorithm AG* is estimated as $2n + n \times \log_2 n$.

In our implementation, *Algorithm AG* returns the range of search depth between *x* and *y* (the number of calls from *x* to *y*), and the range of time difference between x and y, besides the truth value.

4.4   Comprehension and Verification of Multi-Thread Programs

Our implementation, DPA (Dynamic Program Analyzer), integrated call tree mining, visualization and searching. We use an example to demonstrate the capability of DPA. The codes to analyze are a popular java project of open FTP server, JFTP, which can be found on sourceforge[17]. JFTP is designed to support secured FTP, which is able enable server to apply password authentication (represented by function *getStorePasswords*) during each conservation with client (represented by function *init*). The security concern of secured FTP is represented as a weak liveness property:

$$E \ F \ init {\rightarrow} getStorePasswords$$

We use DPA to analyze whether JFTP has successfully implemented the secured FTP. Fig. 6 shows a dynamic call tree of JFTP. Each node is labeled with a function name, prefixed by a thread number and ended with a call number, i.e., *thread_number@function_name#call_number*

Thread number denotes in which thread the function is called. Calling number denotes the temporal order by which the function is called. The *search* tab shows that we input *init* and *getStorePassword* as caller and callee. We could choose *simple* to search function name only, or we could deselect *simple* to include class and path information. The main tab shows the dynamic call graph. All invocations of init are marked in dark color and those of *getStorePasswords* are marked in dark color. The node labeled "*1@getStorePasswords#995*" denotes that function *getStorePasswords* is the 995th function called by thread1, in this program execution. The search results in search tab shows that the weak liveness property is satisfied: *E F init →getStorePasswords (true)*. However, strong liveness property is not satisfied: *A G init→getStorePasswords (false)*.

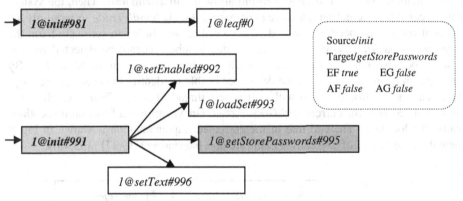

Fig. 6. Dynamic Program Analyzer--JFTP

The search results show that server is able to enable password authentication on certain conversations with a client, which satisfies requirements of a secured FTP server. There are still some conversations where password authentications are not activated, which can be explained as the server provides no-authenticated conversations for anonymous users. Additional information of a function, e.g., class path can also be shown. One can click any node on the call tree to view the additional information.

To further demonstrate the capability of DPA, we use it to analyze an implementation of a Bank Transaction System, whose originally design can be found on IBM website [16]. The bank transaction system allows the user to withdraw any amount of money from one's bank account. Password authentication and balance checking are required before a successful transaction. A user can initiate multiple withdraw actions online or locally. Each time a user starts a withdraw transaction, the system will authenticate his identity and check whether his balance is more than the withdraw amount, which is the critical security requirement. In order to verify this security requirement, we first use DPA to obtain an intuitive understanding of the program through its dynamic call tree. From dynamic call tree, we can identify the key function of withdraw transaction: *withDraw* starts the transaction by initiating a process of authenticating and balance checking, *handleTrade* completes the transaction by writing the after-trade balance into the database. The security requirement can be expressed as liveness property:

$$A\ F\ withDraw{\rightarrow}handleTrade$$

Searching result shows that this property has been satisfied, that is, every call of withdraw has been succeeded by *handleTrade*. However, Fig. 7 shows that one path of calling sequence has the time range of [24, 40], and another path of call sequence has the time range of [25, 41]. The interweaving time span indicates that the transaction represented by the two paths may interfere with each other. To be more specific, a malicious user could exploit this to over withdraw money by starting two transactions simultaneously. The system would allow the first withdraw. Then, the system may receive the second withdraw request before it calls *handleTrade* to write off the account balance reduced by the first withdraw. This loophole can seriously harm the interests of banking systems. Moreover, this loophole cannot be detected by unit testing, because the system has done each withdraw transaction successfully. By studying the dynamic call tree, the developer is able to identify the cause of the loophole: each *withDraw* function will start a new thread of *handleTrade*, which is not synchronized with the current *withDraw* thread. Unsynchronized function invocations cause the loophole. The call tree of the corrected implementation is shown in Fig. 8 here the time ranges of two calling paths are non-interfering ([19,27] , [35,43]).

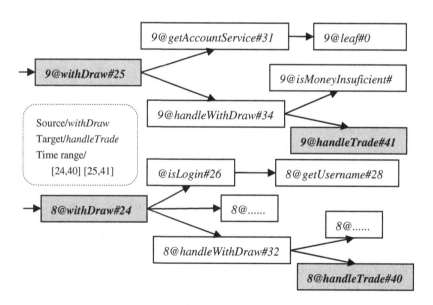

Fig. 7. Bank Transaction System

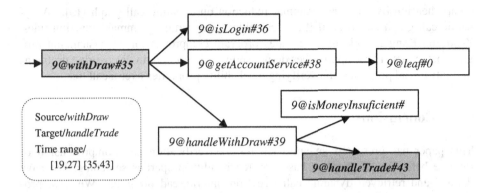

Fig. 8. Bank Transaction System Corrected

5   Related Work

There has been a considerable amount of studies on software verification, graph mining and analysis, formal specification, and aspect oriented programming.

Many studies have successfully used model checking as automated software verification techniques. A variety of model checking tools and specification languages are adopted with respect to the verification object, e.g., program, pieces of detailed designs, or general designs. Dong. has shown that pattern based software designs can be formalized in CCS expressions and subject to automated verification[13][4]. While Dong's work aimed at verifying pieces of software design, other studies have focused on verification of UML diagrams representing detailed software design. Li has advanced a formal semantics of UML sequence diagrams [9]. Lima has proposed a way of specifying a UML sequence diagrams in PROMELA [6], and used SPIN to verifying PROMELA specifications [11].   Although those studies have been proven successfully, they require transforming software system to certain specification language of the model checker.

Graph brings intuitions for program comprehension. Meanwhile, program verifications can be approached by graph grammar reduction or graph searching and matching. AspectJ is a Java implementation for AOP-Aspect Oriented Programming8. Feng has investigated the approaches of using of AspectJ to extract design information, call graph, and values of variables from a program [5]. Li has designed and developed a call graph analysis tool for AOP [10]. Kong has used graph grammar to specify the behavioral aspects of UML diagrams by graph transformations [7]. Obtaining call graph from programs is critical for graph based program verification. Past studies focused on static call graph, which represents dependency of articles, i.e, objects or functions, in programs. Murphy. has proposed a comprehensive study of static call graph exactors [13]. Dynamic call graph represents runtime program behaviors. Many existed studies used dynamic call graph for program verification [15][14]. Yousefi has proposed a way of defect localization based on dynamic call graph mining and matching. Zhao and Kang have advanced an approach for program behavior verification and comprehension by applying grammar reduction on dynamic call graph [15]. A dynamic call graph is correct if it is acceptable under certain grammar reduction rules. Zhao and Kang's work also facilitates program comprehension by visualizing program behaviors. Those approaches do not show their strength in discovering problems of multi-thread programs and verifying reachability properties over a call tree. .

6   Conclusions

This paper has presented an approach for the verification and comprehension of runtime behaviors of program based on reachability properties searching. We have defined and retrieved dynamic call tree from multithread programs. We have proposed tree simplification to reduce loops and recursions. We have formulated reachability properties over a dynamic call tree and defined tree isomorphism. We have proved the soundness of tree simplification and the applicability to transform verification of safety concerns to reachability properties searching. We have designed the algorithms to search reachability properties efficiently. We have shown how the behaviors of multithread programs can be retrieved, comprehended and verified. Moreover, the time range and call depth of reachability properties are analyzed to facilitate verification of multithread program.

Since the dynamic call tree in this paper is actually an ordered tree, we expect to exploit this feature to obtain correct/mistaken patterns from a series of program executions with machine learning techniques.   Then, the patterns can be used to forecast or locate potential characters of other programs with tree matching techniques.

Acknowledgement. This work is supported by the Natural Science Foundation of Beijing (4133088), the Foundation for Doctoral Program by the Ministry of Education (3080036621203), the Foundation for Fundamental Research by Beijing Institute of Technology(20120842004), and the Foundation for Overseas Students of China.

References

1. Behrmann, G., David, A., Larsen, K.G.: A Tutorial on Uppaal: Toolbox for Verification of Realtime System, Department of Computer Science, Aalborg University, Denmark
2. Cleaveland, R., Parrow, J., Steffen, B.: The Concurrency Workbench: A semantics-based tool for the verification of concurrent systems. ACM Trans. on Prog. Lang. and Systems 15(1), 36–72 (1993)
3. Dong, J., Alencar, P., Cowan, D.: A behavioral analysis and verification approach to pattern-based design composition. Software and Systems Modeling 3, 262–272 (2004)
4. Dong, J., Peng, T.: Automated Verfication of Design Pattern Compostiions. Information and Software Technology (IST) 53(3), 274–295 (2010)
5. Feng, X.: Analysis of AspectJ and its Applications in Reverse Engineering. Master Thesis of Software Engineering, Xian Electrical Science and Tehcnology University
6. Gerard, J.: Holzmann, The model checker SPIN. IEEE Transactions on Software Engineering 23(5), 279–295 (1997)

7. Kong, J., Zhang, K., Dong, J., Xua, D.X.: Specifying behavioral semantics of UML diagrams through graph transformations. Journal of Systems and Software 82(2), 292–306 (2009)
8. Kiczales, G., Hilsdale, E., Hugunin, J., Kersten, M., Palm, J., Griswold, W.G.: An Overview of AspectJ. In: Lindskov Knudsen, J. (ed.) ECOOP 2001. LNCS, vol. 2072, pp. 327–353. Springer, Heidelberg (2001)
9. Li, X.S., Liu, Z.M., He, J.F.: A Formal Semantics of UML Sequence Diagrams. In: The Proceedings of ASWEC 2004, Melbourne, Australia, pp. 13–16 (April 2004)
10. Li, N.: Call Graph Analysis Tool for Aspect Oriented Programs, Master Thesis, Shanghai Jiaotong University (2007)
11. Lima, V., Talhi, C., Mouheb, D., Debbabi, M., Wang, L.: Formal Verification and Validation of UML 2.0 Sequence Diagrams using Source and Destination of Messages. Electronic Notes in Theoretical Computer Science 254, 143–160 (2009)
12. Milner, R.: Communication and Concurrency. International Series in Computer Science. Prentice Hall (1989)
13. Murphy, G.C., Notkin, D., Lan, E.: An Empirial Study of Static Call Graph Extractors. In: ICSE (1996)
14. Yousefi, A., Wassyng, A.: A Call Graph Mining and Matching Based Defect Localization Technique. In: 2013 IEEE Sixth International Conference on Software Testing, Verification and Validation Workshops (2013)
15. Zhao, C., Kong, J., Zhang, K.: Program Behavior Discover and Verfication: A Graph Grammar Approach. IEEE Transaction on Software Engineering (2010)
16. UML basics of Sequence Diagrams, IBM developers Work (February 2004), http://www.ibm.com/developerworks/rational/library/3101.html
17. JFtp: a graphical network browser, http://sourceforge.net/projects/j-ftp/files/jftp/

CinHBa: A Secondary Index with Hotscore Caching Policy on Key-Value Data Store

Wei Ge[1,3], Yihua Huang[1], Di Zhao[1], Shengmei Luo[2], Chunfeng Yuan[1],
Wenhui Zhou[1], Yun Tang[1], and Juan Zhou[1]

[1] State Key Laboratory for Novel Software Technology,
Nanjing University,Nanjing, 210046, China
[2] ZTE Corporation, Nanjing, 210012, China
[3] Guangxi Normal University, Guilin, 541000, China
gloria.w.ge@gmail.com, yhuang@nju.edu.cn

Abstract. We are now entering the era of big data. HBase comes out to organize data as key-value pairs and support fast queries on rowkeys, but queries on non-rowkey column are a blind spot of HBase. It is the main topic of this paper to provide high-performance query capability on non-rowkey column. An effective secondary index model is proposed, and the prototype system CinHBa is implemented. Furthermore, a novel caching policy, Hotscore Algorithm, is introduced in CinHBa to cache hottest index data into memory to improve query performance. Experiment evaluation shows that query response time of CinHBa is far less than native HBase without secondary index on 10M records. Besides that, CinHBa has good data scalability.

Keywords: HBase, secondary index, memory cache, caching policy, key-value store.

1 Introduction

We are now entering the era of big data, where the volume of data outgrows the capabilities of relational query processing technology. Querying structured data, an area traditionally dominated by relational databases, are ignored by many emerging applications such as social networks, machine learning, graphics algorithms and scientific computing, which brings the demands of data analytics on PBs(10^{15}) or EBs(10^{18}) of data. For example, Facebook famously ran a proof of concept comparing several parallel relational database vendors before deciding to run their 2.5 petabyte clickstream data using Hadoop instead. Its Hive system ingests 15 terabytes of new data per day in 2009[1]. Hadoop and MapReduce process tasks in a scan-oriented fashion that seems simple. Nevertheless, the outstanding scalability and fault-tolerance of Hadoop gives big data processing unlimited possibility.

HBase is the open-source key-value data store project in Hadoop Ecosystem of Apache Foundation. Data in HBase are composed of much smaller entities in format of key-value pair, and HBase organizes the small records into large

X. Luo, J.X. Yu, and Z. Li (Eds.): ADMA 2014, LNAI 8933, pp. 602–615, 2014.

storage files and offer some sort of indexing as well. Key-value data stores are designed for fast point queries and sequential scans on rowkeys. It is suitable for reading adjacent key-value pairs and is optimized for block disk access operations that can make full use of disk transfer channels. Whereas, in HBase, query on non-rowkey column are supported by filter definition and sequential scan from start to end key. It's an inefficient way of full scans on table, and results in $\mathcal{O}(n)$ time cost. In contrast, the hash location of point query could obtain $\mathcal{O}(\log n)$ time cost or even $\mathcal{O}(1)$ by Bloom Filter in HBase. In relational database, query on non-key attribute is supported by indexing on it. Thus, we devote to building index on query column on HBase.

Increasing capacity and decreasing cost of RAM make memory resident data management an attractive complementation to disk-based solutions. Memory caching is a powerful method to fill the speed gap between CPU and disk. Caching technology is based on the data locality suppose: the recent accessed data are mostly revisited in the near future. On the other hand, data accesses have skew feature in big data scenario. It is commonly accepted that data access is obeyed to Zipf's Law. 80-20 rule is a simple expression of Zipf's Law, that means 80% data accesses focus on 20% hotspot data items. The data that are accessed frequently are called "hot"data, and oppositely, the data with few access frequencies are called "cold"data. It's a popular optimization approach for big data applications to identify hot and cold data and to conduct an effective caching policy for hot data.

In this paper, we propose CinHBa (Cached-index on HBase), which is a hierarchical secondary index prototype system on HBase. The full index is stored in HBase persistently, and the partial hot index is cached into memory to improve query performance. Furthermore, an effective caching policy, Hotscore Algorithm, is provided to keep the hottest data items in cache in nearly real-time fashion during continuous queries, and to maintain cache space fully utilized all the time. CinHBa is fit for both streaming data, that is, data are input as records one by one in streaming pattern and batching data, that is, data are stored as a static dataset. Actually, CinHBa could retrieve arbitrary format of data files by building index on various raw data files, not limited to HBase table.

2   Features of Key-Value Data Store

Key-value storing system stores data with continuous rowkeys orderly, dynamically partitions data horizontally, and distributes data splits on large-scale cluster servers by the system when data split becomes too large. Alternatively, it is possible that data splits are merged to reduce their number and required storage space. The partitioning mechanism of big data storing system allows for fast recovery when a server fails, and fine-grained load balancing since data splits can be moved between servers when the load of server is under pressure, or the server becomes unavailable. Splitting is very fast, close to instantaneous because data are stored with ordered rowkeys. It is easy for key-value data model with partitioning mechanism to provide good scalability by scaling data to petabytes on thousands of nodes.

Key-value storing system is designed for fast point queries and sequential scans on rowkeys. Each data split maintains key-value pairs with ranges of rows sorted by rowkey, so it excels at providing key-based or sequential range access. Point queries are supported by providing a rowkey to find what you are looking for. When faced to non-rowkey query requirement, HBase takes a measurement of sequential scan, that is, scanning data from start to end one by one. Key-value storage is suitable for reading adjacent key-value pairs and is optimized for block disk access operations that can make full use of disk transfer channels. Whereas, sequential scan misses the basic features of key-value retrieval, such as selection of rowkeys based on regular expressions. Thus the approach of non-rowkey query leaves much to be desired. This is just the goal of this paper.

Key-value storing system mostly has a caching mechanism that efficiently retains recently accessed data, and uses its LRU caching policy (one of the simplest and most commonly used replacement policy) to retain data for subsequent reads. LRU-based caching policy assumes data currently visited are more probably revisited in recent future, so it is remarkable when reading the same row more than once or reading data sequentially. Considering the random read, however, LRU, or other simple policy such as FIFO, is not smart enough to make the hot-aware decision to kick out the cold victims. Thus, we propose a novel caching replacement policy to improve the effectiveness of cache.

3   Design of CinHBa

3.1   Secondary Index Model

There are two types of secondary index model on distributed data stores: local index model and global index model. In local index model, such as Hindex[2], index is built on each separate Region Server, faced to local data tables. The index data on multiple compute nodes are independent, so local index model costs much unnecessary computation overhead because query process accesses index table in all Regions, but some Regions returns null result set. In high concurrent circumstance, it would result in the reduction of system throughput.

Global index model, on the other hand, builds index on global data tables. The index tables are partitioned on all the computing nodes in cluster. Query process locates nodes by retrieving index table globally, and then returns result set to client. This mechanism could guarantee only nodes with valid result set participate the query process. Index table are managed by all Region Servers in cluster, providing scalability and fault tolerance. CinHBa is obeyed to this type model. CinHBa selects the query attribute as rowkey, and builds index table on data table globally. In Figure 1, we illustrate CinHBa's system architecture. It could be seen that data tables, index tables and value tables (for range query) are stored in HBase.

Actually, a part of attributes of data table, the more frequent accessed columns, are selected to construct index table because it is helpful to reduce access time of data table. Most queries could be solved in index table, without visit to data table. In CinHBa, rowkey of data table is included in rowkey of index table. It

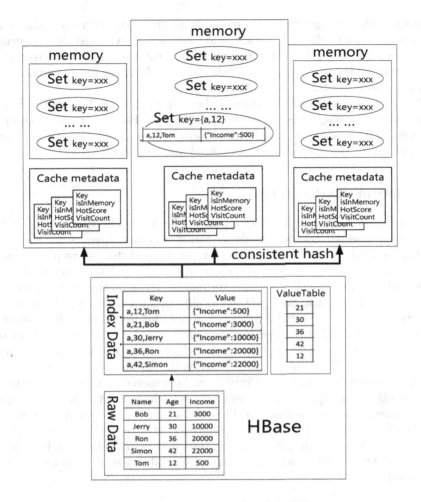

Fig. 1. Hierarchical storage of CinHBa

is a pointer to raw data, which provides a probability to access the source data when the query could not be accomplished in index table.

3.2   Hierachical Storage

CinHBa constructs secondary index hierarchically. Full index table is stored persistently in HBase, and it cache the hottest index data into memory based on our novel Hotscore Algorithm. The hierarchical storage of CinHBa is illustrated in Figure 1.

In CinHBa, rowkey of index table is designed as {index_column_name (abbr.), index_column_value, data_table_rowkey}. Name and value of index column are combined as rowkey of index table, and then query on this column could be solved in index table by rowkey location. The rowkey of data table is included in rowkey of index table, because there are probably several rows with the same index column name and value. In this scenario, the rows are violated the uniqueness of rowkey on HBase table. Thus rowkey of data table are added into rowkey of index table, giving it the uniqueness. On the other hand, as mentioned above, it also provides a pointer to data table records.

Data items of index table are stored in key-value pairs in HBase, and each data item is corresponding to one key-value pair. In CinHBa, we use abbreviation of index column name to save space cost because each non-null data item is stored as <key, value >format, and null is omitted and free of any space.

3.3   Hotscore Caching Policy

As an efficient way to reduce disk access bottleneck, caching technology is the continuous research hotspot on data storing and management all the time. Caching technology stores special data into high-speed storage, such as memory or SSD, so as to improve read performance of the cached data subsequently. The outstanding improvement of caching technology relies on the access speed gap between RAM and disk, and the time locality of data access, which means the data been visited are probably revisited in near future. Since caching space is limited by volume of high-speed storage, some cached data are probably kicked out and some new hot data should be selected into cache while the caching space is full up. This is the caching policy, which is the key factor or performance.

As illustrated in Figure 1, in CinHBa, caching metadata is recorded in memory: $isInMemory$ represents the data item is cached or not; $visitCount$ represents the recent access information. They are served for Hotscore Algorithm, our novel caching policy. In Hotscore Algorithm each set with the key as {index_name, index_value} has its own metadata structure, and $visitCount$ of each set is accumulated periodically and respectively. The evaluation of hotscore could be expressed as:

$$score_n = \alpha \times \frac{visitCount_n}{countPeriod} + (1 - \alpha) \times score_{n-1}$$

where $0 \leq \alpha \leq 1$, $countPeriod$ represents the period of visit count accumulation. CinHBa records the $visitCount$ of each set in every periods, and $score_{n-1}$ represents the history hotscore of the set. History hotscore is accumulated into the current hotscore by $1 - \alpha$ coefficient. That is to say, the nearer visit has higher influential weight, the farther visit has less weight by continuous decay of $1 - \alpha$ coefficient. The variable α is a decay factor that determines the weight given to new hotscore and how quickly to decay old hotscore.

To reduce the computation and update overhead of Hotscore Algorithm, computation and update are conducted periodically. In a period, $visitCount$ is accumulated, and when a period finished (query count comes to $countPeriod$), it is triggered to evaluate hotscore, update metadata, and empty $visitCount$ variable. Then hotscore of all sets are sorted, and TOP-K is chosen to cache into memory. Since the size of each set is not fixed, $HotscoreThreshold$ is computed in the limitation of caching space. That means, the set with hotscore above $HotscoreThreshold$ would be cached.

If we still select the TOP-K hottest sets periodically from scratch, the cost of algorithm's warm-up is high. Whereas in LRU, caching data is obeyed the rule of if-visit-then-insert.When the data item is visited, it would be inserted into the head of cache queue, and the tail of cache queue is the least visited data item. When the cache space is full, the tail of queue should be deleted. Thus, we optimize our Hotscore Algorithm in initial running stage for cache space is free mostly. When cache space is not full, Hotscore Algorithm inserts data into cache space by the rule of if-visit-then-insert. Then after cache space is full, we select the victim to be kicked out from caching by Hotscore evaluation.

Hotscore caching policy considers not only the time distance of data visit, but also the visit frequency, so it's more precise without losing simple.

4   Implementation of CinHBa

4.1   System Infrastructure

Figure 2 shows the system infrastructure of CinHBa. In CinHBa, persistent storing of data tables, index tables and value tables are supported by HBase. Relying on HBase's distributed storing management, raw data and index data are well scalable. Furthermore, hot index data are cached into memory and managed by our Memory Cache Management Module (MCMM). Based on the hierarchical infrastructure, CinHBa could support efficient point query, range query and data update.

There are four modules in CinHBa. They are:

Index Building Module. It manages the metadata of index, such as index table name and structure. It supports two different index building methods oriented to data streaming and batch processing respectively. It also supports the insertion, deletion, update of index tables and value tables.

Persistent Storing and Management Module. It supports the persistent storing of index tables and value tables, and provides scalability and fault tolerance on HBase.

Memory Cache Management Module. It provides the management of hot index's cache storing, cache update, address mapping based on consistent hashing, and the Hotscore caching policy.

Query Engine. It translates query requirements to CinHBa's API. Besides that, it also supports analyzing complex queries, collecting the result set and response to client.

4.2   Index Building Process

There are two index building scenarios: data streaming oriented building and batch processing oriented building. Data streaming oriented building is the main method because most big data application would provide data as streaming

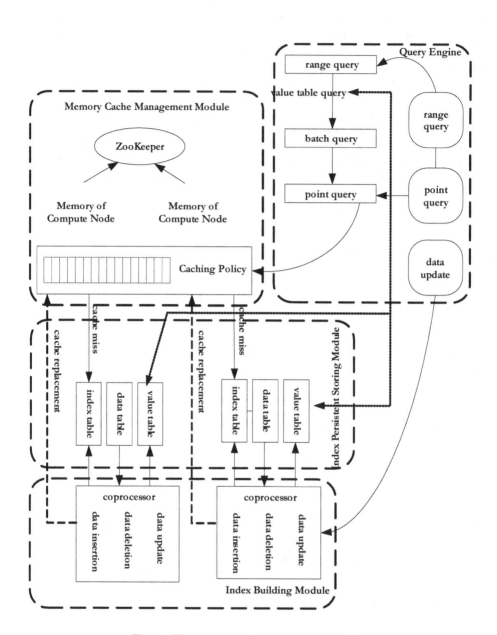

Fig. 2. The system infrastructure of CinHBa

format. Batch processing oriented building could handle a large volume of offline static data. Both have the same process ideas. One data record is put in, and one index record are constructed. Then index data is updated into persistent storage, a metadata item is created if necessary, and value table is updated if necessary.

For data streaming oriented building, CinHBa provides Coprocessor interface to build index table. There are two types of Coprocessor in HBase: Observer and Endpoint. Observer is similar to the trigger of RDBMS, and Endpoint is similar to the storing procedure of RDBMS. CinHBa adopts Observer here. It employs RegionObserver's callback function prePut, which would be triggered when one record is inserted into data table. The method prePut analyzes Put operation. If the record to be put contains index column, index table insertion would be triggered.

For batch processing oriented building, we employ Hadoop MapReduce to execute index building in parallel. The procedure of index building could be described as: input the key-value pair such as <Row, Result>, where Row is rowkey of data table, and Result is the data table record. Then Map task generates index data, and insert it into index table. This operation could be highly paralleled by MapReduce because records in data table are separate each other. Actually, reduce task is not necessary.

4.3   Query Process

CinHBa supports efficient point and range query. For both queries, data in memory cache should be retrieved at first. If cache missed, then query process goes to persistent data store to accomplish the query. Here we describe the query process of CinHBa.

For point query, the process is as follows:

1. Obtain the address of ZooKeeper from configuration, connect to ZooKeeper. Obtain all the MCMM server processes registered on /servers to determine all the locations of MCMM server processes.
2. Send query requirement to MCMM server processes. If cache hit, query is accomplished, and result set is returned.
3. If cache missed, query requirement should be sent to index table on HBase, result set is retrieved and returned.

If a query is hit by cache, the query process has no disk access, so the response time would be reduced in a large degree. This is the original idea of CinHBa. On the other hand, all processes of MCMM servers registered on /servers would be cached in client, so the performance of subsequent queries would be improved further.

In CinHBa, data are distributed on many compute nodes by consistent hashing, which distroys the data's ordering. Thus, for range query, we record the value of index column in value table, so as to obtain all existed values of index column for a range query.

The process of range query is as follows:

1. Obtain the address of ZooKeeper from configuration, connect to ZooKeeper. Obtain all the MCMM server processes registered on /servers to determine all the locations of MCMM server processes.
2. Retrieve value table to obtain all existed values of index column for query.
3. Trigger point query one by one, collect the result and return.

Similar to point query, high cache hit rate would decrease query response time. Value table retrieving is the extra cost, but for each range query, value table is accessed only once, and the execution time is trivial. Thus, the range query method is well performed.

5   Experiments

5.1   Benchmark Setup

In this section, we experimentally evaluate the performance of CinHBa by query response time, hit rate of Hotscore Algorithm (HA) and scalability of CinHBa. Experiments were implemented on Hadoop cluster of 4 compute nodes, the detailed configuration is described in Table 1. We use Brown University's big data benchmark dataset[5], that is proposed for the query performance comparison between RDBMS and Hadoop cluster in [6]. Subsequently, the papers [8,9] test and verify the research work on this big data benchmark dataset. We employ this dataset to generate 10M (10,000,000) data items, that obey uniform distribution. We generate 10,000 query requirements on avgDuration column of Rankings table. They obey scrambled Zipfian distribution according to YCSB Benchmark[7]. Zipfian parameter α is set to 1.0.

The query response time obtained by CinHBa is experimentally evaluated. Three cases are compared: (1) native HBase without index, which conducts non-rowkey query by full scan; (2) Hindex secondary index system; (3) CinHBa secondary index system with Hotscore caching policy. The performance comparison

Table 1. Configuration of Hadoop Cluster

Item	Information or Setting
CPU	4 Core Intel Xeon 2.4GHz × 2
Memory	24 GB
Disk	6 TB, SATA II, 7200RPM
Network Bandwidth	1Gbps
OS	Red Hat Enterprise Linux Server 6.0
JVM Version	Java 1.6.0
Hadoop Version	Hadoop 1.2.1
HBase Version	HBase 0.94.14
ZooKeeper Version	ZooKeeper 3.4.5

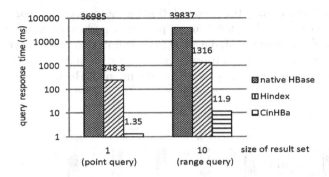

Fig. 3. Query response time of different systems

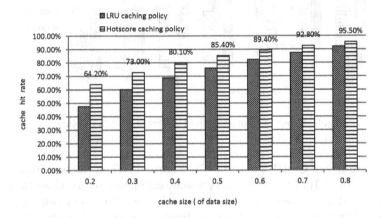

Fig. 4. Cache hit rate of different caching policies

is illustrated in Figure 3. CinHBa greatly outperforms native HBase as it has index on query column. It also outperforms Hindex because CinHBa constructs global index table and further, it caches hot index data into memory.

Cache hit rate is the percentage that query is accomplished by cache hit. High cache hit rate means more queries are solved in cache without disk access. Query performance could be improved in a large degree by cache hit, that is why cache used widely in data management system. Thus, cache hit rate is the most important measurement of caching policy. We give the cache hit rate comparison between LRU and Hotscore caching policy in Figure 4, and query response time in Figure 5.

In Figure 4, cache hit rate is increasing along with the increase of cache size. Especially, when cache size is 0.2 (20% data records are cached), the hit rate of Hotscore exceeds that of LRU for about 16%. Figure 5 illustrates the performance improvement of caching policy. We compare query response time on three cases: no cache, LRU caching policy, and Hotscore caching policy. Query response time is reduced by caching policy in a large degree, and Hotscore is faster than

LRU notably. The reason is the accumulation mechanism of Hotscore Algorithm could evaluate the visit frequency more precisely, and cache more useful data into cache. In big data scenario, data has huge volume, so cache size could never get high percentage. It can be observed from Figure 4 that Hotscore Algorithm is more efficient in low cache size. It is the advantage of Hotscore caching policy.

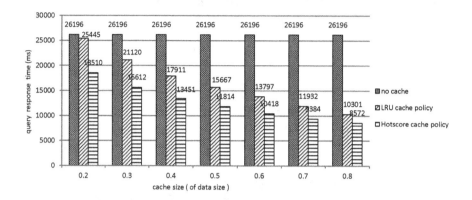

Fig. 5. Query response time of different caching policies

The scalability of CinHBa is verified in Figure 6. When the data record number expands from 10M to 50M, query response time increases close to linearly. Therefore, CinHBa has good data scalability. In CinHBa, query response time is proportional to the size of result set. In Brown University Benchmark dataset, data are distributed uniformly, so the size of result set is proportional to data size.

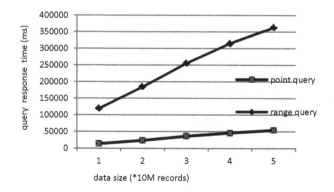

Fig. 6. Data scalability

6   Related Work

Before illustrating our approach to memory-cached HBase index on arbitrary format data files, we demonstrate previous works of improving big data query performance, the encountered problems and inherent tradeoffs. We summarize the works related to our research including efforts to indexes on hadoop big data and caching policies on big data.

6.1   Index on Big Data

Query processing on big data is a challenging and interesting work. A number of database groups concentrate on Hadoop query optimization before HBase release. Earlier Research implements a hybrid system HadoopDB[8]. HadoopDB employs Hadoop as communication layer in high parallel environment, and adopts local RDBMS to process SQL query requirement. Experiments demonstrate the hybrid idea absorbing the advantages of both, but the drawback is also obvious. It needs to extra employment of local RDBMS, and the local RDBMS hides the scalability and fault tolerance of Hadoop totally. Later, Hadoop++[9] has developed by building Trojan Index and the further Hail changes the upload pipeline of local file system to HDFS[10]. Hail fully utilized three replications of HDFS, building three index on different attributes by modifying the layout of raw data. Both Hadoop++ and Hail are similar to RDBMS clustered index to improve its query and join performance. Building index on Hadoop data files is a natural idea to optimize query performance. The effects are remarkable because the native query policy on Hadoop is taking full-scan strategy on massive data. Whereas, the efforts of Hadoop++ and Hail are both batching index data and raw data together physically. Thus the insertion and deletion of data items could result in high overhead.

Hindex [2] is a secondary index on HBase data developed by Huawei. It builds separate index table on each data table split of Regions. Query requirements are sent to Region Server and located in index table. Then index rows are returned as resultset. There is unnecessary computation overhead for Hindex because query process accesses index table in all Regions, but some Regions returns null resultset. Hindex has no awareness of result set's distribution for the index building is oriented local data table, that affects the system throughput and query performance.

Interval Index[11] is an secondary index oriented to range query on HBase. Segment Trees[12] are constructed by MapReduce[13] and stored in memory, and the endpoints of range segments are stored as HBase table to support effective range query. Interval Index has good scalability, but when it faced to point query, Segment Trees are degenerated to binary tree, which is a worse choice for big data query for its high time and space overhead.

6.2   Memory Cached Index

TBF[4] is a SSD-based caching policy in big data scenario. It proposes a space-frugal caching policy by drawing inspiration from CLOCK[3] caching policy and

Bloom Filter. In TBF, the orderly rowkey storage of HBase is a helpful feature, that reduces the space cost of caching metadata. However, TBF has the native drawback of CLOCK, that is, the caching metadata records data items are accessed or not, but not the access frequency. At the same time, the data access information is saved only in one CLOCK period, so there is no accumulation in data access frequency.

In [14], Microsoft proposes an effective method to identify hot data in big data access. It conducts an effective parallel backward algorithm to analyzes data access log in offline mode, and identify hot and cold data without full scan of data log. The effectiveness of this work are remarkable, however, it is fit only for offline analysis because online analysis is not able to preempt the computation early.

7   Conclusion and Future Work

To improve the performance of non-rowkey query on HBase, a hierarchical secondary index model is proposed and the prototype CinHBa is implemented. CinHBa maintains data tables and index tables by persistent storing on HBase, and cache hot index data into memory by Hotscore Algorithm, our novel caching policy. By Hotscore Algorithm, CinHBa obtains higher cache hit rate than LRU remarkably. The query performance of CinHBa is much better than the system with no secondary index, that complete query by full scan. Furthermore, CinHBa has good data scalability.

Acknowledgments. This work is funded in part by China National Science Foundation under Grants 61223003,61362006, Guangxi Natural Science Foundation under Grant 2014GXNSFBA118288. Also supported by Foundation of Key Lab. of Cognitive Radio & Information Processing, the Ministry of Education in China under Grant 2013ZR08, and the ZTE-funded research project.

References

1. DBMS2: DataBase Management System Services,
   http://www.dbms2.com/2009/05/11/facebook-hadoop-and-hive
2. Huawei Hindex, https://github.com/Huawei-Hadoop/hindex
3. Corbato, F.: A Paging Experiment with the Multics System. MIT Project MAC Report MAC-M-384 (1968)
4. Ungureanu, C., Debnath, B., Rago, S., Aranya, A.: TBF: A memory-efficient replacement policy for flash-based caches. In: 29th IEEE International Conference onData Engineering Brisbane (ICDE), pp. 1117–1128. IEEE Press, Brisbane (2013)
5. A Comparison of Approaches to Large-Scale Data Analysis: MapReduce vs. DBMS Benchmarks,
   http://database.cs.brown.edu/projects/mapreduce-vs-dbms
6. Pavlo, A., Paulson, E., Rasin, A., Abadi, D.J., DeWitt, D.J., Madden, S., Stonebraker, M.: A Comparison of Approaches to Large-scale Data Analysis. In: 35th International Conference on Management of Data, New York, pp. 165–178 (2009)

7. Cooper, B.F., Silberstein, A., Tam, E., Ramakrishnan, R., Sears, R.: Benchmarking Cloud Serving Systems with YCSB. In: 1st ACM Symposium on Cloud Computing, Santa Clara, CA, pp. 143–154 (2010)

8. Abouzeid, A., Bajda-Pawlikowski, K., Abadi, D.J., Rasin, A., Silberschatz, A.: HadoopDB: An Architectural Hybrid of MapReduce and DBMS Technologies for Analytical Workloads. In: 35th International Conference on Very Large Data Bases, Lyon, pp. 922–933 (2009)

9. Dittrich, J., Quian-Ruiz, J., Jindal, A., Kargin, Y., Setty, V., Schad, J.: Hadoop++: Making a Yellow Elephant Run Like a Cheetah (WithoutIt Even Noticing). In: 36th International Conference on Very Large Data Bases, Singapore, pp. 518–529 (2010)

10. Dittrich, J., Quian-Ruiz, J., Richter, S., Schuh, S., Jindal, A., Schad, J.: Only Aggressive Elephants are Fast Elephants. In: 38th International Conference on Very Large Data Bases, Istanbul, pp. 1591–1602 (2012)

11. Sfakianakis, G., Patlakas, I., Ntarmos, N., Triantafillou, P.: Interval Indexing and Querying on Key-value Cloud Stores. In: 29th IEEE International Conference on Data Engineering (ICDE), pp. 805–816. IEEE Press, Brisbane (2013)

12. Bentley, J.L.: Solutions to Klee's Rectangle Problem, Technical Report, Carnegie-Mellon University, Pittsburgh (1977)

13. Dean, J., Ghemawat, S.: MapReduce: a Flexible Data Processing Tool. Communications of the ACM 53(1), 72–77 (2010)

14. Levandoski, J.J., Larson, P., Stoica, R.: Identifying Hot and Cold Data in Main-Memory Databases. In: 29th IEEE International Conference on Data Engineering (ICDE), pp. 26–37. IEEE Press, Brisbane (2013)

Towards Ontology-Enhanced Cloud Services Discovery

Abdullah Alfazi[1], Talal H. Noor[2], Quan Z. Sheng[1], and Yong Xu[3]

[1] School of Computer Science
The University of Adelaide, SA 5005, Australia
{abdullah,qsheng}@cs.adelaide.edu.au
[2] College of Computer Science and Engineering
Taibah University, Yanbu, Medinah 46421-7143, Saudi Arabia
tnoor@taibahu.edu.sa
[3] South China University of Technology
Guangzhou 510641, China
xuyong@scut.edu.cn

Abstract. Cloud computing provides high flexibility with regard to on-demand computer resources that are used as services through the Internet. However, searching cloud services still remains a challenge due to their unique characteristics such as dynamic and diverse services offering at different levels, as well as the lack of standardized description languages. In this paper, we propose a novel technique to support cloud services discovery based on a comprehensive ontology. More specifically, our approach has the capability to semi-automatically generate the ontology by mining new concepts from documents related to cloud services. Starting from an initial ontology based on the NIST (US National Institute of Standards and Technology) cloud computing standard, our approach analyzes a real-world cloud service dataset of 5,883 to build the cloud service ontology by identifying and adding new cloud services related concepts. The proposed approaches have been validated by a prototype system and experimental studies.

Keywords: Cloud service, ontology, service discovery, Web Service, concept reasoning.

1 Introduction

Cloud computing is a relatively new computing paradigm that has attracted a considerable attention in the last few years for service delivery on demand. With cloud computing, users are able to deploy their services over a network of a large number of computing resources with practically no capital investment and modest operating cost [1]. Despite active research on addressing various cloud computing challenges such as security and privacy, and trust management, cloud services discovery is still an incipient area of research and development [3,8,7].

Discovering and identifying cloud services is challenging due to a number of reasons. On the one hand, cloud services are provisioned at various levels, not

X. Luo, J.X. Yu, and Z. Li (Eds.): ADMA 2014, LNAI 8933, pp. 616–629, 2014.

only data and business logic, but also infrastructure capabilities. On the other hand, cloud service providers may not follow a standard to describe their services and resources when publishing them [2]. Unlike Web services which use standard languages such as the Web Services Description Language (WSDL) or Unified Service Description Language (USDL) to expose their interfaces and the Universal Description, Discovery and Integration (UDDI) to publish their services to services registries for discovery, the majority of the publicly available cloud services are not based on description standards [7]. In addition, the variety of service level agreements (SLAs) between cloud service users and service providers makes it challenging to identify cloud services [9]. As a result, relying on existing search engines such as Google is of little help in cloud service discovery due to large quantity of irrelevant results (e.g., blogs, news, research papers) returned.

In this paper, we overview the design and the implementation of an ontology-based cloud service search engine. This search engine helps distinguish between cloud services and other services available on the Internet. The main component behind our search engine is a comprehensive ontology for reasoning during cloud service discovery. This cloud service ontology is built semi-automatically based on mining new concepts after analyzing 5,883 real-world cloud services. In a nutshell, the salient features of our ontology-based cloud service search engine are as the following:

- We design and develop a cloud service search engine that achieves more accurate searching results by consulting a comprehensive cloud service ontology for reasoning on the relations of cloud services.
- We develop a novel approach to build the cloud service ontology. We first develop an initial cloud service ontology (i.e., as a roadmap for the ontology builder) based on the NIST cloud computing standard. New cloud service related concepts are then discovered and added to the cloud service ontology by automatically analyzing 5,883 real cloud services.
- We conduct extensive experiments and the results demonstrate the applicability of our approach and show its capability of effectively identifying cloud services on the Internet.

The remainder of our paper is organized as follows. Section 2 overviews the related work. Section 3 briefly presents the ontology-based cloud services search engine's architecture and details the cloud services ontology including the ontology roadmap, cloud service concepts discovery, and the concepts' position determination. Section 4 reports the implementation and experimental results of the proposed approach. Finally, Section 5 provides some concluding remarks.

2   Related Work

Service discovery is considered to be fundamental in several research areas such as ubiquitous computing, mobile ad-hoc networks, peer-to-peer (P2P), and service oriented computing [5,14,6]. Although service discovery is a very active research area, particularly in Web services in the past decade, for cloud services, challenges need to be reconsidered and solutions for effective cloud service discovery are very limited [14,3,7].

Many researchers use an ontology-based approach for service discovery. For example, ArnetMiner [13], a service used to index and search academic social networks, uses researcher profiles and friend-of-a-friend (FOAF) ontology to find researchers and publications. ArnetMiner ontology consists of two main concepts, namely *researcher* and *publication*, as well as 24 properties of publication and two object relationships to link between the authors, their publication and publication properties. Segev and Sheng [12] develop a bootstrapping approach for building Web services ontology. Their approach exploits Term Frequency/Inverse Document Frequency (TF/IDF) and web context generation to automatically build the ontology for describing the Web services functionality. Unlike previous works that use an ontology-based approach to solve the problem of discovery in social networks or Web services, our work focuses on developing an ontology-based approach for cloud services discovery.

There have been several attempts to build an ontology for cloud services. For example, Youseff et al. [16] classify cloud computing based on its components, consisting of five layers: the *applications*, the *software environment*, the *software infrastructure*, the *software kernel*, and the *software hardware*. Each layer can contain one or more services depending on the level of abstraction. Also, each layer relies on computing concepts to measure limitations and strengths. Weinhardt et al. [15] propose to build the ontology based on a cloud business ontology model. This ontology model consists of three layers: the *platform*, the *infrastructure* and the *application*, as well as a content pricing model to help clarify the relationship between cloud service providers and customers. Kang and Sim [3] propose a cloud service discovery system that uses ontology approach to discover cloud services close to users' requirements. However, cloud service providers still need to register at the discovery system in order to publish their cloud services. Furthermore, their work relies on software agents to perform reasoning tasks (e.g., similarity reasoning, equivalent reasoning and numerical reasoning). Zhang et al. [17] propose the Cloud Recommender system to select cloud infrastructure based on their cloud computing ontology called (CoCoO). This ontology defines functional and non-functional concepts of infrastructure services with their attributes and relations. However, the ontology has not provided any details about PaaS and SaaS. In addition, the validation of ontology concepts is only limited for some cloud infrastructure providers such as Amazon, Microsoft Azure, GoGrid,etc. Unlike previous works which fail to develop cloud service ontology automatically and perform cloud services discovery in large-scale such as the Internet, our ontology-based cloud service search engine helps distinguish between cloud services and other services available over the Internet using an automatically-built cloud service ontology.

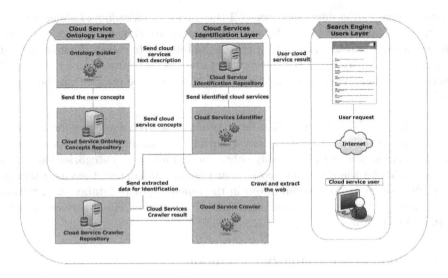

Fig. 1. Architecture of the Cloud Service Search Engine

3   Cloud Service Search Engine and Ontology Building

In this section, we first briefly introduce our cloud service search engine (CSSE), then focus on describing our approach on cloud service ontology generation.

3.1   CSSE Architecture

Figure 1 depicts the architecture of CSSE, which consists of three major layers, namely the *Cloud Service Ontology Layer*, the *Cloud Service Identification Layer*, and the *Search Engine Users Layer*.

Cloud Service Ontology Layer. The cloud services ontology layer is responsible for maintaining the cloud services ontology, which consists of two main parts: the *cloud service ontology repository* and the *ontology builder*. The cloud service ontology repository contains concepts that are generated by the ontology builder. The ontology builder generates concepts semi-automatically using an ontology roadmap and cloud services' metadata. More information on the cloud services ontology generation can be found in Section 3.2.

Cloud Services Identification Layer. The cloud services identification layer contains the *cloud service identification repository* which stores the cloud services that have been identified and the *cloud service identifier* which recognizes cloud services. The cloud service identifier relies on the cloud service ontology that provides a level of understanding and reasoning for cloud services. To judge whether a service is a cloud service, we could simply calculate the ratio between the number of cloud service concepts and the total number of terms appearing in its corresponding document (we call this ratio an *identifier of cloud service*, IC in short) and compare the IC value with a certain threshold. In our system, there is a *cloud service crawler* that crawls the Internet to collect the metadata of potential cloud services. The collected metadata is then processed by the cloud service identifier. More information on the cloud service crawler can be found in [8].

Search Engine User Layer. This layer provides a Web interface for users to search cloud services. A user can simply specify a searching keyword for finding cloud services. She can also specify other constraints (e.g., categories like IaaS) to narrow down the searching scope. Our system will contact the cloud service identification repository and if found in the repository, the detailed information (e.g., access link, categorization, description) of satisfied cloud services will be returned to the user.

3.2 Cloud Service Ontology Generation

Our approach on cloud service ontology generation involves two main steps. Firstly, we create a base ontology by following the US National Institute of Standards and Technology (NIST) cloud computing standards. Since there are very limited concepts in this ontology, we also call it a *cloud service ontology roadmap*. Then, our ontology builder grows the cloud service ontology with new concepts by automatically analyzing the metadata of real-world cloud services. Furthermore, cloud service ontology concepts have two properties, which are the abbreviation and weight of cloud service ontology concepts. The cloud service ontology concepts abbreviation gathers for particular concepts in cloud services ontology roadmap such as SaaS for Software as a Service, PaaS for Platform as a Service, while the weight is computed from the Ontology Builder Algorithm. We will describe each step in details in the sequel.

Cloud Service Ontology Roadmap(*CSOr*): Our cloud service ontology roadmap *CSOr* has been built by following the interpretation of the NIST standards for cloud computing [4] and other published ontology for cloud computing, to obtain a set of concepts that can be used as a roadmap for the cloud services' ontology. These concepts have relationships that can be defined as *is_a* and *is_not_a* cloud service ontology, to enhance the generation of a new concept of the ontology. According to the National Institute of Standards and Technology (NIST), the cloud model comprises of five *essential characteristics*, and three *service models*. The five essential characteristics are as follows:

- *On-demand self-service*: A consumer possesses independent provision of computing capabilities like server time and network storage, whenever necessary without seeking the attention of each service provider.
- *Broad network access*: Capabilities can be accessed on-line and through standard mechanisms which encourage the use of various client platforms such as mobile phones, tablets, laptops and workstations.

- *Resource pooling*: The provider's computing resources are shared among multiple consumers under a multi-tenant model, where dissimilar physical and virtual resources are engaged or disengaged according to consumer preferences. In normal circumstances, consumers do not have the knowledge of the precise location apart from information of a higher level of abstraction (e.g., country, state, or data center), that provide such resources (e.g., storage, processing power, memory and network bandwidth).
- *Rapid elasticity*: Resources are elastically supplied or released automatically or manually on demand. Capabilities as seen by the consumers are often unlimited and are utilized at any amount of time.
- *Measured service*: Cloud services may charge according to a pay-per-use or charge-per-use basis. Resource usage of the service are monitored, controlled and reported, allowing transparency between the provider and the consumers.

The three service models specified by NIST cloud computing standard are as follows:

- *Software as a Service (SaaS)*: The term is referred to a service provided by a cloud application supported on a cloud infrastructure. Consumers can interact the application through a user interface like a Web browser installed on different client devices. It is superfluous for the consumers to manage or manipulate any underlying cloud infrastructure such as the servers or operating systems. However, they may be given the control limited to particular application configuration settings.
- *Platform as a Service (PaaS)*: PaaS is another type of service on which is built upon the cloud infrastructure and involves consumer developed or acquired applications with the help of programming languages, tools, libraries and services powered by the provider. The consumer is not required to manage or configure the cloud infrastructure except possessing the power to control the deployed applications and configuration settings of the hosting environment for which the applications are in.
- *Infrastructure as a Service (IaaS)*: It is defined as the ability to supply the consumer with resources from the ground up - processing, storage, network and other basic resource, in order to allow consumer to develop and run software as well as operating systems and applications. The consumer does not need to manage or configure the cloud infrastructure but has the authority over controlling the operating systems, storage and other mounted applications with the possibility of limited control over selected networking components like the host firewalls settings. Due to space constraints, we only depict part of the IaaS concepts (see Figure 2).

While developing our CSOr based on the interpretation of NISTs cloud computing standards, we determine that cloud computing is the root node to the relationship between is_a and is_not_a. In addition, in CSOr, we consider cloud_service a child of the root node cloud_computing and the parent node for other cloud service concepts. We also define the concepts for Essential

Fig. 2. Infrastructure as a Service (IaaS)

Characteristics and **Service Models**. For example, we treat software as a service (SaaS), platform as a service (PaaS) and infrastructure as a service (IaaS) as three main child nodes of cloud service and parent nodes for other cloud service concept levels. The **Service Models** builds the three main branches in CSOr and additional cloud service concepts are added into the appropriate branches depending on the type of services. For instance, we could add **storage** as a child node of infrastructure as a service (IaaS). In our cloud service ontology roadmap, we consider NIST-interpreted concepts having a higher priority than the available ontology concepts, because the former are more valuable in describing cloud services. As a result, those NIST-interpreted concepts are added at a high level in CSOr.

We also consider is_not_a in our cloud service ontology roadmap, which represents a set of concepts unrelated to cloud services. One example is the concepts such as weather forecast, which may show word "clouds" but does nothing related to cloud services. Another example is the concepts related to research such as reports, articles, and publications. Clearly, social network items can also be considered as is_not_a relations since the word "cloud" may appear in new items, blogs, Twitter or Facebook posts, but not actually a cloud service. Given the large-scale of the Internet, having is_not_a relations in the ontology is very useful for the cloud service crawler to validate and filter out non-cloud services.

Generating Cloud Services Ontology: Since the cloud services ontology roadmap provides very limited number of concepts, it is necessary to generate a more comprehensive ontology with new concepts related to the cloud services. One possible approach in doing so is to analyze known cloud services and mine new concepts from their metadata such as service descriptions. Based on our previous effort where 5,883 cloud services were identified [8], we develop an algorithm (see Algorithm 1) to analyze these cloud services and identify new cloud service concepts to generate our cloud service ontology. There are five steps in our algorithm, which are detailed as the following.

- *Detection*: In this very first step, we detect whether a term should be a candidate concept to be considered in our cloud service ontology. We use the term frequency, a well-known information retrieval technique [10], for this purpose. Term frequency represents the number of times that a term appears in the cloud services text descriptions S. Only those terms that pass a threshold can be considered as candidate concepts, which are then passed to next step for validation.
- *Validation*: The validation step compares the candidate concept with the existing concepts in the cloud services ontology. This eliminates possible repetition and ensures that this candidate concept is not a node or part of a node in the ontology. If the candidate concept has not appeared in CSO, it is considered to be a new cloud service concept, and will be passed to next step for further processing.
- *Balancing*: The purpose of the balancing step is to provide a weight for each new concept by using the popular *TF*IDF* model:

$$TF(nc, C) = \frac{f(nc, s_j)}{f\{t, s_j \, : \, t \in s_j\}} \tag{1}$$

where nc is the new concept, C represents the set of all concepts in CSO, and $f(nc, s_j)$ returns the frequency of nc in s_j.

$$TF^*IDF(nc, C) = tf(nc, C) * \log(\frac{\sum S(s_j, S)}{\sum S\{s_j, s \, : \, nc \in s)\}}) \tag{2}$$

Algorithm 1. Ontlogy Builder Algorithm

Step 1. **Detection:**
/* Determines candidate concepts from cloud services' text descriptions S.*/
for each term $\in$ S **do**
   count its frequency f
   if f < Threshold **then**
     term is not a candidate concept cc
   else
     term is a candidate concept cc and pass cc to validation
   end if
end for

Step 2. **Validation:**
/*Check if the candidate concept $cc \in CSO$.*/
if $cc \in$ CSO **then**
   cc is already at CSO
   Back to Step 1.
else
   cc becomes a new concept nc of CSO
   Pass nc to balancing
end if

Step 3. **Balancing:**
/*Provide a weight for new concept nc */
The balancing step uses $TF*IDF$ technique to give each new concept nc a weight.

Step 4. **Addition:**
/* link the new concept with old concept and determine the new concept position in CSO */
Select new concept nc
for each s $\in$ S **do**
   if nc $\in$ s **then**
     for each CSOc $\in$ CSO **do**
       /* CSOc: cloud service concepts */
       if CSOc in s **then**
         store CSOc in nc Set
         /* The new concept set (nc Set) contains all cloud service ontology concept that
         appear with the new concept */
       end if
     end for
   end if
end for
call maximum frequency concept $max\ nc\text{-}CSOc$ in nc Set
/* this method provide the maximum repeating concept that appear with the new concept nc in nc Set */
call maximum frequency concept $max\ nc\text{-}CSOc$ weight
call $nc\ weight$
if $nc\ weight$ > $max\ nc\text{-}CSOc\ weight$ **then**
   nc add to max $nc\text{-}CSOc$ in CSO as child
else
   nc add to max $nc\text{-}CSOc$ in CSO as siblibng
end if

Step 5. **Updating:**
The CSO ontology updates weights and relation for each concept.

The *TF*IDF* equation has the capability to provide weight for cloud service ontology concepts.

- *Addition*: This step adds a new concept to CSO by using the inverted index and TF*IDF. More specifically, we first link the new concept to the concepts in CSOr using the inverted index, which allows us to determine where the new concept appears in the cloud services text description. We then select the new concept and the documents where the new concept appears. We count the existing CSO concepts that appear in the same cloud services text description. After that, we store the existing CSO concepts and pick up the CSO concept with the maximum appearance and create the link between them.

$$CFS = \langle c_1, c_2 ... c_x \rangle_{|S| \times |C|}. \tag{3}$$

Furthermore, to determine the new concepts position in the cloud services ontology, we need to check the weight of the existing *CSO* concept that has the most appearances. If the weight of the new concept is more than the one of the existing concept, the new concept is inserted to the ontology as the child of this concept. Otherwise, the new concept is added as a sibling.

- *Updating*: The updating step is responsible for upgrading the cloud service ontology after adding the new concept.

4   Experimental Study

To implement the cloud services ontology layer (CSO, see Figure 1), we used 5,883 real-world, valid cloud services [8]. This dataset was chosen because it has been verified and validated. However, the metadata of these cloud services still needed to be processed. In particular, we removed the HTML tags and non-English cloud services from the cloud services metadata. Non-English cloud services were detected using the language detection library[1]. We eventually obtained a set of 5,083 cloud services' text descriptions containing only English language. We implemented the system using Java and JavaServer Pages (JSP).

4.1   Generating Cloud Services Ontology

Based on the collected text descriptions of cloud services, we ran the system to generate our cloud services ontology by using the proposed CSO algorithm. The cloud service text descriptions contained 1,935,185 terms. When setting the threshold frequency as 500, we obtained 654 candidate concepts after the first step. In the validation step, we obtained 105 new concepts for cloud services ontology since some of the candidate concepts were not considered as the CSO concepts or simply stop words (e.g., numbers, dates). To take an example, web_application was found 510 times. Then, we found that the host was the most common concept to appear together with web_application in cloud

---

[1] https://code.google.com/p/language

service text descriptions. Therefore, we linked `web_application` to `host` and further determined `web_application` to be the child for `host` because the TF*IDF weight of `host` is less than the weight of `web_application`. The concepts with greater frequency are at the higher level in the CSO tree.

Table 1. Top 5 High Frequency Concepts in CSO

CSOc	# frequency	# appear CSTD	Percentage CSTD
host	23423	2910	57.7%
server	14151	2500	49.6%
application	6704	2160	42.8%
network	6746	2074	41.1%
Cloud computing	3341	1394	27.7%

Table 1 shows the top 5 high frequency concepts in CSO that appear in cloud service text description (CSTD). From the table we can see that the most common concept is cloud infrastructure services. Moreover, the number of cloud service providers that provide infrastructure services is higher than those that offer platform or software services. From our statistics, it is also interesting to note that some cloud service providers do not use the concept of "cloud service" when advertising their services on the Internet.

4.2   Threshold Identification

It is important to estimate the threshold value in order to identify cloud services. We believe that using a *confidence interval calculator* to estimate the threshold identification [11] in the cloud service text description is a good solution for identifying cloud services because it provides the identifier for cloud services (i.e., *IC*, see Section 3.1) which, is the ratio between the number of cloud service concepts and the total number of terms appearing in its corresponding document. To calculate the estimated value for identifying cloud service, we use the following equation:

$$E\pm = \mu \pm z.\frac{\sigma}{\sqrt{N}} \tag{4}$$

where μ represents the mean of *IC* in the cloud services text description; σ represents the standard deviation of the cloud services text description; and N represents the total number of cloud services text descriptions. In this experiment, we set the value of z to 95% since this is common in research and useful in conducting estimations as well as providing close accuracy of a population set for the estimation equation. We ran the experiment randomly using 2,750 cloud services text descriptions to identify an optimal E threshold for identification. In the first round, we randomly selected cloud services text descriptions, then dynamically increased the number of cloud services text descriptions at each round. Fig. 3 shows the lower bound, the upper bound for *IC* in each set of cloud service text descriptions. We can see from the figure that the lower bound of cloud services ontology is between 7% to 8% of the cloud services text description, whilst the upper bound can be between 11% to 14%. This experiment indicates an interesting fact that most cloud services use 8% to 14% of ontology concepts in their service descriptions.

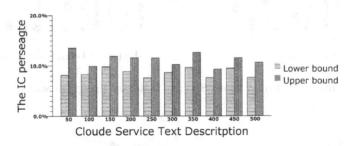

Fig. 3. Identifying E Threshold

4.3 Identifying Cloud Services Ontology

We conducted an experiment to identify cloud services for proving the robustness of the search engine. In this experiment (see Fig. 4), we collected randomly 550 webpage text descriptions which are fake cloud services. Additionally, we used 2,200 cloud service text descriptions to run two experiments using precision, recall and f-measure. The precision represents the percentage rate of distinguishing between real and fake cloud services, whilst recall represents the percentage of cloud services identified and f-measure shows the mean between recall and precision. At the beginning, the experiment started with 40 real cloud service text descriptions and 10 fake cloud services text descriptions. Then, the numbers were increased dynamically. The first experiment used the upper bound of the threshold identification while the second used the lower bound of the threshold identification to compare the results, depending on the threshold. In the experiments, it was concluded that the upper bound threshold identification provides high precision and average recall, which indicates that the upper bound threshold can distinguish text descriptions well. However, it provides an average performance for identifying cloud services. The lower bound threshold identification provides high recall and high average precision, which indicates that the lower bound threshold can effectively find cloud services. However, it can provide more fake cloud services to the search engine.

We conducted another experiment that shows in Fig. 5 to represent the noisy data at the cloud service search engine. The noisy data means the fake cloud service results of the cloud service crawler. In this experiment, we compared the cloud service crawler repository which did not use the cloud service identifier and the one which did use the cloud service identifier. Fig. 5 shows that the percentage of the noisy data decreased to 10% which gives the cloud service identification repository high robustness for cloud service search engine. However,

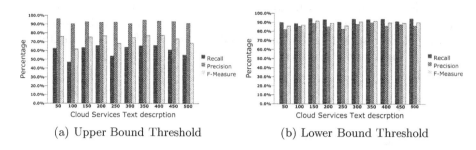

(a) Upper Bound Threshold   (b) Lower Bound Threshold

Fig. 4. Identifying Cloud Services Ontology Using Threshold

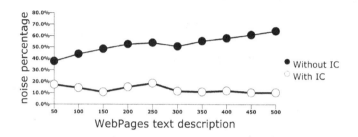

Fig. 5. Noisy Data of Cloud Service Search Engine

from the figure we can see that without using the identifier, with the increase of the number of crawled cloud services, the number of fake cloud services also increases. On the other hand, by using our identifier, fake cloud services can be successfully identified.

5 Conclusion

Effective cloud service discovery remains a challenging issue due to unique characteristics of cloud services such as dynamic, diverse services offering at different levels. In this paper, we have proposed a novel ontology-based approach for cloud service discovery. The contribution of our work includes the ability to effectively identify cloud services on the Internet and an approach to semi-automatically build the cloud service ontology using the metadata of 5,883 real-world cloud services. Our extensive experimental studies demonstrate the applicability of the proposed approach, as well as its capability of effectively identifying cloud services on the Internet. Future work will involve categorization of cloud service based on their service descriptions to enhance the cloud service discovery.

Acknowledgments. Abdullah Alfazi's work is supported by King Abdullahs Postgraduate Scholarships, the Ministry of Higher Education, Kingdom of Saudi Arabia.

References

1. Armbrust, M., Fox, A., Griffith, R., Joseph, A.D., Katz, R., Konwinski, A., Lee, G., Patterson, D., Rabkin, A., Stoica, I., et al.: A View of Cloud Computing. Communications of the ACM 53(4), 50–58 (2010)
2. Garg, S.K., Versteeg, S., Buyya, R.: A Framework for Ranking of Cloud Computing Services. Future Gener. Comput. Syst. 29(4), 1012–1023 (2013)
3. Kang, J., Sim, K.M.: Cloudle: An Ontology-Enhanced Cloud Service Search Engine. In: Chiu, D.K.W., Bellatreche, L., Sasaki, H., Leung, H.-f., Cheung, S.-C., Hu, H., Shao, J. (eds.) WISE Workshops 2010. LNCS, vol. 6724, pp. 416–427. Springer, Heidelberg (2011)
4. Mell, P., Grance, T.: The NIST Definition of Cloud Computing (draft). NIST special publication 800(145), 7 (2011)
5. Meshkova, E., Riihijärvi, J., Petrova, M., Mähönen, P.: A Survey on Resource Discovery Mechanisms, Peer-to-Peer and Service Discovery Frameworks. Computer Networks 52(11), 2097–2128 (2008)
6. Mian, A., Baldoni, R., Beraldi, R.: A Survey of Service Discovery Protocols in Mobile Ad Hoc Networks. IEEE Pervasive Computing 8(1), 66–74 (2009)
7. Noor, T.H., Sheng, Q.Z., Ngu, A.H., Dustdar, S.: Analysis of Web-Scale Cloud Services. IEEE Internet Computing 18(4) (2014)
8. Noor, T., Sheng, Q., Alfazi, A., Ngu, A., Law, J.: CSCE: A Crawler Engine for Cloud Services Discovery on the World Wide Web. In: Proceedings of IEEE 20th International Conference on Web Services (ICWS), pp. 443–450 (2013)
9. Patel, P., Ranabahu, A., Sheth, A.: Service Level Agreement in Cloud Computing. In: Proceedings of Conference on Object Oriented Programming Systems Languages and Applications, pp. 212–217 (2009)
10. Salton, G., Buckley, C.: Term-weighting Approaches in Automatic Text Retrieval. Inf. Process. Manage. 24(5), 513–523 (1988)
11. Samuels, M.L., Witmer, J.A., Schaffner, A.: Statistics for the Life Sciences. Pearson Education (2012)
12. Segev, A., Sheng, Q.: Bootstrapping Ontologies for Web Services. IEEE Transactions on Services Computing 5(1), 33–44 (2012)
13. Tang, J., Zhang, J., Yao, L., Li, J., Zhang, L., Su, Z.: Arnetminer: Extraction and mining of academic social networks. In: Proceedings of the 14th ACM SIGKDD International Conference on Knowledge Discovery and Data Mining, KDD 2008, pp. 990–998. ACM, New York (2008)
14. Wei, Y., Blake, M.B.: Service-Oriented Computing and Cloud Computing: Challenges and Opportunities. IEEE Internet Computing 14(6), 72–75 (2010)
15. Weinhardt, C., Anandasivam, A., Blau, B., Ster, J.: Business Models in the Service World. IT Professional 11(2), 28–33 (2009)
16. Youseff, L., Butrico, M., Da Silva, D.: Toward a Unified Ontology of Cloud Computing. In: Grid Computing Environments Workshop, GCE 2008, pp. 1–10 (2008)
17. Zhang, M., Ranjan, R., Haller, A., Georgakopoulos, D., Menzel, M., Nepal, S.: An ontology-based system for cloud infrastructure services' discovery. In: 2012 8th International Conference on Collaborative Computing: Networking, Applications and Worksharing (CollaborateCom), pp. 524–530 (October 2012)

A Huffman Tree-Based Algorithm
for Clustering Documents

Yaqiong Liu[1,3], Yuzhuo Wen[2], Dingrong Yuan[3,4,*], and Yuwei Cuan[3]

[1] Hubei University for Nationalities, Enshi 445000, China
[2] School of Economics and Management, Guangxi Normal University, Guilin 541004, China
[3] College of Computer Science & IT, Guangxi Normal University, Guilin, 541004, China
[4] Guangxi Key Lab of Multi-source Information Mining & Security
{liuyaqiong1213,cuanyuwei}@163.com,
{dryuan,wenyuzhuo}@mailbox.gxnu.edu.cn

Abstract. Text information processing is one of the important topics in data mining. It involves the techniques of statistics, machine learning, pattern recognition etc. In the age of big data, a huge amount of text data has been accumulated. At present, the most effective text processing way is classifying them before mining. Therefore, it has attracted great interests of scholars and researchers, and many constructive results have been achieved. But along with the increasing of training samples, the shortages of techniques and limits of their application have appeared gradually. In this paper, we propose a new strategy for classifying documents based on Huffman tree. Firstly, we find out all the candidate classifications by generating a Huffman tree, and then we design a quality measure to select the final classification. Our experiment results show that the proposed algorithm is effective and feasible.

Keywords: document classification, document clustering, big data, Huffman tree.

1 Introduction

Text is a traditional information carrier, whether a short one which contains only a few characters or a long one that includes hundreds of pages, all may hide important knowledge, but increasing text documents not only provide us a tremendous wealth of information, but also bring us a significant challenge for mining and managing them. Usually, a text document always includes one or more topics and each topic often described by many documents. Document classification is an effective strategy for managing and mining massive text documents. For example, a given document set includes three topics as computer science, business and mathematics, if we classify these documents into three classes based on different topics in advance, the burden of mining will be greatly reduced.

---

[*] Corresponding author.

X. Luo, J.X. Yu, and Z. Li (Eds.): ADMA 2014, LNAI 8933, pp. 630–640, 2014.
© Springer International Publishing Switzerland 2014

Text classification is one of the key steps of mining text documents, and the quality of the classification result directly affects the following work, thus, an excellent classification method is greatly needed. Many researchers have proposed a series of methods and some of them have obtained good results. The initial classification algorithms for documents are based on document frequency, and the ontology-based methods are proposed subsequently, after that, max entropy, neural networks, SVM, random forest and other classifying strategies have been suggested. Approaches in paper [1-5] need to specify classes in advance, and then classify each new document to one of the pre-specified classes based on some rules. To address the problems caused by large number of training samples and attributes, paper [1,5,6,7] proposed the improved strategies. And many clustering methods have been applied on dealing with documents, such as the algorithms based on KNN, K-means and some improved ways. Some researchers have devoted to mine the useful rules in various kinds of data environment, such as paper [10-15], clustering analysis, as a classic classification technique, has played a major role in the data mining process. In this paper, we propose a new approach based on Huffman tree for clustering text documents. Our approach mainly include two procedures, firstly, abstract the topics of each document based on Latent Dirichlet Allocation(LDA) model, and classify text documents into different classes based on their similarities by a new algorithm, secondly, select the best classification based on our *goodness* measure.

The rest of this paper is organized as follows. In section 2, we introduce the related knowledge about abstracting topic words based on LDA model from documents. Section 3 describes the related concepts and designs a new algorithm for clustering text documents based on Huffman tree. Section 4 gives the experiments, and section 5 is our conclusions.

2   Related Knowledge

Nearly all of the classification algorithms and clustering algorithms are depending on similarity measure, in this paper, we use the similarity between topic words to measure the similarity between documents. Thus, we introduce the related techniques about abstracting topics based on Latent Dirichlet Allocation (LDA) model[9] in this section to lay the groundwork for subsequent work.

2.1   LDA Model

A text document always expresses several topics and each topic is described by a series of words, moreover, a topic or a word often appears in different documents. Therefore, LDA model regards a text as a three level model: "word-topic-document", considers a text is randomly assembled by several topics and a topic is reflected by the probability distribution of words.

Let T is the number of topics, the probability of word w_i in a given document can be expressed as below:

$$p(w_i) = \sum_{j=1}^{T} p(w_i|z_i = j)p(z_i = j) \tag{1}$$

Where w_i represents the i-th word in the document, $p(w_i|z_i =j)$ denotes the probability of w_i belongs to topic j, $p(z_i =j)$ is the probability of topic j belongs to the current document.

Let T, D and W stand for the number of topics, documents and unique words separately. The j-th topic can be expressed as a multinomial distribution of V words in word set $\phi_{w_i}^{j} = p(w_i \mid z_i = j)$. A document can be described as a random mixture of K latent topics $\theta_{w_i}^{j} = p(z_i = j)$. Thus, the probability of word w appears in document d can be denoted as follow:

$$p(w|d) = \sum_{j=1}^{T} \varphi_w^j \cdot \theta_j^d \tag{2}$$

We make a priori probability assumptions of symmetrical *Dirichlet(α)* on θ^d, and make a priori probability assumptions of symmetrical *Dirichlet(χ)* on φ^z.

2.2   Gibbs Sampling

In order to obtain the value of φ and θ, we introduce the Gibbs sampling technique. The key of Gibbs sampling is creating a Markov chain firstly, and then taking the nodes closed to the target probability distribution as the samples.

Let $p(z_i =j| z_{-i}, w_i)$ be the posteriori probability which can be calculated as below:

$$p(z_i|z_{-i}, w_i) = \frac{\dfrac{n_{-i,j}^{w_i} + \chi}{n_{-i,j}^{(\cdot)} + W\chi} \cdot \dfrac{n_{-i,j}^{d_i} + \alpha}{n_{-i,\cdot}^{d_i} + T\alpha}}{\sum_{j=1}^{T} \dfrac{n_{-i,j}^{w_i} + \chi}{n_{-i,j}^{(\cdot)} + W\chi} \cdot \dfrac{n_{-i,j}^{d_i} + \alpha}{n_{-i,\cdot}^{d_i} + T\alpha}} \tag{3}$$

Where $z_i =j$ means the word w_i was assigned to topic j, z_{-i} is the words assignment of all topics z_k, $n_{-i,j}^{w_i}$ expresses the number of words belong to topic j, $n_{-i,j}^{(\cdot)}$ is the number of words belong to topic j, $n_{-i,j}^{d_i}$ stands for the number of words assigned to topic j in d_i, $n_{-i,\cdot}^{d_i}$ means the number of words assigned to some topic in d_i, and all the words numbers are excluding the assignment of $z_i =j$.

The process of Gibbs sampling is as below:

(1) Initial state of Markov Chain: assign each word w_i to one of the T topics randomly.
(2) Obtain the next state: from w_1 to w_N, assign them to topics in turns based on Eq.(3).

(3) Iterate step(2) until the current Markov Chain closes to the target distribution.

For every single sample, estimate the value of φ and θ by the follow formulas.

$$\hat{\phi}_w^{z=j} = \frac{n_j^{(w)} + \chi}{n_j^{(\cdot)} + W\chi}, \quad \hat{\theta}_{z=j}^{(d)} = \frac{n_j^{(d)} + \alpha}{n_{\cdot}^{(d)} + T\alpha} \tag{4}$$

Where, $n_j^{(w)}$ means the times of assigning word w to topic j, $n_j^{(\cdot)}$ stands for the number of words belong to topic j, $n_j^{(d)}$ expresses the number of words that allocated to topic j in document d, $n_{\cdot}^{(d)}$ is the number of words which have been assigned to topics in d.

2.3   Selecting Topic Words

Based on the probability $p(w|d)$ obtained above, we abstract topic words in a document according to their Shannon information.

$$I(w) = -N(w)\ln p(w|d) \tag{5}$$

Where, $N(w)$ is the times that word w occurs in document d, $p(w|d)$ is the probability calculated by Eq.(2). If the Shannon information of a word is high, it is more likely to be treated as a topic word.

In order to improve the precision of topic words, words association is needed to determine the finial topic words after obtain the initial topic words. The process of topic association includes normalize, merger and substitution three steps. J. Shi e.t [9] have given the concrete steps, here we omit the detailed process.

3   Clustering Documents by Huffman Tree-Based Algorithm

In this section, we define some related concepts, and then design a new algorithm for clustering multiple documents based on Huffman tree.

3.1   Related Concepts

In this part we'll discuss the related definitions and the standard for evaluating the clustering results.

3.1.1   Similarity Measure

Let document set $D=\{d_1, d_2,..., d_m\}$, where d_i is the i-th document in D. Let $d_i = \{w_1, w_2,..., w_n\}$, in which w_j is the j-th word in d_i. Some of the words are topic words and some are not, let $W_i =\{ t_1, t_2,..., t_k\}$ be the topic word set, where t_k is the k-th topic word in d_i. Let the supporting of word w in document d as $sup(w,d)=n_{w,d}/n_d$, where $n_{w,d}$ is the times of w occurs in d and n_d is the number of words in d.

Definition 1. The similarity between two documents d_i and d_j can be defined as follow:

$$sim(d_i, d_j) = \frac{\sum_{X \in \{W_i \cap W_j\}} \min\{\sup(X, d_i), \sup(X, d_j)\}}{\sum_{X \in \{W_i \cup W_j\}} \max\{\sup(X, d_i), \sup(X, d_j)\}} \quad (6)$$

Where $\cap$ and $\cup$ detonate the set intersection and set union separately, $X \in \{W_i \cap W_j\}$ expresses X exists as a topic word both in d_i and d_j, $sup(X, d_i)$ is the *support* of X in document d_i.

Definition 2. The similarity between a cluster C_t and a document d_i can be defined as follow.

$$sim(d_i, C_t) = \frac{\sum_{d_j \in C_t} sim(d_i, d_j)}{|C_t|} \quad (7)$$

Where d_j is the j-th document in C_t, $|C_t|$ is the number of documents of C_t.

Definition 3. Let D be a set of documents: $D = \{d_1, d_2, \cdots, d_m\}$, DSM is a $m*m$ matrix, $DSM(i,j)$ is the value of $sim(d_i, d_j)$, we call DSM a similarity matrix.

3.1.2   The Criterion of Measuring the Quality of a Classification

We design a *goodness* measure which considered cohesion, coupling and the number of clusters to evaluate the quality of a classification. We deem a good classification should have high cohesion, low coupling and a relatively small number of classes. Therefore, the classification which has the highest goodness is the goal we are looking for.

We define the intra-class similarity and inter-class similarity as Definition 4 and Definition 5 to evaluate the cohesion and coupling, separately, and then give the *goodness* definition in Definition 6 to measure the quality of a classification.

Definition 4. The intra-class similarity *intra_sim* of a classification C is defined as follow.

$$intra\_sim\ (C) = \frac{1}{n}\sum_{t=1}^{n} in\_sim\ (C_t), \quad for\ |C| = n,$$

$$where, in\_sim\ (C_t) = \begin{cases} 1, & |C_t| = 1, \\ \dfrac{\sum_{d_i, d_j \in C_t} sim\ (d_i, d_j)}{C_{|C_t|}^2}, & |C_t| > 1, \end{cases} \quad (8)$$

Definition 5. The inter-class similarity *inter_sim* of a classification C is defined as below.

$$inter\_sim(C) = \begin{cases} 0, & |C| = 1, \\ \dfrac{\displaystyle\sum_{C_p, C_q \in C} \dfrac{\displaystyle\sum_{d_i \in C_p, d_j \in C_q} sim(d_i, d_j)}{|C_p| * |C_q|}}{C_{|C|}^2}, & |C| > 1, \end{cases} \tag{9}$$

Definition 6. Let $C = \{C_1, C_2, ..., C_n\}$ be a classification of document set $D = \{d_1, d_2, \cdots, d_m\}$. The *goodness* of classification C is defined as follow.

$$goodness(C) = \frac{intra\_sim(C) - inter\_sim(C)}{\ln n} \tag{10}$$

In the light of Definition 6, the higher the goodness value, the better the classification.

3.2   Clustering Algorithm Based on Huffman Tree

3.2.1   Huffman Tree

Huffman tree is an optimal binary search tree which holds the minimum weighted path length. Suppose there are n weights, the corresponding Huffman tree must have n leaf nodes $w_1, w_2, ..., w_n$, the process of generating Huffman tree can be described as follow:

(1) Take the nodes $w_1, w_2, ..., w_n$ as a forest with n independent trees, i.e. treat each node as a single tree.
(2) Choose the two trees with the minimum weights as two sub-trees to build a new tree, and assign the sum of all children's weight to the root of the new tree.
(3) Remove the two trees that have been chosen and add the new tree into forest.
(4) Repeat step (2) to step (3) until there is only one tree in the forest, the final tree with n leaves is the Huffman tree.

Step (2) and (3) are the core of this process, they will be repeated for n times and the number of trees will be decreased by 1 in each cycle. At the end of the process, there would be only one tree which just contains n leaves.

3.2.2   Clustering Algorithm

In this section, we propose a new clustering algorithm for text documents based on Huffman tree. Our algorithm mainly divided into two procedures: firstly, we generate all the candidate classifications based on Huffman tree, and then select the best classification based on the *goodness* measure. The pseudo code is as follows:

Algorithm 1 *HuffmanClustering* algorithm

```
Input: DSM_{m*m}                    //the DSM of m documents
Output: C*={U_1, U_2,…, U_n}        //the best classification
```

Procedure1 Generate the candidate classification set

```
Begin
(1)Let each document as an independent tree: U_i={d_i}.
   Return C_1={ U_1, U_2,…, U_m }.
(2)For  k=2 to m     do{
     while(k<m+1)    do{
       C_k = C_{k-1};
```

$\quad\quad$ If ($\exists\ U_p, U_q \in C_{k-1}$ and $sim(U_p,U_q) = max_{U_x,U_y \in C_{k-1}}\{sim(U_x,U_y)\}$)

```
          do{
                U_{p-q} ← U_p, U_q;
                Remove U_p and U_q from C_k;
                C_k ← U_{p-q; }}
      Return C_k={…, U_{p-q},…}.
      }
```

(3) Return candidate classification set: $SetC=\{C_1,C_2,…,$
$C_k,…,C_m\}$

```
End
```

Procedure2 Select the best classification

```
Begin
```

(1) Calculate the goodness values of all classifications in $SetC$ based on Eq.(10).

(2) If (goodness (C_t)=$max_{t=1\ to\ m}$ { goodness (C_t)})
$\quad\quad\quad$ do{ $C^*= C_t$; }.

(3) Return $C^*=\{U_1, U_2,…, U_n\}$.

```
End
```

In this algorithm, the first procedure is generating candidate classifications based on Huffman tree. We regard all the documents as independent trees in step 1.(1), and then merge the two trees which have the minimum root weights into a new tree in step 1.(2), repeat this process until all the documents has been connected in a single tree. Each loop will generate a new forest, and we treat every forest as a candidate classification. A tree in a forest stands for a cluster which contains several documents. The second procedure of our algorithm is focus on selecting the best classification. We choose the classification based on its *goodness*.

The time complexity of step 1 is $O(m^3)$, where step 1.(1) and step 1.(2) take $O(m)$ and $O(m^3)$ times separately. Step 2 takes $O(m*logm)$ times, in which step 2.(1) takes $O(m)$ times and step 2.(2) takes $O(m*logm)$ times. Thus, the time complexity of our clustering algorithm is $O(m^3)$. But this worst case is not going to happen, because the number of clusters will be decreased after each cycle.

4   Experiments

4.1   Data Preparation

In order to demonstrate the performance of the proposed clustering algorithm, we have taken an experiment using Java Edition 6, and implemented on a computer with 1.6GHz Pentium processor and 2GB of memory. The text classification corpus is the Chinese corpus of Fudan University. This corpus contains 2815 documents which are divided into 10 classes: environment, computer, transport, education, economic, military, sports, medicine, arts and politics. We firstly segment words of documents using ICTCLAS system and remove the stop words based on Hownet dictionary as data preparation. Then we abstract initial topic words based on LDA model and determine the finial topic words through words association. Thus, the similarity between every two documents can be calculated depending on the similarity of topic words.

Table 1 is the result of classifying document corpus by artificial way.

Table 1. The result of artificial classification on document corpus

Class	Environment	Computer	Transport	Education	Economic
Number of documents	200	200	214	220	325

Class	Military	Sports	Medicine	Arts	Politics
Number of documents	249	450	204	248	505

4.2   Performance Evaluation Criteria

We use F1-measure to evaluate the performance of our algorithm. This criterion considers two indicators: precision and recall. Let $cluster_j$ be a cluster of our classification, and $class_i$ is a class belongs to artificial classification. The formulas of precision P_j, recall R_j and F1-measure $F1$ are defined as follows:

$$P_{ij} = \frac{n_{ij}}{n_i}, \quad R_{ij} = \frac{n_{ij}}{n_j},$$

$$F1(i, j) = \frac{2 * P_{ij} * R_{ij}}{P_{ij} + R_{ij}},$$

$$F1_j = \max_i \{F(i, j)\} \tag{11}$$

Where n_{ij} is the number of documents both in $cluster_j$ and $class_i$, n_i and n_j are stand for the size of $cluster_j$ and $class_i$, separately.

In order to evaluate the overall performance of our algorithm on the data set, we introduce a function F based on the F1-measure of all clusters.

$$F = \sum_{j} \frac{n_j}{n} F1_j \qquad (12)$$

Where n is the number of documents in the corpus. The high value of F means more of the documents have been assigned to correctly classes in the corresponding classification.

4.3   Experiment Results

We extract two results from our experiment as Fig.1 and Fig.2. One is the classification C_1 which just has 10 clusters, and the other is the classification C_2 which has the highest *goodness*. Fig.1 shows the number of documents in each class got by artificial classification (n_i), the number of documents in each cluster is obtained through clustering (n_j), the number of documents which are classified correctly (n_ij), precision (P_C_1), recall (R_C_1) and F1-measure $(F1\_C_1)$ of classification C_1. Fig.2 is the same indicators of C_2.

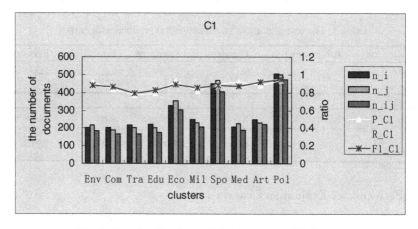

Fig. 1. The classification which just contains 10 clusters

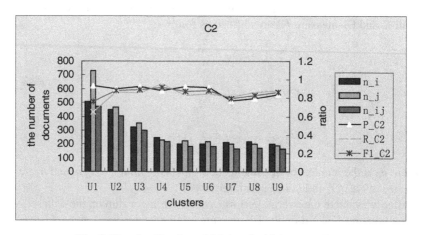

Fig. 2. The classification which has the highest *goodness*

In the two figures above, the horizontal axis stands for the clusters; the vertical axis indicates the number of documents and the secondary vertical axis signifies the ratio. The black bar, light gray bar and dark gray bar represent n_i, n_j and n_ij separately; the yellow line, blue line and purple line express precision, recall and F1-measure respectively.

Fig.1 shows the clustering result which just contains 10 clusters. Through observing the bar distribution, we found the number of documents assigned to each cluster by *HuffmanClustering* algorithm is similar to the result obtained by artificial classification, and the number of documents which have been allocated correctly is considerable. Through observing the line distribution, we discovered the precision, recall and F1-measure are all between 0.7 and 0.9 and the trend is relatively stable, thus, the result is close to the correct one.

Fig.2 is the clustering result which has the highest *goodness*. Through observing the bar distribution, we found 8 of the 9 clusters (except U_1) are the same as 8 clusters in fig.1, while U_1 has been assigned 730 documents. According to preliminary topic analysis, we learned the 8 clusters from U_2 to U_9 in fig.2 correspond to the clusters of sports, economic, arts, medicine, environment, transport, education and computer in fig.1, separately, while cluster U_1 contained most of the military and politics documents. In another words, C_1 is the result of performing the Huffman tree-based clustering algorithm only once based on C_2, in which the clusters of military and politics have been merged into a relatively large cluster. From the line distribution, we found most of the indicators of clusters are about 0.85 in addition to the accuracy of U_1. The main reason of this result is the definitions of the relative standards are not accurate enough, especially for the similarity measure and *goodness* measure.

According to Eq.(12), the F value of C_1 and C_2 are 0.929 and 0.842 respectively. The two relatively high values indicate our algorithm has a certain effect. In the follow-up work, we will improve the accuracy of the measure standards and the performance of our algorithm to achieve more excellent result.

5   Conclusions

Text document, a traditional information carrier, often contains a great amount of valuable knowledge, and the most effective method to mine massive amounts of documents is classifying them before mining, thus, document classification has become an important topic in natural language processing. In this paper, we classify documents with a new clustering method which can be divided into two procedures, firstly, abstract topic words depending on LDA model to obtain the similarity matrix, and design a new clustering algorithm based on Huffman tree to cluster documents, so as to generate the candidate classification collection; secondly, select the best classification from the collection according to our quality evaluation standard. Experiment results proved that our algorithm is effective and feasible.

Acknowledgments. This work was supported partly by NSFC(61462010,613630 36), BaGui scholars team project, the found of university science and technology of Guangxi, and The director found of Guangxi key laboratory.

Many thanks to Guangxi Collaborative Innovation Center of Multi-source Information Integration and Intelligent Processing, and the anonymous referee for comments.

References

1. Yang, Y., Liu, X.: A re-examination of text categorization methods. In: SIGIR 1999 (1999)
2. Salton, G.: Automatic Text Processing: The Transformation, Analysis, and Retrieval of Information by Computer. Addison-Wesley (1989)
3. McCallum, A., Nigam, K.: A comparison of event models for naive bayes text classification. In: AAAI 1998 Workshop on Learning for Text Categorization (1998)
4. Cohen, W.W., Hirsh, H.: Joins that generalize: Text classification using WHIRL. In: Proc. of the Fourth Int'l Conference on Knowledge Discovery and Data Mining (1998)
5. Joachims, T.: Text categorization with support vector machines: Learning with many relevant features. In: Proc. of the European Conference on Machine Learning (1998)
6. Lam, W., Ho, C.Y.: Using a generalized instance set for automatic text categorization. In: SIGIR 1998 (1998)
7. Baker, L., McCallum, A.: Distributional clustering of words for text classification. In: SIGIR 1998 (1998)
8. Blei, D.M., Ng, A.Y., Jordan, M.I.: Latent Dirichlet Allocation. Journal of Machine Learning Research 3, 993–1022 (2003)
9. Shi, J., Fan, M., Li, W.: Topic Analysis Based on LDA Model. Acta Automatica Sinica 35(12), 1586–1592 (2009)
10. Zhang, S., Zhang, C., Yan, X.: Post-mining: maintenance of association rules by weighting. Inf. Syst. 28(7), 691–707 (2003)
11. Zhang, S., Qin, Z., Ling, C.X., Sheng, S.: "Missing Is Useful": Missing Values in Cost-Sensitive Decision Trees. IEEE Trans. Knowl. Data Eng. 17(12), 1689–1693 (2005)
12. Wu, X., Zhang, S.: Synthesizing High-Frequency Rules from Different Data Sources. IEEE Trans. Knowl. Data Eng. 15(2), 353–367 (2003)
13. Wu, X., Zhang, C., Zhang, S.: Efficient mining of both positive and negative association rules. ACM Trans. Inf. Syst. 22(3), 381–405 (2004)
14. Wu, X., Zhang, C., Zhang, S.: Database classification for multi-database mining. Inf. Syst. 30(1), 71–88 (2005)
15. Zhao, Y., Zhang, S.: Generalized Dimension-Reduction Framework for Recent-Biased Time Series Analysis. IEEE Trans. Knowl. Data Eng. 18(2), 231–244 (2006)

Improved Spectral Clustering Algorithm Based on Similarity Measure

Jun Yan[1], Debo Cheng[2,*], Ming Zong[2], and Zhenyun Deng[2]

[1] Geographic Center of Guangxi, Nanning, Guangxi, 530023, China
[2] Guangxi Normal University, Guilin, Guangxi, 541004, China
Cheng7294@foxmail.com

Abstract. Aimed at the Gaussian kernel parameter σ sensitive issue of the traditional spectral clustering algorithm, this paper proposed to utilize the similarity measure based on data density during creating the similarity matrix, inspired by density sensitive similarity measure. Making it increase the distance of the pairs of data in the high density areas, which are located in different spaces. And it can reduce the similarity degree among the pairs of data in the same density region, so as to find the spatial distribution characteristics complex data. According to this point, we designed two similarity measure methods, and both of them didn't introduce Gaussian kernel function parameter σ. The main difference between the two methods is that the first method introduces a shortest path, while the second method doesn't. The second method proved to have better comprehensive performance of similarity measure, experimental verification showed that it improved stability of the entire algorithm. In addition to matching spectral clustering algorithm, the final stage of the algorithm is to use the k-means (or other traditional clustering algorithms) for the selected feature vector to cluster, however the k-means algorithm is sensitive to the initial cluster centers. Therefore, we also designed a simple and effective method to optimize the initial cluster centers leads to improve the k-means algorithm, and applied the improved method to the proposed spectral clustering algorithm. Experimental results on UCI[1] datasets show that the improved k-means clustering algorithm can further make cluster more stable.

Keywords: spectral clustering, Gaussian kernel, density sensitive, similarity measure, shortest path, k-means.

1   Introduction

The clustering is that the sample divided into many different clusters, the obtained clusters after division should meet the samples in the same cluster has a relatively high similarity degree, the similarity of samples in different clusters vary greatly [8, 9, 17].

---

* Corresponding author
[1] University of California Irvine(UCI for short), and the website is
http://archive.ics.uci.edu/ml/

X. Luo, J.X. Yu, and Z. Li (Eds.): ADMA 2014, LNAI 8933, pp. 641–654, 2014.
© Springer International Publishing Switzerland 2014

Cluster analysis technique can be used at all stages of data mining, such as data pre-processing stage, the demand for data, you can have a complex structure of multidimensional data clustering process by making complex structure data standardization, and thus provide data mining pretreatment for other methods of data mining. Cluster analysis can dig out hidden natural classification in the datasets, and the data is divided into different clusters [5,16,18]. Traditional clustering method usually does not require a priori information, it is unsupervised learning. Selecting a suitable method of similarity measure in cluster analysis is crucial, it is used as the basis for division. And a clustering method using different similarity measure is likely to produce different clustering results, even if the same method of similarity measure, when scale parameters are set differently, it would also lead large differences in the clustering results. Because data mining is widely used, its application has been optimistic about the prospects, clustering technology becomes a hot research topic. Therefore, people continue to strengthen the academic study of various clustering algorithms, their work has a very important significance.

Traditional clustering algorithms, such as k-means algorithm [6, 12], EM algorithm [2], which are based on the assumption that the sample space are convex and spherical. When the sample space is not convex, the algorithm is easy to fall into local optimum, so the application of these algorithms is limited. In recent years, spectral clustering algorithm as a new clustering technique attracts the researcher attentions and becomes a hot topic of machine learning, pattern recognition and other areas [19]. It is known that spectral clustering are based on dividing the spectrum theory. Compared with other traditional clustering technology, it can find clustering in the sample space with randomly distributed clustering structure, and eventually converges to the global optimal solution. According to the feature vector data, spectral clustering constructs a more simple data space during the implementation process, it not only reduces the dimension of the sample data, but also makes the distribution structure of the sample data become clearer in the subspace.

Spectral clustering method by solving the Laplacian matrix of feature vectors, and then execute the selected feature vectors division to identify the type of non-convex clusters [15, 20]. The implement of the algorithm is also simple. The efficiency of spectral clustering only concerns the number of data points, rather than the dimension of datum, so the algorithm can avoid the curse of dimensionality caused by high-dimensional. Spectral clustering demonstrated an excellent clustering effect in numerous experiments and applications, and its performance is better than those traditional clustering algorithms, moreover it can handle large data sets. Therefore, the current spectral clustering applications such as image and video segmentation, speech recognition, text mining and so on.

Traditional spectral clustering algorithms typically use Gaussian kernel function as a similarity measure, it needs to introduce a parameter σ. During creating a similarity matrix $\mathbf{W}$, the raw spectral clustering is often based on Euclidean distance, but it is impossible to accurately reflect the complexity of the data distribution [22, 23, 24, 25, 26, 27, 28, 29]. Inspired by the algorithm proposed by Wang Lin [13, 14], this paper makes it to increase the distance of pairs of data in the high density areas which are located in different rooms, while can reduce the similarity degree between the pairs of data in the same density region, so can find the spatial distribution characteristics of complex data. Meanwhile, in order not be affected by parameter σ, the Gaussian kernel function is not introduced in the design of similar matrix, so do not need to set any parameters and it is not affected by parameter σ. Distance can be directly used to calculate the similarity between data points, but also achieve the effect of scaling the similarity, and clustering results obtained are more stable. In addition, we all know that the time complexity of shortest path algorithm is generally $O(n^3)$ when solving figure issues, even though there are a lot improved fast algorithms, the time complexity is still very high.

When the data collection is increasing, the usual computer running the algorithm will not be executed, and data mining is often facing a huge data sets [30, 31, 32, 33]. Based on these reasons, this paper proposed another simple method of similarity measure, which does not need to calculate the shortest path, and does not introduce any parameter. Finally, the last step of traditional spectral clustering algorithm is completed by k-means algorithm. In the experiments, for a certain σ value to do many times experiments, the obtained clustering results (although the difference of clustering results is unapparent) is often different, that is caused by the k-means algorithm. Based on the study k-means algorithm and related improvements, this paper designs a simple and effective method of selecting initial cluster centers. The method is only to do a simple change based on the traditional k-means that is increasing random number (*e.g.* 50) in the step of random initializing the clustering center. For each randomly initialized k cluster centers, computing European cluster each other. Then save and choose the largest distance of k clustering centers as the initial clustering center. Applying the obtained clustering centers in the proposed clustering algorithm, it can make the clustering effect greatly improved, while maintaining the clustering results stable.

The remainder of the paper is arranged as follows. We review the previous work on the applications of clustering in Section 2. We give the detail of the proposed method in Section 3. Furthermore, we analyze the experimental results in Section 4, and give our conclusion in Section 5.

2   Related Work

Clustering analysis method is one of the main analytical methods in data mining, and it has been be studied by many researchers. We introduce the related research work as follows:

Guha proposed a new clustering algorithm called CURE that is more robust to outliers, CURE achieves this by representing each cluster by a certain fixed number of points [3]. Nearest neighbor consistency is a central concept in statistical pattern recognition, Ding and He extended this concept to data clustering, requiring that for any data point in a cluster, its k-nearest neighbors and mutual nearest neighbors should also be in the same cluster. And they proposed kNN and kMN consistency enforcing and improving algorithm that indicates the local consistency information helps the global cluster objective function optimization [1]. Gelbard compared different clustering methods using several datasets, and proposed a novel method called

Binary-Positive clustering, which is a kind of hierarchic cluster algorithm. It adjusted to use binary datasets and developed Binary-Positive similarity measures, and shows that converting raw data into Binary-Positive format will improve clustering accuracy and robustness, especially while using hierarchical algorithms [4]. Zhang Proposed an algorithm for predicting item's long-term popularity through influential users, whose opinions or preferences strongly affect that of the other users. They employed the k-means clustering algorithm and clustering smoothing model for recommender systems has shown extremely performance [21]. Michael present a tabu search based clustering algorithm, to extend the k-means paradigm to categorical domains, and domains with both numeric and categorical values [7]. As is the case with most data clustering algorithms, the algorithm requires a presetting or random selection of initial points (modes) of the clusters. The differences on the initial points often lead to considerable distinct cluster results. Sun present an experimental study on applying Bradley and Fayyad iterative initial-point refinement algorithm to the k-modes clustering to improve the accurate and repetitiveness of the clustering results [10].

3   Method

In this part, we analysis the spectral clustering from our sub experimental results and point out the existed problem, and then introduced the relevant knowledge involved in this paper, such as the similarity measure and k-means, lastly, the improved algorithm we design.

3.1   Spectral Clustering Method

The spectral clustering algorithm is a matching algorithm. The algorithm has three main processes:

The first step: calculate feature values and feature vectors of the sample similarity matrix; the second step: select the appropriate feature vector; the third step: use the k-means algorithm (or other traditional clustering algorithms) to cluster finally feature vector. Mainly, spectral clustering and the k-means clustering algorithm have a large difference is because that the ultimate objects of clustering are not different. Feature vector is selected clustering in spectral clustering, but k-means clustering is performed directly on the sample.

The following sub experiments is used to compare differences in the two algorithms on the clustering effect. The used data are three dedicated individuals working data sets (total six data sets) for studying the spectral clustering. In the Fig 1, (a), (c), (e) is the results of the spectral clustering algorithm, and (b), (d), (f) is the result of using the k-means clustering algorithm. It should be noted that the spectral clustering algorithm parameter σ is set to 0.01 (the value of σ is determined by many experiments). Through the left graphs, they show the perfect clustering results of spectral clustering. That the k-means algorithm have a limitation is because that it is based on a simple distance measure, which easy to fall into local optimal solution. So it is impossible to cluster these three artificial data sets correctly.

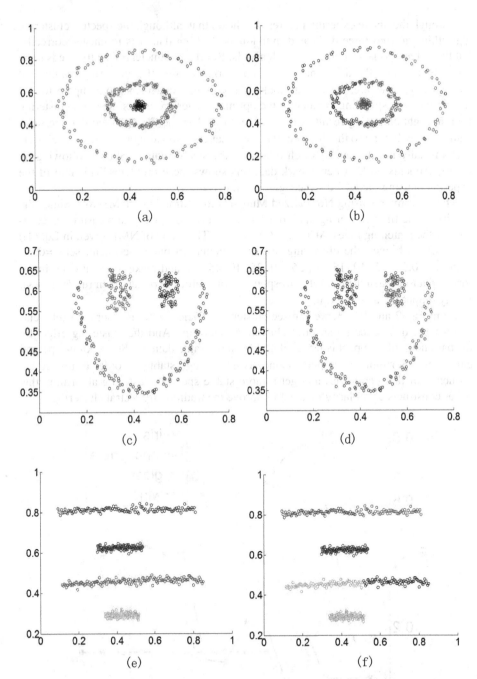

Fig. 1. The result of clustering on Spectral clustering and k-means algorithm

Through the above experimental results shows that, although the spectral clustering algorithm can find three dedicated individuals working data sets to cluster correctly, but the value of σ is needed for high demands. But the parameter σ usually needs to be set manually, and the different values of σ corresponding the clustering results will vary greatly, therefore, we need to make many times experiments to compare for selecting the corresponding σ value of the optimal clustering results. Spectral clustering find the right clustering results when σ is set of 0.01 in the Fig 1. Noted that the same values σ may belong to the different data sets that corresponding to the correct clustering is usually not the same, which related to the size of data attribute. Following by the experiments on UCI benchmark data sets shows these problems. This part of the experiment using the UCI data sets given in the Table 1.

The experiments using Normalized Mutual Information [11] (NMI) to evaluate the quality of the final clustering, and to achieve the level of quality to measure the accuracy of the matching index ACCURACY (ACC). The value of NMI (given in Eq. (4)) and ACC is bigger, the clustering quality is better. In our experiment, set σ =0.01, 0.05, 0.1, 0.5, 1, 5, 10, 15, 20, 25, 30, 35, 40, 45, 50, and make a repeat experiment for the dataset (each dataset set corresponding the values of σ, and then run 50 times), finally output the average value.

From Fig. 2 and Fig. 3, we can see clustering efficiency on the four data sets relatively large difference by spectral clustering algorithm. And the clustering effect by the parameter of σ impact is large relatively on the same data set. So we know spectral clustering algorithm on the performance is not enough stable. In order to reduce the influence of this parameter, and get a more stable spectral clustering algorithm, this paper constructs a similarity matrix to improve the traditional spectral clustering.

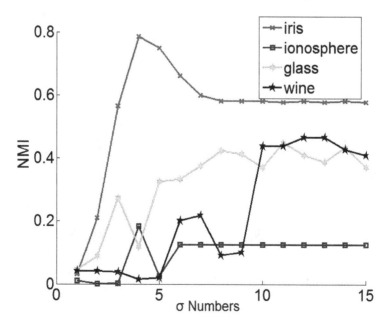

Fig. 2. Impact on the clustering performance parameters σ NMI

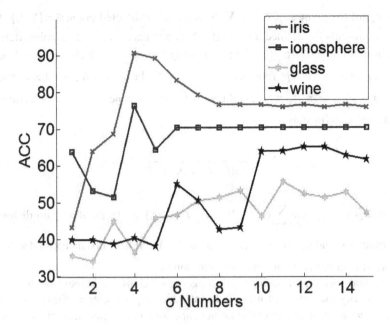

Fig. 3. Impact on the clustering performance parameters σ ACC

3.2   Similarity Measure

In the process of creating a similarity matrix **W**, the traditional spectral clustering usually calculated the similarity by Gaussian kernel based on Euclidean distance, but it can't reflect the complexity of the data distribution accurately. Therefore, this paper proposes that improving the similarity measure based on data density during creating the similarity matrix **W**. Making it to increase the distance of pairs of data in the high density areas which are located in different rooms, while can reduce the similarity degree between the pairs of data in the same density region, so can find the spatial distribution characteristics of complex data. The mainly step is to find the space distribution characteristics. Furthermore, in order to the clustering result is not affected by the parameter σ during design the similarity matrix, we don't introduce the Gaussian kernel function in the new algorithm, therefore, the clustering results are more stable and no affected by the parameter.

Firstly, we need to define the local line length:

$$\mathbf{L}(\mathbf{x}_i, \mathbf{x}_j) = e^{d^2(\mathbf{x}_i, \mathbf{x}_j)} - 1. \tag{1}$$

Where $d(\mathbf{x}_i, \mathbf{x}_j)$ is the Euclidean distance between the data points $\mathbf{x}_i$ and $\mathbf{x}_j$.

We regard the point as a vertex $\mathbf{V}$ in weighted undirected graph $\mathbf{G} = \{\mathbf{V}, \mathbf{E}\}$, the set of edge represents the connection weights between each pair of data points during the calculation. We regard $p \in V^l$ as the path of length $l = |p| - 1$, which connected between the point p_1 and p_2. And p_{ij} represents $(p_k, p_{k+1}) \in \mathbf{E}$ the set of all path that connect the data points of $\{\mathbf{x}_1, \mathbf{x}_2\}$, where $(1 \leq i, j < n)$. And the new Similarity distance between $\mathbf{x}_i$ and $\mathbf{x}_j$ defined as follows:

$$\mathbf{W}_{ij} = \frac{1}{\ln(1 + d_m(\mathbf{x}_i, \mathbf{x}_j)) + 1}. \tag{2}$$

Where $d_m(\mathbf{x}_i, \mathbf{x}_j) = \min\limits_{p \subset p_{i,j}} \sum\limits_{k=1}^{|p|-1} (e^{d^2(p_k, p_{k+1})} - 1)$, $d_m(\mathbf{x}_i, \mathbf{x}_j)$ is the shortest path distance between points $\mathbf{x}_i$ and $\mathbf{x}_j$. And $d(p_k, p_{k+1})$ is the Euclidean distance $\mathbf{x}_i$ and $\mathbf{x}_j$ in graph the shortest path between any two adjacent points.

Traditional spectral clustering algorithms regard Gaussian kernel function as a typically similarity measure, but it needs to introduce a parameter σ. Similarity measure mentioned without introducing the kernel functions in this paper, so there is no need to set any parameters. And not only directly used the distance to calculate the similarity between the data points, but also reached a scaled similarity effect.

It is generally known that the time complexity of the shortest path algorithm to solve the issue of graph is $\mathbf{O}(n^3)$, and the time complexity is very high. When the data collection is increasing, the algorithm will not be executed in PC. Moreover, data mining usually need to face large data sets, so this paper proposed another simple method of similarity measure, without calculate the shortest path.

In Similarity measure, the similarity distance is defined as follows:

$$\mathbf{W}_{ij} = \frac{1}{\ln(1 + d(\mathbf{x}_i, \mathbf{x}_j)) + 1}. \tag{3}$$

Where $d(\mathbf{x}_i, \mathbf{x}_j)$ is the Euclidean distance $\mathbf{x}_i$ to $\mathbf{x}_j$.

3.3   k-Means Method and the Optimization Clustering Method

From the foregoing, the final step of the traditional spectral clustering algorithm is completed by the k-means algorithm to cluster. Do many times experiments of spectral clustering for a certain σ value, they often get different clustering results. Due to the k-means algorithm is dependent on initial cluster centers seriously, however, the initial cluster centers usually randomly generated, so each cluster center is not the same at the start of the basic algorithm, and which may produce the unstable clustering results.

The pseudo of the traditional k-means algorithm as follows:

Algorithm 1: k-means method
Input: Contains n data members to be clustering of the dataset **X** and the number of clusters k.
output: k clusters.

1: Preprocessing the dataset **X**,
2: Choose from the dataset **X** random produce k data members as the initial cluster centers,
3: Calculate the Euclidean distance all data objects in the dataset **X** to each cluster center, the data object is divided into the cluster with a minimum Euclidean distance,
4: Separately calculate the average of in each class all the data objects, we regard these average values as the new cluster center class,
5: Until the criterion function is convergent or the cluster centers is no changed.

The pseudo of improved spectral clustering as follows:

Algorithm 2: The improved spectral clustering
Input: n data points $\{x_i\}_{i=1}^n$.
Output: The divided of data points $C_1, C_2, \cdots C_k$.

1: Construct the matrix $\mathbf{W} \in R^{m \times n}$ by the similarity measure based on data density

similarity matrix w. In matrix, any element of can be represent $\mathbf{W}_{ij} = \dfrac{1}{\ln(1 + d_m(x_i, x_j)) + 1}$

or $\mathbf{W}_{ij} = \dfrac{1}{\ln(1 + d(x_i, x_j)) + 1}$,where the element $\mathbf{W}_{ii} = 0$, $1 \le i, j < n$ on the diagonal.

2: To construct Laplacian matrix $\mathbf{L} = \mathbf{D}^{-1/2} \mathbf{W} \mathbf{D}^{-1/2}$, where **D** is a diagonal matrix,

$$\mathbf{D}_{ii} = \sum_{j=1}^{n} \mathbf{W}_{ij} .$$

3: Seeking the feature vectors $v_1, v_2, \cdots v_k$ corresponding to the k largest feature value in a Laplacian matrix **L**, and to constructing matrix $\mathbf{V} = [v_1, v_2, \cdots v_k] \in R^{n \times k}$, where **V** is the column vector.

4: Unitization the row of **V**, and get matrix **Y** , where $\mathbf{Y}_{ij} = \mathbf{V}_{ij} / (\sum_j \mathbf{V}_{ij}^2)$.

5: Each row of **Y** is regard as a point in R^k , which be clustered into k classes by k-means algorithm to optimize the initial cluster centers.
6: If the i-th row of **Y** belongs to class j, the original data points are classified into class j.

According to study the k-means algorithm and its relevant improved methods, we designed a simple and effective method of selecting initial cluster centers. The method is only to do a simple change based on the traditional k-means that is increasing random number (*e.g.* 50) in the step of random initializing the clustering center. For each randomly initialized k cluster centers, computing European cluster each other. Then save and choose the largest distance of k clustering centers as the initial clustering center. In result, the cluster center made clustering results stable and making the clustering result has been greatly improved.

The core of the improved algorithm as follow: random choose k data objects as the initial cluster centers from the dataset **X**; calculate the Euclidean distance between k clustering centers; and then repeat selected k data objects randomly; calculated Euclidean distance between k clustering centers again. If the distance is larger than the last time, we need to save the k cluster centers and the corresponding distance, or without modification for the next random selection until reach the set random number. Eventually, the algorithm will get a better initial cluster center. The pseudo is given in algorithm 2.

4   Experimental Analysis

We coded the algorithm with MATLAB 7.10.1 (R2010a) in windows 7 system. The used datasets mainly came from UCI, and showed the detail of the datasets in Table 1. In order to evaluate the quality of the clustering by using the Normalized Mutual Information (NMI) , and matching accuracy index of the extent of the clustering quality measure: ACC.

First of all, let **X** and **Y** are respectively the clustering results p_a , and random variables of the predetermined categories results p_b , and **H(X)** and **H(Y)** are **X** and **Y** entropy, let **I(X, Y)** for the mutual information between **X** and **Y**, we can find that **I(X, Y)** is no limit, and the **H(X)** =**I(X, X)**, so in the literature of **I(X, Y)** have been standardized, finally get the NMI as follows:

$$NMI = \frac{I(X, Y)}{\sqrt{H(X)H(Y)}} . \tag{4}$$

Table 1. Detailed information of benchmark data

Dataset	Wine	Iris	Ionosphere	glass
Instances	178	150	351	214
Features	13	4	34	9
classes	3	3	2	6

In order to exhibit the reliability, advantages and disadvantages of the two similarity measures, and the improvement effect of the k-means algorithm, the experiments were divided into three sub experiments, to facilitate comparison.

Experiment 1: algorithm using the similarity measure of Eq. (2), and the step (5) with the traditional k-means algorithm.

Experiment 2: algorithm using the similarity measure of Eq. (3), and the step (5) with the traditional k-means algorithm.

Experiment 3: algorithm adopts the similarity measure of Eq. (3), and the step (5) with the improved k-means algorithm.

Each part of the experiments were repeated 20 times, and then take the average of NMI and ACC. In addition, experiment 1, experiment 2 are recorded the algorithm running time (s) of each time, compared the shortest path effect on the running time of the algorithm.

(1) Experiment 1:

We summarized the clustering quality evaluation indicators results in Table 2. Control of the Table 2 and the Fig 2, Fig 3 in terms of the clustering quality indexes of NMI and ACC, we clearly found that the Fig 2 and Fig 3 reflects the stability of the traditional clustering algorithm exists some shortcomings during the clustering. And our improved method though conducted many times experiments, but, little difference between the results of each experiment, and the experimental result reached the highest value in the stable interval of the traditional spectral clustering algorithm in Fig 2, Fig 3. Especially, the iris dataset of NMI 0.7578 is the maximum among four datasets, the performance is evident.

Table 2. The test results in terms of two evaluation index value

Evaluate indicators	Wine	iris	Ionosphere	glass
NMI	0.4021	**0.7578**	0.1969	0.3553
ACC	71.7977	**90**	72.1368	49.3458
TIME	1510	**634.1531**	45000	3720

(2) Experiment 2:

In experiment 2, we obtained the clustering quality evaluation indicators shown in Table 3. Compared Table 3 with Table 2 in terms of NMI, ACC, TIME, we can easily find TIME in Table 2 is much larger than Table 3, the reason is that the algorithm in experiment 1 needs to calculate the shortest path, and the time complexity of computing for the shortest path algorithm requires too much time. And in terms of NMI and ACC, exception NMI and ACC of the iris dataset is same, the other data sets in Table 3 in terms of the index values were slightly smaller than the Table 2. This shows that the introduction of the shortest path in experiment 1 plays a good role, but not obvious. From the TIME point of view, experiment 2 algorithm is more suitable for practical application.

Table 3. The test results in terms of two evaluation index value

Evaluate index	Wine	iris	Ionosphere	glass
NMI	0.3976	**0.7578**	0.1963	0.3491
ACC	71.6292	**90**	72.0798	48.5514
TIME	0.9878	**0.8096**	5.4177	1.4794

In order to facilitate the experiment, experiment 3 also uses similarity measure, but not introduce the shortest path metric method, and compared with experiment 2.

(3) Experiment 3:

This part experiment, the introduced k-means algorithm to optimize the initial cluster center, and we proposed to use similarity measure methods in Eq. (3), the purpose is to further improve the stability of the improved algorithm. Comparing Table 4 and Table 3, we can see that the evaluation index of NMI and ACC have a small range increase. This is because of the k-means on the initial clustering center is dependent and may cause the clustering quality difference. So the improved k-means algorithm is introduced in this paper greatly, we can get the initial cluster center of high quality, diversity and the clustering result is not significant, so the average index evaluation experiments will be improved. Especially, the glass dataset improved most significantly, the dataset contains the most data categories, thus easily divided class mistake, so the traditional k-means behave more unstable. In Table 4, we can see that the clustering results of the glass data set ACC is least, this also shows that the instability of clustering, the clustering result is relatively more stable after the improved k-means algorithm is introduced than the clustering method in experiment 1 and experiment 2.

Table 4. The test results in terms of two evaluation index value

Evaluate index	Wine	iris	Ionosphere	glass
NMI	0.4122	**0.7632**	0.2135	0.3893
ACC	73.0337	**90.6667**	72.4986	52.8037

5   Conclusion

In view of the density sensitive similarity measure, we designed two methods to measure the similarity to improve the traditional spectral clustering algorithm. The experiments demonstrate the second method of similarity measure has a better performance than the first one. The proposed method solved the sensitive problem of Gauss kernel function parameter σ, the method is a non-parameter method. Lastly, the last stage of spectral clustering is sensitive to the initial clustering center of k-means algorithm to the selected feature vector clustering, so we designed an improved k-means algorithm that is very effective to optimize the initial cluster. The experimental result verified the improved spectral clustering method has better performance than other clustering methods.

References

1. Ding, C., He, X.: k-Nearest-Neighbor consistency in data clustering: Incorporating local information into global optimization. In: ACM Symposium on Applied Computing, pp. 584–589 (2004)
2. Dempster, A., Laird, N., Rubin, D.: Maximum likelihood from incomplete data vis the EM algorithm. Journal of Royal Statistical Society Series B 39(1), 1–38 (1997)
3. Guha, S., Rastogi, R., Shim, K.: CURE: An efficient clustering algorithm for large databases. ACM SIGMOD Record 27(2), 73–84 (1998)

4. Gelbard, R., Goldman, O., Spiegler, I.: Investigating diversity of clustering methods: An empirical comparison. Data & Knowledge Engineering, 155–156 (2007)
5. Huang, Z.: Extensions to the k-means algorithm for clustering large datasets with categorical values. Data Mining and Knowledge Discovery 2, 283–304 (1998)
6. Jain, A.: Data clustering: 50 years beyond k-means. In: ICPR, pp. 651–666 (2010)
7. Michael, K., Joyce, C.: Clustering categorical data sets using tabu search techniques. Pattern Recognition 35, 2783–2790 (2002)
8. Queen, J.M.: Some methods for classification and analysis of multivariate observations. In: Proceedings of the Fifth Berkley Symposium Math. Stat. Prob., vol. 1, pp. 281–297 (1967)
9. Qin, Y., Zhang, S., Zhu, X., Zhang, J., Zhang, C.: Semi-parametric optimization for missing data imputation. Appl. Intell. 27(1), 79–88 (2007)
10. Sun, Y., Zhu, Q., Chen, Z.: An iterative initial-points refinement algorithm for categorical data clustering. Pattern Recognition Letters 23, 875–884 (2002)
11. Strehl, A., Ghosh, J.: Cluster ensembles-a knowledge reuse framework for combining partitioning's. Journal of Machine Learning Research 3, 583–617 (2002)
12. Wagstaff, K., Cardie, C., Rogers, S., Schroedl, S.: Constrained k-means clustering with background knowledge. In: ICML, pp. 577–584 (2001)
13. Wang, L., Bo, L., Jiao, L.: Density-Sensitive Semi-Supervised Spectral Clustering. Journal of Software 18(10), 2412–2422 (2007)
14. Wang, L., Bo, L., Jiao, L.: Density-Sensitive Spectral Clustering. Acta Electronica Sinica 35(8), 1577–1581 (2007)
15. Xiang, T., Gong, S.: Spectral clustering with eigenvector selection. Pattern Recognition 41(3), 1012–1029 (2008)
16. Wu, X., Zhang, S.: Synthesizing High-Frequency Rules from Different Data Sources. IEEE Trans. Knowl. Data Eng. 15(2), 353–367 (2003)
17. Wu, X., Zhang, C., Zhang, S.: Efficient mining of both positive and negative association rules. ACM Trans. Inf. Syst. 22(3), 381–405 (2004)
18. Wu, X., Zhang, C., Zhang, S.: Database classification for multi-database mining. Inf. Syst. 30(1), 71–88 (2005)
19. Zhang, S., Zhang, J., Zhu, X., Qin, Y., Zhang, C.: Missing Value Imputation Based on Data Clustering. Transactions on Computational Science 1, 128–138 (2008)
20. Zhang, S., Chen, F., Wu, X., Zhang, C., Wang, R.: Mining bridging rules between conceptual clusters. Applied Intelligence 36(1), 108–118 (2012)
21. Zhang, J., Zhu, X., Li, X., Zhang, S.: Mining item popularity for recommender systems. In: Motoda, H., Wu, Z., Cao, L., Zaiane, O., Yao, M., Wang, W. (eds.) ADMA 2013, Part II. LNCS (LNAI), vol. 8347, pp. 372–383. Springer, Heidelberg (2013)
22. Zhang, S., Zhang, C., Yan, X.: Post-mining: maintenance of association rules by weighting. Inf. Syst. 28(7), 691–707 (2003)
23. Zhang, S., Qin, Z., Ling, C., Sheng, S.: "Missing Is Useful": Missing Values in Cost-Sensitive Decision Trees. IEEE Trans. Knowl. Data Eng. 17(12), 1689–1693 (2005)
24. Zhao, Y., Zhang, S.: Generalized Dimension-Reduction Framework for Recent-Biased Time Series Analysis. IEEE Trans. Knowl. Data Eng. 18(2), 231–244 (2006)
25. Zhu, X., Zhang, S., Jin, Z., Zhang, Z., Xu, Z.: Missing Value Estimation for Mixed-Attribute Data Sets. IEEE Trans. Knowl. Data Eng. 23(1), 110–121 (2011)
26. Zhu, X., Zhang, L., Huang, Z.: A Sparse Embedding and Least Variance Encoding Approach to Hashing. IEEE Transactions on Image Processing 23(9), 3737–3750 (2014)
27. Zhu, X., Huang, Z., Shen, H., Zhao, X.: Linear cross-modal hashing for efficient multimedia search. In: ACM Multimedia, pp. 143–152 (2013)

28. Zhu, X., Suk, H., Shen, D.: A novel matrix-similarity based loss function for joint regression and classification in AD diagnosis. NeuroImage 100, 91–105 (2014)
29. Zhu, X., Suk, H., Shen, D.: Matrix-Similarity Based Loss Function and Feature Selection for Alzheimer's Disease Diagnosis. In: CVPR, pp. 3089–3096 (2014)
30. Zhu, X., Huang, Z., Yang, Y., Shen, H., Xu, C., Luo, J.: Self-taught dimensionality reduction on the high-dimensional small-sized data. Pattern Recognition 46(1), 215–229 (2013)
31. Zhu, X., Huang, Z., Cui, J., Shen, H.: Video-to-Shot Tag Propagation by Graph Sparse Group Lasso. IEEE Transactions on Multimedia 15(3), 633–646 (2013)
32. Zhu, X., Huang, Z., Cheng, H., Cui, J., Shen, H.: Sparse hashing for fast multimedia search. ACM Trans. Inf. Syst. 31(2), 9 (2013)
33. Zhu, X., Huang, Z., Shen, H., Cheng, J., Xu, C.: Dimensionality reduction by Mixed Kernel Canonical Correlation Analysis. Pattern Recognition 45(8), 3003–3016 (2012)

A Partition Clustering Algorithm for Transaction Database Based on FCM

Meiqin Zhou[1], Yuwei Cuan[1], Dingrong Yuan[1,*], and Yuzhuo Wen[2]

[1] College of Computer Science& IT and Guangxi Key Lab of Multi-source Information Mining & Security at Guangxi Normal University, Guilin, 541004, China
[2] School of Economics and Management, Guangxi Normal University, Guilin, 541004, China
{Mengjhuan,cuanyuwei}@163.com,
dryuan& wenyuzhuo@mailbox.gxnu.edu.cn

Abstract. Large institutions usually have accumulated lots of transaction databases, which refer to multi-database. The effective method for acquiring useful knowledge from the multi-database is to classify them first, and then to mining. Usually, the technology of multi-database classification includes classifying and clustering. This article proposes a partition clustering algorithm to classify multi-database based on FCM. In the algorithm, a membership degree matrix is constructed firstly. And then, in the process of classifying, adjust the matrix to obtain a desired clustering result. Experiments show that our method is reasonable and effective.

Keywords: multi-database mining, membership degree, partition clustering, FCM.

1 Introduction

The development of database has brought a great revolution to the field of information management. At present, large organizations, especially transnational corporations have accumulated a huge amount of database, which is called multi-database. How to acquire useful knowledge from multi-database efficiently is a new challenge in the big data era. Therefore, multi-database mining becomes an important research subject in recent years.

To deal with the large data in multi-database, the effective method for multi-database mining is to classify them first, and then to mining. Database clustering, which is an important technique of multi-database classification, can automatically classify the databases without prior knowledge. Because of the structure of transactions in database is obviously different from text and webpage, the traditional clustering method cannot be directly transferred to the multi-database. Therefore, how to clustering multi-database effectively attracts the interest of scholars and many studies had been conducted.

However, with the demand of application increasing, limitations of existing multi-database clustering techniques have become more apparent. Currently, the clustering

---

* Corresponding author.

X. Luo, J.X. Yu, and Z. Li (Eds.): ADMA 2014, LNAI 8933, pp. 655–666, 2014.

algorithm for multi-database mainly includes partition method and hierarchy method. The partition clustering method can obtain more ideal clustering results, but the time complexity is higher. The hierarchy approach can obtain clustering results in short time, but it may loss better classifications. In this paper, in the light of previous studies, we propose a novel clustering algorithm for multi-database based on FCM. The rest of the paper is organized as follows.

The section 2 addresses the related work of multi-database clustering. In Section 3, the article induces the database clustering problems and related concepts. Following we propose a partition clustering algorithm for multi-database based on FCM in section 4. In section 5 is our experiments, which proves the validity of our strategy. Section 6 is the summary of the paper.

2   Related Work

With the development of network and database technology, large institutions have been accumulated many transaction databases, which means multi-database. Since Ribeiro presented the notion of knowledge discovery from multiple databases [1], many studies about multiple database mining have been conducting.

For instance, Liu et al. proposed an alternative method to search the relevant databases in multi-database [2], but their work was depended on specific applications. Wu and Zhang advocated an approach for identifying interesting patterns hidden in multi-database by weighting [3]. Zhang et al. researched the modes in multi-database and divided them into local, high vote, exceptional, and suggested mode [4].

Database clustering is an important technique in multi-database mining. Unlike text and webpage, structures of transactions in local databases are obviously different, and the traditional clustering method cannot be directly transferred to the multi-database. To solve this problem, Tang et al. measured the similarity between databases based on the itemset of transactions, and proposed an application independent method for classifying multiple-database [5].

Zhang et al. researched the goodness of classification and proposed a partition clustering approach, which can obtain the best classification of multi-database [6]. Furthermore, Animesh Adhikari et al. redefine the similarity measurement by considering the support of itemset in transactions, and presented a novel clustering method for multi-database [7]. Yuan and Fu et al. put forward a hierarchical database clustering approach based on high cohesion and low coupling [8].

In this paper, inspired by the related research, we propose a partition clustering algorithm for multi-database based on FCM. Experiments have proved that our method can obtain satisfactory clustering result in short time.

3   Multi-database Clustering

In this section, we describe the problem of clustering multi-database, and then give the method for measuring similarity between databases.

3.1   Problem Description

In multi-database set, there are many local transaction databases. If mining the multi-database directly, because of the structures of transactions in databases are obviously different, we may not obtain accurate knowledge. The effective method for multi-database mining is to classify them firstly, and then to mining.

In the technology of multi-database classification, database clustering can automatically classify local databases based on the similarity between them. The concepts of multi-database clustering are as follows.

Definition 1: Let D = {D1, D2,...,Dm} be a set of transaction databases. A class of D is defined as:

$$class(D) = \{D_i \mid D_i \in D\}$$

Definition 2: Let D= {D1, D2,...,Dm} be a set of transaction databases. A classification of D is defined as:

$$\pi(D) = \{class_i(D) \mid i = 1, 2, \cdots k\} \quad (k < n)$$

it satisfies

(1) $class_1(D) \bigcup class_2(D) \bigcup \cdots \bigcup class_k(D) = D$

(2) $class_i(D) \bigcap class_j(D) = \varnothing, \quad i, j \in \{1, 2, \cdots, k\}$

Where the k denotes the number of classes

For example, Let D={D1,D2,D3,D4,D5} be a multi-database set which is composed of five transaction databases: D1={a, b, c, d}, D2={b, c, d}, D3={e}, D4={f, h}, and D5={f, g, h}. After comparing the itemset in different databases, we can get a classification of D as: $\pi(D)$={{D1,D2},{D3},{D4,D5}}, and class(D)={D1,D2} is called a class of D.

Given a multi-database set D ={D1,D2,...,Dm}, the number of possible classifications is more than 2m-m. The purpose of multi-database clustering is to select an appropriate classification from them as the final clustering result.

3.2   Similarity Measurement

Similarity among databases is the foundation of database clustering. Because of the difference of structures in transactions, it is difficult to calculate the distance between databases by mapping them into the Euclidean space.

Similar to the category data, we can utilize the Jaccard coefficient to measure the relationship between local databases. However, if we compare the itemset of transactions directly, there might exist interference from abnormal data. In order to avoid such disturbance, we use the frequent itemset to measure the similarity among databases.

Definition 3: Let D={T1,T2,...,Tn} be a set of transaction databases, the itemset of D are Items(D)={I1,I2,...,In}. The support of itemset Ij is Sup (Ij), with the support threshold α, the frequent itemset of D is defined as:

$$FI(D)^{\alpha} = \{I_j \mid I_j \in Items(D), \ Sup(I_j) \geq \alpha\} \tag{1}$$

Where

$$Sup(I_j, D) = |I_j| \Big/ n \quad (j = 1, 2, \cdots, k)$$

For the convenience of description, suppose a transaction database D={(b,c),(a,c),(a)}, we can obtain the support of itemset in D as: {a:0.6,b:0.3,c:0.6,(b,c):0.3,(a,c):0.3}, with the support threshold α=0.5, the frequent itemset of D is FI(D)0.5={a, c}.

Frequent itemset can reflect the characteristic of the database. To measure similarity among databases accurately, we present a novel measurement by considering the difference among itemset of databases as follows.

Definition 4: Let D={D1,D2,...,Dm} be a set of transaction databases, the itemset of D are Items(Di)={I1,I2,...,In}, with the support threshold α. The similarity between Di and Dj is defined as:

$$sim(D_i, D_j)^{\alpha} = \begin{cases} 1 - A/B & , A \leq B \\ 1 - B/A & , A > B \end{cases} \tag{2}$$

Where

$$A = \sum_{X \in (FI(D_1) - FI(D_1) \cap FI(D_2))} Sup(X, D_i)$$

$$B = \sum_{X \in (FI(D_2) - FI(D_1) \cap FI(D_2))} Sup(Y, D_j)$$

For example, supposing a multi-database set D={D1,D2,D3}, with the support threshold α=0.5, the frequent itemset of databases and corresponding support are: FI(D1)0.5={a:0.5,b:1.0,(a,b):0.5}, FI(D2)0.5={a:0.6,c:0.5}.

We have A=1.0+0.5=1.5, B=0.6, and sim(D1,D2)0.5=1-0.6/(1.0+0.5)=0.6.

4   Partition Clustering for Multi-database

In order to classify multiple databases effectively, Zhang et al. presented a clustering methods to obtain the best classification of multi database [6]. On the basis of their studies, this section proposes a partition clustering method for multi-database based on FCM.

4.1   Relate Concepts

FCM algorithm is a partition clustering method which utilizes the degree of membership to determine the class of an object [9]. Based on the idea, we define the notion of membership degree for databases.

Definition 5: Let D = {D1, D2,...,Dm} be a set of transaction databases, classj is a class of D. The membership degree of database Di about classj is defined as μij, which satisfies:

$$\sum\nolimits_{j=1}^{n} \mu_{ij} = 1, \quad \mu_{ij} \in [0,1]$$

The membership degree μij denotes the probability of database Di being allocated in classj, and the whole membership degree of databases form the membership degree matrix of multi-database.

Definition 6: Let D ={D1,D2,...,Dm} be a set of transaction databases, The membership degree matrix of D is defined as:

$$A_{mm} = \begin{bmatrix} \mu_{11} & \mu_{12} & \cdots & \mu_{1m} \\ \mu_{21} & \mu_{22} & \cdots & \mu_{2m} \\ \vdots & \vdots & \ddots & \vdots \\ \mu_{m1} & \mu_{m2} & \cdots & \mu_{mm} \end{bmatrix}$$

Given a multi-database set $D =\{D_1,D_2,...,D_m\}$, and A_{mm} is one membership degree matrix of D. For each local database D_i, if we choose the class which owns the maximum value of membership degree as the true class, then A_{mm} corresponds to certain classifications of D.

For example, there is a multi-database set $D =\{D_1,D_2,D_3,D_4,D_5\}$, one membership degree matrix of D is:

$$A_{32} = \begin{bmatrix} \mu_{11} & \mu_{12} & \mu_{13} \\ \mu_{21} & \mu_{22} & \mu_{23} \\ \mu_{31} & \mu_{32} & \mu_{33} \end{bmatrix} = \begin{bmatrix} 0.6 & 0.3 & 0.1 \\ 0.5 & 0.5 & 0 \\ 0.3 & 0.5 & 0.2 \end{bmatrix}$$

In membership degree matrix A32, the $\mu 12$ stands for that the probability of database D1 being allocated in class2 is 0.3, and the corresponding clustering results of A32 are: $\pi 1=\{\{D1,D2\},\{D3\}\}$, $\pi 2=\{\{D1\},\{D2,D3\}\}$.

4.2   Goodness of Classification

For a multi-database set D = {D1, D2,...,Dm}, there will be a lot of possible classifications. In order to compare the effect of different clustering results, we define the goodness of classification by synthetically considering the inner similarity, outer similarity and amount of classes.

Definition 7: Let D={D1,D2,…,Dm} be a set of transaction databases, and a classification of D is π(D)={class1(D),class2(D),…,classk(D)}. With the support threshold α, the inner similarity of π (D) is defined as:

$$innerSim(\pi(D)) = \frac{1}{k} \cdot \sum_{class_i \in cluster(D)} \frac{\sum_{D_p \in class_i, D_q \in class_i, D_p \neq D_q} sim(D_p, D_q)^{\alpha}}{|class_i| \cdot (|class_i| - 1)/2} \tag{3}$$

The |classi| denotes the number of databases in classi.

Definition 8: Let D={D1,D2,…,Dm} be a set of transaction databases, and a classification of D is π(D)={class1(D),class2(D),…,classk (D)}. With the support threshold α, the outer similarity of π(D) is defined as:

$$outerSim(\pi(D)) = \frac{1}{k(k-1)/2} \cdot \sum_{class_i, class_j \in cluster(D)} \frac{\sum_{D_p \in class_i, D_q \in class_j} sim(D_p, D_q)^{\alpha}}{|class_i| \cdot |class_j|} \tag{4}$$

Where the |classi| and the |classj| denote the number of databases in classi and classj respectively, and the outer similarity equals 1 when k=1.

For a classification of multi-database, the inner similarity reflects the mean similarity among databases in same class, while the outer similarity shows the mean similarity in different classes. Usually, we consider a classification with larger innerSim and smaller outerSim as a better one.

Definition 9: Let D={D1,D2,…,Dm} be a set of transaction databases, and a classification of D is π(D)={class1(D),class2(D),…,classk(D)}. With the support threshold α, the goodness of π (D) is defined as:

$$goodness(\pi(D)) = InnerSim(\pi(D)) - OuterSim(\pi(D)) - \frac{k}{m} \tag{5}$$

Generally, in the possible classifications of a multiple databases set, the best clustering result hold the highest value of goodness.

For example, considering a multi-database set D = {D1, D2, D3, D4}, the similarities of different databases are:

	D_1	D_2	D_3	D_4
D_1	1.000	0.500	0.000	0.200
D_2	0.500	1.000	0.000	0.333
D_3	0.000	0.000	1.000	0.250
D_4	0.200	0.333	0.250	1.000

Suppose there are four classification of D as follows:

$\pi1(D)=\{\{D1\},\{D2\},\{D3\},\{D4\}\}$, $\pi2(D)=\{\{D1\},\{D3\},\{D2,D4\}\}$,
$\pi3(D)=\{\{D1,D2,D4\},\{D3\}\}$, $\pi4(D)=\{\{D1,D2,D3,D4\}\}$.

According definition 9, the goodness of each classification is:

$$goodness(\pi_1(D)) = \frac{1}{4}\times 4 - \frac{2}{4\times 3}\cdot(0.5+0+0.2+0+0.333+0.25) - \frac{4}{4} = -0.214$$

$$goodness(\pi_2(D)) = \frac{1}{3}(1+1+\frac{0.333}{1}) - \frac{2}{6}(\frac{0}{1}+\frac{0.5+0.2}{2}+\frac{0+0.25}{2}) - \frac{3}{4} = -0.131$$

$$goodness(\pi_3(D)) = 0.089, \; goodness(\pi_4(D)) = -1.036$$

Therefore, $\pi3$ (D) = {{D1, D2, D4}, {D3}} is the best classification among them because it has the highest goodness value.

4.3   Partition Clustering Algorithm for Multi-database

In order to classify multiple databases effectively, we design a partition clustering method, which could allocate each database to the nearest class iteratively.

Firstly, we give the definition of mean similarity between the database and class.

Definition 10: Let D={D1,D2,…,Dm} be a set of transaction databases, and a classi-fication of D is $\pi(D)$={class1(D),class2(D),…,classk(D)}. With the support threshold α, the mean similarity between Di and classj is defined as:

$$MeanDis(D_i, class_j) = \frac{1}{|class_j|} \cdot \sum_{D_q \in class_j} sim^\alpha(D_i, D_q) \qquad (6)$$

The |classj| denotes the number of databases in classj.

Given a multi-database set D= {D1, D2,…,Dm}, the processes of partition cluster-ing model are as follows:

(1) With the support threshold α, mine the frequent items of each database, and compute the similarities between different databases.
(2) Assignment the membership degree of each database randomly, and initial the membership degree matrix Amm according to definition 6.
(3) Compare the mean similarity between Di and each class, then find the nearest classj, following the values of μij by a certain step increase, the membership degree of Di about other classes will decrease according to definition 5.
(4) Update the classification according to Amm, and return to step(3). If the class of every database has no change, output the end clustering result.

Procedure 1 is the partition clustering algorithm for multi-database, based on the clustering model above.

```
Procedure 1 PartitionClustering
Input: D={D₁,D₂,…,Dₘ}: m transaction databases.
 α: support threshold of frequent itemset in database.
Output: BestParititon: best clustering result of multiple
databases.
begin
(1) get frenquent itemset of each database:
{FI(Dᵢ)ᵅ|i=1,2,…,m};
construct membership[m][m] to store the membership degree
matrix of multi-database, and initialize it with [0,1];
construct PartitionList[m]={class₁,class₂,…classₘ} to sore
the database in every classes, and initialize it accord-
ing to the membership[m][m].
(2) let flag ← true;
while flag = true do{
  let flag ← false;
  for i:=1 to m do{
    let miniDis ← 1, C ← -1;
      for j:=1 to m do{
        let dis ← 0;
        if classⱼ is empty or only contains Dᵢ do{ let dis
← 0.5;}
        if classⱼ is not empty do{ let dis ←
MeanDis(Dᵢ,classⱼ);}
        if dis < miniDis do{ let miniDis ← dis, C ← j;}
      }
    if membership[i][C]<1 do{
      let membership[i][C] ← membership[i][C] + 0.01;
      for t:=1 to j do{
        if t!=C do{
          let membership[i][t] ← membership[i][t] -
0.01/(j-1);
      }}}
    initialize PartitionList[m] according to the mem-
bership[m][m].
    if every database in PartitionList[m] have no
change do{ let flag ← true;}
  }}}
(3) remove the empty classes in PartitionList[m];
let BestPartition ← PartitionList[m];
return BestPartition;
end
```

In Procedure 1, Step 1 mines the frequent itemset of each database and initializes the membership degree matrix. Step 2 allocates each database to the nearest class and updates the clustering results. Step 3 returns the best classification of multiple databases. The Procedure 1 utilizes the structure of link to store classifications, and the space complexity is O (m2). The time consumption of the algorithm mainly focuses on Step 2, which is O (m3/3+m2/2+m/6). Therefore, the time complexity of Procedure 1 is O (m3).

5   Experiment

In order to prove the efficiency of our algorithm, we have carried out several experiments. One is to research the accuracy of the clustering result obtained by the partition clustering method, and the other experiment is to compare the time consumption of different algorithms. The experiments are implemented on a 1.6GHZ processor with 2GB of memory, and using Java Edition 6 Platform as development tool.

The datasets of our experiment are Iris and Abalone, which are from the UCI [10], and their attributes are shown in Table 1.

Table 1. The attribute of Datasets

Dataset	NI	TA	NA
Iris	150	float	4
Abalone	4,177	integer, float, category	8

Where NI is the number of dataset instances, TA denotes the type of instance attribute, and NA is the amount of instance attribute.

Firstly, we research the accuracy of classifications generated from the clustering algorithm. In the Iris dataset, there are three true classes as Iris Setosa, Iris Versicolour, and Iris Virginica. Based on the true class, we divided the instance of Iris into 10 sub-datasets and made them as the multi-database set D = {D1, D2... D10}. The attribute of each database are shown in Table 2.

In a classification of multi-database, the class of local databases may be different from the true class. In order to measure the effect of classifications, we give the definition of accuracy according to the true class of databases.

Definition 11: Let D={D1,D2,...,Dm} be a set of transaction databases, and a classification of D is $\pi(D)$={class1(D),class2(D),...,classk(D)}. The accuracy of $\pi(D)$ is defined as:

$$Accuracy(\pi(D)) = \frac{n}{m(m-1)/2} \tag{7}$$

Where n denotes the number of databases pair which are allocated in the right classes.

Table 2. The attribute of Databases

Database	NI	Class
D1	20	Iris Setosa
D2	25	Iris Virginica
D3	10	Iris Versicolour
D4	30	Iris Setosa
D5	12	Iris Versicolour
D6	17	Iris Virginica
D7	6	Iris Versicolour
D8	10	Iris Versicolour
D9	8	Iris Virginica
D10	12	Iris Versicolour

With different support threshold α, we mine the frequent itemset of databases, and processe the PartitionClustering in Procedure 1. When the value of α varies, the goodness and accuracy of classifications obtained by the clustering algorithm are shown as Fig.1.

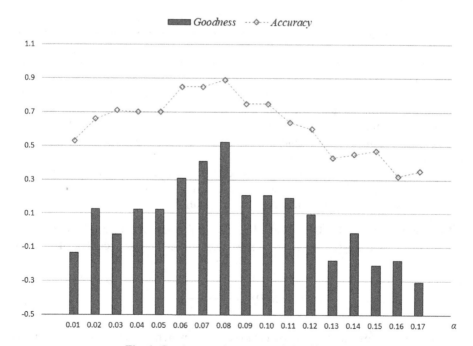

Fig. 1. Goodness and accuracy of classification

In the Figure, the horizontal axis denotes different support threshold α, the column and curve stand for the goodness and accuracy of the clustering results in the experiment respectively.

Fig.1 shows that our method could obtain rational classifications. Moreover, the trend of changes for goodness is almost the same as the accuracy, therefore the measurement in Definition 9 is effective. During the experiment, we also found that when the support threshold α is too small, there would be more disturbance from special items. So it is necessary to mine frequent itemset in multiple databases clustering.

Similar to the Iris dataset, we divide the instance of Abalone into sub-datasets as the multi-database set , and then get the frequent itemset of databases with the support threshold $\alpha=0.10$. When the amount of databases changes, we compare the time consumption of clustering algorithm PartitionClustering, BestClassification[6], and AprioriDatabaseClustering[7], and the result is shown as Fig.2.

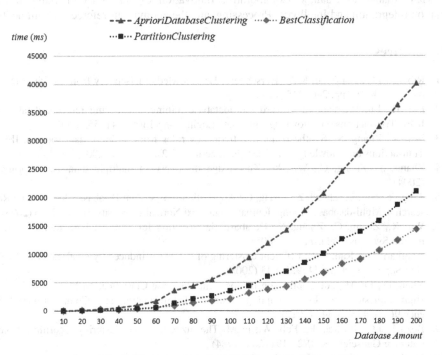

Fig. 2. Time consumption of different algorithms

In the Fig.2, the horizontal axis denotes the number of databases, and the curve stands for the time consumption of clustering algorithm.

We can see that the time consumption of algorithm PartitionClustering and BestClassification rise slowly, with the increaseing number of database, therefore our method is efficient. During the experiment, we found that the clustering result generated from AprioriDatabaseClustering was usually more accurate, and BestClassification could obtain the classification with zero outer similarity.

Consequently, the partition clustering algorithm proposed in this paper can obtain rational classifications of multi-database, and the time consumption is satisfactory.

6 Conclusions

Database clustering is an important technology of Multi-database classification, which can classify large transaction databases effectively[11],[12],[13],[14],[15]. In the light of previous studies, we proposed a partition clustering algorithm based on FCM. Our method optimizes the classes by adjusting the value of membership degree matrix, and then obtains the best classification of multi-database. Experiments have proved that our algorithm is effective and practice.

Acknowledgements. This work was supported partly by NSFC (61462010, 61363036), BaGui scholars team project, the found of university science and technology of Guangxi, The science and technology plan projects of Guilin(20140103-17) and the director found of Guangxi key laboratory.

Many thanks to Guangxi Collaborative Innovation Center of Multi-source Information Integration and Intelligent Processing, and the anonymous referee for comments.

References

1. Ribeiro, J.S., Kaufman, K.A., Kerschberg, L.: Knowledge Discovery from Multiple Databases. In: KDD, pp. 240–245 (1995)
2. Liu, H., Lu, H., Yao, J.: Toward multidatabase mining: Identifying relevant databases. IEEE Transactions on Knowledge and Data Engineering 13(4), 541–553 (2001)
3. Wu, X., Zhang, S.: Synthesizing high-frequency rules from different data sources. IEEE Transactions on Knowledge and Data Engineering 15(2), 353–367 (2003)
4. Zhang, S., Zhang, C., Wu, X.: Knowledge discovery in multiple databases. Springer (2004)
5. Tang, Y., Niu, L., Zhong, Z., Zhang, C.: Application-independent Database Classification Research in Multi-database Mining. Journal of Guangxi Normal University 21(4), 32–36 (2003)
6. Wu, X., Zhang, C., Zhang, S.: Database classification for multidatabase mining. Information Systems 30(1), 71–88 (2005)
7. Adhikaria, A., Rao, P.R.: Efficient clustering of databases induced by local patterns. Decision Support Systems 44, 925–943 (2008)
8. Yuan, D., Fu, H.: An Application-independent Database Classification Method Based on High Cohesion and Low Coupling. Journal of Information and Computational Science 9(15), 4337–4344 (2012)
9. Bezdek, J.C., Ehrlich, R., Full, W.: FCM: The fuzzy c-means clustering algorithm. Computers & Geosciences 10(2), 191–203 (1984)
10. Machine Learning Repository, http://archive.ics.uci.edu/ml
11. Zhang, S., Zhang, C., Yan, X.: Post-mining: Maintenance of association rules by weighting. Inf. Syst. 28(7), 691–707 (2003)
12. Zhang, S., Qin, Z., Ling, C.X., Sheng, S.: "Missing Is Useful': Missing Values in Cost-Sensitive Decision Trees. IEEE Trans. Knowl. Data Eng. 17(12), 1689–1693 (2005)
13. Wu, X., Zhang, S.: Synthesizing High-Frequency Rules from Different Data Sources. IEEE Trans. Knowl. Data Eng. 15(2), 353–367 (2003)
14. Wu, X., Zhang, C., Zhang, S.: Efficient mining of both positive and negative association rules. ACM Trans. Inf. Syst. 22(3), 381–405 (2004)
15. Zhao, Y., Zhang, S.: Generalized Dimension-Reduction Framework for Recent-Biased Time Series Analysis. IEEE Trans. Knowl. Data Eng. 18(2), 231–244 (2006)

Discernibility Matrix Enriching and Boolean-And Algorithm for Attributes Reduction

ZhangYan Xu[*], Ting Wang, JinHu Zhu, and XiaFei Zhang

College of Computer Science and Information Technology,
Guangxi Normal University, Guilin 541004, China
xyzwlx@sina.com.cn, betty0907@sina.cn,
{172861479,980739396}@qq.com

Abstract. discernibility matrix and binary discernibility matrix method is easy to understand and design, which has aroused great concern by many scholar. Research shows that the two methods produce a large number of repeated and useless elements (if A is the subset of B, B is the useless element of A) on the fly. These repeated and useless elements occupy a lot of space and will affect the efficiency of the algorithm. If we delete these elements, the storage is much less than before, and the algorithm will be increased. For this purpose, professor Yang Ming give the definition of enriching discernibility matrix [3],which all the discernibility elements are not repetition and mutually exclusive. Some scholars adopt the method of comparison between every two discernibility elements to get the enriching discernibility matrix. Some present the algorithm, every nonempty entry of a discernibility matrix is stored one path in the enriching tree and a lot of nonempty entries share one path or sub-path. However, these algorithms only delete part of the useless elements in spite of lower storage space. In this paper, we put forward discernibility matrix enriching and Boolean-And algorithm for attributes Reduction. The algorithm is easy to understand and easy to design. The Analysis Experiment and Experimental Comparison show the algorithm is feasible and effective.

Keywords: Rough set, discernibility matrix, enriching discernibility matrix, Boolean-and, attribute reduction.

1 Introduction

As a new analysis of incomplete, inaccurate and inconsistent information system tool, Rough set theory[1,2] has been widely applied in many fields such as data mining, machine learning, artificial neural network, etc.. Attribute reduction is one of the important research content in Rough Set Theory. Under the premise of the certain classification ability unchanged, through the knowledge reduction, deriving problem of decision making or classification rules is the main ideas of attribute reduction. In recent years, some new theories and reduction methods have been developed. However, classical rough set theory is unsuitable for attribution reduction in incomplete decision systems. People are confronted with the incomplete decision table

---

[*] Corresponding author.

X. Luo, J.X. Yu, and Z. Li (Eds.): ADMA 2014, LNAI 8933, pp. 667–676, 2014.
© Springer International Publishing Switzerland 2014

because of the error of measurement data and are limitation on the knowledge acquisition .etc. Owing to the large number of attribute and instance, the efficiency of attribute reduction algorithm is particularly important in the large number data sets of information systems. Attribute reduction algorithm hasn't been a recognized and efficient algorithm until now.

Research shows that the majority reduction methods are based on discernibility matrix which it is simplicity and efficient. Nevertheless, for decision table with N number of object and m number of attributes, the space complexity of Algorithm is $O(|C||U|^2)$ .So if the number N or m is larger; the storage cost is higher based on discernibility matrix.

According to the definition based on discernibility matrix attribute reduction. Many discernibility elements are useless. For this purpose, professor Yang Ming give the definition of enriching discernibility matrix [3],which all the discernibility elements are not repetition and mutually exclusive. Papers [3, 4] give an attribute reduction algorithm based on enriching discernibility matrix. Paper[5]points out that if all the discernibility elements are not repetition and mutually exclusive, the algorithm is complete by the frequency of discrimination elements and proves the discernibility elements are no less than $|C|C_{|C|}^{\lceil|C|/2\rceil}$ .Papers[3,5] adopts method of comparison between every two discernibility elements to get the enriching discernibility matrix. Its time complexity is $O(|C|^2|U|^4)$.

In order to reduce the storage of discernibility matrix, Paper [6] gives attribute reduction algorithm of FP-Tree based on the simple discernibility matrix. Paper [7] gives attribute reduction algorithm based on condensing tree structure(C-Tree), which is an extended order-tree. According to the algorithm, every nonempty entry of a discernibility matrix is stored one path in the C-Tree and a lot of nonempty entries share one path or sub-path, hence the C-Tree has much lower space complexity as compared to discernibility matrix. Paper [8]gives a kind of pattern tree to record discernibility elements, which function is the same as paper [7].Paper [9] points out an algorithm, which aims at static information or static decision table and is unsuitable for dynamic information or dynamic decision Table. Moreover the real world change all the time, and object dynamic and ever changing in the decision table, so that the attribute reduction that has been got might be no longer valid. Therefore, Paper [8]   improves the algorithm based on C-Tree, which not only it gets the valid attribute reduction, but also it reduces cost of storage. In addition, Paper [10]presents a new ordering strategy, according to the descending order of C-Tree improved algorithm, which its design idea is based on FP-Tree. In conclusion, the above literature doesn't present an effective algorithm to solve enriching discernibility matrix from beginning to end.

In order to reduce the time complexity of enriching discernibility matrix, papers [11-16] inspires us to design an algorithm of solving enriching discernibility matrix to delete redundant element. In this paper, firstly, we should deal with elements in binary discernibility matrix by the method of merge sort. Then, the elements have AND operation with each other which are sorted. Enriching discernibility matrix is much higher value by the experiment.

2   Preliminary Concepts of Rough Set Theory

Definition 1[17] In rough set theory, a data set can be formally described using an information system (also called a decision table)An information system is denoted as $S = (U,C,D,V,f)$, where $U = \{x_1, x_2, ..., x_n\}$ is a nonempty finite set of objects or cases, called universe, where $C = \{c_1, c_2, ..., c_n\}$ is the set of conditional(also called conditional attributes in this paper) features, $C \cap D = \varnothing$; $V = \bigcup_{a \in C \cup D} V_a$,and V_a is the domain of the feature a. $f : U \times C \cup D \to V$ is a total function such that $f(x,a) \in V_a$ for every $a \in C \cup D, x \in U$.

If there exists at least $a \in C$ and V_a includes null value (presented *) (recorded $f(x,a) = *$) in the information system, we call it a complete decision Table.

Definition 2[17] given a complete decision table $S = (U,C,D,V,f)$ with $B \subseteq C$, the subset B determines a binary relation, denoted by $T(B)$, which is defined as follows: $T(B) = \{(x, y) \in U \times U | \forall b \in B, f(x,b) = f(y,b) \vee f(x,b) = * \vee f(y,b) = *\}$ It is easy to prove that $T(B)$ is reflexive and symmetric, so is a tolerance relation.

Generally, $T(B)$ denotes the maximal set of objects which are possibly indiscernible by B with object x , Equivalently, $T_B(x) = \{y | (x, y) \in T(B), y \in U\}$.

Definition 3[17] given a complete decision table $S = (U,C,D,V,f)$ with $\forall B \subseteq C \cup D$, by U / T_B we denote the set of all maximal tolerance classes with respect to B . $U / T_B : U / T_B = \{T_B(x) | x \in U\}$ is said to be a full cover of D .

Definition 4[18] given a complete decision table $S = (U,C,D,V,f)$, for a subset $X \subseteq U$ every $B \subseteq C \cup D$, the B-lower and B-upper approximation of X are defined, respectively, as follows:

$$B\_(X) = \{x \in U | T_B(x) \subseteq X\} ; B^-(X) = \{x \in U | T_B(x) \cap X \neq \varnothing\}$$

Definition 5 [13] Let $\forall B \subseteq C$, the B-positive region of D is denoted as $POS_B(D) = \bigcup_{D_i \in U / D} B\_(X)$, simply mark U_{pos} , the B-negative region of D is denoted as $U_{neg} = U - U_{pos}$.

Definition 6[13] given a complete decision Table $S = (U,C,D,V,f)$ with $\forall b \in B \subseteq C$, if $POS_B(D) = POS_{B-\{b\}}(D)$, a is unnecessary for B relative to D .Otherwise, a is necessary for B . $\forall B \subseteq C$, if arbitrary element of B is unnecessary, we call that B is independent with respect to D .

Definition 7[13] given a complete decision table $S = (U,C,D,V,f)$ with $\forall B \subseteq C$, if $POS_B(D) = POS_C(D)$ and B is independent with respect to D .Then B is the attribute reduction.

Definition 8[13] Given a complete decision table $S = (U,C,D,V,f)$, $U = U_{pos} \cup U_{neg}$, a binary discernibility matrix is defined as an $n \times m$ matrix with $(m(i,j),k)$, marked $M = (m(i,j),k)$, given by:

$$m((i,j),k) = \begin{cases} 1 & c_k \in C, f(x_i,c_k) \neq f(x_j,c_k) \wedge f(x_i,c_k) \neq *, \wedge f(x_j,c_k) \neq *, \\ & f(x_i,D) \neq f(x_j,D) \text{ and } x_i,x_j \text{ are in the } U_{pos}; \\ 1 & f(x_i,c_k) \neq f(x_j,c_k) \wedge f(x_i,c_k) \neq *, \wedge f(x_j,c_k) \neq *, \\ & \text{and if } x_i \text{ is in the } U_{pos}, x_j \text{ must be in the } U_{neg}; \\ 0; & \text{otherwise.} \end{cases}$$

Among $k = 1,2,\ldots,r$.

Definition 9[13]   if $M = (m(i,j),k)$ is a simple binary discernibility matrix of a complete decision table. $\forall P \subseteq C$ If P satisfies the following conditions : First, all of the attributes of P corresponding to each column forms the sub matrix of M, called M'. Moreover, the rows what values are not all 0 of M' is equal to M's; second, every $B' \subset B$ doesn't meet the first condition. Then P is the attribute reduction.

Proof 1[13] if $M = (m(i,j),k)$ is a simple binary discernibility matrix of a complete decision table. $\forall P \subseteq C$ If P satisfies the conditions: all of the attributes of P corresponding to each column forms the sub-matrix of M, called M'. Moreover, the rows what values are not all 0 of M' is equal to M's .Then $POS_P(D) = POS_C(D)$ 。

Proof 2 [13] if $M = (m(i,j),k)$ is a simple binary discernibility matrix of a complete decision table. $\forall P \subseteq C$ If there is $POS_P(D) = POS_C(D)$, then, we obtain a conclusion as follows: all of the attributes of P corresponding to each column forms the sub-matrix of M, called M'. Moreover, the rows what values are not all 0 of M' is equal to M's.

Proof 3 [13] Attribute Reduction based on Positive is equal for Attribute Reduction based on simple binary discernibility matrix.

3   Enriching Discernibility Matrix and The Algorithm of Attribute Reduction

The idea of these algorithms describes as follow: firstly, we should deal with elements in binary discernibility matrix by the method of merge sort. Then, the redundancy elements will be deleted while the elements have AND operation with each other which are sorted. There brings in the definition of enriching discernibility matrix in literature [3].

Definition 10 [3] A simple binary discernibility matrix $M = (m(i, j), k)$, its correspond-ing enriching discernibility matrix that all the discernibility elements are not repetition and mutually exclusive. (Marked $IME(M)$ in this paper) is, given by:

$$IME(M) = \{m \mid m(m \neq \varnothing) \in M \ \wedge \text{not exist} \ m'(m' \neq \varnothing) \in M \Rightarrow m' \subset m\}.$$

For example, A simple binary discernibility matrix $M = (m(i, j), k)$ as follow:

$$M = \begin{cases} abc & \varnothing & ac & abcde \\ bc & bd & acd & ace \\ ab & ad & \varnothing & acde \\ ae & \varnothing & bd & bcde \end{cases},$$ its corresponding enriching discernibility matrix is

$$IME(M) = \begin{cases} \varnothing & \varnothing & ac & \varnothing \\ bc & bd & \varnothing & \varnothing \\ ab & ad & \varnothing & \varnothing \\ ae & \varnothing & \varnothing & \varnothing \end{cases}$$

Two matrixes show that simple binary discernibility matrix has 38 discernibility elements and its corresponding enriching discernibility matrix has 12 discernibility elements. That is storage space of enriching discernibility matrix is less than 30% of the original discernibility matrix's. In conclusion, storage space of enriching discernibility matrix is much less than the original discernibility matrix's.

Algorithm: A Boolean-And Algorithm for Attributes Reduction Based on Enriching Discernibility Matrix

Input: An incomplete decision Table $S = (U, C, D, V, f)$, $U = \{x_1, x_2, ..., x_n\}$, $C = \{c_1, c_2, ..., c_n\}$

Output: Enriching Discernibility Matrix IME and reduce(C)

Step1 Calculate $T_{ci}(a)(a \in U)$ with literature [13];

Step2 Calculate $U_{pos} = \{y_1, y_2, ..., y_s\}$, $U_{neg} = \{z_1, z_2, ..., z_t\}$ with literature [13];

Step3 Calculate binary discernibility matrix $M = (m(i, j), c)$ by definition8;

Step4 deal with elements in binary discernibility matrix by the method of merge sort. Marked $M' = (m(i, j), c)$ then store $M' = (m(i, j), c)$ in the array $B[|U|]^2[m]$;

Step5　$for(k = 1; k < |U|^2; k++)$

{

　　　　　$if (B[k] == \{0\})$ //The first k line does not exist of has been deleted
　　　　　$break$;
　　　　　$for(l = k + 1; l < |U|^2; l++)$
　　　　　{ 　$if (B[k] \ \& \ B[l] == B[k])$
　　　　　Delete $B[l]$;
　　　　　$else$
　　　　　$break$; 　}
　　　　　}

Step6 obtain enriching discernibility matrix IME above the three Steps

Step7 $while(IME = (m(i, j), k)! = \varnothing)$

$\{$ $T = \varnothing;$

Step7.1 Judging each row in the IME, if the row has a attribute element recorded 1,then let the attribute element merge into $reduce(C)$ .In the end, delete all the row which the row corresponding attribute element recorded 1;

Step7.2 take free attribute b merge into T in the $C - T - reduce(C)$, then update attribute b corresponding column, that is value is 1 change 0.

Analysis for the time complexity of Algorithm: the time complexity of Step1 is $O(|U|)$;the time complexity of Step2 is $O(K|C||U|)$ ($K = \max\{|T_C(x_i)|, x_i \in U\}$); the time complexity of Step3 is $O(|C||U||U_{pos}|)$; the time complexity of Step4 is $O(|U|^2 lb|U|^2)$;the time complexity of Step5 is $O(|C||U|^4)$; the time complexity of Step6 ignore; the time complexity of Step7 is $O(|C|^2|U2|)$ ($U2$ represents the number of enriching discernibility elements of IME).In conclusion, The worst time complexity of algorithm is $O(|C||U|^4)$.

Analysis for the space complexity of Algorithm: if the number of discernibility elements of IME is $U2$, it is easy to know $U2$ is much less than U^2 while M has deleted the useless elements.

Table 1. Incomplete decision Table

U	c_1	c_2	c_3	c_4	c_5	c_6	c_7	c_8	d
a_1	3	2	1	1	1	0	*	*	0
a_2	2	3	2	0	*	1	2	1	0
a_3	2	3	2	0	1	*	3	1	1
a_4	*	2	*	1	*	2	0	1	1
a_5	*	2	*	1	1	2	0	1	1
a_6	2	3	2	1	3	1	*	1	1
a_7	3	*	*	3	1	0	2	*	0
a_8	*	0	0	*	*	0	2	0	1
a_9	3	2	1	3	1	1	2	1	1
a_{10}	1	*	*	*	1	0	*	0	0
a_{11}	*	2	*	*	1	*	0	1	0
a_{12}	3	2	1	*	*	0	2	3	0

4   Analysis Experiment

First, we use an example to illustrate that the algorithm for computing core is efficient and accurate .The example of incomplete decision Table is described as Table 1[15].

We attain the tolerance classes and positive region of all objects by Step1 and Step2 of Algorithm.

$T_C(a_1) = \{a_1, a_{11}, a_{12}\}$; $T_C(a_2) = \{a_2\}$; $T_C(a_3) = \{a_3\}$;

$T_C(a_4) = \{a_4, a_5, a_{11}\}$; $T_C(a_5) = \{a_5, a_{11}\}$; $T_C(a_6) = \{a_6\}$;

$T_C(a_7) = \{a_7, a_8, a_{12}\}$; $T_C(a_8) = \{a_8, a_7, a_{10}\}$; $T_C(a_9) = \{a_9\}$;

$T_C(a_{10}) = \{a_{10}, a_8, a_{11}\}$; $T_C(a_{11}) = \{a_{11}, a_1, a_4, a_5\}$; $T_C(a_{12}) = \{a_{12}, a_1, a_7\}$;

$U / D = \{\{a_1, a_2, a_7, a_{10}, a_{11}, a_{12}\}, \{a_3, a_4, a_5, a_6, a_8, a_9\}\}$ 。

$U_{pos} = \{a_1, a_2, a_3, a_6, a_9, a_{12}\}$

$U_{neg} = \{a_4, a_5, a_7, a_8, a_{10}, a_{11}\}$

We attain the binary discernibility matrix M by Step4 of Algorithm. Given by:

Table 2. Binary discernibility

m(i,j)	c_1	c_2	c_3	c_4	c_5	c_6	c_7	c_8
m(1,3)	1	1	1	1	0	0	0	0
m(1,6)	1	1	1	0	1	1	0	0
m(1,9)	0	0	0	1	0	1	0	0
m(2,3)	0	0	0	0	0	0	1	0
m(2,6)	0	0	0	1	0	0	0	0
m(2,9)	1	1	1	1	0	0	0	0
m(3,12)	1	1	1	0	0	0	1	1
m(6,12)	1	1	1	0	0	0	1	1
m(9,12)-	0	0	0	0	0	1	0	1
m(1,4)	0	0	0	0	0	1	0	0
m(1,5)	0	0	0	0	0	1	0	0
m(1,7)	0	0	0	1	0	0	0	0
m(1,8)	0	1	1	0	0	0	0	0
m(1,10)	1	0	0	0	0	0	0	0
m(1,11)	0	0	0	0	0	0	0	0
m(2,4)	0	1	0	1	0	1	1	0
m(2,5)	0	1	0	1	0	1	1	0
m(2,7)	1	0	0	1	0	1	0	0
m(2,8)	0	1	1	0	0	1	0	1
m(2,10)	1	0	0	0	0	1	0	1
m(2,11)	0	1	0	0	0	0	1	0
m(3,4)	0	1	0	1	0	0	1	0
m(3,5)	0	1	0	1	0	0	1	0
m(3,7)	1	0	0	1	0	0	1	0

Table 2. (*continued*)

m(3,8)	0	1	1	0	0	0	1	1
m(3,10)	1	0	0	0	0	0	0	1
m(3,11)	0	1	0	0	0	0	1	0
m(6,4)	0	1	0	0	0	1	0	0
m(6,5)	0	1	0	0	1	1	0	0
m(6,7)	1	0	0	1	1	1	0	0
m(6,8)	0	1	1	0	0	1	0	1
m(6,10)	1	0	0	0	1	1	0	1
m(6,11)	0	1	0	0	1	0	0	0
m(9,4)	0	0	0	1	0	1	1	0
m(9,5)	0	0	0	1	0	1	1	0
m(9,7)	0	0	0	0	0	1	0	0
m(9,8)	0	1	1	0	0	1	0	1
m(9,10)	1	0	0	0	0	1	0	1
m(9,11)	0	0	0	0	0	0	1	0
m(12,4)	0	0	0	0	0	1	1	1
m(12,5)	0	0	0	0	0	1	1	1
m(12,7)	0	0	0	0	0	0	0	0
m(12,8)	0	1	1	0	0	0	0	1
m(12,10)	1	0	0	0	0	0	0	1
m(12,11)	0	0	0	0	0	0	1	1

We attain the enriching discernibility matrix *IME* by Step5 of Algorithm. Given by:

Table 3. Enriching discernibility

m(i,j)	c_1	c_2	c_3	c_4	c_5	c_6	c_7	c_8
m(1,10)	1	0	0	0	0	0	0	0
m(2,6)	0	0	0	1	0	0	0	0
m(1,5)	0	0	0	0	0	1	0	0
m(2,3)	0	0	0	0	0	0	1	0
m(1,8)	0	1	1	0	0	0	0	0
m(6,11)	0	1	0	0	1	0	0	0

At first, from Table 3, single attribute merge into *reduce*(C) by Step7.1, that is $reduce(C) = \{c_1, c_4, c_6, c_7\}$ then $C - T - reduce(C) = \{c_2, c_3, c_6, c_8\}$ by Step7.2.

If choose c_2 as attribute b, then $T = \{c_2\}$ .In the end, obtain $reduce(C) = \{c_1, c_4, c_6, c_7, c_3, c_5\}$ by Step7.1.Right now, the matrix IME is null, so the algorithm is end.

If choose c_3 as attribute b, then $T = \{c_3\}$ .In the end, obtain $reduce(C) = \{c_1, c_4, c_6, c_7, c_2\}$ by Step7.1.Right now, the matrix IME is null, so the algorithm is end.

If choose c_5 as attribute b, then $T = \{c_5\}$ .In the end, obtain $reduce(C) = \{c_1, c_4, c_6, c_7, c_3\}$ by Step7.1.Right now, the matrix IME is null, so the algorithm is end.

The number of discernibility elements of original discernibility matrix is 119, and the number of discernibility elements of corresponding enriching discernibility matrix is 8.The rate of enriching is 6.7227%.The operation time of algorithm is 0.371s.

5   Experimental Comparison

Six datasets from the UCI Repository are used in the experiment. Our experimental condition is that the personal computer hardware is CPU with AMD and 2.00memory, and the platform is Visual Stdio2010.The unit of time is seconds. The experimental results take the average of seven times the result of the experiment.

Table 4. The new algorithm comparison on UCI data sets

Data	|U|	|C|	|d|	|MS|	T/s	|PN|	|LN|	R=|LN|/|PN|
H1	155	19	1	167	2.948	37801	184	0.4868%
H2	294	13	1	782	2.577	133854	75	0.056%
S	307	35	1	712	113.659	544639	496	0.0911%
V	435	16	1	392	22.277	412928	13	0.0031%
C	690	15	1	67	6.124	1181600	227	0.0192%
M	8124	22	1	2480	4040	199340696	256	0.0001%

This Table, which H1,H2,S,V,C and M represent Dataset Hepatitis, Dataset Heart, Dataset Soybean-large, Dataset Votes, Dataset Credit and Dataset Mushroom. Which |U|, |C|, |d| and |MS| represent the number of objects, the number of condition attributes, the number of decision attributes and the number of missing value attributes. Which T represents time of algorithm, Which |PN|、|LN|、R=|LN|/|PN|represent the number of original binary discernibility matrix elements, that discernibility elements value is 1,the number of enriched discernibility matrix elements, that discernibility elements value is 1 and the rate of enriching.

Experiment shows that the number of enriched discernibility matrix elements is much less than the number of original binary discernibility matrix elements. Then we argue that enriching discernibility matrix is useful to attribute reduction and certain rule acquisition. In conclusion, enriching discernibility matrix is much higher value.

6   Conclusion

In order to delete repetition and useless elements, in this paper, we put forward Discernibility Matrix Enriching and Boolean-And Algorithm for Attributes Reduction through brining in the idea of enriching discernibility matrix. From Table 4, the larger the dataset of condition attributes, the longer time algorithm takes. In solving enriching discernibility matrix algorithm current situation, the algorithms are feasible and effective, although the time complexity is not ideal. We put forward an enriching discernibility matrix algorithm of low complexity.

Acknowledgements. This work was supported partly by NSFC(61262004, 61363034,60963008); GXNSF(2011GXNSFA018163）;College student innovation fund project (201310602028); BaGui scholars team project, the found of university science and technology of Guangxi, and the director found of Guangxi key laboratory. Many thanks to Guangxi Collaborative Innovation Center of Multi-source Information Integration and Intelligent Processing, and the anonymous referee for comments.

References

1. Pawlak, Z.: Rough sets. Int. J. of Information and Computer Science 11(5), 341–346 (1982)
2. Pawlak, Z.: Rough set approach to multi-attribute decision analysis. European J. of Operational Research 72(3) (1994)
3. Yang, M., Yang, P.: Discernibility Matrix Enriching and Computation for Attributes Reduction. Computer Science 33(9), 181–183 (2006)
4. Yin, Z., Zhang, J.: An attribute algorithm Based on a concentration Boolean matrix. Journal of Harbin Engineering University 30(3), 307–311 (2009)
5. Wang, J.Y., Gao, C.: Improved Algorithm for Attribute Reduction
6. Based on Discernihility Matrix. Computer Engineering 35(3), 66–69 (2009)
7. Huang, L., Xu, Z., Qian, W., Yang, B.: Quick feature reduction algorithm based on improved frequent pattern tree. Computer Engineering and Applications 46(35), 152–191 (2010)
8. Yang, M., Yang, P.: A novel condensing tree structure for rough set feature selection. Neurocomputing 71(4/5/6), 1092–1100 (2008)
9. Ananthanarayana, V.S., Narasimha Murty, M., Subramanian, D.K.: Tree structure for efficient data mining using rough sets. Pattern Recognition Letters 24, 851–862 (2003)
10. Yang, M., Yang, P.: A novel approach to improving, C-Tree for feature selection. Applied Soft Computing 11(2), 1924–1931 (2010)
11. Yang, M., Lv, J.: An incremental updating algorithm for attribute reduction based on C-Tree. Control and Decision 27(12), 1769–1775 (2012)
12. Xu, Z., Yang, B., Song, W.: Qucik Attribute Reduction Algorithm Based on Simple Binary Discernibility Matrix. Computer Science 33(4), 155–158 (2006)
13. Zhi, T., Miao, D.: The Binary Diseernibility Matrix's Transformation and High Efficiency Attributes Reduction Algorithm. Computer Science 29(2), 140–142 (2002)
14. Shu, W., Xu, Z., et al.: Quick Attribute Reduction Algorithm Based on Incomplete Decision Table. Joumal of Chinese Computer Systems 32(9), 103–110 (2011)
15. Xu, Z., Yang, B., Song, W., Hou, W.: Comparative Research of Different Attribute Reduction Definitions. Joumal of Chinese Computer Systems 29(5), 848–853 (2008)
16. Gao, J., Han, Z.: Construction of decision tree with discernibility matrix. Computer Engineering and Application 47(33), 18–21 (2011)
17. Kryszkiewicz, M.: Rough set approach to incomplete information system. Information Science 112(1), 39–49 (1998)
18. Kryszkiewicz, M.: Rules in incomplete information systems. Information Sciences 113(2), 271–292 (1999)

Smart Partitioning for Product DSM Model Based on Improved Genetic Algorithm

Yangjie Zhou, Chao Che, Jianxin Zhang, Qiang Zhang[*], and Xiaopeng Wei

Key Lab of Advanced Design and Intelligent Computing (Dalian University),
Ministry of Education, Dalian 116622, P.R. China
zhangq26@126.com

Abstract. Aiming at the coupling analysis of the design structure matrix(DSM), a new partitioning approach（improved genetic algorithm）is proposed. In the method, the fitness function calculates both the elements in the upper triangular matrix and the ones in the lower triangular matrix. Furthermore, the crossover operator is removed and the mutation rate becomes bigger. Minimized the object function including all elements in DSM, the partitioning accuracy is obvious improved. Through the coupling analysis for the DSM between 'parameter-parameter' of Axial-Flow Turbocharger, our method outperforms the original genetic algorithm. The result of partitioning validates the efficiency and feasibility of the proposed approach.

Keywords: DSM, improved genetic algorithm, coupling analysis, partitioning.

1 Introduction

The product design plays an important part in the whole production process, it directly affects the cost, the life cycle and the production efficiency of products. The coupling and conflicts are inevitable in the design process. The priority for eliminating conflicts is coupling analysis [1], which perform clustering and partitioning for the structure of products. The object of partitioning of product structure is to make the elements close to diagonal position as near as possible by transforming the design structure matrix. For the simple design, the partitioning of DSM can be performed by hand. When the number of elements in DSM increases and the links between elements strengthen, manual manipulation becomes extremely hard. In addition, the results of manual partitioning depend highly on the experience and ability of the operators. The partitioning of DSM should be implemented by smarter method. Whitfield et al [2] presented the objective function which can promote non-zero elements close to diagonal as many as possible. This method does not guarantee that the elements in lower triangular matrix can also achieve the same effect. The accuracy rate is not high. Liu et al [3] applied the genetic algorithm to the partitioning of the DSM. But binary coding need frequent coding and decoding, and the amount of calculation is big. Zhang H et.al [4] applied the particle swarm algorithm to the smart partitioning

---

[*] Corresponding author.

X. Luo, J.X. Yu, and Z. Li (Eds.): ADMA 2014, LNAI 8933, pp. 677–683, 2014.

and clustering of DSM. A mechanical press enterprise was taken as an example to extract typical process routs from the process data, and the effectiveness of proposed method was verified.

However, the objective function presented by Whitfield only calculates the elements in upper triangular matrix. The accuracy rate is not guaranteed. In addition, binary coding needed frequent coding and decoding is complicated. In response to these two issues, this paper uses symbolic coding and presents new objective function. All elements of the DSM model are calculated. It can promote all non-zero elements close to diagonal position as near as possible. The accuracy rate is improved. The symbolic coding method is simple and easy.

2   Smart Partitioning Method for DSM Model

Design structure matrix (DSM) is a straightforward and flexible modeling technique that can be used for designing, developing, and managing complex systems[5].In 1981, the American scholar Dr. Steward firstly proposed the concept of design structure matrix (Design Structure Matrix).It is used to analyze the flow of information [6]. In the 1990s, Eppinger et al further developed the DSM theory. They put forward the notion of the numeric design structure matrix. That is to say, the strength of the link between rows and columns is described by specific values. New approach can not only reorder the elements from DSM model but also simplify the model into smaller modules. It plays an important role in the development of DSM. There are two categories of DSM: Boolean DSM and digital DSM. Boolean DSM is like figure 1. There are only two factors: 'X' and 'BLANK' or '1'and '0'.Digital DSM is as figure 4 in this article. The relationships between two elements are signified by special number.

Coupling analysis based DSM is a quadratic programming problem. It belongs to the combinatorial optimization problems of NP-hard. This type of problem can be solved by genetic algorithm. The genetic algorithm is an optimal probability search algorithm based on genetics and evolution in the natural environment Genetic algorithm derives from the computer simulation studies [7]. The genetic algorithm is suitable for solving complex nonlinear problems which cannot be solved by traditional search methods. Genetic algorithm is widely used in function optimization, combinatorial optimization, robotics, image processing, artificial life, adaptive control, planning and design, artificial intelligence, intelligent manufacturing systems, and other fields. Genetic algorithm has become one of the key technologies in intelligent calculation. There are three main genetic operators: selection, crossover and mutation. Operating parameters of the genetic algorithm are population size, iterations, crossover rate, and mutation rate. The new partitioning approach is specifically addressed in the upcoming time.

2.1   Construction of the Objective Function

The partitioning aims to move the non-zero elements of the matrix to the diagonal position as close as possible. Whitfield et al [2] proposed a method that makes non-zero elements gather the diagonal. The objective function is equation (1).As the DSM is a square, the number of rows and columns equals to 'n'.

$$f = \sum_{i=2}^{n} \sum_{j=i+1}^{n} DM(i, j)|i - j| \tag{1}$$

In the above formula, 'i' stands for different rows. 'j' represents different columns. 'n' is the dimensions of square matrix. '$|i - j|$'indicates the distance from values of elements to diagonal. 'DM (i, j)' represents the value of the element in 'i' row and in 'j' column. Evaluation criterion is the objective function value, the smaller the value, the better.

Since the objective function only calculates the elements in the upper triangular matrix, the method does not guarantee that the elements in lower triangular matrix can also achieve the same effect. Therefore, this paper modifies the objective function to make all the whole non-zero elements in DSM close to diagonal.

$$f = \sum_{i=1}^{n} \sum_{j=1}^{n} DM(i, j)|i - j| \tag{2}$$

2.2   Solving Based on DSM Model Partitioning

Coding Scheme. Optimizing objective function of DSM is to reorder the number sequence of rows and columns. The symbolic coding method is simple and easy. Fig.1 is a simple sample of DSM. (Diagonal elements do not convey any meaning. 'X' signifies existing relations between two parameters 'Blank' describes the independent or weak relations.)[9].

No.	A	B	C	D	E
A		X			
B			X	X	
C					
D		X			X
E			X		

Fig. 1. A simple DSM

Symbol coding is designed as follows. No. A ~ E corresponded to the natural numbers from 1 to 5. That is to say, the chromosome sequence is 12345.

Genetic Operation. The genetic operator consists of Selection, crossover and mutation. When coding scheme is symbol coding, it may give rise to illegal solution during the crossover. As can be seen from Fig.2, there are double 'five' and 'four' in A1, and double 'six' and 'three' in B1. We are not looking forward to see overlapping elements. It is difficult to carry out the following work. Therefore, the issue should be solved. In view of the symbol coding, crossover operator and mutation operator (see Fig.5) are similar. Only mutation operator can also achieve the same function [8]. It may get rid of the crossover operation.

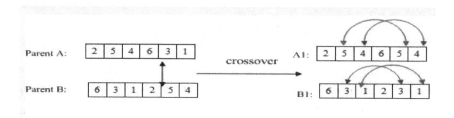

Fig. 2. Crossover operation

The genetic algorithm flow chart of this paper is showed in Fig.3. Comparing to the simple genetic algorithms, the proposed method does not use the crossover operation.

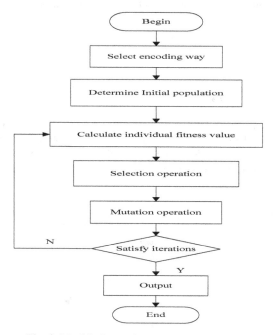

Fig. 3. Modified genetic algorithm flow chart

3   Example and Analysis of Algorithm

In this paper, in order to test the effect of the improved genetic algorithm, the Axial-Flow Turbocharger is used as an example [9]. The original DSM of the Axial-Flow Turbocharger is shown in Fig.4. No. A ~ O represents different parameters. A~O respectively stands for turbine inlet casing., turbine outlet casing, turbocharger support, turbine nozzle ring, compressor impellers, bearing shock absorber, diffuser, turbine rotor blade, intake silencer filter, ball bearing, oil pan, block support, compressor outlet turbine casing, heat screen, spindle.

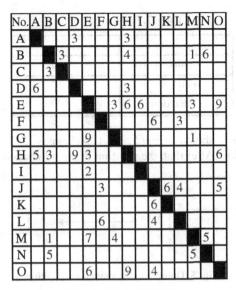

No.	A	B	C	D	E	F	G	H	I	J	K	L	M	N	O
A	■			3			3								
B		■	3				4					1	6		
C		3	■												
D	6			■			3								
E					■	3	6	6				3			9
F						■		6		3					
G				9			■					1			
H	5	3		9	3			■							6
I				2					■						
J					3			6	4	■					5
K								6			■				
L					6			4				■			
M	1			7	4							5	■		
N	5											5		■	
O					6			9	4						■

Fig. 4. Original DSM

3.1 Determining the Fitness Function

In order to improve the accuracy, the 'i' and 'j' begin from one. As the number of the parameters from A to O is 15 in Fig.3, the 'n' is 15. The fitness function equals the objective function.

$$f = \sum_{i=1}^{n} \sum_{j=1}^{n} DM(i, j)|i - j| \quad (n=15) \tag{3}$$

3.2 Coding

As the DSM is a square matrix, the rows and the columns are equal. It is allowed to consider the rows or columns only. The paper calculates the columns. No. A~O corresponded to the natural numbers from 1 to 15. The sequence is random as shown in Tab.1.

Table 1. Symbol coding

1	14	3	4	7	6	2	9	11	10	8	12	5	13	15

3.3 Algorithm Settings

Algorithm settings include population, mutation rate, iterations. As the modified genetic algorithm abandons the crossover operation, mutation rate is up to 0.5. The parameters is shown as Tab.2.

Table 2. Parameters value

title	value
population size	50
mutation rate	0.5
iterations	1000

The mutation style is shown in Fig.5. No.1 exchanges for No.5 in the Parent A. No.2 exchanges for No.3.After that, the position of the numbers is reordered.

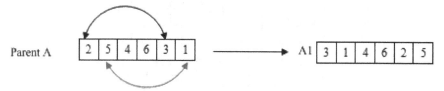

Fig. 5. Double points mutation operation

3.4 Analysis Results

Turbocharger is used to verify the improved genetic algorithm. The optimization value is 445. It is better than the previous value 577 in Figure 6.The final partitioning result is shown in Fig.7.

No.	A	D	H	B	N	C	I	E	G	M	K	J	F	L	O
A	■	3	3												
D	6	■	3												
H	5	9	■	3			3								6
B			4	■	6	3				1					
N			5		■			5							
C			3			■									
I							■	2							
E			6			6		■	3	3					9
G								9	■	1					
M				1	5		7	4		■					
K											■	6			
J										6		■	3	4	5
F												6	■	3	
L												4	6	■	
O		9	6										4		■

Fig. 6. Previous DSM

No.	L	F	J	K	C	B	N	M	G	I	E	H	D	A	O
L	■	6	4												
F	3	■	6												
J	4	3	■					6							5
K			6	■											
C					■	3									
B					3	■	6	1				4			
N					5		■	5							
M						1	5	■	4		7				
G								1	■		9				
I										■	2				
E					6			3	3		■	6			9
H						3				3		■	9	5	6
D												3	■	6	
A												3	3	■	
O		4										6	9		■

Fig. 7. Optimization DSM

We can see from the above two pictures that nine elements are far away from the diagonal in Fig.6, while only four elements keep away from the diagonal in Fig.7. The question is solved by making the non-zero elements of the matrix diagonal position as close as possible to diagonal. However, too many elements are away from the diagonal in Fig.6. It is not good for the following clustering. The elements closed to the diagonal of optimization DSM are more than the previous DSM. The result is obviously improved.

4   Conclusions

In this paper, the new partitioning approach based on modified genetic algorithm and fitness function is proposed. Not only the elements in the upper triangular matrix but also in the lower triangular is involved in calculating. Improved genetic algorithm uses the symbolic coding that is more convenient than the binary coding. Turbocharger is shown as an example. The new partitioning result is better than the original. It is beneficial for the coming clustering. Case study shows that the new method is feasible for the partitioning. However, the problem that emerges the illegal solution during crossover operator is worth further study. It is essential to find better object function.

Acknowledgements. This work is supported by the National Natural Science Foundation of China (No. 61202251, 60875046), the Program for Changjiang Scholars and Innovative Research Team in University (No.IRT1109), the Program for Liaoning Innovative Research Team in University (No. LT2010005), the Natural Science Foundation of Liaoning Province (201102008), and by the Program for Liaoning Excellent Talents in University (LJQ2013132, LJQ2013133).

References

1. Cao, P.B., Xiao, R.B., Ku, Q.: Structure analytical to coupled design in design with axiomatic design. Chinese Journal of Mechanical Engineering 42, 46–55 (2006)
2. Whitfield, R.I., Duffy, A.H.B., Coates, G., et al.: Efficient process optimization. Concurrent Engineering Research and Application 11, 83–92 (2003)
3. Liu, J.G., Wang, N.S., Ye, M.: Decomposition and clustering of product architecture based on genetic algorithm and DSM. Journal of Nanjing University of Aeronautics &Astronautics 38, 454–458 (2006)
4. Zhang, H., Qiu, L.M., Zhang, S.Y., Hu, X.X.: Typical product process route extraction method based on intelligent clustering analysis. Computer Integrated Manufacturing Systems 19, 490–498 (2013)
5. Eppinger, S.D., Browing, T.R.: Design structure matrix methods and applications. MTT Press, Cambrige (2012)
6. Steward, D.V.: The design structure system: A method for managing the design of complex system. IEEE Transactions on Engineering Management 28, 71–74 (1981)
7. Zhou, M., Sun, S.D.: Genetic Algorithm: Theory and Applications. National Defense Industry Press, Beijing (1996)
8. Chen, Y., Teng, H.F.: Comprehensive dependency matrix for heterogeneous modular combination coupling analysis. Chinese Journal of Mechanical Engineering 48, 21–28 (2012)
9. Tang, D.B., Qian, X.M., Liu, J.G.: Product design and development based on the design structure matrix DSM. Science Press, Beijing (2009)

A New Method for Image Understanding and Retrieval Using Text-Mined Knowledge

Jing Tian[1,2], Tinglei Huang[1], Yu Huang[1], Zi Zhang[1], Zhi Guo[1], and Kun Fu[1]

[1] Key Laboratory of Technology in Geo-spatial Information Processing and
Application Systems, Institute of Electronics, Chinese Academy of Sciences,
No.19, North 4th Ring Road West, Haidian District, Beijing 100190, P.R. China
tianjing0303,huangyu23@hotmail.com,tlhuang@mail.ie.ac.cn,
zhangzi@foxmail.com,guozhi@mail.ie.cn,kunfu0519@sohu.com
[2] University of Chinese Academy of Sciences,
No.19A, Yuquan Road, Shijingshan District, Beijing 100049, P.R. China

Abstract. Existing approaches for image annotation generally demand training data with exact image labels or human-generated tags, which are often difficult to obtain. In this paper we present a novel model that utilizes the rich surrounding text of images to perform image annotation. Our work makes two main contributions. First, by integrating the state-of-the-art text analysis methods, words that describe the salient objects in images are extracted. Second, a new probabilistic topic model is built to jointly model image features, extracted words and surrounding text. Our model is demonstrated to be flexible enough to handle multi-modal features and provide better annotation prediction performance than the baseline model proposed in previous research.

Keywords: Image Annotation, Topic Model, Text Analysis, Gibbs Sampling.

1 Introduction

Image annotation is the task of associating text with the semantic content of images, which plays a vital role in reducing the semantic gap. At the same time, as an intermediate step for image retrieval, it allows users to retrieve images using text queries and provides more semantically-related results than traditional content-based image retrieval. The evolution of the Internet has brought many new features to image collections, many of which are in a multi-modal form; for example, images are accompanied by text, including captions, and content descriptions, as well as user generated comments [1]. Generally speaking, the rich surrounding text of images presents challenges while also creating opportunities for automatic image annotation.

The most important part of image annotation is to model the correspondence relationships among different data modalities, including visual features and textual annotations. Previous works either define image annotation as a classification task [2] or attempt to model the correlation between visual features and textual words by estimating the probabilistic relations between different data modalities[3-4,6-8]. One problem of these methods is that they only focus on the generation of image labels using

X. Luo, J.X. Yu, and Z. Li (Eds.): ADMA 2014, LNAI 8933, pp. 684–694, 2014.

annotation models trained with image features and human annotated keywords, but the surrounding texts are seldom considered. In recent years, image annotation methods that leverage the multitude of resources available for natural language processing have attracted much attention [9-12]. These works have practical application in mining and annotating images on the Web, where texts are naturally associated with images and scalability is important.

Inspired by the thriving of Natural Language Processing (NLP), this paper proposes a totally new probabilistic model that integrates the state-of-the-art text mining technologies. We adopt the name Multi-Modal Entity LDA (MME-LDA) to reflect the incorporation of multi-modal data and the specific use of entities. We assume that general words and entity words play different roles when describing an image, thus the novelty of our new approach is the use of entity words [13] and the augmentation of standard LDA to model this feature in a manner analogous to Multi-multinomial LDA (MM-LDA) [14]. In our model, we make several modifications and broaden the definition of "entity" in NLP. We assume that "entity words" in our model are those that can more directly describe the salient objects in the corresponding images when compared to the other words in the surrounding text. We therefore create a special rule to extract such entity words after conducting the traditional text analysis process. We add a parallel path to represent entity words in the original MM-LDA. Built on the standard LDA [5], MME-LDA also supposes the existence of hidden factors which represent the data in a low dimensional way and explicitly model term co-occurrences of different input data types. The latent variables learned through our new model will describe how different data modalities in training data co-occur and offer predictive relations between the image and the corresponding tags. As the exact inference is generally intractable, we derive a collapsed Gibbs sampling algorithm following [15] to perform approximate inference for the proposed model. By using a subset of a PASCAL dataset called UIUC PASCAL Sentence dataset, we demonstrate the power of the proposed model on an image annotation task.

2   Related Work

2.1   Topic Models for Image Annotation

Probabilistic topic models are promising methods in text mining that can automatically organize, understand, search, and summarize large scale text by postulating the existence of latent topics which reveal the semantics of input data. Latent Dirichlet Allocation (LDA) [5] is a seminal work in topic models. It can be represented as a three level hierarchical Bayesian model: each document d of the D documents in a corpus is generated from a mixture over K topics, each of which follows a multinomial distribution $\vec{\theta}$ with a Dirichlet prior $\vec{\alpha}$. Each word n of the d_n words in document d is generated repeatedly from a topic according to the topic distribution $\vec{\theta}$, and a word is selected according to the chosen topic. Through the use of hidden variables, LDA clusters words and models word co-occurrences to obtain a semantic level representation of documents in a way of proportions of topics.

When dealing with annotation tasks, several variants based on LDA have been designed to fit the new requirement of multi-modal situations. The most fundamental principle in realizing image annotation via topic modeling is how to capture the hidden probabilistic correlation between two data modalities. Correspondence LDA(c-LDA) [6] and multi-class-sLDA (mc-sLDA) [7] are two representative works. In c-LDA, each topic consists of two content-specific distributions: a topic-specific distribution over words and a topic-specific distribution over image features. This model builds a language-based correspondence between visual modality and textual modality to achieve simultaneous dimensionality reduction and correspondence modeling between their respective reduced representations. Mc-sLDA combines the basic framework of c-LDA and yields a model which simultaneously classifies images and annotates the individual regions.

All the above methods demonstrate that it is feasible to achieve image annotation through topic modeling, because the topic model has excellent extendibility in handling multi-modal data. However, a common and severe problem of these works is that they depend heavily on training data with exact image labels or human-generated tags, which are often difficult to obtain. In practice, varieties of multi-modal data as mentioned in Section 1 are available, but they are underutilized in most of the image annotation methods. There is a great need to annotate images by mining the surrounding text since data of this type is more common and contains rich information.

2.2   Text Analysis in Image Annotation

The extractive and abstractive caption generation model [9] is a representative work that uses text analysis for image annotation. It focuses on generating descriptions, especially captions for news images. It presents a both extractive and abstractive caption generation manner similar to text summarization, which belongs to the area of nature language processing. Image annotations are achieved through topic modeling and are subsequently used to guide caption generation. The novelty of the model lies in its use of image features when generating summarization. It is also worthwhile to learn from combination of NLP techniques with images. The Corpus-Guided Sentence Generation methods proposed in [10] stress the need to generate descriptions for images. After detecting the objects and scenes in an image, a language model is trained to ground to World Knowledge in the Language Space and the discretional sentences are generated by a HMM model. In both methods above, text analysis is separated from image modeling and they are viewed as two different stages. A clear drawback is error accumulation, as the former stage can influences the performance of the latter. This is too complex to handle in a real world application.

The framework of [11] obtains annotations only on the text surrounding an image without considering the image features. This manner is analogous to keyword extraction and is practical in the Web environment; however, ignoring image features will result a loss of information from images. Furthermore, over-reliance on noisy surrounding text may result in unrelated annotations.

In the next section, we propose a novel annotation model which combines topic modeling and text analysis at the same time. Different from previous works, we jointly model the text analysis results, the surrounding text and the image features into a uniform topic model framework.

3   Multi-modal Entity LDA Model

In this section, we detail the proposed Multi-modal Entity LDA (MME-LDA) model. Our model consists of three stages. First, preprocessing the surrounding text and images provides the input for our model. Second, entity words, surrounding text and image features are jointly modeled, following which a Gibbs sampling algorithm is derived to fit this model. Lastly, the learned latent variable representations are used to predict the corresponding annotations when a new image is presented.

3.1   Data Representation

Following the work in [8], we use the bag-of-words representation for both images and surrounding text. We extract the SIFT feature of images and then quantize the SIFT descriptors using the K-means clustering algorithm to obtain a discrete set of visual terms with size V to form the visual vocabulary. Furthermore, each image I is expressed as a bag-of-words format V -vector. As for the text part, we perform the basic tokenization and remove common stop-words. A dictionary for general words with a dimension of W is then constructed. After processing with the UW Twitter NLP Tool [16] as detailed in the next section, the entity words are generated in the same format as the general words with a vocabulary of E .

3.2   Entity Extraction

We take entity words as descriptions of the more salient objects in the images. Entity extraction detects all the defined entities in the surrounding text according to the salience we need in our task of image annotation. The entity words mentioned in this paper are not all the same as the named entity in NLP [13], which often means the names of persons, organizations, locations, expressions of times, quantities, monetary values, percentages, etc. In contrast, we make some modifications in our model and broaden the definition of "entity" in NLP. We assume that "entity words" in our model are those that can describe the salient objects in the corresponding images more directly when compared to other words in the surrounding text.

Several tools, such as the Stanford Parser toolkit and the Clearnlp have been developed for this task. However, these tools are always trained on particular datasets such as news corpora, and they perform poorly on our short and informal texts. Considering that tweets have almost the same format of the data used in our model, we choose UW Twitter NLP Tools [16] as our processing tool when extracting entities. The output of the UW Twitter NLP Tool contains tokenized and tagged words separated by spaces with tags divided by a forward slash "/" [16]. For example, the input "A city bus driving past a building." will be part-of-speech tagged like this: "A/O/DT/B-NP city/O/NN/I-NP bus/O/NN/I-NP driving/O/VBG/B-VP past/O/JJ/B-ADVP a/O/DT/B-NP building/O/NN/I-NP ./O/./O". The tags of one word refer to chunks, POS and POS with BIO encoding respectively. We subsequently establish a specific rule for choosing our own entity words which are more important for our task of image annotation. We use the second and the third tags to select the entity words. According to our rule, words with tags in the format "B-NP", "B-NP I-NP ...", "B-VP" or "B-VP I-VP ..." are preserved.

3.3   Modeling Images, General Words and Entity Words

The proposed model is a variant of MM-LDA. First proposed in [14], MM-LDA models two types of data by sharing the same topic proportion. It is demonstrated to be useful for multi-modal data and has been well-developed recent years, not only in image annotation but also in video to text summarization [12]. Nevertheless, all these models have been developed on an accurate, manually labeled dataset, so they may run into severe difficulty when applied to real-world image data, which often comes with text descriptions by the way of surrounding text. To tackle these problems, we add another path representing entity words in the original MM-LDA to leverage the vast amount of hidden information in the surrounding text that accompanies the images. Our innovation is that we assume that entity words reveal the more important things in images which better model the correspondence between the visual modality and the textual modality. The input of our model consists of visual words which represent image features, extracted entity words and words in the original surrounding text descriptions, which we call them general words.

The graphical representation of our MME-LDA model is shown in Fig. 1. The total number of documents in the corpus is D. In each document d, there are M_d general words, Q_d entity words and N_d visual words whose vocabulary size is W, E and V respectively. Given K latent topics shared by images, general words and entity words following a Dirichlet distribution with parameter $\vec{\alpha}$, the three parts of the observation separately follow a basic LDA model. The generative process of MME-LDA is given as follows:

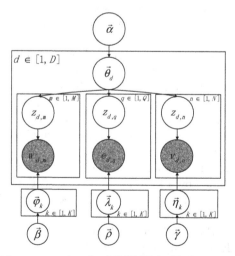

Fig. 1. Graphical model representation for MME-LDA. Nodes represent random variables; edges denote possible dependence between random variables; plates denote replicated structure.

For each document $d \in 1, ..., D$

(a) Sample a Dirichlet random variable as document code $\vec{\theta}_d | \vec{\alpha} \sim Dir(\vec{\alpha})$.

(b) For each observed general textual observation $w_{d,m}$ in d

   (i) Sample a topic indicator $z_{d,m} \sim Mult(\vec{\theta}_d)$

   (ii) Sample a general textual word $w_{d,m} \sim Mult(\vec{\varphi}_{z_{d,m}})$

(c) For each observed general textual observation $e_{d,q}$ in d

   (i) Sample a topic indicator $z_{d,q} \sim Mult(\vec{\theta}_d)$

   (ii) Sample a general textual word $e_{d,q} \sim Mult(\vec{\lambda}_{z_{d,q}})$

(d) For each observed general textual observation $v_{d,n}$ in d

   (i) Sample a topic indicator $z_{d,n} \sim Mult(\vec{\theta}_d)$

   (ii) Sample a general textual word $v_{d,n} \sim Mult(\vec{\eta}_{z_{d,n}})$

where $\vec{\theta}_d$ is the multinomial distribution of topics specific to the document d. $\vec{\varphi}_k$, $\vec{\lambda}_k$ and $\vec{\eta}_k$ respectively represent the multinomial distribution of words of three data modalities specific to the topic k. $\vec{\beta}$, $\vec{\rho}$ and $\vec{\gamma}$ are Dirichlet priors to the corresponding multinomial distributions.

3.4 Inference and Learning

As the exact inference is generally intractable, we derive a collapsed Gibbs sampling algorithm following [15] to perform approximate inference for the proposed model. The Gibbs sampling algorithm is a special application of Markov chain Monte Carlo, which is a procedure for obtaining samples from complicated probability distributions. In the Gibbs sampling procedure, we sequentially sample all latent variables conditioned on the current status of both themselves and the data. Furthermore, as the Dirichlet prior and the multinomial likelihood form a conjugate pair, the derivation of the posterior predictive can be simplified by taking advantage of conjugate priors. In total, we have three variables to sample $p(z_{d,m}|rest)$, $p(z_{d,q}|rest)$ and $p(z_{d,n}|rest)$, which stand for the update rule of general word topic, entity word topic, and visual topic respectively. Due to space limitations, we give only the derived sampling formulas. The "rest" in the formulations represent the current value of both the sampled topic variable and the observed data. The n_d^z represents the number of words in document d assigned to topic z and the n_z^w represents the number of word w assigned to topic z.

Sampling General Word Topic Index $^{z_{d,m}}$. The conditional distribution of the topic index of a general word depends on the likelihood that the specific topic $z_{d,m}$ is assigned to the document d and the likelihood that the specific word $w_{d,m}$ is assigned to the topic $z_{d,m}$.

$$p(z_{d,m}|rest) = \frac{n_d^{(z_{d,m})} + \alpha}{\sum\limits_{z_{d,m}=1}^{K} (n_d^{(z_{d,m})} + \alpha)} \cdot \frac{n_{z_{d,m}}^{(w_{d,m})} + \beta}{\sum\limits_{w_{d,m}=1}^{W} (n_{z_{d,m}}^{(w_{d,m})} + \beta)} \tag{1}$$

Sampling Entity Word Topic Index $z_{d,q}$. As the update rule of general topic $z_{d,q}$, the conditional distribution of the topic index of an entity word depends on the likelihood that the specific topic $z_{d,q}$ is assigned to the document d and the likelihood that the specific word $e_{d,q}$ is assigned to the topic $z_{d,q}$.

$$p(z_{d,q}|rest) = \frac{n_d^{(z_{d,q})} + \alpha}{\sum\limits_{z_{d,q}=1}^{K} (n_d^{(z_{d,q})} + \alpha)} \cdot \frac{n_{z_{d,q}}^{(e_{d,q})} + \rho}{\sum\limits_{e_{d,q}=1}^{E} (n_{z_{d,q}}^{(e_{d,q})} + \rho)} \tag{2}$$

Sampling Visual Topic Index $z_{d,n}$. In the updating of the visual topic index, the conditional distribution is also decided by the likelihood that the specific topic $z_{d,n}$ is assigned to the document d and the likelihood that the specific visual word $v_{d,n}$ is assigned to the topic $z_{d,n}$.

$$p(z_{d,n}|rest) = \frac{n_d^{(z_{d,n})} + \alpha}{\sum\limits_{z_{d,n}=1}^{K} (n_d^{(z_{d,n})} + \alpha)} \cdot \frac{n_{z_{d,n}}^{(v_{d,n})} + \gamma}{\sum\limits_{v_{d,n}=1}^{V} (n_{z_{d,n}}^{(v_{d,n})} + \gamma)} \tag{3}$$

The Gibbs sampling algorithm samples the hidden variables iteratively given the conditional distributions. Samples are collected after the burn in. We can then obtain the multinomial parameter sets $\vec{\varphi}_{train}$, $\vec{\lambda}_{train}$ which are related to annotation prediction during the training process. Since both of them are multinomial distributions with Dirichlet prior, we can apply Bayes' rule and yield:

$$\vec{\varphi}_{train} = \frac{n_{z_{d,m}}^{(w_{d,m})} + \beta}{\sum\limits_{w_{d,m}=1}^{W} (n_{z_{d,m}}^{(w_{d,m})} + \beta)} \tag{4}$$

$$\vec{\lambda}_{train} = \frac{n_{z_{d,q}}^{(e_{d,q})} + \rho}{\sum\limits_{e_{d,q}=1}^{E} (n_{z_{d,q}}^{(e_{d,q})} + \rho)} \tag{5}$$

The procedure for predicting annotations is as follows. We first obtain the topic proportion of new documents $\vec{\theta}_{new}$, and then infer the distribution over predicted general words $p(\vec{w}|I)$ and entity words $p(\vec{e}|I)$ given a new image I by computing the contributions from each topic. Finally, we obtain a W-vector $\vec{w}$ and an E-vector $\vec{e}$ sorted by the conditional probabilities. The formulas are as follows:

$$\vec{\theta}_{new} = \frac{n_d^{(z_{d,n})} + \alpha}{\sum_{z_{d,n}=1}^{K} (n_d^{(z_{d,n})} + \alpha)} \tag{6}$$

$$p(\vec{w}|I) = \vec{\theta}_{new} \cdot \vec{\phi}_{train} = \frac{n_d^{(z_{d,n})} + \alpha}{\sum_{z_{d,n}=1}^{K} (n_d^{(z_{d,n})} + \alpha)} \cdot \frac{n_{z_{d,m}}^{(w_{d,m})} + \beta}{\sum_{w_{d,m}=1}^{W} (n_{z_{d,m}}^{(w_{d,m})} + \beta)} \tag{7}$$

$$p(\vec{e}|I) = \vec{\theta}_{new} \cdot \vec{\lambda}_{train} = \frac{n_d^{(z_{d,n})} + \alpha}{\sum_{z_{d,n}=1}^{K} (n_d^{(z_{d,n})} + \alpha)} \cdot \frac{n_{z_{d,q}}^{(e_{d,q})} + \rho}{\sum_{e_{d,q}=1}^{E} (n_{z_{d,q}}^{(e_{d,q})} + \rho)} \tag{8}$$

4 Experiments

In this section, we conduct our experiments on the PASCAL sentence dataset. First introduced in [17], the dataset contains 1000 images collected from Pascal-VOC, and each image is accompanied by 5 sentences describing the content of the image. We randomly split the 1000 images into 2 sets: 750 images for training and 250 images for test.

4.1 Quantitative Evaluations

Precision and Recall. The purpose of our work is to provide annotations for an entire image and to serve as an intermediate step for image retrieval. The task for each word is thus to predict it for the images where it is a keyword, and not to predict it otherwise. We evaluate our model using precision and recall based on this retrieval aim. Precision is the total number of correct predictions over all images and recall is the total number of correct predictions divided by the number of occurrences as a keyword. Take three single words for example, the precision-recall curve of each word is shown in Fig. 3. It can be observed that MME-LDA gives a better balance between precision and recall and yields higher precision at the same recall values for all the three keywords treated as queries. This shows that more useful correspondence structures are mined by incorporating text analysis and topic modeling.

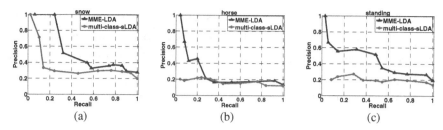

Fig. 2. Precision-recall curve of 3 words as queries: "snow", "horse", "standing"

Table 1. Example topics of general word topics (top panel) , entity word topics (middle panel) of MME-LDA and the caption topics of mc-sLDA (bottom panel)

Topic 1	room, table, living , chairs , couch, dining, wooden, floor, small, chair
Topic 2	car, parked, motorcycle, red , back, lot, truck, next, parking , scooter
Topic 7	kitchen, open, food , cats, paper , full, area, box , sink, various
Topic 5	plane, airplane , sky, small, jet, flying, parked , blue , runway, passenger
Topic 12	horse , riding, bicycle , bike, dirt , person , wearing, rides, racing , road
Topic 15	water, boat, ship, river, lake, large , small , ocean, canoe, crue
Topic 1	man , room , black, cat, sitting, dog, looking , brown, small, blue
Topic 4	room, table, bus , man, street, sitting, kitchen , woman, black, red
Topic 6	water, people, boat, man , small, field, front, brown, large , horse

4.2   Qualitative Evaluations

Example Topics. We compare examples of caption topics learned through mc-sLDA and MME-LDA, including the general word topics, entity word topics of MME-LDA and the caption topics of mc-sLDA. In detail, we employ the 10 most probable words of each topic to represent the topic according to the multinomial parameters. We learn 20 topics and select 3 of them randomly for each kind of the topics mentioned above to examine the capabilities. As shown in Table 1, general word topics and entity word topics are more discriminative. For example, topic 1 of general word topics describes an indoor scene and topic 5 of entity word topics is related to planes and airports. In contrast, the topics learned using mc-sLDA are often meaningless and lack distinctions. It can be seen in Table 1 that the topic 1, 4 and 6 of mc-sLDA contain more general words, such as "man", "woman" and "people". The outstanding performance of MME-LDA illustrates the constraints introduced by entities can well benefit topic modeling.

Example Annotations. Several examples of predicted annotations generated by MME-LDA and mc-sLDA are compared in Fig. 4. We take the five top words of all the predicted words sorted by conditional probability $p(\vec{w}|I)$ and $p(\vec{e}|I)$. It can be seen clearly that the annotation words predicted by MME-LDA are more semantically related to the content of the image and also to the ground truth, while mc-sLDA predicts more general words which are often wrong.

Images			
Ground Truth	A computer on a desk. A computer with a cat screen saver at an office desk. A desktop pc with a close up picture of a cat on the monitor. A picture of cat is displayed on a desktop computer. Worker has a cat as a screen saver at the office.	a bird with bright colors. A close-up of a green bird with a red face and purple tail feathers. A green bird with a red face sitting. A small green bird with a red beak. Tropical bird sitting on wet rock.	Distant view of two lambs perched on top of a rocky plain. Two animals sitting on a rock. Two sheep looking towards the camera while standing on top of a rocky ledge. Two sheep on top of a cliff looking at the camera. Two sheep standing on rocks.
mc-sLDA	white, man, table, room, front	small, yellow, bird, horse, brown	sky, large, front, horse, side
MME-LDA	*Entity Words:* computer, desk, screen, room, table / *General Words:* sitting, table, computer, screen, chair	*Entity Words:* bird, water, animal, close-up, head / *General Words:* animal, wooden, small, bird, water	*Entity Words:* field, standing, sky, cows, hillside / *General Words:* sheep, animal, mountains, stand, field

Fig. 3. Examples of predicted annotations. The red ones are annotated incorrectly.

5   Conclusion

In this paper we have developed a new probabilistic topic model called Multi-Modal Entity LDA model for the task of image annotation. The novelty of the proposed model is the use of rich information mined from the surrounding text and the combination of NLP methods with topic modeling. The extracted entity words by text analysis techniques provide more salient information from the images. We model three data modalities, including general text, specific entity words and visual features in a new topic model framework and derive a collapsed Gibbs sampling algorithm for the framework. Following joint topic modeling of the multi-modal data, the underlying connections between the visual features and text features can be discovered more effectively. Experimental results show that our model outperforms the state-of-the-art image annotation model using the topic model in both annotation and retrieval.

Acknowledgements. This work is supported by the National High Technology Research and Development Program of China (Grant No. 2012AA011005, 2013AA0093).

References

1. Zhang, D., Islam, M.M., Lu, G.: A review on automatic image annotation techniques. Pattern Recognition 45, 346–362 (2012)
2. Carneiro, G., Chan, A.B., Moreno, P.J.: Supervised learning of semantic classes for image annotation and retrieval. IEEE Trans. PAMI 29(3), 394–410 (2007)
3. Hoffman, T.: Unsupervised learning by probabilistic latent semantic analysis. Machine Learning 42(1-2), 177–196 (2001)
4. Barnard, K., Duygulu, P., Forsyth, D., Freitas, N., Blei, D., Jordan, M.: Matching words and pictures. Journal of Machine Learning Research (2003)
5. Blei, D.M., Ng, A.Y., Jordan, M.I.: Latent Dirichlet Allocation. Journal of Machine Learning Research (3), 993–1022 (2003)
6. Blei, D.M., Jordan, M.I.: Modeling annotated data. In: Proceedings of ACM SIGIR, pp. 127–134 (2003)
7. Wang, C., Blei, D.M., Fei-fei, L.: Simultaneous image classification and annotation. In: IEEE Conf. CVPR (2009)
8. Putthividhya, D., Attias, H., Nagarajan, S.: Topic Regression Multi-Modal Latent Dirichlet Allocation for Image Annotation. In: IEEE Conf. CVPR (2010)
9. Feng, Y., Lapata, M.: Automatic Caption Generation for New Image. IEEE Trans. PAMI 35(4), 797–812 (2013)
10. Yang, Y., Teo, C.L., Daum´e, H., Aloimonos, Y.: Corpus-Guided Sentence Generation of Natural Images. In: EMNLP (2011)
11. Leong, C.W., Mihalcea, R., Hassan, S.: Text Mining for Automatic Image Tagging. In: Coling (2010)
12. Das, P., Srihari, R.K., Corso, J.J.: Translating Related Words to Videos and Back through Latent Topics. In: WSDM (2013)
13. Black, W.J., Rinaldi, F., Mowatt, D.: Facile: Description of the NE system used for MUC-7. In: Proceedings of the Seventh Message Understanding Conference (1998)
14. Ramage, D., Heymann, P.: Clustering the Tagged Web. In: WSDM (2009)
15. Neal, R.: Markov Chain Sampling Methods for Dirichlet Process Mixture Models. JCGS 9(2), 249–265 (2000)
16. Ritter, A., Clark, S., Mausam, Etzioni, O.: Named entity recognition in tweets: An experimental study. In: EMNLP (2011)
17. Farhadi, A., Hejrati, M., Sadeghi, M.A., Young, P., Rashtchian, C., Hockenmaier, J., Forsyth, D.: Every picture tells a story: Generating sentences from images. In: Daniilidis, K., Maragos, P., Paragios, N. (eds.) ECCV 2010, Part IV. LNCS, vol. 6314, pp. 15–29. Springer, Heidelberg (2010)

A Web Table Extraction Method Based on Structure and Ontology

Cai Guo[1], Shun Ma[1], and Dingrong Yuan[1,2,*]

[1] College of Computer Science & IT, Guangxi Normal University, Guilin, 541004, China
[2] Guangxi Key Lab of Multi-source Information Mining & Security, Guilin, 541004, China
517385149@qq.com,
dryuan@mailbox.gxnu.edu.cn,
magexiao@foxmail.com

Abstract. The table extraction is an important issue of Webpage information analysis. At present, there are three mainly methods, which is how to construct the wrapper, how to construct the ontology and directly analysis the structure of a table on the webpage. In the process of analysis, usually these methods are applied independently. Aiming at the shortcomings of single method, this paper presents a synthetic method based on the ontology and structure. In this paper, we firstly locates the tables based on heuristic rules, and then analysis the table structure according to the label and the title ontology, at last extract and save the table data on the basis of the obtained characteristics. The experiments show that the introduction of the ontology greatly improved the accuracy of table structure recognition, and the precision and recall of the methods are better.

Keywords: Web table structure, Tidy, heuristic rules, ontology, Data extraction.

1 Introduction

With the double increase of the Webpage number and capacity, a huge of table data is accumulated in the Webpage, how to directly and accurately obtain the information from a table is an important topic of Web information analysis.

Table is different from other data on the Webpage, it not only possesses a certain structure, at the same time also contains certain semantics. For the table on a Webpage, syntactic and semantic concepts are usually mixed in one table, the layout cell with its relative position information contains certain semantic, the syntactic structure is more complex than natural language [1] Thus, how to accurately extract table from webpage becomes a more challenging topic.

At present, the works about extracting table data from a Webpage can be summarized into the following three categories:

(1) Structuring data wrapper based on HTML tag [2].The idea was first proposed by Cohen and Jensen [3] in the wrapper learning. They conclude extraction rules by learning examples, and then according to the rules to extract information. But the method relies on the structure of the Webpage, when the Webpage is changed, we must reconstruct the wrapper. And the user-selected examples determine the accuracy of rules.

---

[*] Corresponding author.

X. Luo, J.X. Yu, and Z. Li (Eds.): ADMA 2014, LNAI 8933, pp. 695–704, 2014.

(2) Extraction method based on ontology. Neches etc consider that ontology specializes the basic terms, relations as well as the rules in a topic area[4]. Thus Cha Songli try to design a framework of ontology by analysis the structure and content characteristics in a specific area, and then obtain the extraction rules[5]. Wang Fang etc proposed another table extraction method based on ontology [6], they defined the Ontology as the object and its relation, the ontology will be perfected gradually by learning, in the same time the ability of ontology's description is enhanced, and the degree of automation is improved step by step. This method has nothing to do with the page format. When the field is changed, we need to redesign the ontology. But it is very cumbersome and complex to design the ontology' framework.

(3)The method based the table structure. The main idea is to transform the table into a logic structure by analyzing HTML table tag [7]. The current main methods are based on Visual cues and tree structure. For example, Gatterbauer made use of the CSS2 Visual Box Model to analysis the Web documents, and extract tabular information according to visual and spatial reasoning [8]. Liu proposed an algorithm of extracting table based on tree edit distance [9]. Zhang ZhiYuan converts the HTML <table> nodes into the DOM tree with semantic information, and then generated the extracting rules based on the Greedy algorithm [10].

This paper presents a new model based on table structure and its title keywords ontology. Firstly, we use a series of heuristic rules to locate the table, and then analyze the table structure, identify its attribute domains and value domains, confirm the way of the table expanding. At last, transform the table into a logic model; thereby the table data are extracted.

2   System Design

Usually a Webpage lacks of specification, so we firstly converse the HTML document to XML document as paper [11], and then denoise on the page [12]. We design the system structure as Fig.1:

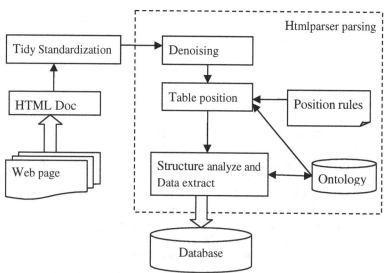

Fig. 1. The logic structure of table extracting System

The principle of the system is described as follow.

2.1   Web Document Standardization

In order to obtain standardized XML document, we check and correct the errors in HTML document by the tool "Tidy". Tidy executes the following tasks:

1. Correct the mismatched element tag.
2. Add the missing element tags, quotes, etc.
3. Report whether uses the proprietary HTML tags.
4. Regulate the code.
5. Cleanup presentation style elements.

2.2   Webpage Clean

Usually a webpage contains much additional information, such as navigation, advertising, video, pictures, etc, which is useless for us. Therefore, we remove the redundant information before extracting the data. In this module, we use the Regular expression to clean the pages. The steps is as following:

1. Clear all contents except the "body" tag.
2. Clear unconcerned contents such as pictures, video, navigation and advertising etc.
3. Recursive-remove the empty tags pair.
4. Use a space instead of a continuous space.

2.3   Table Position

Tag "table" can be used to identify tables, but it is not all. Sometimes, tag "table" is used as layout [13]. So before extracting the tabular data, how to correctly locate the target table is a key step. After analyzing a large number of Web pages, we summarized a series of heuristic rules to identify a table or a layout [14].

Rule 1: the size of a table is at least $2*n$, $n*2$ ($n \in N^*$) 。

Rule 2: If a table tag contains the "caption" or "th" tags, then the table is a data table.

Rule 3: If a table contains a large number of scripts, text, images, and so on, the table is not a data table, it is a layout.

Rule 4: The data "table" tag has following characteristics: "tr" label is not empty and the data type of td" ag is the same as "tr".

Rule 5: For the nested table, we judge it starting from the innermost child node. If one of the children <table> is the data table, its outer layers <table> are non-data tables. Otherwise, pruning the non-data child<table> and then continue judge.

Rule 6: If the number of empty cells in a table exceeds 50%, it is used to control page layout and is not a real data table.

According to the rules, we can eliminate the non-data tables, and then match the contents in the rest data table with the title ontology. If they match each other, the table is the interested one.

2.4   Table Structure Analyze and Contents Extraction

The key steps in the system are to analysis the table structure and extract the content in the table, as well as recognize the way of table expanding, acquire the attribute domain and value domain.

Usually a table consists of rows and columns, which constitute the cells. These cells can be divided into the attribute cell and the value cell. The characteristics of table structure are expressed by the HTML label. For example, <caption> represents the title, <tr> represents the row, <td> represents the cell. For example, the Table 1 can be encoded by HTML as follow:

Table 1. A Non-standard format table

Train number	Price	
	Hard seat	Soft seat
K221	75	125
D223	175	235

```
<table border="1">
<tr align="center">
<td rowspan="2"> Train number </td>
<td colspan="2">Price</td>
</tr>
<tr align="center">
<td> Hard seat </td>
<td> soft seat </td>
….
</table>.
```

In the Table 1, the "price" cell takes up 2 columns, we call the table as non-standard format table. The attribute-value in the table is not easy to be extracted. So the non-standard table needs to be standardized firstly. For the non-standardized table, we firstly remove the span mark, then rewrite the cell's content K-1 times in the below row or column. For example, the table as Table 1 can be converted into the table as Table 2.

Table 2. The standardized format table

Train number	Hard seat Price	Soft seat Price
K221	75	125
D223	175	235

How to correctly identify the attribute domain and value domain in a table and then translate them into a logic model are another critical issue in our extracting system. For this task, we obtain the following rules by learning.

Rule 1: if a row is marked with <th>...</th> then it is the attribute row.

Rule 2: if there are the <font size> tag in the tags <tr> or <td>, then the max is the attribute.

Rule 3: if the structure of a table is not obvious, we match the cell content with the ontology based on Semantic similarity, and judge the attribute row or line by the confidence level.

Based on discuss above, we designed the following algorithm to identify the attribute domain.

Step1 Let C denotes the confidence of a cell belongs to attribute. We define C as follow:

$$C = \begin{cases} 0, \\ \max(sim_i) \end{cases} \tag{1}$$

sim_i represents the semantic similarity between the two words.

Step2 Select the cells that their C is not equal to 0, and compare the dissimilarity between the cell and the other cells in the same column and row. If the dissimilarity is a large value then we increase the value of C, if it is a small one then not increase the value of or just increase a little.

Step3 Calculate the average of C for every row and every column, select the row (column) that C is greater than the threshold S, which is a threshold.

Then, we can determine the way of expanding a table according to the attribute domain. The commonly ways is as shown in Fig.2:

H_1	H_2	....	H_n
C_{11}	C_{12}	...	C_{1n}
...	...	....	
C_{n1}	C_{n2}	...	C_{nn}

(a) Row header table

H_1	C_{11}	...	C_{1n}
H_2	C_{21}	...	C_{2n}
...	...	...	...
H_n	C_{n1}	...	C_{nn}

(b) Column header table

	H_{12}	...	H_{1n}
H_{21}	C_{22}	...	C_{2n}
...	...	...	...
H_{m1}	C_{m2}	...	C_{mn}

(c) Row-Column header table

H_1	H_2	H_3	H_4
...	...	...	...
D_1	D_2	D_3	D_4
...	...	...	...

(d) Multiple-row header table

H_1	...	D_1	...
H_2	...	D_2	...
H_3	...	D_3	...
H_4	...	D_4	...

(e) Multiple-column header table

Fig. 2. Commonly ways of table expanding

For the table (a) (b) in Fig.2, We construct the logic tree model as Fig.3 (a). For the table(c) in Fig.2, We construct of the logic tree model as Fig.3 (b). For the table(d) (e) in Fig.2, we need to split it into a number of tables like table (a) or (b) according to the numbers of its attribute row(column).

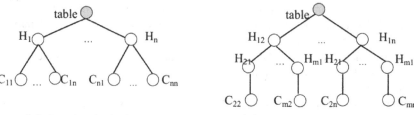

(a) Row header logic tree　　　　　(b) Row-column header logic tree

Fig. 3. Lgical tree model

3   Ontology Structure and Application

3.1   The Title Library

Usually the title of a table is subjectively named by people, the same topic probably possess different title. Therefore, we construct a title library $D=\{a_1,a_2,a_3\ldots\}$ to describe a topic. When we need to match it, first match the cell's content with D, if the content appear in D then it is a title, otherwise, put the content and the keywords which contain the same words into the synonym Ontology in order to calculate the similarity. If the similarity is greater than η, we add the content of the cell to D. So the content of title library will be improved by learning.

3.2   The Synonym Ontology

We describe the synonym Ontology by tree, the nodes represent the words, and the edges represent the semantic relationships between words. Fig.4 is an example.

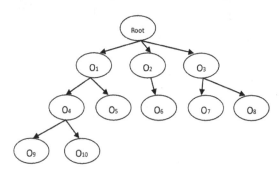

Fig. 4. A simple hierarchical tree

We discuss how to calculate the similarity between the two words:
(1) The semantic distance between two words.
The Semantic distance between two words is the shortest path connected the two nodes in the tree. For the words O_a, O_b, $Sim(O_a, O_b)$ represents their Semantic similarity,

$Dis(O_a, O_b)$ represents their semantic distance then $Dis(O_a, O_b)$ is inversely proportional to $Sim(O_a, O_b)$. When $Dis(O_a, O_b)0$, the $Sim(O_a, O_b)$ is 1, this denote that the two words are exactly the same, when $Dis(O_a, O_b)$ is infinity, it represents the two words are totally dissimilar.

(2) The words' depth in the tree.

In the tree, the words on the top are more abstract, it will be more specific with further down. So for the two nodes have the same semantic distance, the deeper the nodes, the larger semantic similarity they have. For example, the semantic similarity between O_9 and O_{10} is larger than O_4 and O_5. For the different levels of the words the semantic similarity decrease with the increase of increase of depth. For example the semantic similarity between O_9 and O_4 is larger than O_9 and O_1.

(3) Semantic overlap between two words.

The semantic overlap between two words is the number of public nodes, through which the shortest paths between their nearest parent nodes with the root node pass. We define it as $a(x) \cap a(y)$, and the $a(x)$ represents the number of nodes. The semantic overlap shows the same degree between two words, the semantic overlap is greater, the Semantic similarity is larger.

Taking all these factors we define the semantic similarity calculation formula as follows:

$$sim(x, y) = \frac{a(x) \cap a(y)}{(Dis(x, y) + a(x) \cap a(y)) \times \max(\alpha | h_x - h_y |, 1)} \qquad (2)$$

In order to prove the effect and efficiency of proposed table extraction system, we measure the performance of information extraction system based on two main indicators [15]: Recall (Marked R in this paper) and precision (marked P in this paper).

$$R = \frac{\text{the correct results extracted by the system}}{\text{all the correct results should be extracted}} \times 100\% \qquad (3)$$

$$P = \frac{\text{the correct results extract by the system}}{\text{the results exracted by the system}} \times 100\% \qquad (4)$$

However, generally speaking, for the same extracting results, the higher recall, the lower the precision, and the vice versa[17,18,19]. Therefore, we introduce a new measure from paper [16] to evaluate extracting results, the measure is as follow:

$$F = \frac{(\beta^2 + 1.0) \times P \times R}{\beta^2 \times P + R} \times 100\% \qquad (5)$$

β is the relative weight of the precision and recall. In practice, we adjust the β meet our needs. When $\beta = 1$, the precision and recall are equally important, when $\beta > 1$, precision is more important, when $\beta < 1$, recall is more important, in this paper $\beta = 1$.

In the Experiment, We selected five types of Web sites for data extraction. The results are shown in Table 3, and Ef represents correct results extracted by the system,

Et represents all the correct results extracted by the system, Er represents the results extracted by the system, TN represents the number of test tables

Table 3. The expriment result

Site Name	TN	E_f	E_r	E_t	R	P	F
Edu System	50	45	47	45	100	95.7	97.8
Sina.com	60	54	58	60	90.0	93.1	91.5
JD.com	55	51	54	55	92.7	94.4	93.5
Zol.com	145	109	124	115	94.8	87.9	91.2
Library	80	75	78	80	93.6	96.1	94.8

In addition, comparing this method with the method in the paper [16], the results are shown as Fig.5:

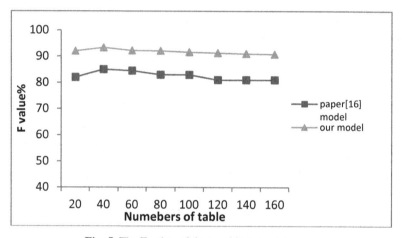

Fig. 5. The F value of the two kinds model

The result shows that the method can locate target table and extract data accurately. The value of F has been increased with the standardizing and the Ontology are been applied in the system before extracting.

4   Conclusion

In this paper we put forward a extraction method based on the structure and the ontology in a table. Experimental data show that the accuracy has been increased, the capacity of identifying the table header has been improved, and the errors of content extraction have been decreased. In the light of discuss above and our experiments, the proposed method is ideal for most web table extraction. But when the Webpage is incomplete and not clear, or its structure is very complex, the performance is poor. Moreover the extraction technology is only for the < table > tag in the table. Not suitable for the < div > tag. So we need to design more perfect heuristic rules for the different type of tables.

Acknowledgements. This work was supported partly by NSFC(61462010,61363036), BaGui scholars team project, the found of university science and technology of Guangxi, and the director found of Guangxi key laboratory.

Many thanks to Guangxi Collaborative Innovation Center of Multi-source Information Integration and Intelligent Processing, and the anonymous referee for comments.

References

1. Gatterbauer, W., Bohunsky, P.: Table extraction using spatial reasoning on the CSS2 visual box model. In: Proceedings of the 21st National Conference on Artificial Intelligence (2006)
2. Amin, M.S., Jamil, H.: FastWrap: An efficient wrapper for tabular data extraction from the web. In: IEEE International Conference on Information Reuse & Integration, IRI 2009, pp. 354–359. IEEE (2009)
3. Garofalakis, M., Gionis, A., Rastogi, R., et al.: XTRACT: A system for extracting document type descriptors from XML documents. ACM SIGMOD Record 29(2), 165–176 (2000)
4. Fensel, D., Van Harmelen, F., Horrocks, I., et al.: OIL: An ontology infrastructure for the semantic web. IEEE Intelligent Systems 16(2), 38–45 (2001)
5. Cha, S., Ma, Z., Cheng, J., et al.: Learning of ontology from the web-table. In: 2011 Eighth International Conference on Fuzzy Systems and Knowledge Discovery (FSKD), vol. 3, pp. 1454–1458. IEEE (2011)
6. Wang, F., Gui, L., Wu, G.W.: Extracting information from Ontology-based WEB Table. MINI-MICRO SYSTEM 24(12), 2142–2146 (2003)
7. Xu, F., Zhang, S.Q., Yao, H.G.: Web Form Data Extraction System Based on Structure. Journal of Xi'an Technological University 29(6), 574–578 (2009)
8. Gatterbauer, W., Bohunsky, P., Herzog, M., et al.: Towards domain-independent information extraction from web tables. In: Proceedings of the 16th International Conference on World Wide Web, pp. 71–80. ACM (2007)
9. Liu, Y., Wu, G.Q., Hu, X.G.: A web table extraction algorithm based on tree edit distance. In: IEEE Conference Anthology, pp. 1–6. IEEE (2013)
10. Zhang, Z.Y., Xu, T., Feng, X.: Auto Generation Technology for Flight Information Extraction Rules. Computer Engineering 37(6), 65–67 (2011)
11. Dalvi, N., Kumar, R., Soliman, M.: Automatic wrappers for large scale web extraction. Proceedings of the VLDB Endowment 4(4), 219–230 (2011)
12. Gultom, R.A.G., Sari, R.F., Budiardjo, B.: Proposing the new Algorithm and Technique Development for Integrating Web Table Extraction and Building a Mashup. Journal of Computer Science 7(2) (2011)
13. Zhao, H., Xiao, H., Xue, D., et al.: A Survey of the Research on Information Extraction over Web Tables. New Technology of Library and Information Service 3, 24–31 (2008)

14. Kuhlins, S., Tredwell, R.: Toolkits for generating wrappers. In: Akşit, M., Mezini, M., Unland, R. (eds.) NODe 2002. LNCS, vol. 2591, pp. 184–198. Springer, Heidelberg (2003)
15. Liu, Y., Bai, K., Mitra, P., et al.: Tableseer: Automatic table metadata extraction and searching in digital libraries. In: Proceedings of the 7th ACM/IEEE-CS Joint Conference on Digital Libraries, pp. 91–100. ACM (2007)
16. Liao, T., Liu, Z.T., Kong, Q.P.: The Design and Implementation of Information Extraction Model on Web Tables. Computer Application and Software 26(4), 72–74 (2009)
17. Zhang, S., Zhang, C., Yan, X.: Post-mining: Maintenance of association rules by weighting. Inf. Syst. 28(7), 691–707 (2003)
18. Zhang, S., Qin, Z., Ling, C.X., Sheng, S.: "Missing Is Useful': Missing Values in Cost-Sensitive Decision Trees. IEEE Trans. Knowl. Data Eng. 17(12), 1689–1693 (2005)
19. Wu, X., Zhang, S.: Synthesizing High-Frequency Rules from Different Data Sources. IEEE Trans. Knowl. Data Eng. 15(2), 353–367 (2003)

Block Nested Join and Sort Merge Join Algorithms: An Empirical Evaluation

Mingxian Chen and Zhi Zhong[*]

Guangxi Teachers Education University, Nanning, Guangxi, 5, China
zhong8662@126.com

Abstract. Both block nested join algorithm and sort merge join algorithm are conventional join algorithms in database systems. To the best of our knowledge, few literature focused on the experimentally comparing these two join algorithms. In this paper, we implement the sort merge join algorithm and the block nested loop join algorithm. And then, experimental results demonstrate the sort merge join algorithm outperforms than the block nested join algorithm on execution time in term of different bytes of page or different number of buffer but with the same result after join.

Keywords: data structure, join algorithm, sort merge join, block nested join.

1 Introduction

Rapid and precise decision-making based on accurate information is a very important issue in real business scenarios [16, 22, 26]. For example, many companies usually encounter the troubles for handling tremendous amounts of complex information. It is a necessary factor for they organize and analyze these data [23,25,28,29].

Data warehousing and online analytical processing (OLAP) applications, as very hot technologies, have been effectively contributed to the application of decision support systems (DSS). However, data warehouses often are very large because of the enormous quantities of information available to companies. For example, when retrievals for aggregations are performed in an OLAP system, the Database management system (DBMS) must search the tables for all records necessary for the process. So it is very important for the OLAP to utilize the high performance joins between tables.

As the tables become larger, the join algorithm becomes increasingly critical. Join algorithms [12,31,34], which were first studied in the context of program analysis, include nested loop join, block nested loop join, sort merge join, hash join and the others [17]. And various techniques have been implemented such as Red Brick's STARjoin TM technology and Bit-Wise technology from Sybase IQ [20,13].

(Luo et al, 2002) in [10] thought common techniques aid join performance, irrespective of the algorithm chosen. In fact, these join techniques use options that can

---

[*] Corresponding author.

X. Luo, J.X. Yu, and Z. Li (Eds.): ADMA 2014, LNAI 8933, pp. 705–715, 2014.

be found on the SQL Properties pane by setting queries. However, selecting a join algorithm is important enough to merit a dedicated topic. In this paper, we review the two traditional join algorithms, i.e., block nested loop join algorithm and sort merge joint algorithm respectively, by all kinds of experiments, and then use the Debug property on the SQL Join Properties pane [1] to run these algorithms. Furthermore, we design query optimizer to make the two join algorithms that yields the best possible performance for the query processing [24,27,30,32,33,35,36].

In the rest of the paper, we review related work in section 2, followed by introduction of the block nested loop join algorithm and sort merge join algorithm in section 3. In section 4, we give all kinds of experiments to evaluate the pros and cons of these two algorithms by design new optimizer. Finally, we conclude the paper and highlight our future work in section 5.

2   Related Work

In relational database systems (RDBMS), join is one of the most fundamental operations, which efficiently retrieves information from two different tables based on a Cartesian product of the two tables [14,7,5,8,6,19,18,3,4]. Meanwhile, join is also one of the most difficult operations to implement efficiently in RDBMS, because in most cases there is no predefined association between tables that can be utilized to facilitate the join processing. Depending on different math operators used in the join condition, there are various types of joins in database systems, including equijoin, natural join, semijoin, outerjoin, and self-join. To process the different joins, there are mainly three sorts of algorithms proposed, namely, nested loop join, sort merge join, and hash join, etc [14,7,5,8,6,9,15,21,19,18,3,2,11,4]. These join algorithms are categorized based on how they partition the tuples from different tables.

Nested loop join is the most straightforward method to process joins [14,6,4]. Specifically, one of the tables being joined is designated as *inner relation*, and the other is the *outer relation*. Then for each tuple of the outer relation, all tuples of the inner relation are fetched from disk and compared with the tuple from the outer relation. Whenever the predefined join condition is satisfied, the two tuples are concatenated and output as a result. Given two tables R of size $|R|$ and S of size $|R|$, the time complexity of nested loop join is in the order of $O(|R|*|S|)$, which is inefficient when R and S are large. The block-oriented implementation of nested loop join, i.e., block nested join, tries to optimize I/O cost by choosing the table with larger cardinality to be the inner relation and the table with smaller cardinality to be the outer relation.

Compared to nested loop join, sort merge join is a more efficient technique which takes advantage of sorted tuples in the tables [7,5]. Specifically, sort merge join consists of two stages. In the first stage, both tables to be joined are sorted on the join attributes. Then, both tables are scanned in the order of the join attributes, and tuples meeting the join condition are concatenated to form a result tuple [14,7]. Sort merge join is superior to nested loop join, in that in sort merge join each table is scanned through only once, because the tuples in the tables are sorted and whenever a tuple in the inner relation does not satisfy the join condition, we need not to examine the rest tuples in the inner relation any more. In terms of time complexity, sort merge join incurs $O(n\log n)$ time, due to the fact that its running time mainly depends on the sorting time. If the two tables are presorted, or the join attributes are indexed, then sort merge join will incur much less computation time.

The main idea of hash join method is that we use some predefined hash function to map all the tuples of one of the tables into a collection of buckets [14,6,15,3]. And tuples mapped to a same bucket have the same hash value on their join attributes. We then scan through the tuples in the other table, using the same hash function to find the bucket that the tuple is mapping to. If the bucket is not empty, we concatenate the tuple with each of the tuples in the bucket, and output the result. Otherwise, we discard the tuple and continue probing the next tuple. The hash join method is one of the most efficient algorithm for join processing, because the hash computation is fast and we only scan each of the tables once, i.e., with a time complexity of $O(|R|+|S|)$. However, the main drawback of hash join algorithm is that it is suitable for equijoin processing, not for non-equijoin processing [14, 7, 5].

3   Block Nested Join and Sort Merge Join

In *Block* algorithm, assuming the number of buffer is n, we let the number of buffer in the left batch be (n-2), one for right batch, and the left one for output batch. In *NEXT()* of the iterator, for each tuple in right batch, *Block* algorithm scans the whole (n-2) batches in left batch to find a join where there are two cases. 1) If there is no a join in the (n-2) batches in the left batches for the tuple in the right tuple, the algorithm continues to read another (n-2) left batches into buffer until all tuples in left file have been read. If there is no a join after scanning the whole left file, the algorithm will read the next tuple in right batch to continue this join process. 2) If there is a join between left batch and right one, then joining them and the joined results are sent into outbatch which will be outputted into the disk while it has been filled with. The algorithm won't stop until the right file is read to the end. So the amount of I/O cost for block algorithm only is the number which is the sum of I/O cost for reading single file in the two files.

Sort algorithm firstly generates two sorted files for joining in *OPEN()* of the iterator. For sorting each file, *Sort* algorithm reads n batches once into buffer. Firstly, we employ quick sort algorithm to sort tuples in one batch, then employing multi-way merge algorithm to sort tuples in (n-1) batches in one buffer. Then the full outbatch with sorted results are sent into the disk saved as the temporary files. After sorting all tuples for one file, and generating m temporary files in the disk, *Sort* algorithm begins to read these temporary files from disk to merge recursive until the file becomes one file, i.e., all temporary files are combined a whole file sorted with non-decreasing ordering. Finally, the sorted file can be got from *OPEN()*.

In *NEXT()* of the iterator, firstly, the algorithm begins to scan left batch to match the first tuple in the right batch till finding a join. The result joined will be sent into outbatch, and read the next tuple in the left batch. After this, there are three cases due to the sorted file with no-decreasing ordering. 1) If the value in the left batch is larger than the right one, then getting a next tuple in the right batch until there is a join between the two batches. 2) If the value in the left batch is smaller than the right one, the continuing to get a next tuple in the left batch until there is a join between the two batches. 3) There exists a join between the two batches, then joining them and outputting the result into *outbatch*. The algorithm continues until finishing scanning all the tuples in the right batch. The pseudo-code of the block nested join and sort merge join algorithm is given in Figure 1 and Figure 2.

```
BlockNestedJoin()
  1. OPEN(){}
  2. NEXT(){
  3        While (!Outbatch.isFull()){
  4                if (lcurs==0 && eosr==true){
  5                    leftbatch=left.next();};
  6                While (eosr==false){
  7                    lefttuple.joinWith(righttuple);}
  8        }
  9        Return outbatch;
 10 }
 11 CLOSE(){};
```

Fig. 1. Pseudo-code of block nested join algorithm

```
SortMergeJoin ()

  1. OPEN(){
  2.    leftIn = sort(left, leftindex);
  3.    rightIn = sort(right, rightindex);
  4. }

  5. NEXT(){
  6.       While(!outbatch.isFull()){
  7.               if(rtuples!=null){
                   lefttuple.joinWith(righttuple);}
  8.       While(lefttuple.dataAt(leftindex)<right){
                   leftbatch=leftIn.readObject();}
  9.       While(lefttuple.dataAt(leftindex)>right){
 10.              rightbatch=rightIn.readObject();}
 11.       While(lefttuple.data.equals(righttuple){
 12.              rightbatch =rightIn.readObject();}
 13.       Return outbatch;
 14. }
 15. CLOSE(){};
```

Fig. 2. Pseudo-code of sort merge join algorithm

4 Experimental Study

We design different experiments to compare the two algorithms implemented in the paper, such as, block nested algorithm (referred to as **Block** in this part) and sort merge join algorithm (referred to as **Sort**), with the nested join algorithm (referred to as **Nested**), in term of the result of the join algorithm and execution time under the different circumstances. The details of datasets generated and designed queries are presented in table 1 and figure 2 respectively, and the experimental results are showed from figure 3 to 4. The worker platform was based on the java applet presented in [1].

Table 1. Datasets used in the experiments

Dataset name	#Tuples	#(bytes/tuple)	#Attributes
Aircraft	10000	24	3
certified	10000	8	2
employee	10000	30	3
flights	10000	80	6
schedules	10000	8	2

Query workload
Query 1: **SELECT * FROM** *SCHEDULE, AIRCRAFTS* **WHERE** *SCHEDULE.aid=AIRCRAFTS.aid*
Query 2: **SELECT * FROM** *EMPLOYEES,CERTIFIED, SCHEDULE* **WHERE** *EMPLOYEES.eid=CERTIFIED.eid*, *CERTIFIED.aid=SCHEDULE.aid*

Fig. 3. Query workload for the two join algorithms

In the designed experiments, we run the three queries and for each query we execute all the three join methods and record the running time for each join method. Also this time keep the buffer size fixed at 20 and change page size progressively and see how the different page size can impact the performance of each join method.

The result is shown in Figure 3, and we can see that as we change page size from 500 from 5000, the running time for the nested loop join is decreasing, but no apparent changes in execution time on block nested loop join and sort merge join. The reason for this is that when the page size become larger, then the number of tuples in one page is become larger, therefore number of times to scan the next relation is become smaller. Hence the I/O cost is decreasing and the running time is decreasing correspondingly. For block nested join and sort merge join we see the similar result

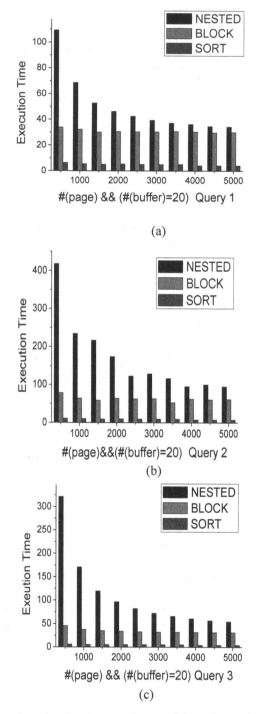

(a)

(b)

(c)

Fig. 4. Comparison of running time between the three join methods with different page size

and the running time decrease for the I/O cost is decreased for the large page size. In Figure 3, we also see that with the relatively large size of relations, the performance between simple nested loop join, block nested join and sort merge join varies a lot. We see that the execution time for running the same query, using the simple nested loop join always means needing much more time than the other two methods. And the sort merge join methods still outperforms block nested join.

Comparing with the experiment (in Figure 3) with relatively smaller relations we did, the difference of running time between sort merge join method and block nested loop join is even larger here. As we can see from Figure 3, the sort merge method is several times faster than the block nested join method. Also we show result of the experiment with a fixed page size and changing buffer size at the bottom part of Figure 3. Still during the starting period when the buffer size increases from a small value, the running time decreases for both block nested loop join and sort merge join. And after the buffer size reaches a threshold value, then running time for both methods do not decrease when we increase the buffer size. The reason here is the buffer is large enough to read all the large outer relation once for block nested join and to sort the relations in one pass. In addition, we also did the comparison of the how the running time changes with the total number of tuple. We see that the simple nested loop join is very sensitive to the size of the data set. And when we increase the size of data set, the running time is increasing faster than the other two join methods. And the increase of running time for block nested increases is larger than for sort merge. As we can see when we the data set twice bigger than the original one, the running time for sort merge slightly increase. So as we see from the scalability of each algorithm, the sort merge join algorithm still outperforms the other two methods.

In practice, we often have a quite large size of relations, and to do the join operations the nested join loop and block nested loop join can't really return the result in a reasonable time. So we can see that sort merge join is a relatively better join method to use especially when now large number of buffer is available for a DBMS. In addition, compared the running time between a random plan and the plan generated by our optimizer can vary by a large range. Due to the same reason, in practice, the little extra time spent on finding the relatively optimal query plan can let us gain great benefit in terms of running time cost. That's why the query plan ordering is used especially when we are dealing with large size of relations. For with larger data sets, different plan cost can vary by several magnitudes. So choosing the relatively smaller cost by query plan ordering is very necessary in practice.

As we can see from Figure 3, the sort merge method is several times faster than the block nested join method. We show the result of the experiment with a fixed page size and change buffer size in Figure 4. And also during the starting period when the buffer size increases from a small value, the running time decreases for both block nested loop join and sort merge join. After the buffer size reaches a threshold value, the running time for both methods do not decrease when we increase the buffer size. The reason is the same as the first experiment presented in Figure 3.

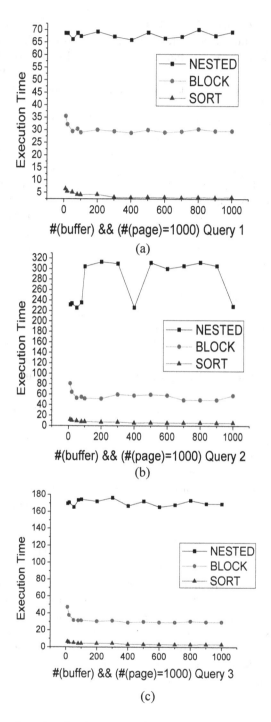

Fig. 5. Comparison of running time between different join methods with different buffer size

5   Conclusion

In this paper, we design experiments to compare the two traditional join algorithm, i.e., blocked nested loop algorithm and sort merge join algorithm respectively, by evaluating different page size and different buffer size. And we have the following findings:

1. The sort-merge join method most of the time out-performs the block nested loop join by several magnitude. Only when the join involve relations with small size or a small size of buffers, the block nested loop join method can outperform sort-merge join. That is because the sort algorithm generated temporary files during the process of merging sorted files so that the algorithm will cost extra time to read and write these files. The I\O cost will be larger than Block nested join when the data set is quite small and the number of join is small also.
2. The cost model we use here gives a good approximation for each query plan of its actual cost in terms of their actual running time.
3. The optimizer implemented by us always gives a plan which is quite close to the optimal solution regarding to the plan's cost.
4. The naïve method for doing the join operation which is nested loop join can only deal with small size of relations and when the relations is relatively large, the time needed to do the join is quite long and sometimes not realistic.
5. The cost difference between different plans with large data sets can vary a lot and query execution plan ordering is necessary for it can let us have a relatively low cost plan instead of some randomly selected plan which will incur much higher cost.

In our future work, we will continue to implement more join algorithms to discover their pros and cons under different settings.

Acknowledgements. This work was supported by the Guangxi Natural Science Foundation under grant 2014GXNSFAA118356.

References

1. http://www.cs.princeton.edu/~appel/modern/java/CUP/
2. Braumandl, R., Claussen, J., Kemper, A., Kossmann, D.: Functional-Join Processing. VLDB Journal 8(3-4), 156–177 (2000)
3. Chen, S., Ailamaki, A., Gibbons, P.B., Mowry, T.C.: Improving Hash Join Performance Through Prefetching. ACM Transactions on Database Systems (TODS) 32(2), 17 (2007)
4. DeWitt, D., Jeffrey, F., Joseph, B.: Nested Loops Revisited. In: Proceedings of the 2nd International Conference on Parallel and Distributed Information Systems, pp. 230–242 (1993)
5. Dittrich, J., Seeger, B., Taylor, D.S., Widmayer, P.: Progressive Merge Join: A Generic and Non-Blocking Sort-Based Join Algorithm. In: Proceedings of the 28th International Conference on Very Large Data Bases (VLDB), pp. 299–310 (2002)

6. Graefe, G., Linville, A., Shapiro, L.: Sort Versus Hash Revisited. IEEE Transactions on Knowledge and Data Engineering (TKDE) 25(2), 73–170 (1993)
7. Harris, E.P., Ramamohanarao, K.: Join Algorithm Costs Revisited, Technical Report. University of Melbourne (1993)
8. Haas, L.M., Carey, M.J., Livny, M., Shukla, A.: Seeking the Truth About Ad Hoc Join Costs. VLDB Journal 6, 241–256 (1997)
9. Ioannidis, Y., Christodoulakis, S.: On the Propagation of Errors in the Size of Join Results. In: Proceedings of ACM International Conference on Management of Data (SIGMOD), pp. 268–277 (1991)
10. Luo, G., Ellmann, C.J., Haas, P.J., Naughton, F.J.: A scalable hash ripple join algorithm. In: Proceedings of the 2002 ACM SIGMOD International Conference on Management of Data, pp. 252–262. ACM, New York (2002)
11. Li, J., Sun, W., Li, Y.: Parallel Join Algorithms based on Parallel B+-trees. In: Proceedings of the 3rd International Symposium on Cooperative Database Systems for Advanced Applications, (CODAS) (2001)
12. Lieberman, M.D., Sankaranarayanan, J., Samet, H.: A Fast Similarity Join Algorithm Using Graphics Processing Units. In: Proceedings of the 24th IEEE International Conference on Data Engineering, Cancun, Mexico, pp. 1111–1120 (April 2008)
13. Mokbe, M.F., Lu, M., Aref, W.G.: Hash-Merge Join: A Non-blocking Join Algorithm for Producing Fast and Early Join Results. In: ICDE (2004)
14. Mishra, P., Eich, M.H.: Join Processing in Relational Databases. ACM Computing Survey 24(1), 63–113 (1992)
15. Patel, J., Carey, M., Vernon, M.: Accurate Modeling of the Hybrid Hash Join Algorithm. In: Proceedings of ACM SIGMETRICS Conference (1994)
16. Qin, Y., Zhang, S., Zhu, X., Zhang, J., Zhang, C.: Semi-parametric optimization for missing data imputation. Appl. Intell. 27(1), 79–88 (2007)
17. Ramakrishnan, R., Gehrke, J.: Database management system, 3rd edn., pp. 452–458 (2002)
18. Ramasamy, K., Patel, J., Naughton, J.F., Kaushik, R.: Set Containment Joins: The Good, The Bad and The Ugly. In: Proceedings of the International Conference on Very Large Data Bases (VLDB), pp. 351–362 (2000)
19. Shekita, E., Carey, M.: A Performance Evaluation of Pointer-based Joins. In: Proceedings of ACM International Conference on Management of Data (SIGMOD) (1990)
20. Toyama, M., Ohara, A.: Hash-Based Symmetric Data Structure and Join Algorithm for OLAP Applications. In: International Database Engineering and Applications Symposium, pp. 231–238 (1999)
21. Valduriez, P.: Join Indices. ACM Transactions on Database Systems (TODS) 12(2), 218–246 (1987)
22. Wu, X., Zhang, S.: Synthesizing High-Frequency Rules from Different Data Sources. IEEE Trans. Knowl. Data Eng. 15(2), 353–367 (2003)
23. Wu, X., Zhang, C., Zhang, S.: Efficient mining of both positive and negative association rules. ACM Trans. Inf. Syst. 22(3), 381–405 (2004)
24. Wu, X., Zhang, C., Zhang, S.: Database classification for multi-database mining. Inf. Syst. 30(1), 71–88 (2005)
25. Zhang, S., Zhang, C., Yan, X.: Post-mining: maintenance of association rules by weighting. Inf. Syst. 28(7), 691–707 (2003)
26. Zhang, S., Qin, Z., Ling, C., Sheng, S.: "Missing Is Useful": Missing Values in Cost-Sensitive Decision Trees. IEEE Trans. Knowl. Data Eng. 17(12), 1689–1693 (2005)
27. Zhao, Y., Zhang, S.: Generalized Dimension-Reduction Framework for Recent-Biased Time Series Analysis. IEEE Trans. Knowl. Data Eng. 18(2), 231–244 (2006)

28. Zhu, X., Zhang, S., Jin, Z., Zhang, Z., Xu, Z.: Missing Value Estimation for Mixed-Attribute Data Sets. IEEE Trans. Knowl. Data Eng. 23(1), 110–121 (2011)

29. Zhu, X., Zhang, L., Huang, Z.: A Sparse Embedding and Least Variance Encoding Approach to Hashing. IEEE Transactions on Image Processing 23(9), 3737–3750 (2014)

30. Zhu, X., Huang, Z., Shen, H., Zhao, X.: Linear cross-modal hashing for efficient multimedia search. In: ACM Multimedia, pp. 143–152 (2013)

31. Zhu, X., Suk, H., Shen, D.: A novel matrix-similarity based loss function for joint regression and classification in AD diagnosis. NeuroImage 100, 91–105 (2014)

32. Zhu, X., Suk, H., Shen, D.: Matrix-Similarity Based Loss Function and Feature Selection for Alzheimer's Disease Diagnosis. In: CVPR, pp. 3089–3096 (2014)

33. Zhu, X., Huang, Z., Yang, Y., Shen, H., Xu, C., Luo, J.: Self-taught dimensionality reduction on the high-dimensional small-sized data. Pattern Recognition 46(1), 215–229 (2013)

34. Zhu, X., Huang, Z., Cui, J., Shen, H.: Video-to-Shot Tag Propagation by Graph Sparse Group Lasso. IEEE Transactions on Multimedia 15(3), 633–646 (2013)

35. Zhu, X., Huang, Z., Cheng, H., Cui, J., Shen, H.: Sparse hashing for fast multimedia search. ACM Trans. Inf. Syst. 31(2), 9 (2013)

36. Zhu, X., Huang, Z., Shen, H., Cheng, J., Xu, C.: Dimensionality reduction by Mixed Kernel Canonical Correlation Analysis. Pattern Recognition 45(8), 3003–3016 (2012)

The Design of Power Security System in Smart Home Based on the Stream Data Mining

Shun Ma[1], Shenglong Fang[1], Dingrong Yuan[1,*], and Xiangchao Wang[2]

[1] College of Computer Science & IT and Guangxi Key Lab of Multi-source Information Mining & Security at Guangxi Normal University, Guangxi Guilin 541004, China
{magexiao,yimenwang}@foxmail.com,
dryuan@mailbox.gxnu.edu.cn
[2] Jian she XI Lu, 109th, Chizhou City Meteorological Bureau,
Anhui Chizhou, China
18905661619@163.com

Abstract. With the improvement of the living standard of people, household electricity consumption has increased significantly. Meanwhile, the accidents in power security occur frequently. In this paper, for ensuring the security of our household electricity appliances, we design a power security system based on stream data mining, which is mainly composed of the intelligent electric outlet, the coordinator and the server etc. In the system, we use the ZigBee module as connector between the traditional power grid and the coordinator, use the intelligent electric outlet to shut or open the power. On and off of the circuit depends on whether the current or voltage is abnormal or not, which is obtained by the technique of stream data mining. In order to test the effectiveness of the proposed system, we have implemented a verification system. The results show that the proposed system is feasible.

Keywords: Data mining, Stream data mining, intelligent electric outlet, ZigBee.

1 Introduction

Nowadays, household appliances have become necessary in our lives, the outlet can be seen everywhere, the security of the outlet is increasingly lightened, even it may be harm our lives. So that the safety precaution, such as remote control, timer switches, power consumption measurement, security and protection, has attracted great attention of academic and industry. Wireless network technology based on ZigBee is provided with the characteristics of high performance, low power, and short range, low transport costs, large network capacity, easy to maintenance and so on. It has been successfully applied in the smart home [1],[2]. But safety precaution of outlet has not been seen. So in this paper, we design a home security power system based on ZigBee technology, which can eliminate the hidden danger and provide strong protection for

---

* Corresponding author.

X. Luo, J.X. Yu, and Z. Li (Eds.): ADMA 2014, LNAI 8933, pp. 716–724, 2014.

household appliances in life. The system include two key parts, one is the ZigBee module, which is used to connect the traditional power grid with and the coordinator, the other is the intelligent electric outlet, which is used to shut and open the power. On or off of the circuit is determined by whether the current or voltage is abnormal or not. This is the issue of stream data mining.

The stream data is a continuous and orderly sequence data in real time. The stream data mining mainly include multiple streams data mining and single stream data mining. The multiple stream data mining is to obtain the relevance between the stream data. The single stream data mining is to obtain the characteristics of Category, the characteristics of frequency and the characteristics of changing. In our proposed system, the current of power obtained by the sensors is continuous and orderly, and it is real time data. So we can regard as stream data, and mining the abnormal in the stream data, to obtain the abnormal data, which is used to determine whether the outlet should be turn on or turn off [10],[11],[12].

2   The Design of the System Function

The traditional power socket is just a connecter used to distribute the electric current to different household appliances. Unlike the traditional power socket, based on ZigBee technology, we design a intelligent outlet possessing the following features:

1. Remote control: The outlet can control switches in the circuit by the commands from the remote terminal, and achieve the purpose of controlling electrical appliance.
2. Monitor power consumption: The socket can collect the value of electric current by the current transformer and the value of electric voltage transformer.
3. Abnormal protection: Any exception, such as current overland, overvoltage, under voltage, short circuit etc, usually occurs in the electric circuit. The path will be immediately disconnected to avoid electrical appliances damaged.

In the system of electric security, the traditional power socket is displaced by the intelligent outlet. The system become secure for its power consumption is monitored by remote control and safeguard by abnormal detection.

3   The Structure of the System

Power security system in smart home is a system composed of household appliances, power line, connector, server, coordinator and remote control etc. In the system, we use the intelligent outlet as connector. So we design the system as Fig.1.

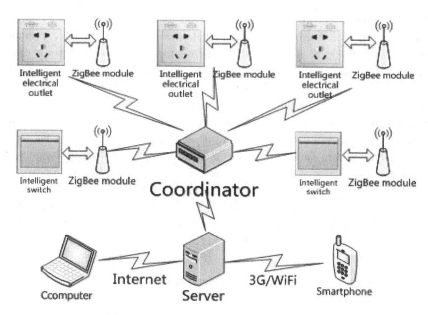

Fig. 1. The overall structure of system

In the Fig.1, the coordinator and the intelligent electrical outlet are linked together by power line. Coordinator is connected to server by the line of communication. The server and the computer or smart phone are linked together by wireless remote or WI-FI, which is the same as the normal network. So to say, the system mainly includes three parts as blow.

1. Server is the heart of the system. It is responsible to collect and analysis the real-time data of the household appliances, intelligent control the electric consume of them, ensure the security of the electricity. It is connector of coordinator and remote terminal such as computer and smart phone.
2. The coordinator is the core of the home electric network. It is responsible to establish and maintain the network, and sent data uploaded by terminal devices to the server through the serial port. It also sends command form the server to the designated terminal devices.
3. Terminal nodes are composed of the ZigBee module and socket. ZigBee modules are embedded in the outlet. It is used to receive a control command from the coordinator, and the same time it collects the real time electric dada and transmits it to the coordinator.

4   The Design and Implementation of the ZigBee Module

In the power security system, ZigBee module is an important component. The coordinator and the terminal node include the ZigBee module. So we introduce the ZigBee module in this section.

4.1   Introduction of the ZigBee Module

In this system, we adopt the chip IC CC2530 made by Texas instruments (TI) company as the ZigBee module. The CC2530 is a true system-on-chip (SoC) solution for ZigBee applications. It is able to establish a strong network node at low cost. The CC2530 combines the excellent performance of a leading RF transceiver with an industry-standard enhanced 8051 MCU, in-system programmable flash memory, 8-KB RAM, and many other powerful features. CC2530 have different operational modes, which makes it especially fit the system of ultra-low power requirements. And the short time of conversion between operational nodes decrease the consumption of energy [3],[4],[5].

The ZigBee network often consists of Coordinator node, Routers and terminal devices (End-Device). The Coordinator is responsible for starting the entire network configuration which is a Full Function Device (FFD) defined by the IEEE 802.15.4. In this system, we firstly select a channel and a network ID, and then start the whole network. The router allows other devices to join the network, and the multi-hop router communicates with its assisting terminal nodes. Terminal nodes, namely Reduced Function Device (RFD), can be in the state of sleep or wake up. Therefore only one battery is enough for the power. The ZigBee module is used to receive the instruction from the coordinator and the real time information from the appliances, and then send the collected data to coordinator [6],[7].

The coordinator, which directly link to the Server and the terminal nodes, is the core of the home power network. It is composed of serial module and ZigBee module. We design the structural of the coordinator as follow.

4.2   The Structural Design of the Coordinator

The coordinator is the core of the wireless network, which is responsible to establish and maintain the network, and sent data uploaded by terminal devices to the server through the serial port. It also sends command form the server to the designated terminal devices. In order to listen to a serial port and receive interrupt, we expend the RS232 serial port on the hardware platform of the coordinator. The concrete structure is shown in Fig.2.

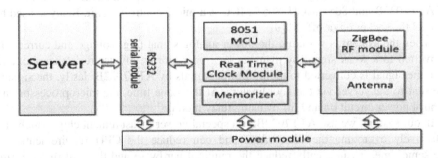

Fig. 2. The structural design of the coordinator

In the Fig.2, ZigBee RF module is used to communicate with the ZigBee in the terminal notes. The 8051 MCU is responsible to collect and analysis the information from the ZigBee RF module. The clock module and the Memorizer are use to record the real time and save the information. The core parts of the coordinator are mainly composed of 8051 MCU and ZigBee RF module. RS322 is used to link the Coordinator and the Server.

4.3   The Design of the Security Outlet

The intelligent outlet is a connector of household appliances and power grid. At the same time, the socket can intelligent control the switches by remote terminal or by an accident. For the reason of security, we design the function of the intelligent outlet based on ZigBee module as below.

1. It can obtain the real-time power included the voltage and current in the circuit.
2. It can control circuit on and off by the relay module.

The structure of security socket is designed as Fig.3.

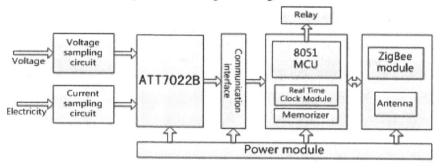

Fig. 3. The design of security outlet

In the Fig.3, the chip of Voltage sampling and Current sampling is used to sample the power signal, which is an analog signal. ATT7022B is a metering chip which is used to measure the voltage and the current. Communication interface is a connector of ATT7022B and 8051 MCU. 8051 MCU is a micro controller unit. Relay is used to control the socket on or off [8],[9].

When the system is working, the power analog signal (e.g. voltage and current) is converted to a weak signal by manganese copper tablets, divider resistance etc, and then the signal is converted to digital power signals by ATT7022B. lastly, the signals are sent to the server via the coordinator . At the same time, the microprocessor can control power circuit on and off by using relay module.

In our system, we use ATT7022B as a special power measurement chip, which can effectively overcome the disturbance and can reduce the CPU requirements. At the same time, it can greatly reduce the required hardware and the cost of hardware circuit.

4.4   The Intelligent Design of the Electrical Outlet

Stream data is a sequence of data items, such as $x_1, x_2, ..., x_n$, where, $x_i, i \in \{1, 2, ..., n\}$ is one data of the sequence data. n is a integer[13],[14].

Let $f(x_i)$ denotes the frequency of the stream data $x_i, i \in \{1, 2, ..., n\}$, $exception\_sup$ denotes the exceptional degree of $x_i, i \in \{1, 2, ..., n\}$, we obtain the abnormal stream data according the method of paper[10], the stream data is the condition of shut of open the power[15],[16].

5   The Design of System Software

In the power security system, the coordinator is the core of the home power network which is responsible to establish and maintain the network, and sent data uploaded by terminal devices to the server through the serial port. It also sends command form the server to the designated terminal devices. So we design the flow chart as Fig.4.

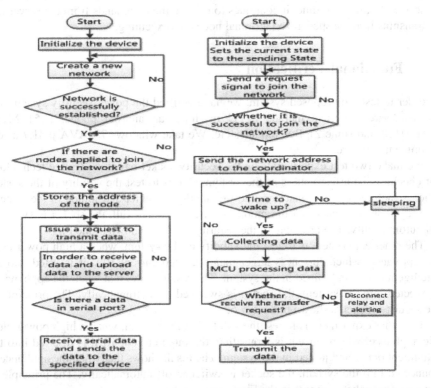

Fig. 4. The coordinator node software flow chart

Fig. 5. The terminal node software flow chart

In the Fig.4, the first step is to establish a power grid based the coordinator and the Server. The second step is to receive the signal from terminal nodes and transmit it to the Server. The third step is to receive the control command from the server and transmit it to the specified terminal node.

In the power security system, the terminal node is another important component, which is responsible to obtain the real time power data in the circuit and transmit it to the coordinator. At the same time, receive and execute the commands from the coordinate. The flow chart of the part is designed as Fig.5.

In the Fig.5, the first step is to link the terminal node to the coordinator. The second step is to obtain the real time power signal and transmit it to the coordinator. The third step is to receive and execute the control command from the coordinate.

The whole workflow of the system is described as follows.

1. Power on the system, start the coordinator and the system software.
2. Initializes the peripheral equipment by function halBoardInit().
3. Initializes power security system, which includes the traditional grid, terminal nodes and the coordinator.
4. After initializing, the coordinator enters the state of circulation working. It continues to monitor the real time signal from the terminal nodes and transmit it to the server. At the same time, it continues to receive the commands from the server and transmits it to the specialized terminal nodes for executing.

6   Functional Verification

In order to test the proposed system, we have verified the power security system. All the hardware materials include a PC machine, an android phone, 8051 MCU, ATT7022B and some ZigBee modules etc. We take windows 7, JAVA jkd1.7 as development software.

We make two test experiments on the security socket based on ZigBee technology for checking the effectiveness of the system. One is to test the validity of the system when a mobile charge is inserted into the intelligent outlet and joined in the system. The result shows that when the mobile phone is charged full, the socket is switched off automatically. The principle of the security socket is shown in the Fig.5.

The other experiment is to test the validity of the system, when a high power electric appliance, which current is larger than the rated current 10A, is inserted into the intelligent outlet and joined in the system. The result shows that when the appliance is connected into the system, the socket is switched off automatically. The principle of the security socket is shown in the Fig.6.

The other experiment is to test the validity of the system, when a high power electric appliance, which current is larger than the rated current 10A, is inserted into the intelligent outlet and joined in the system. The result shows that when the appliance is connected into the system, the socket is switched off automatically. The principle of the security socket is shown in the Fig.6.

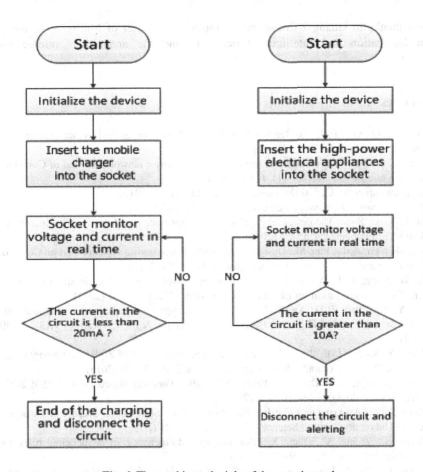

Fig. 6. The working principle of the security socket

7   Conclusions

Based on the current situation of the development of household appliances in the smart home, we constructed a power security system, which can ensure the security when the power in the household appliances is abnormal. The system is composed of the power grid, the coordinator and the server. In order to test the proposed system, we have verified the power security system. The results show that the proposed system has good feasibility.

Acknowledgment. This work was supported partly by NSFC   (61462010,61363 036), the BaGui scholar team project, the found of university science and tech nology of Guangxi, The director found of Guangxi key laboratory. The science and technology plan projects of Guilin(20140103-17). This work was also sup ported by Innovation Project of Guangxi Graduate Education (YCSZ2014092).

Many thanks to Guangxi Collaborative Innovation Center of Multi-source Information Integration and Intelligent Processing, and the anonymous referee for comments.

References

1. Wang, X.-Q., Ou, Y.-J.: ZigBee wireless sensor network design and implementation, pp. 3–6. Chemical Industry Press, Beijing (2012)
2. Li, J.-Z., Gao, H.: The research progress of wireless sensor network. Journal of Computer Research and Development 45(1), 1–15 (2008)
3. Texas Instruments.CC2530 Datasheet [EB/OL] (October 5, 2010), http://www.ti.com.cn/product/cn/cc2530#this
4. Xu, J., Yang, S.-S.: The design of ZigBee coordinator node based on CC2530. Internet of Things Technologies (5), 55–57 (2011)
5. Li, X.-H., Yu, A.-L., Pan, M.: Design of aquaculture monitoring system based on CC2530. Transducer and Microsystem Technologies 32(3), 85–88 (2013)
6. Xu, W., Jiang, Y.-J., Wang, B.: The application of ZigBee technology in smart outlet design. Telecommunication for Electric Power System 32(221), 78–81 (2011)
7. Gao, Y.-W., Zhao, Z.-H., Xu, R.-T.: Design of a fine-grained power monitoring and controlling socket based on ZigBee networks. Microcomputer & Its Applications 31(3), 27–30 (2012)
8. Zhang, Y.-S., Li, H.-B., Jia, J.-Y.: Design and Implementation of ZigBee Gate-way in the Internet of Things. Computer System & Application 22(6), 34–38 (2013)
9. Guo, H., Zhang, S.-J.: Design of Embedded ZigBee Gateway Based on ARM S3C2410. Henan Science 30(8), 1072–1075 (2012)
10. Wang, S., Li, A.: Detecting Out lier Pat terns in Time Series Data. Journal of Guangx i Normal University: Natural Science Edition 24(4), 18–21 (2006)
11. Zhang, S., Zhang, C., Yan, X.: Post-mining: Maintenance of association rules by weighting. Inf. Syst. 28(7), 691–707 (2003)
12. Zhang, S., Qin, Z., Ling, C.X., Sheng, S.: "Missing Is Useful': Missing Values in Cost-Sensitive Decision Trees. IEEE Trans. Knowl. Data Eng. 17(12), 1689–1693 (2005)
13. Wu, X., Zhang, S.: Synthesizing High-Frequency Rules from Different Data Sources. IEEE Trans. Knowl. Data Eng. 15(2), 353–367 (2003)
14. Wu, X., Zhang, C., Zhang, S.: Efficient mining of both positive and nega-tive association rules. ACM Trans. Inf. Syst. 22(3), 381–405 (2004)
15. Wu, X., Zhang, C., Zhang, S.: Database classification for multi-database mining. Inf. Syst. 30(1), 71–88 (2005)
16. Zhao, Y., Zhang, S.: Generalized Dimension-Reduction Framework for Recent-Biased Time Series Analysis. IEEE Trans. Knowl. Data Eng. 18(2), 231–244 (2006)

Agent-Based Privacy Aware Feedback System

Mohammad Shamsul Arefin[1], Rahma Bintey Mufiz Mukta[1],
and Yasuhiko Morimoto[2]

[1] Computer Science and Engineering Department, Chittagong University of
Engineering and Technology, Chittagong-4349, Bangladesh
`sarefin_406@yahoo.com`, `rahmamukta@gmail.com`
[2] Graduate School of Engineering, Hiroshima University, Japan
`morimoto@mis.hiroshima-u.ac.jp`

Abstract. In the field of information technology, a feedback system is
a computer program that receives information from the users and guides
the target audiences in order to achieve the desired outcomes. The Feedback systems can be used as a part of an intervention in organizations to
increase awareness and improve performance. However, due to many factors such as possibility of losing jobs and facing social problems, people of
the organizations are not interested to disclose their identities while providing feedbacks about their organizations. This is the most significant
problem of current feedback systems. Therefore, in this paper, we introduce a framework to provide feedbacks without disclosing individuals'
record's values. In our approach, we introduce an agent-based parallel
computation technique that can collect feedbacks from the users in a
secure environment. We provide an extensive experimental evaluation to
show the effectiveness of our approach.

Keywords: Feedback systems, privacy aware systems, agent-based
computation.

1 Introduction

The economic growth of a country highly depends on proper industrialization.
Due to the high salaries of the employees in the developed countries, many companies of the developed countries are considering rapidly developing countries
like Bangladesh, Thailand, China for the production of their products. As a result, now a days, we can easily find many industries in the developing countries.
As for example, if anyone visits Asulia area near Bangladesh capital Dhaka, he
can easily find many garments and textile industries in that area. These industries produces high volume of garments products for the world markets and play
a vital role for the economic growth of Bangladesh. However, the salaries, working environments, gender equality etc. in these industries are far below than the
world standard. The collapse of Rana Palaza Building that took the life of more
than one thousand people [1] is an example of such an industry. The employees
of these industries are bound to continue their jobs in such horrible situations

X. Luo, J.X. Yu, and Z. Li (Eds.): ADMA 2014, LNAI 8933, pp. 725–738, 2014.

due to their poverty. They cannot even complain about the poor conditions of the garments to the government's garments monitoring authority due to the fear of leaking their identities that may cause losing their jobs and other social harassments. The situations are almost similar in many other industries in the developing countries.

If we can collect feedbacks from the employees of any industry, we can identify the actual scenarios of that industry. From collected feedbacks of the industry, the monitoring authority can find out the shortcomings of the industry. This will help the monitoring authority to take necessary steps against the poorly perform-ing industries. In addition, it will create a competitive environment among the industries. Considering these facts, a feedback system is very much essential for the industries, especially for the industries in the developing countries.

However, an employee of an industry, in general, does not want to disclose her / his identity while providing feedbacks about the industry. This is the most significant problem for developing such a feedback system. Although, in [2], there are some considerations to develop a privacy preserving feedback system for the users, it has both performance limitations and accuracy problems. Therefore, in this paper, we consider an agent-based parallel computation framework to obtain feedbacks from the users. The proposed method solves the privacy problems of feedback systems. The developed system can also ensure performance efficiency and accuracy of the results.

1.1   Motivating Example

The Minimum Wage Board of Bangladesh published *minimum wage 2013 gazette* on 21 November 2013 that would be applicable from 1 December 2013 [3]. The declared minimum wage was BDT 5300 per month that is equivalent to US$75 as the minimum wage. However, according to a survey [4] conducted by Bangladesh Garments Manufacturers and Exporters Association (BGMEA), it was found that nearly 40 percent of the garment factories in Dhaka and its adjacent areas of Bangladesh could not implement the new wage structure.

Now consider that Bangladesh Government's garments monitoring authority wants to identify whether an industry pays its employees according to govern-ment's new wage law. Feedbacks information about the salaries from the employ-ees of the industry can provide very useful information in this regard. Assume the salary information of five employees of an industry as shown in Table 1.

From the information of Table 1, we can see that the salaries of the employees of this industry is far below than the government's declared minimum salary. However, collection of such information is not so easy because garments workers do not want to provide such information to anyone else due to the fear of losing their jobs and social harassments. The situation is almost common in most of the industries, organizations and agencies.

We can easily overcome this problem by developing a feedback system that will preserve individual's privacy while collecting feedbacks. According to our knowledge Hashem et al. [2] consider the fact and propose a theoretical frame-work for privacy preserving feedbacks collection. In their approach, at first an

Table 1. Users' feedbacks example

User ID	Salary
u_1	US$ 40
u_2	US$ 60
u_3	US$ 50
u_4	US$ 35
u_5	US$ 80

user randomly divides each of her/his record's values into several parts and keeps one part for her/him and sends each of the remaining parts to each of remaining users. When all the transactions are completed, the users submit the individual sum of numbers they poses to the servers. Based on the individual sums, the server then computes the average corresponds to each field of the record. However, their approach has several major limitations. First, their system is not scalable well in case of large number of users. As for example, if there are n users and each user's feedback record contains two values then their system needs a transmission of $n(n-1)$ values among the users. Second, there is no consideration about protection of data during transmission in their system. As a result, there is no way to protect data from third party access and modification. Third, their system is highly vulnerable in presence of dishonest users. If there are some dishonest users in the system, they can easily modify the data send to them. As a result, the feedback system will produce wrong output.

In this paper, we provide a framework that can overcome the problems of [2]. Our agent-based computation framework can significantly improve the overall accuracy and computation performance.

The remainder of this paper is organized as follows. Section 2 provides a brief review of related works. In section 3, we detail the computation framework of our proposed approach. Section 4 presents the experimental results. Finally, we conclude and sketch future research directions in Section 5.

2   Related Works

2.1   Privacy Preserving Techniques

In recent years, development of privacy preserving techniques for data mining and location based services attracted great attention among the researchers. As a result, many techniques have been developed for preserving privacy for data mining and location-based services. In the works presented in [5–9] authors propose new techniques based on the randomization approach in order to protect privacy of data. The work in [5] is based on the fact that the probability distribution is sufficient in order to construct data mining models as classifiers. Here, the authors show that the data distribution can be reconstructed with an iterative algorithm. Authors in[6, 7] introduce methods to build a Naive Bayesian classifier over perturbed data. The works in [8, 9] consider users' privacy while mining association rules. The main consideration of [8] is to maximize the privacy of the users and to maintain a high accuracy in the results obtained with the association rule mining. In [9], authors present a privacy preserving framework for mining association rules from randomized data. They propose a class of randomization operators those are more effective than uniform distribution and a data mining approach to recover itemset supports from distorted data.

Sweeney [10] propose the concepts of k-anonymity to preserve privacy of data. In this approach, the author utilizes the concepts of suppression and generalization of some of the data values. However, k-anonymity is not well suited for protecting data privacy due to homogeneous attack and background knowledge attack. To overcome the limitations of k-anonymity, the concept of l-diversity was developed [11]. The main aim of l-diversity is to maintain the diversity of sensitive attributes. In particular, the main idea of this method is that every group of individuals that can be isolated by an attacker should contain at least l well-represented values for a sensitive attribute. Unfortunately, l-diversity is insufficient to protect privacy if overall distribution of data is skewed. In such a situation, the attacker can know the global distribution of the attributes and use it to infer the value of sensitive attributes. To handle such a situation, Li et al. [12] introduce the concept of t-closeness. This technique requires that the distribution of a sensitive attribute in any equivalence class is close to the distribution of the attribute in the overall table. The distance between the two distributions should be no more than a threshold t.

Yarovoy et al. [13] study problem of k-anonymization of moving object databases for the purpose of protecting privacy while publishing the database. In this work, authors propose two approaches that generate anonymity groups satisfying the novel notion of k-anonymity. These approaches are known as Extreme Union and Symmetric Anonymization. In [14], Terrovitis et al. suggest a suppression-based algorithm to protect privacy while publication of trajectories. This work is based on the assumption that different attackers know different and disjoint portions of the trajectories and the data publisher knows the attacker's knowledge. So, the proposed solution considers all the dangerous observations in the database and suppress them all.

2.2   Feedback Systems

Due to the rapid growth of Internet technologies, there are many well known e-commerce companies such as Amazon[15], eBay [16], Olex [17], Agoda [18], TripAdvisor [19], Expedia [20], Rakuten [21], Elance [22] those sell their products and services via Internet. Besides the e-commerce companies, agencies, government organizations, non government organizations provide different types of services via Internet. Most of these companies, agencies and organizations collect feedbacks from the customers / users to improve the quality of their services. Main problem in collecting feedbacks by these companies, agencies or organizations is that customers / users need to disclose their identities while providing

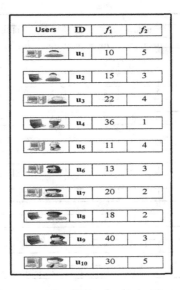

Users	ID	f_1	f_2
	u_1	10	5
	u_2	15	3
	u_3	22	4
	u_4	36	1
	u_5	11	4
	u_6	13	3
	u_7	20	2
	u_8	18	2
	u_9	40	3
	u_{10}	30	5

Fig. 1. Users' feedback information

feedbacks. However, there are many situations where discloser of users' identities can create problems for the users. As for example, if an employee of an organization provides negative feedbacks about the organization, the authority may create financial and social problems for that employee. The same situation is true for almost all companies, organizations and agencies of the developing countries. Here comes the need to develop a framework where customers / users can provide their feedbacks in such a way that it is not necessary to disclose their identities while providing the feedbacks. The main focus of this paper is to develop such a framework.

2.3   Privacy Preserving Feedback Systems

Although privacy of individual's is an important issue in any computation, till now there is very little consideration about preserving individual's privacy in feedback systems. As for the privacy issue, authors in [23] introduce a visualization technique known as *Conversation Votes* to create new backchannels in conversation and augment collocated interaction. In this paper, authors expand the idea of a social mirror to incorporate direct user feedback in the form of anonymous voting. *MyExperience* [24] is a feedback system that captures both objective and subjective in situ data on mobile computing activities. To preserve the privacy of individual's, *MyExperience* uses strong cryptographic hashing, SHA-1, to map personal information. Hashem et al. [2] propose a theoretical framework for privacy preserving feedbacks collection. However, their system is not robust if there are some dishonest users in the system. In addition, the system is not scale well in case of large number of users.

Users	ID	Division of f_1					Division of f_2				
		f_{11}	f_{12}	f_{13}	f_{14}	f_{15}	f_{21}	f_{22}	f_{23}	f_{24}	f_{25}
	u_1	3	8	-3	4	-2	1	2	-1	3	0
	u_2	3	5	2	4	1	5	0	-2	1	-1
	u_3	2	-4	6	7	1	1	2	-2	3	-1
	u_4	7	13	5	-3	14	2	3	-2	-1	1
	u_5	4	1	2	3	1	6	-4	1	2	-1
	u_6	2	1	5	3	2	2	-3	0	3	1
	u_7	8	-3	10	3	2	1	3	-4	-2	4
	u_8	5	3	-3	8	5	4	7	-8	0	-1
	u_9	10	10	-5	15	10	9	4	-7	-4	1
	u_{10}	10	2	3	8	7	1	3	-4	6	-1

Fig. 2. Division of data values of Figure 1

Our agent-based computation framework of this paper can provide accurate feedbacks even in presence of dishonest users. In addition, our system can protect data from third party access and modification. Moreover, due to proper parallelism, the computation time of the propose algorithm is almost independent to the number of users while obtaining feedbacks.

3   Secure Parallel Computation of Feedbacks

We assume that there are m users involve in the system. Let $u_1, u_2, \cdots, u_m$ be the users. Each user provides his feedbacks on features $f_1, f_2, \cdots, f_k$. To preserve the privacy of individual's, instead of publishing the exact feedbacks of each user, we have to publish the feedback results in such a way that the feedbacks information will be accurate while privacy of individual's is preserved. Figure 1 shows the information of ten users. Instead of publishing the information of Figure 1, we utilize a framework to publish aggregated information that will preserve individual's privacy.

3.1   Agent-Based Parallel Computation

We assume there is a coordinator who is responsible for calculating the feedbacks by divide-and-conquer strategy. The coordinator first asks each user within the system to divide each of its data values into s parts in such a way that the sum

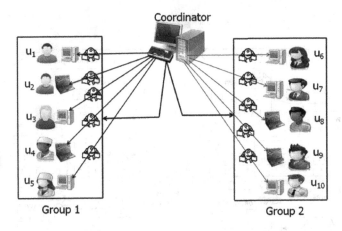

Fig. 3. Example of divide-and-conquer computation

of s parts is equal to the data value. Each user then randomly divides each of its data values into s different parts. As for example, each data value of Figure 1 has been divided into five parts as shown in Figure 2. Based on the number of users involve in feedbacks computation, the coordinator then creates a number of groups and assign s agents to each group. In general, if there are more users in the system more groups are created. For example, the users of Figure 1 has been divided into two groups and there are five agents for each group as shown in Figure 3. Later, the coordinator assigns an unique identifier known as token for each agent within a group. It then sends the agents to the users of the groups. Upon arrival of an agent to an user in a group, the agent asks for data values. The user then sends one part of each data value to the agent. The agent then goes to the next user of the group and ask the user for data values. The user also sends one part of each data value to the agent. At this moment, the agent adds the new values with the values already in the agent. After completing the traversal of all users within the group, the agent contains "local sum" in its data structure and goes back to the coordinator. For each token, the coordinator first computes the "global sum" considering the "local sum" of the groups. Based on the "global sum", the coordinator then computes the average of the users' feedbacks. During the process, agents are used to preserve privacy of users' data. Note that all the groups' computations are performed simultaneously. Now, consider the secure computation of feedbacks from users of Figure 1. For each group, the coordinator computes "local sum" with $s = 5$ having five tokens $T1$, $T2$, $T3$, $T4$, and $T5$.

For each group, the coordinator creates five agents and assign a separate token to each agent of the group. Each agent has an "array data structure" to keep parts of data values. Initially the array contains 0 in each of its index. Each agent travels the users in a circular way pre-defined by the coordinator. Figure 4 shows the computation process in group 1 with token $T1$. When an agent arrives at an user of a group, it asks for a part of each of data values from the user. The user then push a part of each data value to the agent. The agent then adds the

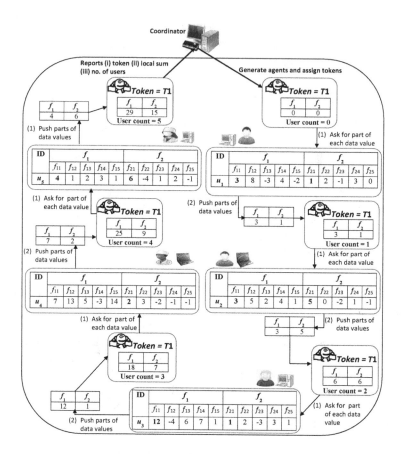

Fig. 4. Computation in group one with token $T1$

values with the values already in the array. The agent then moves to the next user and performs same tasks. Note that during this process, the user cannot see the contents of the array of the agent.

In the example of Figure 4, it is observed that agent with token $T1$ visits the users in the order $(u_1- > u_2- > u_3- > u_4- > u_5)$. From Figure 4, we can see that u_1 pushes $(f_{11}, f_{21}) = (3, 1)$ to the agent. The agent then adds these values with the contents of the array. Next, the agent goes to user u_2. User u_2 then pushes $(f_{11}, f_{21}) = (3, 5)$ to the agent. Here, the contents of the array updates as $((3 + 3) = 6)$ and $((1 + 5) = 6)$. The agent then traverse to users u_3, u_4 and u_5 in a sequential order as defined by the coordinator. During the traversal of any user, the agent performs the same tasks like u_1 and u_2. After visiting all the users in the group, agent with token $T1$ contains $(f_1, f_2) = (19, 15)$ in its array and returns back to the coordinator.

Figure 5 shows similar computation in the group 1 of with token $T2$. Here, in order to minimize idle time, the traversal order of the agent is different. It starts

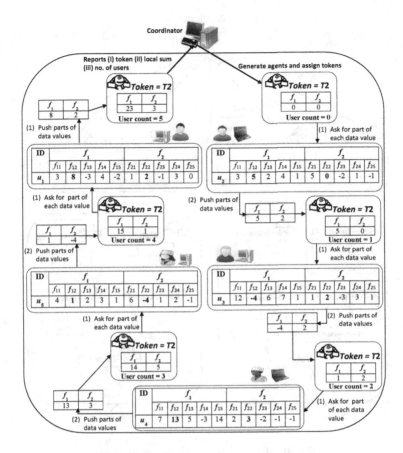

Fig. 5. Computation in group one with token $T2$

its traversal from user u_2 and follows the order $(u_2->u_3->u_4->u_5->u_1)$. After the computation, the agent contains $(f_1, f_2) = (23, 3)$ in its array and goes to the coordinator and reports the results. The agents with tokens $T3, T4$, and $T5$ perform similar computations with traversal order $(u_3->u_4->u_5->u_1->u_2)$, $(u_4->u_5->u_1->u_2->u_3)$, and $(u_5->u_1->u_2->u_3->u_4)$ and reports results $(f_1, f_2) = (12, -7)$, $(f_1, f_2) = (15, 8)$, and $(f_1, f_2) = (15, -2)$, respectively to the coordinator.

During these processes, "local sum" in group 2 is computed simultaneously. After the agent-based computation in two groups, the coordinator first computes the "global sum" for each token. Then, the agents return the average values to the coordinator. Finally, the coordinator performs addition operation among the average values received from the agents to obtain the final feedback results in aggregated form. Figure 6 shows the final computation at the coordinator.

From the example of Figure 6, we get average values $(6.4, 3.2), (3.6, 1.7), (2.2, -3.0), (5.2, 1.1)$, and $(4.1, 0,2)$ for tokens $T1, T2, T3, T4$, and $T5$, respectively.

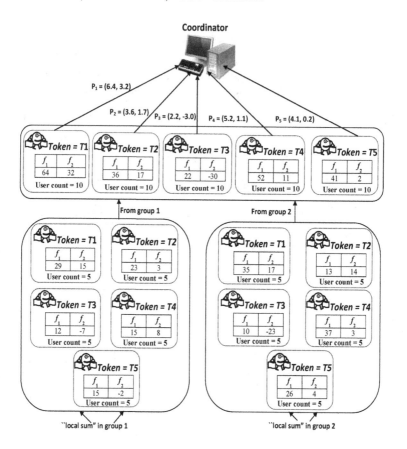

Fig. 6. Merge process at the coordinator

After adding these values, we obtain (21.5, 3.2) as final feedbacks in aggregated form. Note that the obtained results in our system is same as the the results obtained by averaging the values of each attribute of Figure 1 separately. However, in our system there is no discloser of users' feedbacks information.

Note that our system is well scalable in case of large number of users. This is because we have divided users in groups and computation in groups are carried out in parallel that minimizes the time of computation. More over our system protects data from third party access and modification. This due to the fact in our approach we utilize an agent-based computation and our agents can communicate using TLS/SSL that provides strong security while agents moves from one user to another user. In addition, our system is well secure from the modification of feedback information by dishonest users. This is because in our system, we do not need to send parts of data values of an user to other users.

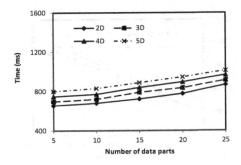

Fig. 7. Time varying number of data parts

3.2 Expansion of the System

We can easily expand our feedback system to collect feedbacks of different companies, agencies and organizations. To do this, we just need to add a super-coordinator. There will be a coordinator for each company / agency / organization. Each coordinator will perform same tasks as describe in subsection 3.1 to obtain feedback record in aggregated form. The coordinator of each company / agency / organization then sends the feedback record to the super-coordinator. Note that each coordinator sends only one feedback record to the super-coordinator. Super-coordinator stores all received feedback records in a table. The super-coordinator can then issue different types of queries such as SQL queries, top-k queries, skyline queries in the table to perform comparative analysis among the companies / agencies / organizations.

4 Experiments

We have implemented our proposed privacy aware feedback system using Java Agent Development Framework. We have performed the experiment in a simulation environment of a PC running on windows OS having an Intel(R) Core i7, 1.73 GHz CPU, and 4 GB main memory. Due to the lack of real data, we evaluate our proposed algorithm using synthetic datasets only.

We first evaluate the effect of data parts s. Figure 7 shows the results when we consider two (2D), three (3D), four (4D), and five (5D) features while distributing 40000 users among twenty groups and each group contains around two thousand users. We observe that with the increases of s, there is very slight increase in computation time. This is because during the computation process each agent in a group follows a different execution order that increase parallelism in computation. In addition, the computation among the groups are also performed in parallel. We can also observed that computation time gradually increases if the number of features increases.

In the next experiment, we evaluate the effect of the number of users involve in the system. In this experiment, we considered 20000, 40000, 60000, 80000, and

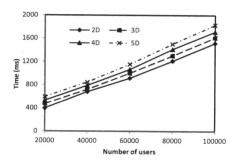

Fig. 8. Time varying number of users

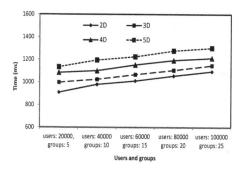

Fig. 9. Time varying number of users and number of groups

100000 users. Same as the previous experiment, users were distributed among twenty groups. In case of 20000 users, each group contains around 1000 users. Similarly, for 20000, 40000, 60000, 80000, and 100000 users, each group contains around 2000, 3000, 4000, and 5000 users. In this experiment, we set s to 10. Figure 8 shows the results. In this experiment, it is observed that in case of fixed number of groups, response time increases with the increase of the number of users. Also, note that there is an increase of response time with the increase in the number of features.

Finally, we conduct the experiment to examine the effects while number of users and number of groups both increases. In this experiment, we distribute 20000, 40000, 60000, 80000, 100000 users among 5, 10, 15, 20, 25 groups, respectively. In this experiment, we set s to 10 and examine 2D, 3D, 4D, and 5D cases. Figure 9 shows the results. From the results, we can find that in case of more users, if we create more groups, there is almost no performance degradation.

5   Conclusion

With the rapid growth of network infrastructure, collecting feedbacks form the users via Internet are becoming popular. In privacy aware environment, users do not want to disclose their identities due to several factors. Therefore, we proposed an agent-based algorithm for computing feedbacks in a parallel manner from the users. The proposed algorithm can efficiently collect users' feedback while preserving individual's privacy. Experimental results demonstrate that the proposed algorithm for feedbacks collection is scalable enough to handle large number of users. The proposed approach can easily expandable to collect and analyze feedbacks for different industries /organizations / agencies.

In this work, we performed different analysis on synthetic data. In future, we aim to show the applicability of our approach in real environment.

References

1. http://en.wikipedia.org/wiki/2013_Savar_building_collapse
2. Hashem, T., Hashem, T.: Privacy preserving feedback and monitoring system. In: Proc. of Workshop on Women Empowerment through ICT: Higher Studies, Research and Career, WE-ICT 2014, pp. 30-31 (2014)
3. http://risebd.com/2013/11/22/minimum-wage-2013-basic-goes-down/
4. http://www.thedailystar.net/40pc-garment-units-fail-to-pay-new-wage-8226
5. Agrawal, R., Srikant., R.: Privacy-preserving data mining. In: Proc. of the 2000 ACM SIGMOD International Conference on Management of Data, SIGMOD 2000, pp. 439–450 (2000)
6. Zhan, J., Matwin, S., Chang, L.: Privacy-preserving collaborative association rule mining. In: Jajodia, S., Wijesekera, D. (eds.) Data and Applications Security 2005. LNCS, vol. 3654, pp. 153–165. Springer, Heidelberg (2005)
7. Zhang, P., Tong, Y., Tang, S.-W., Yang, D.-Q.: Privacy Preserving Naive Bayes Classification. In: Li, X., Wang, S., Dong, Z.Y. (eds.) ADMA 2005. LNCS (LNAI), vol. 3584, pp. 744–752. Springer, Heidelberg (2005)
8. Rizvi, S., Haritsa, J.R.: Maintaining data privacy in association rule mining. In: Proc. of 28th International Conference on Very Large Data Bases, VLDB 2002, pp. 682–693 (2002)
9. Evfimievski, A.V., Srikant, R., Agrawal, R., Gehrke., J.: Privacy preserving mining of association rules. In: Proc. of the 8th ACM SIGKDD International Conference on Knowledge Discovery and Data Mining, SIGKDD 2002, pp. 217–228 (2002)
10. Sweeney., L.: k-anonymity: A model for protecting privacy. International Journal of Uncertainty, Fuzziness, and Knowledge-Based Systems 10(5), 557–570 (2002)
11. Machanavajjhala, A., Gehrke, J., Kifer, D., Venkitasubramaniam, M.: l-diversity: Privacy beyond k-anonymity. In: Proc. of 22nd International Conference on Data Engineering, ICDE 2006, p. 24 (2006)
12. Li, N., Li, T., Venkatasubramanian, S.: t-closeness: Privacy beyond k-anonymity and l-diversity. In: Proc. of 23nd International Conference on Data Engineering, ICDE 2007, pp. 106–115 (2007)
13. Yarovoy, R., Bonchi, F., Lakshmanan, L.V.S., Wang, W.H.: Anonymizing moving objects: How to hide a mob in a crowd? In: Proc. of 12th International Conference on Extending Database Technology, EDBT 2009, pp. 72–83 (2009)
14. Terrovitis, M., Mamoulis, N.: Privacy preservation in the publication of trajectories. In: Proc. of 9th International Conference on Mobile Data Management, MDM 2008, pp. 65–72 (2008)

15. Amazon E-commerce Company, http://www.amazon.com
16. eBay Inc., http://www.ebay.com
17. Olex, http://www.olx.com
18. Agoda Company Pte Ltd., http://www.agoda.com
19. TripAdvisor, http://www.tripadvisor.com/
20. Expedia, http://www.expedia.com
21. Rakuten, http://www.rakuten.com
22. Elance, https://www.elance.com
23. Bergstrom, T., Karahalios, K.: Conversation votes: Enabling anonymous cues. In: Proc. of Computer/Human Interaction, CHI 2007, pp. 2279–2284 (2007)
24. Froehlich, J., Chen, M.Y., Consolvo, S., Harrison, B., Landay, J.A.: MyExperience: A system for in situ tracing and capturing of user feedback on mobile phones. In: Proc. of the 5th International Conference on Mobile Systems, Applications and Services, MobiSys 2007, pp. 57–70 (2007)

Author Index

Author Index